Instructor's Solutions Manual
Volume 2

Sen-Ben Liao
Massachusetts Institute of Technology

FUNDAMENTALS OF PHYSICS

Seventh Edition

David Halliday
University of Pittsburgh

Robert Resnick
Rensselaer Polytechnic Institute

Jearl Walker
Cleveland State University

WILEY

John Wiley & Sons, Inc.

Preface

This book includes the solutions to the PROBLEMS sections of the 7th edition of **Fundamentals of Physics** by Halliday, Resnick and Walker. We have not included solutions or discussions that pertain to the Questions sections. These solutions have been typed using Microsoft Word and MathType™ equation editor. The solution files are available on the Instructor's Resource CD-ROM ISBN:0471470600. Additional information regarding MathType™ can be found at www.mathtype.com

The author has put great time and effort into writing high quality solutions. He welcomes comments and suggestions from readers, please also report any errors that you may find. The author's email address is: senben@ceci.mit.edu

Note to adopters regarding the Instructor's Solutions Manual:

Thank you for adopting **Fundamentals of Physics** 7e by Halliday, Resnick and Walker! We are pleased to provide you with a variety of support material to help you in the teaching of your course.

Please note that all of this material is copyrighted by John Wiley & Sons, Inc. and is explicitly intended for use only at your institution.

Our providing this material does not carry with it permission to distribute it beyond your institution. Before putting any of this material on a website, we ask that you request formal permission from us to do so. Please write to: Permissions Department, John Wiley & Sons, Inc. 111 River St. Hoboken, NJ 07030, or fill out a simple on-line form on our website www.wiley.com/about/permissions. In most cases we will grant such permission PROVIDED THAT THE WEBSITE IS PASSWORD PROTECTED.

Our goal is to prevent students from other campuses from being able to access your materials. We trust that you can understand how that might undermine the efforts of your colleagues at other institutions.

CONTENTS

Chapter 21

1. (a) With a understood to mean the magnitude of acceleration, Newton's second and third laws lead to

$$m_2 a_2 = m_1 a_1 \Rightarrow m_2 = \frac{\left(6.3 \times 10^{-7} \text{kg}\right)\left(7.0 \text{m/s}^2\right)}{9.0 \text{m/s}^2} = 4.9 \times 10^{-7} \text{kg}.$$

(b) The magnitude of the (only) force on particle 1 is

$$F = m_1 a_1 = k \frac{|q_1||q_2|}{r^2} = \left(8.99 \times 10^9\right) \frac{|q|^2}{0.0032^2}.$$

Inserting the values for m_1 and a_1 (see part (a)) we obtain $|q| = 7.1 \times 10^{-11}$ C.

2. The magnitude of the mutual force of attraction at $r = 0.120$ m is

$$F = k \frac{|q_1||q_2|}{r^2} = \left(8.99 \times 10^9\right) \frac{\left(3.00 \times 10^{-6}\right)\left(1.50 \times 10^{-6}\right)}{0.120^2} = 2.81 \text{N}.$$

3. Eq. 21-1 gives Coulomb's Law, $F = k \frac{|q_1||q_2|}{r^2}$, which we solve for the distance:

$$r = \sqrt{\frac{k|q_1||q_2|}{F}} = \sqrt{\frac{\left(8.99 \times 10^9 \text{N} \cdot \text{m}^2/\text{C}^2\right)\left(26.0 \times 10^{-6} \text{C}\right)\left(47.0 \times 10^{-6} \text{C}\right)}{5.70 \text{N}}} = 1.39 \text{m}.$$

4. The fact that the spheres are identical allows us to conclude that when two spheres are in contact, they share equal charge. Therefore, when a charged sphere (q) touches an uncharged one, they will (fairly quickly) each attain half that charge ($q/2$). We start with spheres 1 and 2 each having charge q and experiencing a mutual repulsive force $F = kq^2/r^2$. When the neutral sphere 3 touches sphere 1, sphere 1's charge decreases to $q/2$. Then sphere 3 (now carrying charge $q/2$) is brought into contact with sphere 2, a total amount of $q/2 + q$ becomes shared equally between them. Therefore, the charge of sphere 3 is $3q/4$ in the final situation. The repulsive force between spheres 1 and 2 is finally

$$F' = k \frac{(q/2)(3q/4)}{r^2} = \frac{3}{8} k \frac{q^2}{r^2} = \frac{3}{8} F \quad \Rightarrow \quad \frac{F'}{F} = \frac{3}{8} = 0.375.$$

5. The magnitude of the force of either of the charges on the other is given by

$$F = \frac{1}{4\pi\varepsilon_0} \frac{q(Q-q)}{r^2}$$

where r is the distance between the charges. We want the value of q that maximizes the function $f(q) = q(Q - q)$. Setting the derivative df/dq equal to zero leads to $Q - 2q = 0$, or $q = Q/2$. Thus, $q/Q = 0.500$.

6. For ease of presentation (of the computations below) we assume $Q > 0$ and $q < 0$ (although the final result does not depend on this particular choice).

(a) The x-component of the force experienced by $q_1 = Q$ is

$$F_{1x} = \frac{1}{4\pi\varepsilon_0} \left(-\frac{(Q)(Q)}{\left(\sqrt{2}a\right)^2} \cos 45° + \frac{(|q|)\,(Q)}{a^2} \right) = \frac{Q|q|}{4\pi\varepsilon_0 a^2} \left(-\frac{Q/|q|}{2\sqrt{2}} + 1 \right)$$

which (upon requiring $F_{1x} = 0$) leads to $Q/|q| = 2\sqrt{2}$, or $Q/q = -2\sqrt{2} = -2.83$.

(b) The y-component of the net force on $q_2 = q$ is

$$F_{2y} = \frac{1}{4\pi\varepsilon_0} \left[\frac{|q|^2}{\left(\sqrt{2}a\right)^2} \sin 45° - \frac{(|q|)\,(Q)}{a^2} \right] = \frac{|q|^2}{4\pi\varepsilon_0 a^2} \left(\frac{1}{2\sqrt{2}} - \frac{Q}{|q|} \right)$$

which (if we demand $F_{2y} = 0$) leads to $Q/q = -1/2\sqrt{2}$. The result is inconsistent with that obtained in part (a). Thus, we are unable to construct an equilibrium configuration with this geometry, where the only forces present are given by Eq. 21-1.

7. The force experienced by q_3 is

$$\vec{F}_3 = \vec{F}_{31} + \vec{F}_{32} + \vec{F}_{34} = \frac{1}{4\pi\varepsilon_0} \left(-\frac{|q_3||q_1|}{a^2}\hat{j} + \frac{|q_3||q_2|}{(\sqrt{2}a)^2}(\cos45°\hat{i} + \sin 45°\hat{j}) + \frac{|q_3||q_4|}{a^2}\hat{i} \right)$$

(a) Therefore, the x-component of the resultant force on q_3 is

$$F_{3x} = \frac{|q_3|}{4\pi\varepsilon_0 a^2} \left(\frac{|q_2|}{2\sqrt{2}} + |q_4| \right) = \left(8.99\times10^9\right)\frac{2\left(1.0\times10^{-7}\right)^2}{(0.050)^2}\left(\frac{1}{2\sqrt{2}} + 2\right) = 0.17\text{N}.$$

(b) Similarly, the y-component of the net force on q_3 is

$$F_{3y} = \frac{|q_3|}{4\pi\varepsilon_0 a^2}\left(-|q_1| + \frac{|q_2|}{2\sqrt{2}}\right) = (8.99\times10^9)\frac{2(1.0\times10^{-7})^2}{(0.050)^2}\left(-1 + \frac{1}{2\sqrt{2}}\right) = -0.046\,\text{N}.$$

8. (a) The individual force magnitudes (acting on Q) are, by Eq. 21-1,

$$k\frac{|q_1|Q}{\left(-a-\frac{a}{2}\right)^2} = k\frac{|q_2|Q}{\left(a-\frac{a}{2}\right)^2}$$

which leads to $|q_1| = 9.0\,|q_2|$. Since Q is located between q_1 and q_2, we conclude q_1 and q_2 are like-sign. Consequently, $q_1/q_2 = 9.0$.

(b) Now we have

$$k\frac{|q_1|Q}{\left(-a-\frac{3a}{2}\right)^2} = k\frac{|q_2|Q}{\left(a-\frac{3a}{2}\right)^2}$$

which yields $|q_1| = 25\,|q_2|$. Now, Q is not located between q_1 and q_2, one of them must push and the other must pull. Thus, they are unlike-sign, so $q_1/q_2 = -25$.

9. We assume the spheres are far apart. Then the charge distribution on each of them is spherically symmetric and Coulomb's law can be used. Let q_1 and q_2 be the original charges. We choose the coordinate system so the force on q_2 is positive if it is repelled by q_1. Then, the force on q_2 is

$$F_a = -\frac{1}{4\pi\varepsilon_0}\frac{q_1 q_2}{r^2} = -k\frac{q_1 q_2}{r^2}$$

where $r = 0.500$ m. The negative sign indicates that the spheres attract each other. After the wire is connected, the spheres, being identical, acquire the same charge. Since charge is conserved, the total charge is the same as it was originally. This means the charge on each sphere is $(q_1 + q_2)/2$. The force is now one of repulsion and is given by

$$F_b = \frac{1}{4\pi\varepsilon_0}\frac{\left(\frac{q_1+q_2}{2}\right)\left(\frac{q_1+q_2}{2}\right)}{r^2} = k\frac{(q_1+q_2)^2}{4r^2}.$$

We solve the two force equations simultaneously for q_1 and q_2. The first gives the product

$$q_1 q_2 = -\frac{r^2 F_a}{k} = -\frac{(0.500\,\text{m})^2(0.108\,\text{N})}{8.99\times10^9\,\text{N}\cdot\text{m}^2/\text{C}^2} = -3.00\times10^{-12}\,\text{C}^2,$$

and the second gives the sum

$$q_1 + q_2 = 2r\sqrt{\frac{F_b}{k}} = 2(0.500\,\text{m})\sqrt{\frac{0.0360\,\text{N}}{8.99 \times 10^9\,\text{N} \cdot \text{m}^2/\text{C}^2}} = 2.00 \times 10^{-6}\,\text{C}$$

where we have taken the positive root (which amounts to assuming $q_1 + q_2 \geq 0$). Thus, the product result provides the relation

$$q_2 = \frac{-\left(3.00 \times 10^{-12}\,\text{C}^2\right)}{q_1}$$

which we substitute into the sum result, producing

$$q_1 - \frac{3.00 \times 10^{-12}\,\text{C}^2}{q_1} = 2.00 \times 10^{-6}\,\text{C}.$$

Multiplying by q_1 and rearranging, we obtain a quadratic equation

$$q_1^2 - \left(2.00 \times 10^{-6}\,\text{C}\right)q_1 - 3.00 \times 10^{-12}\,\text{C}^2 = 0.$$

The solutions are

$$q_1 = \frac{2.00 \times 10^{-6}\,\text{C} \pm \sqrt{\left(-2.00 \times 10^{-6}\,\text{C}\right)^2 - 4\left(-3.00 \times 10^{-12}\,\text{C}^2\right)}}{2}.$$

If the positive sign is used, $q_1 = 3.00 \times 10^{-6}$ C, and if the negative sign is used, $q_1 = -1.00 \times 10^{-6}$ C.

(a) Using $q_2 = (-3.00 \times 10^{-12})/q_1$ with $q_1 = 3.00 \times 10^{-6}$ C, we get $q_2 = -1.00 \times 10^{-6}$ C.

(b) If we instead work with the $q_1 = -1.00 \times 10^{-6}$ C root, then we find $q_2 = 3.00 \times 10^{-6}$ C.

Note that since the spheres are identical, the solutions are essentially the same: one sphere originally had charge -1.00×10^{-6} C and the other had charge $+3.00 \times 10^{-6}$ C.

What if we had not made the assumption, above, that $q_1 + q_2 \geq 0$? If the signs of the charges were reversed (so $q_1 + q_2 < 0$), then the forces remain the same, so a charge of $+1.00 \times 10^{-6}$ C on one sphere and a charge of -3.00×10^{-6} C on the other also satisfies the conditions of the problem.

10. With rightwards positive, the net force on q_3 is

$$F_3 = F_{13} + F_{23} = k \frac{q_1 q_3}{\left(L_{12} + L_{23} \right)^2} + k \frac{q_2 q_3}{L_{23}^2} .$$

We note that each term exhibits the proper sign (positive for rightward, negative for leftward) for all possible signs of the charges. For example, the first term (the force exerted on q_3 by q_1) is negative if they are unlike charges, indicating that q_3 is being pulled toward q_1, and it is positive if they are like charges (so q_3 would be repelled from q_1). Setting the net force equal to zero $L_{23} = L_{12}$ and canceling k, q_3 and L_{12} leads to

$$\frac{q_1}{4.00} + q_2 = 0 \quad \Rightarrow \quad \frac{q_1}{q_2} = -4.00.$$

11. (a) Eq. 21-1 gives

$$F_{12} = k \frac{q_1 q_2}{d^2} = \left(8.99 \times 10^9 \, \mathrm{N \cdot m^2 / C^2} \right) \frac{\left(20.0 \times 10^{-6} \, \mathrm{C} \right)^2}{\left(1.50 \, \mathrm{m} \right)^2} = 1.60 \, \mathrm{N}.$$

(b) A force diagram is shown as well as our choice of y axis (the dashed line).

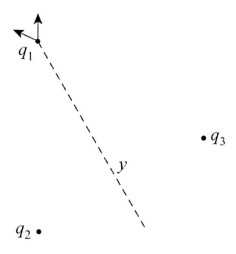

The y axis is meant to bisect the line between q_2 and q_3 in order to make use of the symmetry in the problem (equilateral triangle of side length d, equal-magnitude charges $q_1 = q_2 = q_3 = q$). We see that the resultant force is along this symmetry axis, and we obtain

$$\left| F_y \right| = 2 \left(k \frac{q^2}{d^2} \right) \cos 30° = 2.77 \, \mathrm{N}.$$

12. (a) According to the graph, when q_3 is very close to q_1 (at which point we can consider the force exerted by particle 1 on 3 to dominate) there is a (large) force in the

positive x direction. This is a repulsive force, then, so we conclude q_1 has the same sign as q_3. Thus, q_3 is a positive-valued charge.

(b) Since the graph crosses zero and particle 3 is *between* the others, q_1 must have the same sign as q_2, which means it is also positive-valued. We note that it crosses zero at $r = 0.020$ m (which is a distance $d = 0.060$ m from q_2). Using Coulomb's law at that point, we have

$$\frac{q_1\, q_3}{4\pi\varepsilon_0\, r^2} = \frac{q_3\, q_2}{4\pi\varepsilon_0\, d^2} \quad\Rightarrow\quad q_2 = \left(\frac{d^2}{r^2}\right) q_1 = 9.0\, q_1 \,,$$

or $q_2/q_1 = 9.0$.

13. (a) There is no equilibrium position for q_3 *between* the two fixed charges, because it is being pulled by one and pushed by the other (since q_1 and q_2 have different signs); in this region this means the two force arrows on q_3 are in the same direction and cannot cancel. It should also be clear that off-axis (with the axis defined as that which passes through the two fixed charges) there are no equilibrium positions. On the semi-infinite region of the axis which is nearest q_2 and furthest from q_1 an equilibrium position for q_3 cannot be found because $|q_1| < |q_2|$ and the magnitude of force exerted by q_2 is everywhere (in that region) stronger than that exerted by q_1 on q_3. Thus, we must look in the semi-infinite region of the axis which is nearest q_1 and furthest from q_2, where the net force on q_3 has magnitude

$$\left| k\,\frac{|q_1 q_3|}{x^2} - k\,\frac{|q_2 q_3|}{(d+x)^2} \right|$$

with $d = 10$ cm and x assumed positive. We set this equal to zero, as required by the problem, and cancel k and q_3. Thus, we obtain

$$\frac{|q_1|}{x^2} - \frac{|q_2|}{(d+x)^2} = 0 \Rightarrow \left(\frac{d+x}{x}\right)^2 = \left|\frac{q_2}{q_1}\right| = 3$$

which yields (after taking the square root)

$$\frac{d+x}{x} = \sqrt{3} \Rightarrow x = \frac{d}{\sqrt{3}-1} \approx 14\,\text{cm}$$

for the distance between q_3 and q_1.

(b) As stated above, $y = 0$.

14. Since the forces involved are proportional to q, we see that the essential difference between the two situations is $F_a \propto q_B + q_C$ (when those two charges are on the same side) versus $F_b \propto -q_B + q_C$ (when they are on opposite sides). Setting up ratios, we have

$$\frac{F_a}{F_b} = \frac{q_B + q_C}{-q_B + q_C} \quad \Rightarrow \quad \frac{20.14}{-2.877} = \frac{1+r}{-1+r}$$

where in the last step we have canceled (on the left hand side) 10^{-24} N from the numerator and the denominator, and (on the right hand side) introduced the symbol $r = q_C / q_B$. After noting that the ratio on the left hand side is very close to -7, then, after a couple of algebra steps, we are led to

$$r = \frac{7+1}{7-1} = \frac{8}{6} = 1.333.$$

15. (a) The distance between q_1 and q_2 is

$$r_{12} = \sqrt{(x_2 - x_1)^2 + (y_2 - y_1)^2} = \sqrt{(-0.020 - 0.035)^2 + (0.015 - 0.005)^2} = 0.056 \text{ m}.$$

The magnitude of the force exerted by q_1 on q_2 is

$$F_{21} = k\frac{|q_1 q_2|}{r_{12}^2} = \frac{(8.99 \times 10^9)(3.0 \times 10^{-6})(4.0 \times 10^{-6})}{(0.056)^2} = 35 \text{ N}.$$

(b) The vector $\vec{F}_{21}$ is directed towards q_1 and makes an angle θ with the $+x$ axis, where

$$\theta = \tan^{-1}\left(\frac{y_2 - y_1}{x_2 - x_1}\right) = \tan^{-1}\left(\frac{1.5 - 0.5}{-2.0 - 3.5}\right) = -10.3° \approx -10°.$$

(c) Let the third charge be located at (x_3, y_3), a distance r from q_2. We note that q_1, q_2 and q_3 must be collinear; otherwise, an equilibrium position for any one of them would be impossible to find. Furthermore, we cannot place q_3 on the same side of q_2 where we also find q_1, since in that region both forces (exerted on q_2 by q_3 and q_1) would be in the same direction (since q_2 is attracted to both of them). Thus, in terms of the angle found in part (a), we have $x_3 = x_2 - r \cos\theta$ and $y_3 = y_2 - r \sin\theta$ (which means $y_3 > y_2$ since θ is negative). The magnitude of force exerted on q_2 by q_3 is $F_{23} = k|q_2 q_3|/r^2$, which must equal that of the force exerted on it by q_1 (found in part (a)). Therefore,

$$k\frac{|q_2 q_3|}{r^2} = k\frac{|q_1 q_2|}{r_{12}^2} \Rightarrow r = r_{12}\sqrt{\frac{q_3}{q_1}} = 0.0645 \text{cm}.$$

Consequently, $x_3 = x_2 - r \cos\theta = -2.0$ cm $- (6.45$ cm$) \cos(-10°) = -8.4$ cm,

(d) and $y_3 = y_2 - r \sin\theta = 1.5$ cm $- (6.45$ cm$) \sin(-10°) = 2.7$ cm.

16. (a) For the net force to be in the $+x$ direction, the y components of the individual forces must cancel. The angle of the force exerted by the $q_1 = 40 \ \mu C$ charge on $q_3 = 20 \mu C$ is 45°, and the angle of force exerted on q_3 by Q is at $-\theta$ where

$$\theta = \tan^{-1}\left(\frac{2.0}{3.0}\right) = 33.7°.$$

Therefore, cancellation of y components requires

$$k\frac{q_1 q_3}{\left(0.02\sqrt{2}\right)^2}\sin 45° = k\frac{|Q|q_3}{\left(\sqrt{(0.030)^2 + (0.020)^2}\right)^2}\sin\theta$$

from which we obtain $|Q| = 83 \ \mu C$. Charge Q is "pulling" on q_3, so (since $q_3 > 0$) we conclude $Q = -83 \ \mu C$.

(b) Now, we require that the x components cancel, and we note that in this case, the angle of force on q_3 exerted by Q is $+\theta$ (it is repulsive, and Q is positive-valued). Therefore,

$$k\frac{q_1 q_3}{\left(0.02\sqrt{2}\right)^2}\cos 45° = k\frac{Qq_3}{\left(\sqrt{(0.030)^2 + (0.020)^2}\right)^2}\cos\theta$$

from which we obtain $Q = 55.2 \ \mu C \approx 55 \ \mu C$.

17. (a) If the system of three charges is to be in equilibrium, the force on each charge must be zero. The third charge q_3 must lie between the other two or else the forces acting on it due to the other charges would be in the same direction and q_3 could not be in equilibrium. Suppose q_3 is at a distance x from q, and $L - x$ from $4.00q$. The force acting on it is then given by

$$F_3 = \frac{1}{4\pi\varepsilon_0}\left(\frac{qq_3}{x^2} - \frac{4qq_3}{(L-x)^2}\right)$$

where the positive direction is rightward. We require $F_3 = 0$ and solve for x. Canceling common factors yields $1/x^2 = 4/(L-x)^2$ and taking the square root yields $1/x = 2/(L-x)$. The solution is $x = L/3$.

The force on q is

$$F_q = \frac{-1}{4\pi\varepsilon_0}\left(\frac{qq_3}{x^2} + \frac{4.00q^2}{L^2}\right).$$

The signs are chosen so that a negative force value would cause q to move leftward. We require $F_q = 0$ and solve for q_3:

$$q_3 = -\frac{4qx^2}{L^2} = -\frac{4}{9}q \quad\Rightarrow\quad \frac{q_3}{q} = -\frac{4}{9} = -0.444$$

where $x = L/3$ is used. We may easily verify that the force on $4.00q$ also vanishes:

$$F_{4q} = \frac{1}{4\pi\varepsilon_0}\left(\frac{4q^2}{L^2} + \frac{4qq_0}{(L-x)^2}\right) = \frac{1}{4\pi\varepsilon_0}\left(\frac{4q^2}{L^2} + \frac{4(-4/9)q^2}{(4/9)L^2}\right) = \frac{1}{4\pi\varepsilon_0}\left(\frac{4q^2}{L^2} - \frac{4q^2}{L^2}\right) = 0.$$

(b) As seen above, q_3 is located at $x = L/3$. With $L = 9.00$ cm, we have $x = 3.00$ cm.

(c) Similarly, the y coordinate of q_3 is $y = 0$.

18. (a) We note that $\cos(30°) = \frac{1}{2}\sqrt{3}$, so that the dashed line distance in the figure is $r = 2d/\sqrt{3}$. We net force on q_1 due to the two charges q_3 and q_4 (with $|q_3| = |q_4| = 1.60 \times 10^{-19}$ C) on the y axis has magnitude

$$2\frac{|q_1q_3|}{4\pi\varepsilon_0 r^2}\cos(30°) = \frac{3\sqrt{3}|q_1q_3|}{16\pi\varepsilon_0 d^2}.$$

This must be set equal to the magnitude of the force exerted on q_1 by $q_2 = 8.00 \times 10^{-19}$ C $= 5.00|q_3|$ in order that its net force be zero:

$$\frac{3\sqrt{3}|q_1q_3|}{16\pi\varepsilon_0 d^2} = \frac{|q_1q_2|}{4\pi\varepsilon_0(D+d)^2} \quad\Rightarrow\quad D = d\left(2\sqrt{\frac{5}{3\sqrt{3}}} - 1\right) = 0.9245\,d.$$

Given $d = 2.00$ cm, then this leads to $D = 1.92$ cm.

(b) As the angle decreases, its cosine increases, resulting in a larger contribution from the charges on the y axis. To offset this, the force exerted by q_2 must be made stronger, so that it must be brought closer to q_1 (keep in mind that Coulomb's law is *inversely* proportional to distance-squared). Thus, D must be decreased.

19. The charge dq within a thin shell of thickness dr is $\rho A\,dr$ where $A = 4\pi r^2$. Thus, with $\rho = b/r$, we have

$$q = \int dq = 4\pi b \int_{r_1}^{r_2} r\, dr = 2\pi b\left(r_2^2 - r_1^2\right).$$

With $b = 3.0\ \mu C/m^2$, $r_2 = 0.06$ m and $r_1 = 0.04$ m, we obtain $q = 0.038\ \mu C = 3.8 \times 10^{-8}$ C.

20. If θ is the angle between the force and the x-axis, then

$$\cos\theta = \frac{x}{\sqrt{x^2 + d^2}}.$$

We note that, due to the symmetry in the problem, there is no y component to the net force on the third particle. Thus, F represents the magnitude of force exerted by q_1 or q_2 on q_3. Let $e = +1.60 \times 10^{-19}$ C, then $q_1 = q_2 = +2e$ and $q_3 = 4.0e$ and we have

$$F_{net} = 2F\cos\theta = \frac{2(2e)(4e)}{4\pi\varepsilon_0 (x^2 + d^2)} \frac{x}{\sqrt{x^2 + d^2}} = \frac{4e^2 x}{\pi\varepsilon_0 (x^2 + d^2)^{3/2}}.$$

(a) To find where the force is at an extremum, we can set the derivative of this expression equal to zero and solve for x, but it is good in any case to graph the function for a fuller understanding of its behavior – and as a quick way to see whether an extremum point is a maximum or a miminum. In this way, we find that the value coming from the derivative procedure is a maximum (and will be presented in part (b)) and that the minimum is found at the lower limit of the interval. Thus, the net force is found to be zero at $x = 0$, which is the smallest value of the net force in the interval $5.0\ m \geq x \geq 0$.

(b) The maximum is found to be at $x = d/\sqrt{2}$ or roughly 12 cm.

(c) The value of the net force at $x = 0$ is $F_{net} = 0$.

(d) The value of the net force at $x = d/\sqrt{2}$ is $F_{net} = 4.9 \times 10^{-26}$ N.

21. (a) The magnitude of the force between the (positive) ions is given by

$$F = \frac{(q)(q)}{4\pi\varepsilon_0 r^2} = k\frac{q^2}{r^2}$$

where q is the charge on either of them and r is the distance between them. We solve for the charge:

$$q = r\sqrt{\frac{F}{k}} = \left(5.0 \times 10^{-10}\ m\right)\sqrt{\frac{3.7 \times 10^{-9}\ N}{8.99 \times 10^9\ N\cdot m^2/C^2}} = 3.2 \times 10^{-19}\ C.$$

(b) Let N be the number of electrons missing from each ion. Then, $Ne = q$, or

$$N = \frac{q}{e} = \frac{3.2 \times 10^{-9}\,\text{C}}{1.6 \times 10^{-19}\,\text{C}} = 2.$$

22. The magnitude of the force is

$$F = k\frac{e^2}{r^2} = \left(8.99 \times 10^9\,\frac{\text{N}\cdot\text{m}^2}{\text{C}^2}\right)\frac{\left(1.60 \times 10^{-19}\,\text{C}\right)^2}{\left(2.82 \times 10^{-10}\,\text{m}\right)^2} = 2.89 \times 10^{-9}\,\text{N}.$$

23. Eq. 21-11 (in absolute value) gives

$$n = \frac{|q|}{e} = \frac{1.0 \times 10^{-7}\,\text{C}}{1.6 \times 10^{-19}\,\text{C}} = 6.3 \times 10^{11}.$$

24. (a) Eq. 21-1 gives

$$F = \frac{\left(8.99 \times 10^9\,\text{N}\cdot\text{m}^2/\text{C}^2\right)\left(1.00 \times 10^{-16}\,\text{C}\right)^2}{\left(1.00 \times 10^{-2}\,\text{m}\right)^2} = 8.99 \times 10^{-19}\,\text{N}.$$

(b) If n is the number of excess electrons (of charge $-e$ each) on each drop then

$$n = -\frac{q}{e} = -\frac{-1.00 \times 10^{-16}\,\text{C}}{1.60 \times 10^{-19}\,\text{C}} = 625.$$

25. The unit Ampere is discussed in §21-4. The proton flux is given as 1500 protons per square meter per second, where each proton provides a charge of $q = +e$. The current through the spherical area $4\pi R^2 = 4\pi\,(6.37 \times 10^6\,\text{m})^2 = 5.1 \times 10^{14}\,\text{m}^2$ would be

$$i = \left(5.1 \times 10^{14}\,\text{m}^2\right)\left(1500\,\frac{\text{protons}}{\text{s}\cdot\text{m}^2}\right)\left(1.6 \times 10^{-19}\,\text{C/proton}\right) = 0.122\,\text{A}.$$

26. The volume of 250 cm^3 corresponds to a mass of 250 g since the density of water is 1.0 g/cm^3. This mass corresponds to 250/18 = 14 moles since the molar mass of water is 18. There are ten protons (each with charge $q = +e$) in each molecule of H_2O, so

$$Q = 14 N_A q = 14\left(6.02 \times 10^{23}\right)(10)\left(1.60 \times 10^{-19}\,\text{C}\right) = 1.3 \times 10^7\,\text{C}.$$

27. Since the graph crosses zero, q_1 must be positive-valued: $q_1 = +8.00e$. We note that it crosses zero at $r = 0.40$ m. Now the asymptotic value of the force yields the magnitude and sign of q_2:

$$\frac{q_1 q_2}{4\pi\varepsilon_0 r^2} = F \quad \Rightarrow \quad q_2 = \left(\frac{1.5 \times 10^{-25}}{k\, q_1}\right) r^2 = 2.086 \times 10^{-18}\, C = 13e.$$

28. Let d be the vertical distance from the coordinate origin to $q_3 = -q$ and $q_4 = -q$ on the $+y$ axis, where the symbol q is assumed to be a positive value. Similarly, d is the (positive) distance from the origin $q_4 = -$ on the $-y$ axis. If we take each angle θ in the figure to be positive, then we have $\tan\theta = d/R$ and $\cos\theta = R/r$ (where r is the dashed line distance shown in the figure). The problem asks us to consider θ to be a variable in the sense that, once the charges on the x axis are fixed in place (which determines R), d can then be arranged to some multiple of R, since $d = R \tan\theta$. The aim of this exploration is to show that if q is bounded then θ (and thus d) is also bounded.

From symmetry, we see that there is no net force in the vertical direction on $q_2 = -e$ sitting at a distance R to the left of the coordinate origin. We note that the net x force caused by q_3 and q_4 on the y axis will have a magnitude equal to

$$2\frac{q\,e}{4\pi\varepsilon_0 r^2}\cos(\theta) = \frac{2\,q\,e\cos(\theta)}{4\pi\varepsilon_0\,(R/\cos(\theta))^2} = \frac{2\,q\,e\cos^3(\theta)}{4\pi\varepsilon_0\,R^2}.$$

Consequently, to achieve a zero net force along the x axis, the above expression must equal the magnitude of the repulsive force exerted on q_2 by $q_1 = -e$. Thus,

$$\frac{2\,q\,e\cos^3(\theta)}{4\pi\varepsilon_0\,R^2} = \frac{e^2}{4\pi\varepsilon_0\,R^2} \quad \Rightarrow \quad q = \frac{e}{2\cos^3(\theta)}.$$

Below we plot q/e as a function of the angle (in degrees):

The graph suggests that $q/e < 5$ for $\theta < 60°$, roughly. We can be more precise by solving the above equation. The requirement that $q \le 5e$ leads to

$$\frac{e}{2\cos^3(\theta)} \le 5e \quad \Rightarrow \quad \frac{1}{(10)^{1/3}} \le \cos\theta$$

which yields $\theta \leq 62.34°$. The problem asks for "physically possible values," and it is reasonable to suppose that only positive-integer-multiple values of e are allowed for q. If we let $q = Ne$, for $N = 1 \dots 5$, then θ_N will be found by taking the inverse cosine of the cube root of $(1/2N)$.

(a) The smallest value of angle is $\theta_1 = 37.5°$ (or 0.654 rad).

(b) The second smallest value of angle is $\theta_2 = 50.95°$ (or 0.889 rad).

(c) The third smallest value of angle is $\theta_3 = 56.6°$ (or 0.988 rad).

29. (a) Every cesium ion at a corner of the cube exerts a force of the same magnitude on the chlorine ion at the cube center. Each force is a force of attraction and is directed toward the cesium ion that exerts it, along the body diagonal of the cube. We can pair every cesium ion with another, diametrically positioned at the opposite corner of the cube. Since the two ions in such a pair exert forces that have the same magnitude but are oppositely directed, the two forces sum to zero and, since every cesium ion can be paired in this way, the total force on the chlorine ion is zero.

(b) Rather than remove a cesium ion, we superpose charge $-e$ at the position of one cesium ion. This neutralizes the ion, and as far as the electrical force on the chlorine ion is concerned, it is equivalent to removing the ion. The forces of the eight cesium ions at the cube corners sum to zero, so the only force on the chlorine ion is the force of the added charge.

The length of a body diagonal of a cube is $\sqrt{3}a$, where a is the length of a cube edge. Thus, the distance from the center of the cube to a corner is $d = \left(\sqrt{3}/2\right)a$. The force has magnitude

$$F = k\frac{e^2}{d^2} = \frac{ke^2}{(3/4)a^2} = \frac{\left(8.99 \times 10^9 \text{ N} \cdot \text{m}^2/\text{C}^2\right)\left(1.60 \times 10^{-19}\text{C}\right)^2}{(3/4)\left(0.40 \times 10^{-9}\text{ m}\right)^2} = 1.9 \times 10^{-9}\text{N}.$$

Since both the added charge and the chlorine ion are negative, the force is one of repulsion. The chlorine ion is pushed away from the site of the missing cesium ion.

30. (a) Since the proton is positively charged, the emitted particle must be a positron (as opposed to the negatively charged electron) in accordance with the law of charge conservation.

(b) In this case, the initial state had zero charge (the neutron is neutral), so the sum of charges in the final state must be zero. Since there is a proton in the final state, there should also be an electron (as opposed to a positron) so that $\Sigma q = 0$.

31. None of the reactions given include a beta decay, so the number of protons, the number of neutrons, and the number of electrons are each conserved. Atomic numbers (numbers of protons and numbers of electrons) and molar masses (combined numbers of protons and neutrons) can be found in Appendix F of the text.

(a) ^{1}H has 1 proton, 1 electron, and 0 neutrons and ^{9}Be has 4 protons, 4 electrons, and 9 − 4 = 5 neutrons, so X has 1 + 4 = 5 protons, 1 + 4 = 5 electrons, and 0 + 5 − 1 = 4 neutrons. One of the neutrons is freed in the reaction. X must be boron with a molar mass of 5 + 4 = 9 g/mol: ^{9}B.

(b) ^{12}C has 6 protons, 6 electrons, and 12 − 6 = 6 neutrons and ^{1}H has 1 proton, 1 electron, and 0 neutrons, so X has 6 + 1 = 7 protons, 6 + 1 = 7 electrons, and 6 + 0 = 6 neutrons. It must be nitrogen with a molar mass of 7 + 6 = 13 g/mol: ^{13}N.

(c) ^{15}N has 7 protons, 7 electrons, and 15 − 7 = 8 neutrons; ^{1}H has 1 proton, 1 electron, and 0 neutrons; and ^{4}He has 2 protons, 2 electrons, and 4 − 2 = 2 neutrons; so X has 7 + 1 − 2 = 6 protons, 6 electrons, and 8 + 0 − 2 = 6 neutrons. It must be carbon with a molar mass of 6 + 6 = 12: ^{12}C.

32. We note that the problem is examining the force <u>on</u> charge A, so that the respective distances (involved in the Coulomb force expressions) between B and A, and between C and A, do not change as particle B is moved along its circular path. We focus on the endpoints ($\theta = 0°$ and $180°$) of each graph, since they represent cases where the forces (on A) due to B and C are either parallel or antiparallel (yielding maximum or minimum force magnitudes, respectively). We note, too, that since Coulomb's law is inversely proportional to r^2 then the (if, say, the charges were all the same) force due to C would be one-fourth as big as that due to B (since C is twice as far away from A). The charges, it turns out, are not the same, so there is also a factor of the charge ratio ξ (the charge of C divided by the charge of B), as well as the aforementioned ¼ factor. That is, the force exerted by C is, by Coulomb's law equal to ±¼ξ multiplied by the force exerted by B.

(a) The maximum force is $2F_0$ and occurs when $\theta = 180°$ (B is to the left of A, while C is the right of A). We choose the minus sign and write

$$2 F_0 = (1 - ¼\xi) F_0 \quad \Rightarrow \quad \xi = -4 .$$

One way to think of the minus sign choice is $\cos(180°) = -1$. This is certainly consistent with the minimum force ratio (zero) at $\theta = 0°$ since that would also imply

$$0 = 1 + ¼\xi \quad \Rightarrow \quad \xi = -4 .$$

(b) The ratio of maximum to minimum forces is 1.25/0.75 = 5/3 in this case, which implies

$$\frac{5}{3} = \frac{1 + ¼\xi}{1 - ¼\xi} \quad \Rightarrow \quad \xi = 16 .$$

Of course, this could also be figured as illustrated in part (a), looking at the maximum force ratio by itself and solving, or looking at the minimum force ratio ($\frac{3}{4}$) at $\theta = 180°$ and solving for ξ.

33. We note that, as result of the fact that the Coulomb force is inversely proportional to r^2, a particle of charge Q which is distance d from the origin will exert a force on some charge q_0 at the origin of equal strength as a particle of charge $4Q$ at distance $2d$ would exert on q_0. Therefore, $q_6 = +8e$ on the $-y$ axis could be replaced with a $+2e$ closer to the origin (at half the distance); this would add to the $q_5 = +2e$ already there and produce $+4e$ below the origin which exactly cancels the force due to $q_2 = +4e$ above the origin.

Similarly, $q_4 = +4e$ to the far right could be replaced by a $+e$ at half the distance, which would add to $q_3 = +e$ already there to produce a $+2e$ at distance d to the right of the central charge q_7. The horizontal force due to this $+2e$ is cancelled exactly by that of $q_1 = +2e$ on the $-x$ axis, so that the net force on q_7 is zero.

34. For the Coulomb force to be sufficient for circular motion at that distance (where $r = 0.200$ m and the acceleration needed for circular motion is $a = v^2/r$) the following equality is required:

$$\frac{Qq}{4\pi\varepsilon_0 r^2} = -\frac{mv^2}{r} .$$

With $q = 4.00 \times 10^{-6}$ C, $m = 0.000800$ kg, $v = 50.0$ m/s, this leads to $Q = -1.11 \times 10^{-5}$ C.

35. Let $\vec{F}_{12}$ denotes the force on q_1 exerted by q_2 and F_{12} be its magnitude.

(a) We consider the net force on q_1. $\vec{F}_{12}$ points in the $+x$ direction since q_1 is attracted to q_2. $\vec{F}_{13}$ and $\vec{F}_{14}$ both point in the $-x$ direction since q_1 is repelled by q_3 and q_4. Thus, using $d = 0.0200$ m, the net force is

$$F_1 = F_{12} - F_{13} - F_{14} = \frac{2e|-e|}{4\pi\varepsilon_0 d^2} - \frac{(2e)(e)}{4\pi\varepsilon_0 (2d)^2} - \frac{(2e)(4e)}{4\pi\varepsilon_0(3d)^2} = +3.52 \times 10^{-25}\,\text{N} ,$$

or $\vec{F}_1 = (3.52 \times 10^{-25}\,\text{N})\hat{\text{i}}$.

(b) We now consider the net force on q_2. We note that $\vec{F}_{21} = -\vec{F}_{12}$ points in the $-x$ direction, and $\vec{F}_{23}$ and $\vec{F}_{24}$ both point in the $+x$ direction. The net force is

$$F_{23} + F_{24} - F_{21} = \frac{4e|-e|}{4\pi\varepsilon_0 (2d)^2} + \frac{e|-e|}{4\pi\varepsilon_0 d^2} - \frac{2e|-e|}{4\pi\varepsilon_0 d^2} = 0$$

36. As a result of the first action, both sphere W and sphere A possess charge $\frac{1}{2}q_A$, where q_A is the initial charge of sphere A. As a result of the second action, sphere W has charge

$$\frac{1}{2}\left(\frac{1}{2}q_A - 32e\right).$$

As a result of the final action, sphere W now has charge equal to

$$\frac{1}{2}\left[\frac{1}{2}\left(\frac{1}{2}q_A - 32e\right) + 48e\right].$$

Setting this final expression equal to $+18e$ as required by the problem leads (after a couple of algebra steps) to the answer: $q_A = +16e$.

37. If θ is the angle between the force and the x axis, then

$$\cos\theta = \frac{d_2}{\sqrt{d_1^2 + d_2^2}}.$$

Thus, using Coulomb's law for F, we have

$$F_x = F\cos\theta = \frac{q_1\,q_2}{4\pi\varepsilon_0\,(d_1^2 + d_2^2)}\frac{d_2}{\sqrt{d_1^2 + d_2^2}} = 1.31 \times 10^{-22}\,\text{N}.$$

38. (a) We note that $\tan(30°) = 1/\sqrt{3}$. In the initial (highly symmetrical) configuration, the net force on the central bead is in the $-y$ direction and has magnitude $3F$ where F is the Coulomb's law force of one bead on another at distance $d = 10$ cm. This is due to the fact that the forces exerted on the central bead (in the initial situation) by the beads on the x axis cancel each other; also, the force exerted "downward" by bead 4 on the central bead is four times larger than the "upward" force exerted by bead 2. This net force along the y axis does not change as bead 1 is now moved, though there is now a nonzero x-component F_x. The components are now related by

$$\tan(30°) = \frac{F_x}{F_y} \quad\Rightarrow\quad \frac{1}{\sqrt{3}} = \frac{F_x}{3F}$$

which implies $F_x = \sqrt{3}\,F$. Now, bead 3 exerts a "leftward" force of magnitude F on the central bead, while bead 1 exerts a "rightward" force of magnitude F'. Therefore,

$$F' - F = \sqrt{3}\,F. \quad\Rightarrow\quad F' = (\sqrt{3} + 1)\,F.$$

The fact that Coulomb's law depends inversely on distance-squared then implies

$$r^2 = \frac{d^2}{\sqrt{3} + 1} \quad\Rightarrow\quad r = \frac{d}{\sqrt{\sqrt{3} + 1}}$$

where r is the distance between bead 1 and the central bead. Thus $r = 6.05$ cm.

(b) To regain the condition of high symmetry (in particular, the cancellation of x-components) bead 3 must be moved closer to the central bead so that it, too, is the distance r (as calculated in part(a)) away from it.

39. (a) Charge $Q_1 = +80 \times 10^{-9}$ C is on the y axis at $y = 0.003$ m, and charge $Q_2 = +80 \times 10^{-9}$ C is on the y axis at $y = -0.003$ m. The force on particle 3 (which has a charge of $q = +18 \times 10^{-9}$ C) is due to the vector sum of the repulsive forces from Q_1 and Q_2. In symbols, $\vec{F}_{31} + \vec{F}_{32} = \vec{F}_3$, where

$$|\vec{F}_{31}| = k\frac{q_3|q_1|}{r_{31}^2} \quad \text{and} \quad |\vec{F}_{32}| = k\frac{q_3 q_2}{r_{32}^2}.$$

Using the Pythagorean theorem, we have $r_{31} = r_{32} = 0.005$ m. In magnitude-angle notation (particularly convenient if one uses a vector-capable calculator in polar mode), the indicated vector addition becomes

$$\vec{F}_3 = (0.518\angle-37°) + (0.518\angle 37°) = (0.829\angle 0°).$$

Therefore, the net force is $\vec{F}_3 = (0.829 \text{ N})\hat{i}$.

(b) Switching the sign of Q_2 amounts to reversing the direction of its force on q. Consequently, we have

$$\vec{F}_3 = (0.518\angle-37°) + (0.518\angle-143°) = (0.621\angle-90°).$$

Therefore, the net force is $\vec{F}_3 = -(0.621 \text{ N})\hat{j}$.

40. (a) Let x be the distance between particle 1 and particle 3. Thus, the distance between particle 3 and particle 2 is $L - x$. Both particles exert leftward forces on q_3 (so long as it is on the line <u>between</u> them), so the magnitude of the net force on q_3 is

$$F_{net} = |\vec{F}_{13}| + |\vec{F}_{23}| = \frac{|q_1 q_3|}{4\pi\varepsilon_0 x^2} + \frac{|q_2 q_3|}{4\pi\varepsilon_0 (L-x)^2} = \frac{e^2}{\pi\varepsilon_0}\left(\frac{1}{x^2} + \frac{27}{(L-x)^2}\right)$$

with the values of the charges (stated in the problem) plugged in. Finding the value of x which minimizes this expression leads to $x = ¼ L$. Thus, $x = 2.00$ cm.

(b) Substituting $x = ¼ L$ back into the expression for the net force magnitude and using the standard value for e leads to $F_{net} = 9.21 \times 10^{-24}$ N.

41. The individual force magnitudes are found using Eq. 21-1, with SI units (so $a = 0.02$ m) and k as in Eq. 21-5. We use magnitude-angle notation (convenient if one uses a vector-capable calculator in polar mode), listing the forces due to $+4.00q$, $+2.00q$, and $-2.00q$ charges:

$$\left(4.60 \times 10^{-24} \angle 180°\right) + \left(2.30 \times 10^{-24} \angle -90°\right) + \left(1.02 \times 10^{-24} \angle -145°\right) = \left(6.16 \times 10^{-24} \angle -152°\right)$$

(a) Therefore, the net force has magnitude 6.16×10^{-24} N.

(b) The direction of the net force is at an angle of $-152°$ (or $208°$ measured counterclockwise from the $+x$ axis).

42. The charge dq within a thin section of the rod (of thickness dx) is $\rho A \, dx$ where $A = 4.00 \times 10^{-4}$ m^2 and ρ is the charge per unit volume. The number of (excess) electrons in the rod (of length $L = 2.00$ m) is $N = q/(-e)$ where e is given in Eq. 21-14.

(a) In the case where $\rho = -4.00 \times 10^{-6}$ C/m^3, we have

$$N = \frac{q}{-e} = \frac{\rho A}{-e} \int_0^L dx = \frac{|\rho| A L}{e} = 2.00 \times 10^{10}.$$

(b) With $\rho = bx^2$ ($b = -2.00 \times 10^{-6}$ C/m^5) we obtain

$$N = \frac{b A}{-e} \int_0^L x^2 \, dx = \frac{|b| A L^3}{3e} = 1.33 \times 10^{10}.$$

43. The magnitude of the net force on the $q = 42 \times 10^{-6}$ C charge is

$$k \frac{q_1 q}{0.28^2} + k \frac{|q_2| q}{0.44^2}$$

where $q_1 = 30 \times 10^{-9}$ C and $|q_2| = 40 \times 10^{-9}$ C. This yields 0.22 N. Using Newton's second law, we obtain

$$m = \frac{F}{a} = \frac{0.22 \, \text{N}}{100 \times 10^3 \, \text{m/s}^2} = 2.2 \times 10^{-6} \, \text{kg}.$$

44. Let q_1 be the charge of one part and q_2 that of the other part; thus, $q_1 + q_2 = Q = 6.0$ μC. The repulsive force between them is given by Coulomb's law:

$$F = \frac{q_1 q_2}{4\pi\varepsilon_0 r^2} = \frac{q_1 (Q - q_1)}{4\pi\varepsilon_0 r^2} .$$

If we maximize this expression by taking the derivative with respect to q_1 and setting equal to zero, we find $q_1 = Q/2$, which might have been anticipated (based on symmetry arguments). This implies $q_2 = Q/2$ also. With $r = 0.0030$ m and $Q = 6.0 \times 10^{-6}$ C, we find

$$F = \frac{(Q/2)(Q/2)}{4\pi\varepsilon_0 r^2} \approx 9.0 \times 10^3 \text{ N}.$$

45. For the net force on $q_1 = +Q$ to vanish, the x force component due to $q_2 = q$ must exactly cancel the force of attraction caused by $q_4 = -2Q$. Consequently,

$$\frac{Qq}{4\pi\varepsilon_0 a^2} = \frac{Q|2Q|}{4\pi\varepsilon_0(\sqrt{2}a)^2}\cos 45° = \frac{Q^2}{4\pi\varepsilon_0\sqrt{2}a^2}$$

or $q = Q/\sqrt{2}$. This implies that $q/Q = 1/\sqrt{2} = 0.707$.

46. We are looking for a charge q which, when placed at the origin, experiences $\vec{F}_{net} = 0$, where

$$\vec{F}_{net} = \vec{F}_1 + \vec{F}_2 + \vec{F}_3.$$

The magnitude of these individual forces are given by Coulomb's law, Eq. 21-1, and without loss of generality we assume $q > 0$. The charges q_1 (+6 μC), q_2 (-4 μC), and q_3 (unknown), are located on the +x axis, so that we know $\vec{F}_1$ points towards –x, $\vec{F}_2$ points towards +x, and $\vec{F}_3$ points towards –x if $q_3 > 0$ and points towards +x if $q_3 < 0$. Therefore, with $r_1 = 8$ m, $r_2 = 16$ m and $r_3 = 24$ m, we have

$$0 = -k\frac{q_1 q}{r_1^2} + k\frac{|q_2|q}{r_2^2} - k\frac{q_3 q}{r_3^2}.$$

Simplifying, this becomes

$$0 = -\frac{6}{8^2} + \frac{4}{16^2} - \frac{q_3}{24^2}$$

where q_3 is now understood to be in μC. Thus, we obtain $q_3 = -45$ μC.

47. There are two protons (each with charge $q = +e$) in each molecule, so

$$Q = N_A q = (6.02 \times 10^{23})(2)(1.60 \times 10^{-19} \text{ C}) = 1.9 \times 10^5 \text{ C} = 0.19 \text{ MC}.$$

48. (a) Since the rod is in equilibrium, the net force acting on it is zero, and the net torque about any point is also zero. We write an expression for the net torque about the bearing, equate it to zero, and solve for x. The charge Q on the left exerts an upward force of

magnitude $(1/4\pi\varepsilon_0)$ (qQ/h^2), at a distance $L/2$ from the bearing. We take the torque to be negative. The attached weight exerts a downward force of magnitude W, at a distance $x - L/2$ from the bearing. This torque is also negative. The charge Q on the right exerts an upward force of magnitude $(1/4\pi\varepsilon_0)$ $(2qQ/h^2)$, at a distance $L/2$ from the bearing. This torque is positive. The equation for rotational equilibrium is

$$\frac{-1}{4\pi\varepsilon_0}\frac{qQ}{h^2}\frac{L}{2} - W\left(x - \frac{L}{2}\right) + \frac{1}{4\pi\varepsilon_0}\frac{2qQ}{h^2}\frac{L}{2} = 0.$$

The solution for x is

$$x = \frac{L}{2}\left(1 + \frac{1}{4\pi\varepsilon_0}\frac{qQ}{h^2 W}\right).$$

(b) If F_N is the magnitude of the upward force exerted by the bearing, then Newton's second law (with zero acceleration) gives

$$W - \frac{1}{4\pi\varepsilon_0}\frac{qQ}{h^2} - \frac{1}{4\pi\varepsilon_0}\frac{2qQ}{h^2} - F_N = 0.$$

We solve for h so that $F_N = 0$. The result is

$$h = \sqrt{\frac{1}{4\pi\varepsilon_0}\frac{3qQ}{W}}.$$

49. Charge $q_1 = -80 \times 10^{-6}$ C is at the origin, and charge $q_2 = +40 \times 10^{-6}$ C is at $x = 0.20$ m. The force on $q_3 = +20 \times 10^{-6}$ C is due to the attractive and repulsive forces from q_1 and q_2, respectively. In symbols, $\vec{F}_{3\,net} = \vec{F}_{31} + \vec{F}_{32}$, where

$$\left|\vec{F}_{31}\right| = k\frac{q_3|q_1|}{r_{31}^2} \quad \text{and} \quad |\vec{F}_{32}| = k\frac{q_3 q_2}{r_{32}^2}.$$

(a) In this case $r_{31} = 0.40$ m and $r_{32} = 0.20$ m, with $\vec{F}_{31}$ directed towards $-x$ and $\vec{F}_{32}$ directed in the $+x$ direction. Using the value of k in Eq. 21-5, we obtain $\vec{F}_{3\,net} = (89.9 \text{ N})\hat{i}$.

(b) In this case $r_{31} = 0.80$ m and $r_{32} = 0.60$ m, with $\vec{F}_{31}$ directed towards $-x$ and $\vec{F}_{32}$ towards $+x$. Now we obtain $\vec{F}_{3\,net} = (-2.50 \text{ N})\hat{i}$.

(c) Between the locations treated in parts (a) and (b), there must be one where $\vec{F}_{3\,net} = 0$. Writing $r_{31} = x$ and $r_{32} = x - 0.20$ m, we equate $\left|\vec{F}_{31}\right|$ and $\left|\vec{F}_{32}\right|$, and after canceling common factors, arrive at

$$\frac{|q_1|}{x^2} = \frac{q_2}{(x-0.2)^2}.$$

This can be further simplified to

$$\frac{(x-0.2)^2}{x^2} = \frac{q_2}{|q_1|} = \frac{1}{2}.$$

Taking the (positive) square root and solving, we obtain $x = 0.683$ m. If one takes the negative root and 'solves', one finds the location where the net force *would* be zero *if* q_1 and q_2 were of like sign (which is not the case here).

(d) From the above, we see that $y = 0$.

50. We are concerned with the charges in the nucleus (not the "orbiting" electrons, if there are any). The nucleus of Helium has 2 protons and that of Thorium has 90.

(a) Eq. 21-1 gives

$$F = k\frac{q^2}{r^2} = \frac{\left(8.99\times10^9 \text{ N}\cdot\text{m}^2/\text{C}^2\right)(2(1.60\times10^{-19}\text{ C}))(90(1.60\times10^{-19}\text{C}))}{(9.0\times10^{-15}\text{ m})^2} = 5.1\times10^2\,\text{N}.$$

(b) Estimating the helium nucleus mass as that of 4 protons (actually, that of 2 protons and 2 neutrons, but the neutrons have approximately the same mass), Newton's second law leads to

$$a = \frac{F}{m} = \frac{5.1\times10^2\text{ N}}{4\left(1.67\times10^{-27}\text{ kg}\right)} = 7.7\times10^{28}\,\text{m/s}^2.$$

51. Coulomb's law gives

$$F = \frac{|q|\cdot|q|}{4\pi\varepsilon_0 r^2} = \frac{k(e/3)^2}{r^2} = \frac{\left(8.99\times10^9 \text{ N}\cdot\text{m}^2/\text{C}^2\right)(1.60\times10^{-19}\text{ C})^2}{9(2.6\times10^{-15}\text{m})^2} = 3.8\,\text{N}.$$

52. (a) Since $q_A = -2.00$ nC and $q_C = +8.00$ nC Eq. 21-4 leads to

$$|\vec{F}_{AC}| = \frac{|q_A q_C|}{4\pi\varepsilon_0 d^2} = \frac{|(8.99 \times 10^9 \text{ N} \cdot \text{m}^2/\text{C}^2)(-2.00 \times 10^{-9} \text{ C})(8.00 \times 10^{-9} \text{ C})|}{(0.200 \text{ m})^2} = 3.60 \times 10^{-6} \text{ N}.$$

(b) After making contact with each other, both A and B have a charge of

$$\frac{q_A + q_B}{2} = \left(\frac{-2.00 + (-4.00)}{2} \right) \text{ nC} = -3.00 \text{ nC}.$$

When B is grounded its charge is zero. After making contact with C, which has a charge of +8.00 nC, B acquires a charge of $[0 + (-8.00 \text{ nC})]/2 = -4.00$ nC, which charge C has as well. Finally, we have $Q_A = -3.00$ nC and $Q_B = Q_C = -4.00$ nC. Therefore,

$$|\vec{F}_{AC}| = \frac{|q_A q_C|}{4\pi\varepsilon_0 d^2} = \frac{|(8.99 \times 10^9 \text{ N} \cdot \text{m}^2/\text{C}^2)(-3.00 \times 10^{-9} \text{ C})(-4.00 \times 10^{-9} \text{ C})|}{(0.200 \text{ m})^2} = 2.70 \times 10^{-6} \text{ N}.$$

(c) We also obtain

$$|\vec{F}_{BC}| = \frac{|q_B q_C|}{4\pi\varepsilon_0 d^2} = \frac{|(8.99 \times 10^9 \text{ N} \cdot \text{m}^2/\text{C}^2)(-4.00 \times 10^{-9} \text{ C})(-4.00 \times 10^{-9} \text{ C})|}{(0.200 \text{ m})^2} = 3.60 \times 10^{-6} \text{ N}.$$

53. Let the two charges be q_1 and q_2. Then $q_1 + q_2 = Q = 5.0 \times 10^{-5}$ C. We use Eq. 21-1:

$$1.0\text{N} = \frac{\left(8.99 \times 10^9 \text{ N} \cdot \text{m}^2/\text{C}^2 \right) q_1 q_2}{\left(2.0\text{m} \right)^2}.$$

We substitute $q_2 = Q - q_1$ and solve for q_1 using the quadratic formula. The two roots obtained are the values of q_1 and q_2, since it does not matter which is which. We get 1.2×10^{-5} C and 3.8×10^{-5} C. Thus, the charge on the sphere with the smaller charge is 1.2×10^{-5} C.

54. The unit Ampere is discussed in §21-4. Using i for current, the charge transferred is

$$q = it = \left(2.5 \times 10^4 \text{ A} \right) \left(20 \times 10^{-6} \text{s} \right) = 0.50 \text{ C}.$$

55. (a) Using Coulomb's law, we obtain

$$F = \frac{q_1 q_2}{4\pi\varepsilon_0 r^2} = \frac{kq^2}{r^2} = \frac{\left(8.99 \times 10^9 \text{ N} \cdot \text{m}^2/\text{C}^2 \right) \left(1.00\text{C} \right)^2}{\left(1.00\text{m} \right)^2} = 8.99 \times 10^9 \text{ N}.$$

(b) If $r = 1000$ m, then

$$F = \frac{q_1 q_2}{4\pi\varepsilon_0 r^2} = \frac{kq^2}{r^2} = \frac{\left(8.99\times10^9\ \text{N}\cdot\text{m}^2/\text{C}^2\right)\left(1.00\text{C}\right)^2}{\left(1.00\times10^3\text{m}\right)^2} = 8.99\times10^3\ \text{N}.$$

56. Keeping in mind that an Ampere is a Coulomb per second, and that a minute is 60 seconds, the charge (in absolute value) that passes through the chest is

$$|q| = (\,0.300\ \tfrac{\text{Coulomb}}{\text{second}}\,)(\,120\ \text{seconds}\,) = 36.0\ \text{Coulombs}.$$

This charge consists of a number N of electrons (each of which has an absolute value of charge equal to e). Thus,

$$N = \frac{|q|}{e} = \frac{36.0\ \text{C}}{1.60 \times 10^{-19}\text{C}} = 2.25 \times 10^{20}.$$

57. When sphere C touches sphere A, they divide up their total charge ($Q/2$ plus Q) equally between them. Thus, sphere A now has charge $3Q/4$, and the magnitude of the force of attraction between A and B becomes

$$F = k\frac{(3Q/4)(Q/4)}{d^2} = 4.68\times10^{-19}\ \text{N}.$$

58. In experiment 1, sphere C first touches sphere A, and they divided up their total charge ($Q/2$ plus Q) equally between them. Thus, sphere A and sphere C each acquired charge $3Q/4$. Then, sphere C touches B and those spheres split up their total charge ($3Q/4$ plus $-Q/4$) so that B ends up with charge equal to $Q/4$. The force of repulsion between A and B is therefore

$$F_1 = k\frac{(3Q/4)(Q/4)}{d^2}$$

at the end of experiment 1. Now, in experiment 2, sphere C first touches B which leaves each of them with charge $Q/8$. When C next touches A, sphere A is left with charge $9Q/16$. Consequently, the force of repulsion between A and B is

$$F_2 = k\frac{(9Q/16)(Q/8)}{d^2}$$

at the end of experiment 2. The ratio is

$$\frac{F_2}{F_1} = \frac{(9/16)(1/8)}{(3/4)(1/4)} = 0.375.$$

59. If the relative difference between the proton and electron charges (in absolute value) were

$$\frac{q_p - |q_e|}{e} = 0.0000010$$

then the actual difference would be $q_p - |q_e| = 1.6 \times 10^{-25}$ C. Amplified by a factor of $29 \times 3 \times 10^{22}$ as indicated in the problem, this amounts to a deviation from perfect neutrality of

$$\Delta q = (29 \times 3 \times 10^{22})(1.6 \times 10^{-25} \text{ C}) = 0.14 \text{ C}$$

in a copper penny. Two such pennies, at $r = 1.0$ m, would therefore experience a very large force. Eq. 21-1 gives

$$F = k \frac{(\Delta q)^2}{r^2} = 1.7 \times 10^8 \text{ N}.$$

60. With $F = m_e g$, Eq. 21-1 leads to

$$y^2 = \frac{ke^2}{m_e g} = \frac{(8.99 \times 10^9 \text{ N} \cdot \text{m}^2/\text{C}^2)(1.60 \times 10^{-19} \text{ C})^2}{(9.11 \times 10^{-31} \text{ kg})(9.8 \text{ m/s}^2)}$$

which leads to $y = \pm 5.1$ m. We choose $y = -5.1$ m since the second electron must be below the first one, so that the repulsive force (acting on the first) is in the direction opposite to the pull of Earth's gravity.

61. Letting $kq^2/r^2 = mg$, we get

$$r = q \sqrt{\frac{k}{mg}} = (1.60 \times 10^{-19} \text{ C}) \sqrt{\frac{8.99 \times 10^9 \text{ N} \cdot \text{m}^2/\text{C}^2}{(1.67 \times 10^{-27} \text{ kg})(9.8 \text{ m/s}^2)}} = 0.119 \text{m}.$$

62. The net charge carried by John whose mass is m is roughly

$$q = (0.0001) \frac{m N_A Z e}{M}$$

$$= (0.0001) \frac{(90 \text{kg})(6.02 \times 10^{23} \text{ molecules/mol})(18 \text{ electron proton pairs/molecule})(1.6 \times 10^{-19} \text{C})}{0.018 \text{ kg/mol}}$$

$$= 8.7 \times 10^5 \text{C},$$

and the net charge carried by Mary is half of that. So the electrostatic force between them is estimated to be

$$F \approx k\frac{q(q/2)}{d^2} = (8.99\times10^9\,\text{N}\cdot\text{m}^2/\text{C}^2)\frac{(8.7\times10^5\,\text{C})^2}{2(30\text{m})^2} \approx 4\times10^{18}\,\text{N}.$$

Thus, the order of magnitude of the electrostatic force is 10^{18} N .

63. (a) Eq. 21-11 (in absolute value) gives

$$n = \frac{|q|}{e} = \frac{2.00\times10^{-6}\,\text{C}}{1.60\times10^{-19}\,\text{C}} = 1.25\times10^{13}\ \text{electrons}.$$

(b) Since you have the excess electrons (and electrons are lighter and more mobile than protons) then the electrons "leap" from you to the faucet instead of protons moving from the faucet to you (in the process of neutralizing your body).

(c) Unlike charges attract, and the faucet (which is grounded and is able to gain or lose any number of electrons due to its contact with Earth's large reservoir of mobile charges) becomes positively charged, especially in the region closest to your (negatively charged) hand, just before the spark.

(d) The cat is positively charged (before the spark), and by the reasoning given in part (b) the flow of charge (electrons) is from the faucet to the cat.

(e) If we think of the nose as a conducting sphere, then the side of the sphere closest to the fur is of one sign (of charge) and the side furthest from the fur is of the opposite sign (which, additionally, is oppositely charged from your bare hand which had stroked the cat's fur). The charges in your hand and those of the furthest side of the "sphere" therefore attract each other, and when close enough, manage to neutralize (due to the "jump" made by the electrons) in a painful spark.

64. The two charges are $q = \alpha Q$ (where α is a pure number presumably less than 1 and greater than zero) and $Q - q = (1 - \alpha)Q$. Thus, Eq. 21-4 gives

$$F = \frac{1}{4\pi\varepsilon_0}\frac{(\alpha Q)((1-\alpha)Q)}{d^2} = \frac{Q^2\alpha(1-\alpha)}{4\pi\varepsilon_0 d^2}.$$

The graph below, of F versus α, has been scaled so that the maximum is 1. In actuality, the maximum value of the force is $F_{max} = Q^2/16\pi\varepsilon_0 d^2$.

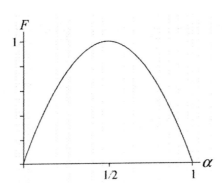

(a) It is clear that $\alpha = \dfrac{1}{2} = 0.5$ gives the maximum value of F.

(b) Seeking the half-height points on the graph is difficult without grid lines or some of the special tracing features found in a variety of modern calculators. It is not difficult to algebraically solve for the half-height points (this involves the use of the quadratic formula). The results are

$$\alpha_1 = \frac{1}{2}\left(1 - \frac{1}{\sqrt{2}}\right) \approx 0.15 \quad \text{and} \quad \alpha_2 = \frac{1}{2}\left(1 + \frac{1}{\sqrt{2}}\right) \approx 0.85.$$

Thus, the smaller value of α is $\alpha_1 = 0.15$,

(c) and the larger value of α is $\alpha_2 = 0.85$.

65. (a) The magnitudes of the gravitational and electrical forces must be the same:

$$\frac{1}{4\pi\varepsilon_0}\frac{q^2}{r^2} = G\frac{mM}{r^2}$$

where q is the charge on either body, r is the center-to-center separation of Earth and Moon, G is the universal gravitational constant, M is the mass of Earth, and m is the mass of the Moon. We solve for q:

$$q = \sqrt{4\pi\varepsilon_0 GmM}.$$

According to Appendix C of the text, $M = 5.98 \times 10^{24}$ kg, and $m = 7.36 \times 10^{22}$ kg, so (using $4\pi\varepsilon_0 = 1/k$) the charge is

$$q = \sqrt{\frac{\left(6.67\times10^{-11}\,\text{N}\cdot\text{m}^2/\text{kg}^2\right)\left(7.36\times10^{22}\,\text{kg}\right)\left(5.98\times10^{24}\,\text{kg}\right)}{8.99\times10^{9}\,\text{N}\cdot\text{m}^2/\text{C}^2}} = 5.7\times10^{13}\,\text{C}.$$

(b) The distance r cancels because both the electric and gravitational forces are proportional to $1/r^2$.

(c) The charge on a hydrogen ion is $e = 1.60 \times 10^{-19}$ C, so there must be

$$\frac{q}{e} = \frac{5.7 \times 10^{13} \text{ C}}{1.6 \times 10^{-19} \text{ C}} = 3.6 \times 10^{32} \text{ ions.}$$

Each ion has a mass of 1.67×10^{-27} kg, so the total mass needed is

$$\left(3.6 \times 10^{32}\right)\left(1.67 \times 10^{-27} \text{ kg}\right) = 6.0 \times 10^5 \text{ kg.}$$

66. (a) A force diagram for one of the balls is shown below. The force of gravity $m\vec{g}$ acts downward, the electrical force $\vec{F}_e$ of the other ball acts to the left, and the tension in the thread acts along the thread, at the angle θ to the vertical. The ball is in equilibrium, so its acceleration is zero. The y component of Newton's second law yields $T \cos\theta - mg = 0$ and the x component yields $T \sin\theta - F_e = 0$. We solve the first equation for T and obtain $T = mg/\cos\theta$. We substitute the result into the second to obtain $mg \tan\theta - F_e = 0$.

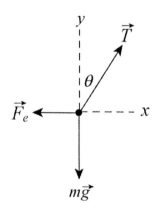

Examination of the geometry of Figure 21-43 leads to

$$\tan\theta = \frac{x/2}{\sqrt{L^2 - (x/2)^2}}.$$

If L is much larger than x (which is the case if θ is very small), we may neglect $x/2$ in the denominator and write $\tan\theta \approx x/2L$. This is equivalent to approximating $\tan\theta$ by $\sin\theta$. The magnitude of the electrical force of one ball on the other is

$$F_e = \frac{q^2}{4\pi\varepsilon_0 x^2}$$

by Eq. 21-4. When these two expressions are used in the equation $mg \tan\theta = F_e$, we obtain

$$\frac{mgx}{2L} \approx \frac{1}{4\pi\varepsilon_0}\frac{q^2}{x^2} \Rightarrow x \approx \left(\frac{q^2 L}{2\pi\varepsilon_0 mg}\right)^{1/3}.$$

(b) We solve $x^3 = 2kq^2 L/mg$) for the charge (using Eq. 21-5):

$$q = \sqrt{\frac{mgx^3}{2kL}} = \sqrt{\frac{(0.010\,\text{kg})(9.8\,\text{m/s}^2)(0.050\,\text{m})^3}{2(8.99\times10^9\,\text{N}\cdot\text{m}^2/\text{C}^2)(1.20\,\text{m})}} = \pm2.4\times10^{-8}\,\text{C}.$$

Thus, the magnitude is $|q| = 2.4\times10^{-8}\,\text{C}$.

67. (a) If one of them is discharged, there would no electrostatic repulsion between the two balls and they would both come to the position $\theta = 0$, making contact with each other.

(b) A redistribution of the remaining charge would then occur, with each of the balls getting $q/2$. Then they would again be separated due to electrostatic repulsion, which results in the new equilibrium separation

$$x' = \left[\frac{(q/2)^2 L}{2\pi\varepsilon_0 mg}\right]^{1/3} = \left(\frac{1}{4}\right)^{1/3} x = \left(\frac{1}{4}\right)^{1/3}(5.0\,\text{cm}) = 3.1\,\text{cm}.$$

68. Regarding the forces on q_3 exerted by q_1 and q_2, one must "push" and the other must "pull" in order that the net force is zero; hence, q_1 and q_2 have opposite signs. For individual forces to cancel, their magnitudes must be equal:

$$k\frac{|q_1||q_3|}{(L_{12}+L_{23})^2} = k\frac{|q_2||q_3|}{(L_{23})^2}.$$

With $L_{23} = 2.00 L_{12}$, the above expression simplifies to $\dfrac{|q_1|}{9} = \dfrac{|q_2|}{4}$. Therefore, $q_1 = -9q_2/4$, or $q_1/q_2 = -2.25$.

69. (a) The charge q placed at the origin is a distance r from Q (which is the positive charge on which the forces are being evaluated), and the charge q placed at $x = d$ is a distance r' from Q. Depending on what region Q is located in, the relation between r, r' and d will be either

$$r' = r + d \qquad \text{if } Q \text{ is along the } -x \text{ axis (region } \mathbf{A})$$
$$r' = d - r \qquad \text{if } Q \text{ is between the charges (region } \mathbf{B})$$

$$r' = r - d \qquad \text{if } Q \text{ is at } x > d \text{ (region C).}$$

Since all charges in this problem are taken to be positive, then the net force in region **A** will in the $-x$ direction; its magnitude will consist of the individual force magnitudes *added* together. In region **C** the net force will be in the $+x$ direction and will consist again of the individual force magnitudes *added* together. It is in region **B** where the individual force magnitudes must be *subtracted*, and in order for the result to exhibit the correct sign (positive when the net force $\vec{F}$ should point in the $+x$ direction, and so forth), we must write

$$\vec{F}_{\text{B}} = \frac{qQ}{4\pi\varepsilon_0 r^2} - \frac{qQ}{4\pi\varepsilon_0 r'^2} = \frac{qQ}{4\pi\varepsilon_0 r^2} - \frac{qQ}{4\pi\varepsilon_0 (d-r)^2} .$$

If we further adopt the notation suggested in the problem, then $r = \alpha d$ in regions **B** and **C**, and $r = -\alpha d$ in region **A**.(since r must by definition be a positive number, yet α is negative-valued in region **A**). Using this notation, too, it is clear that we can factor out a common $qQ/4\pi\varepsilon_0 d^2$ from our expressions. For brevity we will use the notation

$$J = \frac{qQ}{4\pi\varepsilon_0 d^2} .$$

Then, using the observations noted above, we are able to write down the expressions for the force in each region:

$$\vec{F}_{\text{A}} = -J\left(\frac{1}{\alpha^2} + \frac{1}{(1-\alpha)^2}\right)$$
$$\vec{F}_{\text{B}} = J\left(\frac{1}{\alpha^2} - \frac{1}{(1-\alpha)^2}\right)$$
$$\vec{F}_{\text{C}} = J\left(\frac{1}{\alpha^2} + \frac{1}{(\alpha-1)^2}\right)$$

(b) We set $J = 1$ in our plot of the force, below.

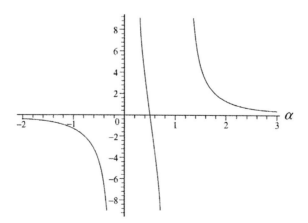

70. The mass of an electron is $m = 9.11 \times 10^{-31}$ kg, so the number of electrons in a collection with total mass $M = 75.0$ kg is

$$N = \frac{M}{m} = \frac{75.0\,\text{kg}}{9.11 \times 10^{-31}\,\text{kg}} = 8.23 \times 10^{31}\,\text{electrons.}$$

The total charge of the collection is

$$q = -\,Ne = -\left(8.23 \times 10^{31}\right)\left(1.60 \times 10^{-19}\,\text{C}\right) = -1.32 \times 10^{13}\,\text{C.}$$

71. (a) If a (negative) charged particle is placed a distance x to the right of the $+2q$ particle, then its attraction to the $+2q$ particle will be exactly balanced by its repulsion from the $-5q$ particle is we require

$$\frac{5}{(L+x)^2} = \frac{2}{x^2}$$

which is obtained by equating the Coulomb force magnitudes and then canceling common factors. Cross-multiplying and taking the square root, we obtain

$$\frac{x}{L+x} = \sqrt{\frac{2}{5}}$$

which can be rearranged to produce

$$x = \frac{L}{\sqrt{\dfrac{2}{5}} - 1} \approx 1.72\,L$$

(b) The y coordinate of particle 3 is $y = 0$.

Chapter 22

1. We note that the symbol q_2 is used in the problem statement to mean the absolute value of the negative charge which resides on the larger shell. The following sketch is for $q_1 = q_2$.

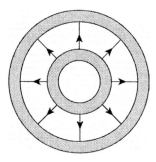

The following two sketches are for the cases $q_1 > q_2$ (left figure) and $q_1 < q_2$ (right figure).

2. (a) We note that the electric field points leftward at both points. Using $\vec{F} = q_0 \vec{E}$, and orienting our x axis rightward (so $\hat{i}$ points right in the figure), we find

$$\vec{F} = \left(+1.6 \times 10^{-19} \, \text{C} \right)\left(-40\frac{\text{N}}{\text{C}}\, \hat{i} \right) = -6.4 \times 10^{-18}\,\text{N}\,\hat{i}$$

which means the magnitude of the force on the proton is 6.4×10^{-18} N and its direction $(-\hat{i})$ is leftward.

(b) As the discussion in §22-2 makes clear, the field strength is proportional to the "crowdedness" of the field lines. It is seen that the lines are twice as crowded at A than at B, so we conclude that $E_A = 2E_B$. Thus, $E_B = 20$ N/C.

3. The following diagram is an edge view of the disk and shows the field lines above it. Near the disk, the lines are perpendicular to the surface and since the disk is uniformly charged, the lines are uniformly distributed over the surface. Far away from the disk, the

lines are like those of a single point charge (the charge on the disk). Extended back to the disk (along the dotted lines of the diagram) they intersect at the center of the disk.

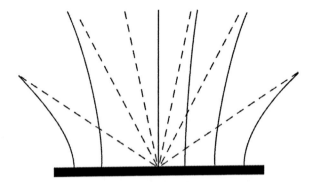

If the disk is positively charged, the lines are directed outward from the disk. If the disk is negatively charged, they are directed inward toward the disk. A similar set of lines is associated with the region below the disk.

4. We find the charge magnitude $|q|$ from $E = |q|/4\pi\varepsilon_0 r^2$:

$$q = 4\pi\varepsilon_0 E r^2 = \frac{(1.00\,\text{N/C})(1.00\,\text{m})^2}{8.99\times10^9\,\text{N}\cdot\text{m}^2/\text{C}^2} = 1.11\times10^{-10}\,\text{C}.$$

5. Since the magnitude of the electric field produced by a point charge q is given by $E = |q|/4\pi\varepsilon_0 r^2$, where r is the distance from the charge to the point where the field has magnitude E, the magnitude of the charge is

$$|q| = 4\pi\varepsilon_0 r^2 E = \frac{(0.50\,\text{m})^2\,(2.0\,\text{N/C})}{8.99\times10^9\,\text{N}\cdot\text{m}^2/\text{C}^2} = 5.6\times10^{-11}\,\text{C}.$$

6. With $x_1 = 6.00$ cm and $x_2 = 21.00$ cm, the point midway between the two charges is located at $x = 13.5$ cm. The values of the charge are $q_1 = -q_2 = -2.00\times10^{-7}$ C, and the magnitudes and directions of the individual fields are given by:

$$\vec{E}_1 = -\frac{|q_1|}{4\pi\varepsilon_0(x-x_1)^2}\hat{\text{i}} = -(3.196\times10^5\,\text{N/C})\hat{\text{i}}$$

$$\vec{E}_2 = -\frac{q_2}{4\pi\varepsilon_0(x-x_1)^2}\hat{\text{i}} = -(3.196\times10^5\,\text{N/C})\hat{\text{i}}$$

Thus, the net electric field is

$$\vec{E}_{\text{net}} = \vec{E}_1 + \vec{E}_2 = -(6.39\times10^5\,\text{N/C})\hat{\text{i}}$$

7. Since the charge is uniformly distributed throughout a sphere, the electric field at the surface is exactly the same as it would be if the charge were all at the center. That is, the magnitude of the field is

$$E = \frac{q}{4\pi\varepsilon_0 R^2}$$

where q is the magnitude of the total charge and R is the sphere radius.

(a) The magnitude of the total charge is Ze, so

$$E = \frac{Ze}{4\pi\varepsilon_0 R^2} = \frac{\left(8.99\times10^9\,\text{N}\cdot\text{m}^2/\text{C}^2\right)(94)\left(1.60\times10^{-19}\,\text{C}\right)}{\left(6.64\times10^{-15}\,\text{m}\right)^2} = 3.07\times10^{21}\,\text{N/C}.$$

(b) The field is normal to the surface and since the charge is positive, it points outward from the surface.

8. (a) The individual magnitudes $\left|\vec{E}_1\right|$ and $\left|\vec{E}_2\right|$ are figured from Eq. 22-3, where the absolute value signs for q_2 are unnecessary since this charge is positive. Whether we add the magnitudes or subtract them depends on if $\vec{E}_1$ is in the same, or opposite, direction as $\vec{E}_2$. At points left of q_1 (on the $-x$ axis) the fields point in opposite directions, but there is no possibility of cancellation (zero net field) since $\left|\vec{E}_1\right|$ is everywhere bigger than $\left|\vec{E}_2\right|$ in this region. In the region between the charges ($0 < x < L$) both fields point leftward and there is no possibility of cancellation. At points to the right of q_2 (where $x > L$), $\vec{E}_1$ points leftward and $\vec{E}_2$ points rightward so the net field in this range is

$$\vec{E}_{\text{net}} = \left(|\vec{E}_2| - |\vec{E}_1|\right)\hat{\text{i}}.$$

Although $|q_1| > q_2$ there is the possibility of $\vec{E}_{\text{net}} = 0$ since these points are closer to q_2 than to q_1. Thus, we look for the zero net field point in the $x > L$ region:

$$|\vec{E}_1| = |\vec{E}_2| \quad \Rightarrow \quad \frac{1}{4\pi\varepsilon_0}\frac{|q_1|}{x^2} = \frac{1}{4\pi\varepsilon_0}\frac{q_2}{(x-L)^2}$$

which leads to

$$\frac{x-L}{x} = \sqrt{\frac{q_2}{|q_1|}} = \sqrt{\frac{2}{5}}.$$

Thus, we obtain $x = \dfrac{L}{1 - \sqrt{2/5}} \approx 2.72L$.

(b) A sketch of the field lines is shown in the figure below:

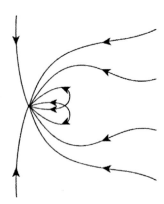

9. At points between the charges, the individual electric fields are in the same direction and do not cancel. Since charge $q_2 = -4.00\, q_1$ located at $x_2 = 70$ cm has a greater magnitude than $q_1 = 2.1 \times 10^{-8}$ C located at $x_1 = 20$ cm, a point of zero field must be closer to q_1 than to q_2. It must be to the left of q_1.

Let x be the coordinate of P, the point where the field vanishes. Then, the total electric field at P is given by

$$E = \frac{1}{4\pi\varepsilon_0}\left(\frac{|q_2|}{(x - x_2)^2} - \frac{|q_1|}{(x - x_1)^2}\right).$$

If the field is to vanish, then

$$\frac{|q_2|}{(x - x_2)^2} = \frac{|q_1|}{(x - x_1)^2} \quad \Rightarrow \quad \frac{|q_2|}{|q_1|} = \frac{(x - x_2)^2}{(x - x_1)^2}.$$

Taking the square root of both sides, noting that $|q_2|/|q_1| = 4$, we obtain

$$\frac{x - 70}{x - 20} = \pm 2.0\,.$$

Choosing -2.0 for consistency, the value of x is found to be $x = -30$ cm.

10. We place the origin of our coordinate system at point P and orient our y axis in the direction of the $q_4 = -12q$ charge (passing through the $q_3 = +3q$ charge). The x axis is perpendicular to the y axis, and thus passes through the identical $q_1 = q_2 = +5q$ charges. The individual magnitudes $|\vec{E}_1|, |\vec{E}_2|, |\vec{E}_3|,$ and $|\vec{E}_4|$ are figured from Eq. 22-3, where the

absolute value signs for q_1, q_2, and q_3 are unnecessary since those charges are positive (assuming $q > 0$). We note that the contribution from q_1 cancels that of q_2 (that is, $|\vec{E}_1| = |\vec{E}_2|$), and the net field (if there is any) should be along the y axis, with magnitude equal to

$$\vec{E}_{net} = \frac{1}{4\pi\varepsilon_0}\left(\frac{|q_4|}{(2d)^2} - \frac{q_3}{d^2}\right)\hat{j} = \frac{1}{4\pi\varepsilon_0}\left(\frac{12q}{4d^2} - \frac{3q}{d^2}\right)\hat{j}$$

which is seen to be zero. A rough sketch of the field lines is shown below:

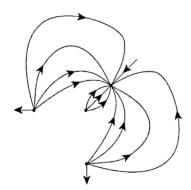

11. The x component of the electric field at the center of the square is given by

$$E_x = \frac{1}{4\pi\varepsilon_0}\left[\frac{|q_1|}{(a/\sqrt{2})^2} + \frac{|q_2|}{(a/\sqrt{2})^2} - \frac{|q_3|}{(a/\sqrt{2})^2} - \frac{|q_4|}{(a/\sqrt{2})^2}\right]\cos 45°$$

$$= \frac{1}{4\pi\varepsilon_0}\frac{1}{a^2/2}(|q_1| + |q_2| - |q_3| - |q_4|)\frac{1}{\sqrt{2}}$$

$$= 0.$$

Similarly, the y component of the electric field is

$$E_y = \frac{1}{4\pi\varepsilon_0}\left[-\frac{|q_1|}{(a/\sqrt{2})^2} + \frac{|q_2|}{(a/\sqrt{2})^2} + \frac{|q_3|}{(a/\sqrt{2})^2} - \frac{|q_4|}{(a/\sqrt{2})^2}\right]\cos 45°$$

$$= \frac{1}{4\pi\varepsilon_0}\frac{1}{a^2/2}(-|q_1| + |q_2| + |q_3| - |q_4|)\frac{1}{\sqrt{2}}$$

$$= \frac{(8.99\times10^9\ \text{N}\cdot\text{m}^2/\text{C}^2)(2.0\times10^{-8}\ \text{C})}{(0.050\ \text{m})^2/2}\frac{1}{\sqrt{2}} = 1.02\times10^5\ \text{N/C}.$$

Thus, the electric field at the center of the square is $\vec{E} = E_y\hat{j} = (1.02\times10^5\ \text{N/C})\hat{j}$.

12. By symmetry we see the contributions from the two charges $q_1 = q_2 = +e$ cancel each other, and we simply use Eq. 22-3 to compute magnitude of the field due to $q_3 = +2e$.

(a) The magnitude of the net electric field is

$$| \vec{E}_{net} | = \frac{1}{4\pi\varepsilon_0} \frac{2e}{r^2} = \frac{1}{4\pi\varepsilon_0} \frac{2e}{(a/\sqrt{2})^2} = \frac{1}{4\pi\varepsilon_0} \frac{4e}{a^2} = (8.99\times10^9) \frac{4(1.60\times10^{-19})}{(6.00\times10^{-6})^2} = 160 \text{ N/C.}$$

(b) This field points at 45.0°, counterclockwise from the *x* axis.

13. (a) The vertical components of the individual fields (due to the two charges) cancel, by symmetry. Using $d = 3.00$ m, the horizontal components (both pointing to the –*x* direction) add to give a magnitude of

$$E_{x,\,net} = \frac{2\,q\,d}{4\pi\varepsilon_0\,(d^2 + y^2)^{3/2}} = 1.38 \times 10^{-10} \text{ N/C .}$$

(b) The net electric field points in the –*x* direction, or 180° counterclockwise from the +*x* axis.

14. For it to be possible for the net field to vanish at some *x* > 0, the two individual fields (caused by q_1 and q_2) must point in opposite directions for *x* > 0. Given their locations in the figure, we conclude they are therefore oppositely charged. Further, since the net field points more strongly leftward for the small positive *x* (where it is very close to q_2) then we conclude that q_2 is the negative-valued charge. Thus, q_1 is a positive-valued charge. We write each charge as a multiple of some positive number ξ (not determined at this point). Since the problem states the absolute value of their ratio, and we have already inferred their signs, we have $q_1 = 4\,\xi$ and $q_2 = -\xi$. Using Eq. 22-3 for the individual fields, we find

$$E_{net} = E_1 + E_2 = \frac{4\,\xi}{4\pi\varepsilon_0\,(L + x)^2} - \frac{\xi}{4\pi\varepsilon_0\,x^2}$$

for points along the positive *x* axis. Setting $E_{net} = 0$ at *x* = 20 cm (see graph) immediately leads to *L* = 20 cm.

(a) If we differentiate E_{net} with respect to *x* and set equal to zero (in order to find where it is maximum), we obtain (after some simplification) that location:

$$x = \left(\frac{2}{3} \sqrt[3]{2} + \frac{1}{3} \sqrt[3]{4} + \frac{1}{3} \right) L = 34 \text{ cm.}$$

We note that the result for part (a) does not depend on the particular value of ξ.

(b) Now we are asked to set $\xi = 3e$, where $e = 1.60 \times 10^{-19}$ C, and evaluate E_{net} at the value of *x* (converted to meters) found in part (a). The result is 2.2×10^{-8} N/C .

15. The field of each charge has magnitude

$$E = k \frac{e}{(0.020\,\text{m})^2} = 3.6 \times 10^{-6}\,\text{N/C}.$$

The directions are indicated in standard format below. We use the magnitude-angle notation (convenient if one is using a vector-capable calculator in polar mode) and write (starting with the proton on the left and moving around clockwise) the contributions to $\vec{E}_{net}$ as follows:

$$\left(E\angle -20°\right) + \left(E\angle 130°\right) + \left(E\angle -100°\right) + \left(E\angle -150°\right) + \left(E\angle 0°\right).$$

This yields $\left(3.93 \times 10^{-6}\angle -76.4°\right)$, with the N/C unit understood.

(a) The result above shows that the magnitude of the net electric field is $|\vec{E}_{net}| = 3.93 \times 10^{-6}$ N/C.

(b) Similarly, the direction of $\vec{E}_{net}$ is $-76.4°$ from the x axis.

16. The net field components along the x and y axes are

$$E_{net\,x} = \frac{q_1}{4\pi\varepsilon_0 R^2} - \frac{q_2 \cos\theta}{4\pi\varepsilon_0 R^2}, \qquad E_{net\,y} = -\frac{q_2 \sin\theta}{4\pi\varepsilon_0 R^2}.$$

The magnitude is the square root of the sum of the components-squared. Setting the magnitude equal to $E = 2.00 \times 10^5$ N/C, squaring and simplifying, we obtain

$$E^2 = \frac{q_1^2 + q_2^2 - 2\,q_1\,q_2 \cos\theta}{16\pi^2\varepsilon_0^2\,R^4}.$$

With $R = 0.500$ m, $q_1 = 2.00 \times 10^{-6}$ C and $q_2 = 6.00 \times 10^{-6}$ C, we can solve this expression for $\cos\theta$ and then take the inverse cosine to find the angle. There are two answers.

(a) The positive value of angle is $\theta = 67.8°$.

(b) The positive value of angle is $\theta = -67.8°$.

17. The magnitude of the dipole moment is given by $p = qd$, where q is the positive charge in the dipole and d is the separation of the charges. For the dipole described in the problem,

$$p = \left(1.60 \times 10^{-19}\text{ C}\right)\left(4.30 \times 10^{-9}\text{ m}\right) = 6.88 \times 10^{-28}\text{ C}\cdot\text{m}.$$

The dipole moment is a vector that points from the negative toward the positive charge.

18. According to the problem statement, E_{act} is Eq. 22-5 (with $z = 5d$)

$$\frac{q}{4\pi\varepsilon_0 (4.5d)^2} - \frac{q}{4\pi\varepsilon_0 (5.5d)^2} = \frac{40\,q}{9801\,\pi\varepsilon_0\,d^2}$$

and E_{approx} is

$$\frac{q\,d}{2\pi\varepsilon_0 (5d)^3} = \frac{q}{250\pi\varepsilon_0\,d^2}.$$

The ratio is therefore

$$\frac{E_{approx}}{E_{act}} = 0.9801 \approx 0.98.$$

19. Consider the figure below.

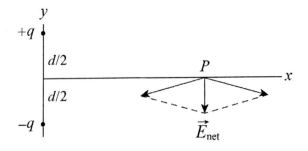

(a) The magnitude of the net electric field at point P is

$$\left|\vec{E}_{net}\right| = 2E_1 \sin\theta = 2\left[\frac{1}{4\pi\varepsilon_0}\frac{q}{(d/2)^2 + r^2}\right]\frac{d/2}{\sqrt{(d/2)^2 + r^2}} = \frac{1}{4\pi\varepsilon_0}\frac{qd}{\left[(d/2)^2 + r^2\right]^{3/2}}$$

For $r \gg d$, we write $[(d/2)^2 + r^2]^{3/2} \approx r^3$ so the expression above reduces to

$$|\vec{E}_{net}| \approx \frac{1}{4\pi\varepsilon_0}\frac{qd}{r^3}.$$

(b) From the figure, it is clear that the net electric field at point P points in the $-\hat{j}$ direction, or $-90°$ from the $+x$ axis.

20. Referring to Eq. 22-6, we use the binomial expansion (see Appendix E) but keeping higher order terms than are shown in Eq. 22-7:

$$E = \frac{q}{4\pi\varepsilon_0 z^2}\left(\left(1 + \frac{d}{z} + \frac{3}{4}\frac{d^2}{z^2} + \frac{1}{2}\frac{d^3}{z^3} + \ldots\right) - \left(1 - \frac{d}{z} + \frac{3}{4}\frac{d^2}{z^2} - \frac{1}{2}\frac{d^3}{z^3} + \ldots\right)\right)$$

$$= \frac{q\,d}{2\pi\varepsilon_0 z^3} + \frac{q\,d^3}{4\pi\varepsilon_0 z^5} + \ldots$$

Therefore, in the terminology of the problem, $E_{next} = q\,d^3/4\pi\varepsilon_0 z^5$.

21. Think of the quadrupole as composed of two dipoles, each with dipole moment of magnitude $p = qd$. The moments point in opposite directions and produce fields in opposite directions at points on the quadrupole axis. Consider the point P on the axis, a distance z to the right of the quadrupole center and take a rightward pointing field to be positive. Then, the field produced by the right dipole of the pair is $qd/2\pi\varepsilon_0(z - d/2)^3$ and the field produced by the left dipole is $-qd/2\pi\varepsilon_0(z + d/2)^3$. Use the binomial expansions $(z - d/2)^{-3} \approx z^{-3} - 3z^{-4}(-d/2)$ and $(z + d/2)^{-3} \approx z^{-3} - 3z^{-4}(d/2)$ to obtain

$$E = \frac{qd}{2\pi\varepsilon_0}\left[\frac{1}{z^3} + \frac{3d}{2z^4} - \frac{1}{z^3} + \frac{3d}{2z^4}\right] = \frac{6qd^2}{4\pi\varepsilon_0 z^4}.$$

Let $Q = 2qd^2$. Then,

$$E = \frac{3Q}{4\pi\varepsilon_0 z^4}.$$

22. We use Eq. 22-3, assuming both charges are positive. At P, we have

$$E_{left\ ring} = E_{right\ ring} \quad \Rightarrow \quad \frac{q_1 R}{4\pi\varepsilon_0\left(R^2 + R^2\right)^{3/2}} = \frac{q_2(2R)}{4\pi\varepsilon_0[(2R)^2 + R^2]^{3/2}}$$

Simplifying, we obtain

$$\frac{q_1}{q_2} = 2\left(\frac{2}{5}\right)^{3/2} \approx 0.506.$$

23. (a) We use the usual notation for the linear charge density: $\lambda = q/L$. The arc length is $L = r\theta$ if θ is expressed in radians. Thus,

$$L = (0.0400\ \text{m})(0.698\ \text{rad}) = 0.0279\ \text{m}.$$

With $q = -300(1.602 \times 10^{-19}\ \text{C})$, we obtain $\lambda = -1.72 \times 10^{-15}\ \text{C/m}$.

(b) We consider the same charge distributed over an area $A = \pi r^2 = \pi(0.0200 \text{ m})^2$ and obtain $\sigma = q/A = -3.82 \times 10^{-14} \text{ C/m}^2$.

(c) Now the area is four times larger than in the previous part $(A_{\text{sphere}} = 4\pi r^2)$ and thus obtain an answer that is one-fourth as big:

$$\sigma = q/A_{\text{sphere}} = -9.56 \times 10^{-15} \text{ C/m}^2.$$

(d) Finally, we consider that same charge spread throughout a volume of $4\pi r^3/3$ and obtain the charge density $\rho = \text{charge/volume} = -1.43 \times 10^{-12} \text{ C/m}^3$.

24. From symmetry, we see that the net field at P is twice the field caused by the upper semicircular charge $+q = \lambda \cdot \pi R$ (and that it points downward). Adapting the steps leading to Eq. 22-21, we find

$$\vec{E}_{\text{net}} = 2\left(-\hat{\jmath}\right)\frac{\lambda}{4\pi\varepsilon_0 R}\sin\theta \Big|_{-90°}^{90°} = -\frac{q}{\varepsilon_0\pi^2 R^2}\hat{\jmath}.$$

(a) With $R = 8.50 \times 10^{-2}$ m and $q = 1.50 \times 10^{-8}$ C, $|\vec{E}_{\text{net}}| = 23.8$ N/C.

(b) The net electric field $\vec{E}_{\text{net}}$ points in the $-\hat{\jmath}$ direction, or $-90°$ counterclockwise from the $+x$ axis.

25. Studying Sample Problem 22-4, we see that the field evaluated at the center of curvature due to a charged distribution on a circular arc is given by

$$\vec{E} = \frac{\lambda}{4\pi\varepsilon_0 r}[\sin\theta]_{-\theta/2}^{\theta/2} \qquad \text{along the symmetry axis}$$

where $\lambda = q/r\theta$ with θ in radians. In this problem, each charged quarter-circle produces a field of magnitude

$$|\vec{E}| = \frac{|q|}{r\pi/2}\frac{1}{4\pi\varepsilon_0 r}[\sin\theta]_{-\pi/4}^{\pi/4} = \frac{1}{4\pi\varepsilon_0}\frac{2\sqrt{2}|q|}{\pi r^2}.$$

That produced by the positive quarter-circle points at $-45°$, and that of the negative quarter-circle points at $+45°$.

(a) The magnitude of the net field is

$$E_{net,x} = 2\left(\frac{1}{4\pi\varepsilon_0}\frac{2\sqrt{2}|q|}{\pi r^2}\right)\cos 45° = \frac{1}{4\pi\varepsilon_0}\frac{4|q|}{\pi r^2} = \frac{(8.99\times10^9)4(4.50\times10^{-12})}{\pi(5.00\times10^{-2})^2} = 20.6 \text{ N/C.}$$

(b) By symmetry, the net field points vertically downward in the $-\hat{j}$ direction, or $-90°$ counterclockwise from the $+x$ axis.

26. We find the maximum by differentiating Eq. 22-16 and setting the result equal to zero.

$$\frac{d}{dz}\left(\frac{qz}{4\pi\varepsilon_0(z^2+R^2)^{3/2}}\right) = \frac{q}{4\pi\varepsilon_0}\frac{R^2-2z^2}{(z^2+R^2)^{5/2}} = 0$$

which leads to $z = R/\sqrt{2}$. With $R = 2.40$ cm, we have $z = 1.70$ cm.

27. (a) The linear charge density is the charge per unit length of rod. Since the charge is uniformly distributed on the rod,

$$\lambda = \frac{-q}{L} = \frac{-4.23\times10^{-15} \text{ C}}{0.0815 \text{ m}} = -5.19\times10^{-14} \text{ C/m.}.$$

(b) We position the x axis along the rod with the origin at the left end of the rod, as shown in the diagram.

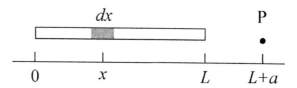

Let dx be an infinitesimal length of rod at x. The charge in this segment is $dq = \lambda\,dx$. The charge dq may be considered to be a point charge. The electric field it produces at point P has only an x component and this component is given by

$$dE_x = \frac{1}{4\pi\varepsilon_0}\frac{\lambda\,dx}{(L+a-x)^2}.$$

The total electric field produced at P by the whole rod is the integral

$$E_x = \frac{\lambda}{4\pi\varepsilon_0}\int_0^L \frac{dx}{(L+a-x)^2} = \frac{\lambda}{4\pi\varepsilon_0}\frac{1}{L+a-x}\Big|_0^L = \frac{\lambda}{4\pi\varepsilon_0}\left(\frac{1}{a}-\frac{1}{L+a}\right)$$

$$= \frac{\lambda}{4\pi\varepsilon_0}\frac{L}{a(L+a)} = -\frac{1}{4\pi\varepsilon_0}\frac{q}{a(L+a)},$$

upon substituting $-q = \lambda L$. With $q = 4.23 \times 10^{-15}$ C, $L = 0.0815$ m and $a = 0.120$ m, we obtain $E_x = -1.57 \times 10^{-3}$ N/C.

(c) The negative sign indicates that the field points in the $-x$ direction, or $-180°$ counterclockwise form the $+x$ axis.

(d) If a is much larger than L, the quantity $L + a$ in the denominator can be approximated by a and the expression for the electric field becomes

$$E_x = -\frac{q}{4\pi\varepsilon_0 a^2}.$$

Since $a = 50$ m $\gg L = 0.0815$ m, the above approximation applies and we have $E_x = -1.52 \times 10^{-8}$ N/C, or $|E_x| = 1.52 \times 10^{-8}$ N/C.

(e) For a particle of charge $-q = -4.23 \times 10^{-15}$ C, the electric field at a distance $a = 50$ m away has a magnitude $|E_x| = 1.52 \times 10^{-8}$ N/C.

28. First, we need a formula for the field due to the arc. We use the notation λ for the charge density, $\lambda = Q/L$. Sample Problem 22-4 illustrates the simplest approach to circular arc field problems. Following the steps leading to Eq. 22-21, we see that the general result (for arcs that subtend angle θ) is

$$E_{arc} = \frac{\lambda}{4\pi\varepsilon_0 r} [\sin(\theta/2) - \sin(-\theta/2)] = \frac{\lambda \sin(\theta/2)}{2\pi\varepsilon_0 r}.$$

Now, the arc length is $L = r\theta$ if θ is expressed in radians. Thus, using R instead of r, we obtain

$$E_{arc} = \frac{Q/L \sin(\theta/2)}{2\pi\varepsilon_0 R} = \frac{Q \sin(\theta/2)}{2\pi\varepsilon_0 \theta R^2}.$$

Thus, with $\theta = \pi$, the problem asks for the ratio $E_{particle} / E_{arc}$ where $E_{particle}$ is given by Eq. 22-3. We obtain

$$\frac{Q/4\pi\varepsilon_0 R^2}{Q\sin(\theta/2)/2\pi\varepsilon_0 R^2} = \frac{\pi}{2} \approx 1.57.$$

29. We assume $q > 0$. Using the notation $\lambda = q/L$ we note that the (infinitesimal) charge on an element dx of the rod contains charge $dq = \lambda \, dx$. By symmetry, we conclude that all horizontal field components (due to the dq's) cancel and we need only "sum" (integrate) the vertical components. Symmetry also allows us to integrate these contributions over

only half the rod ($0 \leq x \leq L/2$) and then simply double the result. In that regard we note that $\sin \theta = R/r$ where $r = \sqrt{x^2 + R^2}$.

(a) Using Eq. 22-3 (with the 2 and $\sin \theta$ factors just discussed) the magnitude is

$$|\vec{E}| = 2 \int_0^{L/2} \left(\frac{dq}{4\pi\varepsilon_0 r^2} \right) \sin \theta = \frac{2}{4\pi\varepsilon_0} \int_0^{L/2} \left(\frac{\lambda\, dx}{x^2 + R^2} \right) \left(\frac{y}{\sqrt{x^2 + R^2}} \right)$$

$$= \frac{\lambda R}{2\pi\varepsilon_0} \int_0^{L/2} \frac{dx}{\left(x^2 + R^2 \right)^{3/2}} = \frac{(q/L)R}{2\pi\varepsilon_0} \left[\frac{x}{R^2 \sqrt{x^2 + R^2}} \right]_0^{L/2}$$

$$= \frac{q}{2\pi\varepsilon_0 LR} \frac{L/2}{\sqrt{(L/2)^2 + R^2}} = \frac{q}{2\pi\varepsilon_0 R} \frac{1}{\sqrt{L^2 + 4R^2}}$$

where the integral may be evaluated by elementary means or looked up in Appendix E (item #19 in the list of integrals). With $q = 7.81 \times 10^{-12}$ C, $L = 0.145$ m and $R = 0.0600$ m, we have $|\vec{E}| = 12.4$ N/C.

(b) As noted above, the electric field $\vec{E}$ points in the $+y$ direction, or $+90°$ counterclockwise from the $+x$ axis.

30. From Eq. 22-26

$$E = \frac{\sigma}{2\varepsilon_0} \left(1 - \frac{z}{\sqrt{z^2 + R^2}} \right) = \frac{5.3 \times 10^{-6}\ \text{C/m}^2}{2 \left(8.85 \times 10^{-12} \dfrac{\text{C}^2}{\text{N} \cdot \text{m}^2} \right)} \left[1 - \frac{12\,\text{cm}}{\sqrt{(12\,\text{cm})^2 + (2.5\,\text{cm})^2}} \right] = 6.3 \times 10^3\ \text{N/C}.$$

31. At a point on the axis of a uniformly charged disk a distance z above the center of the disk, the magnitude of the electric field is

$$E = \frac{\sigma}{2\varepsilon_0} \left[1 - \frac{z}{\sqrt{z^2 + R^2}} \right]$$

where R is the radius of the disk and σ is the surface charge density on the disk. See Eq. 22-26. The magnitude of the field at the center of the disk ($z = 0$) is $E_c = \sigma/2\varepsilon_0$. We want to solve for the value of z such that $E/E_c = 1/2$. This means

$$1 - \frac{z}{\sqrt{z^2 + R^2}} = \frac{1}{2} \Rightarrow \frac{z}{\sqrt{z^2 + R^2}} = \frac{1}{2}.$$

Squaring both sides, then multiplying them by $z^2 + R^2$, we obtain $z^2 = (z^2/4) + (R^2/4)$. Thus, $z^2 = R^2/3$, or $z = R/\sqrt{3}$. With $R = 0.600$ m, we have $z = 0.346$ m.

32. We write Eq. 22-26 as

$$\frac{E}{E_{max}} = 1 - \frac{z}{(z^2 + R^2)^{1/2}}$$

and note that this ratio is $\frac{1}{2}$ (according to the graph shown in the figure) when $z = 4.0$ cm. Solving this for R we obtain $R = z\sqrt{3} = 6.9$ cm.

33. We use Eq. 22-26, noting that the disk in figure (*b*) is effectively equivalent to the disk in figure (*a*) plus a concentric smaller disk (of radius $R/2$) with the opposite value of σ. That is,

$$E_{(b)} = E_{(a)} - \frac{\sigma}{2\varepsilon_0}\left(1 - \frac{2R}{\sqrt{(2R)^2 + (R/2)^2}}\right)$$

where

$$E_{(a)} = \frac{\sigma}{2\varepsilon_0}\left(1 - \frac{2R}{\sqrt{(2R)^2 + R^2}}\right).$$

We find the relative difference and simplify:

$$\frac{E_{(a)} - E_{(b)}}{E_{(a)}} = \frac{1 - \dfrac{2}{\sqrt{4 + \frac{1}{4}}}}{1 - \dfrac{2}{\sqrt{4 + 1}}} = 0.283$$

or approximately 28%.

34. (a) Vertical equilibrium of forces leads to the equality

$$q\left|\vec{E}\right| = mg \Rightarrow \left|\vec{E}\right| = \frac{mg}{2e}.$$

Using the mass given in the problem, we obtain $\left|\vec{E}\right| = 2.03 \times 10^{-7}$ N/C.

(b) Since the force of gravity is downward, then $q\vec{E}$ must point upward. Since $q > 0$ in this situation, this implies $\vec{E}$ must itself point upward.

35. The magnitude of the force acting on the electron is $F = eE$, where E is the magnitude of the electric field at its location. The acceleration of the electron is given by Newton's second law:

$$a = \frac{F}{m} = \frac{eE}{m} = \frac{(1.60 \times 10^{-19}\,\text{C})(2.00 \times 10^{4}\,\text{N/C})}{9.11 \times 10^{-31}\,\text{kg}} = 3.51 \times 10^{15}\,\text{m/s}^2.$$

36. Eq. 22-28 gives

$$\vec{E} = \frac{\vec{F}}{q} = \frac{m\vec{a}}{(-e)} = -\left(\frac{m}{e}\right)\vec{a}$$

using Newton's second law.

(a) With *east* being the $\hat{i}$ direction, we have

$$\vec{E} = -\left(\frac{9.11 \times 10^{-31}\,\text{kg}}{1.60 \times 10^{-19}\,\text{C}}\right)\left(1.80 \times 10^{9}\,\text{m/s}^2\,\hat{i}\right) = -0.0102\,\text{N/C}\,\hat{i}$$

which means the field has a magnitude of 0.0102 N/C

(b) The result shows that the field $\vec{E}$ is directed in the $-x$ direction, or westward.

37. We combine Eq. 22-9 and Eq. 22-28 (in absolute values).

$$F = |q|E = |q|\left(\frac{p}{2\pi\varepsilon_0 z^3}\right) = \frac{2kep}{z^3}$$

where we have used Eq. 21-5 for the constant k in the last step. Thus, we obtain

$$F = \frac{2(8.99 \times 10^{9}\,\text{N}\cdot\text{m}^2/\text{C}^2)(1.60 \times 10^{-19}\,\text{C})(3.6 \times 10^{-29}\,\text{C}\cdot\text{m})}{(25 \times 10^{-9}\,\text{m})^3}$$

which yields a force of magnitude 6.6×10^{-15} N. If the dipole is oriented such that $\vec{p}$ is in the $+z$ direction, then $\vec{F}$ points in the $-z$ direction.

38. (a) $F_e = Ee = (3.0 \times 10^{6}\,\text{N/C})(1.6 \times 10^{-19}\,\text{C}) = 4.8 \times 10^{-13}$ N.

(b) $F_i = Eq_{ion} = Ee = 4.8 \times 10^{-13}$ N.

39. (a) The magnitude of the force on the particle is given by $F = qE$, where q is the magnitude of the charge carried by the particle and E is the magnitude of the electric field at the location of the particle. Thus,

$$E = \frac{F}{q} = \frac{3.0 \times 10^{-6} \text{ N}}{2.0 \times 10^{-9} \text{ C}} = 1.5 \times 10^{3} \text{ N/C}.$$

The force points downward and the charge is negative, so the field points upward.

(b) The magnitude of the electrostatic force on a proton is

$$F_{el} = eE = \left(1.60 \times 10^{-19} \text{ C}\right)\left(1.5 \times 10^{3} \text{ N/C}\right) = 2.4 \times 10^{-16} \text{N}.$$

(c) A proton is positively charged, so the force is in the same direction as the field, upward.

(d) The magnitude of the gravitational force on the proton is

$$F_{g} = mg = \left(1.67 \times 10^{-27} \text{ kg}\right)\left(9.8 \text{ m/s}^{2}\right) = 1.6 \times 10^{-26} \text{N}.$$

The force is downward.

(e) The ratio of the forces is

$$\frac{F_{el}}{F_{g}} = \frac{2.4 \times 10^{-16} \text{ N}}{1.64 \times 10^{-26} \text{ N}} = 1.5 \times 10^{10}.$$

40. (a) The initial direction of motion is taken to be the $+x$ direction (this is also the direction of $\vec{E}$). We use $v_{f}^{2} - v_{i}^{2} = 2a\Delta x$ with $v_{f} = 0$ and $\vec{a} = \vec{F}/m = -e\vec{E}/m_{e}$ to solve for distance Δx:

$$\Delta x = \frac{-v_{i}^{2}}{2a} = \frac{-m_{e}v_{i}^{2}}{-2eE} = \frac{-\left(9.11 \times 10^{-31} \text{ kg}\right)\left(5.00 \times 10^{6} \text{ m/s}\right)^{2}}{-2\left(1.60 \times 10^{-19} \text{ C}\right)\left(1.00 \times 10^{3} \text{ N/C}\right)} = 7.12 \times 10^{-2} \text{ m}.$$

(b) Eq. 2-17 leads to

$$t = \frac{\Delta x}{v_{avg}} = \frac{2\Delta x}{v_{i}} = \frac{2\left(7.12 \times 10^{-2} \text{ m}\right)}{5.00 \times 10^{6} \text{ m/s}} = 2.85 \times 10^{-8} \text{ s}.$$

(c) Using $\Delta v^{2} = 2a\Delta x$ with the new value of Δx, we find

$$\frac{\Delta K}{K_i} = \frac{\Delta\left(\frac{1}{2}m_e v^2\right)}{\frac{1}{2}m_e v_i^2} = \frac{\Delta v^2}{v_i^2} = \frac{2a\Delta x}{v_i^2} = \frac{-2eE\Delta x}{m_e v_i^2}$$

$$= \frac{-2\left(1.60\times10^{-19}\,\text{C}\right)\left(1.00\times10^3\,\text{N/C}\right)\left(8.00\times10^{-3}\,\text{m}\right)}{\left(9.11\times10^{-31}\,\text{kg}\right)\left(5.00\times10^6\,\text{m/s}\right)^2} = -0.112.$$

Thus, the fraction of the initial kinetic energy lost in the region is 0.112 or 11.2%.

41. (a) The magnitude of the force acting on the proton is $F = eE$, where E is the magnitude of the electric field. According to Newton's second law, the acceleration of the proton is $a = F/m = eE/m$, where m is the mass of the proton. Thus,

$$a = \frac{\left(1.60\times10^{-19}\,\text{C}\right)\left(2.00\times10^4\,\text{N/C}\right)}{1.67\times10^{-27}\,\text{kg}} = 1.92\times10^{12}\,\text{m/s}^2.$$

(b) We assume the proton starts from rest and use the kinematic equation $v^2 = v_0^2 + 2ax$ (or else $x = \frac{1}{2}at^2$ and $v = at$) to show that

$$v = \sqrt{2ax} = \sqrt{2\left(1.92\times10^{12}\,\text{m/s}^2\right)\left(0.0100\,\text{m}\right)} = 1.96\times10^5\,\text{m/s}.$$

42. When the drop is in equilibrium, the force of gravity is balanced by the force of the electric field: $mg = -qE$, where m is the mass of the drop, q is the charge on the drop, and E is the magnitude of the electric field. The mass of the drop is given by $m = (4\pi/3)r^3\rho$, where r is its radius and ρ is its mass density. Thus,

$$q = -\frac{mg}{E} = -\frac{4\pi r^3 \rho g}{3E} = -\frac{4\pi\left(1.64\times10^{-6}\,\text{m}\right)^3\left(851\,\text{kg/m}^3\right)\left(9.8\,\text{m/s}^2\right)}{3\left(1.92\times10^5\,\text{N/C}\right)} = -8.0\times10^{-19}\,\text{C}$$

and $q/e = (-8.0\times10^{-19}\,\text{C})/(1.60\times10^{-19}\,\text{C}) = -5$, or $q = -5e$.

43. (a) We use $\Delta x = v_{\text{avg}}t = vt/2$:

$$v = \frac{2\Delta x}{t} = \frac{2\left(2.0\times10^{-2}\,\text{m}\right)}{1.5\times10^{-8}\,\text{s}} = 2.7\times10^6\,\text{m/s}.$$

(b) We use $\Delta x = \frac{1}{2}at^2$ and $E = F/e = ma/e$:

$$E = \frac{ma}{e} = \frac{2\Delta x m}{et^2} = \frac{2(2.0 \times 10^{-2}\,\text{m})(9.11 \times 10^{-31}\,\text{kg})}{(1.60 \times 10^{-19}\,\text{C})(1.5 \times 10^{-8}\,\text{s})^2} = 1.0 \times 10^3\,\text{N/C}.$$

44. We assume there are no forces or force-components along the x direction. We combine Eq. 22-28 with Newton's second law, then use Eq. 4-21 to determine time t followed by Eq. 4-23 to determine the final velocity (with $-g$ replaced by the a_y of this problem); for these purposes, the velocity components *given* in the problem statement are re-labeled as v_{0x} and v_{0y} respectively.

(a) We have $\vec{a} = q\vec{E}/m = -(e/m)\vec{E}$ which leads to

$$\vec{a} = -\left(\frac{1.60 \times 10^{-19}\,\text{C}}{9.11 \times 10^{-31}\,\text{kg}}\right)\left(120\,\frac{\text{N}}{\text{C}}\right)\hat{j} = -(2.1 \times 10^{13}\,\text{m/s}^2)\,\hat{j}.$$

(b) Since $v_x = v_{0x}$ in this problem (that is, $a_x = 0$), we obtain

$$t = \frac{\Delta x}{v_{0x}} = \frac{0.020\,\text{m}}{1.5 \times 10^5\,\text{m/s}} = 1.3 \times 10^{-7}\,\text{s}$$

$$v_y = v_{0y} + a_y t = 3.0 \times 10^3\,\text{m/s} + (-2.1 \times 10^{13}\,\text{m/s}^2)(1.3 \times 10^{-7}\,\text{s})$$

which leads to $v_y = -2.8 \times 10^6$ m/s. Therefore, the final velocity is

$$\vec{v} = (1.5 \times 10^5\,\text{m/s})\,\hat{i} - (2.8 \times 10^6\,\text{m/s})\,\hat{j}.$$

45. We take the positive direction to be to the right in the figure. The acceleration of the proton is $a_p = eE/m_p$ and the acceleration of the electron is $a_e = -eE/m_e$, where E is the magnitude of the electric field, m_p is the mass of the proton, and m_e is the mass of the electron. We take the origin to be at the initial position of the proton. Then, the coordinate of the proton at time t is $x = \frac{1}{2}a_p t^2$ and the coordinate of the electron is $x = L + \frac{1}{2}a_e t^2$. They pass each other when their coordinates are the same, or $\frac{1}{2}a_p t^2 = L + \frac{1}{2}a_e t^2$. This means $t^2 = 2L/(a_p - a_e)$ and

$$x = \frac{a_p}{a_p - a_e}L = \frac{eE/m_p}{(eE/m_p) + (eE/m_e)}L = \frac{m_e}{m_e + m_p}L$$

$$= \frac{9.11 \times 10^{-31}\,\text{kg}}{9.11 \times 10^{-31}\,\text{kg} + 1.67 \times 10^{-27}\,\text{kg}}(0.050\text{m})$$

$$= 2.7 \times 10^{-5}\,\text{m}.$$

46. Due to the fact that the electron is negatively charged, then (as a consequence of Eq. 22-28 and Newton's second law) the field $\vec{E}$ pointing in the $+y$ direction (which we will

call "upward") leads to a downward acceleration. This is exactly like a projectile motion problem as treated in Chapter 4 (but with g replaced with $a = eE/m = 8.78 \times 10^{11}$ m/s^2). Thus, Eq. 4-21 gives

$$t = \frac{x}{v_0 \cos 40°} = \frac{3.00 \text{ m}}{1.53 \times 10^7 \text{ m/s}} = 1.96 \times 10^{-6} \text{ s}.$$

This leads (using Eq. 4-23) to

$$v_y = v_0 \sin 40° - a\,t = -4.34 \times 10^5 \text{ m/s}.$$

Since the x component of velocity does not change, then the final velocity is

$$\vec{v} = (1.53 \times 10^6 \text{ m/s})\,\hat{i} - (4.34 \times 10^5 \text{ m/s})\,\hat{j}.$$

47. (a) Using Eq. 22-28, we find

$$\vec{F} = (8.00 \times 10^{-5} \text{C})(3.00 \times 10^3 \text{ N/C})\hat{i} + (8.00 \times 10^{-5} \text{C})(-600 \text{ N/C})\hat{j}$$
$$= (0.240 \text{N})\hat{i} - (0.0480 \text{N})\hat{j}.$$

Therefore, the force has magnitude equal to

$$F = \sqrt{(0.240 \text{N})^2 + (0.0480 \text{N})^2} = 0.245 \text{N}.$$

(b) The angle the force $\vec{F}$ makes with the $+x$ axis is

$$\theta = \tan^{-1}\left(\frac{F_y}{F_x}\right) = \tan^{-1}\left(\frac{-0.0480 \text{N}}{0.240 \text{N}}\right) = -11.3°$$

measured counterclockwise from the $+x$ axis.

(c) With $m = 0.0100$ kg, the (x, y) coordinates at $t = 3.00$ s can be found by combining Newton's second law with the kinematics equations of Chapters 2–4. The x coordinate is

$$x = \frac{1}{2}a_x t^2 = \frac{F_x t^2}{2m} = \frac{(0.240)(3.00)^2}{2(0.0100)} = 108 \text{m}.$$

(d) Similarly, the y coordinate is

$$y = \frac{1}{2}a_y t^2 = \frac{F_y t^2}{2m} = \frac{(-0.0480)(3.00)^2}{2(0.0100)} = -21.6 \text{m}.$$

48. We are given $\sigma = 4.00 \times 10^{-6}\,C/m^2$ and various values of z (in the notation of Eq. 22-26 which specifies the field E of the charged disk). Using this with $F = eE$ (the magnitude of Eq. 22-28 applied to the electron) and $F = ma$, we obtain

(a) The magnitude of the acceleration at a distance R is

$$a = \frac{e\,\sigma\,(2 - \sqrt{2}\,)}{4\,m\,\varepsilon_0} = 1.16 \times 10^{16}\ m/s^2\ .$$

(b) At a distance $R/100$, $a = \dfrac{e\,\sigma\,(10001 - \sqrt{10001}\,)}{20002\,m\,\varepsilon_0} = 3.94 \times 10^{16}\,m/s^2\ .$

(c) At a distance $R/1000$, $a = \dfrac{e\,\sigma\,(1000001 - \sqrt{1000001}\,)}{2000002\,m\,\varepsilon_0} = 3.97 \times 10^{16}\,m/s^2\ .$

(d) The field due to the disk becomes more uniform as the electron nears the center point. One way to view this is to consider the forces exerted on the electron by the charges near the edge of the disk; the net force on the electron caused by those charges will decrease due to the fact that their contributions come closer to canceling out as the electron approaches the middle of the disk.

49. (a) Due to the fact that the electron is negatively charged, then (as a consequence of Eq. 22-28 and Newton's second law) the field $\vec{E}$ pointing in the same direction as the velocity leads to deceleration. Thus, with $t = 1.5 \times 10^{-9}\,s$, we find

$$v = v_0 - |a|\,t = v_0 - \frac{eE}{m}t = 2.7 \times 10^4\ m/s\ .$$

(b) The displacement is equal to the distance since the electron does not change its direction of motion. The field is uniform, which implies the acceleration is constant. Thus,

$$d = \frac{v + v_0}{2}t = 5.0 \times 10^{-5}\,m.$$

50. (a) Eq. 22-33 leads to $\tau = pE\sin 0° = 0.$

(b) With $\theta = 90°$, the equation gives

$$\tau = pE = \left(2\left(1.6 \times 10^{-19}\,C\right)\left(0.78 \times 10^{-9}\,m\right)\right)\left(3.4 \times 10^6\,N/C\right) = 8.5 \times 10^{-22}\ N\cdot m.$$

(c) Now the equation gives $\tau = pE\sin 180° = 0.$

51. (a) The magnitude of the dipole moment is

$$p = qd = \left(1.50 \times 10^{-9}\,\text{C}\right)\left(6.20 \times 10^{-6}\,\text{m}\right) = 9.30 \times 10^{-15}\,\text{C} \cdot \text{m}.$$

(b) Following the solution to part (c) of Sample Problem 22-6, we find

$$U(180°) - U(0) = 2pE = 2\left(9.30 \times 10^{-15}\right)(1100) = 2.05 \times 10^{-11}\,\text{J}.$$

52. Using Eq. 22-35, considering θ as a variable, we note that it reaches its maximum value when $\theta = -90°$: $\tau_{max} = pE$. Thus, with $E = 40$ N/C and $\tau_{max} = 100 \times 10^{-28}$ N·m (determined from the graph), we obtain the dipole moment: $p = 2.5 \times 10^{-28}$ C·m.

53. Following the solution to part (c) of Sample Problem 22-6, we find

$$W = U\left(\theta_0 + \pi\right) - U\left(\theta_0\right) = -pE\left(\cos\left(\theta_0 + \pi\right) - \cos\left(\theta_0\right)\right) = 2pE\cos\theta_0$$
$$= 2(3.02 \times 10^{-25}\,\text{C} \cdot \text{m})(46.0\,\text{N/C})\cos 64.0°$$
$$= 1.22 \times 10^{-23}\,\text{J}.$$

54. We make the assumption that bead 2 is in the lower half of the circle, partly because it would be awkward for bead 1 to "slide through" bead 2 if it were in the path of bead 1 (which is the upper half of the circle) and partly to eliminate a second solution to the problem (which would have opposite angle and charge for bead 2). We note that the net y component of the electric field evaluated at the origin is negative (points *down*) for all positions of bead 1, which implies (with our assumption in the previous sentence) that bead 2 is a negative charge.

(a) When bead 1 is on the $+y$ axis, there is no x component of the net electric field, which implies bead 2 is on the $-y$ axis, so its angle is $-90°$.

(b) Since the downward component of the net field, when bead 1 is on the $+y$ axis, is of largest magnitude, then bead 1 must be a positive charge (so that its field is in the same direction as that of bead 2, in that situation). Comparing the values of E_y at $0°$ and at $90°$ we see that the absolute values of the charges on beads 1 and 2 must be in the ratio of 5 to 4. This checks with the $180°$ value from the E_x graph, which further confirms our belief that bead 1 is positively charged. In fact, the $180°$ value from the E_x graph allows us to solve for its charge (using Eq. 22-3):

$$q_1 = 4\pi\varepsilon_0 r^2 E = 4\pi(\,8.854 \times 10^{-12}\,\tfrac{\text{C}^2}{\text{N m}^2})(0.60\,\text{m})^2\,(5.0 \times 10^4\,\tfrac{\text{N}}{\text{C}}) = 2.0 \times 10^{-6}\,\text{C}\,.$$

(c) Similarly, the $0°$ value from the E_y graph allows us to solve for the charge of bead 2:

$$q_2 = 4\pi\varepsilon_0 r^2 E = 4\pi(\,8.854 \times 10^{-12}\,\tfrac{\text{C}^2}{\text{N m}^2})(0.60\,\text{m})^2\,(-4.0 \times 10^4\,\tfrac{\text{N}}{\text{C}}) = -1.6 \times 10^{-6}\,\text{C}\,.$$

55. Consider an infinitesimal section of the rod of length dx, a distance x from the left end, as shown in the following diagram.

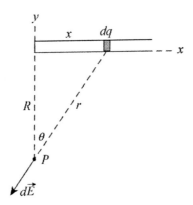

It contains charge $dq = \lambda\, dx$ and is a distance r from P. The magnitude of the field it produces at P is given by

$$dE = \frac{1}{4\pi\varepsilon_0}\frac{\lambda\, dx}{r^2}.$$

The x and the y components are

$$dE_x = -\frac{1}{4\pi\varepsilon_0}\frac{\lambda\, dx}{r^2}\sin\theta$$

and

$$dE_y = -\frac{1}{4\pi\varepsilon_0}\frac{\lambda\, dx}{r^2}\cos\theta,$$

respectively. We use θ as the variable of integration and substitute $r = R/\cos\theta$, $x = R\tan\theta$ and $dx = (R/\cos^2\theta)\, d\theta$. The limits of integration are 0 and $\pi/2$ rad. Thus,

$$E_x = -\frac{\lambda}{4\pi\varepsilon_0 R}\int_0^{\pi/2}\sin\theta d\theta = \frac{\lambda}{4\pi\varepsilon_0 R}\cos\theta\Big|_0^{\pi/2} = -\frac{\lambda}{4\pi\varepsilon_0 R}$$

and

$$E_y = -\frac{\lambda}{4\pi\varepsilon_0 R}\int_0^{\pi/2}\cos\theta d\theta = -\frac{\lambda}{4\pi\varepsilon_0 R}\sin\theta\Big|_0^{\pi/2} = -\frac{\lambda}{4\pi\varepsilon_0 R}.$$

We notice that $E_x = E_y$ no matter what the value of R. Thus, $\vec{E}$ makes an angle of $45°$ with the rod for all values of R.

56. From $dA = 2\pi r\, dr$ (which can be thought of as the differential of $A = \pi r^2$) and $dq = \sigma\, dA$ (from the definition of the surface charge density σ), we have

$$dq = \left(\frac{Q}{\pi R^2}\right) 2\pi r\, dr$$

where we have used the fact that the disk is uniformly charged to set the surface charge density equal to the total charge (Q) divided by the total area (πR^2). We next set $r = 0.0050$ m and make the approximation $dr \approx 30 \times 10^{-6}$ m. Thus we get $dq \approx 2.4 \times 10^{-16}$ C.

57. Our approach (based on Eq. 22-29) consists of several steps. The first is to find an *approximate* value of e by taking differences between all the given data. The smallest difference is between the fifth and sixth values:

$$18.08 \times 10^{-19}\, \text{C} - 16.48 \times 10^{-19}\, \text{C} = 1.60 \times 10^{-19}\, \text{C}$$

which we denote e_{approx}. The goal at this point is to assign integers n using this approximate value of e:

datum1	$\dfrac{6.563 \times 10^{-19}\text{C}}{e_{\text{approx}}} = 4.10 \Rightarrow n_1 = 4$	datum6	$\dfrac{18.08 \times 10^{-19}\text{C}}{e_{\text{appeox}}} = 11.30 \Rightarrow n_6 = 11$
datum2	$\dfrac{8.204 \times 10^{-19}\text{C}}{e_{\text{approx}}} = 5.13 \Rightarrow n_2 = 5$	datum7	$\dfrac{19.71 \times 10^{-19}\text{C}}{e_{\text{approx}}} = 12.32 \Rightarrow n_7 = 12$
datum3	$\dfrac{11.50 \times 10^{-19}\text{C}}{e_{\text{approx}}} = 7.19 \Rightarrow n_3 = 7$	datum8	$\dfrac{22.89 \times 10^{-19}\text{C}}{e_{\text{approx}}} = 14.31 \Rightarrow n_8 = 14$
datum4	$\dfrac{13.13 \times 10^{-19}\text{C}}{e_{\text{approx}}} = 8.21 \Rightarrow n_4 = 8$	datum9	$\dfrac{26.13 \times 10^{-19}\text{C}}{e_{\text{approx}}} = 16.33 \Rightarrow n_9 = 16$
datum5	$\dfrac{16.48 \times 10^{-19}\text{C}}{e_{\text{approx}}} = 10.30 \Rightarrow n_5 = 10$		

Next, we construct a new data set (e_1, e_2, e_3 ...) by dividing the given data by the respective exact integers n_i (for $i = 1, 2, 3$ ...):

$$(e_1, e_2, e_3 ...) = \left(\frac{6.563 \times 10^{-19}\, \text{C}}{n_1}, \frac{8.204 \times 10^{-19}\, \text{C}}{n_2}, \frac{11.50 \times 10^{-19}\, \text{C}}{n_3} ...\right)$$

which gives (carrying a few more figures than are significant)

$$\left(1.64075 \times 10^{-19}\, \text{C},\ 1.6408 \times 10^{-19}\, \text{C},\ 1.64286 \times 10^{-19}\, \text{C} ...\right)$$

as the new data set (our experimental values for e). We compute the average and standard deviation of this set, obtaining

$$e_{\text{exptal}} = e_{\text{avg}} \pm \Delta e = \left(1.641 \pm 0.004\right) \times 10^{-19}\,\text{C}$$

which does not agree (to within one standard deviation) with the modern accepted value for e. The lower bound on this spread is $e_{\text{avg}} - \Delta e = 1.637 \times 10^{-19}$ C which is still about 2% too high.

58. (a) It is clear from symmetry (also from Eq. 22-16) that the field vanishes at the center.

(b) The result ($E = 0$) for points infinitely far away can be reasoned directly from Eq. 22-16 (it goes as $1/z^2$ as $z \to \infty$) or by recalling the starting point of its derivation (Eq. 22-11, which makes it clearer that the field strength decreases as $1/r^2$ at distant points).

(c) Differentiating Eq. 22-16 and setting equal to zero (to obtain the location where it is maximum) leads to

$$\frac{dE}{dz} = \frac{q(R^2 - 2z^2)}{4\pi\varepsilon_0\,(R^2 + z^2)^{5/2}} = 0 \quad \Rightarrow \quad z = +\frac{R}{\sqrt{2}} = 0.707\,R.$$

(d) Plugging this value back into Eq. 22-16 with the values stated in the problem, we find $E_{\max} = 3.46 \times 10^7\,\text{N/C}$.

59. The distance from Q to P is $5a$, and the distance from q to P is $3a$. Therefore, the magnitudes of the individual electric fields are, using Eq. 22-3 (writing $1/4\pi\varepsilon_0 = k$),

$$\left|\vec{E}_Q\right| = \frac{k|Q|}{25\,a^2}, \quad \left|\vec{E}_q\right| = \frac{k|q|}{9\,a^2}.$$

We note that $\vec{E}_q$ is along the y axis (directed towards $\pm y$ in accordance with the sign of q), and $\vec{E}_Q$ has x and y components, with $\vec{E}_{Qx} = \pm\frac{4}{5}\left|\vec{E}_Q\right|$ and $\vec{E}_{Qy} = \pm\frac{3}{5}\left|\vec{E}_Q\right|$ (signs corresponding to the sign of Q). Consequently, we can write the addition of components in a simple way (basically, by dropping the absolute values):

$$\vec{E}_{\text{net }x} = \frac{4\,k\,Q}{125\,a^2}$$

$$\vec{E}_{\text{net }y} = \frac{3\,k\,Q}{125\,a^2} + \frac{k\,q}{9\,a^2}$$

(a) Equating $\vec{E}_{\text{net }x}$ and $\vec{E}_{\text{net }y}$, it is straightforward to solve for the relation between Q and q. We obtain $Q/q = 125/9 \approx 14$.

(b) We set $\vec{E}_{net\,y} = 0$ and find the necessary relation between Q and q. We obtain $Q/q = -125/27 \approx -4.6$.

60. First, we need a formula for the field due to the arc. We use the notation λ for the charge density, $\lambda = Q/L$. Sample Problem 22-4 illustrates the simplest approach to circular arc field problems. Following the steps leading to Eq. 22-21, we see that the general result (for arcs that subtend angle θ) is

$$E_{arc} = \frac{\lambda}{4\pi\varepsilon_0 r}[\sin(\theta/2) - \sin(-\theta/2)] = \frac{\lambda\,\sin(\theta/2)}{2\pi\varepsilon_0 r}\,.$$

Now, the arc length is $L = r\theta$ if θ is expressed in <u>radians</u>. Thus, using R instead of r, we obtain

$$E_{arc} = \frac{Q/L\,\sin(\theta/2)}{2\pi\varepsilon_0 R} = \frac{Q\,\sin(\theta/2)}{2\pi\varepsilon_0\,\theta\,R^2}\,.$$

Thus, the problem requires $E_{arc} = \frac{1}{2}E_{particle}$ where $E_{particle}$ is given by Eq. 22-3. Hence,

$$\frac{Q\,\sin(\theta/2)}{2\pi\varepsilon_0\,\theta\,R^2} = \frac{1}{2}\frac{Q}{4\pi\varepsilon_0\,R^2} \quad\Rightarrow\quad \sin\left(\frac{\theta}{2}\right) = \frac{\theta}{4}$$

where we note, again, that the angle is in radians. The approximate solution to this equation is $\theta = 3.791$ rad $\approx 217°$.

61. Most of the individual fields, caused by diametrically opposite charges, will cancel, except for the pair that lie on the x axis passing through the center. This pair of charges produces a field pointing to the right

$$\vec{E} = \frac{3q}{4\pi\varepsilon_0\,d^2}\,\hat{i} = \frac{3e}{4\pi\varepsilon_0\,(0.020\text{ m})^2} = (1.08 \times 10^{-5}\text{ N/C})\hat{i}\,.$$

62. We use Eq. 22-16, with "q" denoting the charge on the larger ring:

$$\frac{Q\,z}{4\pi\varepsilon_0(z^2 + R^2)^{3/2}} + \frac{q\,z}{4\pi\varepsilon_0(z^2 + (3R)^2)^{3/2}} = 0 \quad\Rightarrow\quad q = -Q\left(\frac{13\sqrt{13}}{5\sqrt{5}}\right)$$

which gives $q \approx -4.19Q$. Note: we set $z = 2R$ in the above calculation.

63. (a) We refer to the same figure to which problem 63 refers (but without "q"). From symmetry, we see the net field component along the x axis is zero; the net field component along the y axis points upward. With $\theta = 60°$,

$$E_{\text{net } y} = 2 \frac{Q \sin \theta}{4\pi\varepsilon_0 a^2} .$$

Since $\sin(60°) = \sqrt{3}/2$, we can write this as $E_{\text{net}} = kQ\sqrt{3}/a^2$ (using the notation of the constant k defined in Eq. 21-5). Numerically, this gives roughly 47 N/C.

(b) From symmetry, we see in this case that the net field component along the y axis is zero; the net field component along the x axis points rightward. With $\theta = 60°$,

$$E_{\text{net } x} = 2 \frac{Q \cos \theta}{4\pi\varepsilon_0 a^2} .$$

Since $\cos(60°) = 1/2$, we can write this as $E_{\text{net}} = kQ/a^2$ (using the notation of Eq. 21-5). Thus, $E_{\text{net}} \approx 27$ N/C.

64. The smallest arc is of length $L_1 = \pi r_1/2 = \pi R/2$; the middle-sized arc has length $L_2 = \pi r_2/2 = \pi(2R)/2 = \pi R$; and, the largest arc has $L_3 = \pi(3R)/2$. The charge per unit length for each arc is $\lambda = q/L$ where each charge q is specified in the figure. Following the steps that lead to Eq. 22-21 in Sample Problem 22-4, we find

$$E_{\text{net}} = \frac{\lambda_1 [2 \sin(45°)]}{4\pi\varepsilon_0 r_1} + \frac{\lambda_2 [2 \sin(45°)]}{4\pi\varepsilon_0 r_2} + \frac{\lambda_3 [2 \sin(45°)]}{4\pi\varepsilon_0 r_3} = \frac{Q}{\sqrt{2}\, \pi^2 \varepsilon_0 R^2}$$

which yields $E_{\text{net}} = 1.62 \times 10^6$ N/C .

(b) The direction is $-45°$, measured counterclockwise from the $+x$ axis.

65. (a) Since the two charges in question are of the same sign, the point $x = 2.0$ mm should be located in between them (so that the field vectors point in the opposite direction). Let the coordinate of the second particle be x' ($x' > 0$). Then, the magnitude of the field due to the charge $-q_1$ evaluated at x is given by $E = q_1/4\pi\varepsilon_0 x^2$, while that due to the second charge $-4q_1$ is $E' = 4q_1/4\pi\varepsilon_0(x'-x)^2$. We set the net field equal to zero:

$$\vec{E}_{\text{net}} = 0 \implies E = E'$$

so that

$$\frac{q_1}{4\pi\varepsilon_0 x^2} = \frac{4q_1}{4\pi\varepsilon_0 (x'-x)^2} .$$

Thus, we obtain $x' = 3x = 3(2.0 \text{ mm}) = 6.0$ mm.

(b) In this case, with the second charge now positive, the electric field vectors produced by both charges are in the negative x direction, when evaluated at $x = 2.0$ mm. Therefore, the net field points in the negative x direction, or 180°, measured counterclockwise from the $+x$ axis.

66. (a) The electron e_c is a distance $r = z = 0.020$ meter away. Thus,

$$E_c = \frac{e}{4\pi\varepsilon_0 r^2} = 3.60 \times 10^{-6}\,\text{N/C}.$$

(b) The horizontal components of the individual fields (due to the two e_s charges) cancel, and the vertical components add to give

$$E_{s,\,net} = \frac{2\,e\,z}{4\pi\varepsilon_0 (R^2 + z^2)^{3/2}} = 2.55 \times 10^{-6}\,\text{N/C}.$$

(c) Calculation similar to that shown in part (a) now leads to a stronger field $E_c = 3.60 \times 10^{-4}$ N/C from the central charge.

(d) The field due to the side charges may be obtained from calculation similar to that shown in part (b). The result is $E_{s,\,net} = 7.09 \times 10^{-7}$ N/C.

(e) Since E_c is inversely proportional to z^2, this is a simple result of the fact that z is now much smaller than in part (a). For the net effect due to the side charges, it is the "trigonometric factor" for the y component (here expressed as $z/\sqrt{r}$) which shrinks almost linearly (as z decreases) for very small z, plus the fact that the x components cancel, which leads to the decreasing value of $E_{s,\,net}$.

67. We interpret the linear charge density, $\lambda = |Q|/L$, to indicate a positive quantity (so we can relate it to the magnitude of the field). Sample Problem 22-4 illustrates the simplest approach to circular arc field problems. Following the steps leading to Eq. 22-21, we see that the general result (for arcs that subtend angle θ) is

$$E = \frac{\lambda}{4\pi\varepsilon_0 r}[\sin(\theta/2) - \sin(-\theta/2)] = \frac{\lambda \sin(\theta/2)}{2\pi\varepsilon_0 r}.$$

Now, the arc length is $L = r\theta$ if θ is expressed in radians. Thus, using R instead of r, we obtain

$$E = \frac{|Q|/L \sin(\theta/2)}{2\pi\varepsilon_0 R} = \frac{|Q| \sin(\theta/2)}{2\pi\varepsilon_0 \theta R^2}$$

With $|Q| = 6.25 \times 10^{-12}$ C, $\theta = 2.40$ rad $= 137.5°$ and $R = 9.00 \times 10^{-2}$ m, the magnitude of the electric field is $E = 5.39$ N/C.

68. Examining the lowest value on the graph, we have (using Eq. 22-38)

$$U = -\vec{p} \cdot \vec{E} = -1.00 \times 10^{-28} \text{ J.}$$

If $E = 20$ N/C, we find $p = 5.0 \times 10^{-28}$ C·m.

69. From symmetry, we see the net force component along the y axis is zero; the net force component along the x axis points rightward. With $\theta = 60°$,

$$F_3 = 2\frac{q_3 q_1 \cos\theta}{4\pi\varepsilon_0 a^2}.$$

Since $\cos(60°) = 1/2$, we can write this as

$$F_3 = \frac{kq_3 q_1}{a^2} = \frac{(8.99\times10^9)(5.00\times10^{-12})(2.00\times10^{-12})}{(0.0950)^2} = 9.96\times10^{-12} \text{ N.}$$

70. The two closest charges produce fields at the midpoint which cancel each other out. Thus, the only significant contribution is from the furthest charge, which is a distance $r = \sqrt{3}d/2$ away from that midpoint. Plugging this into Eq. 22-3 immediately gives the result:

$$E = \frac{Q}{4\pi\varepsilon_0 r^2} = \frac{Q}{3\pi\varepsilon_0 d^2}.$$

71. From the second measurement (at (2.0, 0)) we see that the charge must be somewhere on the x axis. A line passing through (3.0, 3.0) with slope $\tan^{-1}(3/4)$ will intersect the x axis at $x = -1.0$. Thus, the location of the particle is specified by the coordinates (in cm): $(-1.0, 0)$.

(a) Thus, the x coordinate is $x = -1.0$ cm.

(b) Similarly, the y coordinate is $y = 0$.

(c) Using $k = 1/4\pi\varepsilon_0$, the field magnitude measured at (2.0, 0) (which is $r = 0.030$ m from the charge) is

$$\left|\vec{E}\right| = k\frac{q}{r^2} = 100 \text{ N/C.}$$

Therefore, $q = 1.0 \times 10^{-11}$ C.

72. We denote the electron with subscript e and the proton with p. From the figure below we see that

$$\left|\vec{E}_e\right| = \left|\vec{E}_p\right| = \frac{e}{4\pi\varepsilon_0 d^2}$$

where $d = 2.0 \times 10^{-6}$ m. We note that the components along the y axis cancel during the vector summation. With $k = 1/4\pi\varepsilon_0$ and $\theta = 60°$, the magnitude of the net electric field is obtained as follows:

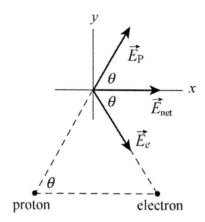

$$\left|\vec{E}_{net}\right| = E_x = 2E_e\cos\theta = 2\left(\frac{e}{4\pi\varepsilon_0 d^2}\right)\cos\theta = 2k\left[\frac{e}{d^2}\right]\cos\theta$$

$$= 2\left(8.99\times10^9 \ \frac{\text{N}\cdot\text{m}^2}{\text{C}^2}\right)\left[\frac{\left(1.6\times10^{-19} \ \text{C}\right)}{\left(2.0\times10^{-6} \ \text{m}\right)^2}\right]\cos 60°$$

$$= 3.6\times10^2 \ \text{N/C}.$$

73. On the one hand, the conclusion (that $Q = +1.00 \ \mu C$) is clear from symmetry. If a more in-depth justification is desired, one should use Eq. 22-3 for the electric field magnitudes of the three charges (each at the same distance $r = a/\sqrt{3}$ from C) and then find field components along suitably chosen axes, requiring each component-sum to be zero. If the y axis is vertical, then (assuming $Q > 0$) the component-sum along that axis leads to $2kq\sin 30°/r^2 = kQ/r^2$ where q refers to either of the charges at the bottom corners. This yields $Q = 2q\sin 30° = q$ and thus to the conclusion mentioned above.

74. (a) Let $E = \sigma/2\varepsilon_0 = 3 \times 10^6$ N/C. With $\sigma = |q|/A$, this leads to

$$|q| = \pi R^2\sigma = 2\pi\varepsilon_0 R^2 E = \frac{R^2 E}{2k} = \frac{\left(2.5\times10^{-2} \ \text{m}\right)^2\left(3.0\times10^6 \ \text{N/C}\right)}{2\left(8.99\times10^9 \ \frac{\text{N}\cdot\text{m}^2}{\text{C}^2}\right)} = 1.0\times10^{-7} \ \text{C}.$$

(b) Setting up a simple proportionality (with the areas), the number of atoms is estimated to be

$$N = \frac{\pi \left(2.5 \times 10^{-2}\,\text{m}\right)^2}{0.015 \times 10^{-18}\,\text{m}^2} = 1.3 \times 10^{17}.$$

(c) Therefore, the fraction is

$$\frac{q}{Ne} = \frac{1.0 \times 10^{-7}\,\text{C}}{\left(1.3 \times 10^{17}\right)\left(1.6 \times 10^{-19}\,\text{C}\right)} \approx 5.0 \times 10^{-6}.$$

75. (a) Using the density of water ($\rho = 1000$ kg/m^3), the weight mg of the spherical drop (of radius $r = 6.0 \times 10^{-7}$ m) is

$$W = \rho V g = \left(1000\,\text{kg/m}^3\right)\left(\frac{4\pi}{3}\left(6.0 \times 10^{-7}\,\text{m}\right)^3\right)\left(9.8\,\text{m/s}^2\right) = 8.87 \times 10^{-15}\,\text{N}.$$

(b) Vertical equilibrium of forces leads to $mg = qE = neE$, which we solve for n, the number of excess electrons:

$$n = \frac{mg}{eE} = \frac{8.87 \times 10^{-15}\,\text{N}}{\left(1.60 \times 10^{-19}\,\text{C}\right)\left(462\,\text{N/C}\right)} = 120.$$

76. Eq. 22-38 gives $U = -\vec{p} \cdot \vec{E} = -pE\cos\theta$. We note that $\theta_i = 110°$ and $\theta_f = 70.0°$. Therefore,

$$\Delta U = -pE\left(\cos 70.0° - \cos 110°\right) = -3.28 \times 10^{-21}\,\text{J}.$$

77. A small section of the distribution that has charge dq is $\lambda\,dx$, where $\lambda = 9.0 \times 10^{-9}$ C/m. Its contribution to the field at $x_P = 4.0$ m is

$$d\vec{E} = \frac{dq}{4\pi\varepsilon_0 \left(x - x_P\right)^2}$$

pointing in the $+x$ direction. Thus, we have

$$\vec{E} = \int_0^{3.0\,\text{m}} \frac{\lambda\,dx}{4\pi\varepsilon_0 \left(x - x_P\right)^2}\,\hat{\text{i}}$$

which becomes, using the substitution $u = x - x_P$,

$$\vec{E} = \frac{\lambda}{4\pi\varepsilon_0} \int_{-4.0\text{m}}^{-1.0\text{m}} \frac{du}{u^2}\,\hat{\text{i}} = \frac{\lambda}{4\pi\varepsilon_0}\left(\frac{-1}{-1.0\,\text{m}} - \frac{-1}{-4.0\,\text{m}}\right)\hat{\text{i}}$$

which yields 61 N/C in the +x direction.

78. Studying Sample Problem 22-4, we see that the field evaluated at the center of curvature due to a charged distribution on a circular arc is given by

$$\vec{E} = \frac{\lambda}{4\pi\varepsilon_0 r}[\sin\theta]_{-\theta/2}^{\theta/2} \quad \text{along the symmetry axis}$$

where $\lambda = q/\ell = q/r\theta$ with θ in radians. Here ℓ is the length of the arc, given as $\ell = 4.0\,\text{m}$. Therefore, $\theta = \ell/r = 4.0/2.0 = 2.0\,\text{rad}$. Thus, with $q = 20 \times 10^{-9}$ C, we obtain

$$\left|\vec{E}\right| = \frac{q}{\ell}\frac{1}{4\pi\varepsilon_0 r}[\sin\theta]_{-1.0\,\text{rad}}^{1.0\,\text{rad}} = 38\,\text{N/C}.$$

79. (a) We combine Eq. 22-28 (in absolute value) with Newton's second law:

$$a = \frac{|q|E}{m} = \left(\frac{1.60 \times 10^{-19}\,\text{C}}{9.11 \times 10^{-31}\,\text{kg}}\right)\left(1.40 \times 10^6\,\frac{\text{N}}{\text{C}}\right) = 2.46 \times 10^{17}\,\text{m/s}^2.$$

(b) With $v = \dfrac{c}{10} = 3.00 \times 10^7$ m/s, we use Eq. 2-11 to find

$$t = \frac{v - v_0}{a} = \frac{3.00 \times 10^7}{2.46 \times 10^{17}} = 1.22 \times 10^{-10}\,\text{s}.$$

(c) Eq. 2-16 gives

$$\Delta x = \frac{v^2 - v_0^2}{2a} = \frac{\left(3.00 \times 10^7\right)^2}{2\left(2.46 \times 10^{17}\right)} = 1.83 \times 10^{-3}\,\text{m}.$$

80. Let q_1 denote the charge at $y = d$ and q_2 denote the charge at $y = -d$. The individual magnitudes $\left|\vec{E}_1\right|$ and $\left|\vec{E}_2\right|$ are figured from Eq. 22-3, where the absolute value signs for q are unnecessary since these charges are both positive. The distance from q_1 to a point on the x axis is the same as the distance from q_2 to a point on the x axis: $r = \sqrt{x^2 + d^2}$. By symmetry, the y component of the net field along the x axis is zero. The x component of the net field, evaluated at points on the positive x axis, is

$$E_x = 2\left(\frac{1}{4\pi\varepsilon_0}\right)\left(\frac{q}{x^2 + d^2}\right)\left(\frac{x}{\sqrt{x^2 + d^2}}\right)$$

where the last factor is $\cos\theta = x/r$ with θ being the angle for each individual field as measured from the *x* axis.

(a) If we simplify the above expression, and plug in $x = \alpha d$, we obtain

$$E_x = \frac{q}{2\pi\varepsilon_0 d^2}\left(\frac{\alpha}{\left(\alpha^2 + 1\right)^{3/2}}\right).$$

(b) The graph of $E = E_x$ versus α is shown below. For the purposes of graphing, we set $d = 1$ m and $q = 5.56 \times 10^{-11}$ C.

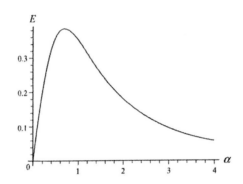

(c) From the graph, we estimate E_{max} occurs at about $\alpha = 0.71$. More accurate computation shows that the maximum occurs at $\alpha = 1/\sqrt{2}$.

(d) The graph suggests that "half-height" points occur at $\alpha \approx 0.2$ and $\alpha \approx 2.0$. Further numerical exploration leads to the values: $\alpha = 0.2047$ and $\alpha = 1.9864$.

81. (a) From Eq. 22-38 (and the facts that $\hat{i}\cdot\hat{i} = 1$ and $\hat{j}\cdot\hat{i} = 0$), the potential energy is

$$U = -\vec{p}\cdot\vec{E} = -\left[\left(3.00\hat{i} + 4.00\hat{j}\right)\left(1.24\times10^{-30}\,\text{C}\cdot\text{m}\right)\right]\cdot\left[\left(4000\,\text{N/C}\right)\hat{i}\right]$$

$$= -1.49\times10^{-26}\,\text{J}.$$

(b) From Eq. 22-34 (and the facts that $\hat{i}\times\hat{i} = 0$ and $\hat{j}\times\hat{i} = -\hat{k}$), the torque is

$$\vec{\tau} = \vec{p}\times\vec{E} = \left[\left(3.00\hat{i} + 4.00\hat{j}\right)\left(1.24\times10^{-30}\,\text{C}\cdot\text{m}\right)\right]\times\left[\left(4000\,\text{N/C}\right)\hat{i}\right]$$

$$= \left(-1.98\times10^{-26}\,\text{N}\cdot\text{m}\right)\hat{k}.$$

(c) The work done is

$$W = \Delta U = \Delta\left(-\vec{p}\cdot\vec{E}\right) = \left(\vec{p}_i - \vec{p}_f\right)\cdot\vec{E}$$
$$= \left[\left(3.00\hat{i} + 4.00\hat{j}\right) - \left(-4.00\hat{i} + 3.00\hat{j}\right)\right]\left(1.24\times10^{-30}\,C\cdot m\right)\cdot\left[\left(4000\,N/C\right)\hat{i}\right]$$
$$= 3.47\times10^{-26}\,J.$$

82. We consider pairs of diametrically opposed charges. The net field due to just the charges in the one o'clock ($-q$) and seven o'clock ($-7q$) positions is clearly equivalent to that of a single $-6q$ charge sitting at the seven o'clock position. Similarly, the net field due to just the charges in the six o'clock ($-6q$) and twelve o'clock ($-12q$) positions is the same as that due to a single $-6q$ charge sitting at the twelve o'clock position. Continuing with this line of reasoning, we see that there are six equal-magnitude electric field vectors pointing at the seven o'clock, eight o'clock ... twelve o'clock positions. Thus, the resultant field of all of these points, by symmetry, is directed toward the position midway between seven and twelve o'clock. Therefore, $\vec{E}_{\text{resultant}}$ points towards the nine-thirty position.

83. (a) For point A, we have (in SI units)

$$\vec{E}_A = \left[\frac{q_1}{4\pi\varepsilon_0 r_1^2} + \frac{q_2}{4\pi\varepsilon_0 r_2^2}\right]\left(-\hat{i}\right)$$
$$= \frac{\left(8.99\times10^9\right)\left(1.00\times10^{-12}\,C\right)}{\left(5.00\times10^{-2}\right)^2}\left(-\hat{i}\right) + \frac{\left(8.99\times10^9\right)|-2.00\times10^{-12}\,C|}{\left(2\times5.00\times10^{-2}\right)^2}\left(\hat{i}\right).$$
$$= \left(-1.80\,N/C\right)\hat{i}.$$

(b) Similar considerations leads to

$$\vec{E}_B = \left[\frac{q_1}{4\pi\varepsilon_0 r_1^2} + \frac{|q_2|}{4\pi\varepsilon_0 r_2^2}\right]\hat{i} = \frac{\left(8.99\times10^9\right)\left(1.00\times10^{-12}\,C\right)}{\left(0.500\times5.00\times10^{-2}\right)^2}\hat{i} + \frac{\left(8.99\times10^9\right)|-2.00\times10^{-12}\,C|}{\left(0.500\times5.00\times10^{-2}\right)^2}\hat{i}$$
$$= \left(43.2\,N/C\right)\hat{i}.$$

(c) For point C, we have

$$\vec{E}_C = \left[\frac{q_1}{4\pi\varepsilon_0 r_1^2} - \frac{|q_2|}{4\pi\varepsilon_0 r_2^2}\right]\hat{i} = \frac{\left(8.99\times10^9\right)\left(1.00\times10^{-12}\,C\right)}{\left(2.00\times5.00\times10^{-2}\right)^2}\hat{i} - \frac{\left(8.99\times10^9\right)|-2.00\times10^{-12}\,C|}{\left(5.00\times10^{-2}\right)^2}\hat{i}$$
$$= -\left(6.29\,N/C\right)\hat{i}.$$

(d) Although a sketch is not shown here, it would be somewhat similar to Fig. 22-5 in the textbook except that there would be twice as many field lines "coming into" the negative charge (which would destroy the simple up/down symmetry seen in Fig. 22-5).

84. The electric field at a point on the axis of a uniformly charged ring, a distance z from the ring center, is given by

$$E = \frac{qz}{4\pi\varepsilon_0\left(z^2 + R^2\right)^{3/2}}$$

where q is the charge on the ring and R is the radius of the ring (see Eq. 22-16). For q positive, the field points upward at points above the ring and downward at points below the ring. We take the positive direction to be upward. Then, the force acting on an electron on the axis is

$$F = -\frac{eqz}{4\pi\varepsilon_0\left(z^2 + R^2\right)^{3/2}}.$$

For small amplitude oscillations $z \ll R$ and z can be neglected in the denominator. Thus,

$$F = -\frac{eqz}{4\pi\varepsilon_0 R^3}.$$

The force is a restoring force: it pulls the electron toward the equilibrium point $z = 0$. Furthermore, the magnitude of the force is proportional to z, just as if the electron were attached to a spring with spring constant $k = eq/4\pi\varepsilon_0 R^3$. The electron moves in simple harmonic motion with an angular frequency given by

$$\omega = \sqrt{\frac{k}{m}} = \sqrt{\frac{eq}{4\pi\varepsilon_0 mR^3}}$$

where m is the mass of the electron.

85. (a) Since $\vec{E}$ points down and we need an upward electric force (to cancel the downward pull of gravity), then we require the charge of the sphere to be negative. The magnitude of the charge is found by working with the absolute value of Eq. 22-28:

$$|q| = \frac{F}{E} = \frac{mg}{E} = \frac{4.4\text{N}}{150 \text{ N/C}} = 0.029\text{C},$$

or $q = -0.029$ C.

(b) The feasibility of this experiment may be studied by using Eq. 22-3 (using k for $1/4\pi\varepsilon_0$). We have $E = k|q|/r^2$ with

$$\rho_{sulfur}\left(\frac{4}{3}\pi r^3\right) = m_{sphere}$$

Since the mass of the sphere is $4.4/9.8 \approx 0.45$ kg and the density of sulfur is about 2.1×10^3 kg/m^3 (see Appendix F), then we obtain

$$r = \left(\frac{3m_{sphere}}{4\pi\rho_{sulfur}}\right)^{1/3} = 0.037\,\text{m} \Rightarrow E = k\frac{|q|}{r^2} \approx 2 \times 10^{11}\,\text{N/C}$$

which is much too large a field to maintain in air.

86. (a) The electric field is upward in the diagram and the charge is negative, so the force of the field on it is downward. The magnitude of the acceleration is $a = eE/m$, where E is the magnitude of the field and m is the mass of the electron. Its numerical value is

$$a = \frac{\left(1.60 \times 10^{-19}\,\text{C}\right)\left(2.00 \times 10^3\,\text{N/C}\right)}{9.11 \times 10^{-31}\,\text{kg}} = 3.51 \times 10^{14}\,\text{m/s}^2.$$

We put the origin of a coordinate system at the initial position of the electron. We take the x axis to be horizontal and positive to the right; take the y axis to be vertical and positive toward the top of the page. The kinematic equations are

$$x = v_0 t \cos\theta, \quad y = v_0 t \sin\theta - \frac{1}{2}at^2, \quad \text{and} \quad v_y = v_0 \sin\theta - at.$$

First, we find the greatest y coordinate attained by the electron. If it is less than d, the electron does not hit the upper plate. If it is greater than d, it will hit the upper plate if the corresponding x coordinate is less than L. The greatest y coordinate occurs when $v_y = 0$. This means $v_0 \sin\theta - at = 0$ or $t = (v_0/a)\sin\theta$ and

$$y_{max} = \frac{v_0^2 \sin^2\theta}{a} - \frac{1}{2}a\frac{v_0^2 \sin^2\theta}{a^2} = \frac{1}{2}\frac{v_0^2 \sin^2\theta}{a} = \frac{\left(6.00 \times 10^6\,\text{m/s}\right)^2 \sin^2 45°}{2\left(3.51 \times 10^{14}\,\text{m/s}^2\right)} = 2.56 \times 10^{-2}\,\text{m}.$$

Since this is greater than $d = 2.00$ cm, the electron might hit the upper plate.

(b) Now, we find the x coordinate of the position of the electron when $y = d$. Since

$$v_0 \sin\theta = \left(6.00 \times 10^6\,\text{m/s}\right)\sin 45° = 4.24 \times 10^6\,\text{m/s}$$

and

$$2ad = 2\left(3.51 \times 10^{14} \text{ m/s}^2\right)\left(0.0200\,\text{m}\right) = 1.40 \times 10^{13} \text{ m}^2/\text{s}^2$$

the solution to $d = v_0 t \sin\theta - \frac{1}{2}at^2$ is

$$t = \frac{v_0 \sin\theta - \sqrt{v_0^2 \sin^2\theta - 2ad}}{a} = \frac{4.24 \times 10^6 \text{ m/s} - \sqrt{\left(4.24 \times 10^6 \text{ m/s}\right)^2 - 1.40 \times 10^{13} \text{ m}^2/\text{s}^2}}{3.51 \times 10^{14} \text{ m/s}^2}$$

$$= 6.43 \times 10^{-9}\,\text{s}.$$

The negative root was used because we want the *earliest* time for which $y = d$. The x coordinate is

$$x = v_0 t \cos\theta = \left(6.00 \times 10^6 \text{ m/s}\right)\left(6.43 \times 10^{-9}\,\text{s}\right)\cos 45° = 2.72 \times 10^{-2}\,\text{m}.$$

This is less than L so the electron hits the upper plate at $x = 2.72$ cm.

87. Eq. 22-35 $\left(\tau = -pE \sin\theta\right)$ captures the sense as well as the magnitude of the effect. That is, this is a restoring torque, trying to bring the tilted dipole back to its aligned equilibrium position. If the amplitude of the motion is small, we may replace $\sin\theta$ with θ in radians. Thus, $\tau \approx -pE\theta$. Since this exhibits a simple negative proportionality to the angle of rotation, the dipole oscillates in simple harmonic motion, like a torsional pendulum with torsion constant $\kappa = pE$. The angular frequency ω is given by

$$\omega^2 = \frac{\kappa}{I} = \frac{pE}{I}$$

where I is the rotational inertia of the dipole. The frequency of oscillation is

$$f = \frac{\omega}{2\pi} = \frac{1}{2\pi}\sqrt{\frac{pE}{I}}.$$

88. (a) If we subtract each value from the next larger value in the table, we find a set of numbers which are suggestive of a basic unit of charge: 1.64×10^{-19}, 3.3×10^{-19}, 1.63×10^{-19}, 3.35×10^{-19}, 1.6×10^{-19}, 1.63×10^{-19}, 3.18×10^{-19}, 3.24×10^{-19}, where the SI unit Coulomb is understood. These values are either close to a common $e \approx 1.6 \times 10^{-19}$ C value or are double that. Taking this, then, as a crude approximation to our experimental e we divide it into all the values in the original data set and round to the nearest integer, obtaining $n = 4, 5, 7, 8, 10, 11, 12, 14$, and 16.

(b) When we perform a least squares fit of the original data set versus these values for n we obtain the linear equation:

$$q = 7.18 \times 10^{-21} + 1.633 \times 10^{-19} n .$$

If we dismiss the constant term as unphysical (representing, say, systematic errors in our measurements) then we obtain $e = 1.63 \times 10^{-19}$ when we set $n = 1$ in this equation.

89. (a) Using $k = 1/4\pi\varepsilon_0$, we estimate the field at $r = 0.02$ m using Eq. 22-3:

$$E = k\frac{q}{r^2} = \left(8.99 \times 10^9 \ \frac{\text{N} \cdot \text{m}^2}{\text{C}^2}\right) \frac{45 \times 10^{-12} \ \text{C}}{(0.02 \ \text{m})^2} \approx 1 \times 10^3 \ \text{N/C}.$$

(b) The field described by Eq. 22-3 is nonuniform.

(c) As the positively charged bee approaches the grain, a concentration of negative charge is induced on the closest side of the grain, leading to a force of attraction which makes the grain jump to the bee. Although in physical contact, it is not in electrical contact with the bee, or else it would acquire a net positive charge causing it to be repelled from the bee. As the bee (with grain) approaches the stigma, a concentration of negative charge is induced on the closest side of the stigma which is presumably highly nonuniform. In some configurations, the field from the stigma (acting on the positive side of the grain) will overcome the field from the bee acting on the negative side, and the grain will jump to the stigma.

90. Since both charges are positive (and aligned along the z axis) we have

$$\left|\vec{E}_{net}\right| = \frac{1}{4\pi\varepsilon_0}\left[\frac{q}{(z-d/2)^2} + \frac{q}{(z+d/2)^2}\right].$$

For $z \gg d$ we have $(z \pm d/2)^{-2} \approx z^{-2}$, so

$$\left|\vec{E}_{net}\right| \approx \frac{1}{4\pi\varepsilon_0}\left(\frac{q}{z^2} + \frac{q}{z^2}\right) = \frac{2q}{4\pi\varepsilon_0 z^2}.$$

91. (a) Suppose the pendulum is at the angle θ with the vertical. The force diagram is shown below. $\vec{T}$ is the tension in the thread, mg is the magnitude of the force of gravity, and qE is the magnitude of the electric force. The field points upward and the charge is positive, so the force is upward. Taking the angle shown to be positive, then the torque on the sphere about the point where the thread is attached to the upper plate is $\tau = -(mg - qE)L\sin\theta$. If $mg > qE$ then the torque is a restoring torque; it tends to pull the pendulum back to its equilibrium position.

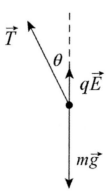

If the amplitude of the oscillation is small, $\sin\theta$ can be replaced by θ in radians and the torque is $\tau = -(mg - qE) L\theta$. The torque is proportional to the angular displacement and the pendulum moves in simple harmonic motion. Its angular frequency is $\omega = \sqrt{(mg - qE) L/I}$, where I is the rotational inertia of the pendulum. Since $I = mL^2$ for a simple pendulum,

$$\omega = \sqrt{\frac{(mg - qE) L}{mL^2}} = \sqrt{\frac{g - qE/m}{L}}$$

and the period is

$$T = \frac{2\pi}{\omega} = 2\pi\sqrt{\frac{L}{g - qE/m}}.$$

If $qE > mg$ the torque is not a restoring torque and the pendulum does not oscillate.

(b) The force of the electric field is now downward and the torque on the pendulum is $\tau = -(mg + qE)L\theta$ if the angular displacement is small. The period of oscillation is

$$T = 2\pi\sqrt{\frac{L}{g + qE/m}}.$$

Chapter 23

1. The vector area $\vec{A}$ and the electric field $\vec{E}$ are shown on the diagram below. The angle θ between them is $180° - 35° = 145°$, so the electric flux through the area is

$$\Phi = \vec{E} \cdot \vec{A} = EA \cos\theta = (1800 \text{ N/C})(3.2 \times 10^{-3} \text{ m})^2 \cos 145° = -1.5 \times 10^{-2} \text{ N} \cdot \text{m}^2/\text{C}.$$

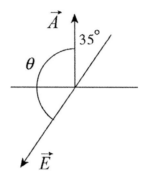

2. We use $\Phi = \vec{E} \cdot \vec{A}$, where $\vec{A} = A\hat{j} = (1.40\text{m})^2 \hat{j}$.

(a) $\Phi = (6.00 \text{ N/C})\hat{i} \cdot (1.40 \text{ m})^2 \hat{j} = 0.$

(b) $\Phi = (-2.00 \text{ N/C})\hat{j} \cdot (1.40 \text{ m})^2 \hat{j} = -3.92 \text{ N} \cdot \text{m}^2/\text{C}.$

(c) $\Phi = \left[(-3.00 \text{ N/C})\hat{i} + (400 \text{ N/C})\hat{k} \right] \cdot (1.40 \text{ m})^2 \hat{j} = 0.$

(d) The total flux of a uniform field through a closed surface is always zero.

3. We use $\Phi = \int \vec{E} \cdot d\vec{A}$ and note that the side length of the cube is $(3.0 \text{ m} - 1.0 \text{ m}) = 2.0 \text{ m}$.

(a) On the top face of the cube $y = 2.0$ m and $d\vec{A} = (dA)\hat{j}$. Therefore, we have
$\vec{E} = 4\hat{i} - 3((2.0)^2 + 2)\hat{j} = 4\hat{i} - 18\hat{j}$. Thus the flux is

$$\Phi = \int_{\text{top}} \vec{E} \cdot d\vec{A} = \int_{\text{top}} (4\hat{i} - 18\hat{j}) \cdot (dA)\hat{j} = -18 \int_{\text{top}} dA = (-18)(2.0)^2 \text{ N} \cdot \text{m}^2/\text{C} = -72 \text{ N} \cdot \text{m}^2/\text{C}.$$

(b) On the bottom face of the cube $y = 0$ and $d\vec{A} = (dA)(-\hat{j})$. Therefore, we have

$E = 4\hat{i} - 3(0^2 + 2)\hat{j} = 4\hat{i} - 6\hat{j}$. Thus, the flux is

$$\Phi = \int_{bottom} \vec{E} \cdot d\vec{A} = \int_{bottom} (4\hat{i} - 6\hat{j}) \cdot (dA)(-\hat{j}) = 6 \int_{bottom} dA = 6(2.0)^2 \text{ N} \cdot \text{m}^2/\text{C} = +24 \text{ N} \cdot \text{m}^2/\text{C}.$$

(c) On the left face of the cube $d\vec{A} = (dA)(-\hat{i})$. So

$$\Phi = \int_{left} \hat{E} \cdot d\vec{A} = \int_{left} (4\hat{i} + E_y\hat{j}) \cdot (dA)(-\hat{i}) = -4 \int_{bottom} dA = -4(2.0)^2 \text{ N} \cdot \text{m}^2/\text{C} = -16 \text{ N} \cdot \text{m}^2/\text{C}.$$

(d) On the back face of the cube $d\vec{A} = (dA)(-\hat{k})$. But since $\vec{E}$ has no z component $\vec{E} \cdot d\vec{A} = 0$. Thus, $\Phi = 0$.

(e) We now have to add the flux through all six faces. One can easily verify that the flux through the front face is zero, while that through the right face is the opposite of that through the left one, or +16 N·m²/C. Thus the net flux through the cube is

$$\Phi = (-72 + 24 - 16 + 0 + 0 + 16) \text{ N} \cdot \text{m}^2/\text{C} = -48 \text{ N} \cdot \text{m}^2/\text{C}.$$

4. The flux through the flat surface encircled by the rim is given by $\Phi = \pi a^2 E$. Thus, the flux through the netting is

$$\Phi' = -\Phi = -\pi a^2 E = -\pi(0.11 \text{ m})^2 (3.0 \times 10^{-3} \text{ N/C}) = -1.1 \times 10^{-4} \text{ N} \cdot \text{m}^2/\text{C}.$$

5. We use Gauss' law: $\varepsilon_0 \Phi = q$, where Φ is the total flux through the cube surface and q is the net charge inside the cube. Thus,

$$\Phi = \frac{q}{\varepsilon_0} = \frac{1.8 \times 10^{-6} \text{ C}}{8.85 \times 10^{-12} \text{ C}^2/\text{N} \cdot \text{m}^2} = 2.0 \times 10^5 \text{ N} \cdot \text{m}^2/\text{C}.$$

6. There is no flux through the sides, so we have two "inward" contributions to the flux, one from the top (of magnitude $(34)(3.0)^2$) and one from the bottom (of magnitude $(20)(3.0)^2$). With "inward" flux being negative, the result is $\Phi = -486$ N·m²/C. Gauss' law then leads to $q_{enc} = \varepsilon_0 \Phi = -4.3 \times 10^{-9}$ C.

7. To exploit the symmetry of the situation, we imagine a closed Gaussian surface in the shape of a cube, of edge length d, with a proton of charge $q = +1.6 \times 10^{-19}$ C situated at the inside center of the cube. The cube has six faces, and we expect an equal amount of flux through each face. The total amount of flux is $\Phi_{net} = q/\varepsilon_0$, and we conclude that the flux through the square is one-sixth of that. Thus, $\Phi = q/6\varepsilon_0 = 3.01 \times 10^{-9}$ N·m²/C.

8. (a) The total surface area bounding the bathroom is

$$A = 2(2.5 \times 3.0) + 2(3.0 \times 2.0) + 2(2.0 \times 2.5) = 37 \text{ m}^2.$$

The absolute value of the total electric flux, with the assumptions stated in the problem, is

$$|\Phi| = \left| \sum \vec{E} \cdot \vec{A} \right| = |\vec{E}| A = (600)(37) = 22 \times 10^3 \text{ N} \cdot \text{m}^2 / \text{C}.$$

By Gauss' law, we conclude that the enclosed charge (in absolute value) is $|q_{enc}| = \varepsilon_0 |\Phi| = 2.0 \times 10^{-7}$ C. Therefore, with volume $V = 15$ m^3, and recognizing that we are dealing with negative charges (see problem), the charge density is $q_{enc}/V = -1.3 \times 10^{-8}$ C/m^3.

(b) We find $(|q_{enc}|/e)/V = (2.0 \times 10^{-7}/1.6 \times 10^{-19})/15 = 8.2 \times 10^{10}$ excess electrons per cubic meter.

9. Let A be the area of one face of the cube, E_u be the magnitude of the electric field at the upper face, and E_l be the magnitude of the field at the lower face. Since the field is downward, the flux through the upper face is negative and the flux through the lower face is positive. The flux through the other faces is zero, so the total flux through the cube surface is $\Phi = A(E_l - E_u)$. The net charge inside the cube is given by Gauss' law:

$$q = \varepsilon_0 \Phi = \varepsilon_0 A(E_l - E_u) = (8.85 \times 10^{-12} \text{ C}^2 / \text{N} \cdot \text{m}^2)(100 \text{ m})^2 (100 \text{ N/C} - 60.0 \text{ N/C})$$
$$= 3.54 \times 10^{-6} \text{ C} = 3.54 \ \mu\text{C}.$$

10. We note that only the smaller shell contributes a (non-zero) field at the designated point, since the point is inside the radius of the large sphere (and $E = 0$ inside of a spherical charge), and the field points towards the $-x$ direction. Thus,

$$\vec{E} = E(-\hat{j}) = -\frac{q}{4\pi\varepsilon_0 r^2} \hat{j} = -\frac{\sigma_2 \, 4\pi R^2}{4\pi\varepsilon_0 (L-x)^2} \hat{j} = -(2.8 \times 10^4 \text{ N/C}) \hat{j},$$

where $R = 0.020$ m (the radius of the smaller shell), $d = 0.10$ m and $x = 0.020$ m.

11. The total flux through any surface that completely surrounds the point charge is q/ε_0.

(a) If we stack identical cubes side by side and directly on top of each other, we will find that eight cubes meet at any corner. Thus, one-eighth of the field lines emanating from the point charge pass through a cube with a corner at the charge, and the total flux through the surface of such a cube is $q/8\varepsilon_0$. Now the field lines are radial, so at each of the three cube faces that meet at the charge, the lines are parallel to the face and the flux through the face is zero.

(b) The fluxes through each of the other three faces are the same, so the flux through each of them is one-third of the total. That is, the flux through each of these faces is $(1/3)(q/8\varepsilon_0)$ $= q/24\varepsilon_0$. Thus, the multiple is $1/24 = 0.0417$.

12. Eq. 23-6 (Gauss' law) gives $\varepsilon_0 \Phi = q_{\text{enclosed}}$.

(a) Thus, the value $\Phi = 2.0 \times 10^5$ (in SI units) for small r leads to $q_{\text{central}} = +1.77 \times 10^{-6}\,\text{C}$ or roughly 1.8 μC.

(b) The next value that Φ takes is -4.0×10^5 (in SI units), which implies $q_{\text{enc}} = -3.54 \times 10^{-6}\,\text{C}$. But we have already accounted for some of that charge in part (a), so the result for part (b) is $q_A = q_{\text{enc}} - q_{\text{central}} = -5.3 \times 10^{-6}\,\text{C}$.

(c) Finally, the large r value for Φ is 6.0×10^5 (in SI units), which implies $q_{\text{total enc}} = 5.31 \times 10^{-6}\,\text{C}$. Considering what we have already found, then the result is $q_{\text{total enc}} - q_A - q_{\text{central}} = +8.9\,\mu\text{C}$.

13. (a) Let $A = (1.40\ \text{m})^2$. Then

$$\Phi = \left(3.00 y\,\hat{j}\right) \cdot \left(-A\,\hat{j}\right)\Big|_{y=0} + \left(3.00 y\,\hat{j}\right) \cdot \left(A\,\hat{j}\right)\Big|_{y=1.40} = (3.00)(1.40)(1.40)^2 = 8.23\ \text{N} \cdot \text{m}^2/\text{C}.$$

(b) The charge is given by

$$q_{\text{enc}} = \varepsilon_0 \Phi = \left(8.85 \times 10^{-12}\,\text{C}^2/\text{N} \cdot \text{m}^2\right)\left(8.23\ \text{N} \cdot \text{m}^2/\text{C}\right) = 7.29 \times 10^{-11}\ \text{C}.$$

(c) The electric field can be re-written as $\vec{E} = 3.00 y\,\hat{j} + \vec{E}_0$, where $\vec{E}_0 = -4.00\,\hat{i} + 6.00\,\hat{j}$ is a constant field which does not contribute to the net flux through the cube. Thus Φ is still 8.23 N·m²/C.

(d) The charge is again given by

$$q_{\text{enc}} = \varepsilon_0 \Phi = \left(8.85 \times 10^{-12}\,\text{C}^2/\text{N} \cdot \text{m}^2\right)\left(8.23\ \text{N} \cdot \text{m}^2/\text{C}\right) = 7.29 \times 10^{-11}\ \text{C}.$$

14. The total electric flux through the cube is $\Phi = \oint \vec{E} \cdot d\vec{A}$. The net flux through the two faces parallel to the yz plane is

$$\Phi_{yz} = \iint [E_x(x=x_2) - E_x(x=x_1)]dydz = \int_{y_1=0}^{y_2=1} dy \int_{z_1=1}^{z_2=3} dz[10 + 2(4) - 10 - 2(1)]$$

$$= 6 \int_{y_1=0}^{y_2=1} dy \int_{z_1=1}^{z_2=3} dz = 6(1)(2) = 12.$$

Similarly, the net flux through the two faces parallel to the *xz* plane is

$$\Phi_{xz} = \iint [E_y(y=y_2) - E_y(y=y_1)]dxdz = \int_{x_1=1}^{x_2=4} dy \int_{z_1=1}^{z_2=3} dz[-3-(-3)] = 0,$$

and the net flux through the two faces parallel to the *xy* plane is

$$\Phi_{xy} = \iint [E_z(z=z_2) - E_z(z=z_1)]dxdy = \int_{x_1=1}^{x_2=4} dx \int_{y_1=0}^{y_2=1} dy(3b-b) = 2b(3)(1) = 6b.$$

Applying Gauss' law, we obtain

$$q_{enc} = \varepsilon_0 \Phi = \varepsilon_0 (\Phi_{xy} + \Phi_{xz} + \Phi_{yz}) = \varepsilon_0 (6.00b + 0 + 12.0) = 24.0\varepsilon_0$$

which implies that $b = 2.00$ N/C·m .

15. (a) The charge on the surface of the sphere is the product of the surface charge density σ and the surface area of the sphere (which is $4\pi r^2$, where r is the radius). Thus,

$$q = 4\pi r^2 \sigma = 4\pi \left(\frac{1.2 \text{ m}}{2}\right)^2 (8.1\times10^{-6} \text{ C/m}^2) = 3.7\times10^{-5} \text{ C}.$$

(b) We choose a Gaussian surface in the form of a sphere, concentric with the conducting sphere and with a slightly larger radius. The flux is given by Gauss's law:

$$\Phi = \frac{q}{\varepsilon_0} = \frac{3.66\times10^{-5} \text{ C}}{8.85\times10^{-12} \text{ C}^2/\text{N}\cdot\text{m}^2} = 4.1\times10^6 \text{ N}\cdot\text{m}^2/\text{C} .$$

16. Using Eq. 23-11, the surface charge density is

$$\sigma = E\varepsilon_0 = (2.3\times10^5 \text{ N/C})(8.85\times10^{-12}\text{C}^2/\text{N}\cdot\text{m}^2) = 2.0\times10^{-6} \text{ C/m}^2.$$

17. (a) The area of a sphere may be written $4\pi R^2 = \pi D^2$. Thus,

$$\sigma = \frac{q}{\pi D^2} = \frac{2.4\times10^{-6} \text{ C}}{\pi(1.3 \text{ m})^2} = 4.5\times10^{-7} \text{ C/m}^2.$$

(b) Eq. 23-11 gives

$$E = \frac{\sigma}{\varepsilon_0} = \frac{4.5\times10^{-7} \text{ C/m}^2}{8.85\times10^{-12} \text{ C}^2/\text{N.m}^2} = 5.1\times10^4 \text{ N/C}.$$

18. Eq. 23-6 (Gauss' law) gives $\varepsilon_0 \Phi = q_{enc}$.

(a) The value $\Phi = -9.0 \times 10^5$ (in SI units) for small r leads to $q_{central} = -7.97 \times 10^{-6}$ C or roughly -8.0 μC.

(b) The next (non-zero) value that Φ takes is $+4.0 \times 10^5$ (in SI units), which implies $q_{enc} = 3.54 \times 10^{-6}$ C. But we have already accounted for some of that charge in part (a), so the result is

$$q_A = q_{enc} - q_{central} = 11.5 \times 10^{-6}\,\text{C} \approx 12\ \mu\text{C}.$$

(c) Finally, the large r value for Φ is -2.0×10^5 (in SI units), which implies $q_{\text{total enc}} = -1.77 \times 10^{-6}$ C. Considering what we have already found, then the result is

$$q_{\text{total enc}} - q_A - q_{central} = -5.3\ \mu\text{C}.$$

19. (a) Consider a Gaussian surface that is completely within the conductor and surrounds the cavity. Since the electric field is zero everywhere on the surface, the net charge it encloses is zero. The net charge is the sum of the charge q in the cavity and the charge q_w on the cavity wall, so $q + q_w = 0$ and $q_w = -q = -3.0 \times 10^{-6}$C.

(b) The net charge Q of the conductor is the sum of the charge on the cavity wall and the charge q_s on the outer surface of the conductor, so $Q = q_w + q_s$ and

$$q_s = Q - q_\omega = \left(10 \times 10^{-6}\ \text{C}\right) - \left(-3.0 \times 10^{-6}\ \text{C}\right) = +1.3 \times 10^{-5}\ \text{C}.$$

20. We imagine a cylindrical Gaussian surface A of radius r and unit length concentric with the metal tube. Then by symmetry $\oint_A \vec{E} \cdot d\vec{A} = 2\pi r E = \dfrac{q_{enc}}{\varepsilon_0}$.

(a) For $r < R$, $q_{enc} = 0$, so $E = 0$.

(b) For $r > R$, $q_{enc} = \lambda$, so $E(r) = \lambda / 2\pi r \varepsilon_0$. With $\lambda = 2.00 \times 10^{-8}$ C/m and $r = 2.00R = 0.0600$ m, we obtain

$$E = \frac{\left(2.0 \times 10^{-8}\ \text{C/m}\right)}{2\pi \left(0.0600\ \text{m}\right)\left(8.85 \times 10^{-12}\ \text{C}^2 / \text{N} \cdot \text{m}^2\right)} = 5.99 \times 10^3\ \text{N/C}.$$

(c) The plot of E vs. r is shown below.

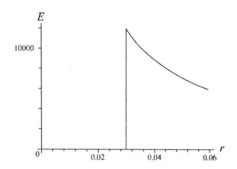

Here, the maximum value is

$$E_{max} = \frac{\lambda}{2\pi r \varepsilon_0} = \frac{\left(2.0\times 10^{-8} \text{ C/m}\right)}{2\pi\left(0.030 \text{ m}\right)\left(8.85\times 10^{-12} \text{ C}^2 / \text{N}\cdot\text{m}^2\right)} = 1.2\times 10^4 \text{ N/C.}$$

21. The magnitude of the electric field produced by a uniformly charged infinite line is $E = \lambda/2\pi\varepsilon_0 r$, where λ is the linear charge density and r is the distance from the line to the point where the field is measured. See Eq. 23-12. Thus,

$$\lambda = 2\pi\varepsilon_0 E r = 2\pi\left(8.85\times 10^{-12} \text{ C}^2 / \text{N}\cdot\text{m}^2\right)\left(4.5\times 10^4 \text{ N/C}\right)\left(2.0 \text{ m}\right) = 5.0\times 10^{-6} \text{ C/m.}$$

22. We combine Newton's second law ($F = ma$) with the definition of electric field ($F = qE$) and with Eq. 23-12 (for the field due to a line of charge). In terms of magnitudes, we have (if $r = 0.080$ m and $\lambda = 6.0 \times 10^{-6}$ C/m)

$$ma = eE = \frac{e\lambda}{2\pi\varepsilon_0 r} \quad\Rightarrow\quad a = \frac{e\lambda}{2\pi\varepsilon_0 r \, m} = 2.1 \times 10^{17} \text{ m/s}^2 .$$

23. (a) The side surface area A for the drum of diameter D and length h is given by $A = \pi D h$. Thus

$$q = \sigma A = \sigma\pi D h = \pi\varepsilon_0 E D h = \pi\left(8.85\times 10^{-12} \frac{\text{C}^2}{\text{N}\cdot\text{m}^2}\right)\left(2.3\times 10^5 \text{ N/C}\right)\left(0.12 \text{ m}\right)\left(0.42 \text{ m}\right)$$

$$= 3.2\times 10^{-7} \text{ C.}$$

(b) The new charge is

$$q' = q\left(\frac{A'}{A}\right) = q\left(\frac{\pi D' h'}{\pi D h}\right) = \left(3.2\times 10^{-7} \text{C}\right)\left[\frac{\left(8.0 \text{ cm}\right)\left(28 \text{ cm}\right)}{\left(12 \text{ cm}\right)\left(42 \text{ cm}\right)}\right] = 1.4\times 10^{-7} \text{ C.}$$

24. We reason that point P (the point on the x axis where the net electric field is zero) cannot be between the lines of charge (since their charges have opposite sign). We reason further that P is not to the left of "line 1" since its magnitude of charge (per unit

length) exceeds that of "line 2"; thus, we look in the region to the right of "line 2" for P. Using Eq. 23-12, we have

$$E_{net} = E_1 + E_2 = \frac{\lambda_1}{2\pi\varepsilon_0(x + L/2)} + \frac{\lambda_2}{2\pi\varepsilon_0(x - L/2)} \, .$$

Setting this equal to zero and solving for x we find

$$x = \frac{\lambda_1 - \lambda_2}{\lambda_1 + \lambda_2} \frac{L}{2}$$

which, for the values given in the problem, yields $x = 8.0$ cm.

25. We denote the inner and outer cylinders with subscripts i and o, respectively.

(a) Since $r_i < r = 4.0$ cm $< r_o$,

$$E(r) = \frac{\lambda_i}{2\pi\varepsilon_0 r} = \frac{5.0 \times 10^{-6} \text{ C/m}}{2\pi(8.85 \times 10^{-12} \text{ C}^2/\text{N} \cdot \text{m}^2)(4.0 \times 10^{-2} \text{ m})} = 2.3 \times 10^6 \text{ N/C}.$$

(b) The electric field $\vec{E}(r)$ points radially outward.

(c) Since $r > r_o$,

$$E(r) = \frac{\lambda_i + \lambda_o}{2\pi\varepsilon_0 r} = \frac{5.0 \times 10^{-6} \text{ C/m} - 7.0 \times 10^{-6} \text{ C/m}}{2\pi(8.85 \times 10^{-12} \text{ C}^2/\text{N} \cdot \text{m}^2)(8.0 \times 10^{-2} \text{ m})} = -4.5 \times 10^5 \text{ N/C},$$

or $|E(r)| = 4.5 \times 10^5$ N/C.

(d) The minus sign indicates that $\vec{E}(r)$ points radially inward.

26. As we approach $r = 3.5$ cm from the inside, we have

$$E_{internal} = \frac{\lambda}{2\pi\varepsilon_0 r} = 1000 \text{ N/C} .$$

And as we approach $r = 3.5$ cm from the outside, we have

$$E_{external} = \frac{\lambda}{2\pi\varepsilon_0 r} + \frac{\lambda'}{2\pi\varepsilon_0 r} = -3000 \text{ N/C} .$$

Considering the difference ($E_{external} - E_{internal}$) allows us to find λ' (the charge per unit length on the larger cylinder). Using $r = 0.035$ m, we obtain $\lambda' = -5.8 \times 10^{-9}$ C/m.

27. We assume the charge density of both the conducting cylinder and the shell are uniform, and we neglect fringing effect. Symmetry can be used to show that the electric field is radial, both between the cylinder and the shell and outside the shell. It is zero, of course, inside the cylinder and inside the shell.

(a) We take the Gaussian surface to be a cylinder of length L, coaxial with the given cylinders and of larger radius r than either of them. The flux through this surface is $\Phi = 2\pi rLE$, where E is the magnitude of the field at the Gaussian surface. We may ignore any flux through the ends. Now, the charge enclosed by the Gaussian surface is $q_{enc} = Q1 + Q_2 = -Q_1 = -3.40 \times 10^{-12}$ C. Consequently, Gauss' law yields $2\pi r\varepsilon_0 LE = q_{enc}$, or

$$E = \frac{q_{enc}}{2\pi\varepsilon_0 Lr} = \frac{-3.40 \times 10^{-12} \text{ C}}{2\pi(8.85 \times 10^{-12} \text{ C}^2/\text{N} \cdot \text{m}^2)(11.0 \text{ m})(20.0 \times 1.30 \times 10^{-3}\text{m})} = -0.214 \text{ N/C},$$

or $|E| = 0.214$ N/C.

(b) The negative sign in E indicates that the field points inward.

(c) Next, for $r = 5.00 R_1$, the charge enclosed by the Gaussian surface is $q_{enc} = Q_1 = 3.40 \times 10^{-12}$ C. Consequently, Gauss' law yields $2\pi r\varepsilon_0 LE = q_{enc}$, or

$$E = \frac{q_{enc}}{2\pi\varepsilon_0 Lr} = \frac{3.40 \times 10^{-12} \text{ C}}{2\pi(8.85 \times 10^{-12} \text{ C}^2/\text{N} \cdot \text{m}^2)(11.0 \text{ m})(5.00 \times 1.30 \times 10^{-3}\text{m})} = 0.855 \text{ N/C}.$$

(d) The positive sign indicates that the field points outward.

(e) we consider a cylindrical Gaussian surface whose radius places it within the shell itself. The electric field is zero at all points on the surface since any field within a conducting material would lead to current flow (and thus to a situation other than the electrostatic ones being considered here), so the total electric flux through the Gaussian surface is zero and the net charge within it is zero (by Gauss' law). Since the central rod has charge Q_1, the inner surface of the shell must have charge $Q_{in} = -Q_1 = -3.40 \times 10^{-12}$ C.

(f) Since the shell is known to have total charge $Q_2 = -2.00Q_1$, it must have charge $Q_{out} = Q_2 - Q_{in} = -Q_1 = -3.40 \times 10^{-12}$ C on its outer surface.

28. (a) In Eq. 23-12, $\lambda = q/L$ where q is the net charge enclosed by a cylindrical Gaussian surface of radius r. The field is being measured outside the system (the charged rod coaxial with the neutral cylinder) so that the net enclosed charge is only that which is on the rod. Consequently,

$$\left|\vec{E}\right| = \frac{\lambda}{2\pi\varepsilon_0\, r} = \frac{2.0\times10^{-9}}{2\pi\varepsilon_0(0.15)} = 2.4\times10^2 \text{ N/C}.$$

(b) Since the field is zero inside the conductor (in an electrostatic configuration), then there resides on the inner surface charge $-q$, and on the outer surface, charge $+q$ (where q is the charge on the rod at the center). Therefore, with $r_i = 0.05$ m, the surface density of charge is

$$\sigma_{inner} = \frac{-q}{2\pi r_i L} = -\frac{\lambda}{2\pi r_i} = -6.4\times10^{-9} \text{ C/m}^2$$

for the inner surface.

(c) With $r_o = 0.10$ m, the surface charge density of the outer surface is

$$\sigma_{outer} = \frac{+q}{2\pi r_o L} = \frac{\lambda}{2\pi r_o} = +3.2\times10^{-9} \text{ C/m}^2.$$

29. We denote the radius of the thin cylinder as $R = 0.015$ m. Using Eq. 23-12, the net electric field for $r > R$ is given by

$$E_{net} = E_{wire} + E_{cylinder} = \frac{-\lambda}{2\pi\varepsilon_0 r} + \frac{\lambda'}{2\pi\varepsilon_0 r}$$

where $-\lambda = -3.6$ nC/m is the linear charge density of the wire and λ' is the linear charge density of the thin cylinder. We note that the surface and linear charge densities of the thin cylinder are related by

$$q_{cylinder} = \lambda'\, L = \sigma(2\pi R L) \Rightarrow \lambda' = \sigma(2\pi R).$$

Now, E_{net} outside the cylinder will equal zero, provided that $2\pi R\sigma = \lambda$, or

$$\sigma = \frac{\lambda}{2\pi R} = \frac{3.6\times10^{-6} \text{ C/m}}{(2\pi)(0.015 \text{ m})} = 3.8\times10^{-8} \text{ C/m}^2.$$

30. To evaluate the field using Gauss' law, we employ a cylindrical surface of area $2\pi r L$ where L is very large (large enough that contributions from the ends of the cylinder become irrelevant to the calculation). The volume within this surface is $V = \pi r^2 L$, or expressed more appropriate to our needs: $dV = 2\pi r L\, dr$. The charge enclosed is, with $A = 2.5\times10^{-6}\,\text{C/m}^5$,

$$q_{enc} = \int_0^r Ar^2\, 2\pi r\, L\, dr = \frac{\pi}{2} A L r^4.$$

By Gauss' law, we find $\Phi = |\vec{E}|(2\pi r L) = q_{enc}/\varepsilon_0$; we thus obtain $|\vec{E}| = \dfrac{A r^3}{4\varepsilon_0}$.

(a) With $r = 0.030$ m, we find $|\vec{E}| = 1.9$ N/C.

(b) Once outside the cylinder, Eq. 23-12 is obeyed. To find $\lambda = q/L$ we must find the total charge q. Therefore,

$$\frac{q}{L} = \frac{1}{L}\int_0^{0.04} A r^2\, 2\pi r\, L\, dr = 1.0 \times 10^{-11} \text{ C/m}.$$

And the result, for $r = 0.050$ m, is $|\vec{E}| = \lambda/2\pi\varepsilon_0 r = 3.6$ N/C.

31. (a) To calculate the electric field at a point very close to the center of a large, uniformly charged conducting plate, we may replace the finite plate with an infinite plate with the same area charge density and take the magnitude of the field to be $E = \sigma/\varepsilon_0$, where σ is the area charge density for the surface just under the point. The charge is distributed uniformly over both sides of the original plate, with half being on the side near the field point. Thus,

$$\sigma = \frac{q}{2A} = \frac{6.0 \times 10^{-6} \text{ C}}{2(0.080 \text{ m})^2} = 4.69 \times 10^{-4} \text{ C/m}^2.$$

The magnitude of the field is

$$E = \frac{4.69 \times 10^{-4} \text{ C/m}^2}{8.85 \times 10^{-12} \text{ C}^2/\text{N} \cdot \text{m}^2} = 5.3 \times 10^{7} \text{ N/C}.$$

The field is normal to the plate and since the charge on the plate is positive, it points away from the plate.

(b) At a point far away from the plate, the electric field is nearly that of a point particle with charge equal to the total charge on the plate. The magnitude of the field is $E = q/4\pi\varepsilon_0 r^2 = kq/r^2$, where r is the distance from the plate. Thus,

$$E = \frac{(8.99 \times 10^{9} \text{ N} \cdot \text{m}^2/\text{C}^2)(6.0 \times 10^{-6} \text{ C})}{(30 \text{ m})^2} = 60 \text{ N/C}.$$

32. According to Eq. 23-13 the electric field due to either sheet of charge with surface charge density $\sigma = 1.77 \times 10^{-22}$ C/m^2 is perpendicular to the plane of the sheet (pointing *away* from the sheet if the charge is positive) and has magnitude $E = \sigma/2\varepsilon_0$. Using the superposition principle, we conclude:

(a) $E = \sigma/\varepsilon_0 = (1.77 \times 10^{-22})/(8.85 \times 10^{-12}) = 2.00 \times 10^{-11}$ N/C, pointing in the upward direction, or $\vec{E} = (2.00 \times 10^{-11}$ N/C$)\hat{j}$.

(b) $E = 0$;

(c) and, $E = \sigma/\varepsilon_0$, pointing down, or $\vec{E} = -(2.00 \times 10^{-11}$ N/C$)\hat{j}$.

33. In the region between sheets 1 and 2, the net field is $E_1 - E_2 + E_3 = 2.0 \times 10^5$ N/C.

In the region between sheets 2 and 3, the net field is at its greatest value:

$$E_1 + E_2 + E_3 = 6.0 \times 10^5 \text{ N/C}.$$

The net field vanishes in the region to the right of sheet 3, where $E_1 + E_2 = E_3$. We note the implication that σ_3 is negative (and is the largest surface-density, in magnitude). These three conditions are sufficient for finding the fields:

$$E_1 = 1.0 \times 10^5 \text{ N/C}, \quad E_2 = 2.0 \times 10^5 \text{ N/C}, \quad E_3 = 3.0 \times 10^5 \text{ N/C}.$$

From Eq. 23-13, we infer (from these values of E)

$$\frac{|\sigma_3|}{|\sigma_2|} = \frac{3.0 \times 10^5 \text{ N/C}}{2.0 \times 10^5 \text{ N/C}} = 1.5.$$

Recalling our observation, above, about σ_3, we conclude $\dfrac{\sigma_3}{\sigma_2} = -1.5$.

34. The charge distribution in this problem is equivalent to that of an infinite sheet of charge with surface charge density $\sigma = 4.50 \times 10^{-12}$ C/m^2 plus a small circular pad of radius $R = 1.80$ cm located at the middle of the sheet with charge density $-\sigma$. We denote the electric fields produced by the sheet and the pad with subscripts 1 and 2, respectively. Using Eq. 22-26 for $\vec{E}_2$, the net electric field $\vec{E}$ at a distance $z = 2.56$ cm along the central axis is then

$$\vec{E} = \vec{E}_1 + \vec{E}_2 = \left(\frac{\sigma}{2\varepsilon_0}\right)\hat{k} + \frac{(-\sigma)}{2\varepsilon_0}\left(1 - \frac{z}{\sqrt{z^2 + R^2}}\right)\hat{k} = \frac{\sigma z}{2\varepsilon_0\sqrt{z^2 + R^2}}\hat{k}$$

$$= \frac{(4.50 \times 10^{-12})(2.56 \times 10^{-2})}{2(8.85 \times 10^{-12})\sqrt{(2.56 \times 10^{-2})^2 + (1.80 \times 10^{-2})^2}}\hat{k} = (0.208 \text{ N/C})\,\hat{k}$$

35. We use Eq. 23-13.

(a) To the left of the plates:

$$\vec{E} = (\sigma / 2\varepsilon_0)(-\hat{i}) \text{ (from the right plate) } + (\sigma / 2\varepsilon_0)\hat{i} \text{ (from the left one) } = 0.$$

(b) To the right of the plates:

$$\vec{E} = (\sigma / 2\varepsilon_0)\hat{i} \text{ (from the right plate)} + (\sigma / 2\varepsilon_0)(-\hat{i}) \text{ (from the left one) } = 0.$$

(c) Between the plates:

$$\vec{E} = \left(\frac{\sigma}{2\varepsilon_0}\right)(-\hat{i}) + \left(\frac{\sigma}{2\varepsilon_0}\right)(-\hat{i}) = \left(\frac{\sigma}{\varepsilon_0}\right)(-\hat{i}) = -\left(\frac{7.00 \times 10^{-22} \text{ C/m}^2}{8.85 \times 10^{-12} \dfrac{\text{N} \cdot \text{m}^2}{\text{C}^2}}\right)\hat{i} = \left(-7.91 \times 10^{-11} \text{ N/C}\right)\hat{i}.$$

36. The field due to the sheet is $E = \dfrac{\sigma}{2\varepsilon_0}$. The force (in magnitude) on the electron (due to that field) is $F = eE$, and assuming it's the *only* force then the acceleration is

$$a = \frac{e\sigma}{2\varepsilon_0 m} = \text{slope of the graph } \left(= 2.0 \times 10^5 \text{ m/s divided by } 7.0 \times 10^{-12} \text{ s}\right) \, .$$

Thus we obtain $\sigma = 2.9 \times 10^{-6} \text{ C/m}^2$.

37. The charge on the metal plate, which is negative, exerts a force of repulsion on the electron and stops it. First find an expression for the acceleration of the electron, then use kinematics to find the stopping distance. We take the initial direction of motion of the electron to be positive. Then, the electric field is given by $E = \sigma/\varepsilon_0$, where σ is the surface charge density on the plate. The force on the electron is $F = -eE = -e\sigma/\varepsilon_0$ and the acceleration is

$$a = \frac{F}{m} = -\frac{e\sigma}{\varepsilon_0 m}$$

where m is the mass of the electron. The force is constant, so we use constant acceleration kinematics. If v_0 is the initial velocity of the electron, v is the final velocity, and x is the distance traveled between the initial and final positions, then $v^2 - v_0^2 = 2ax$. Set $v = 0$ and replace a with $-e\sigma/\varepsilon_0 m$, then solve for x. We find

$$x = -\frac{v_0^2}{2a} = \frac{\varepsilon_0 m v_0^2}{2e\sigma}.$$

Now $\frac{1}{2}mv_0^2$ is the initial kinetic energy K_0, so

$$x = \frac{\varepsilon_0 K_0}{e\sigma} = \frac{\left(8.85 \times 10^{-12} \ \mathrm{C^2 / N \cdot m^2}\right)\left(1.60 \times 10^{-17} \ \mathrm{J}\right)}{\left(1.60 \times 10^{-19} \ \mathrm{C}\right)\left(2.0 \times 10^{-6} \ \mathrm{C/m^2}\right)} = 4.4 \times 10^{-4} \ \mathrm{m}.$$

38. We use the result of part (c) of problem 35 to obtain the surface charge density.

$$E = \sigma / \varepsilon_0 \Rightarrow \sigma = \varepsilon_0 E = \left(8.85 \times 10^{-12} \ \frac{\mathrm{C^2}}{\mathrm{N \cdot m^2}}\right)(55 \ \mathrm{N/C}) = 4.9 \times 10^{-10} \ \mathrm{C/m^2}.$$

Since the area of the plates is $A = 1.0 \ \mathrm{m^2}$, the magnitude of the charge on the plate is $Q = \sigma A = 4.9 \times 10^{-10}$ C.

39. The forces acting on the ball are shown in the diagram below. The gravitational force has magnitude mg, where m is the mass of the ball; the electrical force has magnitude qE, where q is the charge on the ball and E is the magnitude of the electric field at the position of the ball; and, the tension in the thread is denoted by T. The electric field produced by the plate is normal to the plate and points to the right. Since the ball is positively charged, the electric force on it also points to the right. The tension in the thread makes the angle $\theta (= 30°)$ with the vertical.

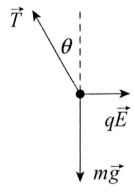

Since the ball is in equilibrium the net force on it vanishes. The sum of the horizontal components yields $qE - T \sin \theta = 0$ and the sum of the vertical components yields $T \cos \theta - mg = 0$. The expression $T = qE/\sin \theta$, from the first equation, is substituted into the second to obtain $qE = mg \tan \theta$. The electric field produced by a large uniform plane of charge is given by $E = \sigma/2\varepsilon_0$, where σ is the surface charge density. Thus,

$$\frac{q\sigma}{2\varepsilon_0} = mg \tan \theta$$

and

$$\sigma = \frac{2\varepsilon_0 mg \tan\theta}{q} = \frac{2\left(8.85\times10^{-12} \ C^2/N.m^2\right)\left(1.0\times10^{-6} \ kg\right)\left(9.8 \ m/s^2\right)\tan 30°}{2.0\times10^{-8} \ C}$$
$$= 5.0\times10^{-9} \ C/m^2.$$

40. The point where the individual fields cancel cannot be in the region between the sheet and the particle ($-d < x < 0$) since the sheet and the particle have opposite-signed charges. The point(s) could be in the region to the right of the particle ($x > 0$) and in the region to the left of the sheet ($x < d$); this is where the condition

$$\frac{|\sigma|}{2\varepsilon_0} = \frac{Q}{4\pi\varepsilon_0 r^2}$$

must hold. Solving this with the given values, we find $r = x = \pm\sqrt{3/2\pi} \approx \pm 0.691$ m.

If $d = 0.20$ m (which is less than the magnitude of r found above), then neither of the points ($x \approx \pm 0.691$ m) is in the "forbidden region" between the particle and the sheet. Thus, both values are allowed. Thus, we have

(a) $x = 0.691$ m on the positive axis, and

(b) $x = -0.691$ m on the negative axis.

(c) If, however, $d = 0.80$ m (greater than the magnitude of r found above), then one of the points ($x \approx -0.691$ m) is in the "forbidden region" between the particle and the sheet and is disallowed. In this part, the fields cancel only at the point $x \approx +0.691$ m.

41. We use a Gaussian surface in the form of a box with rectangular sides. The cross section is shown with dashed lines in the diagram below.

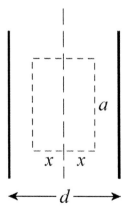

It is centered at the central plane of the slab, so the left and right faces are each a distance x from the central plane. We take the thickness of the rectangular solid to be a, the same as its length, so the left and right faces are squares. The electric field is normal to the left and right faces and is uniform over them. Since $\rho = 5.80$ fC/m^3 is positive, it points outward at both faces: toward the left at the left face and toward the right at the right face. Furthermore, the magnitude is the same at both faces. The electric flux through each of these faces is Ea^2. The field is parallel to the other faces of the Gaussian surface and the flux through them is zero. The total flux through the Gaussian surface is $\Phi = 2Ea^2$. The volume enclosed by the Gaussian surface is $2a^2x$ and the charge contained within it is $q = 2a^2x\rho$. Gauss' law yields $2\varepsilon_0 Ea^2 = 2a^2x\rho$. We solve for the magnitude of the electric field: $E = \rho x / \varepsilon_0$.

(a) For $x = 0$, $E = 0$.

(b) For $x = 2.00$ mm $= 2.00 \times 10^{-3}$ m,

$$E = (5.80 \times 10^{-15})(2.00 \times 10^{-3})/(8.85 \times 10^{-12}) = 1.31 \times 10^{-6} \text{ N/C}.$$

(c) For $x = d/2 = 4.70$ mm $= 4.70 \times 10^{-3}$ m,

$$E = (5.80 \times 10^{-15})(4.70 \times 10^{-3})/(8.85 \times 10^{-12}) = 3.08 \times 10^{-6} \text{ N/C}.$$

(d) For $x = 26.0$ mm $= 2.60 \times 10^{-2}$ m, we take a Gaussian surface of the same shape and orientation, but with $x > d/2$, so the left and right faces are outside the slab. The total flux through the surface is again $\Phi = 2Ea^2$ but the charge enclosed is now $q = a^2 d\rho$. Gauss' law yields $2\varepsilon_0 Ea^2 = a^2 d\rho$, so

$$E = \frac{\rho d}{2\varepsilon_0} = \frac{(5.80 \times 10^{-15})(9.40 \times 10^{-3})}{2(8.85 \times 10^{-12})} = 3.08 \times 10^{-6} \text{ N/C}.$$

42. We determine the (total) charge on the ball by examining the maximum value ($E = 5.0 \times 10^7$ N/C) shown in the graph (which occurs at $r = 0.020$ m). Thus,

$$E = \frac{q}{4\pi\varepsilon_0 r^2} \quad \Rightarrow \quad q = 2.2 \times 10^{-6} \text{C}.$$

43. Charge is distributed uniformly over the surface of the sphere and the electric field it produces at points outside the sphere is like the field of a point particle with charge equal to the net charge on the sphere. That is, the magnitude of the field is given by $E = q/4\pi\varepsilon_0 r^2$, where q is the magnitude of the charge on the sphere and r is the distance from the center of the sphere to the point where the field is measured. Thus,

$$q = 4\pi\varepsilon_0 r^2 E = \frac{(0.15 \text{ m})^2 (3.0 \times 10^3 \text{ N/C})}{8.99 \times 10^9 \text{ N} \cdot \text{m}^2 / \text{C}^2} = 7.5 \times 10^{-9} \text{ C}.$$

The field points inward, toward the sphere center, so the charge is negative: -7.5×10^{-9} C.

44. (a) The flux is still $-750 \text{ N} \cdot \text{m}^2/\text{C}$, since it depends only on the amount of charge enclosed.

(b) We use $\Phi = q / \varepsilon_0$ to obtain the charge q:

$$q = \varepsilon_0 \Phi = \left(8.85 \times 10^{-12} \frac{\text{C}^2}{\text{N} \cdot \text{m}^2} \right) \left(-750 \text{ N} \cdot \text{m}^2 / \text{C} \right) = -6.64 \times 10^{-9} \text{ C}.$$

45. (a) Since $r_1 = 10.0 \text{ cm} < r = 12.0 \text{ cm} < r_2 = 15.0 \text{ cm}$,

$$E(r) = \frac{1}{4\pi\varepsilon_0} \frac{q_1}{r^2} = \frac{\left(8.99 \times 10^9 \text{ N} \cdot \text{m}^2 \text{C}^2 \right) \left(4.00 \times 10^{-8} \text{ C} \right)}{\left(0.120 \text{ m} \right)^2} = 2.50 \times 10^4 \text{ N/C}.$$

(b) Since $r_1 < r_2 < r = 20.0 \text{ cm}$,

$$E(r) = \frac{1}{4\pi\varepsilon_0} \frac{q_1 + q_2}{r^2} = \frac{\left(8.99 \times 10^9 \text{ N} \cdot \text{m}^2 / \text{C}^2 \right) \left(4.00 + 2.00 \right) \left(1 \times 10^{-8} \text{ C} \right)}{\left(0.200 \text{ m}^2 \right)} = 1.35 \times 10^4 \text{ N/C}.$$

46. The point where the individual fields cancel cannot be in the region between the shells since the shells have opposite-signed charges. It cannot be inside the radius R of one of the shells since there is only one field contribution there (which would not be canceled by another field contribution and thus would not lead to zero net field). We note shell 2 has greater magnitude of charge ($|\sigma_2|A_2$) than shell 1, which implies the point is not to the right of shell 2 (any such point would always be closer to the larger charge and thus no possibility for cancellation of equal-magnitude fields could occur). Consequently, the point should be in the region to the left of shell 1 (at a distance $r > R_1$ from its center); this is where the condition

$$E_1 = E_2 \Rightarrow \frac{|q_1|}{4\pi\varepsilon_0 r^2} = \frac{|q_2|}{4\pi\varepsilon_0(r + L)^2}$$

or

$$\frac{\sigma_1 A_1}{4\pi\varepsilon_0 r^2} = \frac{|\sigma_2| A_2}{4\pi\varepsilon_0(r + L)^2} .$$

Using the fact that the area of a sphere is $A = 4\pi R^2$, this condition simplifies to

$$r = \frac{L}{(R_2/R_1)\sqrt{|\sigma_2|/\sigma_1} - 1} = 3.3 \text{ cm} .$$

We note that this value satisfies the requirement $r > R_1$. The answer, then, is that the net field vanishes at $x = -r = -3.3 \text{ cm}$.

47. To find an expression for the electric field inside the shell in terms of A and the distance from the center of the shell, select A so the field does not depend on the distance. We use a Gaussian surface in the form of a sphere with radius r_g, concentric with the spherical shell and within it ($a < r_g < b$). Gauss' law will be used to find the magnitude of the electric field a distance r_g from the shell center. The charge that is both in the shell and within the Gaussian sphere is given by the integral $q_s = \int \rho \, dV$ over the portion of the shell within the Gaussian surface. Since the charge distribution has spherical symmetry, we may take dV to be the volume of a spherical shell with radius r and infinitesimal thickness dr: $dV = 4\pi r^2 \, dr$. Thus,

$$q_s = 4\pi \int_a^{r_g} \rho r^2 dr = 4\pi \int_a^{r_g} \frac{A}{r} r^2 dr = 4\pi A \int_a^{r_g} r \, dr = 2\pi A \left(r_g^2 - a^2 \right).$$

The total charge inside the Gaussian surface is $q + q_s = q + 2\pi A \left(r_g^2 - a^2 \right)$. The electric field is radial, so the flux through the Gaussian surface is $\Phi = 4\pi r_g^2 E$, where E is the magnitude of the field. Gauss' law yields $4\pi \varepsilon_0 E r_g^2 = q + 2\pi A \left(r_g^2 - a^2 \right)$. We solve for E:

$$E = \frac{1}{4\pi\varepsilon_0} \left[\frac{q}{r_g^2} + 2\pi A - \frac{2\pi A a^2}{r_g^2} \right].$$

For the field to be uniform, the first and last terms in the brackets must cancel. They do if $q - 2\pi A a^2 = 0$ or $A = q/2\pi a^2$. With $a = 2.00 \times 10^{-2}$ m and $q = 45.0 \times 10^{-15}$ C, we have $A = 1.79 \times 10^{-11}$ C/m^2.

48. Let E_A designate the magnitude of the field at $r = 2.4$ cm. Thus $E_A = 2.0 \times 10^7$ N/C, and is totally due to the particle. Since

$$E_{\text{particle}} = \frac{q}{4\pi\varepsilon_0 r^2}$$

then the field due to the particle at any other point will relate to E_A by a ratio of distances squared. Now, we note that at $r = 3.0$ cm the total contribution (from particle and shell) is 8.0×10^7 N/C. Therefore,

$$E_{\text{shell}} + E_{\text{particle}} = E_{\text{shell}} + (2.4/3)^2 E_A = 8.0 \times 10^7 \text{ N/C}$$

Using the value for E_A noted above, we find $E_{\text{shell}} = 6.6 \times 10^7$ N/C. Thus, with $r = 0.030$ m, we find the charge Q using

$$E_{\text{shell}} = \frac{Q}{4\pi\varepsilon_0 r^2} \quad \Rightarrow \quad Q = 6.6 \times 10^{-6} \text{ C}.$$

49. At all points where there is an electric field, it is radially outward. For each part of the problem, use a Gaussian surface in the form of a sphere that is concentric with the sphere of charge and passes through the point where the electric field is to be found. The field is uniform on the surface, so $\oint \vec{E} \cdot d\vec{A} = 4\pi r^2 E$, where r is the radius of the Gaussian surface.

For $r < a$, the charge enclosed by the Gaussian surface is $q_1(r/a)^3$. Gauss' law yields

$$4\pi r^2 E = \left(\frac{q_1}{\varepsilon_0}\right)\left(\frac{r}{a}\right)^3 \Rightarrow E = \frac{q_1 r}{4\pi\varepsilon_0 a^3}.$$

(a) For $r = 0$, the above equation implies $E = 0$.

(b) For $r = a/2$, we have

$$E = \frac{q_1(a/2)}{4\pi\varepsilon_0 a^3} = \frac{(8.99\times10^9\,\text{N}\cdot\text{m}^2/\text{C}^2)(5.00\times10^{-15}\,\text{C})}{2(2.00\times10^{-2}\,\text{m})^2} = 5.62\times10^{-2}\ \text{N/C}.$$

(c) For $r = a$, we have

$$E = \frac{q_1}{4\pi\varepsilon_0 a^2} = \frac{(8.99\times10^9\,\text{N}\cdot\text{m}^2/\text{C}^2)(5.00\times10^{-15}\,\text{C})}{(2.00\times10^{-2}\,\text{m})^2} = 0.112\ \text{N/C}.$$

In the case where $a < r < b$, the charge enclosed by the Gaussian surface is q_1, so Gauss' law leads to

$$4\pi r^2 E = \frac{q_1}{\varepsilon_0} \Rightarrow E = \frac{q_1}{4\pi\varepsilon_0 r^2}.$$

(d) For $r = 1.50a$, we have

$$E = \frac{q_1}{4\pi\varepsilon_0 r^2} = \frac{(8.99\times10^9\,\text{N}\cdot\text{m}^2/\text{C}^2)(5.00\times10^{-15}\,\text{C})}{(1.50\times2.00\times10^{-2}\,\text{m})^2} = 0.0499\ \text{N/C}.$$

(e) In the region $b < r < c$, since the shell is conducting, the electric field is zero. Thus, for $r = 2.30a$, we have $E = 0$.

(f) For $r > c$, the charge enclosed by the Gaussian surface is zero. Gauss' law yields $4\pi r^2 E = 0 \Rightarrow E = 0$. Thus, $E = 0$ at $r = 3.50a$.

(g) Consider a Gaussian surface that lies completely within the conducting shell. Since the electric field is everywhere zero on the surface, $\oint \vec{E} \cdot d\vec{A} = 0$ and, according to Gauss' law, the net charge enclosed by the surface is zero. If Q_i is the charge on the inner surface of the shell, then $q_1 + Q_i = 0$ and $Q_i = -q_1 = -5.00$ fC.

(h) Let Q_o be the charge on the outer surface of the shell. Since the net charge on the shell is $-q$, $Q_i + Q_o = -q_1$. This means $Q_o = -q_1 - Q_i = -q_1 - (-q_1) = 0$.

50. The field is zero for $0 \le r \le a$ as a result of Eq. 23-16. Thus,

(a) $E = 0$ at $r = 0$,

(b) $E = 0$ at $r = a/2.00$, and

(c) $E = 0$ at $r = a$.

For $a \le r \le b$ the enclosed charge q_{enc} (for $a \le r \le b$) is related to the volume by

$$q_{enc} = \rho \left(\frac{4\pi r^3}{3} - \frac{4\pi a^3}{3} \right).$$

Therefore, the electric field is

$$E = \frac{1}{4\pi\varepsilon_0} \frac{q_{enc}}{r^2} = \frac{\rho}{4\pi\varepsilon_0 r^2} \left(\frac{4\pi r^3}{3} - \frac{4\pi a^3}{3} \right) = \frac{\rho}{3\varepsilon_0} \frac{r^3 - a^3}{r^2}$$

for $a \le r \le b$.

(d) For $r = 1.50a$, we have

$$E = \frac{\rho}{3\varepsilon_0} \frac{(1.50a)^3 - a^3}{(1.50a)^2} = \frac{\rho a}{3\varepsilon_0} \frac{2.375}{2.25} = \frac{(1.84\times10^{-9})(0.100)}{3(8.85\times10^{-12})} \frac{2.375}{2.25} = 7.32 \text{ N/C.}$$

(e) For $r = b = 2.00a$, the electric field is

$$E = \frac{\rho}{3\varepsilon_0} \frac{(2.00a)^3 - a^3}{(2.00a)^2} = \frac{\rho a}{3\varepsilon_0} \frac{7}{4} = \frac{(1.84\times10^{-9})(0.100)}{3(8.85\times10^{-12})} \frac{7}{4} = 12.1 \text{ N/C.}$$

(f) For $r \ge b$ we have $E = q_{total} / 4\pi\varepsilon_0 r^2$ or

$$E = \frac{\rho}{3\varepsilon_0} \frac{b^3 - a^3}{r^2}.$$

Thus, for $r = 3.00b = 6.00a$, the electric field is

$$E = \frac{\rho}{3\varepsilon_0} \frac{(2.00a)^3 - a^3}{(6.00a)^2} = \frac{\rho a}{3\varepsilon_0} \frac{7}{36} = \frac{(1.84 \times 10^{-9})(0.100)}{3(8.85 \times 10^{-12})} \frac{7}{4} = 1.35 \text{ N/C}.$$

51. (a) We integrate the volume charge density over the volume and require the result be equal to the total charge:

$$\int dx \int dy \int dz \, \rho = 4\pi \int_0^R dr \, r^2 \, \rho = Q.$$

Substituting the expression $\rho = \rho_s r / R$, with $\rho_s = 14.1$ pC/m^3, and performing the integration leads to

$$4\pi \left(\frac{\rho_s}{R} \right) \left(\frac{R^4}{4} \right) = Q$$

or

$$Q = \pi \rho_s R^3 = \pi (14.1 \times 10^{-12} \text{ C/m}^3)(0.0560 \text{ m})^3 = 7.78 \times 10^{-15} \text{ C}.$$

(b) At $r = 0$, the electric field is zero ($E = 0$) since the enclosed charge is zero.

At a certain point within the sphere, at some distance r from the center, the field (see Eq. 23-8 through Eq. 23-10) is given by Gauss' law:

$$E = \frac{1}{4\pi\varepsilon_0} \frac{q_{enc}}{r^2}$$

where q_{enc} is given by an integral similar to that worked in part (a):

$$q_{enc} = 4\pi \int_0^r dr \, r^2 \rho = 4\pi \left(\frac{\rho_s}{R} \right) \left(\frac{r^4}{4} \right).$$

Therefore,

$$E = \frac{1}{4\pi\varepsilon_0} \frac{\pi \rho_s r^4}{R r^2} = \frac{1}{4\pi\varepsilon_0} \frac{\pi \rho_s r^2}{R}.$$

(c) For $r = R/2.00$, where $R = 5.60$ cm, the electric field is

$$E = \frac{1}{4\pi\varepsilon_0} \frac{\pi \rho_s (R/2.00)^2}{R} = \frac{1}{4\pi\varepsilon_0} \frac{\pi \rho_s R}{4.00} = \frac{(8.99 \times 10^9)\pi(14.1 \times 10^{-12})(0.0560)}{4.00} = 5.58 \times 10^{-3} \text{ N/C}.$$

(d) For $r = R$, the electric field is

$$E = \frac{1}{4\pi\varepsilon_0}\frac{\pi\rho_s R^2}{R} = \frac{\pi\rho_s R}{4\pi\varepsilon_0} = (8.99\times10^9)\pi(14.1\times10^{-12})(0.0560) = 2.23\times10^{-2}\,\text{N/C}.$$

(e) The electric field strength as a function of r is depicted below:

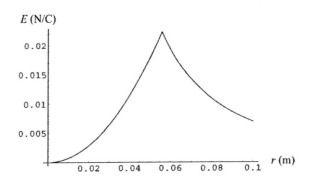

52. Applying Eq. 23-20, we have

$$E_1 = \frac{|q_1|}{4\pi\varepsilon_0 R^3}r_1 = \frac{|q_1|}{4\pi\varepsilon_0 R^3}\left(\frac{R}{2}\right) = \frac{|q_1|}{8\pi\varepsilon_0 R^2}.$$

Also, outside sphere 2 we have

$$E_2 = \frac{|q_2|}{4\pi\varepsilon_0 r^2} = \frac{|q_2|}{4\pi\varepsilon_0(1.5\,R)^2}.$$

Equating these and solving for the ratio of charges, we arrive at $\dfrac{q_2}{q_1} = \dfrac{9}{8} = 1.125.$

53. We use

$$E(r) = \frac{q_{\text{encl}}}{4\pi\varepsilon_0 r^2} = \frac{1}{4\pi\varepsilon_0 r^2}\int_0^r \rho(r)4\pi r^2 dr$$

to solve for $\rho(r)$:

$$\rho(r) = \frac{\varepsilon_0}{r^2}\frac{d}{dr}\left[r^2 E(r)\right] = \frac{\varepsilon_0}{r^2}\frac{d}{dr}\left(Kr^6\right) = 6K\varepsilon_0 r^3.$$

54. (a) We consider the radial field produced at points within a uniform cylindrical distribution of charge. The volume enclosed by a Gaussian surface in this case is $L\pi r^2$. Thus, Gauss' law leads to

$$E = \frac{|q_{enc}|}{\varepsilon_0 A_{cylinder}} = \frac{|\rho|(L\pi r^2)}{\varepsilon_0 (2\pi r L)} = \frac{|\rho| r}{2\varepsilon_0}.$$

(b) We note from the above expression that the magnitude of the radial field grows with r.

(c) Since the charged powder is negative, the field points radially inward.

(d) The largest value of r which encloses charged material is $r_{max} = R$. Therefore, with $|\rho| = 0.0011$ C/m^3 and $R = 0.050$ m, we obtain

$$E_{max} = \frac{|\rho| R}{2\varepsilon_0} = 3.1 \times 10^6 \text{ N/C}.$$

(e) According to condition 1 mentioned in the problem, the field is high enough to produce an electrical discharge (at $r = R$).

55. (a) The cube is totally within the spherical volume, so the charge enclosed is

$$q_{enc} = \rho V_{cube} = (500 \times 10^{-9})(0.0400)^3 = 3.20 \times 10^{-11} \text{ C}.$$

By Gauss' law, we find $\Phi = q_{enc}/\varepsilon_0 = 3.62$ N·m^2/C.

(b) Now the sphere is totally contained within the cube (note that the radius of the sphere is less than half the side-length of the cube). Thus, the total charge is

$$q_{enc} = \rho V_{sphere} = 4.5 \times 10^{-10} \text{ C}.$$

By Gauss' law, we find $\Phi = q_{enc}/\varepsilon_0 = 51.1$ N·m^2/C.

56. (a) Since the volume contained within a radius of $\frac{1}{2} R$ is one-eighth the volume contained within a radius of R, so the charge at $0 < r < R/2$ is $Q/8$. The fraction is $1/8 = 0.125$.

(b) At $r = R/2$, the magnitude of the field is

$$E = \frac{Q/8}{4\pi\varepsilon_0 (R/2)^2} = \frac{Q}{8\pi\varepsilon_0 R^2}$$

and is equivalent to *half* the field at the surface. Thus, the ratio is 0.500.

57. (a) We use $m_e g = eE = e\sigma/\varepsilon_0$ to obtain the surface charge density.

$$\sigma = \frac{m_e g \varepsilon_0}{e} = \frac{(9.11 \times 10^{-31} \text{kg})(9.8 \text{ m/s})(8.85 \times 10^{-12} \frac{C^2}{N \cdot m^2})}{1.60 \times 10^{-19} \text{ C}} = 4.9 \times 10^{-22} \text{ C/m}^2 .$$

(b) Downward (since the electric force exerted on the electron must be upward).

58. None of the constant terms will result in a nonzero contribution to the flux (see Eq. 23-4 and Eq. 23-7), so we focus on the x dependent term only. In Si units, we have

$$E_{\text{non-constant}} = 3x \, \hat{i} .$$

The face of the cube located at $x = 0$ (in the yz plane) has area $A = 4$ m^2 (and it "faces" the $+\hat{i}$ direction) and has a "contribution" to the flux equal to $E_{\text{non-constant}} A = (3)(0)(4) = 0$. The face of the cube located at $x = -2$ m has the same area A (and this one "faces" the $-\hat{i}$ direction) and a contribution to the flux: $-E_{\text{non-constant}} A = -(3)(-2)(4) = 24$ (in SI units). Thus, the net flux is $\Phi = 0 + 24 = 24$ N·m/C^2. According to Gauss' law, we therefore have $q_{\text{enc}} = \varepsilon_0 \Phi = 2.13 \times 10^{-10}$ C .

59. None of the constant terms will result in a nonzero contribution to the flux (see Eq. 23-4 and Eq. 23-7), so we focus on the x dependent term only:

$$E_{\text{non-constant}} = (-4.00y^2) \, \hat{i} \text{ (in SI units)} .$$

The face of the cube located at $y = 4.00$ has area $A = 4.00$ m^2 (and it "faces" the $+\hat{j}$ direction) and has a "contribution" to the flux equal to $E_{\text{non-constant}} A = (-4)(4^2)(4) = -256$ (in SI units). The face of the cube located at $y = 2.00$ m has the same area A (and this one "faces" the $-\hat{j}$ direction) and a contribution to the flux: $-E_{\text{non-constant}} A = -(-4)(2^2)(4) = 64$ (in SI units). Thus, the net flux is $\Phi = -256 + 64 = -192$ N·m/C^2. According to Gauss's law, we therefore have $q_{\text{enc}} = \varepsilon_0 \Phi = -1.70 \times 10^{-9}$ C .

60. (a) The field maximum occurs at the outer surface:

$$E_{\text{max}} = \left(\frac{|q|}{4\pi\varepsilon_0 r^2} \right)_{\text{at } r = R} = \frac{|q|}{4\pi\varepsilon_0 R^2}$$

Applying Eq. 23-20, we have

$$E_{\text{internal}} = \frac{|q|}{4\pi\varepsilon_0 R^3} r = \frac{1}{4} E_{\text{max}} \quad \Rightarrow \quad r = \frac{R}{4} = 0.25 R .$$

(b) Outside sphere 2 we have

$$E_{\text{external}} = \frac{|q|}{4\pi\varepsilon_0 r^2} = \frac{1}{4} E_{\text{max}} \quad \Rightarrow \quad r = 2.0R .$$

61. The initial field (evaluated "just outside the outer surface" which means it is evaluated at $r = 0.20$ m) is related to the charge q on the hollow conductor by Eq. 23-15. After the point charge Q is placed at the geometric center of the hollow conductor, the final field at that point is a combination of the initial and that due to Q (determined by Eq. 22-3).

(a) $q = 4\pi\varepsilon_0 r^2 E_{initial} = +2.0 \times 10^{-9}$ C.

(b) $Q = 4\pi\varepsilon_0 r^2 (E_{final} - E_{initial}) = -1.2 \times 10^{-9}$ C.

(c) In order to cancel the field (due to Q) within the conducting material, there must be an amount of charge equal to $-Q$ distributed uniformly on the inner surface. Thus, the answer is $+1.2 \times 10^{-9}$ C.

(d) Since the total excess charge on the conductor is q and is located on the surfaces, then the outer surface charge must equal the total minus the inner surface charge. Thus, the answer is 2.0×10^{-9} C $- 1.2 \times 10^{-9}$ C $= +0.80 \times 10^{-9}$ C.

62. Since the charge distribution is uniform, we can find the total charge q by multiplying ρ by the spherical volume ($\frac{4}{3}\pi r^3$) with $r = R = 0.050$ m. This gives $q = 1.68$ nC.

(a) Applying Eq. 23-20 with $r = 0.035$ m, we have

$$E_{internal} = \frac{|q|}{4\pi\varepsilon_0 R^3} r = 4.2 \times 10^3 \text{ N/C.}$$

(b) Outside the sphere we have (with $r = 0.080$ m)

$$E_{external} = \frac{|q|}{4\pi\varepsilon_0 r^2} = 2.4 \times 10^3 \text{ N/C.}$$

63. (a) In order to have net charge -10 μC when -14 μC is known to be on the outer surface, then there must be $+4.0$ μC on the inner surface (since charges reside on the surfaces of a conductor in electrostatic situations).

(b) In order to cancel the electric field inside the conducting material, the contribution from the $+4$ μC on the inner surface must be canceled by that of the charged particle in the hollow. Thus, the particle's charge is -4.0 μC.

64. The field at the proton's location (but not caused by the proton) has magnitude E. The proton's charge is e. The ball's charge has <u>magnitude</u> q. Thus, as long as the proton is at $r \geq R$ then the force on the proton (caused by the ball) has magnitude

$$F = eE = e\left(\frac{q}{4\pi\varepsilon_0 r^2}\right) = \frac{e\,q}{4\pi\varepsilon_0 r^2}$$

where r is measured from the center of the ball (to the proton).This agrees with Coulomb's law from Chapter 22.   We note that if $r = R$ then this expression becomes

$$F_R = \frac{e\,q}{4\pi\varepsilon_0\,R^2}.$$

(a) If we require $F = \frac{1}{2}F_R$, and solve for r, we obtain $r = \sqrt{2}\,R$. Since the problem asks for the measurement from the surface then the answer is $\sqrt{2}\,R - R = 0.41R$.

(b) Now we require $F_{inside} = \frac{1}{2}F_R$ where $F_{inside} = eE_{inside}$ and E_{inside} is given by Eq. 23-20. Thus,

$$e\left(\frac{q}{4\pi\varepsilon_0\,R^2}\right)r = \frac{1}{2}\frac{e\,q}{4\pi\varepsilon_0\,R^2} \qquad \Rightarrow \qquad r = \frac{1}{2}R = 0.50\ R.$$

65. (a) At $x = 0.040$ m, the net field has a rightward $(+x)$ contribution (computed using Eq. 23-13) from the charge lying between $x = -0.050$ m and $x = 0.040$ m, and a leftward $(-x)$ contribution (again computed using Eq. 23-13) from the charge in the region from $x = 0.040$ m to $x = 0.050$ m. Thus, since $\sigma = q/A = \rho V/A = \rho \Delta x$ in this situation, we have

$$\left|\vec{E}\right| = \frac{\rho(0.090\,\text{m})}{2\varepsilon_0} - \frac{\rho(0.010\,\text{m})}{2\varepsilon_0} = 5.4\,\text{N/C}.$$

(b) In this case, the field contributions from all layers of charge point rightward, and we obtain

$$\left|\vec{E}\right| = \frac{\rho(0.100\,\text{m})}{2\varepsilon_0} = 6.8\,\text{N/C}.$$

66. From Gauss's law, we have

$$\Phi = \frac{q_{enclosed}}{\varepsilon_0} = \frac{\sigma\,\pi r^2}{\varepsilon_0} = \frac{(8.0 \times 10^{-9}\,\text{C/m}^2)\,\pi\,(0.050\,\text{m})^2}{8.85 \times 10^{-12}\,\text{C}^2/\text{N·m}^2} = 7.1\ \text{N·m}^2/\text{C}.$$

67. (a) For $r < R$, $E = 0$ (see Eq. 23-16).

(b) For r slightly greater than R,

$$E_R = \frac{1}{4\pi\varepsilon_0}\frac{q}{r^2} \approx \frac{q}{4\pi\varepsilon_0 R^2} = \frac{\left(8.99\times10^9\ \text{N·m}^2/\text{C}^2\right)\left(2.00\times10^{-7}\,\text{C}\right)}{(0.250\,\text{m})^2} = 2.88\times10^4\ \text{N/C}.$$

(c) For $r > R$,

$$E = \frac{1}{4\pi\varepsilon_0}\frac{q}{r^2} = E_R\left(\frac{R}{r}\right)^2 = (2.88\times10^4 \text{ N/C})\left(\frac{0.250 \text{ m}}{3.00 \text{ m}}\right)^2 = 200 \text{ N/C}.$$

68. (a) There is no flux through the sides, so we have two contributions to the flux, one from the $x = 2$ end (with $\Phi_2 = +(2 + 2)(\pi (0.20)^2) = 0.50$ N·m²/C) and one from the $x = 0$ end (with $\Phi_0 = -(2)(\pi (0.20)^2)$).

(b) By Gauss' law we have $q_{enc} = \varepsilon_0 (\Phi_2 + \Phi_0) = 2.2 \times 10^{-12}$ C.

69. (a) Outside the sphere, we use Eq. 23-15 and obtain $E = \frac{1}{4\pi\varepsilon_0}\frac{q}{r^2} = 15.0 \text{ N/C}.$

(b) With $q = +6.00 \times 10^{-12}$ C, Eq. 23-20 leads to $E = 25.3 \text{ N/C}$.

70. Since the fields involved are uniform, the precise location of P is not relevant; what is important is it is above the three sheets, with the positively charged sheets contributing upward fields and the negatively charged sheet contributing a downward field, which conveniently conforms to usual conventions (of upward as positive and downward as negative). The net field is directed upward $(+\hat{j})$, and (from Eq. 23-13) its magnitude is

$$|\vec{E}| = \frac{\sigma_1}{2\varepsilon_0} + \frac{\sigma_2}{2\varepsilon_0} + \frac{\sigma_3}{2\varepsilon_0} = \frac{1.0\times10^{-6}}{2\times8.85\times10^{-12}} = 5.65\times10^4 \text{ N/C}.$$

In unit-vector notation, we have $\vec{E} = (5.65\times10^4 \text{ N/C})\hat{j}$.

71. Let $\Phi_0 = 10^3 \text{ N}\cdot\text{m}^2/\text{C}$. The net flux through the entire surface of the dice is given by

$$\Phi = \sum_{n=1}^{6}\Phi_n = \sum_{n=1}^{6}(-1)^n n\Phi_0 = \Phi_0(-1+2-3+4-5+6) = 3\Phi_0.$$

Thus, the net charge enclosed is

$$q = \varepsilon_0\Phi = 3\varepsilon_0\Phi_0 = 3\left(8.85\times10^{-12}\frac{\text{C}^2}{\text{N}\cdot\text{m}^2}\right)(10^3\text{N}\cdot\text{m}^2\text{/C}) = 2.66\times10^{-8}\text{C}.$$

72. (a) From Gauss' law,

$$\vec{E}(\vec{r}) = \frac{1}{4\pi\varepsilon_0}\frac{q_{encl}}{r^3}\vec{r} = \frac{1}{4\pi\varepsilon_0}\frac{(4\pi\rho r^3/3)\vec{r}}{r^3} = \frac{\rho\vec{r}}{3\varepsilon_0}.$$

(b) The charge distribution in this case is equivalent to that of a whole sphere of charge density ρ plus a smaller sphere of charge density $-\rho$ which fills the void. By superposition

$$\vec{E}(\vec{r}) = \frac{\rho\vec{r}}{3\varepsilon_0} + \frac{(-\rho)(\vec{r}-\vec{a})}{3\varepsilon_0} = \frac{\rho\vec{a}}{3\varepsilon_0}.$$

73. We choose a coordinate system whose origin is at the center of the flat base, such that the base is in the xy plane and the rest of the hemisphere is in the $z > 0$ half space.

(a) $\Phi = \pi R^2 \left(-\hat{k}\right) \cdot E\hat{k} = -\pi R^2 E = -\pi(0.0568\ \text{m})^2(2.50\ \text{N/C}) = -0.0253\ \text{N}\cdot\text{m}^2/\text{C}.$

(b) Since the flux through the entire hemisphere is zero, the flux through the curved surface is $\vec{\Phi}_c = -\Phi_{\text{base}} = \pi R^2 E = 0.0253\ \text{N}\cdot\text{m}^2/\text{C}.$

74. (a) The direction of the electric field at P_1 is away from q_1 and its magnitude is

$$\left|\vec{E}\right| = \frac{q}{4\pi\varepsilon_0 r_1^2} = \frac{(8.99\times10^9\ \text{N}\cdot\text{m}^2/\text{C}^2)(1.0\times10^{-7}\ \text{C})}{(0.015\,\text{m})^2} = 4.0\times10^6\ \text{N/C}.$$

(b) $\vec{E} = 0$, since P_2 is inside the metal.

75. The field due to a sheet of charge is given by Eq. 23-13. Both sheets are horizontal (parallel to the xy plane), producing vertical fields (parallel to the z axis). At points above the $z = 0$ sheet (sheet A), its field points upward (towards $+z$); at points above the $z = 2.0$ sheet (sheet B), its field does likewise. However, below the $z = 2.0$ sheet, its field is oriented downward.

(a) The magnitude of the net field in the region between the sheets is

$$|\vec{E}| = \frac{\sigma_A}{2\varepsilon_0} + \frac{\sigma_B}{2\varepsilon_0} = 2.82\times10^2\ \text{N/C}.$$

(b) The magnitude of the net field at points above both sheets is

$$|\vec{E}| = \frac{\sigma_A}{2\varepsilon_0} + \frac{\sigma_B}{2\varepsilon_0} = 6.21\times10^2\ \text{N/C}.$$

76. Since the fields involved are uniform, the precise location of P is not relevant. Since the sheets are oppositely charged (though not equally so), the field contributions are additive (since P is between them). Using Eq. 23-13, we obtain

$$\vec{E} = \frac{\sigma_1}{2\varepsilon_0} + \frac{3\sigma_1}{2\varepsilon_0} = \frac{2\sigma_1}{\varepsilon_0}$$

directed towards the negatively charged sheet. The multiple is 2.00.

77. We use Eqs. 23-15, 23-16 and the superposition principle.

(a) $E = 0$ in the region inside the shell.

(b) $E = q_a / 4\pi\varepsilon_0 r^2$.

(c) $E = (q_a + q_b) / 4\pi\varepsilon_0 r^2$.

(d) Since $E = 0$ for $r < a$ the charge on the inner surface of the inner shell is always zero. The charge on the outer surface of the inner shell is therefore q_a. Since $E = 0$ inside the metallic outer shell the net charge enclosed in a Gaussian surface that lies in between the inner and outer surfaces of the outer shell is zero. Thus the inner surface of the outer shell must carry a charge $-q_a$, leaving the charge on the outer surface of the outer shell to be $q_b + q_a$.

78. The net enclosed charge q is given by

$$q = \varepsilon_0 \Phi = \left(8.85 \times 10^{-12} \frac{C^2}{N \cdot m^2}\right)\left(-48 \, N \cdot m^2/C\right) = -4.2 \times 10^{-10} C.$$

79. (a) At A, the only field contribution is from the $+5.00$ pC particle in the hollow (this follows from Gauss' law — it is the only charge enclosed by a Gaussian spherical surface passing through point A, concentric with the shell). Thus, using k for $1/4\pi\varepsilon_0$, we have
$|\vec{E}| = k(5.00 \times 10^{-12})/(0.5)^2 = 0.180$.

(b) The direction is radially outward.

(c) Point B is in the conducting material, where the field must be zero in any electrostatic situation.

(d) Point C is outside the sphere where the net charge at smaller values of radius is $(-3.00 \, pC + 5.00 \, pC) = 2.00 \, pC$. Therefore, we have

$$|\vec{E}| = k(2.00 \times 10^{-12})/(2)^2 = 4.50 \times 10^{-3} \, N/C$$

directed radially outward.

80. We can express Eq. 23-17 in terms of the charge density ρ as follows:

$$E = \frac{q}{4\pi\varepsilon_0 r^2} = \frac{\rho \frac{4}{3}\pi R^3}{4\pi\varepsilon_0 r^2} = \frac{\rho R^3}{3\varepsilon_0 r^2}.$$

Thus, at $r = 2R$, we have (when the ball is solid)

$$E_1 = \frac{\rho R^3}{3\varepsilon_0 (2R)^2} = \frac{\rho}{12\varepsilon_0 R}.$$

Now, with the hollow core of radius $R/2$, we have a similar field but without the contribution from those charges that would have been in that core:

$$E_{new} = E_1 - \left(\frac{\rho (R/2)^3}{3\varepsilon_0 r^2}\right)_{at\ r=2R} = \frac{\rho}{12\varepsilon_0 R} - \frac{\rho}{96\varepsilon_0 R} = \frac{7\rho}{96\varepsilon_0 R}$$

which is equivalent to $\frac{7}{8}E_1$. Thus, the fraction is $7/8 = 0.875$.

81. The proton is in uniform circular motion, with the electrical force of the sphere on the proton providing the centripetal force. According to Newton's second law, $F = mv^2/r$, where F is the magnitude of the force, v is the speed of the proton, and r is the radius of its orbit, essentially the same as the radius of the sphere. The magnitude of the force on the proton is $F = eq/4\pi\varepsilon_0 r^2$, where q is the magnitude of the charge on the sphere. Thus,

$$\frac{1}{4\pi\varepsilon_0}\frac{eq}{r^2} = \frac{mv^2}{r}$$

so

$$q = \frac{4\pi\varepsilon_0 mv^2 r}{e} = \frac{\left(1.67\times10^{-27}\ kg\right)\left(3.00\times10^5\ m/s\right)^2 \left(0.0100\ m\right)}{\left(8.99\times10^9\ N\cdot m^2/C^2\right)\left(1.60\times10^{-9}\ C\right)} = 1.04\times10^{-9}\ C.$$

The force must be inward, toward the center of the sphere, and since the proton is positively charged, the electric field must also be inward. The charge on the sphere is negative: $q = -1.04 \times 10^{-9}\ C.$

82. We interpret the question as referring to the field *just* outside the sphere (that is, at locations roughly equal to the radius r of the sphere). Since the area of a sphere is $A = 4\pi r^2$ and the surface charge density is $\sigma = q/A$ (where we assume q is positive for brevity), then

$$E = \frac{\sigma}{\varepsilon_0} = \frac{1}{\varepsilon_0}\left(\frac{q}{4\pi r^2}\right) = \frac{1}{4\pi\varepsilon_0}\frac{q}{r^2}$$

which we recognize as the field of a point charge (see Eq. 22-3).

83. The field is radially outward and takes on equal magnitude-values over the surface of any sphere centered at the atom's center. We take the Gaussian surface to be such a sphere (of radius r). If E is the magnitude of the field, then the total flux through the Gaussian sphere is $\Phi = 4\pi r^2 E$. The charge enclosed by the Gaussian surface is the positive charge at the center of the atom plus that portion of the negative charge within the surface. Since the negative charge is uniformly distributed throughout the large sphere of radius R, we can compute the charge inside the Gaussian sphere using a ratio of volumes. That is, the negative charge inside is $-Zer^3/R^3$. Thus, the total charge enclosed is $Ze - Zer^3/R^3$ for $r \leq R$. Gauss' law now leads to

$$4\pi\varepsilon_0 r^2 E = Ze\left(1 - \frac{r^3}{R^3}\right) \Rightarrow E = \frac{Ze}{4\pi\varepsilon_0}\left(\frac{1}{r^2} - \frac{r}{R^3}\right).$$

84. The electric field is radially outward from the central wire. We want to find its magnitude in the region between the wire and the cylinder as a function of the distance r from the wire. Since the magnitude of the field at the cylinder wall is known, we take the Gaussian surface to coincide with the wall. Thus, the Gaussian surface is a cylinder with radius R and length L, coaxial with the wire. Only the charge on the wire is actually enclosed by the Gaussian surface; we denote it by q. The area of the Gaussian surface is $2\pi RL$, and the flux through it is $\Phi = 2\pi RLE$. We assume there is no flux through the ends of the cylinder, so this Φ is the total flux. Gauss' law yields $q = 2\pi\varepsilon_0 RLE$. Thus,

$$q = 2\pi\left(8.85\times10^{-12} \ \frac{\text{C}^2}{\text{N}\cdot\text{m}^2}\right)(0.014 \text{ m})(0.16 \text{ m})\,(2.9\times10^4 \text{ N/C}) = 3.6\times10^{-9} \text{ C}.$$

85. (a) The diagram below shows a cross section (or, perhaps more appropriately, "end view") of the charged cylinder (solid circle).

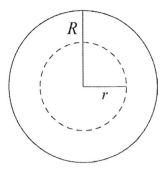

Consider a Gaussian surface in the form of a cylinder with radius r and length ℓ, coaxial with the charged cylinder. An "end view" of the Gaussian surface is shown as a dotted circle. The charge enclosed by it is $q = \rho V = \pi r^2 \ell\rho$, where $V = \pi r^2\ell$ is the volume of the cylinder.

If ρ is positive, the electric field lines are radially outward, normal to the Gaussian surface and distributed uniformly along it. Thus, the total flux through the Gaussian cylinder is $\Phi = EA_{\text{cylinder}} = E(2\pi r\ell)$. Now, Gauss' law leads to

$$2\pi\varepsilon_0 r\ell E = \pi r^2 \ell\rho \Rightarrow E = \frac{\rho r}{2\varepsilon_0}.$$

(b) Next, we consider a cylindrical Gaussian surface of radius $r > R$. If the external field E_{ext} then the flux is $\Phi = 2\pi r\ell E_{\text{ext}}$. The charge enclosed is the total charge in a section of the charged cylinder with length ℓ. That is, $q = \pi R^2 \ell\rho$. In this case, Gauss' law yields

$$2\pi\varepsilon_0 r\ell E_{\text{ext}} = \pi R^2 \ell\rho \Rightarrow E_{\text{ext}} = \frac{R^2\rho}{2\varepsilon_0 r}.$$

86. (a) The mass flux is $wd\rho v = (3.22 \text{ m}) (1.04 \text{ m}) (1000 \text{ kg/m}^3) (0.207 \text{ m/s}) = 693 \text{ kg/s}$.

(b) Since water flows only through area wd, the flux through the larger area is still 693 kg/s.

(c) Now the mass flux is $(wd/2)\rho v = (693 \text{ kg/s})/2 = 347 \text{ kg/s}$.

(d) Since the water flows through an area $(wd/2)$, the flux is 347 kg/s.

(e) Now the flux is $(wd\cos\theta)\rho v = (693 \text{ kg/s})(\cos 34°) = 575 \text{ kg/s}$.

87. (a) We note that the symbol "e" stands for the elementary charge in the manipulations below. From

$$-e = \int_0^\infty \rho(r) 4\pi r^2 dr = \int_0^\infty A \exp(-2r/a_0) 4\pi r^2 \, dr = \pi a_0^3 A$$

we get $A = -e/\pi a_0^3$.

(b) The magnitude of the field is

$$E = \frac{q_{\text{encl}}}{4\pi\varepsilon_0 a_0^2} = \frac{1}{4\pi\varepsilon_0 a_0^2}\left(e + \int_0^{a_0} \rho(r) 4\pi r^2 dr\right) = \frac{e}{4\pi\varepsilon_0 a_0^2}\left(1 - \frac{4}{a_0^3}\int_0^{a_0} \exp(-2r/a_0) r^2 dr\right)$$
$$= \frac{5 e \exp(-2)}{4\pi\varepsilon_0 a_0^2}.$$

We note that $\vec{E}$ points radially outward.

Chapter 24

1. (a) An Ampere is a Coulomb per second, so

$$84 \; A \cdot h = \left(84 \frac{C \cdot h}{s}\right)\left(3600 \frac{s}{h}\right) = 3.0 \times 10^5 \; C.$$

(b) The change in potential energy is $\Delta U = q\Delta V = (3.0 \times 10^5 \; C)(12 \; V) = 3.6 \times 10^6 \; J.$

2. The magnitude is $\Delta U = e\Delta V = 1.2 \times 10^9 \; eV = 1.2 \; GeV.$

3. The electric field produced by an infinite sheet of charge has magnitude $E = \sigma/2\varepsilon_0$, where σ is the surface charge density. The field is normal to the sheet and is uniform. Place the origin of a coordinate system at the sheet and take the x axis to be parallel to the field and positive in the direction of the field. Then the electric potential is

$$V = V_s - \int_0^x E \; dx = V_s - Ex,$$

where V_s is the potential at the sheet. The equipotential surfaces are surfaces of constant x; that is, they are planes that are parallel to the plane of charge. If two surfaces are separated by Δx then their potentials differ in magnitude by $\Delta V = E\Delta x = (\sigma/2\varepsilon_0)\Delta x$. Thus,

$$\Delta x = \frac{2\varepsilon_0 \Delta V}{\sigma} = \frac{2(8.85 \times 10^{-12} \; C^2/N \cdot m^2)(50 \; V)}{0.10 \times 10^{-6} \; C/m^2} = 8.8 \times 10^{-3} \; m.$$

4. (a) $V_B - V_A = \Delta U/q = -W/(-e) = -(3.94 \times 10^{-19} \; J)/(-1.60 \times 10^{-19} \; C) = 2.46 \; V.$

(b) $V_C - V_A = V_B - V_A = 2.46 \; V.$

(c) $V_C - V_B = 0$ (Since C and B are on the same equipotential line).

5. (a) $E = F/e = (3.9 \times 10^{-15} \; N)/(1.60 \times 10^{-19} \; C) = 2.4 \times 10^4 \; N/C.$

(b) $\Delta V = E\Delta s = (2.4 \times 10^4 \; N/C)(0.12 \; m) = 2.9 \times 10^3 \; V.$

6. (a) By Eq. 24-18, the change in potential is the negative of the "area" under the curve. Thus, using the area-of-a-triangle formula, we have

$$V - 10 = -\int_0^{x=2} \vec{E} \cdot d\vec{s} = \frac{1}{2}(2)(20)$$

which yields $V = 30$ V.

(b) For any region within $0 < x < 3$ m, $-\int \vec{E} \cdot d\vec{s}$ is positive, but for any region for which $x > 3$ m it is negative. Therefore, $V = V_{max}$ occurs at $x = 3$ m.

$$V - 10 = -\int_0^{x=3} \vec{E} \cdot d\vec{s} = \frac{1}{2}(3)(20)$$

which yields $V_{max} = 40$ V.

(c) In view of our result in part (b), we see that now (to find $V = 0$) we are looking for some $X > 3$ m such that the "area" from $x = 3$ m to $x = X$ is 40 V. Using the formula for a triangle ($3 < x < 4$) and a rectangle ($4 < x < X$), we require

$$\frac{1}{2}(1)(20) + (X - 4)(20) = 40.$$

Therefore, $X = 5.5$ m.

7. (a) The work done by the electric field is (in SI units)

$$W = \int_i^f q_0 \vec{E} \cdot d\vec{s} = \frac{q_0 \sigma}{2\varepsilon_0} \int_0^d dz = \frac{q_0 \sigma d}{2\varepsilon_0} = \frac{(1.60 \times 10^{-19})(5.80 \times 10^{-12})(0.0356)}{2(8.85 \times 10^{-12})} = 1.87 \times 10^{-21} \text{ J}.$$

(b) Since $V - V_0 = -W/q_0 = -\sigma z/2\varepsilon_0$, with V_0 set to be zero on the sheet, the electric potential at P is (in SI units)

$$V = -\frac{\sigma z}{2\varepsilon_0} = -\frac{(5.80 \times 10^{-12})(0.0356)}{2(8.85 \times 10^{-12})} = -1.17 \times 10^{-2} \text{ V}.$$

8. We connect A to the origin with a line along the y axis, along which there is no change of potential (Eq. 24-18: $\int \vec{E} \cdot d\vec{s} = 0$). Then, we connect the origin to B with a line along the x axis, along which the change in potential is

$$\Delta V = -\int_0^{x=4} \vec{E} \cdot d\vec{s} = -4.00 \int_0^4 x \, dx = -4.00 \left(\frac{4^2}{2} \right)$$

which yields $V_B - V_A = -32.0$ V.

9. (a) The potential as a function of r is (in SI units)

$$V(r) = V(0) - \int_0^r E(r)\,dr = 0 - \int_0^r \frac{qr}{4\pi\varepsilon_0 R^3}\,dr = -\frac{qr^2}{8\pi\varepsilon_0 R^3}$$

$$= -\frac{(8.99\times10^9)(3.50\times10^{-15})(0.0145)^2}{2(0.0231)^3} = -2.68\times10^{-4} \text{ V}.$$

(b) Since $\Delta V = V(0) - V(R) = q/8\pi\varepsilon_0 R$, we have (in SI units)

$$V(R) = -\frac{q}{8\pi\varepsilon_0 R} = -\frac{(8.99\times10^9)(3.50\times10^{-15})}{2(0.0231)} = -6.81\times10^{-4} \text{ V}.$$

10. The charge is

$$q = 4\pi\varepsilon_0 RV = \frac{(10\text{m})(-1.0\text{V})}{8.99\times10^9 \text{ N}\cdot\text{m}^2/\text{C}^2} = -1.1\times10^{-9}\text{C}.$$

11. (a) The charge on the sphere is

$$q = 4\pi\varepsilon_0 VR = \frac{(200\text{ V})(0.15\text{ m})}{8.99\times10^9 \dfrac{\text{N}\cdot\text{m}^2}{\text{C}^2}} = 3.3\times10^{-9} \text{ C}.$$

(b) The (uniform) surface charge density (charge divided by the area of the sphere) is

$$\sigma = \frac{q}{4\pi R^2} = \frac{3.3\times10^{-9}\text{ C}}{4\pi(0.15\text{ m})^2} = 1.2\times10^{-8} \text{ C}/\text{m}^2.$$

12. (a) The potential difference is

$$V_A - V_B = \frac{q}{4\pi\varepsilon_0 r_A} - \frac{q}{4\pi\varepsilon_0 r_B} = (1.0\times10^{-6}\text{ C})\left(8.99\times10^9 \frac{\text{N}\cdot\text{m}^2}{\text{C}^2}\right)\left(\frac{1}{2.0\text{ m}} - \frac{1}{1.0\text{ m}}\right)$$

$$= -4.5\times10^3 \text{ V}.$$

(b) Since $V(r)$ depends only on the magnitude of $\vec{r}$, the result is unchanged.

13. First, we observe that $V(x)$ cannot be equal to zero for $x > d$. In fact $V(x)$ is always negative for $x > d$. Now we consider the two remaining regions on the x axis: $x < 0$ and $0 < x < d$.

(a) For $0 < x < d$ we have $d_1 = x$ and $d_2 = d - x$. Let

$$V(x) = k\left(\frac{q_1}{d_1} + \frac{q_2}{d_2}\right) = \frac{q}{4\pi\varepsilon_0}\left(\frac{1}{x} + \frac{-3}{d-x}\right) = 0$$

and solve: $x = d/4$. With $d = 24.0$ cm, we have $x = 6.00$ cm.

(b) Similarly, for $x < 0$ the separation between q_1 and a point on the x axis whose coordinate is x is given by $d_1 = -x$; while the corresponding separation for q_2 is $d_2 = d - x$. We set

$$V(x) = k\left(\frac{q_1}{d_1} + \frac{q_2}{d_2}\right) = \frac{q}{4\pi\varepsilon_0}\left(\frac{1}{-x} + \frac{-3}{d-x}\right) = 0$$

to obtain $x = -d/2$. With $d = 24.0$ cm, we have $x = -12.0$ cm.

14. Since according to the problem statement there is a point in between the two charges on the x axis where the net electric field is zero, the fields at that point due to q_1 and q_2 must be directed opposite to each other. This means that q_1 and q_2 must have the same sign (i.e., either both are positive or both negative). Thus, the potentials due to either of them must be of the same sign. Therefore, the net electric potential cannot possibly be zero anywhere except at infinity.

15. A charge $-5q$ is a distance $2d$ from P, a charge $-5q$ is a distance d from P, and two charges $+5q$ are each a distance d from P, so the electric potential at P is (in SI units)

$$V = \frac{q}{4\pi\varepsilon_0}\left[-\frac{1}{2d} - \frac{1}{d} + \frac{1}{d} + \frac{1}{d}\right] = \frac{q}{8\pi\varepsilon_0 d} = \frac{(8.99\times10^9)(5.00\times10^{-15})}{2(4.00\times10^{-2})} = 5.62\times10^{-4}\,\text{V}.$$

The zero of the electric potential was taken to be at infinity.

16. In applying Eq. 24-27, we are assuming $V \to 0$ as $r \to \infty$. All corner particles are equidistant from the center, and since their total charge is

$$2q_1 - 3q_1 + 2\,q_1 - q_1 = 0,$$

then their contribution to Eq. 24-27 vanishes. The net potential is due, then, to the two $+4q_2$ particles, each of which is a distance of $a/2$ from the center. In SI units, it is

$$V = \frac{1}{4\pi\varepsilon_0}\frac{4q_2}{a/2} + \frac{1}{4\pi\varepsilon_0}\frac{4q_2}{a/2} = \frac{16q_2}{4\pi\varepsilon_0 a} = \frac{16(8.99\times10^9)(6.00\times10^{-12})}{0.39} = 2.21\,\text{V}.$$

17. (a) The electric potential V at the surface of the drop, the charge q on the drop, and the radius R of the drop are related by $V = q/4\pi\varepsilon_0 R$. Thus

$$R = \frac{q}{4\pi\varepsilon_0 V} = \frac{\left(8.99 \times 10^9 \ \text{N} \cdot \text{m}^2 / \text{C}^2\right)\left(30 \times 10^{-12} \ \text{C}\right)}{500 \ \text{V}} = 5.4 \times 10^{-4} \ \text{m}.$$

(b) After the drops combine the total volume is twice the volume of an original drop, so the radius R' of the combined drop is given by $(R')^3 = 2R^3$ and $R' = 2^{1/3}R$. The charge is twice the charge of original drop: $q' = 2q$. Thus,

$$V' = \frac{1}{4\pi\varepsilon_0}\frac{q'}{R'} = \frac{1}{4\pi\varepsilon_0}\frac{2q}{2^{1/3} R} = 2^{2/3}V = 2^{2/3}(500 \ \text{V}) \approx 790 \ \text{V}.$$

18. When the charge q_2 is infinitely far away, the potential at the origin is due only to the charge q_1 :

$$V_1 = \frac{q_1}{4\pi\varepsilon_0 d} = 5.76 \times 10^{-7} \ \text{V}.$$

Thus, $q_1/d = 6.41 \times 10^{-17}$ C/m. Next, we note that when q_2 is located at $x = 0.080$ m, the net potential vanishes ($V_1 + V_2 = 0$). Therefore,

$$0 = \frac{kq_2}{0.08 \ \text{m}} + \frac{kq_1}{d}$$

Thus, we find $q_2 = -(q_1/d)(0.08 \ \text{m}) = -5.13 \times 10^{-18} \ \text{C} = -32 \ e.$

19. We use Eq. 24-20:

$$V = \frac{1}{4\pi\varepsilon_0}\frac{p}{r^2} = \frac{\left(8.99 \times 10^9 \ \frac{\text{N} \cdot \text{m}^2}{\text{C}^2}\right)\left(1.47 \times 3.34 \times 10^{-30} \ \text{C} \cdot \text{m}\right)}{\left(52.0 \times 10^{-9} \ \text{m}\right)^2} = 1.63 \times 10^{-5} \ \text{V}.$$

20. From Eq. 24-30 and Eq. 24-14, we have (for $\theta_i = 0°$)

$$W_a = q\Delta V = e\left(\frac{p\cos\theta}{4\pi\varepsilon_0 r^2} - \frac{p\cos\theta_i}{4\pi\varepsilon_0 r^2}\right) = \frac{e\,p}{4\pi\varepsilon_0 r^2}(\cos\theta - 1).$$

where $r = 20 \times 10^{-9}$ m. For $\theta = 180°$ the graph indicates $W_a = -4.0 \times 10^{-30}$ J, from which we can determine p. The magnitude of the dipole moment is therefore 5.6×10^{-37} C·m.

21. (a) From Eq. 24-35, in SI units,

$$V = 2 \frac{\lambda}{4\pi\varepsilon_0} \ln\left[\frac{L/2 + \sqrt{(L^2/4) + d^2}}{d}\right]$$

$$= 2(8.99\times10^9)(3.68\times10^{-12})\ln\left[\frac{(0.06/2) + \sqrt{(0.06)^2/4 + (0.08)^2}}{0.08}\right]$$

$$= 2.43\times10^{-2} \text{ V.}$$

(b) The potential at P is $V = 0$ due to superposition.

22. The potential is (in SI units)

$$V_P = \frac{1}{4\pi\varepsilon_0}\int_{rod}\frac{dq}{R} = \frac{1}{4\pi\varepsilon_0 R}\int_{rod}dq = \frac{-Q}{4\pi\varepsilon_0 R} = -\frac{(8.99\times10^9)(25.6\times10^{-12})}{3.71\times10^{-2}} = -6.20 \text{ V.}$$

We note that the result is exactly what one would expect for a point-charge $-Q$ at a distance R. This "coincidence" is due, in part, to the fact that V is a scalar quantity.

23. (a) All the charge is the same distance R from C, so the electric potential at C is (in SI units)

$$V = \frac{1}{4\pi\varepsilon_0}\left[\frac{Q_1}{R} - \frac{6Q_1}{R}\right] = -\frac{5Q_1}{4\pi\varepsilon_0 R} = -\frac{5(8.99\times10^9)(4.20\times10^{-12})}{8.20\times10^{-2}} = -2.30 \text{ V,}$$

where the zero was taken to be at infinity.

(b) All the charge is the same distance from P. That distance is $\sqrt{R^2 + D^2}$, so the electric potential at P is (in SI units)

$$V = \frac{1}{4\pi\varepsilon_0}\left[\frac{Q_1}{\sqrt{R^2 + D^2}} - \frac{6Q_1}{\sqrt{R^2 + D^2}}\right] = -\frac{5Q_1}{4\pi\varepsilon_0\sqrt{R^2 + D^2}}$$

$$= -\frac{5(8.99\times10^9)(4.20\times10^{-12})}{\sqrt{(8.20\times10^{-2})^2 + (6.71\times10^{-2})^2}} = -1.78 \text{ V.}$$

24. Since the charge distribution on the arc is equidistant from the point where V is evaluated, its contribution is identical to that of a point charge at that distance. We assume $V \to 0$ as $r \to \infty$ and apply Eq. 24-27:

$$V = \frac{1}{4\pi\varepsilon_0}\frac{+Q_1}{R} + \frac{1}{4\pi\varepsilon_0}\frac{+4Q_1}{2R} + \frac{1}{4\pi\varepsilon_0}\frac{-2Q_1}{R} = \frac{1}{4\pi\varepsilon_0}\frac{Q_1}{R}$$

$$= \frac{(8.99\times10^9)(7.21\times10^{-12})}{2.00} = 3.24\times10^{-2}\ \text{V}.$$

25. The disk is uniformly charged. This means that when the full disk is present each quadrant contributes equally to the electric potential at P, so the potential at P due to a single quadrant is one-fourth the potential due to the entire disk. First find an expression for the potential at P due to the entire disk. We consider a ring of charge with radius r and (infinitesimal) width dr. Its area is $2\pi r\ dr$ and it contains charge $dq = 2\pi\sigma r\ dr$. All the charge in it is a distance $\sqrt{r^2 + D^2}$ from P, so the potential it produces at P is

$$dV = \frac{1}{4\pi\varepsilon_0}\frac{2\pi\sigma r dr}{\sqrt{r^2 + D^2}} = \frac{\sigma r dr}{2\varepsilon_0\sqrt{r^2 + D^2}}.$$

The total potential at P is

$$V = \frac{\sigma}{2\varepsilon_0}\int_0^R \frac{r dr}{\sqrt{r^2 + D^2}} = \frac{\sigma}{2\varepsilon_0}\sqrt{r^2 + D^2}\Big|_0^R = \frac{\sigma}{2\varepsilon_0}\left[\sqrt{R^2 + D^2} - D\right].$$

The potential V_{sq} at P due to a single quadrant is (in SI units)

$$V_{sq} = \frac{V}{4} = \frac{\sigma}{8\varepsilon_0}\left[\sqrt{R^2 + D^2} - D\right] = \frac{(7.73\times10^{-15})}{8(8.85\times10^{-12})}\left[\sqrt{(0.640)^2 + (0.259)^2} - 0.259\right]$$

$$= 4.71\times10^{-5}\ \text{V}.$$

26. The dipole potential is given by Eq. 24-30 (with $\theta = 90°$ in this case)

$$V = \frac{p\cos\theta}{4\pi\varepsilon_0 r^2} = 0$$

since $\cos(90°) = 0$. The potential due to the short arc is $q_1/4\pi\varepsilon_0 r_1$ and that caused by the long arc is $q_2/4\pi\varepsilon_0 r_2$. Since $q_1 = +2\ \mu\text{C}$, $r_1 = 4.0$ cm, $q_2 = -3\ \mu\text{C}$, and $r_2 = 6.0$ cm, the potentials of the arcs cancel. The result is zero.

27. Letting d denote 0.010 m, we have (in SI units)

$$V = \frac{Q_1}{4\pi\varepsilon_0 d} + \frac{3Q_1}{8\pi\varepsilon_0 d} - \frac{3Q_1}{16\pi\varepsilon_0 d} = \frac{Q_1}{8\pi\varepsilon_0 d} = \frac{(8.99\times10^9)(30\times10^{-9})}{2(0.01)} = 1.3\times10^4\ \text{V}.$$

28. Consider an infinitesimal segment of the rod, located between x and $x + dx$. It has length dx and contains charge $dq = \lambda\, dx$, where $\lambda = Q/L$ is the linear charge density of the rod. Its distance from P_1 is $d + x$ and the potential it creates at P_1 is

$$dV = \frac{1}{4\pi\varepsilon_0}\frac{dq}{d+x} = \frac{1}{4\pi\varepsilon_0}\frac{\lambda dx}{d+x}.$$

To find the total potential at P_1, we integrate over the length of the rod and obtain (in SI units):

$$V = \frac{\lambda}{4\pi\varepsilon_0}\int_0^L \frac{dx}{d+x} = \frac{\lambda}{4\pi\varepsilon_0}\ln(d+x)\Big|_0^L = \frac{Q}{4\pi\varepsilon_0 L}\ln\left(1+\frac{L}{d}\right)$$

$$= \frac{(8.99\times10^9)(56.1\times10^{-15})}{0.12}\ln\left(1+\frac{0.12}{0.025}\right) = 7.39\times10^{-3}\ \text{V}.$$

29. Consider an infinitesimal segment of the rod, located between x and $x + dx$. It has length dx and contains charge $dq = \lambda\, dx = cx\, dx$. Its distance from P_1 is $d + x$ and the potential it creates at P_1 is

$$dV = \frac{1}{4\pi\varepsilon_0}\frac{dq}{d+x} = \frac{1}{4\pi\varepsilon_0}\frac{cx\, dx}{d+x}.$$

To find the total potential at P_1, we integrate over the length of the rod and obtain (in SI units):

$$V = \frac{c}{4\pi\varepsilon_0}\int_0^L \frac{x\, dx}{d+x} = \frac{c}{4\pi\varepsilon_0}[x - d\ln(x+d)]\Big|_0^L = \frac{c}{4\pi\varepsilon_0}\left[L - d\ln\left(1+\frac{L}{d}\right)\right]$$

$$= (8.99\times10^9)(28.9\times10^{-12})\left[0.12 - (0.03)\ln\left(1+\frac{0.12}{0.03}\right)\right] = 1.86\times10^{-2}\ \text{V}.$$

30. The magnitude of the electric field is given by

$$|E| = \left|-\frac{\Delta V}{\Delta x}\right| = \frac{2(5.0\,\text{V})}{0.015\text{m}} = 6.7\times10^2\ \text{V/m}.$$

At any point in the region between the plates, $\vec{E}$ points away from the positively charged plate, directly towards the negatively charged one.

31. We use Eq. 24-41:

$$E_x(x,y) = -\frac{\partial V}{\partial x} = -\frac{\partial}{\partial x}\left((2.0\,\text{V}/\text{m}^2)x^2 - 3.0\,\text{V}/\text{m}^2)y^2\right) = -2(2.0\,\text{V}/\text{m}^2)x;$$

$$E_y(x,y) = -\frac{\partial V}{\partial y} = -\frac{\partial}{\partial y}\left((2.0\,\text{V}/\text{m}^2)x^2 - 3.0\,\text{V}/\text{m}^2)y^2\right) = 2(3.0\,\text{V}/\text{m}^2)y.$$

We evaluate at $x = 3.0$ m and $y = 2.0$ m to obtain

$$\vec{E} = (-12\ \text{V/m})\hat{i} + (12\ \text{V/m})\hat{j}.$$

32. We use Eq. 24-41. This is an ordinary derivative since the potential is a function of only one variable.

$$\vec{E} = -\left(\frac{dV}{dx}\right)\hat{i} = -\frac{d}{dx}(1500x^2)\hat{i} = (-3000x)\hat{i}$$
$$= (-3000\,\text{V}/\text{m}^2)(0.0130\,\text{m})\hat{i} = (-39\,\text{V}/\text{m})\hat{i}.$$

(a) Thus, the magnitude of the electric field is $E = 39$ V/m.

(b) The direction of $\vec{E}$ is $-\hat{i}$, or toward plate 1.

33. We apply Eq. 24-41:

$$E_x = -\frac{\partial V}{\partial x} = -2.00\,yz^2$$

$$E_y = -\frac{\partial V}{\partial y} = -2.00\,xz^2$$

$$E_z = -\frac{\partial V}{\partial z} = -4.00\,xyz$$

which, at $(x, y, z) = (3.00, -2.00, 4.00)$, gives $(E_x, E_y, E_z) = (64.0, -96.0, 96.0)$ in SI units. The magnitude of the field is therefore

$$\left|\vec{E}\right| = \sqrt{E_x^2 + E_y^2 + E_z^2} = 150\,\text{V/m} = 150\,\text{N/C}.$$

34. (a) According to the result of problem 28, the electric potential at a point with coordinate x is given by

$$V = \frac{Q}{4\pi\varepsilon_0 L}\ln\left(\frac{x-L}{x}\right).$$

At $x = -d$ we obtain (in SI units)

$$V = \frac{Q}{4\pi\varepsilon_0 L}\ln\left(\frac{d+L}{d}\right) = \frac{(8.99\times10^9)(43.6\times10^{-15})}{0.135}\ln\left(1+\frac{0.135}{d}\right)$$

$$= (2.90\times10^{-3}\text{ V})\ln\left(1+\frac{0.135}{d}\right).$$

(b) We differentiate the potential with respect to x to find the x component of the electric field (in SI units):

$$E_x = -\frac{\partial V}{\partial x} = -\frac{Q}{4\pi\varepsilon_0 L}\frac{\partial}{\partial x}\ln\left(\frac{x-L}{x}\right) = -\frac{Q}{4\pi\varepsilon_0 L}\frac{x}{x-L}\left(\frac{1}{x}-\frac{x-L}{x^2}\right) = -\frac{Q}{4\pi\varepsilon_0 x(x-L)}$$

$$= -\frac{(8.99\times10^9)(43.6\times10^{-15})}{x(x+0.135)} = -\frac{(3.92\times10^{-4})}{x(x+0.135)},$$

or

$$|E_x| = \frac{(3.92\times10^{-4})}{x(x+0.135)}.$$

(c) Since $E_x < 0$, its direction relative to the positive x axis is $180°$.

(d) At $x = -d$ we obtain (in SI units)

$$|E_x| = \frac{(3.92\times10^{-4})}{(0.0620)(0.0620+0.135)} = 0.0321\text{ N/C}.$$

(e) Consider two points an equal infinitesimal distance on either side of P_1, along a line that is perpendicular to the x axis. The difference in the electric potential divided by their separation gives the transverse component of the electric field. Since the two points are situated symmetrically with respect to the rod, their potentials are the same and the potential difference is zero. Thus, the transverse component of the electric field E_y is zero.

35. The electric field (along some axis) is the (negative of the) derivative of the potential V with respect to the corresponding coordinate. In this case, the derivatives can be read off of the graphs as slopes (since the graphs are of straight lines). Thus,

$$E_x = -\frac{dV}{dx} = -\left(\frac{-500\text{ V}}{0.20\text{ m}}\right) = 2500\text{ V/m} = 2500\text{ N/C}$$

$$E_y = -\frac{dV}{dy} = -\left(\frac{300\text{ V}}{0.30\text{ m}}\right) = -1000\text{ V/m} = -1000\text{ N/C}.$$

These components imply the electric field has a magnitude of 2693 N/C and a direction of $-21.8°$ (with respect to the positive x axis). The force on the electron is given by $\vec{F} = q\vec{E}$ where $q = -e$. The minus sign associated with the value of q has the implication that $\vec{F}$ points in the opposite direction from $\vec{E}$ (which is to say that its angle is found by adding $180°$ to that of $\vec{E}$). With $e = 1.60 \times 10^{-19}$ C, we obtain

$$\vec{F} = (-1.60 \times 10^{-19} \text{C})[(2500 \text{ N/C})\hat{i} - (1000 \text{ N/C})\hat{j}] = (-4.0 \times 10^{-16} \text{N})\hat{i} + (1.60 \times 10^{-16} \text{ N})\hat{j}.$$

36. (a) Consider an infinitesimal segment of the rod from x to $x + dx$. Its contribution to the potential at point P_2 is

$$dV = \frac{1}{4\pi\varepsilon_0} \frac{\lambda(x)dx}{\sqrt{x^2 + y^2}} = \frac{1}{4\pi\varepsilon_0} \frac{cx}{\sqrt{x^2 + y^2}} dx.$$

Thus, (in SI units)

$$V = \int_{\text{rod}} dV_P = \frac{c}{4\pi\varepsilon_0} \int_0^L \frac{x}{\sqrt{x^2 + y^2}} dx = \frac{c}{4\pi\varepsilon_0}\left(\sqrt{L^2 + y^2} - y\right)$$

$$= (8.99 \times 10^9)(49.9 \times 10^{-12})\left(\sqrt{(0.100)^2 + (0.0356)^2} - 0.0356\right)$$

$$= 3.16 \times 10^{-2} \text{ V}.$$

(b) The y component of the field there is

$$E_y = -\frac{\partial V_P}{\partial y} = -\frac{c}{4\pi\varepsilon_0} \frac{d}{dy}\left(\sqrt{L^2 + y^2} - y\right) = \frac{c}{4\pi\varepsilon_0}\left(1 - \frac{y}{\sqrt{L^2 + y^2}}\right).$$

$$= (8.99 \times 10^9)(49.9 \times 10^{-12})\left(1 - \frac{0.0356}{\sqrt{(0.100)^2 + (0.0356)^2}}\right)$$

$$= 0.298 \text{ N/C}.$$

(c) We obtained above the value of the potential at any point P strictly on the y-axis. In order to obtain $E_x(x, y)$ we need to first calculate $V(x, y)$. That is, we must find the potential for an arbitrary point located at (x, y). Then $E_x(x, y)$ can be obtained from $E_x(x, y) = -\partial V(x, y)/\partial x$.

37. We choose the zero of electric potential to be at infinity. The initial electric potential energy U_i of the system before the particles are brought together is therefore zero. After the system is set up the final potential energy is

$$U_f = \frac{q^2}{4\pi\varepsilon_0}\left(-\frac{1}{a}-\frac{1}{a}+\frac{1}{\sqrt{2}a}-\frac{1}{a}-\frac{1}{a}+\frac{1}{\sqrt{2}a}\right)=\frac{2q^2}{4\pi\varepsilon_0 a}\left(\frac{1}{\sqrt{2}}-2\right).$$

Thus the amount of work required to set up the system is given by (in SI units)

$$W = \Delta U = U_f - U_i = U_f = \frac{2q^2}{4\pi\varepsilon_0 a}\left(\frac{1}{\sqrt{2}}-2\right)=\frac{2(8.99\times10^9)(2.30\times10^{-12})^2}{(0.640)}\left(\frac{1}{\sqrt{2}}-2\right)$$

$$=-1.92\times10^{-13}\,\text{J}.$$

38. The work done must equal the change in the electric potential energy. From Eq. 24-14 and Eq. 24-26, we find (with $r = 0.020$ m)

$$W = \frac{(3e)\,(7e)}{4\pi\varepsilon_0 r} = 2.1\times10^{-25}\,\text{J}\,.$$

39. (a) We use Eq. 24-43 with $q_1 = q_2 = -e$ and $r = 2.00$ nm:

$$U = k\frac{q_1 q_2}{r} = k\frac{e^2}{r} = \frac{\left(8.99\times10^9\,\frac{\text{N·m}^2}{\text{C}^2}\right)(1.60\times10^{-19}\,\text{C})^2}{2.00\times10^{-9}\,\text{m}} = 1.15\times10^{-19}\,\text{J}.$$

(b) Since $U > 0$ and $U \propto r^{-1}$ the potential energy U decreases as r increases.

40. The work required is

$$W = \Delta U = \frac{1}{4\pi\varepsilon_0}\left[\frac{q_1 Q}{2d}+\frac{q_2 Q}{d}\right]=\frac{1}{4\pi\varepsilon_0}\left[\frac{q_1 Q}{2d}+\frac{(-q_1/2)Q}{d}\right]=0.$$

41. (a) Let $\ell = 0.15$ m be the length of the rectangle and $w = 0.050$ m be its width. Charge q_1 is a distance ℓ from point A and charge q_2 is a distance w, so the electric potential at A is

$$V_A = \frac{1}{4\pi\varepsilon_0}\left[\frac{q_1}{\ell}+\frac{q_2}{w}\right]=(8.99\times10^9\,\text{N·m}^2/\text{C}^2)\left[\frac{-5.0\times10^{-6}\,\text{C}}{0.15\,\text{m}}+\frac{2.0\times10^{-6}\,\text{C}}{0.050\,\text{m}}\right]$$

$$= 6.0\times10^4\,\text{V}.$$

(b) Charge q_1 is a distance w from point b and charge q_2 is a distance ℓ, so the electric potential at B is

$$V_B = \frac{1}{4\pi\varepsilon_0}\left[\frac{q_1}{w}+\frac{q_2}{\ell}\right]=(8.99\times10^9\,\text{N·m}^2/\text{C}^2)\left[\frac{-5.0\times10^{-6}\,\text{C}}{0.050\,\text{m}}+\frac{2.0\times10^{-6}\,\text{C}}{0.15\,\text{m}}\right]$$

$$=-7.8\times10^5\,\text{V}.$$

(c) Since the kinetic energy is zero at the beginning and end of the trip, the work done by an external agent equals the change in the potential energy of the system. The potential energy is the product of the charge q_3 and the electric potential. If U_A is the potential energy when q_3 is at A and U_B is the potential energy when q_3 is at B, then the work done in moving the charge from B to A is

$$W = U_A - U_B = q_3(V_A - V_B) = (3.0 \times 10^{-6}\ \text{C})(6.0 \times 10^4\ \text{V} + 7.8 \times 10^5\ \text{V}) = 2.5\ \text{J}.$$

(d) The work done by the external agent is positive, so the energy of the three-charge system increases.

(e) and (f) The electrostatic force is conservative, so the work is the same no matter which path is used.

42. Let $r = 1.5$ m, $x = 3.0$ m, $q_1 = -9.0$ nC, and $q_2 = -6.0$ pC. The work done by an external agent is given by

$$W = \Delta U = \frac{q_1 q_2}{4\pi\varepsilon_0}\left(\frac{1}{r} - \frac{1}{\sqrt{r^2 + x^2}}\right)$$

$$= (-9.0 \times 10^{-9}\ \text{C})(-6.0 \times 10^{-12}\text{C})\left(8.99 \times 10^9\ \frac{\text{N} \cdot \text{m}^2}{\text{C}^2}\right) \cdot \left[\frac{1}{1.5\ \text{m}} - \frac{1}{\sqrt{(1.5\ \text{m})^2 + (3.0\ \text{m})^2}}\right]$$

$$= 1.8 \times 10^{-10}\ \text{J}.$$

43. We use the conservation of energy principle. The initial potential energy is $U_i = q^2/4\pi\varepsilon_0 r_1$, the initial kinetic energy is $K_i = 0$, the final potential energy is $U_f = q^2/4\pi\varepsilon_0 r_2$, and the final kinetic energy is $K_f = \frac{1}{2}mv^2$, where v is the final speed of the particle. Conservation of energy yields

$$\frac{q^2}{4\pi\varepsilon_0 r_1} = \frac{q^2}{4\pi\varepsilon_0 r_2} + \frac{1}{2}mv^2.$$

The solution for v is

$$v = \sqrt{\frac{2q^2}{4\pi\varepsilon_0 m}\left(\frac{1}{r_1} - \frac{1}{r_2}\right)} = \sqrt{\frac{(8.99 \times 10^9\ \text{N} \cdot \text{m}^2/\text{C}^2)(2)(3.1 \times 10^{-6}\text{C})^2}{20 \times 10^{-6}\text{kg}}\left(\frac{1}{0.90 \times 10^{-3}\text{m}} - \frac{1}{2.5 \times 10^{-3}\text{m}}\right)}$$

$$= 2.5 \times 10^3\ \text{m/s}.$$

44. The change in electric potential energy of the electron-shell system as the electron starts from its initial position and just reaches the shell is $\Delta U = (-e)(-V) = eV$. Thus from $\Delta U = K = \frac{1}{2}m_e v_i^2$ we find the initial electron speed to be (in SI units)

$$v_i = \sqrt{\frac{2\Delta U}{m_e}} = \sqrt{\frac{2eV}{m_e}} = \sqrt{\frac{2(1.6\times10^{-19})(125)}{9.11\times10^{-31}}} = 6.63\times10^6 \text{ m/s.}$$

45. We use conservation of energy, taking the potential energy to be zero when the moving electron is far away from the fixed electrons. The final potential energy is then $U_f = 2e^2/4\pi\varepsilon_0 d$, where d is half the distance between the fixed electrons. The initial kinetic energy is $K_i = \frac{1}{2}mv^2$, where m is the mass of an electron and v is the initial speed of the moving electron. The final kinetic energy is zero. Thus $K_i = U_f$ or $\frac{1}{2}mv^2 = 2e^2/4\pi\varepsilon_0 d$. Hence

$$v = \sqrt{\frac{4e^2}{4\pi\varepsilon_0 dm}} = \sqrt{\frac{\left(8.99\times10^9 \text{ N}\cdot\text{m}^2/\text{C}^2\right)(4)\left(1.60\times10^{-19} \text{ C}\right)^2}{(0.010 \text{ m})\left(9.11\times10^{-31} \text{ kg}\right)}} = 3.2\times10^2 \text{ m/s.}$$

46. (a) The electric field between the plates is leftward in Fig, 24-50 since it points towards lower values of potential. The force (associated with the field, by Eq. 23-28) is evidently leftward, from the problem description (indicating deceleration of the rightward moving particle), so that $q > 0$ (ensuring that $\vec{F}$ is parallel to $\vec{E}$); it is a proton.

(b) We use conservation of energy:

$$K_0 + U_0 = K + U \implies \frac{1}{2}m_p v_0^2 + qV_1 = \frac{1}{2}m_p v^2 + qV_2 .$$

Using $q = +1.6\times10^{-19}$ C, $m_p = 1.67\times10^{-27}$ kg, $v_0 = 90\times10^3$ m/s, $V_1 = -70$ V and $V_2 = -50$ V, we obtain the final speed $v = 6.53\times10^4$ m/s. We note that the value of d is not used in the solution.

47. Let the distance in question be r. The initial kinetic energy of the electron is $K_i = \frac{1}{2}m_e v_i^2$, where $v_i = 3.2\times10^5$ m/s. As the speed doubles, K becomes $4K_i$. Thus

$$\Delta U = \frac{-e^2}{4\pi\varepsilon_0 r} = -\Delta K = -(4K_i - K_i) = -3K_i = -\frac{3}{2}m_e v_i^2 ,$$

or

$$r = \frac{2e^2}{3(4\pi\varepsilon_0)m_e v_i^2} = \frac{2\left(1.6\times10^{-19} \text{ C}\right)^2\left(8.99\times10^9 \text{ } \tfrac{\text{N}\cdot\text{m}^2}{\text{C}^2}\right)}{3\left(9.11\times10^{-19} \text{ kg}\right)\left(3.2\times10^5 \text{ m/s}\right)^2} = 1.6\times10^{-9}\text{m.}$$

48. When particle 3 is at $x = 0.10$ m, the total potential energy vanishes. Using Eq. 24-43, we have (with meters understood at the length unit)

$$0 = \frac{q_1 q_2}{4\pi\varepsilon_0 d} + \frac{q_1 q_3}{4\pi\varepsilon_0 (d + 0.10)} + \frac{q_3 q_2}{4\pi\varepsilon_0 (0.10)} .$$

This leads to

$$q_3 \left(\frac{q_1}{(d + 0.10)} + \frac{q_2}{0.10} \right) = - \frac{q_1 q_2}{d}$$

which yields $q_3 = -5.7$ μC.

49. We apply conservation of energy for particle 3 (with $q' = -15 \times 10^{-6}$ C):

$$K_0 + U_0 = K_f + U_f$$

where (letting $x = \pm 3$ m and $q_1 = q_2 = 50 \times 10^{-6}$ C $= q$)

$$U = \frac{q_1 q'}{4\pi\varepsilon_0 \sqrt{x^2 + y^2}} + \frac{q_2 q'}{4\pi\varepsilon_0 \sqrt{x^2 + y^2}} = \frac{q q'}{2\pi\varepsilon_0 \sqrt{x^2 + y^2}} .$$

(a) We solve for K_f (with $y_0 = 4$ m):

$$K_f = K_0 + U_0 - U_f = 1.2 \text{ J} + \frac{q q'}{2\pi\varepsilon_0} + \left(\frac{1}{\sqrt{x^2 + y_0^2}} - \frac{1}{|x|} \right) = 3.0 \text{ J} .$$

(b) We set $K_f = 0$ and solve for y (choosing the negative root, as indicated in the problem statement):

$$K_0 + U_0 = U_f \Rightarrow 1.2 \text{ J} + \frac{q q'}{2\pi\varepsilon_0 \sqrt{x^2 + y_0^2}} = \frac{q q'}{2\pi\varepsilon_0 \sqrt{x^2 + y^2}}$$

This yields $y = -8.5$ m.

50. From Eq. 24-30 and Eq. 24-7, we have (for $\theta = 180°$)

$$U = qV = -e \left(\frac{p \cos \theta}{4\pi\varepsilon_0 r^2} \right) = \frac{e p}{4\pi\varepsilon_0 r^2}$$

where $r = 0.020$ m. Appealing to energy conservation, we set this expression equal to 100 eV and solve for p. The magnitude of the dipole moment is therefore 4.5×10^{-12} C·m.

51. (a) Using $U = qV$ we can "translate" the graph of voltage into a potential energy graph (in eV units). From the information in the problem, we can calculate its kinetic

energy (which is its total energy at $x = 0$) in those units: $K_i = 284$ eV. This is less than the "height" of the potential energy "barrier" (500 eV high once we've translated the graph as indicated above). Thus, it must reach a turning point and then reverse its motion.

(b) Its final velocity, then, is in the negative x direction with a magnitude equal to that of its initial velocity. That is, its speed (upon leaving this region) is 1.0×10^7 m/s.

52. (a) The work done results in a potential energy gain:

$$W = q\,\Delta V = (-e)\left(\frac{Q}{4\pi\varepsilon_0 R}\right) = +2.16 \times 10^{-13}\,\text{J}.$$

With $R = 0.0800$ m, we find $Q = -1.20 \times 10^{-5}$ C.

(b) The work is the same, so the increase in the potential energy is $\Delta U = +2.16 \times 10^{-13}$ J.

53. If the electric potential is zero at infinity, then the potential at the surface of the sphere is given by $V = q/4\pi\varepsilon_0 r$, where q is the charge on the sphere and r is its radius. Thus

$$q = 4\pi\varepsilon_0 rV = \frac{(0.15\,\text{m})(1500\,\text{V})}{8.99\times10^9\,\text{N}\cdot\text{m}^2/\text{C}^2} = 2.5\times10^{-8}\,\text{C}.$$

54. Since the electric potential throughout the entire conductor is a constant, the electric potential at its center is also +400 V.

55. (a) The electric potential is the sum of the contributions of the individual spheres. Let q_1 be the charge on one, q_2 be the charge on the other, and d be their separation. The point halfway between them is the same distance $d/2$ (= 1.0 m) from the center of each sphere, so the potential at the halfway point is

$$V = \frac{q_1+q_2}{4\pi\varepsilon_0\,d/2} = \frac{\left(8.99\times10^9\,\text{N}\cdot\text{m}^2/\text{C}^2\right)\left(1.0\times10^{-8}\text{C}-3.0\times10^{-8}\text{C}\right)}{1.0\,\text{m}} = -1.8\times10^2\,\text{V}.$$

(b) The distance from the center of one sphere to the surface of the other is $d - R$, where R is the radius of either sphere. The potential of either one of the spheres is due to the charge on that sphere and the charge on the other sphere. The potential at the surface of sphere 1 is

$$V_1 = \frac{1}{4\pi\varepsilon_0}\left[\frac{q_1}{R}+\frac{q_2}{d-R}\right] = \left(8.99\times10^9\text{N}\cdot\text{m}^2/\text{C}^2\right)\left[\frac{1.0\times10^{-8}\text{C}}{0.030\,\text{m}} - \frac{3.0\times10^{-8}\text{C}}{2.0\,\text{m}-0.030\,\text{m}}\right] = 2.9\times10^3\,\text{V}.$$

(c) The potential at the surface of sphere 2 is

$$V_2 = \frac{1}{4\pi\varepsilon_0}\left[\frac{q_1}{d-R} + \frac{q_2}{R}\right] = \left(8.99\times10^9\,\text{N}\cdot\text{m}^2/\text{C}^2\right)\left[\frac{1.0\times10^{-8}\,\text{C}}{2.0\,\text{m} - 0.030\,\text{m}} - \frac{3.0\times10^{-8}\,\text{C}}{0.030\,\text{m}}\right] = -8.9\times10^3\,\text{V}.$$

56. (a) Since the two conductors are connected V_1 and V_2 must be equal to each other.

Let $V_1 = q_1/4\pi\varepsilon_0 R_1 = V_2 = q_2/4\pi\varepsilon_0 R_2$ and note that $q_1 + q_2 = q$ and $R_2 = 2R_1$. We solve for q_1 and q_2: $q_1 = q/3$, $q_2 = 2q/3$, or

(b) $q_1/q = 1/3 = 0.333$,

(c) and $q_2/q = 2/3 = 0.667$.

(d) The ratio of surface charge densities is

$$\frac{\sigma_1}{\sigma_2} = \frac{q_1/4\pi R_1^2}{q_2/4\pi R_2^2} = \left(\frac{q_1}{q_2}\right)\left(\frac{R_2}{R_1}\right)^2 = 2.00.$$

57. (a) The magnitude of the electric field is

$$E = \frac{\sigma}{\varepsilon_0} = \frac{q}{4\pi\varepsilon_0 R^2} = \frac{\left(3.0\times10^{-8}\,\text{C}\right)\left(8.99\times10^9\,\dfrac{\text{N}\cdot\text{m}^2}{\text{C}^2}\right)}{\left(0.15\,\text{m}\right)^2} = 1.2\times10^4\,\text{N/C}.$$

(b) $V = RE = (0.15\,\text{m})(1.2\times10^4\,\text{N/C}) = 1.8\times10^3\,\text{V}$.

(c) Let the distance be x. Then

$$\Delta V = V(x) - V = \frac{q}{4\pi\varepsilon_0}\left(\frac{1}{R+x} - \frac{1}{R}\right) = -500\,\text{V},$$

which gives

$$x = \frac{R\Delta V}{-V - \Delta V} = \frac{(0.15\,\text{m})(-500\,\text{V})}{-1800\,\text{V} + 500\,\text{V}} = 5.8\times10^{-2}\,\text{m}.$$

58. Since the charge distribution is spherically symmetric we may write

$$E(r) = \frac{1}{4\pi\varepsilon_0}\frac{q_{encl}}{r},$$

where q_{encl} is the charge enclosed in a sphere of radius r centered at the origin.

(a) For $r = 4.00$ m, $R_2 = 1.00$ m and $R_1 = 0.500$ m, with $r > R_2 > R_1$ we have (in SI units)

$$E(r) = \frac{q_1 + q_2}{4\pi\varepsilon_0 r^2} = \frac{(8.99\times10^9)(2.00\times10^{-6} + 1.00\times10^{-6})}{(4.00)^2} = 1.69\times10^3 \text{ V/m}.$$

(b) For $R_2 > r = 0.700$ m $> R_2$

$$E(r) = \frac{q_1}{4\pi\varepsilon_0 r^2} = \frac{(8.99\times10^9)(2.00\times10^{-6})}{(0.700)^2} = 3.67\times10^4 \text{ V/m}.$$

(c) For $R_2 > R_1 > r$, the enclosed charge is zero. Thus, $E = 0$.

The electric potential may be obtained using Eq. 24-18: $V(r) - V(r') = \int_r^{r'} E(r)dr$.

(d) For $r = 4.00$ m $> R_2 > R_1$, we have

$$V(r) = \frac{q_1 + q_2}{4\pi\varepsilon_0 r} = \frac{(8.99\times10^9)(2.00\times10^{-6} + 1.00\times10^{-6})}{(4.00)} = 6.74\times10^3 \text{ V}.$$

(e) For $r = 1.00$ m $= R_2 > R_1$, we have

$$V(r) = \frac{q_1 + q_2}{4\pi\varepsilon_0 r} = \frac{(8.99\times10^9)(2.00\times10^{-6} + 1.00\times10^{-6})}{(1.00)} = 2.70\times10^4 \text{ V}.$$

(f) For $R_2 > r = 0.700$ m $> R_2$,

$$V(r) = \frac{1}{4\pi\varepsilon_0}\left(\frac{q_1}{r} + \frac{q_2}{R_2}\right) = (8.99\times10^9)\left(\frac{2.00\times10^{-6}}{0.700} + \frac{1.00\times10^{-6}}{1.00}\right) = 3.47\times10^4 \text{ V}.$$

(g) For $R_2 > r = 0.500$ m $= R_2$,

$$V(r) = \frac{1}{4\pi\varepsilon_0}\left(\frac{q_1}{r} + \frac{q_2}{R_2}\right) = (8.99\times10^9)\left(\frac{2.00\times10^{-6}}{0.500} + \frac{1.00\times10^{-6}}{1.00}\right) = 4.50\times10^4 \text{ V}.$$

(h) For $R_2 > R_1 > r$,

$$V = \frac{1}{4\pi\varepsilon_0}\left(\frac{q_1}{R_1} + \frac{q_2}{R_2}\right) = (8.99\times10^9)\left(\frac{2.00\times10^{-6}}{0.500} + \frac{1.00\times10^{-6}}{1.00}\right) = 4.50\times10^4 \text{ V}.$$

(i) At $r = 0$, the potential remains constant, $V = 4.50\times10^4$ V.

(j) The electric field and the potential as a function of r are depicted below:

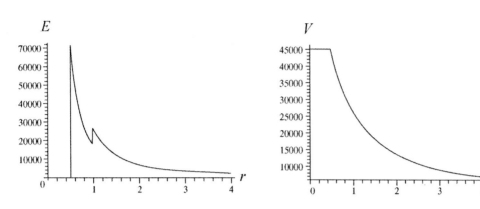

59. Using Gauss' law, $q = \varepsilon_0\Phi = +495.8$ nC. Consequently, $V = \dfrac{q}{4\pi\varepsilon_0 r} = 37.1$ kV.

60. (a) We use Eq. 24-18 to find the potential: $V_{wall} - V = -\int_r^R E\,dr$, or

$$0 - V = -\int_r^R \left(\frac{\rho r}{2\varepsilon_0}\right) \quad\Rightarrow\quad -V = -\frac{\rho}{4\varepsilon_0}\left(R^2 - r^2\right).$$

Consequently, $V = \rho(R^2 - r^2)/4\varepsilon_0$.

(b) The value at $r = 0$ is

$$V_{center} = \frac{-1.1\times 10^{-3}\,\text{C}/\text{m}^3}{4\left(8.85\times 10^{-12}\,\text{C}/\text{V}\cdot\text{m}\right)}\left((0.05\,\text{m})^2 - 0\right) = -7.8\times 10^4\,\text{V}.$$

Thus, the difference is $|V_{center}| = 7.8\times 10^4\,\text{V}$.

61. The electric potential energy in the presence of the dipole is

$$U = qV_{dipole} = \frac{qp\cos\theta}{4\pi\varepsilon_0 r^2} = \frac{(-e)(e\,d)\cos\theta}{4\pi\varepsilon_0 r^2}.$$

Noting that $\theta_i = \theta_f = 0°$, conservation of energy leads to

$$K_f + U_f = K_i + U_i \quad\Rightarrow\quad v = \sqrt{\frac{2\,e^2}{4\pi\varepsilon_0\,m\,d}\left(\frac{1}{25} - \frac{1}{49}\right)} = 7.0\times 10^5\,\text{m/s}.$$

62. (a) When the proton is released, its energy is $K + U = 4.0$ eV $+ 3.0$ eV (the latter value is inferred from the graph). This implies that if we draw a horizontal line at the 7.0

Volt "height" in the graph and find where it intersects the voltage plot, then we can determine the turning point. Interpolating in the region between 1.0 cm and 3.0 cm, we find the turning point is at roughly $x = 1.7$ cm.

(b) There is no turning point towards the right, so the speed there is nonzero, and is given by energy conservation:

$$v = \sqrt{\frac{2(7.0 \text{ eV})}{m}} = \sqrt{\frac{2(7.0 \text{ eV})(1.6 \times 10^{-19} \text{ J/eV})}{1.67 \times 10^{-27} \text{ kg}}} = 20 \text{ km/s}.$$

(c) The electric field at any point P is the (negative of the) slope of the voltage graph evaluated at P. Once we know the electric field, the force on the proton follows immediately from $\vec{F} = q\vec{E}$, where $q = +e$ for the proton. In the region just to the left of $x = 3.0$ cm, the field is $\vec{E} = (+300 \text{ V/m})\hat{i}$ and the force is $F = +4.8 \times 10^{-17} \text{ N}$.

(d) The force $\vec{F}$ points in the $+x$ direction, as the electric field $\vec{E}$.

(e) In the region just to the right of $x = 5.0$ cm, the field is $\vec{E} = (-200 \text{ V/m})\hat{i}$ and the magnitude of the force is $F = 3.2 \times 10^{-17} \text{ N}$.

(f) The force $\vec{F}$ points in the $-x$ direction, as the electric field $\vec{E}$.

63. Eq. 24-32 applies with $dq = \lambda \, dx = bx \, dx$ (along $0 \leq x \leq 0.20$ m).

(a) Here $r = x > 0$, so that

$$V = \frac{1}{4\pi\varepsilon_0} \int_0^{0.20} \frac{bx \, dx}{x} = \frac{b(0.20)}{4\pi\varepsilon_0} = 36 \text{ V}.$$

(b) Now $r = \sqrt{x^2 + d^2}$ where $d = 0.15$ m, so that

$$V = \frac{1}{4\pi\varepsilon_0} \int_0^{0.20} \frac{bx dx}{\sqrt{x^2 + d^2}} = \frac{b}{4\pi\varepsilon_0} \left(\sqrt{x^2 + d^2} \right) \Big|_0^{0.20} = 18 \text{ V}.$$

64. (a) When the electron is released, its energy is $K + U = 3.0 \text{ eV} - 6.0 \text{ eV}$ (the latter value is inferred from the graph along with the fact that $U = qV$ and $q = -e$). Because of the minus sign (of the charge) it is convenient to imagine the graph multiplied by a minus sign so that it represents potential energy in eV. Thus, the 2 V value shown at $x = 0$ would become -2 eV, and the 6 V value at $x = 4.5$ cm becomes -6 eV, and so on. The total energy (-3.0 eV) is constant and can then be represented on our (imagined) graph as a horizontal line at -3.0 V. This intersects the potential energy plot at a point we recognize as the turning point. Interpolating in the region between 1.0 cm and 4.0 cm, we find the turning point is at $x = 1.75$ cm ≈ 1.8 cm.

(b) There is no turning point towards the right, so the speed there is nonzero. Noting that the kinetic energy at $x = 7.0$ cm is -3.0 eV $-(-5.0$ eV$) = 2.0$ eV, we find the speed using energy conservation:

$$v = \sqrt{\frac{2(2.0 \text{ eV})}{m}} = \sqrt{\frac{2(2.0 \text{ eV})(1.6 \times 10^{-19} \text{ J/eV})}{9.11 \times 10^{-31} \text{ kg}}} = 8.4 \times 10^5 \text{ m/s}.$$

(c) The electric field at any point P is the (negative of the) slope of the voltage graph evaluated at P. Once we know the electric field, the force on the electron follows immediately from $\vec{F} = q\vec{E}$, where $q = -e$ for the electron. In the region just to the left of $x = 4.0$ cm, the field is $\vec{E} = (-133 \text{ V/m})\,\hat{\text{i}}$ and the magnitude of the force is $F = 2.1 \times 10^{-17}$ N.

(d) The force points in the $+x$ direction.

(e) In the region just to the right of $x = 5.0$ cm, the field is $\vec{E} = +100$ V/m $\hat{\text{i}}$ and the force is $\vec{F} = (-1.6 \times 10^{-17} \text{N})\,\hat{\text{i}}$. Thus, the magnitude of the force is $F = 1.6 \times 10^{-17}$ N.

(f) The minus sign indicates that $\vec{F}$ points in the $-x$ direction.

65. We treat the system as a superposition of a disk of surface charge density σ and radius R and a smaller, oppositely charged, disk of surface charge density $-\sigma$ and radius r. For each of these, Eq 24-37 applies (for $z > 0$)

$$V = \frac{\sigma}{2\varepsilon_0}\left(\sqrt{z^2 + R^2} - z\right) + \frac{-\sigma}{2\varepsilon_0}\left(\sqrt{z^2 + r^2} - z\right).$$

This expression does vanish as $r \to \infty$, as the problem requires. Substituting $r = 0.200R$ and $z = 2.00R$ and simplifying, we obtain

$$V = \frac{\sigma R}{\varepsilon_0}\left(\frac{5\sqrt{5} - \sqrt{101}}{10}\right) = \frac{(6.20 \times 10^{-12})(0.130)}{8.85 \times 10^{-12}}\left(\frac{5\sqrt{5} - \sqrt{101}}{10}\right) = 1.03 \times 10^{-2} \text{ V}.$$

66. Since the electric potential energy is not changed by the introduction of the third particle, we conclude that the net electric potential evaluated at P caused by the original two particles must be zero:

$$\frac{q_1}{4\pi\varepsilon_0 r_1} + \frac{q_2}{4\pi\varepsilon_0 r_2} = 0.$$

Setting $r_1 = 5d/2$ and $r_2 = 3d/2$ we obtain $q_1 = -5q_2/3$, or $q_1/q_2 = -5/3 \approx -1.7$.

67. The electric field throughout the conducting volume is zero, which implies that the potential there is constant and equal to the value it has on the surface of the charged sphere:

$$V_A = V_S = \frac{q}{4\pi\varepsilon_0 \, R}$$

where $q = 30 \times 10^{-9}$ C and $R = 0.030$ m. For points beyond the surface of the sphere, the potential follows Eq. 24-26:

$$V_B = \frac{q}{4\pi\varepsilon_0 \, r}$$

where $r = 0.050$ m.

(a) We see that

$$V_S - V_B = \frac{q}{4\pi\varepsilon_0} \left(\frac{1}{R} - \frac{1}{r} \right) = 3.6 \times 10^3 \text{ V}.$$

(b) Similarly,

$$V_A - V_B = \frac{q}{4\pi\varepsilon_0} \left(\frac{1}{R} - \frac{1}{r} \right) = 3.6 \times 10^3 \text{ V}.$$

68. The *escape speed* may be calculated from the requirement that the initial kinetic energy (of *launch*) be equal to the absolute value of the initial potential energy (compare with the gravitational case in chapter 14). Thus,

$$\frac{1}{2} m v^2 = \frac{e \, q}{4\pi\varepsilon_0 \, r}$$

where $m = 9.11 \times 10^{-31}$ kg, $e = 1.60 \times 10^{-19}$ C, $q = 10000e$, and $r = 0.010$ m. This yields the answer $v = 22490$ m/s $\approx 2.2 \times 10^4$ m/s.

69. We apply conservation of energy for the particle with $q = 7.5 \times 10^{-6}$ C (which has zero initial kinetic energy):

$$U_0 = K_f + U_f \qquad \text{where } U = \frac{q \, Q}{4\pi\varepsilon_0 r} \quad .$$

(a) The initial value of r is 0.60 m and the final value is $(0.6 + 0.4)$ m $= 1.0$ m (since the particles repel each other). Conservation of energy, then, leads to $K_f = 0.90$ J.

(b) Now the particles attract each other so that the final value of r is $0.60 - 0.40 = 0.20$ m. Use of energy conservation yields $K_f = 4.5$ J in this case.

70. (a) Using $d = 2$ m, we find the potential at P:

$$V_P = \frac{1}{4\pi\varepsilon_0}\frac{+2e}{d} + \frac{1}{4\pi\varepsilon_0}\frac{-2e}{2d} = \frac{1}{4\pi\varepsilon_0}\frac{e}{d} \; .$$

Thus, with $e = 1.60 \times 10^{-19}$ C, we find $V_P = 7.19 \times 10^{-10}$ V. Note that we are implicitly assuming that $V \to 0$ as $r \to \infty$.

(b) Since $U = qV$, then the movable particle's contribution of the potential energy when it is at $r = \infty$ is zero, and its contribution to U_{system} when it is at P is $(2e)V_P = 2.30 \times 10^{-28}$ J. Thus, we obtain $W_{app} = 2.30 \times 10^{-28}$ J.

(c) Now, combining the contribution to U_{system} from part (b) and from the original pair of fixed charges

$$U_{fixed} = \frac{1}{4\pi\varepsilon_0}\frac{(2e)(-2e)}{\sqrt{4^2 + 2^2}} = -2.1 \times 10^{-28}\text{ J} \; ,$$

we obtain

$$U_{system} = U_{part\,(b)} + U_{fixed} = 2.43 \times 10^{-29}\text{ J} \; .$$

71. The derivation is shown in the book (Eq. 24-33 through Eq. 24-35) except for the change in the lower limit of integration (which is now $x = D$ instead of $x = 0$). The result is therefore (cf. Eq. 24-35)

$$V = \frac{\lambda}{4\pi\varepsilon_0}\ln\left(\frac{L + \sqrt{L^2 + d^2}}{D + \sqrt{D^2 + d^2}}\right) = \frac{2.0 \times 10^{-6}}{4\pi\varepsilon_0}\ln\left(\frac{4 + \sqrt{17}}{1 + \sqrt{2}}\right) = 2.18 \times 10^4\text{ V.}$$

72. Using Eq. 24-18, we have

$$\Delta V = -\int_2^3 \frac{A}{r^4}\,dr = \frac{A}{3}\left(\frac{1}{2^3} - \frac{1}{3^3}\right) = A(0.029/\text{m}^3).$$

73. The work done results in a change of potential energy:

$$W = \Delta U = \frac{2(0.12\text{ C})^2}{4\pi\varepsilon_0\left(\dfrac{1.7\text{ m}}{2}\right)} - \frac{2(0.12\text{ C})^2}{4\pi\varepsilon_0\,(1.7\text{ m})} = 1.5 \times 10^8\text{ J} \; .$$

At a rate of $P = 0.83 \times 10^3$ Joules per second, it would take $W/P = 1.8 \times 10^5$ seconds or about 2.1 days to do this amount of work.

74. The charges are equidistant from the point where we are evaluating the potential — which is computed using Eq. 24-27 (or its integral equivalent). Eq. 24-27 implicitly assumes $V \to 0$ as $r \to \infty$. Thus, we have

$$V = \frac{1}{4\pi\varepsilon_0}\frac{+Q_1}{R} + \frac{1}{4\pi\varepsilon_0}\frac{-2Q_1}{R} + \frac{1}{4\pi\varepsilon_0}\frac{+3Q_1}{R} = \frac{1}{4\pi\varepsilon_0}\frac{2Q_1}{R} = \frac{2(8.99\times10^9)(4.52\times10^{-12})}{0.0850}$$

$$= 0.956 \text{ V}.$$

75. The radius of the cylinder (0.020 m, the same as r_B) is denoted R, and the field magnitude there (160 N/C) is denoted E_B. The electric field beyond the surface of the sphere follows Eq. 23-12, which expresses inverse proportionality with r:

$$\frac{|\vec{E}|}{E_B} = \frac{R}{r} \quad \text{for } r \geq R .$$

(a) Thus, if $r = r_C = 0.050$ m, we obtain $|\vec{E}| = (160)(0.020)/(0.050) = 64 \text{ N/C}$.

(b) Integrating the above expression (where the variable to be integrated, r, is now denoted ρ) gives the potential difference between V_B and V_C.

$$V_B - V_C = \int_R^r \frac{E_B R}{\rho} d\rho = E_B R \ln\left(\frac{r}{R}\right) = 2.9 \text{ V} .$$

(c) The electric field throughout the conducting volume is zero, which implies that the potential there is constant and equal to the value it has on the surface of the charged cylinder: $V_A - V_B = 0$.

76. We note that for two points on a circle, separated by angle θ (in radians), the direct-line distance between them is $r = 2R \sin(\theta/2)$. Using this fact, distinguishing between the cases where $N =$ odd and $N =$ even, and counting the pair-wise interactions very carefully, we arrive at the following results for the total potential energies. We use $k = 1/4\pi\varepsilon_0$. For configuration 1 (where all N electrons are on the circle), we have

$$U_{1,N=\text{even}} = \frac{Nke^2}{2R}\left(\sum_{j=1}^{\frac{N}{2}-1}\frac{1}{\sin(j\theta/2)} + \frac{1}{2}\right), \quad U_{1,N=\text{odd}} = \frac{Nke^2}{2R}\left(\sum_{j=1}^{\frac{N-1}{2}}\frac{1}{\sin(j\theta/2)}\right)$$

where $\theta = \frac{2\pi}{N}$. For configuration 2, we find

$$U_{2,N=\text{even}} = \frac{(N-1)ke^2}{2R}\left(\sum_{j=1}^{\frac{N}{2}-1}\frac{1}{\sin(j\theta'/2)}+2\right)$$

$$U_{2,N=\text{odd}} = \frac{(N-1)ke^2}{2R}\left(\sum_{j=1}^{\frac{N-3}{2}}\frac{1}{\sin(j\theta'/2)}+\frac{5}{2}\right)$$

where $\theta' = \dfrac{2\pi}{N-1}$. The results are all of the form

$$U_{1\text{or}2}\frac{ke^2}{2R}\times\text{a pure number.}$$

In our table, below, we have the results for those "pure numbers" as they depend on N and on which configuration we are considering. The values listed in the U rows are the potential energies divided by $ke^2/2R$.

N	4	5	6	7	8	9	10	11	12	13	14	15
U_1	3.83	6.88	10.96	16.13	22.44	29.92	38.62	48.58	59.81	72.35	86.22	101.5
U_2	4.73	7.83	11.88	16.96	23.13	30.44	39.92	48.62	59.58	71.81	85.35	100.2

We see that the potential energy for configuration 2 is greater than that for configuration 1 for $N < 12$, but for $N \geq 12$ it is configuration 1 that has the greatest potential energy.

(a) $N = 12$ is the smallest value such that $U_2 < U_1$.

(b) For $N = 12$, configuration 2 consists of 11 electrons distributed at equal distances around the circle, and one electron at the center. A specific electron e_0 on the circle is R distance from the one in the center, and is

$$r = 2R\sin\left(\frac{\pi}{11}\right) \approx 0.56R$$

distance away from its nearest neighbors on the circle (of which there are two — one on each side). Beyond the nearest neighbors, the next nearest electron on the circle is

$$r = 2R\sin\left(\frac{2\pi}{11}\right) \approx 1.1R$$

distance away from e_0. Thus, we see that there are only two electrons closer to e_0 than the one in the center.

77. We note that the net potential (due to the "fixed" charges) is zero at the first location ("at ∞") being considered for the movable charge q (where $q = +2e$). Thus, the work required is equal to the potential energy in the final configuration: qV where

$$V = \frac{1}{4\pi\varepsilon_0}\frac{(+2e)}{2D} + \frac{1}{4\pi\varepsilon_0}\frac{+e}{D} \ .$$

Using $D = 4.00$ m and $e = 1.60 \times 10^{-19}$ C, we obtain

$$W_{app} = qV = (2e)(7.20 \times 10^{-10} \text{ V}) = 2.30 \times 10^{-28} \text{ J}.$$

78. Since the electric potential is a scalar quantity, this calculation is far simpler than it would be for the electric field. We are able to simply take half the contribution that would be obtained from a complete (whole) sphere. If it were a whole sphere (of the same density) then its charge would be $q_{whole} = 8.00$ µC. Then

$$V = \frac{1}{2}V_{whole} = \frac{1}{2}\frac{q_{whole}}{4\pi\varepsilon_0\, r} = \frac{1}{2}\frac{8.00 \times 10^{-6}\, \text{C}}{4\pi\varepsilon_0(0.15\text{ m})} = 2.40 \times 10^5 \text{ V} \ .$$

79. The net potential at point P (the place where we are to place the third electron) due to the fixed charges is computed using Eq. 24-27 (which assumes $V \to 0$ as $r \to \infty$):

$$V_P = \frac{1}{4\pi\varepsilon_0}\frac{-e}{d} + \frac{1}{4\pi\varepsilon_0}\frac{-e}{d} = \frac{-e}{2\pi\varepsilon_0 d} \ .$$

Thus, with $d = 2.00 \times 10^{-6}$ m and $e = 1.60 \times 10^{-19}$ C, we find $V_P = -1.438 \times 10^{-3}$ V. Then the required "applied" work is, by Eq. 24-14,

$$W_{app} = (-e)\, V_P = 2.30 \times 10^{-22} \text{ J} \ .$$

80. The work done is equal to the change in the (total) electric potential energy U of the system, where

$$U = \frac{q_1\, q_2}{4\pi\varepsilon_0\, r_{12}} + \frac{q_3\, q_2}{4\pi\varepsilon_0\, r_{23}} + \frac{q_1\, q_3}{4\pi\varepsilon_0\, r_{13}}$$

and the notation r_{13} indicates the distance between q_1 and q_3 (similar definitions apply to r_{12} and r_{23}).

(a) We consider the difference in U where initially $r_{12} = b$ and $r_{23} = a$, and finally $r_{12} = a$ and $r_{23} = b$ (r_{13} doesn't change). Converting the values given in the problem to SI units (µC to C, cm to m), we obtain $\Delta U = -24$ J.

(b) Now we consider the difference in U where initially $r_{23} = a$ and $r_{13} = a$, and finally r_{23} is again equal to a and r_{13} is also again equal to a (and of course, r_{12} doesn't change in this case). Thus, we obtain $\Delta U = 0$.

81. (a) Clearly, the net voltage

$$V = \frac{q}{4\pi\varepsilon_0 |x|} + \frac{2q}{4\pi\varepsilon_0 |d - x|}$$

is not zero for any finite value of x.

(b) The electric field cancels at a point between the charges:

$$\frac{q}{4\pi\varepsilon_0 x^2} = \frac{2q}{4\pi\varepsilon_0 (d - x)^2}$$

which has the solution: $x = (\sqrt{2} - 1)d = 0.41$ m.

82. (a) The potential on the surface is

$$V = \frac{q}{4\pi\varepsilon_0 R} = \frac{\left(4.0 \times 10^{-6}\,\mathrm{C}\right)\left(8.99 \times 10^9\,\frac{\mathrm{N \cdot m^2}}{\mathrm{C^2}}\right)}{0.10\,\mathrm{m}} = 3.6 \times 10^5\,\mathrm{V}\ .$$

(b) The field just outside the sphere would be

$$E = \frac{q}{4\pi\varepsilon_0 R^2} = \frac{V}{R} = \frac{3.6 \times 10^5\,\mathrm{V}}{0.10\,\mathrm{m}} = 3.6 \times 10^6\,\mathrm{V/m}\ ,$$

which would have exceeded 3.0 MV/m. So this situation cannot occur.

83. This can be approached more than one way, but the simplest is to observe that the net potential (using Eq. 24-27) due to $q_1 = +2e$ and $q_3 = -2e$ is zero at both the initial and final positions of the movable charge $q_2 = +5q$. This implies that no work is necessary to effect its change of position, which, in turn, implies there is no resulting change in potential energy of the configuration. Hence, the ratio is unity.

84. We use $E_x = -dV/dx$, where dV/dx is the local slope of the V vs. x curve depicted in Fig. 24-54. The results are:

(a) $E_x(ab) = -6.0$ V/m,

(b) $E_x(bc) = 0$,

(c) $E_x(cd)$ 3.0 V/m,

(d) $E_x(de) = 3.0$ V/m,

(e) $E_x(ef) = 15$ V/m,

(f) $E_x(fg) = 0$,

(g) $E_x(gh) = -3.0$ V/m.

Since these values are constant during their respective time-intervals, their graph consists of several disconnected line-segments (horizontal) and is not shown here.

85. (a) We denote the surface charge density of the disk as σ_1 for $0 < r < R/2$, and as σ_2 for $R/2 < r < R$. Thus the total charge on the disk is given by

$$q = \int_{disk} dq = \int_0^{R/2} 2\pi\sigma_1 r\, dr + \int_{R/2}^R 2\pi\sigma_2 r\, dr = \frac{\pi}{4} R^2 (\sigma_1 + 3\sigma_2)$$

$$= \frac{\pi}{4}\left(2.20\times10^{-2}\text{ m}\right)^2 \left[1.50\times10^{-6}\text{ C/m}^2 + 3\left(8.00\times10^{-7}\text{ C/m}^2\right)\right]$$

$$= 1.48\times10^{-9}\text{ C}.$$

(b) We use Eq. 24-36:

$$V(z) = \int_{disk} dV = k\left[\int_0^{R/2} \frac{\sigma_1(2\pi R')dR'}{\sqrt{z^2+R'^2}} + \int_{R/2}^R \frac{\sigma_2(2\pi R')dR'}{\sqrt{z^2+R'^2}}\right]$$

$$= \frac{\sigma_1}{2\varepsilon_0}\left(\sqrt{z^2+\frac{R^2}{4}} - z\right) + \frac{\sigma_2}{2\varepsilon_0}\left(\sqrt{z^2+R^2} - \sqrt{z^2+\frac{R^2}{4}}\right).$$

Substituting the numerical values of σ_1, σ_2, R and z, we obtain $V(z) = 7.95 \times 10^2$ V.

86. The net potential (at point A or B) is computed using Eq. 24-27. Thus, using k for $1/4\pi\varepsilon_0$, the difference is

$$V_A - V_B = \left(\frac{ke}{d} + \frac{k(-5e)}{5d}\right) - \left(\frac{ke}{2d} + \frac{k(-5e)}{2d}\right) = \frac{2ke}{d} = \frac{2(8.99\times10^9)(1.6\times10^{-19})}{5.60\times10^{-6}} = 5.14\times10^{-4}\text{ V}.$$

87. We denote $q = 25 \times 10^{-9}$ C, $y = 0.6$ m, $x = 0.8$ m, with $V =$ the net potential (assuming $V \to 0$ as $r \to \infty$). Then,

$$V_A = \frac{1}{4\pi\varepsilon_0}\frac{q}{y} + \frac{1}{4\pi\varepsilon_0}\frac{(-q)}{x}$$

$$V_B = \frac{1}{4\pi\varepsilon_0}\frac{q}{x} + \frac{1}{4\pi\varepsilon_0}\frac{(-q)}{y}$$

leads to

$$V_B - V_A = \frac{2}{4\pi\varepsilon_0}\frac{q}{x} - \frac{2}{4\pi\varepsilon_0}\frac{q}{y} = \frac{q}{2\pi\varepsilon_0}\left(\frac{1}{x} - \frac{1}{y}\right)$$

which yields $\Delta V = -187\,\text{V}$.

88. In the "inside" region between the plates, the individual fields (given by Eq. 24-13) are in the same direction $(-\hat{i})$:

$$\vec{E}_{in} = -\left(\frac{50\times 10^{-9}}{2\varepsilon_0} + \frac{25\times 10^{-9}}{2\varepsilon_0}\right)\hat{i} = -4.2\times 10^3\hat{i}$$

in SI units (N/C or V/m). And in the "outside" region where $x > 0.5$ m, the individual fields point in opposite directions:

$$\vec{E}_{out} = -\frac{50\times 10^{-9}}{2\varepsilon_0}\hat{i} + \frac{25\times 10^{-9}}{2\varepsilon_0}\hat{i} = -1.4\times 10^3\hat{i}.$$

Therefore, by Eq. 24-18, we have

$$\Delta V = -\int_0^{0.8}\vec{E}\cdot d\vec{s} = -\int_0^{0.5}\left|\vec{E}\right|_{in}dx - \int_{0.5}^{0.8}\left|\vec{E}\right|_{out}dx = -\left(4.2\times 10^3\right)(0.5) - \left(1.4\times 10^3\right)(0.3)$$

$$= 2.5\times 10^3\,\text{V}.$$

89. (a) The charges are equal and are the same distance from C. We use the Pythagorean theorem to find the distance $r = \sqrt{(d/2)^2 + (d/2)^2} = d/\sqrt{2}$. The electric potential at C is the sum of the potential due to the individual charges but since they produce the same potential, it is twice that of either one:

$$V = \frac{2q}{4\pi\varepsilon_0}\frac{\sqrt{2}}{d} = \frac{2\sqrt{2}q}{4\pi\varepsilon_0 d} = \frac{\left(8.99\times 10^9\,\text{N}\cdot\text{m}^2/\text{C}^2\right)(2)\sqrt{2}\left(2.0\times 10^{-6}\,\text{C}\right)}{0.020\,\text{m}} = 2.5\times 10^6\,\text{V}.$$

(b) As you move the charge into position from far away the potential energy changes from zero to qV, where V is the electric potential at the final location of the charge. The change in the potential energy equals the work you must do to bring the charge in:

$$W = qV = \left(2.0 \times 10^{-6}\,\text{C}\right)\left(2.54 \times 10^{6}\,\text{V}\right) = 5.1\ \text{J}.$$

(c) The work calculated in part (b) represents the potential energy of the interactions between the charge brought in from infinity and the other two charges. To find the total potential energy of the three-charge system you must add the potential energy of the interaction between the fixed charges. Their separation is d so this potential energy is $q^2/4\pi\varepsilon_0 d$. The total potential energy is

$$U = W + \frac{q^2}{4\pi\varepsilon_0 d} = 5.1\ \text{J} + \frac{\left(8.99 \times 10^{9}\,\text{N} \cdot \text{m}^2/\text{C}^2\right)\left(2.0 \times 10^{-6}\,\text{C}\right)^2}{0.020\,\text{m}} = 6.9\ \text{J}.$$

90. The potential energy of the two-charge system is

$$U = \frac{1}{4\pi\varepsilon_0}\left[\frac{q_1 q_2}{\sqrt{(x_1 - x_2)^2 + (y_1 - y_2)^2}}\right] = \frac{\left(8.99 \times 10^{9}\,\dfrac{\text{N} \cdot \text{m}^2}{\text{C}^2}\right)\left(3.00 \times 10^{-6}\,\text{C}\right)\left(-4.00 \times 10^{-6}\,\text{C}\right)}{\sqrt{(3.50 + 2.00)^2 + (0.500 - 1.50)^2}\ \text{cm}}$$

$$= -1.93\ \text{J}.$$

Thus, -1.93 J of work is needed.

91. For a point on the axis of the ring the potential (assuming $V \to 0$ as $r \to \infty$) is

$$V = \frac{q}{4\pi\varepsilon_0 \sqrt{z^2 + R^2}}$$

where $q = 16 \times 10^{-6}$ C and $R = 0.0300$ m. Therefore,

$$V_B - V_A = \frac{q}{4\pi\varepsilon_0}\left(\frac{1}{\sqrt{z_B^2 + R^2}} - \frac{1}{R}\right)$$

where $z_B = 0.040$ m. The result is -1.92×10^{6} V.

92. The initial speed v_i of the electron satisfies $K_i = \frac{1}{2}m_e v_i^2 = e\Delta V$, which gives

$$v_i = \sqrt{\frac{2e\Delta V}{m_e}} = \sqrt{\frac{2\left(1.60 \times 10^{-19}\,\text{J}\right)\left(625\ \text{V}\right)}{9.11 \times 10^{-31}\,\text{kg}}} = 1.48 \times 10^{7}\ \text{m/s}.$$

93. (a) The potential energy is

$$U = \frac{q^2}{4\pi\varepsilon_0 d} = \frac{(8.99\times 10^9 \text{ N}\cdot\text{m}^2/\text{C}^2)(5.0\times 10^{-6} \text{ C})^2}{1.00 \text{ m}} = 0.225 \text{ J}$$

relative to the potential energy at infinite separation.

(b) Each sphere repels the other with a force that has magnitude

$$F = \frac{q^2}{4\pi\varepsilon_0 d^2} = \frac{(8.99\times 10^9 \text{ N}\cdot\text{m}^2/\text{C}^2)(5.0\times 10^{-6} \text{ C})^2}{(1.00 \text{ m})^2} = 0.225 \text{ N}.$$

According to Newton's second law the acceleration of each sphere is the force divided by the mass of the sphere. Let m_A and m_B be the masses of the spheres. The acceleration of sphere A is

$$a_A = \frac{F}{m_A} = \frac{0.225 \text{ N}}{5.0\times 10^{-3} \text{ kg}} = 45.0 \text{ m/s}^2$$

and the acceleration of sphere B is

$$a_B = \frac{F}{m_B} = \frac{0.225 \text{ N}}{10\times 10^{-3} \text{ kg}} = 22.5 \text{ m/s}^2 .$$

(c) Energy is conserved. The initial potential energy is $U = 0.225$ J, as calculated in part (a). The initial kinetic energy is zero since the spheres start from rest. The final potential energy is zero since the spheres are then far apart. The final kinetic energy is $\frac{1}{2}m_A v_A^2 + \frac{1}{2}m_B v_B^2$, where v_A and v_B are the final velocities. Thus,

$$U = \frac{1}{2}m_A v_A^2 + \frac{1}{2}m_B v_B^2.$$

Momentum is also conserved, so

$$0 = m_A v_A + m_B v_B.$$

These equations may be solved simultaneously for v_A and v_B. Substituting $v_B = -(m_A/m_B)v_A$, from the momentum equation into the energy equation, and collecting terms, we obtain $U = \frac{1}{2}(m_A/m_B)(m_A+m_B)v_A^2$. Thus,

$$v_A = \sqrt{\frac{2Um_B}{m_A(m_A+m_B)}} = \sqrt{\frac{2(0.225 \text{ J})(10\times 10^{-3} \text{ kg})}{(5.0\times 10^{-3} \text{ kg})(5.0\times 10^{-3} \text{ kg}+10\times 10^{-3} \text{ kg})}} = 7.75 \text{ m/s}.$$

We thus obtain

$$v_B = -\frac{m_A}{m_B} v_A = -\left(\frac{5.0 \times 10^{-3} \text{ kg}}{10 \times 10^{-3} \text{ kg}}\right)(7.75 \text{ m/s}) = -3.87 \text{ m/s},$$

or $|v_B| = 3.87$ m/s.

94. The particle with charge $-q$ has both potential and kinetic energy, and both of these change when the radius of the orbit is changed. We first find an expression for the total energy in terms of the orbit radius r. Q provides the centripetal force required for $-q$ to move in uniform circular motion. The magnitude of the force is $F = Qq/4\pi\varepsilon_0 r^2$. The acceleration of $-q$ is v^2/r, where v is its speed. Newton's second law yields

$$\frac{Q_q}{4\pi\varepsilon_0 r^2} = \frac{mv^2}{r} \Rightarrow mv^2 = \frac{Qq}{4\pi\varepsilon_0 r},$$

and the kinetic energy is $K = \frac{1}{2}mv^2 = Qq/8\pi\varepsilon_0 r$. The potential energy is $U = -Qq/4\pi\varepsilon_0 r$, and the total energy is

$$E = K + U = \frac{Qq}{8\pi\varepsilon_0 r} - \frac{Qq}{4\pi\varepsilon_0 r} = -\frac{Qq}{8\pi\varepsilon_0 r}.$$

When the orbit radius is r_1 the energy is $E_1 = -Qq/8\pi\varepsilon_0 r_1$ and when it is r_2 the energy is $E_2 = -Qq/8\pi\varepsilon_0 r_2$. The difference $E_2 - E_1$ is the work W done by an external agent to change the radius:

$$W = E_2 - E_1 = -\frac{Qq}{8\pi\varepsilon_0}\left(\frac{1}{r_2} - \frac{1}{r_1}\right) = \frac{Qq}{8\pi\varepsilon_0}\left(\frac{1}{r_1} - \frac{1}{r_2}\right).$$

95. (a) The total electric potential energy consists of three equal terms:

$$U = \frac{q_1 q_2}{4\pi\varepsilon_0 r} + \frac{q_2 q_3}{4\pi\varepsilon_0 r} + \frac{q_1 q_3}{4\pi\varepsilon_0 r}$$

where $q_1 = q_2 = q_3 = -\frac{e}{3}$, and r as given in the problem. The result is $U = 2.72 \times 10^{-14}$ J.

(b) Dividing by the square of the speed of light (roughly 3.0×10^8 m/s), we obtain a value in kilograms (about a third of the correct electron mass value): 3.02×10^{-31} kg.

96. A positive charge q is a distance $r - d$ from P, another positive charge q is a distance r from P, and a negative charge $-q$ is a distance $r + d$ from P. Sum the individual electric potentials created at P to find the total:

$$V = \frac{q}{4\pi\varepsilon_0}\left[\frac{1}{r-d}+\frac{1}{r}-\frac{1}{r+d}\right].$$

We use the binomial theorem to approximate $1/(r-d)$ for r much larger than d:

$$\frac{1}{r-d} = (r-d)^{-1} \approx (r)^{-1} - (r)^{-2}(-d) = \frac{1}{r}+\frac{d}{r^2}.$$

Similarly,

$$\frac{1}{r+d} \approx \frac{1}{r}-\frac{d}{r^2}.$$

Only the first two terms of each expansion were retained. Thus,

$$V \approx \frac{q}{4\pi\varepsilon_0}\left[\frac{1}{r}+\frac{d}{r^2}+\frac{1}{r}-\frac{1}{r}+\frac{d}{r^2}\right] = \frac{q}{4\pi\varepsilon_0}\left[\frac{1}{r}+\frac{2d}{r^2}\right] = \frac{q}{4\pi\varepsilon_0 r}\left[1+\frac{2d}{r}\right].$$

97. Assume the charge on Earth is distributed with spherical symmetry. If the electric potential is zero at infinity then at the surface of Earth it is $V = q/4\pi\varepsilon_0 R$, where q is the charge on Earth and $R = 6.37 \times 10^6$ m is the radius of Earth. The magnitude of the electric field at the surface is $E = q/4\pi\varepsilon_0 R^2$, so

$$V = ER = (100 \text{ V/m}) (6.37 \times 10^6 \text{ m}) = 6.4 \times 10^8 \text{ V}.$$

98. The net electric potential at point P is the sum of those due to the six charges:

$$V_P = \sum_{i=1}^{6}V_{Pi} = \sum_{i=1}^{6}\frac{q_i}{4\pi\varepsilon_0 r_i} = \frac{10^{-15}}{4\pi\varepsilon_0}\left[\frac{5.00}{\sqrt{d^2+(d/2)^2}}+\frac{-2.00}{d/2}+\frac{-3.00}{\sqrt{d^2+(d/2)^2}}\right.$$

$$\left.+\frac{3.00}{\sqrt{d^2+(d/2)^2}}+\frac{-2.00}{d/2}+\frac{+5.00}{\sqrt{d^2+(d/2)^2}}\right] = \frac{9.4\times10^{-16}}{4\pi\varepsilon_0(2.54\times10^{-2})} = 3.34\times10^{-4} \text{ V}.$$

99. In the sketches shown next, the lines with the arrows are field lines and those without are the equipotentials (which become more circular the closer one gets to the individual charges). In all pictures, q_2 is on the left and q_1 is on the right (which is reversed from the way it is shown in the textbook).

(a)

(b)

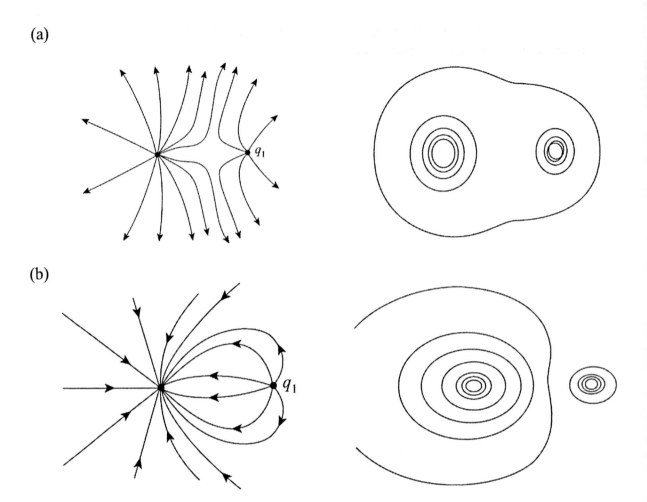

100. (a) We use Gauss' law to find expressions for the electric field inside and outside the spherical charge distribution. Since the field is radial the electric potential can be written as an integral of the field along a sphere radius, extended to infinity. Since different expressions for the field apply in different regions the integral must be split into two parts, one from infinity to the surface of the distribution and one from the surface to a point inside. Outside the charge distribution the magnitude of the field is $E = q/4\pi\varepsilon_0 r^2$ and the potential is $V = q/4\pi\varepsilon_0 r$, where r is the distance from the center of the distribution. This is the same as the field and potential of a point charge at the center of the spherical distribution. To find an expression for the magnitude of the field inside the charge distribution, we use a Gaussian surface in the form of a sphere with radius r, concentric with the distribution. The field is normal to the Gaussian surface and its magnitude is uniform over it, so the electric flux through the surface is $4\pi r^2 E$. The charge enclosed is qr^3/R^3. Gauss' law becomes

$$4\pi\varepsilon_0 r^2 E = \frac{qr^3}{R^3},$$

so

$$E = \frac{qr}{4\pi\varepsilon_0 R^3}.$$

If V_s is the potential at the surface of the distribution ($r = R$) then the potential at a point inside, a distance r from the center, is

$$V = V_s - \int_R^r E\,dr = V_s - \frac{q}{4\pi\varepsilon_0 R^3}\int_R^r r\,dr = V_s - \frac{qr^2}{8\pi\varepsilon_0 R^3} + \frac{q}{8\pi\varepsilon_0 R}.$$

The potential at the surface can be found by replacing r with R in the expression for the potential at points outside the distribution. It is $V_s = q/4\pi\varepsilon_0 R$. Thus,

$$V = \frac{q}{4\pi\varepsilon_0}\left[\frac{1}{R} - \frac{r^2}{2R^3} + \frac{1}{2R}\right] = \frac{q}{8\pi\varepsilon_0 R^3}\left(3R^2 - r^2\right).$$

(b) The potential difference is

$$\Delta V = V_s - V_c = \frac{2q}{8\pi\varepsilon_0 R} - \frac{3q}{8\pi\varepsilon_0 R} = -\frac{q}{8\pi\varepsilon_0 R},$$

or $|\Delta V| = q/8\pi\varepsilon_0 R$.

101. (a) For $r > r_2$ the field is like that of a point charge and

$$V = \frac{1}{4\pi\varepsilon_0}\frac{Q}{r},$$

where the zero of potential was taken to be at infinity.

(b) To find the potential in the region $r_1 < r < r_2$, first use Gauss's law to find an expression for the electric field, then integrate along a radial path from r_2 to r. The Gaussian surface is a sphere of radius r, concentric with the shell. The field is radial and therefore normal to the surface. Its magnitude is uniform over the surface, so the flux through the surface is $\Phi = 4\pi r^2 E$. The volume of the shell is $(4\pi/3)(r_2^3 - r_1^3)$, so the charge density is

$$\rho = \frac{3Q}{4\pi\left(r_2^3 - r_1^3\right)},$$

and the charge enclosed by the Gaussian surface is

$$q = \left(\frac{4\pi}{3}\right)\left(r^3 - r_1^3\right)\rho = Q\left(\frac{r^3 - r_1^3}{r_2^3 - r_1^3}\right).$$

Gauss' law yields

$$4\pi\varepsilon_0 r^2 E = Q\left(\frac{r^3 - r_1^3}{r_2^3 - r_1^3}\right) \Rightarrow E = \frac{Q}{4\pi\varepsilon_0}\frac{r^3 - r_1^3}{r^2\left(r_2^3 - r_1^3\right)}.$$

If V_s is the electric potential at the outer surface of the shell ($r = r_2$) then the potential a distance r from the center is given by

$$V = V_s - \int_{r_2}^{r} E\,dr = V_s - \frac{Q}{4\pi\varepsilon_0}\frac{1}{r_2^3 - r_1^3}\int_{r_2}^{r}\left(r - \frac{r_1^3}{r^2}\right)dr$$

$$= V_s - \frac{Q}{4\pi\varepsilon_0}\frac{1}{r_2^3 - r_1^3}\left(\frac{r^2}{2} - \frac{r_2^2}{2} + \frac{r_1^3}{r} - \frac{r_1^3}{r_2}\right).$$

The potential at the outer surface is found by placing $r = r_2$ in the expression found in part (a). It is $V_s = Q/4\pi\varepsilon_0 r_2$. We make this substitution and collect terms to find

$$V = \frac{Q}{4\pi\varepsilon_0}\frac{1}{r_2^3 - r_1^3}\left(\frac{3r_2^2}{2} - \frac{r^2}{2} - \frac{r_1^3}{r}\right).$$

Since $\rho = 3Q/4\pi\left(r_2^3 - r_1^3\right)$ this can also be written

$$V = \frac{\rho}{3\varepsilon_0}\left(\frac{3r_2^2}{2} - \frac{r^2}{2} - \frac{r_1^3}{r}\right).$$

(c) The electric field vanishes in the cavity, so the potential is everywhere the same inside and has the same value as at a point on the inside surface of the shell. We put $r = r_1$ in the result of part (b). After collecting terms the result is

$$V = \frac{Q}{4\pi\varepsilon_0}\frac{3\left(r_2^2 - r_1^2\right)}{2\left(r_2^3 - r_1^3\right)},$$

or in terms of the charge density $V = \frac{\rho}{2\varepsilon_0}\left(r_2^2 - r_1^2\right)$.

(d) The solutions agree at $r = r_1$ and at $r = r_2$.

102. The distance r being looked for is that where the alpha particle has (momentarily) zero kinetic energy. Thus, energy conservation leads to

$$K_0 + U_0 = K + U \Rightarrow (0.48 \times 10^{-12}\text{ J}) + \frac{(2e)(92e)}{4\pi\varepsilon_0 r_0} = 0 + \frac{(2e)(92e)}{4\pi\varepsilon_0 r}.$$

If we set $r_0 = \infty$ (so $U_0 = 0$) then we obtain $r = 8.8 \times 10^{-14}$ m.

103. (a) The net potential is

$$V = V_1 + V_2 = \frac{q_1}{4\pi\varepsilon_0 r_1} + \frac{q_2}{4\pi\varepsilon_0 r_2}$$

where $r_1 = \sqrt{x^2 + y^2}$ and $r_2 = \sqrt{(x-d)^2 + y^2}$. The distance d is 8.6 nm. To find the locus of points resulting in $V = 0$, we set V_1 equal to the (absolute value of) V_2 and square both sides. After simplifying and rearranging we arrive at an equation for a circle:

$$y^2 + \left(x + \frac{9d}{16}\right)^2 = \frac{225}{256}\, d^2.$$

From this form, we recognize that the center of the circle is $-9d/16 = -4.8$ nm.

(b) Also from this form, we identify the radius as the square root of the right-hand side: $R = 15d/16 = 8.1$ nm.

(c) If one uses a graphing program with "implicitplot" features, it is certainly possible to set $V = 5$ volts in the expression (shown in part (a)) and find its (or one of its) equipotential curves in the xy plane. In fact, it will look very much like a circle. Algebraically, attempts to put the expression into any standard form for a circle will fail, but that can be a frustrating endeavor. Perhaps the easiest way to show that it is not truly a circle is to find where its "horizontal diameter" D_x and its "vertical diameter" D_y (not hard to do); we find $D_x = 2.582$ nm and $D_y = 2.598$ nm. The fact that $D_x \neq D_y$ is evidence that it is not a true circle.

104. The electric field (along the radial axis) is the (negative of the) derivative of the voltage with respect to r. There are no other components of $\vec{E}$ in this case, so (noting that the derivative of a constant is zero) we conclude that the magnitude of the field is

$$E = -\frac{dV}{dr} = -\frac{Ze}{4\pi\varepsilon_0}\left(\frac{dr^{-1}}{dr} + 0 + \frac{1}{2R^3}\frac{dr^2}{dr}\right) = \frac{Ze}{4\pi\varepsilon_0}\left(\frac{1}{r^2} - \frac{r}{R^3}\right)$$

for $r \leq R$. This agrees with the Rutherford field expression shown in exercise 37 (in the textbook). We note that he has designed his voltage expression to be zero at $r = R$. Since the zero point for the voltage of this system (in an otherwise empty space) is arbitrary, then choosing $V = 0$ at $r = R$ is certainly permissible.

105. If the electric potential is zero at infinity then at the surface of a uniformly charged sphere it is $V = q/4\pi\varepsilon_0 R$, where q is the charge on the sphere and R is the sphere radius. Thus $q = 4\pi\varepsilon_0 RV$ and the number of electrons is

$$N = \frac{|q|}{e} = \frac{4\pi\varepsilon_0 R|V|}{e} = \frac{(1.0\times10^{-6}\,\text{m})(400\,\text{V})}{(8.99\times10^9\,\text{N}\cdot\text{m}^2/\text{C}^2)(1.60\times10^{-19}\,\text{C})} = 2.8\times10^5 .$$

106. We imagine moving all the charges on the surface of the sphere to the center of the the sphere. Using Gauss' law, we see that this would not change the electric field *outside* the sphere. The magnitude of the electric field E of the uniformly charged sphere as a function of r, the distance from the center of the sphere, is thus given by $E(r) = q/(4\pi\varepsilon_0 r^2)$ for $r > R$. Here R is the radius of the sphere. Thus, the potential V at the surface of the sphere (where $r = R$) is given by

$$V(R) = V\big|_{r=\infty} + \int_R^\infty E(r)\,dr = \int_\infty^R \frac{q}{4\pi\varepsilon_0 r^2}\,dr = \frac{q}{4\pi\varepsilon_0 R} = \frac{\left(8.99\times10^9\,\frac{\text{N}\cdot\text{m}^2}{\text{C}^2}\right)\left(1.50\times10^8\,\text{C}\right)}{0.160\,\text{m}}$$

$$= 8.43\times10^2\,\text{V}.$$

107. On the dipole axis $\theta = 0$ or π, so $|\cos\theta| = 1$. Therefore, magnitude of the electric field is

$$|E(r)| = \left|-\frac{\partial V}{\partial r}\right| = \frac{p}{4\pi\varepsilon_0}\left|\frac{d}{dr}\left(\frac{1}{r^2}\right)\right| = \frac{p}{2\pi\varepsilon_0 r^3} .$$

108. The potential difference is

$$\Delta V = E\Delta s = (1.92\times10^5\,\text{N/C})(0.0150\,\text{m}) = 2.90\times10^3\,\text{V}.$$

109. (a) Using Eq. 24-26, we calculate the radius r of the sphere representing the 30 V equipotential surface:

$$r = \frac{q}{4\pi\varepsilon_0 V} = 4.5\,\text{m}.$$

(b) If the potential were a linear function of r then it would have equally spaced equipotentials, but since $V \propto 1/r$ they are spaced more and more widely apart as r increases.

110. (a) Let the quark-quark separation be r. To "naturally" obtain the eV unit, we only plug in for one of the e values involved in the computation:

$$U_{\text{up–up}} = \frac{1}{4\pi\varepsilon_0}\frac{\left(\dfrac{2e}{3}\right)\left(\dfrac{2e}{3}\right)}{r} = \frac{4ke}{9r}e = \frac{4\left(8.99\times10^9\,\dfrac{\text{N}\cdot\text{m}^2}{\text{C}^2}\right)\left(1.60\times10^{-19}\,\text{C}\right)}{9\left(1.32\times10^{-15}\,\text{m}\right)}e$$

$$= 4.84\times10^5\,\text{eV} = 0.484\,\text{MeV}.$$

(b) The total consists of all pair-wise terms:

$$U = \frac{1}{4\pi\varepsilon_0}\left[\frac{\left(\frac{2e}{3}\right)\left(\frac{2e}{3}\right)}{r} + \frac{\left(\frac{-e}{3}\right)\left(\frac{2e}{3}\right)}{r} + \frac{\left(\frac{-e}{3}\right)\left(\frac{2e}{3}\right)}{r}\right] = 0.$$

111. (a) At the smallest center-to-center separation d_p the initial kinetic energy K_i of the proton is entirely converted to the electric potential energy between the proton and the nucleus. Thus,

$$K_i = \frac{1}{4\pi\varepsilon_0}\frac{eq_{\text{lead}}}{d_p} = \frac{82e^2}{4\pi\varepsilon_0 d_p}.$$

In solving for d_p using the eV unit, we note that a factor of e cancels in the middle line:

$$d_p = \frac{82e^2}{4\pi\varepsilon_0 K_i} = k\frac{82e^2}{4.80\times10^6\,eV} = \left(8.99\times10^9\,\frac{\text{N}\cdot\text{m}^2}{\text{C}^2}\right)\frac{82\left(1.6\times10^{-19}\,\text{C}\right)}{4.80\times10^6\,\text{V}}$$

$$= 2.5\times10^{-14}\,\text{m} = 25\,\text{fm}\,.$$

It is worth recalling that a volt is a newton·meter/coulomb, in making sense of the above manipulations.

(b) An alpha particle has 2 protons (as well as 2 neutrons). Therefore, using $r'_{\min}$ for the new separation, we find

$$K_i = \frac{1}{4\pi\varepsilon_0}\frac{q_\alpha q_{\text{lead}}}{d_\alpha} = 2\left(\frac{82e^2}{4\pi\varepsilon_0 d_\alpha}\right) = \frac{82e^2}{4\pi\varepsilon_0 d_p}$$

which leads to $d_\alpha / d_p = 2.00$.

112. (a) The potential would be

$$V_e = \frac{Q_e}{4\pi\varepsilon_0 R_e} = \frac{4\pi R_e^2 \sigma_e}{4\pi\varepsilon_0 R_e} = 4\pi R_e \sigma_e k$$

$$= 4\pi\left(6.37\times10^6\,\text{m}\right)\left(1.0\,\text{electron/m}^2\right)\left(-1.6\times10^{-9}\,\text{C/electron}\right)\left(8.99\times10^9\,\frac{\text{N}\cdot\text{m}^2}{\text{C}^2}\right)$$

$$= -0.12\,\text{V}.$$

(b) The electric field is

$$E = \frac{\sigma_e}{\varepsilon_0} = \frac{V_e}{R_e} = -\frac{0.12\,\text{V}}{6.37 \times 10^6\,\text{m}} = -1.8 \times 10^{-8}\,\text{N/C},$$

or $|E| = 1.8 \times 10^{-8}$ N/C.

(c) The minus sign in E indicates that $\vec{E}$ is radially inward.

113. The electric potential energy is

$$U = k\sum_{i \neq j} \frac{q_i q_j}{r_{ij}} = \frac{1}{4\pi\varepsilon_0 d}\left(q_1 q_2 + q_1 q_3 + q_2 q_4 + q_3 q_4 + \frac{q_1 q_4}{\sqrt{2}} + \frac{q_2 q_3}{\sqrt{2}}\right)$$

$$= \frac{(8.99 \times 10^9)}{1.3}\left[(12)(-24) + (12)(31) + (-24)(17) + (31)(17) + \frac{(12)(17)}{\sqrt{2}} + \frac{(-24)(31)}{\sqrt{2}}\right](10^{-19})^2$$

$$= -1.2 \times 10^{-6}\,\text{J}.$$

114. (a) The charge on every part of the ring is the same distance from any point P on the axis. This distance is $r = \sqrt{z^2 + R^2}$, where R is the radius of the ring and z is the distance from the center of the ring to P. The electric potential at P is

$$V = \frac{1}{4\pi\varepsilon_0}\int\frac{dq}{r} = \frac{1}{4\pi\varepsilon_0}\int\frac{dq}{\sqrt{z^2 + R^2}} = \frac{1}{4\pi\varepsilon_0}\frac{1}{\sqrt{z^2 + R^2}}\int dq = \frac{1}{4\pi\varepsilon_0}\frac{q}{\sqrt{z^2 + R^2}}.$$

(b) The electric field is along the axis and its component is given by

$$E = -\frac{\partial V}{\partial z} = -\frac{q}{4\pi\varepsilon_0}\frac{\partial}{\partial z}(z^2 + R^2)^{-1/2} = \frac{q}{4\pi\varepsilon_0}\left(\frac{1}{2}\right)(z^2 + R^2)^{-3/2}(2z) = \frac{q}{4\pi\varepsilon_0}\frac{z}{(z^2 + R^2)^{3/2}}.$$

This agrees with Eq. 23-16.

115. From the previous chapter, we know that the radial field due to an infinite line-source is

$$E = \frac{\lambda}{2\pi\varepsilon_0 r}$$

which integrates, using Eq. 24-18, to obtain

$$V_i = V_f + \frac{\lambda}{2\pi\varepsilon_0}\int_{r_i}^{r_f}\frac{dr}{r} = V_f + \frac{\lambda}{2\pi\varepsilon_0}\ln\left(\frac{r_f}{r_i}\right).$$

The subscripts i and f are somewhat arbitrary designations, and we let $V_i = V$ be the potential of some point P at a distance $r_i = r$ from the wire and $V_f = V_o$ be the potential along some reference axis (which intersects the plane of our figure, shown next, at the xy coordinate origin, placed midway between the bottom two line charges — that is, the midpoint of the bottom side of the equilateral triangle) at a distance $r_f = a$ from each of the bottom wires (and a distance $a\sqrt{3}$ from the topmost wire). Thus, each side of the triangle is of length $2a$. Skipping some steps, we arrive at an expression for the net potential created by the three wires (where we have set $V_o = 0$):

$$V_{net} = \frac{\lambda}{4\pi\varepsilon_0} \ln\left(\frac{\left(x^2 + \left(y - a\sqrt{3}\right)^2\right)^2}{\left((x+a)^2 + y^2\right)\left((x-a)^2 + y^2\right)} \right)$$

which forms the basis of our contour plot shown below. On the same plot we have shown four electric field lines, which have been sketched (as opposed to rigorously calculated) and are not meant to be as accurate as the equipotentials. The $\pm 2\lambda$ by the top wire in our figure should be -2λ (the $\pm$ typo is an artifact of our plotting routine).

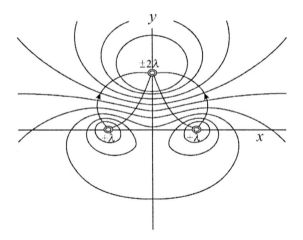

116. From the previous chapter, we know that the radial field due to an infinite line-source is

$$E = \frac{\lambda}{2\pi\varepsilon_0 r}$$

which integrates, using Eq. 24-18, to obtain

$$V_i = V_f + \frac{\lambda}{2\pi\varepsilon_0} \int_{r_i}^{r_f} \frac{dr}{r} = V_f + \frac{\lambda}{2\pi\varepsilon_0} \ln\left(\frac{r_f}{r_i}\right).$$

The subscripts i and f are somewhat arbitrary designations, and we let $V_i = V$ be the potential of some point P at a distance $r_i = r$ from the wire and $V_f = V_o$ be the potential along some reference axis (which will be the z axis described in this problem) at a distance $r_f = a$ from the wire. In the "end-view" presented here, the wires and the z axis appear as points as they intersect the xy plane. The potential due to the wire on the left (intersecting the plane at $x = -a$) is

$$V_{\text{negative wire}} = V_o + \frac{(-\lambda)}{2\pi\varepsilon_0}\ln\left(\frac{a}{\sqrt{(x+a)^2+y^2}}\right),$$

and the potential due to the wire on the right (intersecting the plane at $x = +a$) is

$$V_{\text{positive wire}} = V_o + \frac{(+\lambda)}{2\pi\varepsilon_0}\ln\left(\frac{a}{\sqrt{(x-a)^2+y^2}}\right).$$

Since potential is a scalar quantity, the net potential at point P is the addition of $V_{-\lambda}$ and $V_{+\lambda}$ which simplifies to

$$V_{\text{net}} = 2V_0 + \frac{\lambda}{2\pi\varepsilon_0}\left(\ln\left(\frac{a}{\sqrt{(x-a)^2+y^2}}\right) - \ln\left(\frac{a}{\sqrt{(x+a)^2+y^2}}\right)\right) = \frac{\lambda}{4\pi\varepsilon_0}\ln\left(\frac{(x+a)^2+y^2}{(x-a)^2+y^2}\right)$$

where we have set the potential along the z axis equal to zero ($V_o = 0$) in the last step (which we are free to do). This is the expression used to obtain the equipotentials shown next. The center dot in the figure is the intersection of the z axis with the xy plane, and the dots on either side are the intersections of the wires with the plane.

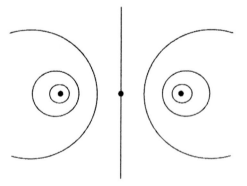

117. (a) With $V = 1000$ V, we solve

$$V = \frac{q}{4\pi\varepsilon_0 R} \qquad \text{where } R = 0.010 \text{ m}$$

for the net charge on the sphere, and find $q = 1.1 \times 10^{-9}$ C. Dividing this by e yields 6.95 $\times 10^9$ electrons that entered the copper sphere. Now, half of the 3.7×10^8 decays per second resulted in electrons entering the sphere, so the time required is

$$\frac{6.95 \times 10^9}{\frac{1}{2}(3.7 \times 10^8)} = 38 \text{ seconds.}$$

(b) We note that 100 keV is 1.6×10^{-14} J (per electron that entered the sphere). Using the given heat capacity, we note that a temperature increase of $\Delta T = 5.0$ K $= 5.0$ C° required 71.5 J of energy. Dividing this by 1.6×10^{-14} J, we find the number of electrons needed to enter the sphere (in order to achieve that temperature change); since this is half the number of decays, we multiply to 2 and find

$$N = 8.94 \times 10^{15} \text{ decays.}$$

We divide N by 3.7×10^8 to obtain the number of seconds. Converting to days, this becomes roughly 280 days.

118. The (implicit) equation for the pair (x,y) in terms of a specific V is

$$V = \frac{q_1}{4\pi\varepsilon_0\sqrt{x^2 + y^2}} + \frac{q_2}{4\pi\varepsilon_0\sqrt{x^2 + (y - d)^2}}$$

where $d = 0.50$ m. The values of q_1 and q_2 are given in the problem.

(a) We set $V = 5.0$ V and plotted (using MAPLE's implicit plotting routine) those points in the xy plane which (when plugged into the above expression for V) yield 5.0 volts. The result is

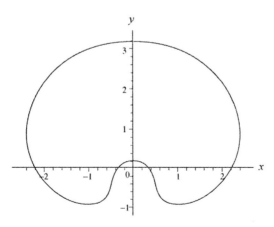

(b) In this case, the same procedure yields these two equipotential lines:

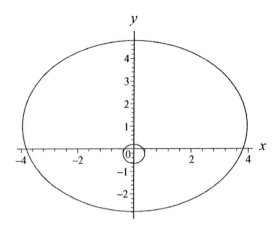

(c) One way to search for the "crossover" case (from a single equipotential line, to two) is to "solve" for a point on the y axis (chosen here to be an absolute distance ξ below q_1 – that is, the point is at a negative value of y, specifically at $y = -\xi$) in terms of V (or more conveniently, in terms of the parameter $\eta = 4\pi\varepsilon_o V \times 10^{10}$). Thus, the above expression for V becomes simply

$$\eta = \frac{-12}{\xi} + \frac{25}{d+\xi} .$$

This leads to a quadratic equation with the (formal) solution

$$\xi = \frac{13 - d\eta \pm \sqrt{d^2 \eta^2 + 169 - 74\, d\eta}}{2\,\eta} .$$

Clearly there is the possibility of having two solutions (implying two intersections of equipotential lines with the $-y$ axis) when the square root term is nonzero. This suggests that we explore the special case where the square root term is zero; that is,

$$\sqrt{d^2 \eta^2 + 169 - 74\, d\eta} = 0 .$$

Squaring both sides, using the fact that $d = 0.50$ m and recalling how we have defined the parameter η, this leads to a "critical value" of the potential (corresponding to the crossover case, between one and two equipotentials):

$$\eta_{\text{critical}} = \frac{37 - 20\sqrt{3}}{d} \quad \Rightarrow \quad V_{\text{critical}} = \frac{\eta_{\text{critical}}}{4\pi\varepsilon_o \times 10^{10}} = 4.2 \text{ V}.$$

Chapter 25

1. Charge flows until the potential difference across the capacitor is the same as the potential difference across the battery. The charge on the capacitor is then $q = CV$, and this is the same as the total charge that has passed through the battery. Thus,

$$q = (25 \times 10^{-6} \text{ F})(120 \text{ V}) = 3.0 \times 10^{-3} \text{ C}.$$

2. (a) The capacitance of the system is

$$C = \frac{q}{\Delta V} = \frac{70 \text{ pC}}{20 \text{ V}} = 3.5 \text{ pF}.$$

(b) The capacitance is independent of q; it is still 3.5 pF.

(c) The potential difference becomes

$$\Delta V = \frac{q}{C} = \frac{200 \text{ pC}}{3.5 \text{ pF}} = 57 \text{ V}.$$

3. (a) The capacitance of a parallel-plate capacitor is given by $C = \varepsilon_0 A / d$, where A is the area of each plate and d is the plate separation. Since the plates are circular, the plate area is $A = \pi R^2$, where R is the radius of a plate. Thus,

$$C = \frac{\varepsilon_0 \pi R^2}{d} = \frac{\left(8.85 \times 10^{-12} \text{ F/m}\right) \pi \left(8.2 \times 10^{-2} \text{ m}\right)^2}{1.3 \times 10^{-3} \text{ m}} = 1.44 \times 10^{-10} \text{ F} = 144 \text{ pF}.$$

(b) The charge on the positive plate is given by $q = CV$, where V is the potential difference across the plates. Thus,

$$q = (1.44 \times 10^{-10} \text{ F})(120 \text{ V}) = 1.73 \times 10^{-8} \text{ C} = 17.3 \text{ nC}.$$

4. We use $C = A\varepsilon_0 / d$.

(a) Thus,

$$d = \frac{A\varepsilon_0}{C} = \frac{\left(1.00 \text{ m}^2\right)\left(8.85 \times 10^{-12} \frac{\text{C}^2}{\text{N} \cdot \text{m}^2}\right)}{1.00 \text{ F}} = 8.85 \times 10^{-12} \text{ m}.$$

(b) Since d is much less than the size of an atom ($\sim 10^{-10}$ m), this capacitor cannot be constructed.

5. Assuming conservation of volume, we find the radius of the combined spheres, then use $C = 4\pi\varepsilon_0 R$ to find the capacitance. When the drops combine, the volume is doubled. It is then $V = 2(4\pi/3)R^3$. The new radius R' is given by

$$\frac{4\pi}{3}\left(R'\right)^3 = 2\frac{4\pi}{3}R^3 \quad \Rightarrow \quad R' = 2^{1/3}R.$$

The new capacitance is

$$C' = 4\pi\varepsilon_0 R' = 4\pi\varepsilon_0 2^{1/3} R = 5.04\pi\varepsilon_0 R.$$

With $R = 2.00$ mm, we obtain $C = 5.04\pi\left(8.85\times10^{-12}\ \text{F/m}\right)\left(2.00\times10^{-3}\text{m}\right) = 2.80\times10^{-13}\text{F}$.

6. (a) We use Eq. 25-17:

$$C = 4\pi\varepsilon_0 \frac{ab}{b-a} = \frac{\left(40.0\,\text{mm}\right)\left(38.0\,\text{mm}\right)}{\left(8.99\times10^9\ \frac{\text{N·m}^2}{\text{C}^2}\right)\left(40.0\,\text{mm} - 38.0\,\text{mm}\right)} = 84.5\,\text{pF}.$$

(b) Let the area required be A. Then $C = \varepsilon_0 A/(b - a)$, or

$$A = \frac{C\left(b-a\right)}{\varepsilon_0} = \frac{\left(84.5\,\text{pF}\right)\left(40.0\,\text{mm} - 38.0\,\text{mm}\right)}{\left(8.85\times10^{-12}\ \frac{\text{C}^2}{\text{N·m}^2}\right)} = 191\,\text{cm}^2.$$

7. The equivalent capacitance is given by $C_{eq} = q/V$, where q is the total charge on all the capacitors and V is the potential difference across any one of them. For N identical capacitors in parallel, $C_{eq} = NC$, where C is the capacitance of one of them. Thus, $NC = q/V$ and

$$N = \frac{q}{VC} = \frac{1.00\,\text{C}}{\left(110\text{V}\right)\left(1.00\times10^{-6}\text{F}\right)} = 9.09\times10^3.$$

8. The equivalent capacitance is

$$C_{eq} = C_3 + \frac{C_1 C_2}{C_1 + C_2} = 4.00\,\mu\text{F} + \frac{\left(10.0\,\mu\text{F}\right)\left(5.00\,\mu\text{F}\right)}{10.0\,\mu\text{F} + 5.00\,\mu\text{F}} = 7.33\,\mu\text{F}.$$

9. The equivalent capacitance is

$$C_{eq} = \frac{(C_1 + C_2)C_3}{C_1 + C_2 + C_3} = \frac{(10.0\,\mu\text{F} + 5.00\,\mu\text{F})(4.00\,\mu\text{F})}{10.0\,\mu\text{F} + 5.00\,\mu\text{F} + 4.00\,\mu\text{F}} = 3.16\,\mu\text{F}.$$

10. The charge that passes through meter A is

$$q = C_{eq}V = 3CV = 3(25.0\,\mu\text{F})(4200\,\text{V}) = 0.315\,\text{C}.$$

11. (a) and (b) The original potential difference V_1 across C_1 is

$$V_1 = \frac{C_{eq}V}{C_1 + C_2} = \frac{(3.16\,\mu\text{F})(100.0\,\text{V})}{10.0\,\mu\text{F} + 5.00\,\mu\text{F}} = 21.1\,\text{V}.$$

Thus $\Delta V_1 = 100.0\,\text{V} - 21.1\,\text{V} = 78.9\,\text{V}$ and

$$\Delta q_1 = C_1 \Delta V_1 = (10.0\,\mu\text{F})(78.9\,\text{V}) = 7.89 \times 10^{-4}\,\text{C}.$$

12. (a) The potential difference across C_1 is $V_1 = 10.0\,\text{V}$. Thus,

$$q_1 = C_1 V_1 = (10.0\,\mu\text{F})(10.0\,\text{V}) = 1.00 \times 10^{-4}\,\text{C}.$$

(b) Let $C = 10.0\,\mu\text{F}$. We first consider the three-capacitor combination consisting of C_2 and its two closest neighbors, each of capacitance C. The equivalent capacitance of this combination is

$$C_{eq} = C + \frac{C_2 C}{C + C_2} = 1.50\,C.$$

Also, the voltage drop across this combination is

$$V = \frac{CV_1}{C + C_{eq}} = \frac{CV_1}{C + 1.50\,C} = 0.40V_1.$$

Since this voltage difference is divided equally between C_2 and the one connected in series with it, the voltage difference across C_2 satisfies $V_2 = V/2 = V_1/5$. Thus

$$q_2 = C_2 V_2 = (10.0\,\mu\text{F})\left(\frac{10.0\,\text{V}}{5}\right) = 2.00 \times 10^{-5}\,\text{C}.$$

13. The charge initially on the charged capacitor is given by $q = C_1 V_0$, where $C_1 = 100\,\text{pF}$ is the capacitance and $V_0 = 50\,\text{V}$ is the initial potential difference. After the battery is disconnected and the second capacitor wired in parallel to the first, the charge on the first capacitor is $q_1 = C_1 V$, where $V = 35\,\text{V}$ is the new potential difference. Since charge is conserved in the process, the charge on the second capacitor is $q_2 = q - q_1$, where C_2 is

the capacitance of the second capacitor. Substituting C_1V_0 for q and C_1V for q_1, we obtain $q_2 = C_1 (V_0 - V)$. The potential difference across the second capacitor is also V, so the capacitance is

$$C_2 = \frac{q_2}{V} = \frac{V_0 - V}{V}C_1 = \frac{50\,\text{V} - 35\,\text{V}}{35\,\text{V}}(100\,\text{pF}) = 43\,\text{pF}.$$

14. The two 6.0 μF capacitors are in parallel and are consequently equivalent to $C_{eq} = 12\ \mu$F. Thus, the total charge stored (before the squeezing) is

$$q_{total} = C_{eq}V_{battery} = 120\ \mu\text{C}.$$

(a) and (b) As a result of the squeezing, one of the capacitors is now 12 μF (due to the inverse proportionality between C and d in Eq. 25-9) which represents an increase of 6.0 μF and thus a charge increase of

$$\Delta q_{total} = \Delta C_{eq}V_{battery} = (6.0\ \mu\text{F})(10\ \text{V}) = 60\ \mu\text{C}.$$

15. (a) First, the equivalent capacitance of the two 4.00 μF capacitors connected in series is given by 4.00 μF/2 = 2.00 μF. This combination is then connected in parallel with two other 2.00-μF capacitors (one on each side), resulting in an equivalent capacitance C = 3(2.00 μF) = 6.00 μF. This is now seen to be in series with another combination, which consists of the two 3.0-μF capacitors connected in parallel (which are themselves equivalent to $C' = 2(3.00\ \mu$F) = 6.00 μF). Thus, the equivalent capacitance of the circuit is

$$C_{eq} = \frac{CC'}{C+C'} = \frac{(6.00\,\mu\text{F})\,(6.00\,\mu\text{F})}{6.00\,\mu\text{F} + 6.00\,\mu\text{F}} = 3.00\,\mu\text{F}.$$

(b) Let $V = 20.0$ V be the potential difference supplied by the battery. Then

$$q = C_{eq}V = (3.00\ \mu\text{F})(20.0\ \text{V}) = 6.00 \times 10^{-5}\ \text{C}.$$

(c) The potential difference across C_1 is given by

$$V_1 = \frac{CV}{C+C'} = \frac{(6.00\,\mu\text{F})\,(20.0\,\text{V})}{6.00\,\mu\text{F} + 6.00\,\mu\text{F}} = 10.0\,\text{V}.$$

(d) The charge carried by C_1 is $q_1 = C_1V_1 = (3.00\ \mu$F)(10.0 V) = 3.00×10^{-5} C.

(e) The potential difference across C_2 is given by $V_2 = V - V_1 = 20.0$ V $- 10.0$ V $= 10.0$ V.

(f) The charge carried by C_2 is $q_2 = C_2V_2 = (2.00\ \mu$F)(10.0 V) = 2.00×10^{-5} C.

(g) Since this voltage difference V_2 is divided equally between C_3 and the other 4.00-μF capacitors connected in series with it, the voltage difference across C_3 is given by $V_3 = V_2/2 = 10.0\ \text{V}/2 = 5.00\ \text{V}$.

(h) Thus, $q_3 = C_3 V_3 = (4.00\ \mu\text{F})(5.00\ \text{V}) = 2.00 \times 10^{-5}\ \text{C}$.

16. We determine each capacitance from the slope of the appropriate line in the graph. Thus, $C_1 = (12\ \mu\text{C})/(2.0\ \text{V}) = 6.0\ \mu\text{F}$. Similarly, $C_2 = 4.0\ \mu\text{F}$ and $C_3 = 2.0\ \mu\text{F}$. The total equivalent capacitance is

$$C_{123} = ((C_1)^{-1} + (C_3 + C_2)^{-1})^{-1} = 3.0\ \mu\text{F}.$$

This implies that the charge on capacitor 1 is $(3.0\ \mu\text{F})(6.0\ \text{V}) = 18\ \mu\text{C}$. The voltage across capacitor 1 is therefore $(18\ \mu\text{C})/(6.0\ \mu\text{F}) = 3.0\ \text{V}$. From the discussion in section 25-4, we conclude that the voltage across capacitor 2 must be $6.0\ \text{V} - 3.0\ \text{V} = 3.0\ \text{V}$. Consequently, the charge on capacitor 2 is $(4.0\ \mu\text{F})(3.0\ \text{V}) = 12\ \mu\text{C}$.

17. (a) After the switches are closed, the potential differences across the capacitors are the same and the two capacitors are in parallel. The potential difference from a to b is given by $V_{ab} = Q/C_{eq}$, where Q is the net charge on the combination and C_{eq} is the equivalent capacitance. The equivalent capacitance is $C_{eq} = C_1 + C_2 = 4.0 \times 10^{-6}\ \text{F}$. The total charge on the combination is the net charge on either pair of connected plates. The charge on capacitor 1 is

$$q_1 = C_1 V = \left(1.0 \times 10^{-6}\ \text{F}\right)(100\ \text{V}) = 1.0 \times 10^{-4}\ \text{C}$$

and the charge on capacitor 2 is

$$q_2 = C_2 V = \left(3.0 \times 10^{-6}\ \text{F}\right)(100\ \text{V}) = 3.0 \times 10^{-4}\ \text{C},$$

so the net charge on the combination is $3.0 \times 10^{-4}\ \text{C} - 1.0 \times 10^{-4}\ \text{C} = 2.0 \times 10^{-4}\ \text{C}$. The potential difference is

$$V_{ab} = \frac{2.0 \times 10^{-4}\ \text{C}}{4.0 \times 10^{-6}\ \text{F}} = 50\ \text{V}.$$

(b) The charge on capacitor 1 is now $q_1 = C_1 V_{ab} = (1.0 \times 10^{-6}\ \text{F})(50\ \text{V}) = 5.0 \times 10^{-5}\ \text{C}$.

(c) The charge on capacitor 2 is now $q_2 = C_2 V_{ab} = (3.0 \times 10^{-6}\ \text{F})(50\ \text{V}) = 1.5 \times 10^{-4}\ \text{C}$.

18. Eq. 23-14 applies to each of these capacitors. Bearing in mind that $\sigma = q/A$, we find the total charge to be

$$q_{total} = q_1 + q_2 = \sigma_1 A_1 + \sigma_2 A_2 = \varepsilon_o E_1 A_1 + \varepsilon_o E_2 A_2 = 3.6 \text{ pC}$$

where we have been careful to convert cm^2 to m^2 by dividing by 10^4.

19. (a) and (b) We note that the charge on C_3 is $q_3 = 12 \mu C - 8.0 \mu C = 4.0 \mu C$. Since the charge on C_4 is $q_4 = 8.0 \mu C$, then the voltage across it is $q_4/C_4 = 2.0$ V. Consequently, the voltage V_3 across C_3 is 2.0 V $\Rightarrow C_3 = q_3/V_3 = 2.0 \mu F$.

Now C_3 and C_4 are in parallel and are thus equivalent to 6 μF capacitor which would then be in series with C_2; thus, Eq 25-20 leads to an equivalence of 2.0 μF which is to be thought of as being in series with the unknown C_1. We know that the total effective capacitance of the circuit (in the sense of what the battery "sees" when it is hooked up) is $(12 \mu C)/V_{battery} = 4\mu F/3$. Using Eq 25-20 again, we find

$$\frac{1}{2 \mu F} + \frac{1}{C_1} = \frac{3}{4 \mu F} \quad \Rightarrow \quad C_1 = 4.0 \mu F .$$

20. We note that the total equivalent capacitance is $C_{123} = [(C_3)^{-1} + (C_1 + C_2)^{-1}]^{-1} = 6 \mu F$.

(a) Thus, the charge that passed point a is $C_{123} V_{batt} = (6 \mu F)(12 \text{ V}) = 72 \mu C$. Dividing this by the value $e = 1.60 \times 10^{-19}$ C gives the number of electrons: 4.5×10^{14}, which travel to the left – towards the positive terminal of the battery.

(b) The equivalent capacitance of the parallel pair is $C_{12} = C_1 + C_2 = 12 \mu F$. Thus, the voltage across the pair (which is the same as the voltage across C_1 and C_2 individually) is

$$\frac{72 \mu C}{12 \mu F} = 6 \text{ V} .$$

Thus, the charge on C_1 is $(4 \mu F)(6 \text{ V}) = 24 \mu C$, and dividing this by e gives the number of electrons (1.5×10^{14}) which have passed (upward) though point b.

(c) Similarly, the charge on C_2 is $(8 \mu F)(6 \text{ V}) = 48 \mu C$, and dividing this by e gives the number of electrons (3.0×10^{14}) which have passed (upward) though point c.

(d) Finally, since C_3 is in series with the battery, its charge is the same that passed through the battery (the same as passed through the switch). Thus, 4.5×10^{14} electrons passed rightward though point d. By leaving the rightmost plate of C_3, that plate is then the positive plate of the fully charged capacitor – making its leftmost plate (the one closest to the negative terminal of the battery) the negative plate, as it should be.

(e) As stated in (b), the electrons travel up through point b.

(f) As stated in (c), the electrons travel up through point c.

21. The charges on capacitors 2 and 3 are the same, so these capacitors may be replaced by an equivalent capacitance determined from

$$\frac{1}{C_{eq}} = \frac{1}{C_2} + \frac{1}{C_3} = \frac{C_2 + C_3}{C_2 C_3}.$$

Thus, $C_{eq} = C_2 C_3/(C_2 + C_3)$. The charge on the equivalent capacitor is the same as the charge on either of the two capacitors in the combination and the potential difference across the equivalent capacitor is given by q_2/C_{eq}. The potential difference across capacitor 1 is q_1/C_1, where q_1 is the charge on this capacitor. The potential difference across the combination of capacitors 2 and 3 must be the same as the potential difference across capacitor 1, so $q_1/C_1 = q_2/C_{eq}$. Now some of the charge originally on capacitor 1 flows to the combination of 2 and 3. If q_0 is the original charge, conservation of charge yields $q_1 + q_2 = q_0 = C_1 V_0$, where V_0 is the original potential difference across capacitor 1.

(a) Solving the two equations

$$\frac{q_1}{C_1} = \frac{q_2}{C_{eq}} \qquad \text{and} \qquad q_1 + q_2 = C_1 V_0$$

for q_1 and q_2, we obtain

$$q_1 = \frac{C_1^2 V_0}{C_{eq} + C_1} = \frac{C_1^2 V_0}{\dfrac{C_2 C_3}{C_2 + C_3} + C_1} = \frac{C_1^2 (C_2 + C_3) V_0}{C_1 C_2 + C_1 C_3 + C_2 C_3}.$$

With $V_0 = 12.0$ V, $C_1 = 4.00 \ \mu$F, $C_2 = 6.00 \ \mu$F and $C_3 = 3.00 \ \mu$F, we find $C_{eq} = 2.00 \ \mu$F and $q_1 = 32.0 \ \mu$C.

(b) The charge on capacitors 2 is

$$q_2 = C_1 V_0 - q_1 = (4.00 \mu F)(12.0 V) - 32.0 \mu F = 16.0 \mu F$$

(c) The charge on capacitor 3 is the same as that on capacitor 2:

$$q_3 = C_1 V_0 - q_1 = (4.00 \mu F)(12.0 V) - 32.0 \mu F = 16.0 \mu F$$

22. Initially the capacitors C_1, C_2, and C_3 form a combination equivalent to a single capacitor which we denote C_{123}. This obeys the equation

$$\frac{1}{C_1} + \frac{1}{C_2 + C_3} = \frac{1}{C_{123}}.$$

Hence, using $q = C_{123} V$ and the fact that $q = q_1 = C_1 V_1$, we arrive at

$$V_1 = \frac{C_2 + C_3}{C_1 + C_2 + C_3} V.$$

(a) As $C_3 \rightarrow \infty$ this expression becomes $V_1 = V$. Since the problem states that V_1 approaches 10 volts in this limit, so we conclude $V = 10$ V.

(b) and (c)   At $C_3 = 0$, the graph indicates $V_1 = 2.0$ V. The above expression consequently implies $C_1 = 4C_2$. Next we note that the graph shows that, at $C_3 = 6.0$ μF, the voltage across C_1 is exactly half of the battery voltage. Thus,

$$\frac{1}{2} = \frac{C_2 + 6.0 \text{ μF}}{C_1 + C_2 + 6.0 \text{ μF}} = \frac{C_2 + 6.0 \text{ μF}}{4C_2 + C_2 + 6.0 \text{ μF}}$$

which leads to $C_2 = 2.0$ μF. We conclude, too, that $C_1 = 8.0$ μF.

23. (a) In this situation, capacitors 1 and 3 are in series, which means their charges are necessarily the same:

$$q_1 = q_3 = \frac{C_1 C_3 V}{C_1 + C_3} = \frac{(1.00 \mu F) (3.00 \mu F) (12.0 V)}{1.00 \mu F + 3.00 \mu F} = 9.00 \mu C.$$

(b) Capacitors 2 and 4 are also in series:

$$q_2 = q_4 = \frac{C_2 C_4 V}{C_2 + C_4} = \frac{(2.00 \mu F) (4.00 \mu F) (12.0 V)}{2.00 \mu F + 4.00 \mu F} = 16.0 \mu C.$$

(c) $q_3 = q_1 = 9.00 \mu C.$

(d) $q_4 = q_2 = 16.0 \mu C.$

(e) With switch 2 also closed, the potential difference V_1 across C_1 must equal the potential difference across C_2 and is

$$V_1 = \frac{C_3 + C_4}{C_1 + C_2 + C_3 + C_4} V = \frac{(3.00 \mu F + 4.00 \mu F)(12.0 V)}{1.00 \mu F + 2.00 \mu F + 3.00 \mu F + 4.00 \mu F} = 8.40 V.$$

Thus, $q_1 = C_1 V_1 = (1.00$ μF$)(8.40$ V$) = 8.40$ μC.

(f) Similarly, $q_2 = C_2 V_1 = (2.00$ μF$)(8.40$ V$) = 16.8$ μC.

(g) $q_3 = C_3(V - V_1) = (3.00$ μF$)(12.0$ V $- 8.40$ V$) = 10.8$ μC.

(h) $q_4 = C_4(V - V_1) = (4.00\ \mu F)(12.0\ V - 8.40\ V) = 14.4\ \mu C.$

24. Let $\mathcal{V} = 1.00\ m^3$. Using Eq. 25-25, the energy stored is

$$U = u\mathcal{V} = \frac{1}{2}\varepsilon_0 E^2 \mathcal{V} = \frac{1}{2}\left(8.85\times 10^{-12}\ \frac{C^2}{N\cdot m^2}\right)(150\ V/m)^2\,(1.00\ m^3) = 9.96\times 10^{-8}\ J.$$

25. The energy stored by a capacitor is given by $U = \frac{1}{2}CV^2$, where V is the potential difference across its plates. We convert the given value of the energy to Joules. Since a Joule is a watt·second, we multiply by $(10^3$ W/kW) (3600 s/h) to obtain $10\ kW\cdot h = 3.6\times 10^7\ J$. Thus,

$$C = \frac{2U}{V^2} = \frac{2(3.6\times 10^7\ J)}{(1000\ V)^2} = 72\ F.$$

26. (a) The capacitance is

$$C = \frac{\varepsilon_0 A}{d} = \frac{\left(8.85\times 10^{-12}\ \frac{C^2}{N\cdot m^2}\right)(40\times 10^{-4}\ m^2)}{1.0\times 10^{-3}\ m} = 3.5\times 10^{-11}\ F = 35\ pF.$$

(b) $q = CV = (35\ pF)(600\ V) = 2.1\times 10^{-8}\ C = 21\ nC.$

(c) $U = \frac{1}{2}CV^2 = \frac{1}{2}(35\ pF)(21\ nC)^2 = 6.3\times 10^{-6}\ J = 6.3\ \mu J.$

(d) $E = V/d = 600\ V/1.0\times 10^{-3}\ m = 6.0\times 10^5\ V/m.$

(e) The energy density (energy per unit volume) is

$$u = \frac{U}{Ad} = \frac{6.3\times 10^{-6}\ J}{(40\times 10^{-4}\ m^2)(1.0\times 10^{-3}\ m)} = 1.6\ J/m^3.$$

27. The total energy is the sum of the energies stored in the individual capacitors. Since they are connected in parallel, the potential difference V across the capacitors is the same and the total energy is

$$U = \frac{1}{2}(C_1 + C_2)V^2 = \frac{1}{2}(2.0\times 10^{-6}\ F + 4.0\times 10^{-6}\ F)(300\ V)^2 = 0.27\ J.$$

28. (a) The potential difference across C_1 (the same as across C_2) is given by

$$V_1 = V_2 = \frac{C_3 V}{C_1 + C_2 + C_3} = \frac{(15.0\,\mu\text{F})(100\,\text{V})}{10.0\,\mu\text{F} + 5.00\,\mu\text{F} + 15.0\,\mu\text{F}} = 50.0\,\text{V}.$$

Also, $V_3 = V - V_1 = V - V_2 = 100\,\text{V} - 50.0\,\text{V} = 50.0\,\text{V}$. Thus,

$$q_1 = C_1 V_1 = (10.0\,\mu\text{F})(50.0\,\text{V}) = 5.00 \times 10^{-4}\,\text{C}$$
$$q_2 = C_2 V_2 = (5.00\,\mu\text{F})(50.0\,\text{V}) = 2.50 \times 10^{-4}\,\text{C}$$
$$q_3 = q_1 + q_2 = 5.00 \times 10^{-4}\,\text{C} + 2.50 \times 10^{-4}\,\text{C} = 7.50 \times 10^{-4}\,\text{C}.$$

(b) The potential difference V_3 was found in the course of solving for the charges in part (a). Its value is $V_3 = 50.0\,\text{V}$.

(c) The energy stored in C_3 is

$$U_3 = \frac{1}{2} C_3 V_3^2 = \frac{1}{2}(15.0\,\mu\text{F})(50.0\,\text{V})^2 = 1.88 \times 10^{-2}\,\text{J}.$$

(d) From part (a), we have $q_1 = 5.00 \times 10^{-4}\,\text{C}$, and

(e) $V_1 = 50.0\,\text{V}$.

(f) The energy stored in C_1 is

$$U_1 = \frac{1}{2} C_1 V_1^2 = \frac{1}{2}(10.0\,\mu\text{F})(50.0\,\text{V})^2 = 1.25 \times 10^{-2}\,\text{J}.$$

(g) Again, from part (a), $q_2 = 2.50 \times 10^{-4}\,\text{C}$, and

(h) $V_2 = 50.0\,\text{V}$.

(i) The energy stored in C_2 is

$$U_2 = \frac{1}{2} C_2 V_2^2 = \frac{1}{2}(5.00\,\mu\text{F})(50.0\,\text{V})^2 = 6.25 \times 10^{-3}\,\text{J}.$$

29. The energy per unit volume is

$$u = \frac{1}{2}\varepsilon_0 E^2 = \frac{1}{2}\varepsilon_0 \left(\frac{e}{4\pi\varepsilon_0 r^2}\right)^2 = \frac{e^2}{32\pi^2\varepsilon_0 r^4}.$$

(a) At $r = 1.00 \times 10^{-3}\,\text{m}$, with $e = 1.60 \times 10^{-19}\,\text{C}$ and $\varepsilon_0 = 8.85 \times 10^{-12}\,\text{C}^2/\text{N}\cdot\text{m}^2$, we have $u = 9.16 \times 10^{-18}\,\text{J/m}^3$.

(b) Similarly, at $r = 1.00 \times 10^{-6}$ m, $u = 9.16 \times 10^{-6}$ J/m^3,

(c) at $r = 1.00 \times 10^{-9}$ m, $u = 9.16 \times 10^{6}$ J/m^3, and

(d) at $r = 1.00 \times 10^{-12}$ m, $u = 9.16 \times 10^{18}$ J/m^3.

(e) From the expression above $u \propto r^{-4}$. Thus, for $r \to 0$, the energy density $u \to \infty$.

30. (a) The charge q_3 in the Figure is $q_3 = C_3 V = (4.00\,\mu\text{F})(100\text{ V}) = 4.00 \times 10^{-4}$ C.

(b) $V_3 = V = 100$ V.

(c) Using $U_i = \frac{1}{2} C_i V_i^2$, we have $U_3 = \frac{1}{2} C_3 V_3^2 = 2.00 \times 10^{-2}$ J.

(d) From the Figure,

$$q_1 = q_2 = \frac{C_1 C_2 V}{C_1 + C_2} = \frac{(10.0\,\mu\text{F})(5.00\,\mu\text{F})(100\text{ V})}{10.0\,\mu\text{F} + 5.00\,\mu\text{F}} = 3.33 \times 10^{-4}\text{C}.$$

(e) $V_1 = q_1/C_1 = 3.33 \times 10^{-4}$ C/10.0 μF = 33.3 V.

(f) $U_1 = \frac{1}{2} C_1 V_1^2 = 5.55 \times 10^{-3}$ J.

(g) From part (d), we have $q_2 = q_1 = 3.33 \times 10^{-4}\text{C}$.

(h) $V_2 = V - V_1 = 100$ V $-$ 33.3 V $=$ 66.7 V.

(i) $U_2 = \frac{1}{2} C_2 V_2^2 = 1.11 \times 10^{-2}$ J.

31. (a) Let q be the charge on the positive plate. Since the capacitance of a parallel-plate capacitor is given by $\varepsilon_0 A/d_i$, the charge is $q = CV = \varepsilon_0 AV_i/d_i$. After the plates are pulled apart, their separation is d_f and the potential difference is V_f. Then $q = \varepsilon_0 AV_f/2d_f$ and

$$V_f = \frac{d_f}{\varepsilon_0 A} q = \frac{d_f}{\varepsilon_0 A} \frac{\varepsilon_0 A}{d_i} V_i = \frac{d_f}{d_i} V_i.$$

With $d_i = 3.00 \times 10^{-3}$ m, $V_i = 6.00$ V and $d_f = 8.00 \times 10^{-3}$ m, we have $V_f = 16.0$ V.

(b) The initial energy stored in the capacitor is (in SI units)

$$U_i = \frac{1}{2}CV_i^2 = \frac{\varepsilon_0 A V_i^2}{2d_i} = \frac{(8.85\times10^{-12})(8.50\times10^{-4})(6.00)^2}{2(3.00\times10^{-3})} = 4.51\times10^{-11} \text{ J}.$$

(c) The final energy stored is

$$U_f = \frac{1}{2}\frac{\varepsilon_0 A}{d_f}V_f^2 = \frac{1}{2}\frac{\varepsilon_0 A}{d_f}\left(\frac{d_f}{d_i}V_i\right)^2 = \frac{d_f}{d_i}\left(\frac{\varepsilon_0 A V_i^2}{d_i}\right) = \frac{d_f}{d_i}U_i.$$

With $d_f/d_i = 8.00/3.00$, we have $U_f = 1.20\times10^{-10}$ J.

(d) The work done to pull the plates apart is the difference in the energy:

$$W = U_f - U_i = 7.52\times10^{-11} \text{ J}.$$

32. We use $E = q/4\pi\varepsilon_0 R^2 = V/R$. Thus

$$u = \frac{1}{2}\varepsilon_0 E^2 = \frac{1}{2}\varepsilon_0\left(\frac{V}{R}\right)^2 = \frac{1}{2}\left(8.85\times10^{-12}\frac{\text{C}^2}{\text{N}\cdot\text{m}^2}\right)\left(\frac{8000\text{ V}}{0.050\text{ m}}\right)^2 = 0.11 \text{ J/m}^3.$$

33. (a) They each store the same charge, so the maximum voltage is across the smallest capacitor. With 100 V across 10 μF, then the voltage across the 20 μF capacitor is 50 V and the voltage across the 25 μF capacitor is 40 V. Therefore, the voltage across the arrangement is 190 V.

(b) Using Eq. 25-21 or Eq. 25-22, we sum the energies on the capacitors and obtain U_{total} = 0.095 J.

34. If the original capacitance is given by $C = \varepsilon_0 A/d$, then the new capacitance is $C' = \varepsilon_0 \kappa A/2d$. Thus $C'/C = \kappa/2$ or

$$\kappa = 2C'/C = 2(2.6\text{ pF}/1.3\text{ pF}) = 4.0.$$

35. The capacitance with the dielectric in place is given by $C = \kappa C_0$, where C_0 is the capacitance before the dielectric is inserted. The energy stored is given by $U = \frac{1}{2}CV^2 = \frac{1}{2}\kappa C_0 V^2$, so

$$\kappa = \frac{2U}{C_0 V^2} = \frac{2(7.4\times10^{-6}\text{ J})}{(7.4\times10^{-12}\text{ F})(652\text{V})^2} = 4.7.$$

According to Table 25-1, you should use Pyrex.

36. (a) We use $C = \varepsilon_0 A/d$ to solve for d:

$$d = \frac{\varepsilon_0 A}{C} = \frac{\left(8.85 \times 10^{-12} \frac{C^2}{N \cdot m^2}\right)(0.35 \text{ m}^2)}{50 \times 10^{-12} \text{ F}} = 6.2 \times 10^{-2} \text{ m}.$$

(b) We use $C \propto \kappa$. The new capacitance is $C' = C(\kappa/\kappa_{air}) = (50 \text{ pf})(5.6/1.0) = 2.8 \times 10^2 \text{ pF}$.

37. The capacitance of a cylindrical capacitor is given by

$$C = \kappa C_0 = \frac{2\pi\kappa\varepsilon_0 L}{\ln(b/a)},$$

where C_0 is the capacitance without the dielectric, κ is the dielectric constant, L is the length, a is the inner radius, and b is the outer radius. The capacitance per unit length of the cable is

$$\frac{C}{L} = \frac{2\pi\kappa\varepsilon_0}{\ln(b/a)} = \frac{2\pi(2.6)(8.85 \times 10^{-12} \text{ F/m})}{\ln[(0.60 \text{ mm})/(0.10 \text{ mm})]} = 8.1 \times 10^{-11} \text{F/m} = 81 \text{ pF/m}.$$

38. Each capacitor has 12.0 V across it, so Eq. 25-1 yields the charge values once we know C_1 and C_2. From Eq. 25-9,

$$C_2 = \frac{\varepsilon_0 A}{d} = 2.21 \times 10^{-11} \text{ F},$$

and from Eq. 25-27,

$$C_1 = \frac{\kappa\varepsilon_0 A}{d} = 6.64 \times 10^{-11} \text{ F}.$$

This leads to $q_1 = C_1 V_1 = 8.00 \times 10^{-10}$ C and $q_2 = C_2 V_2 = 2.66 \times 10^{-10}$ C. The addition of these gives the desired result: $q_{tot} = 1.06 \times 10^{-9}$ C. Alternatively, the circuit could be reduced to find the q_{tot}.

39. The capacitance is given by $C = \kappa C_0 = \kappa\varepsilon_0 A/d$, where C_0 is the capacitance without the dielectric, κ is the dielectric constant, A is the plate area, and d is the plate separation. The electric field between the plates is given by $E = V/d$, where V is the potential difference between the plates. Thus, $d = V/E$ and $C = \kappa\varepsilon_0 AE/V$. Thus,

$$A = \frac{CV}{\kappa\varepsilon_0 E}.$$

For the area to be a minimum, the electric field must be the greatest it can be without breakdown occurring. That is,

$$A = \frac{(7.0 \times 10^{-8} \text{F})(4.0 \times 10^{3} \text{V})}{2.8(8.85 \times 10^{-12} \text{F/m})(18 \times 10^{6} \text{ V/m})} = 0.63 \text{ m}^{2}.$$

40. (a) We use Eq. 25-14:

$$C = 2\pi\varepsilon_{0}\kappa \frac{L}{\ln(b/a)} = \frac{(4.7)(0.15 \text{ m})}{2\left(8.99 \times 10^{9} \frac{\text{N}\cdot\text{m}^{2}}{\text{C}^{2}}\right)\ln(3.8 \text{ cm}/3.6 \text{ cm})} = 0.73 \text{ nF}.$$

(b) The breakdown potential is (14 kV/mm) (3.8 cm – 3.6 cm) = 28 kV.

41. Using Eq. 25-29, with $\sigma = q/A$, we have

$$\left|\vec{E}\right| = \frac{q}{\kappa\varepsilon_{0}A} = 200 \times 10^{3} \text{ N/C}$$

which yields $q = 3.3 \times 10^{-7}$ C. Eq. 25-21 and Eq. 25-27 therefore lead to

$$U = \frac{q^{2}}{2C} = \frac{q^{2}d}{2\kappa\varepsilon_{0}A} = 6.6 \times 10^{-5} \text{ J} .$$

42. The capacitor can be viewed as two capacitors C_1 and C_2 in parallel, each with surface area $A/2$ and plate separation d, filled with dielectric materials with dielectric constants κ_1 and κ_2, respectively. Thus, (in SI units),

$$C = C_{1} + C_{2} = \frac{\varepsilon_{0}(A/2)\kappa_{1}}{d} + \frac{\varepsilon_{0}(A/2)\kappa_{2}}{d} = \frac{\varepsilon_{0}A}{d}\left(\frac{\kappa_{1} + \kappa_{2}}{2}\right)$$

$$= \frac{(8.85 \times 10^{-12})(5.56 \times 10^{-4})}{5.56 \times 10^{-3}} \frac{7.00 + 12.00}{2} = 8.41 \times 10^{-12} \text{ F}.$$

43. We assume there is charge q on one plate and charge $-q$ on the other. The electric field in the lower half of the region between the plates is

$$E_{1} = \frac{q}{\kappa_{1}\varepsilon_{0}A},$$

where A is the plate area. The electric field in the upper half is

$$E_{2} = \frac{q}{\kappa_{2}\varepsilon_{0}A}.$$

Let $d/2$ be the thickness of each dielectric. Since the field is uniform in each region, the potential difference between the plates is

$$V = \frac{E_1 d}{2} + \frac{E_2 d}{2} = \frac{qd}{2\varepsilon_0 A}\left[\frac{1}{\kappa_1} + \frac{1}{\kappa_2}\right] = \frac{qd}{2\varepsilon_0 A}\frac{\kappa_1 + \kappa_2}{\kappa_1 \kappa_2},$$

so

$$C = \frac{q}{V} = \frac{2\varepsilon_0 A}{d}\frac{\kappa_1 \kappa_2}{\kappa_1 + \kappa_2}.$$

This expression is exactly the same as that for C_{eq} of two capacitors in series, one with dielectric constant κ_1 and the other with dielectric constant κ_2. Each has plate area A and plate separation $d/2$. Also we note that if $\kappa_1 = \kappa_2$, the expression reduces to $C = \kappa_1\varepsilon_0 A/d$, the correct result for a parallel-plate capacitor with plate area A, plate separation d, and dielectric constant κ_1.

With $A = 7.89\times10^{-4}\,\text{m}^2$, $d = 4.62\times10^{-3}\,\text{m}$, $\kappa_1 = 11.0$ and $\kappa_2 = 12.0$, the capacitance is, (in SI units)

$$C = \frac{2(8.85\times10^{-12})(7.89\times10^{-4})}{4.62\times10^{-3}}\frac{(11.0)(12.0)}{11.0+12.0} = 1.73\times10^{-11}\,\text{F}.$$

44. Let $C_1 = \varepsilon_0(A/2)\kappa_1/2d = \varepsilon_0 A\kappa_1/4d$, $C_2 = \varepsilon_0(A/2)\kappa_2/d = \varepsilon_0 A\kappa_2/2d$, and $C_3 = \varepsilon_0 A\kappa_3/2d$. Note that C_2 and C_3 are effectively connected in series, while C_1 is effectively connected in parallel with the C_2-C_3 combination. Thus,

$$C = C_1 + \frac{C_2 C_3}{C_2 + C_3} = \frac{\varepsilon_0 A\kappa_1}{4d} + \frac{(\varepsilon_0 A/d)(\kappa_2/2)(\kappa_3/2)}{\kappa_2/2 + \kappa_3/2} = \frac{\varepsilon_0 A}{4d}\left(\kappa_1 + \frac{2\kappa_2\kappa_3}{\kappa_2 + \kappa_3}\right).$$

With $A = 1.05\times10^{-3}\,\text{m}^2$, $d = 3.56\times10^{-3}\,\text{m}$, $\kappa_1 = 21.0$, $\kappa_2 = 42.0$ and $\kappa_3 = 58.0$, the capacitance is, (in SI units)

$$C = \frac{(8.85\times10^{-12})(1.05\times10^{-3})}{4(3.56)\times10^{-3}}\left(21.0 + \frac{2(42.0)(58.0)}{42.0+58.0}\right) = 4.55\times10^{-11}\,\text{F}.$$

45. (a) The electric field in the region between the plates is given by $E = V/d$, where V is the potential difference between the plates and d is the plate separation. The capacitance is given by $C = \kappa\varepsilon_0 A/d$, where A is the plate area and κ is the dielectric constant, so $d = \kappa\varepsilon_0 A/C$ and

$$E = \frac{VC}{\kappa \varepsilon_0 A} = \frac{(50 \text{ V})(100 \times 10^{-12} \text{ F})}{5.4(8.85 \times 10^{-12} \text{ F/m})(100 \times 10^{-4} \text{ m}^2)} = 1.0 \times 10^4 \text{ V/m}.$$

(b) The free charge on the plates is $q_f = CV = (100 \times 10^{-12} \text{ F})(50 \text{ V}) = 5.0 \times 10^{-9} \text{ C}.$

(c) The electric field is produced by both the free and induced charge. Since the field of a large uniform layer of charge is $q/2\varepsilon_0 A$, the field between the plates is

$$E = \frac{q_f}{2\varepsilon_0 A} + \frac{q_f}{2\varepsilon_0 A} - \frac{q_i}{2\varepsilon_0 A} - \frac{q_i}{2\varepsilon_0 A},$$

where the first term is due to the positive free charge on one plate, the second is due to the negative free charge on the other plate, the third is due to the positive induced charge on one dielectric surface, and the fourth is due to the negative induced charge on the other dielectric surface. Note that the field due to the induced charge is opposite the field due to the free charge, so they tend to cancel. The induced charge is therefore

$$q_i = q_f - \varepsilon_0 AE = 5.0 \times 10^{-9} \text{ C} - (8.85 \times 10^{-12} \text{ F/m})(100 \times 10^{-4} \text{ m}^2)(1.0 \times 10^4 \text{ V/m})$$
$$= 4.1 \times 10^{-9} \text{ C} = 4.1 \text{ nC}.$$

46. (a) The electric field E_1 in the free space between the two plates is $E_1 = q/\varepsilon_0 A$ while that inside the slab is $E_2 = E_1/\kappa = q/\kappa \varepsilon_0 A$. Thus,

$$V_0 = E_1(d-b) + E_2 b = \left(\frac{q}{\varepsilon_0 A}\right)\left(d - b + \frac{b}{\kappa}\right),$$

and the capacitance is

$$C = \frac{q}{V_0} = \frac{\varepsilon_0 A\kappa}{\kappa(d-b)+b} = \frac{(8.85 \times 10^{-12} \frac{\text{C}^2}{\text{N·m}^2})(115 \times 10^{-4} \text{ m}^2)(2.61)}{(2.61)(0.0124 \text{ m} - 0.00780 \text{ m}) + (0.00780 \text{ m})} = 13.4 \text{ pF}.$$

(b) $q = CV = (13.4 \times 10^{-12} \text{ F})(85.5 \text{ V}) = 1.15 \text{ nC}.$

(c) The magnitude of the electric field in the gap is

$$E_1 = \frac{q}{\varepsilon_0 A} = \frac{1.15 \times 10^{-9} \text{ C}}{(8.85 \times 10^{-12} \frac{\text{C}^2}{\text{N·m}^2})(115 \times 10^{-4} \text{ m}^2)} = 1.13 \times 10^4 \text{ N/C}.$$

(d) Using Eq. 25-34, we obtain

$$E_2 = \frac{E_1}{\kappa} = \frac{1.13 \times 10^4 \text{ N/C}}{2.61} = 4.33 \times 10^3 \text{ N/C}.$$

47. (a) According to Eq. 25-17 the capacitance of an air-filled spherical capacitor is given by

$$C_0 = 4\pi\varepsilon_0 \left(\frac{ab}{b-a} \right).$$

When the dielectric is inserted between the plates the capacitance is greater by a factor of the dielectric constant κ. Consequently, the new capacitance is

$$C = 4\pi\kappa\varepsilon_0 \left(\frac{ab}{b-a} \right) = \frac{23.5}{8.99 \times 10^9} \frac{(0.0120)(0.0170)}{0.0170 - 0.0120} = 0.107 \text{ nF}.$$

(b) The charge on the positive plate is

$$q = CV = (0.107 \text{ nF})(73.0 \text{ V}) = 7.79 \text{ nC}.$$

(c) Let the charge on the inner conductor be $-q$. Immediately adjacent to it is the induced charge q'. Since the electric field is less by a factor $1/\kappa$ than the field when no dielectric is present, then $-q + q' = -q/\kappa$. Thus,

$$q' = \frac{\kappa-1}{\kappa} q = 4\pi(\kappa-1)\varepsilon_0 \frac{ab}{b-a} V$$
$$= \left(\frac{23.5 - 1.00}{23.5} \right)(7.79 \text{ nC}) = 7.45 \text{ nC}.$$

48. (a) We apply Gauss's law with dielectric: $q/\varepsilon_0 = \kappa EA$, and solve for κ:

$$\kappa = \frac{q}{\varepsilon_0 EA} = \frac{8.9 \times 10^{-7} \text{ C}}{\left(8.85 \times 10^{-12} \frac{C^2}{N \cdot m^2}\right)\left(1.4 \times 10^{-6} \text{ V/m}\right)\left(100 \times 10^{-4} \text{ m}^2\right)} = 7.2.$$

(b) The charge induced is

$$q' = q\left(1 - \frac{1}{\kappa}\right) = (8.9 \times 10^{-7} \text{ C})\left(1 - \frac{1}{7.2}\right) = 7.7 \times 10^{-7} \text{ C}.$$

49. (a) Initially, the capacitance is

$$C_0 = \frac{\varepsilon_0 A}{d} = \frac{\left(8.85 \times 10^{-12} \frac{C^2}{N \cdot m^2}\right)(0.12 \text{ m}^2)}{1.2 \times 10^{-2} \text{ m}} = 89 \text{ pF}.$$

(b) Working through Sample Problem 25-7 algebraically, we find:

$$C = \frac{\varepsilon_0 A \kappa}{\kappa(d-b)+b} = \frac{\left(8.85 \times 10^{-12} \, \frac{C^2}{N \cdot m^2}\right)(0.12 \, m^2)(4.8)}{(4.8)(1.2-0.40)(10^{-2} \, m)+(4.0 \times 10^{-3} \, m)} = 1.2 \times 10^2 \, pF.$$

(c) Before the insertion, $q = C_0 V$ (89 pF)(120 V) = 11 nC.

(d) Since the battery is disconnected, q will remain the same after the insertion of the slab, with q = 11 nC.

(e) $E = q / \varepsilon_0 A = 11 \times 10^{-9} \, C / (8.85 \times 10^{-12} \, \frac{C^2}{N \cdot m^2})(0.12 \, m^2) = 10 \, kV / m.$

(f) $E' = E/\kappa = (10 \, kV/m)/4.8 = 2.1 \, kV/m.$

(g) $V = E(d-b) + E'b = (10 \, kV/m)(0.012 \, m - 0.0040 \, m) + (2.1 \, kV/m)(0.40 \times 10^{-3} \, m) = 88 \, V.$

(h) The work done is

$$W_{ext} = \Delta U = \frac{q^2}{2}\left(\frac{1}{C}-\frac{1}{C_0}\right) = \frac{(11 \times 10^{-9} \, C)^2}{2}\left(\frac{1}{89 \times 10^{-12} \, F}-\frac{1}{120 \times 10^{-12} \, F}\right) = -1.7 \times 10^{-7} \, J.$$

50. (a) Eq. 25-22 yields

$$U = \frac{1}{2}CV^2 = \frac{1}{2}\left(200 \times 10^{-12} \, F\right)\left(7.0 \times 10^3 \, V\right)^2 = 4.9 \times 10^{-3} \, J.$$

(b) Our result from part (a) is much less than the required 150 mJ, so such a spark should not have set off an explosion.

51. One way to approach this is to note that – since they are identical – the voltage is evenly divided between them. That is, the voltage across each capacitor is $V = (10/n)$ volt. With $C = 2.0 \times 10^{-6}$ F, the electric energy stored by each capacitor is $\frac{1}{2}CV^2$. The total energy stored by the capacitors is n times that value, and the problem requires the total be equal to 25×10^{-6} J. Thus,

$$\frac{n}{2}(2.0 \times 10^{-6})\left(\frac{10}{n}\right)^2 = 25 \times 10^{-6}$$

leads to $n = 4$.

52. Initially the capacitors C_1, C_2, and C_3 form a series combination equivalent to a single capacitor which we denote C_{123}. Solving the equation

$$\frac{1}{C_1} + \frac{1}{C_2} + \frac{1}{C_3} = \frac{1}{C_{123}},$$

we obtain $C_{123} = 2.40$ μF. With $V = 12.0$ V, we then obtain $q = C_{123}V = 28.8$ μC. In the final situation, C_2 and C_4 are in parallel and are thus effectively equivalent to $C_{24} = 12.0$ μF. Similar to the previous computation, we use

$$\frac{1}{C_1} + \frac{1}{C_{24}} + \frac{1}{C_3} = \frac{1}{C_{1234}}$$

and find $C_{1234} = 3.00$ μF. Therefore, the final charge is $q = C_{1234}V = 36.0$ μC.

(a) This represents a change (relative to the initial charge) of $\Delta q = 7.20$ μC.

(b) The capacitor C_{24} which we imagined to replace the parallel pair C_2 and C_4 is in series with C_1 and C_3 and thus also has the final charge $q = 36.0$ μC found above. The voltage across C_{24} would be $V_{24} = q/C_{24} = 36.0/12.0 = 3.00$ V. This is the same voltage across each of the parallel pair. In particular, $V_4 = 3.00$ V implies that $q_4 = C_4 V_4 = 18.0$ μC.

(c) The battery supplies charges only to the plates where it is connected. The charges on the rest of the plates are due to electron transfers between them, in accord with the new distribution of voltages across the capacitors. So, the battery does not directly supply the charge on capacitor 4.

53. In series, their equivalent capacitance (and thus their total energy stored) is smaller than either one individually (by Eq. 25-20). In parallel, their equivalent capacitance (and thus their total energy stored) is larger than either one individually (by Eq. 25-19). Thus, the middle two values quoted in the problem must correspond to the individual capacitors. We use Eq. 25-22 and find

(a) 100 $\mu J = \frac{1}{2}C_1 (10 \text{ V})^2$ $\Rightarrow$ $C_1 = 2.0$ μF

(b) 300 $\mu J = \frac{1}{2}C_2 (10 \text{ V})^2$ $\Rightarrow$ $C_2 = 6.0$ μF .

54. We note that the voltage across C_3 is $V_3 = (12 \text{ V} - 2 \text{ V} - 5 \text{ V}) = 5$ V. Thus, its charge is $q_3 = C_3 V_3 = 4$ μC.

(a) Therefore, since C_1, C_2 and C_3 are in series (so they have the same charge), then

$$C_1 = \frac{4 \ \mu C}{2 \text{ V}} = 2.0 \ \mu F .$$

(b) Similarly, $C_2 = 4/5 = 0.80$ μF.

55. (a) The number of (conduction) electrons per cubic meter is $n = 8.49 \times 10^{28}$ m^3. The volume in question is the face area multiplied by the depth: $A \cdot d$. The total number of electrons which have moved to the face is

$$N = \frac{-3.0 \times 10^{-6}\,\text{C}}{-1.6 \times 10^{-19}\,\text{C}} \approx 1.9 \times 10^{13}\ .$$

Using the relation $N = nAd$, we obtain $d = 1.1 \times 10^{-12}$ m, a remarkably small distance!

56. Initially, the total equivalent capacitance is $C_{12} = [(C_1)^{-1} + (C_2)^{-1}]^{-1} = 3.0$ μF, and the charge on the positive plate of each one is $(3.0\ \mu\text{F})(10\ \text{V}) = 30\ \mu\text{C}$. Next, the capacitor (call is C_1) is squeezed as described in the problem, with the effect that the new value of C_1 is 12 μF (see Eq. 25-9). The new total equivalent capacitance then becomes

$$C_{12} = [(C_1)^{-1} + (C_2)^{-1}]^{-1} = 4.0\ \mu\text{F},$$

and the new charge on the positive plate of each one is $(4.0\ \mu\text{F})(10\ \text{V}) = 40\ \mu\text{C}$.

(a) Thus we see that the charge transferred from the battery as a result of the squeezing is $40\ \mu\text{C} - 30\ \mu\text{C} = 10\ \mu\text{C}$.

(b) The total increase in positive charge (on the respective positive plates) stored on the capacitors is twice the value found in part (a) (since we are dealing with two capacitors in series): 20 μC.

57. (a) Put five such capacitors in series. Then, the equivalent capacitance is 2.0 μF/5 = 0.40 μF. With each capacitor taking a 200-V potential difference, the equivalent capacitor can withstand 1000 V.

(b) As one possibility, you can take three identical arrays of capacitors, each array being a five-capacitor combination described in part (a) above, and hook up the arrays in parallel. The equivalent capacitance is now $C_{eq} = 3(0.40\ \mu\text{F}) = 1.2\ \mu\text{F}$. With each capacitor taking a 200-V potential difference the equivalent capacitor can withstand 1000 V.

58. Equation 25-14 leads to $C_1 = 2.53$ pF and $C_1 = 2.17$ pF. Initially, the total equivalent capacitance is

$$C_{12} = [(C_1)^{-1} + (C_2)^{-1}]^{-1} = 1.488\ \text{pF},$$

and the charge on the positive plate of each one is (1.488 pF)(10 V) = 14.88 pC. Next, capacitor 2 is modified as described in the problem, with the effect that the new value of C_2 is 2.17 pF (again using Eq. 25-14). The new total equivalent capacitance is

$$C_{12} = [(C_1)^{-1} + (C_2)^{-1}]^{-1} = 1.170 \text{ pF},$$

and the new charge on the positive plate of each one is (1.170 pF)(10 V) = 11.70 pC. Thus we see that the charge transferred from the battery (considered in absolute value) as a result of the modification is 14.88 pC – 11.70 pC = 3.18 pC.

(a) This charge, divided by e gives the number of electrons that pass point P. Thus,

$$\frac{3.18 \times 10^{-12}}{1.6 \times 10^{-19}} = 2.0 \times 10^7 .$$

(b) These electrons move rightwards in the figure (that is, away from the battery) since the positive plates (the ones closest to point P) of the capacitors have suffered a *decease* in their positive charges. The usual reason for a metal plate to be positive is that it has more protons than electrons. Thus, in this problem some electrons have "returned" to the positive plates (making them less positive).

59. (a) We do not employ energy conservation since, in reaching equilibrium, some energy is dissipated either as heat or radio waves. Charge is conserved; therefore, if $Q = C_1 V_{\text{bat}} = 40$ μC, and q_1 and q_2 are the charges on C_1 and C_2 after the switch is thrown to the right and equilibrium is reached, then

$$Q = q_1 + q_2.$$

Reducing the right portion of the circuit (the C_3, C_4 parallel pair which are in series with C_2) we have an equivalent capacitance of $C' = 8.0$ μF which has charge $q' = q_2$ and potential difference equal to that of C_1. Thus, $V_1 = V'$, or

$$\frac{q_1}{C_1} = \frac{q_2}{C'}$$

which yields $4q_1 = q_2$. Therefore, $Q = q_1 + 4q_1$. This leads to $q_1 = 8.0$ μC and consequently to $q_2 = 32$ μC.

(b) From Eq. 25-1, we have $V_2 = (32 \text{ μC})(16 \text{ μF}) = 2.0$ V.

60. (a) We calculate the charged surface area of the cylindrical volume as follows:

$$A = 2\pi r h + \pi r^2 = 2\pi(0.20 \text{ m})(0.10 \text{ m}) + \pi(0.20 \text{ m})^2 = 0.25 \text{ m}^2$$

where we note from the figure that although the bottom is charged, the top is not. Therefore, the charge is $q = \sigma A = -0.50 \ \mu C$ on the exterior surface, and consequently (according to the assumptions in the problem) that same charge q is induced in the interior of the fluid.

(b) By Eq. 25-21, the energy stored is

$$U = \frac{q^2}{2C} = \frac{(5.0 \times 10^{-7} \ C)^2}{2(35 \times 10^{-12} \ F)} = 3.6 \times 10^{-3} \ J.$$

(c) Our result is within a factor of three of that needed to cause a spark. Our conclusion is that it will probably not cause a spark; however, there is not enough of a safety factor to be sure.

61. (a) In the top right portion of the circuit is a pair of 4.00 μF which we reduce to a single 8.00 μF capacitor (which is then in series with the bottom capacitor that the problem is asking about). The further reduction with the bottom 4.00 μF capacitor results in an equivalence of $\frac{8}{3}$ μF, which clearly has the battery voltage across it -- and therefore has charge ($\frac{8}{3}$ μF)(9.00 V) = 24.0 μC. This is seen to be the same as the charge on the bottom capacitor.

(b) The voltage across the bottom capacitor is

$$V = \frac{q}{C} = \frac{24.0 \ \mu C}{4.00 \ \mu F} = 6.00 \ V \quad .$$

62. We do not employ energy conservation since, in reaching equilibrium, some energy is dissipated either as heat or radio waves. Charge is conserved; therefore, if $Q = C_1 V_{bat} = 100 \ \mu C$, and q_1, q_2 and q_3 are the charges on C_1, C_2 and C_3 after the switch is thrown to the right and equilibrium is reached, then

$$Q = q_1 + q_2 + q_3 \quad .$$

Since the parallel pair C_2 and C_3 are identical, it is clear that $q_2 = q_3$. They are in parallel with C_1 so that $V_1 = V_3$, or

$$\frac{q_1}{C_1} = \frac{q_3}{C_3}$$

which leads to $q_1 = q_3/2$. Therefore,

$$Q = \left(\frac{1}{2} q_3\right) + q_3 + q_3$$

which yields $q_3 = 40 \ \mu C$ and consequently $q_1 = 20 \ \mu C$.

63. The pair C_3 and C_4 are in parallel and consequently equivalent to 30 μF. Since this numerical value is identical to that of the others (with which it is in series, with the battery), we observe that each has one-third the battery voltage across it. Hence, 3.0 V is across C_4, producing a charge

$$q_4 = C_4 V_4 = (15 \ \mu F)(3.0 \ V) = 45 \ \mu C \ .$$

64. (a) We reduce the parallel group C_2, C_3 and C_4, and the parallel pair C_5 and C_6, obtaining equivalent values $C' = 12 \ \mu F$ and $C'' = 12 \ \mu F$, respectively. We then reduce the series group C_1, C' and C'' to obtain an equivalent capacitance of $C_{eq} = 3 \ \mu F$ hooked to the battery. Thus, the charge stored in the system is

$$q_{sys} = C_{eq} V_{bat} = 36 \ \mu C \ .$$

(b) Since $q_{sys} = q_1$ then the voltage across C_1 is

$$V_1 = \frac{q_1}{C_1} = \frac{36 \ \mu C}{6.0 \ \mu F} = 6.0 \ V \ .$$

The voltage across the series-pair C' and C'' is consequently $V_{bat} - V_1 = 6.0$ V. Since $C' = C''$, we infer $V' = V'' = 6.0/2 = 3.0$ V, which, in turn, is equal to V_4, the potential across C_4. Therefore,

$$q_4 = C_4 V_4 = (4.0 \ \mu F)(3.0 \ V) = 12 \ \mu C \ .$$

65. (a) The potential across C_1 is 10 V, so the charge on it is

$$q_1 = C_1 V_1 = (10.0 \ \mu F)(10.0 \ V) = 100 \ \mu C.$$

(b) Reducing the right portion of the circuit produces an equivalence equal to 6.00 μF, with 10.0 V across it. Thus, a charge of 60.0 μC is on it -- and consequently also on the bottom right capacitor. The bottom right capacitor has, as a result, a potential across it equal to

$$V = \frac{q}{C} = \frac{60 \ \mu C}{10 \ \mu F} = 6.00 \ V$$

which leaves 10.0 V − 6.00 V = 4.00 V across the group of capacitors in the upper right portion of the circuit. Inspection of the arrangement (and capacitance values) of that group reveals that this 4.00 V must be equally divided by C_2 and the capacitor directly below it (in series with it). Therefore, with 2.00 V across C_2 we find

$$q_2 = C_2 V_2 = (10.0 \ \mu F)(2.00 \ V) = 20.0 \ \mu C \ .$$

66. The pair C_1 and C_2 are in parallel, as are the pair C_3 and C_4; they reduce to equivalent values 6.0 µF and 3.0 µF, respectively. These are now in series and reduce to 2.0 µF, across which we have the battery voltage. Consequently, the charge on the 2.0 µF equivalence is (2.0 µF)(12 V) = 24 µC. This charge on the 3.0 µF equivalence (of C_3 and C_4) has a voltage of

$$V = \frac{q}{C} = \frac{24 \text{ µC}}{3 \text{ µF}} = 8.0 \text{ V} \quad .$$

Finally, this voltage on capacitor C_4 produces a charge (2.0 µF)(8.0 V) = 16 µC.

67. For maximum capacitance the two groups of plates must face each other with maximum area. In this case the whole capacitor consists of $(n - 1)$ identical single capacitors connected in parallel. Each capacitor has surface area A and plate separation d so its capacitance is given by $C_0 = \varepsilon_0 A/d$. Thus, the total capacitance of the combination is (in SI units)

$$C = (n-1)C_0 = \frac{(n-1)\varepsilon_0 A}{d} = \frac{(8-1)(8.85 \times 10^{-12})(1.25 \times 10^{-4})}{3.40 \times 10^{-3}} = 2.28 \times 10^{-12} \text{F}.$$

68. (a) Here D is not attached to anything, so that the $6C$ and $4C$ capacitors are in series (equivalent to $2.4C$). This is then in parallel with the $2C$ capacitor, which produces an equivalence of $4.4C$. Finally the $4.4C$ is in series with C and we obtain

$$C_{eq} = \frac{(C)(4.4\,C)}{C + 4.4\,C} = 0.82\,C = 41\,\mu F$$

where we have used the fact that $C = 50\ \mu F$.

(b) Now, B is the point which is not attached to anything, so that the $6C$ and $2C$ capacitors are now in series (equivalent to $1.5C$), which is then in parallel with the $4C$ capacitor (and thus equivalent to $5.5C$). The $5.5C$ is then in series with the C capacitor; consequently,

$$C_{eq} = \frac{(C)(5.5C)}{C + 5.5C} = 0.85C = 42\,\mu F \quad .$$

69. (a) In the first case the two capacitors are effectively connected in series, so the output potential difference is $V_{out} = CV_{in}/2C = V_{in}/2 = 50.0$ V.

(b) In the second case the lower diode acts as a wire so $V_{out} = 0$.

70. The voltage across capacitor 1 is

$$V_1 = \frac{q_1}{C_1} = \frac{30\,\mu C}{10\,\mu F} = 3.0\ V\ .$$

Since $V_1 = V_2$, the total charge on capacitor 2 is

$$q_2 = C_2 V_2 = (20\,\mu F)(2\ V) = 60\,\mu C\ ,$$

which means a total of 90 μC of charge is on the pair of capacitors C_1 and C_2. This implies there is a total of 90 μC of charge also on the C_3 and C_4 pair. Since $C_3 = C_4$, the charge divides equally between them, so $q_3 = q_4 = 45\ \mu C$. Thus, the voltage across capacitor 3 is

$$V_3 = \frac{q_3}{C_3} = \frac{45\,\mu C}{20\,\mu F} = 2.3\ V\ .$$

Therefore, $|V_A - V_B| = V_1 + V_3 = 5.3\ V$.

71. (a) The equivalent capacitance is

$$C_{eq} = \frac{C_1 C_2}{C_1 + C_2} = \frac{(6.00\,\mu F)(4.00\,\mu F)}{6.00\,\mu F + 4.00\,\mu F} = 2.40\,\mu F\ .$$

(b) $q_1 = C_{eq}V = (2.40\ \mu F)(200\ V) = 4.80 \times 10^{-4}\ C$.

(c) $V_1 = q_1/C_1 = 4.80 \times 10^{-4}\ C/6.00\ \mu F = 80.0\ V$.

(d) $q_2 = q_1 = 4.80 \times 10^{-4}\ C$.

(e) $V_2 = V - V_1 = 200\ V - 80.0\ V = 120\ V$.

72. (a) Now $C_{eq} = C_1 + C_2 = 6.00\ \mu F + 4.00\ \mu F = 10.0\ \mu F$.

(b) $q_1 = C_1 V = (6.00\ \mu F)(200\ V) = 1.20 \times 10^{-3}\ C$.

(c) $V_1 = 200\ V$.

(d) $q_2 = C_2 V = (4.00\ \mu F)(200\ V) = 8.00 \times 10^{-4}\ C$.

(e) $V_2 = V_1 = 200\ V$.

73. We cannot expect simple energy conservation to hold since energy is presumably dissipated either as heat in the hookup wires or as radio waves while the charge oscillates in the course of the system "settling down" to its final state (of having 40 V across the parallel pair of capacitors C and 60 μF). We do expect charge to be conserved. Thus, if

Q is the charge originally stored on C and q_1, q_2 are the charges on the parallel pair after "settling down," then

$$Q = q_1 + q_2$$
$$C(100\,\text{V}) = C(40\,\text{V}) + (60\,\mu\text{F})(40\,\text{V})$$

which leads to the solution $C = 40\ \mu\text{F}$.

74. We first need to find an expression for the energy stored in a cylinder of radius R and length L, whose surface lies between the inner and outer cylinders of the capacitor ($a < R < b$). The energy density at any point is given by $u = \frac{1}{2}\varepsilon_0 E^2$, where E is the magnitude of the electric field at that point. If q is the charge on the surface of the inner cylinder, then the magnitude of the electric field at a point a distance r from the cylinder axis is given by

$$E = \frac{q}{2\pi\varepsilon_0 Lr}$$

(see Eq. 25-12), and the energy density at that point is given by

$$u = \frac{1}{2}\varepsilon_0 E^2 = \frac{q^2}{8\pi^2\varepsilon_0 L^2 r^2}.$$

The energy in the cylinder is the volume integral

$$U_R = \int u\,dV.$$

Now, $dV = 2\pi rL\,dr$, so

$$U_R = \int_a^R \frac{q^2}{8\pi^2\varepsilon_0 L^2 r^2} 2\pi rL\,dr = \frac{q^2}{4\pi\varepsilon_0 L}\int_a^R \frac{dr}{r} = \frac{q^2}{4\pi\varepsilon_0 L}\ln\frac{R}{a}.$$

To find an expression for the total energy stored in the capacitor, we replace R with b:

$$U_b = \frac{q^2}{4\pi\varepsilon_0 L}\ln\frac{b}{a}.$$

We want the ratio U_R/U_b to be 1/2, so

$$\ln\frac{R}{a} = \frac{1}{2}\ln\frac{b}{a}$$

or, since $\frac{1}{2}\ln(b/a) = \ln(\sqrt{b/a})$, $\ln(R/a) = \ln(\sqrt{b/a})$. This means $R/a = \sqrt{b/a}$ or $R = \sqrt{ab}$.

75. (a) Since the field is constant and the capacitors are in parallel (each with 600 V across them) with identical distances ($d = 0.00300$ m) between the plates, then the field in A is equal to the field in B:

$$\left|\vec{E}\right| = \frac{V}{d} = 2.00 \times 10^5 \text{ V/m} .$$

(b) $|\vec{E}| = 2.00 \times 10^5$ V/m. See the note in part (a).

(c) For the air-filled capacitor, Eq. 25-4 leads to

$$\sigma = \frac{q}{A} = \varepsilon_0 \left|\vec{E}\right| = 1.77 \times 10^{-6} \text{ C/m}^2 .$$

(d) For the dielectric-filled capacitor, we use Eq. 25-29:

$$\sigma = \kappa\varepsilon_0 \left|\vec{E}\right| = 4.60 \times 10^{-6} \text{ C/m}^2 .$$

(e) Although the discussion in the textbook (§25-8) is in terms of the charge being held fixed (while a dielectric is inserted), it is readily adapted to this situation (where comparison is made of two capacitors which have the same *voltage* and are identical except for the fact that one has a dielectric). The fact that capacitor B has a relatively large charge but only produces the field that A produces (with its smaller charge) is in line with the point being made (in the text) with Eq. 25-34 and in the material that follows. Adapting Eq. 25-35 to this problem, we see that the difference in charge densities between parts (c) and (d) is due, in part, to the (negative) layer of charge at the top surface of the dielectric; consequently,

$$\sigma' = (1.77 \times 10^{-6}) - (4.60 \times 10^{-6}) = -2.83 \times 10^{-6} \text{ C/m}^2 .$$

76. (a) The equivalent capacitance is $C_{eq} = C_1 C_2/(C_1 + C_2)$. Thus the charge q on each capacitor is

$$q = q_1 = q_2 = C_{eq}V = \frac{C_1 C_2 V}{C_1 + C_2} = \frac{(2.00\,\mu\text{F})(8.00\,\mu\text{F})(300\,\text{V})}{2.00\,\mu\text{F} + 8.00\,\mu\text{F}} = 4.80 \times 10^{-4}\text{C}.$$

(b) The potential difference is $V_1 = q/C_1 = 4.80 \times 10^{-4}$ C/2.0 μF = 240 V.

(c) As noted in part (a), $q_2 = q_1 = 4.80 \times 10^{-4}$ C.

(d) $V_2 = V - V_1 = 300\ \text{V} - 240\ \text{V} = 60.0\ \text{V}$.

Now we have $q'_1/C_1 = q'_2/C_2 = V'$ (V' being the new potential difference across each capacitor) and $q'_1 + q'_2 = 2q$. We solve for q'_1, q'_2 and V:

(e) $q'_1 = \dfrac{2C_1 q}{C_1 + C_2} = \dfrac{2(2.00\,\mu\text{F})(4.80\times10^{-4}\text{C})}{2.00\,\mu\text{F} + 8.00\,\mu\text{F}} = 1.92\times10^{-4}\text{C}$.

(f) $V'_1 = \dfrac{q'_1}{C_1} = \dfrac{1.92\times10^{-4}\text{C}}{2.00\,\mu\text{F}} = 96.0\ \text{V}$.

(g) $q'_2 = 2q - q_1 = 7.68\times10^{-4}\text{C}$.

(h) $V'_2 = V'_1 = 96.0\ \text{V}$.

(i) In this circumstance, the capacitors will simply discharge themselves, leaving $q_1 = 0$,

(j) $V_1 = 0$,

(k) $q_2 = 0$,

(l) and $V_2 = V_1 = 0$.

77. We use $U = \tfrac{1}{2}CV^2$. As V is increased by ΔV, the energy stored in the capacitor increases correspondingly from U to $U + \Delta U$: $U + \Delta U = \tfrac{1}{2}C(V + \Delta V)^2$. Thus, $(1 + \Delta V/V)^2 = 1 + \Delta U/U$, or

$$\frac{\Delta V}{V} = \sqrt{1 + \frac{\Delta U}{U}} - 1 = \sqrt{1 + 10\%} - 1 = 4.9\%\ .$$

78. (a) The voltage across C_1 is 12 V, so the charge is

$$q_1 = C_1 V_1 = 24\,\mu\text{C}\ .$$

(b) We reduce the circuit, starting with C_4 and C_3 (in parallel) which are equivalent to 4 μF. This is then in series with C_2, resulting in an equivalence equal to $\tfrac{4}{3}\,\mu$F which would have 12 V across it. The charge on this $\tfrac{4}{3}\,\mu$F capacitor (and therefore on C_2) is $(\tfrac{4}{3}\,\mu\text{F})(12\,\text{V}) = 16\,\mu\text{C}$. Consequently, the voltage across C_2 is

$$V_2 = \frac{q_2}{C_2} = \frac{16\,\mu\text{C}}{2\,\mu\text{F}} = 8\ \text{V}.$$

This leaves $12 - 8 = 4$ V across C_4 (similarly for C_3).

79. We reduce the circuit, starting with C_1 and C_2 (in series) which are equivalent to 4 μF. This is then parallel to C_3 and results in a total of 8 μF, which is now in series with C_4 and can be further reduced. However, the final step in the reduction is not necessary, as we observe that the 8 μF equivalence from the top 3 capacitors has the same capacitance as C_4 and therefore the same voltage; since they are in series, that voltage is then $12/2 = 6.0$ V.

80. We use $C = \varepsilon_0 \kappa A / d \propto \kappa / d$. To maximize C we need to choose the material with the greatest value of κ / d. It follows that the mica sheet should be chosen.

81. We may think of this as two capacitors in series C_1 and C_2, the former with the $\kappa_1 = 3.00$ material and the latter with the $\kappa_2 = 4.00$ material. Upon using Eq. 25-9, Eq. 25-27 and then reducing C_1 and C_2 to an equivalent capacitance (connected directly to the battery) with Eq. 25-20, we obtain

$$C_{eq} = \left(\frac{\kappa_1 \kappa_2}{\kappa_1 + \kappa_2} \right) \frac{\varepsilon_0 A}{d} = 1.52 \times 10^{-10} \text{ F} \quad .$$

Therefore, $q = C_{eq} V = 1.06 \times 10^{-9}$ C.

82. (a) The length d is effectively shortened by b so $C' = \varepsilon_0 A / (d - b) = 0.708$ pF.

(b) The energy before, divided by the energy after inserting the slab is

$$\frac{U}{U'} = \frac{q^2 / 2C}{q^2 / 2C'} = \frac{C'}{C} = \frac{\varepsilon_0 A / (d - b)}{\varepsilon_0 A / d} = \frac{d}{d - b} = \frac{5.00}{5.00 - 2.00} = 1.67.$$

(c) The work done is

$$W = \Delta U = U' - U = \frac{q^2}{2} \left(\frac{1}{C'} - \frac{1}{C} \right) = \frac{q^2}{2 \varepsilon_0 A} (d - b - d) = -\frac{q^2 b}{2 \varepsilon_0 A} = -5.44 \text{ J}.$$

(d) Since $W < 0$ the slab is sucked in.

83. (a) $C' = \varepsilon_0 A / (d - b) = 0.708$ pF, the same as part (a) in problem 82.

(b) Now,

$$\frac{U}{U'} = \frac{\frac{1}{2} C V^2}{\frac{1}{2} C' V^2} = \frac{C}{C'} = \frac{\varepsilon_0 A / d}{\varepsilon_0 A / (d - b)} = \frac{d - b}{d} = \frac{5.00 - 2.00}{5.00} = 0.600.$$

(c) The work done is

$$W = \Delta U = U' - U = \frac{1}{2}(C' - C)V^2 = \frac{\varepsilon_0 A}{2}\left(\frac{1}{d-b} - \frac{1}{d}\right)V^2 = \frac{\varepsilon_0 A b V^2}{2d(d-b)} = 1.02 \times 10^{-9}\,\text{J}.$$

(d) In Problem 82 where the capacitor is disconnected from the battery and the slab is sucked in, F is certainly given by $-dU/dx$. However, that relation does not hold when the battery is left attached because the force on the slab is not conservative. The charge distribution in the slab causes the slab to be sucked into the gap by the charge distribution on the plates. This action causes an increase in the potential energy stored by the battery in the capacitor.

84. We do not employ energy conservation since, in reaching equilibrium, some energy is dissipated either as heat or radio waves. Charge is conserved; therefore, if $Q = 48\ \mu\text{C}$, and q_1 and q_3 are the charges on C_1 and C_3 after the switch is thrown to the right (and equilibrium is reached), then

$$Q = q_1 + q_3.$$

We note that $V_{1\text{ and }2} = V_3$ because of the parallel arrangement, and $V_1 = \frac{1}{2}V_{1\text{ and }2}$ since they are identical capacitors. This leads to

$$2V_1 = V_3$$
$$2\frac{q_1}{C_1} = \frac{q_3}{C_3}$$
$$2q_1 = q_3$$

where the last step follows from multiplying both sides by 2.00 μF. Therefore,

$$Q = q_1 + (2q_1)$$

which yields $q_1 = 16.0\ \mu\text{C}$ and $q_3 = 32.0\ \mu\text{C}$.

Chapter 26

1. (a) The charge that passes through any cross section is the product of the current and time. Since 4.0 min = (4.0 min)(60 s/min) = 240 s, $q = it = (5.0 \text{ A})(240 \text{ s}) = 1.2 \times 10^3$ C.

(b) The number of electrons N is given by $q = Ne$, where e is the magnitude of the charge on an electron. Thus,

$$N = q/e = (1200 \text{ C})/(1.60 \times 10^{-19} \text{ C}) = 7.5 \times 10^{21}.$$

2. We adapt the discussion in the text to a moving two-dimensional collection of charges. Using σ for the charge per unit area and w for the belt width, we can see that the transport of charge is expressed in the relationship $i = \sigma v w$, which leads to

$$\sigma = \frac{i}{vw} = \frac{100 \times 10^{-6} \text{ A}}{(30 \text{ m/s})(50 \times 10^{-2} \text{ m})} = 6.7 \times 10^{-6} \text{ C/m}^2.$$

3. Suppose the charge on the sphere increases by Δq in time Δt. Then, in that time its potential increases by

$$\Delta V = \frac{\Delta q}{4\pi\varepsilon_0 r},$$

where r is the radius of the sphere. This means

$$\Delta q = 4\pi\varepsilon_0 r \, \Delta V.$$

Now, $\Delta q = (i_{in} - i_{out}) \, \Delta t$, where i_{in} is the current entering the sphere and i_{out} is the current leaving. Thus,

$$\Delta t = \frac{\Delta q}{i_{in} - i_{out}} = \frac{4\pi\varepsilon_0 r \, \Delta V}{i_{in} - i_{out}}$$

$$= \frac{(0.10 \text{ m})(1000 \text{ V})}{(8.99 \times 10^9 \text{ F/m})(1.0000020 \text{ A} - 1.0000000 \text{ A})} = 5.6 \times 10^{-3} \text{ s}.$$

4. We express the magnitude of the current density vector in SI units by converting the diameter values in mils to inches (by dividing by 1000) and then converting to meters (by multiplying by 0.0254) and finally using

$$J = \frac{i}{A} = \frac{i}{\pi R^2} = \frac{4i}{\pi D^2}.$$

For example, the gauge 14 wire with $D = 64$ mil $= 0.0016$ m is found to have a (maximum safe) current density of $J = 7.2 \times 10^6$ A/m^2. In fact, this is the wire with the largest value of J allowed by the given data. The values of J in SI units are plotted below as a function of their diameters in mils.

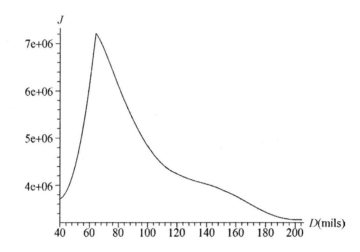

5. (a) The magnitude of the current density is given by $J = nqv_d$, where n is the number of particles per unit volume, q is the charge on each particle, and v_d is the drift speed of the particles. The particle concentration is $n = 2.0 \times 10^8$/cm$^3 = 2.0 \times 10^{14}$ m^{-3}, the charge is

$$q = 2e = 2(1.60 \times 10^{-19} \text{ C}) = 3.20 \times 10^{-19} \text{ C},$$

and the drift speed is 1.0×10^5 m/s. Thus,

$$J = (2 \times 10^{14} / \text{m})(3.2 \times 10^{-19} \text{ C})(1.0 \times 10^5 \text{ m/s}) = 6.4 \text{ A} / \text{m}^2.$$

(b) Since the particles are positively charged the current density is in the same direction as their motion, to the north.

(c) The current cannot be calculated unless the cross-sectional area of the beam is known. Then $i = JA$ can be used.

6. (a) The magnitude of the current density vector is

$$J = \frac{i}{A} = \frac{i}{\pi d^2 / 4} = \frac{4(1.2 \times 10^{-10} \text{ A})}{\pi(2.5 \times 10^{-3} \text{ m})^2} = 2.4 \times 10^{-5} \text{ A} / \text{m}^2.$$

(b) The drift speed of the current-carrying electrons is

$$v_d = \frac{J}{ne} = \frac{2.4 \times 10^{-5} \text{ A/m}^2}{(8.47 \times 10^{28} / \text{m}^3)(1.60 \times 10^{-19} \text{ C})} = 1.8 \times 10^{-15} \text{ m/s}.$$

7. The cross-sectional area of wire is given by $A = \pi r^2$, where r is its radius (half its thickness). The magnitude of the current density vector is $J = i / A = i / \pi r^2$, so

$$r = \sqrt{\frac{i}{\pi J}} = \sqrt{\frac{0.50 \text{ A}}{\pi (440 \times 10^4 \text{ A/m}^2)}} = 1.9 \times 10^{-4} \text{ m}.$$

The diameter of the wire is therefore $d = 2r = 2(1.9 \times 10^{-4} \text{ m}) = 3.8 \times 10^{-4} \text{ m}.$

8. (a) Since $1 \text{ cm}^3 = 10^{-6} \text{ m}^3$, the magnitude of the current density vector is

$$J = nev = \left(\frac{8.70}{10^{-6} \text{ m}^3}\right)(1.60 \times 10^{-19} \text{ C})(470 \times 10^3 \text{ m/s}) = 6.54 \times 10^{-7} \text{ A/m}^2.$$

(b) Although the total surface area of Earth is $4\pi R_E^2$ (that of a sphere), the area to be used in a computation of how many protons in an approximately unidirectional beam (the solar wind) will be captured by Earth is its projected area. In other words, for the beam, the encounter is with a "target" of circular area πR_E^2. The rate of charge transport implied by the influx of protons is

$$i = AJ = \pi R_E^2 J = \pi (6.37 \times 10^6 \text{ m})^2 (6.54 \times 10^{-7} \text{ A/m}^2) = 8.34 \times 10^7 \text{ A}.$$

9. We use $v_d = J/ne = i/Ane$. Thus,

$$t = \frac{L}{v_d} = \frac{L}{i/Ane} = \frac{LAne}{i} = \frac{(0.85 \text{ m})(0.21 \times 10^{-14} \text{ m}^2)(8.47 \times 10^{28}/\text{m}^3)(1.60 \times 10^{-19} \text{ C})}{300 \text{ A}}$$
$$= 8.1 \times 10^2 \text{ s} = 13 \text{ min}.$$

10. We note that the radial width $\Delta r = 10 \text{ μm}$ is small enough (compared to $r = 1.20$ mm) that we can make the approximation

$$\int Br\, 2\pi r\, dr \approx Br\, 2\pi r\, \Delta r.$$

Thus, the enclosed current is $2\pi Br^2 \Delta r = 18.1$ μA. Performing the integral gives the same answer.

11. (a) The current resulting from this non-uniform current density is

$$i = \int_{\text{cylinder}} J_a \, dA = \frac{J_0}{R} \int_0^R r \cdot 2\pi r \, dr = \frac{2}{3} \pi R^2 J_0 = \frac{2}{3} \pi (3.40 \times 10^{-3} \, \text{m})^2 (5.50 \times 10^4 \, \text{A/m}^2) = 1.33 \, \text{A} \, .$$

(b) In this case,

$$i = \int_{\text{cylinder}} J_b \, dA = \int_0^R J_0 \left(1 - \frac{r}{R} \right) 2\pi r \, dr = \frac{1}{3} \pi R^2 J_0 = \frac{1}{3} \pi (3.40 \times 10^{-3} \, \text{m})^2 (5.50 \times 10^4 \, \text{A/m}^2)$$

$$= 0.666 \, \text{A} \, .$$

(c) The result is different from that in part (a) because J_b is higher near the center of the cylinder (where the area is smaller for the same radial interval) and lower outward, resulting in a lower average current density over the cross section and consequently a lower current than that in part (a). So, J_a has its maximum value near the surface of the wire.

12. Assuming $\vec{J}$ is directed along the wire (with no radial flow) we integrate, starting with Eq. 26-4,

$$i = \int |\vec{J}| \, dA = \int_{9R/10}^R (kr^2) 2\pi r \, dr = \frac{1}{2} k\pi \left(R^4 - 0.656 R^4 \right)$$

where $k = 3.0 \times 10^8$ and SI units understood. Therefore, if $R = 0.00200$ m, we obtain $i = 2.59 \times 10^{-3} \, \text{A}$.

13. We find the conductivity of Nichrome (the reciprocal of its resistivity) as follows:

$$\sigma = \frac{1}{\rho} = \frac{L}{RA} = \frac{L}{(V/i)A} = \frac{Li}{VA} = \frac{(1.0 \, \text{m})(4.0 \, \text{A})}{(2.0 \, \text{V})(1.0 \times 10^{-6} \, \text{m}^2)} = 2.0 \times 10^6 \, / \Omega \cdot \text{m}.$$

14. We use $R/L = \rho/A = 0.150$ Ω/km.

(a) For copper $J = i/A = (60.0 \, \text{A})(0.150 \, \Omega/\text{km})/(1.69 \times 10^{-8} \, \Omega \cdot \text{m}) = 5.32 \times 10^5 \, \text{A/m}^2$.

(b) We denote the mass densities as ρ_m. For copper,

$$(m/L)_c = (\rho_m A)_c = (8960 \, \text{kg/m}^3) \, (1.69 \times 10^{-8} \, \Omega \cdot \text{m})/(0.150 \, \Omega/\text{km}) = 1.01 \, \text{kg/m}.$$

(c) For aluminum $J = (60.0 \, \text{A})(0.150 \, \Omega/\text{km})/(2.75 \times 10^{-8} \, \Omega \cdot \text{m}) = 3.27 \times 10^5 \, \text{A/m}^2$.

(d) The mass density of aluminum is

$$(m/L)_a = (\rho_m A)_a = (2700 \, \text{kg/m}^3)(2.75 \times 10^{-8} \, \Omega \cdot \text{m})/(0.150 \, \Omega/\text{km}) = 0.495 \, \text{kg/m}.$$

15. The resistance of the wire is given by $R = \rho L / A$, where ρ is the resistivity of the material, L is the length of the wire, and A is its cross-sectional area. In this case,

$$A = \pi r^2 = \pi \left(0.50 \times 10^{-3} \text{ m}\right)^2 = 7.85 \times 10^{-7} \text{ m}^2.$$

Thus,

$$\rho = \frac{RA}{L} = \frac{\left(50 \times 10^{-3} \Omega\right)\left(7.85 \times 10^{-7} \text{ m}^2\right)}{2.0 \text{ m}} = 2.0 \times 10^{-8} \Omega \cdot \text{m}.$$

16. (a) $i = V/R = 23.0 \text{ V}/15.0 \times 10^{-3} \Omega = 1.53 \times 10^3 \text{ A}$.

(b) The cross-sectional area is $A = \pi r^2 = \frac{1}{4}\pi D^2$. Thus, the magnitude of the current density vector is

$$J = \frac{i}{A} = \frac{4i}{\pi D^2} = \frac{4\left(1.53 \times 10^{-3} \text{ A}\right)}{\pi \left(6.00 \times 10^{-3} \text{ m}\right)^2} = 5.41 \times 10^7 \text{ A}/\text{m}^2.$$

(c) The resistivity is

$$\rho = RA/L = \left(15.0 \times 10^{-3} \Omega\right)(\pi)\left(6.00 \times 10^3 \text{ m}\right)^2 / [4(4.00 \text{ m})] = 10.6 \times 10^{-8} \Omega \cdot \text{m}.$$

(d) The material is platinum.

17. Since the potential difference V and current i are related by $V = iR$, where R is the resistance of the electrician, the fatal voltage is $V = (50 \times 10^{-3} \text{ A})(2000 \Omega) = 100 \text{ V}$.

18. The thickness (diameter) of the wire is denoted by D. We use $R \propto L/A$ (Eq. 26-16) and note that $A = \frac{1}{4}\pi D^2 \propto D^2$. The resistance of the second wire is given by

$$R_2 = R\left(\frac{A_1}{A_2}\right)\left(\frac{L_2}{L_1}\right) = R\left(\frac{D_1}{D_2}\right)^2\left(\frac{L_2}{L_1}\right) = R(2)^2\left(\frac{1}{2}\right) = 2R.$$

19. The resistance of the coil is given by $R = \rho L/A$, where L is the length of the wire, ρ is the resistivity of copper, and A is the cross-sectional area of the wire. Since each turn of wire has length $2\pi r$, where r is the radius of the coil, then

$$L = (250)2\pi r = (250)(2\pi)(0.12 \text{ m}) = 188.5 \text{ m}.$$

If r_w is the radius of the wire itself, then its cross-sectional area is $A = \pi r^2 w = \pi(0.65 \times 10^{-3}$ m$)^2 = 1.33 \times 10^{-6}$ m^2. According to Table 26-1, the resistivity of copper is $1.69 \times 10^{-8}\,\Omega \cdot$m. Thus,

$$R = \frac{\rho L}{A} = \frac{\left(1.69 \times 10^{-8}\,\Omega \cdot \text{m}\right)\left(188.5 \text{ m}\right)}{1.33 \times 10^{-6} \text{ m}^2} = 2.4\,\Omega.$$

20. (a) Since the material is the same, the resistivity ρ is the same, which implies (by Eq. 26-11) that the electric fields (in the various rods) are directly proportional to their current-densities. Thus, J_1: J_2: J_3 are in the ratio 2.5/4/1.5 (see Fig. 26-24). Now the currents in the rods must be the same (they are "in series") so

$$J_1 A_1 = J_3 A_3 \quad \text{and} \quad J_2 A_2 = J_3 A_3 \; .$$

Since $A = \pi r^2$ this leads (in view of the aforementioned ratios) to

$$4r_2^2 = 1.5 r_3^2 \quad \text{and} \quad 2.5 r_1^2 = 1.5 r_3^2 \; .$$

Thus, with $r_3 = 2$ mm, the latter relation leads to $r_1 = 1.55$ mm.

(b) The $4r_2^2 = 1.5 r_3^2$ relation leads to $r_2 = 1.22$ mm.

21. Since the mass density of the material do not change, the volume remains the same. If L_0 is the original length, L is the new length, A_0 is the original cross-sectional area, and A is the new cross-sectional area, then $L_0 A_0 = LA$ and $A = L_0 A_0/L = L_0 A_0/3L_0 = A_0/3$. The new resistance is

$$R = \frac{\rho L}{A} = \frac{\rho 3 L_0}{A_0 / 3} = 9 \frac{\rho L_0}{A_0} = 9 R_0,$$

where R_0 is the original resistance. Thus, $R = 9(6.0\ \Omega) = 54\ \Omega$.

22. The cross-sectional area is $A = \pi r^2 = \pi(0.002$ m$)^2$. The resistivity from Table 26-1 is $\rho = 1.69 \times 10^{-8}\,\Omega$·m. Thus, with $L = 3$ m, Ohm's Law leads to $V = iR = i\rho L/A$, or

$$12 \times 10^{-6}\,\text{V} = i(1.69 \times 10^{-8}\,\Omega\text{·m})(3.0\text{ m})/\,\pi(0.002\text{ m})^2$$

which yields $i = 0.00297$ A or roughly 3.0 mA.

23. The resistance of conductor A is given by

$$R_A = \frac{\rho L}{\pi r_A^2},$$

where r_A is the radius of the conductor. If r_o is the outside diameter of conductor B and r_i is its inside diameter, then its cross-sectional area is $\pi(r_o^2 - r_i^2)$, and its resistance is

$$R_B = \frac{\rho L}{\pi(r_o^2 - r_i^2)}.$$

The ratio is

$$\frac{R_A}{R_B} = \frac{r_o^2 - r_i^2}{r_A^2} = \frac{(1.0\,\text{mm})^2 - (0.50\,\text{mm})^2}{(0.50\,\text{mm})^2} = 3.0.$$

24. The absolute values of the slopes (for the straight-line segments shown in the graph of Fig. 26-26(b)) are equal to the respective electric field magnitudes. Thus, applying Eq. 26-5 and Eq. 26-13 to the three sections of the resistive strip, we have

$$J_1 = \frac{i}{A} = \sigma_1 E_1 = \sigma_1 (0.50 \times 10^3 \text{ V/m})$$

$$J_2 = \frac{i}{A} = \sigma_2 E_2 = \sigma_2 (4.0 \times 10^3 \text{ V/m})$$

$$J_3 = \frac{i}{A} = \sigma_3 E_3 = \sigma_3 (1.0 \times 10^3 \text{ V/m}) \ .$$

We note that the current densities are the same since the values of i and A are the same (see the problem statement) in the three sections, so $J_1 = J_2 = J_3$.

(a) Thus we see that

$$\sigma_1 = 2\sigma_3 = 2\ (3.00 \times 10^7 (\Omega \cdot \text{m})^{-1}) = 6.00 \times 10^7 \ (\Omega \cdot \text{m})^{-1}.$$

(b) Similarly, $\sigma_2 = \sigma_3/4 = (3.00 \times 10^7 (\Omega \cdot \text{m})^{-1})/4 = 7.50 \times 10^6 \ (\Omega \cdot \text{m})^{-1}$.

25. The resistance at operating temperature T is $R = V/i = 2.9$ V/0.30 A = 9.67 Ω. Thus, from $R - R_0 = R_0 \alpha\ (T - T_0)$, we find

$$T = T_0 + \frac{1}{\alpha}\left(\frac{R}{R_0} - 1\right) = 20°\text{C} + \left(\frac{1}{4.5 \times 10^{-3}/\text{K}}\right)\left(\frac{9.67\,\Omega}{1.1\,\Omega} - 1\right) = 1.9 \times 10^3 \ °\text{C}.$$

Since a change in Celsius is equivalent to a change on the Kelvin temperature scale, the value of α used in this calculation is not inconsistent with the other units involved. Table 26-1 has been used.

26. First we find $R = \rho L/A = 2.69 \times 10^{-5}\,\Omega$. Then $i = V/R = 1.115 \times 10^{-4}\,A$. Finally, $\Delta Q = i\,\Delta t = 3.35 \times 10^{-7}\,C$.

27. We use $J = E/\rho$, where E is the magnitude of the (uniform) electric field in the wire, J is the magnitude of the current density, and ρ is the resistivity of the material. The electric field is given by $E = V/L$, where V is the potential difference along the wire and L is the length of the wire. Thus $J = V/L\rho$ and

$$\rho = \frac{V}{LJ} = \frac{115\ V}{(10\ \text{m})(1.4 \times 10^4\ \text{A/m}^2)} = 8.2 \times 10^{-4}\,\Omega \cdot \text{m}.$$

28. We use $J = \sigma E = (n_+ + n_-)ev_d$, which combines Eq. 26-13 and Eq. 26-7.

(a) $J = \sigma E = (2.70 \times 10^{-14}\,/\,\Omega \cdot \text{m})\,(120\ \text{V/m}) = 3.24 \times 10^{-12}\ \text{A/m}^2$.

(b) The drift velocity is

$$v_d = \frac{\sigma E}{(n_+ + n_-)e} = \frac{(2.70 \times 10^{-14}/\Omega \cdot \text{m})(120\ \text{V/m})}{[(620 + 550)/\text{cm}^3](1.60 \times 10^{-19}\ \text{C})} = 1.73\ \text{cm/s}.$$

29. (a) $i = V/R = 35.8\ \text{V}/935\ \Omega = 3.83 \times 10^{-2}\ \text{A}$.

(b) $J = i/A = (3.83 \times 10^{-2}\ \text{A})/(3.50 \times 10^{-4}\ \text{m}^2) = 109\ \text{A/m}^2$.

(c) $v_d = J/ne = (109\ \text{A/m}^2)/[(5.33 \times 10^{22}/\text{m}^3)\,(1.60 \times 10^{-19}\ \text{C})] = 1.28 \times 10^{-2}\ \text{m/s}$.

(d) $E = V/L = 35.8\ \text{V}/0.158\ \text{m} = 227\ \text{V/m}$.

30. We use $R \propto L/A$. The diameter of a 22-gauge wire is 1/4 that of a 10-gauge wire. Thus from $R = \rho L/A$ we find the resistance of 25 ft of 22-gauge copper wire to be

$$R = (1.00\ \Omega)\,(25\ \text{ft}/1000\ \text{ft})(4)^2 = 0.40\ \Omega.$$

31. (a) The current in each strand is $i = 0.750\ \text{A}/125 = 6.00 \times 10^{-3}\ \text{A}$.

(b) The potential difference is $V = iR = (6.00 \times 10^{-3}\ \text{A})\,(2.65 \times 10^{-6}\ \Omega) = 1.59 \times 10^{-8}\ \text{V}$.

(c) The resistance is $R_{\text{total}} = 2.65 \times 10^{-6}\ \Omega/125 = 2.12 \times 10^{-8}\ \Omega$.

32. The number density of conduction electrons in copper is $n = 8.49 \times 10^{28}\,/\text{m}^3$. The electric field in section 2 is $(10.0\ \mu\text{V})/(2.00\ \text{m}) = 5.00\ \mu\text{V/m}$. Since $\rho = 1.69 \times 10^{-8}\,\Omega \cdot \text{m}$ for copper (see Table 26-1) then Eq. 26-10 leads to a current density vector of magnitude

$J_2 = (5.00 \ \mu V/m)/(1.69 \times 10^{-8} \ \Omega \cdot m) = 296 \ A/m^2$ in section 2. Conservation of electric current from section 1 into section 2 implies

$$J_1 A_1 = J_2 A_2$$

$$J_1 (4\pi R^2) = J_2 (\pi R^2)$$

(see Eq. 26-5). This leads to $J_1 = 74 \ A/m^2$. Now, Eq. 26-7 immediately yields

$$v_d = \frac{J_1}{ne} = 5.44 \times 10^{-9} \ m/s$$

for the drift speed of conduction-electrons in section 1.

33. (a) The current i is shown in Fig. 26-29 entering the truncated cone at the left end and leaving at the right. This is our choice of positive x direction. We make the assumption that the current density J at each value of x may be found by taking the ratio i/A where $A = \pi r^2$ is the cone's cross-section area at that particular value of x. The direction of $\vec{J}$ is identical to that shown in the figure for i (our $+x$ direction). Using Eq. 26-11, we then find an expression for the electric field at each value of x, and next find the potential difference V by integrating the field along the x axis, in accordance with the ideas of Chapter 25. Finally, the resistance of the cone is given by $R = V/i$. Thus,

$$J = \frac{i}{\pi r^2} = \frac{E}{\rho}$$

where we must deduce how r depends on x in order to proceed. We note that the radius increases linearly with x, so (with c_1 and c_2 to be determined later) we may write

$$r = c_1 + c_2 x.$$

Choosing the origin at the left end of the truncated cone, the coefficient c_1 is chosen so that $r = a$ (when $x = 0$); therefore, $c_1 = a$. Also, the coefficient c_2 must be chosen so that (at the right end of the truncated cone) we have $r = b$ (when $x = L$); therefore, $c_2 = (b - a)/L$. Our expression, then, becomes

$$r = a + \left(\frac{b-a}{L}\right) x.$$

Substituting this into our previous statement and solving for the field, we find

$$E = \frac{i\rho}{\pi} \left(a + \frac{b-a}{L} x\right)^{-2}.$$

Consequently, the potential difference between the faces of the cone is

$$V = -\int_0^L E\,dx = -\frac{i\rho}{\pi}\int_0^L \left(a + \frac{b-a}{L}x\right)^{-2} dx = \frac{i\rho}{\pi}\frac{L}{b-a}\left(a + \frac{b-a}{L}x\right)^{-1}\Bigg|_0^L$$

$$= \frac{i\rho}{\pi}\frac{L}{b-a}\left(\frac{1}{a} - \frac{1}{b}\right) = \frac{i\rho}{\pi}\frac{L}{b-a}\frac{b-a}{ab} = \frac{i\rho L}{\pi ab}.$$

The resistance is therefore

$$R = \frac{V}{i} = \frac{\rho L}{\pi ab} = \frac{(731\ \Omega\cdot\text{m})(1.94\times10^{-2}\ \text{m})}{\pi(2.00\times10^{-3}\ \text{m})(2.30\times10^{-3}\ \text{m})} = 9.81\times10^5\ \Omega$$

Note that if $b = a$, then $R = \rho L/\pi a^2 = \rho L/A$, where $A = \pi a^2$ is the cross-sectional area of the cylinder.

34. From Eq. 26-25, $\rho \propto \tau^{-1} \propto v_{\text{eff}}$. The connection with v_{eff} is indicated in part (b) of Sample Problem 26-6, which contains useful insight regarding the problem we are working now. According to Chapter 20, $v_{\text{eff}} \propto \sqrt{T}$. Thus, we may conclude that $\rho \propto \sqrt{T}$.

35. (a) Electrical energy is converted to heat at a rate given by

$$P = \frac{V^2}{R},$$

where V is the potential difference across the heater and R is the resistance of the heater. Thus,

$$P = \frac{(120\ \text{V})^2}{14\ \Omega} = 1.0\times10^3\ \text{W} = 1.0\ \text{kW}.$$

(b) The cost is given by $(1.0\,\text{kW})(5.0\text{h})(5.0\text{cents/kW}\cdot\text{h}) = \text{US}\$0.25$.

36. Since $P = iV$, $q = it = Pt/V = (7.0\ \text{W})\ (5.0\ \text{h})\ (3600\ \text{s/h})/9.0\ \text{V} = 1.4 \times 10^4\ \text{C}$.

37. The relation $P = V^2/R$ implies $P \propto V^2$. Consequently, the power dissipated in the second case is

$$P = \left(\frac{1.50\ \text{V}}{3.00\ \text{V}}\right)^2 (0.540\ \text{W}) = 0.135\ \text{W}.$$

38. The resistance is $R = P/i^2 = (100\ \text{W})/(3.00\ \text{A})^2 = 11.1\ \Omega$.

39. (a) The power dissipated, the current in the heater, and the potential difference across the heater are related by $P = iV$. Therefore,

$$i = \frac{P}{V} = \frac{1250 \text{ W}}{115 \text{ V}} = 10.9 \text{ A}.$$

(b) Ohm's law states $V = iR$, where R is the resistance of the heater. Thus,

$$R = \frac{V}{i} = \frac{115 \text{ V}}{10.9 \text{ A}} = 10.6 \text{ }\Omega.$$

(c) The thermal energy E generated by the heater in time $t = 1.0 \text{ h} = 3600 \text{ s}$ is

$$E = Pt = (1250 \text{ W})(3600 \text{ s}) = 4.50 \times 10^6 \text{ J}.$$

40. (a) From $P = V^2/R$ we find $R = V^2/P = (120 \text{ V})^2/500 \text{ W} = 28.8 \text{ }\Omega$.

(b) Since $i = P/V$, the rate of electron transport is

$$\frac{i}{e} = \frac{P}{eV} = \frac{500 \text{ W}}{(1.60 \times 10^{-19} \text{ C})(120 \text{ V})} = 2.60 \times 10^{19} \text{ / s}.$$

41. (a) From $P = V^2/R = AV^2 / \rho L$, we solve for the length:

$$L = \frac{AV^2}{\rho P} = \frac{(2.60 \times 10^{-6} \text{ m}^2)(75.0 \text{ V})^2}{(5.00 \times 10^{-7} \text{ }\Omega \cdot \text{m})(500 \text{ W})} = 5.85 \text{ m}.$$

(b) Since $L \propto V^2$ the new length should be

$$L' = L\left(\frac{V'}{V}\right)^2 = (5.85 \text{ m})\left(\frac{100 \text{ V}}{75.0 \text{ V}}\right)^2 = 10.4 \text{ m}.$$

42. The slopes of the lines yield $P_1 = 8 \text{ mW}$ and $P_2 = 4 \text{ mW}$. Their sum (by energy conservation) must be equal to that supplied by the battery: $P_{batt} = (8 + 4) \text{ mW} = 12 \text{ mW}$.

43. (a) The monthly cost is $(100 \text{ W}) (24 \text{ h/day}) (31 \text{ day/month}) (6 \text{ cents/kW} \cdot \text{h}) = 446 \text{ cents} = \text{US\$}4.46$, assuming a 31-day month.

(b) $R = V^2/P = (120 \text{ V})^2/100 \text{ W} = 144 \text{ }\Omega$.

(c) $i = P/V = 100 \text{ W}/120 \text{ V} = 0.833 \text{ A}$.

44. Assuming the current is along the wire (not radial) we find the current from Eq. 26-4:

$$i = \int |\vec{J}| \, dA = \int_0^R kr^2 \, 2\pi r \, dr = \frac{1}{2} k\pi R^4 = 3.50 \text{ A}$$

where $k = 2.75 \times 10^{10}$ A/m^4 and $R = 0.00300$ m. The rate of thermal energy generation is found from Eq. 26-26: $P = iV = 210$ W. Assuming a steady rate, the thermal energy generated in 40 s is $(210 \text{ J/s})(3600 \text{ s}) = 7.56 \times 10^5$ J.

45. (a) Using Table 26-1 and Eq. 26-10 (or Eq. 26-11), we have

$$|\vec{E}| = \rho |\vec{J}| = \left(1.69 \times 10^{-8} \, \Omega \cdot \text{m}\right) \left(\frac{2.00 \text{ A}}{2.00 \times 10^{-6} \text{ m}^2} \right) = 1.69 \times 10^{-2} \text{ V/m}.$$

(b) Using $L = 4.0$ m, the resistance is found from Eq. 26-16: $R = \rho L/A = 0.0338$ Ω. The rate of thermal energy generation is found from Eq. 26-27:

$$P = i^2 R = (2.00 \text{ A})^2 (0.0338 \text{ } \Omega) = 0.135 \text{ W}.$$

Assuming a steady rate, the thermal energy generated in 30 minutes is $(0.135 \text{ J/s})(30 \times 60\text{s}) = 2.43 \times 10^2$ J.

46. From $P = V^2/R$, we have $R = (5.0 \text{ V})^2/(200 \text{ W}) = 0.125$ Ω. To meet the conditions of the problem statement, we must therefore set

$$\int_0^L 5.00x \, dx = 0.125 \text{ } \Omega$$

Thus,

$$\frac{5}{2} L^2 = 0.125 \qquad \Rightarrow \qquad L = 0.224 \text{ m}.$$

47. (a) We use Eq. 26-16 to compute the resistances in SI units:

$$R_C = \rho_C \frac{L_C}{\pi r_C^2} = (2 \times 10^{-6}) \frac{1}{\pi (0.0005)^2} = 2.5 \text{ } \Omega.$$

The voltage follows from Ohm's law:

$$|V_1 - V_2| = V_C = iR_C = 5.1 \text{V}.$$

(b) Similarly,

$$R_D = \rho_D \frac{L_D}{\pi r_D^2} = (1 \times 10^{-6}) \frac{1}{\pi (0.00025)^2} = 5.1\,\Omega$$

and $|V_2 - V_3| = V_D = iR_D = 10\,\text{V}$.

(c) The power is calculated from Eq. 26-27: $P_C = i^2 R_C = 10\,\text{W}$.

(d) Similarly, $P_D = i^2 R_D = 20\,\text{W}$.

48. (a) Current is the transport of charge; here it is being transported "in bulk" due to the volume rate of flow of the powder. From Chapter 14, we recall that the volume rate of flow is the product of the cross-sectional area (of the stream) and the (average) stream velocity. Thus, $i = \rho A v$ where ρ is the charge per unit volume. If the cross-section is that of a circle, then $i = \rho \pi R^2 v$.

(b) Recalling that a Coulomb per second is an Ampere, we obtain

$$i = (1.1 \times 10^{-3}\,\text{C/m}^3)\pi(0.050\,\text{m})^2(2.0\,\text{m/s}) = 1.7 \times 10^{-5}\,\text{A}.$$

(c) The motion of charge is not in the same direction as the potential difference computed in problem 57 of Chapter 25. It might be useful to think of (by analogy) Eq. 7-48; there, the scalar (dot) product in $P = \vec{F} \cdot \vec{v}$ makes it clear that $P = 0$ if $\vec{F} \perp \vec{v}$. This suggests that a radial potential difference and an axial flow of charge will not together produce the needed transfer of energy (into the form of a spark).

(d) With the assumption that there is (at least) a voltage equal to that computed in problem 60 of Chapter 24, in the proper direction to enable the transference of energy (into a spark), then we use our result from that problem in Eq. 26-26:

$$P = iV = (1.7 \times 10^{-5}\,\text{A})(7.8 \times 10^4\,\text{V}) = 1.3\,\text{W}.$$

(e) Recalling that a Joule per second is a Watt, we obtain (1.3 W)(0.20 s) = 0.27 J for the energy that can be transferred at the exit of the pipe.

(f) This result is greater than the 0.15 J needed for a spark, so we conclude that the spark was likely to have occurred at the exit of the pipe, going into the silo.

49. (a) We are told that $r_B = \frac{1}{2} r_A$ and $L_B = 2L_A$. Thus, using Eq. 26-16,

$$R_B = \rho \frac{L_B}{\pi r_B^2} = \rho \frac{2LA}{\frac{1}{4}\pi r_A^2} = 8R_A = 64\,\Omega.$$

(b) The current densities are assumed uniform.

$$\frac{J_A}{J_B} = \frac{i/\pi r_A^2}{i/\pi r_B^2} = \frac{i/\pi r_A^2}{4i/\pi r_A^2} = 0.25.$$

50. (a) Circular area depends, of course, on r^2, so the horizontal axis of the graph in Fig. 26-33(b) is effectively the same as the area (enclosed at variable radius values), except for a factor of π. The fact that the current increases linearly in the graph means that $i/A = J$ = constant. Thus, the answer is "yes, the current density is uniform."

(b) We find $i/(\pi r^2) = (0.005\ A)/(\pi \times 4 \times 10^{-6}\ m^2) = 398 \approx 4.0 \times 10^2\ A/m^2$.

51. Using $A = \pi r^2$ with $r = 5 \times 10^{-4}$ m with Eq. 26-5 yields

$$|\vec{J}| = \frac{i}{A} = 2.5 \times 10^6\ A/m^2.$$

Then, with $|\vec{E}| = 5.3\ V/m$, Eq. 26-10 leads to

$$\rho = \frac{5.3\ V/m}{2.5 \times 10^6\ A/m^2} = 2.1 \times 10^{-6}\ \Omega \cdot m.$$

52. We find the drift speed from Eq. 26-7:

$$v_d = \frac{|\vec{J}|}{ne} = 1.5 \times 10^{-4}\ m/s.$$

At this (average) rate, the time required to travel $L = 5.0$ m is

$$t = \frac{L}{v_d} = 3.4 \times 10^4\ s.$$

53. (a) Referring to Fig. 26-34, the electric field would point down (towards the bottom of the page) in the strip, which means the current density vector would point down, too (by Eq. 26-11). This implies (since electrons are negatively charged) that the conduction-electrons would be "drifting" upward in the strip.

(b) Eq. 24-6 immediately gives 12 eV, or (using $e = 1.60 \times 10^{-19}$ C) 1.9×10^{-18} J for the work done by the field (which equals, in magnitude, the potential energy change of the electron).

(c) Since the electrons don't (on average) gain kinetic energy as a result of this work done, it is generally dissipated as heat. The answer is as in part (b): 12 eV or 1.9×10^{-18} J.

54. Since 100 cm = 1 m, then $10^4\ cm^2 = 1\ m^2$. Thus,

$$R = \frac{\rho L}{A} = \frac{\left(3.00 \times 10^{-7}\,\Omega \cdot m\right)\left(10.0 \times 10^3\ m\right)}{56.0 \times 10^{-4}\ m^2} = 0.536\ \Omega.$$

55. (a) The charge that strikes the surface in time Δt is given by $\Delta q = i\,\Delta t$, where i is the current. Since each particle carries charge $2e$, the number of particles that strike the surface is

$$N = \frac{\Delta q}{2e} = \frac{i\Delta t}{2e} = \frac{\left(0.25 \times 10^{-6}\,A\right)\left(3.0\,s\right)}{2\left(1.6 \times 10^{-19}\,C\right)} = 2.3 \times 10^{12}.$$

(b) Now let N be the number of particles in a length L of the beam. They will all pass through the beam cross section at one end in time $t = L/v$, where v is the particle speed. The current is the charge that moves through the cross section per unit time. That is, $i = 2eN/t = 2eNv/L$. Thus $N = iL/2ev$. To find the particle speed, we note the kinetic energy of a particle is

$$K = 20\,\text{MeV} = \left(20 \times 10^6\,eV\right)\left(1.60 \times 10^{-19}\,J\,/\,eV\right) = 3.2 \times 10^{-12}\,J.$$

Since $K = \frac{1}{2}mv^2$, then the speed is $v = \sqrt{2\,K/m}$. The mass of an alpha particle is (very nearly) 4 times the mass of a proton, or $m = 4(1.67 \times 10^{-27}\ \text{kg}) = 6.68 \times 10^{-27}\ \text{kg}$, so

$$v = \sqrt{\frac{2\left(3.2 \times 10^{-12}\,J\right)}{6.68 \times 10^{-27}\ \text{kg}}} = 3.1 \times 10^7\ m/s$$

and

$$N = \frac{iL}{2ev} = \frac{\left(0.25 \times 10^{-6}\right)\left(20 \times 10^{-2}\ m\right)}{2\left(1.60 \times 10^{-19}\ C\right)\left(3.1 \times 10^7\ m/s\right)} = 5.0 \times 10^3.$$

(c) We use conservation of energy, where the initial kinetic energy is zero and the final kinetic energy is 20 MeV $= 3.2 \times 10^{-12}$ J. We note, too, that the initial potential energy is $U_i = qV = 2eV$, and the final potential energy is zero. Here V is the electric potential through which the particles are accelerated. Consequently,

$$K_f = U_i = 2eV \quad \Rightarrow \quad V = \frac{K_f}{2e} = \frac{3.2 \times 10^{-12}\,J}{2\left(1.60 \times 10^{-19}\,C\right)} = 1.0 \times 10^7\ V.$$

56. (a) Since $P = i^2 R = J^2 A^2 R$, the current density is

$$J = \frac{1}{A}\sqrt{\frac{P}{R}} = \frac{1}{A}\sqrt{\frac{P}{\rho L / A}} = \sqrt{\frac{P}{\rho L A}}$$

$$= \sqrt{\frac{1.0\,\text{W}}{\pi(3.5\times10^{-5}\,\Omega\cdot\text{m})(2.0\times10^{-2}\,\text{m})(5.0\times10^{-3}\,\text{m})^2}} = 1.3\times10^5\,\text{A}/\text{m}^2.$$

(b) From $P = iV = JAV$ we get

$$V = \frac{P}{AJ} = \frac{P}{\pi r^2 J} = \frac{1.0\,\text{W}}{\pi(5.0\times10^{-3}\,\text{m})^2(1.3\times10^5\,\text{A}/\text{m}^2)} = 9.4\times10^{-2}\,\text{V}.$$

57. Let R_H be the resistance at the higher temperature (800°C) and let R_L be the resistance at the lower temperature (200°C). Since the potential difference is the same for the two temperatures, the power dissipated at the lower temperature is $P_L = V^2/R_L$, and the power dissipated at the higher temperature is $P_H = V^2 / R_H$, so $P_L = (R_H/R_L)P_H$. Now $R_L = R_H + \alpha R_H \Delta T$, where ΔT is the temperature difference $T_L - T_H = -600$ C° = −600 K. Thus,

$$P_L = \frac{R_H}{R_H + \alpha R_H \Delta T}P_H = \frac{P_H}{1 + \alpha \Delta T} = \frac{500\,\text{W}}{1 + (4.0\times10^{-4}/\text{K})(-600\,\text{K})} = 660\,\text{W}.$$

58. (a) We use $P = V^2/R \propto V^2$, which gives $\Delta P \propto \Delta V^2 \approx 2V\,\Delta V$. The percentage change is roughly

$$\Delta P/P = 2\Delta V/V = 2(110 - 115)/115 = -8.6\%.$$

(b) A drop in V causes a drop in P, which in turn lowers the temperature of the resistor in the coil. At a lower temperature R is also decreased. Since $P \propto R^{-1}$ a decrease in R will result in an increase in P, which partially offsets the decrease in P due to the drop in V. Thus, the actual drop in P will be smaller when the temperature dependency of the resistance is taken into consideration.

59. (a) The current is

$$i = \frac{V}{R} = \frac{V}{\rho L / A} = \frac{\pi V d^2}{4\rho L}$$

$$= \frac{\pi(1.20\,\text{V})[(0.0400\,\text{in.})(2.54\times10^{-2}\,\text{m/in.})]^2}{4(1.69\times10^{-8}\,\Omega\cdot\text{m})(33.0\,\text{m})} = 1.74\,\text{A}.$$

(b) The magnitude of the current density vector is

$$|\vec{J}| = \frac{i}{A} = \frac{4i}{\pi d^2} = \frac{4(1.74\ \text{A})}{\pi[(0.0400\ \text{in.})(2.54 \times 10^{-2}\ \text{m/in.})]^2} = 2.15 \times 10^6\ \text{A/m}^2.$$

(c) $E = V/L = 1.20\ \text{V}/33.0\ \text{m} = 3.63 \times 10^{-2}\ \text{V/m}.$

(d) $P = Vi = (1.20\ \text{V})(1.74\ \text{A}) = 2.09\ \text{W}.$

60. (a) Since

$$\rho = RA/L = \pi Rd^2/4L = \pi(1.09 \times 10^{-3}\ \Omega)(5.50 \times 10^{-3}\ \text{m})^2/[4(1.60\ \text{m})] = 1.62 \times 10^{-8}\ \Omega\cdot\text{m},$$

the material is silver.

(b) The resistance of the round disk is

$$R = \rho \frac{L}{A} = \frac{4\rho L}{\pi d^2} = \frac{4(1.62 \times 10^{-8}\ \Omega\cdot\text{m})(1.00 \times 10^{-3}\ \text{m})}{\pi(2.00 \times 10^{-2}\ \text{m})^2} = 5.16 \times 10^{-8}\ \Omega.$$

61. Eq. 26-26 gives the rate of thermal energy production:

$$P = iV = (10.0\,\text{A})(120\,\text{V}) = 1.20\,\text{kW}.$$

Dividing this into the 180 kJ necessary to cook the three hot-dogs leads to the result $t = 150$ s.

62. (a) We denote the copper wire with subscript c and the aluminum wire with subscript a.

$$R = \rho_a \frac{L}{A} = \frac{(2.75 \times 10^{-8}\ \Omega\cdot\text{m})(1.3\,\text{m})}{(5.2 \times 10^{-3}\ \text{m})^2} = 1.3 \times 10^{-3}\ \Omega.$$

(b) Let $R = \rho_c L/(\pi d^2/4)$ and solve for the diameter d of the copper wire:

$$d = \sqrt{\frac{4\rho_c L}{\pi R}} = \sqrt{\frac{4(1.69 \times 10^{-8}\ \Omega\cdot\text{m})(1.3\ \text{m})}{\pi(1.3 \times 10^{-3}\ \Omega)}} = 4.6 \times 10^{-3}\ \text{m}.$$

63. We use $P = i^2 R = i^2 \rho L/A$, or $L/A = P/i^2\rho$.

(a) The new values of L and A satisfy

$$\left(\frac{L}{A}\right)_{\text{new}} = \left(\frac{P}{i^2\rho}\right)_{\text{new}} = \frac{30}{4^2}\left(\frac{P}{i^2\rho}\right)_{\text{old}} = \frac{30}{16}\left(\frac{L}{A}\right)_{\text{old}}.$$

Consequently, $(L/A)_{new} = 1.875(L/A)_{old}$, and

$$L_{new} = \sqrt{1.875}L_{old} = 1.37L_{old} \quad\Rightarrow\quad \frac{L_{new}}{L_{old}} = 1.37 .$$

(b) Similarly, we note that $(LA)_{new} = (LA)_{old}$, and

$$A_{new} = \sqrt{1/1.875}A_{old} = 0.730A_{old} \quad\Rightarrow\quad \frac{A_{new}}{A_{old}} = 0.730 .$$

64. The horsepower required is

$$P = \frac{iV}{0.80} = \frac{(10A)(12\ V)}{(0.80)(746\ W/hp)} = 0.20\ hp.$$

65. We find the current from Eq. 26-26: $i = P/V = 2.00$ A. Then, from Eq. 26-1 (with constant current), we obtain

$$\Delta q = i\Delta t = 2.88 \times 10^4\ C .$$

66. We find the resistance from $A = \pi r^2$ and Eq. 26-16:

$$R = \rho\frac{L}{A} = (1.69 \times 10^{-8})\frac{45}{1.3 \times 10^{-5}} = 0.061\ \Omega .$$

Then the rate of thermal energy generation is found from Eq. 26-28: $P = V^2/R = 2.4$ kW. Assuming a steady rate, the thermal energy generated in 40 s is $(2.4)(40) = 95$ kJ.

67. We find the rate of energy consumption from Eq. 26-28:

$$P = \frac{V^2}{R} = \frac{90^2}{400} = 20.3\ W .$$

Assuming a steady rate, the energy consumed is $(20.3\ J/s)(2.00 \times 3600\ s) = 1.46 \times 10^5$ J.

68. We use Eq. 26-28:

$$R = \frac{V^2}{P} = \frac{200^2}{3000} = 13.3\ \Omega .$$

69. The slope of the graph is $P = 5.0 \times 10^{-4}$ W. Using this in the $P = V^2/R$ relation leads to $V = 0.10$ Vs.

70. The rate at which heat is being supplied is $P = iV = (5.2 \text{ A})(12 \text{ V}) = 62.4 \text{ W}$. Considered on a one-second time-frame, this means 62.4 J of heat are absorbed the liquid each second. Using Eq. 18-16, we find the heat of transformation to be

$$L = \frac{Q}{m} = \frac{62.4 \text{ J}}{21 \times 10^{-6} \text{ kg}} = 3.0 \times 10^{6} \text{ J/kg} \ .$$

71. (a) The current is $4.2 \times 10^{18} \, e$ divided by 1 second. Using $e = 1.60 \times 10^{-19} \text{ C}$ we obtain 0.67 A for the current.

(b) Since the electric field points away from the positive terminal (high potential) and towards the negative terminal (low potential), then the current density vector (by Eq. 26-11) must also point towards the negative terminal.

72. Combining Eq. 26-28 with Eq. 26-16 demonstrates that the power is inversely proportional to the length (when the voltage is held constant, as in this case). Thus, a new length equal to 7/8 of its original value leads to

$$P = \frac{8}{7} \, (2.0 \text{ kW}) = 2.4 \text{ kW} \ \ .$$

73. (a) Since the field is considered to be uniform inside the wire, then its magnitude is, by Eq. 24-42,

$$|\vec{E}| = \frac{|\Delta V|}{L} = \frac{50}{200} = 0.25 \text{ V / m}.$$

Using Eq. 26-11, with $\rho = 1.7 \times 10^{-8} \ \Omega \cdot \text{m}$, we obtain

$$\vec{E} = \rho \vec{J} \Rightarrow \vec{J} = 1.5 \times 10^{7} \ \hat{\text{i}}$$

in SI units (A/m^2).

(b) The electric field points towards lower values of potential (see Eq. 24-40) so $\vec{E}$ is directed towards point B (which we take to be the $\hat{\text{i}}$ direction in our calculation).

74. We use Eq. 26-17: $\rho - \rho_0 = \rho \alpha (T - T_0)$, and solve for T:

$$T = T_0 + \frac{1}{\alpha}\left(\frac{\rho}{\rho_0} - 1 \right) = 20^{\circ}\text{C} + \frac{1}{4.3 \times 10^{-3} \text{ / K}}\left(\frac{58\Omega}{50\Omega} - 1 \right) = 57^{\circ}\text{C}.$$

We are assuming that $\rho / \rho_0 = R/R_0$.

75. (a) With $\rho = 1.69 \times 10^{-8} \ \Omega \cdot \text{m}$ (from Table 26-1) and $L = 1000$ m, Eq. 26-16 leads to

$$A = \rho \frac{L}{R} = (1.69 \times 10^{-8}) \frac{1000}{33} = 5.1 \times 10^{-7} \, \text{m}^2 \quad .$$

Then, $A = \pi r^2$ yields $r = 4.0 \times 10^{-4}$ m; doubling that gives the diameter as 8.1×10^{-4} m.

(b) Repeating the calculation in part (a) with $\rho = 2.75 \times 10^{-8}$ Ω·m leads to a diameter of 1.0×10^{-3} m.

76. (a) The charge q that flows past any cross section of the beam in time Δt is given by $q = i\Delta t$, and the number of electrons is $N = q/e = (i/e) \, \Delta t$. This is the number of electrons that are accelerated. Thus

$$N = \frac{(0.50 \, \text{A})(0.10 \times 10^{-6} \, \text{s})}{1.60 \times 10^{-19} \, \text{C}} = 3.1 \times 10^{11}.$$

(b) Over a long time t the total charge is $Q = nqt$, where n is the number of pulses per unit time and q is the charge in one pulse. The average current is given by $i_{avg} = Q/t = nq$. Now $q = i\Delta t = (0.50 \, \text{A}) (0.10 \times 10^{-6} \, \text{s}) = 5.0 \times 10^{-8}$ C, so

$$i_{avg} = (500 / \text{s})(5.0 \times 10^{-8} \, \text{C}) = 2.5 \times 10^{-5} \, \text{A}.$$

(c) The accelerating potential difference is $V = K/e$, where K is the final kinetic energy of an electron. Since $K = 50$ MeV, the accelerating potential is $V = 50$ kV $= 5.0 \times 10^7$ V. The average power is

$$P_{avg} = i_{avg}V = (2.5 \times 10^{-5} \, \text{A})(5.0 \times 10^7 \, \text{V}) = 1.3 \times 10^3 \, \text{W}.$$

(d) During a pulse the power output is

$$P = iV = (0.50 \, \text{A})(5.0 \times 10^7 \, \text{V}) = 2.5 \times 10^7 \, \text{W}.$$

This is the peak power.

77. The power dissipated is given by the product of the current and the potential difference: $P = iV = (7.0 \times 10^{-3} \, \text{A})(80 \times 10^3 \, \text{V}) = 560$ W.

78. (a) Let ΔT be the change in temperature and κ be the coefficient of linear expansion for copper. Then $\Delta L = \kappa L \, \Delta T$ and

$$\frac{\Delta L}{L} = \kappa \Delta T = (1.7 \times 10^{-5} / \text{K})(1.0°\text{C}) = 1.7 \times 10^{-5}.$$

This is equivalent to 0.0017%. Since a change in Celsius is equivalent to a change on the Kelvin temperature scale, the value of κ used in this calculation is not inconsistent with the other units involved. Incorporating a factor of 2 for the two-dimensional nature of A, the fractional change in area is

$$\frac{\Delta A}{A} = 2\kappa\Delta T = 2(1.7 \times 10^{-5} / \text{K})(1.0^\circ\text{C}) = 3.4 \times 10^{-5}$$

which is 0.0034%. For small changes in the resistivity ρ, length L, and area A of a wire, the change in the resistance is given by

$$\Delta R = \frac{\partial R}{\partial \rho}\Delta\rho + \frac{\partial R}{\partial L}\Delta L + \frac{\partial R}{\partial A}\Delta A.$$

Since $R = \rho L/A$, $\partial R/\partial\rho = L/A = R/\rho$, $\partial R/\partial L = \rho/A = R/L$, and $\partial R/\partial A = -\rho L/A^2 = -R/A$. Furthermore, $\Delta\rho/\rho = \alpha\Delta T$, where α is the temperature coefficient of resistivity for copper (4.3×10^{-3}/K $= 4.3 \times 10^{-3}$/C°, according to Table 26-1). Thus,

$$\frac{\Delta R}{R} = \frac{\Delta\rho}{\rho} + \frac{\Delta L}{L} - \frac{\Delta A}{A} = (\alpha + \kappa - 2\kappa)\Delta T = (\alpha - \kappa)\Delta T$$
$$= (4.3 \times 10^{-3} / \text{C}^\circ - 1.7 \times 10^{-5} / \text{C}^\circ)(1.0\text{C}^\circ) = 4.3 \times 10^{-3}.$$

This is 0.43%, which we note (for the purposes of the next part) is primarily determined by the $\Delta\rho/\rho$ term in the above calculation.

(b) As shown in part (a), the percentage change in L is 0.0017%.

(c) As shown in part (a), the percentage change in A is 0.0034%.

(d) The fractional change in resistivity is much larger than the fractional change in length and area. Changes in length and area affect the resistance much less than changes in resistivity.

79. (a) In Eq. 26-17, we let $\rho = 2\rho_0$ where ρ_0 is the resistivity at $T_0 = 20^\circ\text{C}$:

$$\rho - \rho_0 = 2\rho_0 - \rho_0 = \rho_0\alpha(T - T_0),$$

and solve for the temperature T:

$$T = T_0 + \frac{1}{\alpha} = 20^\circ\text{C} + \frac{1}{4.3 \times 10^{-3} / \text{K}} \approx 250^\circ\text{C}.$$

(b) Since a change in Celsius is equivalent to a change on the Kelvin temperature scale, the value of α used in this calculation is not inconsistent with the other units involved. It is worth noting that this agrees well with Fig. 26-10.

80. Since values from the referred-to graph can only be crudely estimated, we do not present a graph here, but rather indicate a few values. Since $R = V/i$ then we see $R = \infty$ when $i = 0$ (which the graph seems to show throughout the range $-\infty < V < 2$ V) and $V \neq 0$. For voltages values larger than 2 V, the resistance changes rapidly according to the ratio V/i. For instance, $R \approx 3.1/0.002 = 1550 \ \Omega$ when $V = 3.1$ V, and $R \approx 3.8/0.006 = 633 \ \Omega$ when $V = 3.8$ V.

81. (a) $V = iR = i\rho\dfrac{L}{A} = \dfrac{(12 \ \text{A})(1.69 \times 10^{-8} \ \Omega \cdot \text{m})(4.0 \times 10^{-2} \ \text{m})}{\pi(5.2 \times 10^{-3} \ \text{m}/2)^2} = 3.8 \times 10^{-4} \ \text{V}.$

(b) Since it moves in the direction of the electron drift which is against the direction of the current, its tail is negative compared to its head.

(c) The time of travel relates to the drift speed:

$$t = \frac{L}{v_d} = \frac{lAne}{i} = \frac{\pi L d^2 ne}{4i}$$

$$= \frac{\pi(1.0 \times 10^{-2} \ \text{m})(5.2 \times 10^{-3} \ \text{m})^2(8.47 \times 10^{28} / \text{m}^3)(1.60 \times 10^{-19} \ \text{C})}{4(12 \ \text{A})}$$

$$= 238 \ \text{s} = 3 \ \text{min} \ 58 \ \text{s}.$$

82. Using Eq. 7-48 and Eq. 26-27, the rate of change of mechanical energy of the piston-Earth system, mgv, must be equal to the rate at which heat is generated from the coil: $mgv = i^2 R$. Thus

$$v = \frac{i^2 R}{mg} = \frac{(0.240 \ \text{A})^2(550 \Omega)}{(12 \ \text{kg})(9.8 \ \text{m/s}^2)} = 0.27 \ \text{m/s}.$$

83. (a) $i = (n_h + n_e)e = (2.25 \times 10^{15}/\text{s} + 3.50 \times 10^{15}/\text{s}) (1.60 \times 10^{-19} \ \text{C}) = 9.20 \times 10^{-4} \ \text{A}.$

(b) The magnitude of the current density vector is

$$|\vec{J}| = \frac{i}{A} = \frac{9.20 \times 10^{-4} \ \text{A}}{\pi(0.165 \times 10^{-3} \ \text{m})^2} = 1.08 \times 10^4 \ \text{A/m}^2.$$

Chapter 27

1. (a) The cost is $(100 \text{ W} \cdot 8.0 \text{ h}/2.0 \text{ W} \cdot \text{h})$ ($\$0.80$) $= \$3.2 \times 10^2$.

(b) The cost is $(100 \text{ W} \cdot 8.0 \text{ h}/10^3 \text{ W} \cdot \text{h})$ ($\$0.06$) $= \$0.048 = 4.8$ cents.

2. The chemical energy of the battery is reduced by $\Delta E = q\varepsilon$, where q is the charge that passes through in time $\Delta t = 6.0$ min, and ε is the emf of the battery. If i is the current, then $q = i \, \Delta t$ and

$$\Delta E = i\varepsilon \, \Delta t = (5.0 \text{ A})(6.0 \text{ V}) \, (6.0 \text{ min}) \, (60 \text{ s/min}) = 1.1 \times 10^4 \text{ J}.$$

We note the conversion of time from minutes to seconds.

3. If P is the rate at which the battery delivers energy and Δt is the time, then $\Delta E = P \, \Delta t$ is the energy delivered in time Δt. If q is the charge that passes through the battery in time Δt and ε is the emf of the battery, then $\Delta E = q\varepsilon$. Equating the two expressions for ΔE and solving for Δt, we obtain

$$\Delta t = \frac{q\varepsilon}{P} = \frac{(120 \text{ A} \cdot \text{h})(12.0 \text{ V})}{100 \text{ W}} = 14.4 \text{ h}.$$

4. (a) The energy transferred is

$$U = Pt = \frac{\varepsilon^2 t}{r + R} = \frac{(2.0 \text{ V})^2 (2.0 \text{ min}) (60 \text{ s/min})}{1.0\Omega + 5.0\Omega} = 80 \text{ J}.$$

(b) The amount of thermal energy generated is

$$U' = i^2 Rt = \left(\frac{\varepsilon}{r + R}\right)^2 Rt = \left(\frac{2.0 \text{ V}}{1.0\Omega + 5.0\Omega}\right)^2 (5.0\Omega)(2.0 \text{ min})(60 \text{ s/min}) = 67 \text{ J}.$$

(c) The difference between U and U', which is equal to 13 J, is the thermal energy that is generated in the battery due to its internal resistance.

5. (a) Let i be the current in the circuit and take it to be positive if it is to the left in R_1. We use Kirchhoff's loop rule: $\varepsilon_1 - iR_2 - iR_1 - \varepsilon_2 = 0$. We solve for i:

$$i = \frac{\varepsilon_1 - \varepsilon_2}{R_1 + R_2} = \frac{12 \text{ V} - 6.0 \text{ V}}{4.0\Omega + 8.0\Omega} = 0.50 \text{ A}.$$

A positive value is obtained, so the current is counterclockwise around the circuit.

If i is the current in a resistor R, then the power dissipated by that resistor is given by $P = i^2 R$.

(b) For R_1, $P_1 = (0.50 \text{ A})^2 (4.0 \text{ }\Omega) = 1.0$ W,

(c) and for R_2, $P_2 = (0.50 \text{ A})^2 (8.0 \text{ }\Omega) = 2.0$ W.

If i is the current in a battery with emf ε, then the battery supplies energy at the rate $P = i\varepsilon$ provided the current and emf are in the same direction. The battery absorbs energy at the rate $P = i\varepsilon$ if the current and emf are in opposite directions.

(d) For ε_1, $P_1 = (0.50 \text{ A})(12 \text{ V}) = 6.0$ W

(e) and for ε_2, $P_2 = (0.50 \text{ A})(6.0 \text{ V}) = 3.0$ W.

(f) In battery 1 the current is in the same direction as the emf. Therefore, this battery supplies energy to the circuit; the battery is discharging.

(g) The current in battery 2 is opposite the direction of the emf, so this battery absorbs energy from the circuit. It is charging.

6. The current in the circuit is

$$i = (150 \text{ V} - 50 \text{ V})/(3.0 \text{ }\Omega + 2.0 \text{ }\Omega) = 20 \text{ A}.$$

So from $V_Q + 150 \text{ V} - (2.0 \text{ }\Omega)i = V_P$, we get $V_Q = 100 \text{ V} + (2.0 \text{ }\Omega)(20 \text{ A}) - 150 \text{ V} = -10$ V.

7. (a) The potential difference is $V = \varepsilon + ir = 12 \text{ V} + (0.040 \text{ }\Omega)(50 \text{ A}) = 14$ V.

(b) $P = i^2 r = (50 \text{ A})^2 (0.040 \text{ }\Omega) = 1.0 \times 10^2$ W.

(c) $P' = iV = (50 \text{ A})(12 \text{ V}) = 6.0 \times 10^2$ W.

(d) In this case $V = \varepsilon - ir = 12 \text{ V} - (0.040 \text{ }\Omega)(50 \text{ A}) = 10$ V.

(e) $P_r = i^2 r = 1.0 \times 10^2$ W.

8. (a) We solve $i = (\varepsilon_2 - \varepsilon_1)/(r_1 + r_2 + R)$ for R:

$$R = \frac{\varepsilon_2 - \varepsilon_1}{i} - r_1 - r_2 = \frac{3.0 \text{ V} - 2.0 \text{ V}}{1.0 \times 10^{-3} \text{ A}} - 3.0\Omega - 3.0\Omega = 9.9 \times 10^2 \text{ }\Omega.$$

(b) $P = i^2R = (1.0 \times 10^{-3} \text{ A})^2(9.9 \times 10^2 \text{ } \Omega) = 9.9 \times 10^{-4} \text{ W.}$

9. (a) If i is the current and ΔV is the potential difference, then the power absorbed is given by $P = i\,\Delta V$. Thus,

$$\Delta V = \frac{P}{i} = \frac{50 \text{ W}}{1.0 \text{ A}} = 50 \text{ V.}$$

Since the energy of the charge decreases, point A is at a higher potential than point B; that is, $V_A - V_B = 50 \text{ V.}$

(b) The end-to-end potential difference is given by $V_A - V_B = +iR + \varepsilon$, where ε is the emf of element C and is taken to be positive if it is to the left in the diagram. Thus,

$$\varepsilon = V_A - V_B - iR = 50 \text{ V} - (1.0 \text{ A})(2.0 \text{ } \Omega) = 48 \text{ V.}$$

(c) A positive value was obtained for ε, so it is toward the left. The negative terminal is at B.

10. (a) For each wire, $R_{\text{wire}} = \rho L/A$ where $A = \pi r^2$. Consequently, we have

$$R_{\text{wire}} = (1.69 \times 10^{-8})(0.200)/\pi(0.00100)^2 = 0.0011 \text{ } \Omega.$$

The total resistive load on the battery is therefore $2R_{\text{wire}} + 6.00 \text{ } \Omega$. Dividing this into the battery emf gives the current $i = 1.9993$ A. The voltage across the $6.00 \text{ } \Omega$ resistor is therefore $(1.9993 \text{ A})(6.00 \text{ } \Omega) = 11.996 \text{ V} \approx 12 \text{ V.}$

(b) Similarly, we find the voltage-drop across each wire to be 2.15 mV.

(c) $P = i^2R = (1.9993 \text{ A})(6 \text{ } \Omega)^2 = 23.98 \text{ W} \approx 24.0 \text{ W.}$

(d) Similarly, we find the power dissipated in each wire to be 4.30 mW.

11. Let the emf be V. Then $V = iR = i'(R + R')$, where $i = 5.0$ A, $i' = 4.0$ A and $R' = 2.0 \text{ } \Omega$. We solve for R:

$$R = \frac{i'R'}{i - i'} = \frac{(4.0)(2.0)}{5.0 - 4.0} = 8.0 \text{ } \Omega.$$

12. (a) Here we denote the battery emf's as V_1 and V_2. The loop rule gives

$$V_2 - ir_2 + V_1 - ir_1 - iR = 0 \quad \Rightarrow \quad i = \frac{V_2 + V_1}{r_1 + r_2 + R} \quad .$$

The terminal voltage of battery 1 is V_{1T} and (see Fig. 27-4(a)) is easily seen to be equal to $V_1 - ir_1$; similarly for battery 2. Thus,

$$V_{1T} = V_1 - \frac{r_1(V_2 + V_1)}{r_1 + r_2 + R} \quad \text{and} \quad V_{2T} = V_2 - \frac{r_1(V_2 + V_1)}{r_1 + r_2 + R} \; .$$

The problem tells us that V_1 and V_2 each equal 1.20 V. From the graph in Fig. 27-30(b) we see that $V_{2T} = 0$ and $V_{1T} = 0.40$ V for $R = 0.10\ \Omega$. This supplies us (in view of the above relations for terminal voltages) with simultaneous equations, which, when solved, lead to $r_1 = 0.20\ \Omega$.

(b) The simultaneous solution also gives $r_2 = 0.30\ \Omega$.

13. To be as general as possible, we refer to the individual emf's as ε_1 and ε_2 and wait until the latter steps to equate them ($\varepsilon_1 = \varepsilon_2 = \varepsilon$). The batteries are placed in series in such a way that their voltages add; that is, they do not "oppose" each other. The total resistance in the circuit is therefore $R_{total} = R + r_1 + r_2$ (where the problem tells us $r_1 > r_2$), and the "net emf" in the circuit is $\varepsilon_1 + \varepsilon_2$. Since battery 1 has the higher internal resistance, it is the one capable of having a zero terminal voltage, as the computation in part (a) shows.

(a) The current in the circuit is

$$i = \frac{\varepsilon_1 + \varepsilon_2}{r_1 + r_2 + R},$$

and the requirement of zero terminal voltage leads to

$$\varepsilon_1 = ir_1 \Rightarrow R = \frac{\varepsilon_2 r_1 - \varepsilon_1 r_2}{\varepsilon_1} = \frac{(12.0)(0.016) - (12.0)(0.012)}{12.0} = 0.004\ \Omega$$

Note that $R = r_1 - r_2$ when we set $\varepsilon_1 = \varepsilon_2$.

(b) As mentioned above, this occurs in battery 1.

14. (a) Let the emf of the solar cell be ε and the output voltage be V. Thus,

$$V = \varepsilon - ir = \varepsilon - \left(\frac{V}{R}\right)r$$

for both cases. Numerically, we get

$$0.10\ \text{V} = \varepsilon - (0.10\ \text{V}/500\ \Omega)r$$
$$0.15\ \text{V} = \varepsilon - (0.15\ \text{V}/1000\ \Omega)r.$$

We solve for ε and r.

(a) $r = 1.0 \times 10^3$ Ω.

(b) $\varepsilon = 0.30$ V.

(c) The efficiency is

$$\frac{V^2/R}{P_{\text{received}}} = \frac{0.15\,\text{V}}{(1000\,\Omega)\,(5.0\,\text{cm}^2)\,(2.0\times10^{-3}\,\text{W/cm}^2)} = 2.3\times10^{-3} = 0.23\%.$$

15. The potential difference across each resistor is $V = 25.0$ V. Since the resistors are identical, the current in each one is $i = V/R = (25.0\,\text{V})/(18.0\,\Omega) = 1.39$ A. The total current through the battery is then $i_{\text{total}} = 4(1.39\,\text{A}) = 5.56$ A. One might alternatively use the idea of equivalent resistance; for four identical resistors in parallel the equivalent resistance is given by

$$\frac{1}{R_{\text{eq}}} = \sum \frac{1}{R} = \frac{4}{R}.$$

When a potential difference of 25.0 V is applied to the equivalent resistor, the current through it is the same as the total current through the four resistors in parallel. Thus

$$i_{\text{total}} = V/R_{\text{eq}} = 4V/R = 4(25.0\,\text{V})/(18.0\,\Omega) = 5.56\,\text{A}.$$

16. Let the resistances of the two resistors be R_1 and R_2, with $R_1 < R_2$. From the statements of the problem, we have

$$R_1 R_2 / (R_1 + R_2) = 3.0\,\Omega \text{ and } R_1 + R_2 = 16\,\Omega.$$

So R_1 and R_2 must be 4.0 Ω and 12 Ω, respectively.

(a) The smaller resistance is $R_1 = 4.0$ Ω.

(b) The larger resistance is $R_2 = 12$ Ω.

17. We note that two resistors in parallel, R_1 and R_2, are equivalent to

$$R_{12} = \frac{1}{\dfrac{1}{R_1} + \dfrac{1}{R_2}} = \frac{R_1 R_2}{R_1 + R_2}.$$

This situation (Figure 27-32) consists of a parallel pair which are then in series with a single $R_3 = 2.50 \ \Omega$ resistor. Thus, the situation has an equivalent resistance of

$$R_{eq} = R_3 + R_{12} = 2.50 \Omega + \frac{(4.00\Omega)(4.00\Omega)}{4.00\Omega + 4.00\Omega} = 4.50\Omega.$$

18. (a) R_{eq} (FH) = $(10.0 \ \Omega)(10.0 \ \Omega)(5.00 \ \Omega)/[(10.0 \ \Omega)(10.0 \ \Omega) + 2(10.0 \ \Omega)(5.00 \ \Omega)]$ = $2.50 \ \Omega$.

(b) R_{eq} (FG) = $(5.00 \ \Omega) \ R/(R + 5.00 \ \Omega)$, where

$$R = 5.00 \ \Omega + (5.00 \ \Omega)(10.0 \ \Omega)/(5.00 \ \Omega + 10.0 \ \Omega) = 8.33 \ \Omega.$$

So R_{eq} (FG) = $(5.00 \ \Omega)(8.33 \ \Omega)/(5.00 \ \Omega + 8.33 \ \Omega) = 3.13 \ \Omega$.

19. Let i_1 be the current in R_1 and take it to be positive if it is to the right. Let i_2 be the current in R_2 and take it to be positive if it is upward.

(a) When the loop rule is applied to the lower loop, the result is

$$\varepsilon_2 - i_1 R_1 = 0$$

The equation yields

$$i_1 = \frac{\varepsilon_2}{R_1} = \frac{5.0 \ \text{V}}{100 \Omega} = 0.050 \ \text{A}.$$

(b) When it is applied to the upper loop, the result is

$$\varepsilon_1 - \varepsilon_2 - \varepsilon_3 - i_2 R_2 = 0 \ .$$

The equation yields

$$i_2 = \frac{\varepsilon_1 - \varepsilon_2 - \varepsilon_3}{R_2} = \frac{6.0 \ \text{V} - 5.0 \ \text{V} - 4.0 \ \text{V}}{50 \Omega} = -0.060 \ \text{A} \ ,$$

or $|i_2| = 0.060$ A. The negative sign indicates that the current in R_2 is actually downward.

(c) If V_b is the potential at point b, then the potential at point a is $V_a = V_b + \varepsilon_3 + \varepsilon_2$, so $V_a - V_b = \varepsilon_3 + \varepsilon_2 = 4.0 \ \text{V} + 5.0 \ \text{V} = 9.0 \ \text{V}$.

20. The currents i_1, i_2 and i_3 are obtained from Eqs. 27-18 through 27-20:

$$i_1 = \frac{\varepsilon_1(R_2 + R_3) - \varepsilon_2 R_3}{R_1 R_2 + R_2 R_3 + R_1 R_3} = \frac{(4.0 \text{ V})(10\,\Omega + 5.0\,\Omega) - (1.0 \text{ V})(5.0\,\Omega)}{(10\,\Omega)(10\,\Omega) + (10\,\Omega)(5.0\,\Omega) + (10\,\Omega)(5.0\,\Omega)} = 0.275 \text{ A},$$

$$i_2 = \frac{\varepsilon_1 R_3 - \varepsilon_2(R_1 + R_2)}{R_1 R_2 + R_2 R_3 + R_1 R_3} = \frac{(4.0 \text{ V})(5.0\,\Omega) - (1.0 \text{ V})(10\,\Omega + 5.0\,\Omega)}{(10\,\Omega)(10\,\Omega) + (10\,\Omega)(5.0\,\Omega) + (10\,\Omega)(5.0\,\Omega)} = 0.025 \text{ A},$$

$$i_3 = i_2 - i_1 = 0.025 \text{A} - 0.275 \text{A} = -0.250 \text{A}.$$

$V_d - V_c$ can now be calculated by taking various paths. Two examples: from $V_d - i_2 R_2 = V_c$ we get

$$V_d - V_c = i_2 R_2 = (0.0250 \text{ A})(10\ \Omega) = +0.25 \text{ V};$$

from $V_d + i_3 R_3 + \varepsilon_2 = V_c$ we get

$$V_d - V_c = i_3 R_3 - \varepsilon_2 = -(-0.250 \text{ A})(5.0\ \Omega) - 1.0 \text{ V} = +0.25 \text{ V}.$$

21. Let r be the resistance of each of the narrow wires. Since they are in parallel the resistance R of the composite is given by

$$\frac{1}{R} = \frac{9}{r},$$

or $R = r/9$. Now $r = 4\rho\ell / \pi d^2$ and $R = 4\rho\ell / \pi D^2$, where ρ is the resistivity of copper. $A = \pi d^2/4$ was used for the cross-sectional area of a single wire, and a similar expression was used for the cross-sectional area of the thick wire. Since the single thick wire is to have the same resistance as the composite,

$$\frac{4\rho\ell}{\pi D^2} = \frac{4\rho\ell}{9\pi d^2} \Rightarrow D = 3d.$$

22. Using the junction rule ($i_3 = i_1 + i_2$) we write two loop rule equations:

$$10.0 \text{ V} - i_1 R_1 - (i_1 + i_2) R_3 = 0$$

$$5.00 \text{ V} - i_2 R_2 - (i_1 + i_2) R_3 = 0.$$

(a) Solving, we find $i_2 = 0$, and

(b) $i_3 = i_1 + i_2 = 1.25$ A (downward, as was assumed in writing the equations as we did).

23. First, we note V_4, that the voltage across R_4 is equal to the sum of the voltages across R_5 and R_6:

$$V_4 = i_6(R_5 + R_6) = (1.40 \text{ A})(8.00 \text{ } \Omega + 4.00 \text{ } \Omega) = 16.8 \text{ V}.$$

The current through R_4 is then equal to $i_4 = V_4/R_4 = 16.8 \text{ V}/(16.0 \text{ } \Omega) = 1.05 \text{ A}$.

By the junction rule, the current in R_2 is $i_2 = i_4 + i_6 = 1.05 \text{ A} + 1.40 \text{ A} = 2.45 \text{ A}$, so its voltage is $V_2 = (2.00 \text{ } \Omega)(2.45 \text{ A}) = 4.90 \text{ V}$.

The loop rule tells us the voltage across R_3 is $V_3 = V_2 + V_4 = 21.7 \text{ V}$ (implying that the current through it is $i_3 = V_3/(2.00 \text{ } \Omega) = 10.85 \text{ A}$).

The junction rule now gives the current in R_1 as $i_1 = i_2 + i_3 = 2.45 \text{ A} + 10.85 \text{ A} = 13.3 \text{ A}$, implying that the voltage across it is $V_1 = (13.3 \text{ A})(2.00 \text{ } \Omega) = 26.6 \text{ V}$. Therefore, by the loop rule,

$$\varepsilon = V_1 + V_3 = 26.6 \text{ V} + 21.7 \text{ V} = 48.3 \text{ V}.$$

24. (a) By the loop rule, it remains the same. This question is aimed at student conceptualization of voltage; many students apparently confuse the concepts of voltage and current and speak of "voltage going through" a resistor – which would be difficult to rectify with the conclusion of this problem.

(b) The loop rule still applies, of course, but (by the junction rule and Ohm's law) the voltages across R_1 and R_3 (which were the same when the switch was open) are no longer equal. More current is now being supplied by the battery which means more current is in R_3, implying its voltage-drop has increased (in magnitude). Thus, by the loop rule (since the battery voltage has not changed) the voltage across R_1 has decreased a corresponding amount. When the switch was open, the voltage across R_1 was 6.0 V (easily seen from symmetry considerations). With the switch closed, R_1 and R_2 are equivalent (by Eq. 27-24) to 3.0 Ω, which means the total load on the battery is 9.0 Ω. The current therefore is 1.33 A which implies the voltage-drop across R_3 is 8.0 V. The loop rule then tells us that voltage-drop across R_1 is 12 V – 8.0 V = 4.0 V. This is a decrease of 2.0 volts from the value it had when the switch was open.

25. The voltage difference across R_3 is $V_3 = \varepsilon R' /(R' + 2.00 \text{ } \Omega)$, where

$$R' = (5.00 \text{ } \Omega R)/(5.00 \text{ } \Omega + R_3).$$

Thus,

$$P_3 = \frac{V_3^2}{R_3} = \frac{1}{R_3}\left(\frac{\varepsilon R'}{R' + 2.00 \text{ } \Omega}\right)^2 = \frac{1}{R_3}\left(\frac{\varepsilon}{1 + 2.00 \text{ } \Omega/R'}\right)^2 = \frac{\varepsilon^2}{R_3}\left[1 + \frac{(2.00 \text{ } \Omega)(5.00 \text{ } \Omega + R)}{(5.00 \text{ } \Omega) R_3}\right]^{-2}$$

$$\equiv \frac{\varepsilon^2}{f(R_3)}$$

where we use the equivalence symbol $\equiv$ to define the expression $f(R_3)$. To maximize P_3 we need to minimize the expression $f(R_3)$. We set

$$\frac{df(R_3)}{dR_3} = -\frac{4.00\,\Omega^2}{R_3^2} + \frac{49}{25} = 0$$

to obtain $R_3 = \sqrt{(4.00\,\Omega^2)(25)/49} = 1.43\,\Omega$.

26. (a) The voltage across $R_3 = 6.0\,\Omega$ is $V_3 iR_3 = (6.0\,\text{A})\,(6.0\,\Omega) = 36\,\text{V}$. Now, the voltage across $R_1 = 2.0\,\Omega$ is

$$(V_A - V_B) - V_3 = 78 - 36 = 42\,\text{V},$$

which implies the current is $i_1 = (42\,\text{V})/(2.0\,\Omega) = 21\,\text{A}$. By the junction rule, then, the current in $R_2 = 4.0\,\Omega$ is $i_2 = i_1 - i = 21 - 6.0 = 15\,\text{A}$. The total power dissipated by the resistors is (using Eq. 26-27)

$$i_1^2\,(2.0\,\Omega) + i_2^2\,(4.0\,\Omega) + i^2\,(6.0\,\Omega) = 1998\,\text{W} \approx 2.0\,\text{kW} \ .$$

By contrast, the power supplied (externally) to this section is $P_A = i_A\,(V_A - V_B)$ where $i_A = i_1 = 21\,\text{A}$. Thus, $P_A = 1638\,\text{W}$. Therefore, the "Box" must be providing energy.

(b) The rate of supplying energy is $(1998 - 1638)\,\text{W} = 3.6 \times 10^2\,\text{W}$.

27. (a) We note that the R_1 resistors occur in series pairs, contributing net resistance $2R_1$ in each branch where they appear. Since $\varepsilon_2 = \varepsilon_3$ and $R_2 = 2R_1$, from symmetry we know that the currents through ε_2 and ε_3 are the same: $i_2 = i_3 = i$. Therefore, the current through ε_1 is $i_1 = 2i$. Then from $V_b - V_a = \varepsilon_2 - iR_2 = \varepsilon_1 + (2R_1)(2i)$ we get

$$i = \frac{\varepsilon_2 - \varepsilon_1}{4R_1 + R_2} = \frac{4.0\,\text{V} - 2.0\,\text{V}}{4(1.0\,\Omega) + 2.0\,\Omega} = 0.33\,\text{A}.$$

Therefore, the current through ε_1 is $i_1 = 2i = 0.67\,\text{A}$.

(b) The direction of i_1 is downward.

(c) The current through ε_2 is $i_2 = 0.33\,\text{A}$.

(d) The direction of i_2 is upward.

(e) From part (a), we have $i_3 = i_2 = 0.33\,\text{A}$.

(f) The direction of i_3 is also upward.

(g) $V_a - V_b = -iR_2 + \varepsilon_2 = -(0.333 \text{ A})(2.0 \text{ }\Omega) + 4.0 \text{ V} = 3.3 \text{ V}$.

28. (a) For typing convenience, we denote the emf of battery 2 as V_2 and the emf of battery 1 as V_1. The loop rule (examining the left-hand loop) gives $V_2 + i_1 R_1 - V_1 = 0$. Since V_1 is held constant while V_2 and i_1 vary, we see that this expression (for large enough V_2) will result in a negative value for i_1 – so the downward sloping line (the line that is dashed in Fig. 27-41(b)) must represent i_1. It appears to be zero when $V_2 = 6$ V. With $i_1 = 0$, our loop rule gives $V_1 = V_2$ which implies that $V_1 = 6.0$ V.

(b) At $V_2 = 2$ V (in the graph) it appears that $i_1 = 0.2$ A. Now our loop rule equation (with the conclusion about V_1 found in part (a)) gives $R_1 = 20 \text{ }\Omega$.

(c) Looking at the point where the upward-sloping i_2 line crosses the axis (at $V_2 = 4$ V), we note that $i_1 = 0.1$ A there and that the loop rule around the right-hand loop should give

$$V_1 - i_1 R_1 = i_1 R_2 \quad \text{when} \quad i_1 = 0.1 \text{ A} \quad \text{and} \quad i_2 = 0.$$

This leads directly to $R_2 = 40 \text{ }\Omega$.

29. Let the resistors be divided into groups of n resistors each, with all the resistors in the same group connected in series. Suppose there are m such groups that are connected in parallel with each other. Let R be the resistance of any one of the resistors. Then the equivalent resistance of any group is nR, and R_{eq}, the equivalent resistance of the whole array, satisfies

$$\frac{1}{R_{\text{eq}}} = \sum_1^m \frac{1}{nR} = \frac{m}{nR}.$$

Since the problem requires $R_{\text{eq}} = 10 \text{ }\Omega = R$, we must select $n = m$. Next we make use of Eq. 27-16. We note that the current is the same in every resistor and there are $n \cdot m = n^2$ resistors, so the maximum total power that can be dissipated is $P_{\text{total}} = n^2 P$, where $P = 1.0$ W is the maximum power that can be dissipated by any one of the resistors. The problem demands $P_{\text{total}} \geq 5.0P$, so n^2 must be at least as large as 5.0. Since n must be an integer, the smallest it can be is 3. The least number of resistors is $n^2 = 9$.

30. (a) R_2, R_3 and R_4 are in parallel. By finding a common denominator and simplifying, the equation $1/R = 1/R_2 + 1/R_3 + 1/R_4$ gives an equivalent resistance of

$$R = \frac{R_2 R_3 R_4}{R_2 R_3 + R_2 R_4 + R_3 R_4} = \frac{(50.0\Omega)(50.0\Omega)(75.0\Omega)}{(50.0\Omega)(50.0\Omega) + (50.0\Omega)(75.0\Omega) + (50.0\Omega)(75.0\Omega)}$$
$$= 18.8\Omega.$$

Thus, considering the series contribution of resistor R_1, the equivalent resistance for the network is $R_{eq} = R_1 + R = 100\ \Omega + 18.8\ \Omega = 118.8\ \Omega \approx 119\ \Omega$.

(b) $i_1 = \varepsilon/R_{eq} = 6.0\ \text{V}/(118.8\ \Omega) = 5.05 \times 10^{-2}\ \text{A}$.

(c) $i_2 = (\varepsilon - V_1)/R_2 = (\varepsilon - i_1R_1)/R_2 = [6.0\text{V} - (5.05 \times 10^{-2}\ \text{A})(100\Omega)]/50\ \Omega = 1.90 \times 10^{-2}\ \text{A}$.

(d) $i_3 = (\varepsilon - V_1)/R_3 = i_2R_2/R_3 = (1.90 \times 10^{-2}\ \text{A})(50.0\ \Omega/50.0\ \Omega) = 1.90 \times 10^{-2}\ \text{A}$.

(e) $i_4 = i_1 - i_2 - i_3 = 5.05 \times 10^{-2}\ \text{A} - 2(1.90 \times 10^{-2}\ \text{A}) = 1.25 \times 10^{-2}\ \text{A}$.

31. (a) The batteries are identical and, because they are connected in parallel, the potential differences across them are the same. This means the currents in them are the same. Let i be the current in either battery and take it to be positive to the left. According to the junction rule the current in R is $2i$ and it is positive to the right. The loop rule applied to either loop containing a battery and R yields

$$\varepsilon - ir - 2iR = 0 \Rightarrow i = \frac{\varepsilon}{r + 2R}.$$

The power dissipated in R is

$$P = (2i)^2 R = \frac{4\varepsilon^2 R}{(r + 2R)^2}.$$

We find the maximum by setting the derivative with respect to R equal to zero. The derivative is

$$\frac{dP}{dR} = \frac{4\varepsilon^2}{(r + 2R)^3} - \frac{16\varepsilon^2 R}{(r + 2R)^3} = \frac{4\varepsilon^2(r - 2R)}{(r + 2R)^3}.$$

The derivative vanishes (and P is a maximum) if $R = r/2$. With $r = 0.300\ \Omega$, we have $R = 0.150\ \Omega$.

(b) We substitute $R = r/2$ into $P = 4\varepsilon^2 R/(r + 2R)^2$ to obtain

$$P_{max} = \frac{4\varepsilon^2(r/2)}{[r + 2(r/2)]^2} = \frac{\varepsilon^2}{2r} = \frac{(12.0\ \text{V})^2}{2(0.300\ \Omega)} = 240\ \text{W}.$$

32. (a) By symmetry, when the two batteries are connected in parallel the current i going through either one is the same. So from $\varepsilon = ir + (2i)R$ with $r = 0.200\ \Omega$ and $R = 2.00r$, we get

$$i_R = 2i = \frac{2\varepsilon}{r+2R} = \frac{2(12.0\text{V})}{0.200\Omega + 2(0.400\Omega)} = 24.0 \text{ A.}$$

(b) When connected in series $2\varepsilon - i_R r - i_R r - i_R R = 0$, or $i_R = 2\varepsilon/(2r + R)$.

$$i_R = 2i = \frac{2\varepsilon}{2r+R} = \frac{2(12.0\text{V})}{2(0.200\Omega) + 0.400\Omega} = 30.0 \text{ A.}$$

(c) In series, since $R > r$.

(d) If $R = r/2.00$, then for parallel connection,

$$i_R = 2i = \frac{2\varepsilon}{r+2R} = \frac{2(12.0\text{V})}{0.200\Omega + 2(0.100\Omega)} = 60.0 \text{ A.}$$

(e) For series connection, we have

$$i_R = 2i = \frac{2\varepsilon}{2r+R} = \frac{2(12.0\text{V})}{2(0.200\Omega) + 0.100\Omega} = 48.0 \text{ A.}$$

(f) In parallel, since $R < r$.

33. (a) We first find the currents. Let i_1 be the current in R_1 and take it to be positive if it is to the right. Let i_2 be the current in R_2 and take it to be positive if it is to the left. Let i_3 be the current in R_3 and take it to be positive if it is upward. The junction rule produces

$$i_1 + i_2 + i_3 = 0 \ .$$

The loop rule applied to the left-hand loop produces

$$\varepsilon_1 - i_1 R_1 + i_3 R_3 = 0$$

and applied to the right-hand loop produces

$$\varepsilon_2 - i_2 R_2 + i_3 R_3 = 0.$$

We substitute $i_3 = -i_2 - i_1$, from the first equation, into the other two to obtain

$$\varepsilon_1 - i_1 R_1 - i_2 R_3 - i_1 R_3 = 0$$

and

$$\varepsilon_2 - i_2 R_2 - i_2 R_3 - i_1 R_3 = 0.$$

Solving the above equations yield

$$i_1 = \frac{\varepsilon_1(R_2 + R_3) - \varepsilon_2 R_3}{R_1 R_2 + R_1 R_3 + R_2 R_3} = \frac{(3.00 \text{ V})(2.00 \,\Omega + 5.00 \,\Omega) - (1.00 \text{ V})(5.00 \,\Omega)}{(4.00 \,\Omega)(2.00 \,\Omega) + (4.00 \,\Omega)(5.00\Omega) + (2.00 \,\Omega)(5.00 \,\Omega)} = 0.421 \text{ A}.$$

$$i_2 = \frac{\varepsilon_2(R_1 + R_3) - \varepsilon_1 R_3}{R_1 R_2 + R_1 R_3 + R_2 R_3} = \frac{(1.00 \text{ V})(4.00 \,\Omega + 5.00 \,\Omega) - (3.00 \text{ V})(5.00 \,\Omega)}{(4.00 \,\Omega)(2.00 \,\Omega) + (4.00 \,\Omega)(5.00\Omega) + (2.00 \,\Omega)(5.00 \,\Omega)} = -0.158 \text{ A}.$$

$$i_3 = -\frac{\varepsilon_2 R_1 + \varepsilon_1 R_2}{R_1 R_2 + R_1 R_3 + R_2 R_3} = -\frac{(1.00 \text{ V})(4.00 \,\Omega) + (3.00 \text{ V})(2.00 \,\Omega)}{(4.00 \,\Omega)(2.00 \,\Omega) + (4.00 \,\Omega)(5.00\Omega) + (2.00 \,\Omega)(5.00 \,\Omega)} = -0.263 \text{ A}.$$

Note that the current i_3 in R_3 is actually downward and the current i_2 in R_2 is to the right. The current i_1 in R_1 is to the right.

(a) The power dissipated in R_1 is $P_1 = i_1^2 R_1 = (0.421 \text{A})^2 (4.00 \,\Omega) = 0.709 \text{ W}$.

(b) The power dissipated in R_2 is $P_2 = i_2^2 R_2 = (-0.158 \text{A})^2 (2.00 \,\Omega) = 0.0499 \text{ W} \approx 0.050 \text{ W}$.

(c) The power dissipated in R_3 is $P_3 = i_3^2 R_3 = (-0.263 \text{A})^2 (5.00 \,\Omega) = 0.346 \text{ W}$.

(d) The power supplied by ε_1 is $i_3 \varepsilon_1 = (0.421 \text{ A})(3.00 \text{ V}) = 1.26 \text{ W}$.

(e) The power "supplied" by ε_2 is $i_2 \varepsilon_2 = (-0.158 \text{ A})(1.00 \text{ V}) = -0.158 \text{ W}$. The negative sign indicates that ε_2 is actually absorbing energy from the circuit.

34. (a) When $R_3 = 0$ all the current passes through R_1 and R_3 and avoids R_2 altogether. Since that value of the current (through the battery) is 0.006 A (see Fig. 27-45(b)) for $R_3 = 0$ then (using Ohm's law) $R_1 = (12 \text{ V})/(0.006 \text{ A}) = 2.0 \times 10^3 \,\Omega$.

(b) When $R_3 = \infty$ all the current passes through R_1 and R_2 and avoids R_3 altogether. Since that value of the current (through the battery) is 0.002 A (stated in problem) for $R_3 = \infty$ then (using Ohm's law) $R_2 = (12 \text{ V})/(0.002 \text{ A}) - R_1 = 4.0 \times 10^3 \,\Omega$.

35. (a) The copper wire and the aluminum sheath are connected in parallel, so the potential difference is the same for them. Since the potential difference is the product of the current and the resistance, $i_C R_C = i_A R_A$, where i_C is the current in the copper, i_A is the current in the aluminum, R_C is the resistance of the copper, and R_A is the resistance of the aluminum. The resistance of either component is given by $R = \rho L/A$, where ρ is the resistivity, L is the length, and A is the cross-sectional area. The resistance of the copper wire is $R_C = \rho_C L/\pi a^2$, and the resistance of the aluminum sheath is $R_A = \rho_A L/\pi (b^2 - a^2)$.

We substitute these expressions into $i_C R_C = i_A R_A$, and cancel the common factors L and π to obtain

$$\frac{i_C \rho_C}{a^2} = \frac{i_A \rho_A}{b^2 - a^2}.$$

We solve this equation simultaneously with $i = i_C + i_A$, where i is the total current. We find

$$i_C = \frac{r_C^2 \rho_C i}{\left(r_A^2 - r_C^2\right)\rho_C + r_C^2 \rho_A}$$

and

$$i_A = \frac{\left(r_A^2 - r_C^2\right)\rho_C i}{\left(r_A^2 - r_C^2\right)\rho_C + r_C^2 \rho_A}.$$

The denominators are the same and each has the value

$$\left(b^2 - a^2\right)\rho_C + a^2 \rho_A = \left[\left(0.380 \times 10^{-3}\ \text{m}\right)^2 - \left(0.250 \times 10^{-3}\ \text{m}\right)^2\right]\left(1.69 \times 10^{-8}\ \Omega\cdot\text{m}\right)$$

$$+ \left(0.250 \times 10^{-3}\ \text{m}\right)^2 \left(2.75 \times 10^{-8}\ \Omega\cdot\text{m}\right)$$

$$= 3.10 \times 10^{-15}\ \Omega\cdot\text{m}^3.$$

Thus,

$$i_C = \frac{\left(0.250 \times 10^{-3}\ \text{m}\right)^2 \left(2.75 \times 10^{-8}\ \Omega\cdot\text{m}\right)\left(2.00\ \text{A}\right)}{3.10 \times 10^{-15}\ \Omega\cdot\text{m}^3} = 1.11\ \text{A}$$

(b) and

$$i_A = \frac{\left[\left(0.380 \times 10^{-3}\ \text{m}\right)^2 - \left(0.250 \times 10^{-3}\ \text{m}\right)^2\right]\left(1.69 \times 10^{-8}\ \Omega\cdot\text{m}\right)\left(2.00\ \text{A}\right)}{3.10 \times 10^{-15}\ \Omega\cdot\text{m}^3} = 0.893\ \text{A}.$$

(c) Consider the copper wire. If V is the potential difference, then the current is given by $V = i_C R_C = i_C \rho_C L / \pi\, a^2$, so

$$L = \frac{\pi a^2 V}{i_C \rho_C} = \frac{(\pi)\left(0.250 \times 10^{-3}\ \text{m}\right)^2 \left(12.0\ \text{V}\right)}{\left(1.11\ \text{A}\right)\left(1.69 \times 10^{-8}\ \Omega\cdot\text{m}\right)} = 126\ \text{m}.$$

36. (a) We use $P = \varepsilon^2/R_{eq}$, where

$$R_{eq} = 7.00\,\Omega + \frac{(12.0\,\Omega)(4.00\,\Omega)R}{(12.0\,\Omega)(4.0\,\Omega)+(12.0\,\Omega)R+(4.00\,\Omega)R}.$$

Put $P = 60.0$ W and $\varepsilon = 24.0$ V and solve for R: $R = 19.5\,\Omega$.

(b) Since $P \propto R_{eq}$, we must minimize R_{eq}, which means $R = 0$.

(c) Now we must maximize R_{eq}, or set $R = \infty$.

(d) Since $R_{eq,\,min} = 7.00\,\Omega$, $P_{max} = \varepsilon^2/R_{eq,\,min} = (24.0\text{ V})^2/7.00\,\Omega = 82.3$ W.

(e) Since

$$R_{eq,\,max} = 7.00\,\Omega + (12.0\,\Omega)(4.00\,\Omega)/(12.0\,\Omega + 4.00\,\Omega) = 10.0\,\Omega,$$

$P_{min} = \varepsilon^2/R_{eq,\,max} = (24.0\text{ V})^2/10.0\,\Omega = 57.6$ W.

37. (a) The current in R_1 is given by

$$i_1 = \frac{\varepsilon}{R_1 + R_2 R_3/(R_2 + R_3)} = \frac{5.0\text{ V}}{2.0\Omega+(4.0\Omega)(6.0\Omega)/(4.0\Omega+6.0\Omega)} = 1.14\text{ A}.$$

Thus

$$i_3 = \frac{\varepsilon - V_1}{R_3} = \frac{\varepsilon - i_1 R_1}{R_3} = \frac{5.0\text{ V} - (1.14\text{ A})(2.0\Omega)}{6.0\Omega} = 0.45\text{ A}.$$

(b) We simply interchange subscripts 1 and 3 in the equation above. Now

$$i_3 = \frac{\varepsilon}{R_3 + (R_2 R_1/(R_2 + R_1))} = \frac{5.0\text{ V}}{6.0\Omega + ((2.0\Omega)(4.0\Omega)/(2.0\Omega + 4.0\Omega))} = 0.6818\text{ A}$$

and

$$i_1 = \frac{5.0\text{ V} - (0.6818\text{ A})(6.0\Omega)}{2.0\Omega} = 0.45\text{ A},$$

the same as before.

38. (a) Since $i = \varepsilon/(r + R_{ext})$ and $i_{max} = \varepsilon/r$, we have $R_{ext} = R(i_{max}/i - 1)$ where $r = 1.50$ V/1.00 mA $= 1.50 \times 10^3 \ \Omega$. Thus,

$$R_{ext} = (1.5 \times 10^3 \ \Omega)(1/0.100 - 1) = 1.35 \times 10^4 \ \Omega.$$

(b) $R_{ext} = (1.5 \times 10^3 \ \Omega)(1/0.500 - 1) = 1.50 \times 10^3 \ \Omega.$

(c) $R_{ext} = (1.5 \times 10^3 \ \Omega)(1/0.900 - 1) = 167 \ \Omega.$

(d) Since $r = 20.0 \ \Omega + R$, $R = 1.50 \times 10^3 \ \Omega - 20.0 \ \Omega = 1.48 \times 10^3 \ \Omega.$

39. The current in R_2 is i. Let i_1 be the current in R_1 and take it to be downward. According to the junction rule the current in the voltmeter is $i - i_1$ and it is downward. We apply the loop rule to the left-hand loop to obtain

$$\varepsilon - iR_2 - i_1R_1 - ir = 0.$$

We apply the loop rule to the right-hand loop to obtain

$$i_1R_1 - (i - i_1)R_V = 0.$$

The second equation yields

$$i = \frac{R_1 + R_V}{R_V} i_1.$$

We substitute this into the first equation to obtain

$$\varepsilon - \frac{(R_2 + r)(R_1 + R_V)}{R_V} i_1 + R_1 i_1 = 0.$$

This has the solution

$$i_1 = \frac{\varepsilon R_V}{(R_2 + r)(R_1 + R_V) + R_1 R_V}.$$

The reading on the voltmeter is

$$i_1 R_1 = \frac{\varepsilon R_V R_1}{(R_2 + r)(R_1 + R_V) + R_1 R_V} = \frac{(3.0\text{V})(5.0 \times 10^3 \ \Omega)(250\Omega)}{(300\Omega + 100\Omega)(250\Omega + 5.0 \times 10^3 \ \Omega) + (250\Omega)(5.0 \times 10^3 \ \Omega)}$$

$$= 1.12\text{V}.$$

The current in the absence of the voltmeter can be obtained by taking the limit as R_V becomes infinitely large. Then

$$i_1 R_1 = \frac{\varepsilon R_1}{R_1 + R_2 + r} = \frac{(3.0\,\text{V})(250\,\Omega)}{250\,\Omega + 300\,\Omega + 100\,\Omega} = 1.15\,\text{V}.$$

The fractional error is $(1.12 - 1.15)/(1.15) = -0.030$, or -3.0%.

40. (a) $\varepsilon = V + ir = 12\,\text{V} + (10.0\,\text{A})(0.0500\,\Omega) = 12.5\,\text{V}$.

(b) Now $\varepsilon = V' + (i_{\text{motor}} + 8.00\,\text{A})r$, where $V' = i'_A R_{\text{light}} = (8.00\,\text{A})(12.0\,\text{V}/10\,\text{A}) = 9.60\,\text{V}$. Therefore,

$$i_{\text{motor}} = \frac{\varepsilon - V'}{r} - 8.00\,\text{A} = \frac{12.5\,\text{V} - 9.60\,\text{V}}{0.0500\,\Omega} - 8.00\,\text{A} = 50.0\,\text{A}.$$

41. Since the current in the ammeter is i, the voltmeter reading is $V' = V + i\,R_A = i\,(R + R_A)$, or $R = V'/i - R_A = R' - R_A$, where $R' = V'/i$ is the apparent reading of the resistance.

Now, from the lower loop of the circuit diagram, the current through the voltmeter is $i_V = \varepsilon/(R_{\text{eq}} + R_0)$, where

$$\frac{1}{R_{\text{eq}}} = \frac{1}{R_V} + \frac{1}{R_A + R} \quad\Rightarrow\quad R_{\text{eq}} = \frac{R_V(R + R_A)}{R_V + R + R_A} = \frac{(300\,\Omega)(85.0\,\Omega + 3.00\,\Omega)}{300\,\Omega + 85.0\,\Omega + 3.00\,\Omega} = 68.0\,\Omega.$$

The voltmeter reading is then

$$V' = i_V R_{\text{eq}} = \frac{\varepsilon R_{\text{eq}}}{R_{\text{eq}} + R_0} = \frac{(12.0\,\text{V})(68.0\,\Omega)}{68.0\,\Omega + 100\,\Omega} = 4.86\,\text{V}.$$

(a) The ammeter reading is

$$i = \frac{V'}{R + R_A} = \frac{4.86\,\text{V}}{85.0\,\Omega + 3.00\,\Omega} = 0.0552\,\text{A}.$$

(b) As shown above, the voltmeter reading is $V' = 4.86\,\text{V}$.

(c) $R' = V'/i = 4.86\,\text{V}/(5.52 \times 10^{-2}\,\text{A}) = 88.0\,\Omega$.

(d) Since $R = R' - R_A$, if R_A is decreased, the difference between R' and R decreases. In fact, when $R_A = 0$, $R' = R$.

42. The currents in R and R_V are i and $i' - i$, respectively. Since $V = iR = (i' - i)R_V$ we have, by dividing both sides by V, $1 = (i'/V - i/V)R_V = (1/R' - 1/R)R_V$. Thus,

$$\frac{1}{R} = \frac{1}{R'} - \frac{1}{R_V} \quad \Rightarrow R' = \frac{RR_V}{R+R_V}.$$

The equivalent resistance of the circuit is $R_{eq} = R_A + R_0 + R' = R_A + R_0 + \dfrac{RR_V}{R+R_V}$.

(a) The ammeter reading is

$$i' = \frac{\varepsilon}{R_{eq}} = \frac{\varepsilon}{R_A + R_0 + R_V\, R/(R+R_V)} = \frac{12.0\,\text{V}}{3.00\Omega + 100\Omega + (300\Omega)\,(85.0\Omega)/(300\Omega + 85.0\Omega)}$$

$$= 7.09 \times 10^{-2}\,\text{A}.$$

(b) The voltmeter reading is

$$V = \varepsilon - i'\,(R_A + R_0) = 12.0\,\text{V} - (0.0709\,\text{A})\,(103.00\,\Omega) = 4.70\,\text{V}.$$

(c) The apparent resistance is $R' = V/i' = 4.70\,\text{V}/(7.09 \times 10^{-2}\,\text{A}) = 66.3\,\Omega$.

(d) If R_V is increased, the difference between R and R' decreases. In fact, $R' \to R$ as $R_V \to \infty$.

43. Let i_1 be the current in R_1 and R_2, and take it to be positive if it is toward point a in R_1. Let i_2 be the current in R_s and R_x, and take it to be positive if it is toward b in R_s. The loop rule yields $(R_1 + R_2)i_1 - (R_x + R_s)i_2 = 0$. Since points a and b are at the same potential, $i_1 R_1 = i_2 R_s$. The second equation gives $i_2 = i_1 R_1/R_s$, which is substituted into the first equation to obtain

$$(R_1 + R_2)i_1 = (R_x + R_s)\frac{R_1}{R_s}i_1 \Rightarrow R_x = \frac{R_2 R_s}{R_1}.$$

44. (a) We use $q = q_0 e^{-t/\tau}$, or $t = \tau \ln (q_0/q)$, where $\tau = RC$ is the capacitive time constant. Thus, $t_{1/3} = \tau \ln[q_0/(2q_0/3)] = \tau \ln(3/2) = 0.41\,\tau$, or $t_{1/3}/\tau = 0.41$.

(b) $t_{2/3} = \tau \ln[q_0/(q_0/3)] = \tau \ln 3 = 1.1\,\tau$, or $t_{2/3}/\tau = 1.1$.

45. During charging, the charge on the positive plate of the capacitor is given by

$$q = C\varepsilon\left(1 - e^{-t/\tau}\right),$$

where C is the capacitance, ε is applied emf, and $\tau = RC$ is the capacitive time constant. The equilibrium charge is $q_{eq} = C\varepsilon$. We require $q = 0.99 q_{eq} = 0.99 C\varepsilon$, so

$$0.99 = 1 - e^{-t/\tau}.$$

Thus, $e^{-t/\tau} = 0.01$. Taking the natural logarithm of both sides, we obtain $t/\tau = -\ln 0.01 = 4.61$ or $t = 4.61\,\tau$.

46. (a) $\tau = RC = (1.40 \times 10^6\,\Omega)(1.80 \times 10^{-6}\,\text{F}) = 2.52$ s.

(b) $q_o = \varepsilon C = (12.0\,\text{V})(1.80\,\mu\text{F}) = 21.6\,\mu\text{C}$.

(c) The time t satisfies $q = q_0(1 - e^{-t/RC})$, or

$$t = RC \ln\left(\frac{q_0}{q_0 - q}\right) = (2.52\,\text{s}) \ln\left(\frac{21.6\,\mu\text{C}}{21.6\,\mu\text{C} - 16.0\,\mu\text{C}}\right) = 3.40\,\text{s}.$$

47. (a) The voltage difference V across the capacitor is $V(t) = \varepsilon(1 - e^{-t/RC})$. At $t = 1.30\,\mu\text{s}$ we have $V(t) = 5.00$ V, so $5.00\,\text{V} = (12.0\,\text{V})(1 - e^{-1.30\,\mu\text{s}/RC})$, which gives

$$\tau = (1.30\,\mu\text{s})/\ln(12/7) = 2.41\,\mu\text{s}.$$

(b) $C = \tau/R = 2.41\,\mu\text{s}/15.0\,\text{k}\Omega = 161$ pF.

48. Here we denote the battery emf as V. Then the requirement stated in the problem that the resistor voltage be equal to the capacitor voltage becomes $iR = V_{\text{cap}}$, or

$$Ve^{-t/RC} = V(1 - e^{-t/RC})$$

where Eqs. 27-34 and 27-35 have been used. This leads to $t = RC \ln 2$, or $t = 0.208$ ms.

49. (a) The potential difference V across the plates of a capacitor is related to the charge q on the positive plate by $V = q/C$, where C is capacitance. Since the charge on a discharging capacitor is given by $q = q_0\,e^{-t/\tau}$, this means $V = V_0\,e^{-t/\tau}$ where V_0 is the initial potential difference. We solve for the time constant τ by dividing by V_0 and taking the natural logarithm:

$$\tau = -\frac{t}{\ln(V/V_0)} = -\frac{10.0\,\text{s}}{\ln[(1.00\,\text{V})/(100\,\text{V})]} = 2.17\,\text{s}.$$

(b) At $t = 17.0$ s, $t/\tau = (17.0\,\text{s})/(2.17\,\text{s}) = 7.83$, so

$$V = V_0 e^{-t/\tau} = (100\,\text{V})e^{-7.83} = 3.96 \times 10^{-2}\,\text{V}.$$

50. The potential difference across the capacitor varies as a function of time t as $V(t) = V_0 e^{-t/RC}$. Using $V = V_0/4$ at $t = 2.0$ s, we find

$$R = \frac{t}{C \ln(V_0/V)} = \frac{2.0\,\mathrm{s}}{(2.0 \times 10^{-6}\,\mathrm{F}) \ln 4} = 7.2 \times 10^5\,\Omega.$$

51. (a) The initial energy stored in a capacitor is given by $U_C = q_0^2 / 2C$, where C is the capacitance and q_0 is the initial charge on one plate. Thus

$$q_0 = \sqrt{2CU_C} = \sqrt{2(1.0 \times 10^{-6}\,\mathrm{F})(0.50\,\mathrm{J})} = 1.0 \times 10^{-3}\,\mathrm{C}.$$

(b) The charge as a function of time is given by $q = q_0 e^{-t/\tau}$, where τ is the capacitive time constant. The current is the derivative of the charge

$$i = -\frac{dq}{dt} = \frac{q_0}{\tau} e^{-t/\tau},$$

and the initial current is $i_0 = q_0/\tau$. The time constant is

$$\tau = RC = (1.0 \times 10^{-6}\,\mathrm{F})(1.0 \times 10^6\,\Omega) = 1.0\,\mathrm{s}.$$

Thus $i_0 = (1.0 \times 10^{-3}\,\mathrm{C})/(1.0\,\mathrm{s}) = 1.0 \times 10^{-3}\,\mathrm{A}$.

(c) We substitute $q = q_0 e^{-t/\tau}$ into $V_C = q/C$ to obtain

$$V_C = \frac{q_0}{C} e^{-t/\tau} = \left(\frac{1.0 \times 10^{-3}\,\mathrm{C}}{1.0 \times 10^{-6}\,\mathrm{F}} \right) e^{-t/1.0\,\mathrm{s}} = (1.0 \times 10^3\,\mathrm{V}) e^{-1.0t},$$

where t is measured in seconds.

(d) We substitute $i = (q_0/\tau) e^{-t/\tau}$ into $V_R = iR$ to obtain

$$V_R = \frac{q_0 R}{\tau} e^{-t/\tau} = \frac{(1.0 \times 10^{-3}\,\mathrm{C})(1.0 \times 10^6\,\Omega)}{1.0\,\mathrm{s}} e^{-t/1.0\,\mathrm{s}} = (1.0 \times 10^3\,\mathrm{V}) e^{-1.0t},$$

where t is measured in seconds.

(e) We substitute $i = (q_0/\tau) e^{-t/\tau}$ into $P = i^2 R$ to obtain

$$P = \frac{q_0^2 R}{\tau^2} e^{-2t/\tau} = \frac{(1.0 \times 10^{-3}\,\mathrm{C})^2 (1.0 \times 10^6\,\Omega)}{(1.0\,\mathrm{s})^2} e^{-2t/1.0\,\mathrm{s}} = (1.0\,\mathrm{W}) e^{-2.0t},$$

where t is again measured in seconds.

52. The time it takes for the voltage difference across the capacitor to reach V_L is given by $V_L = \varepsilon\left(1 - e^{-t/RC}\right)$. We solve for R:

$$R = \frac{t}{C\ln\left[\varepsilon/(\varepsilon - V_L)\right]} = \frac{0.500\,\mathrm{s}}{\left(0.150 \times 10^{-6}\,\mathrm{F}\right)\ln\left[95.0\,\mathrm{V}/(95.0\,\mathrm{V} - 72.0\,\mathrm{V})\right]} = 2.35 \times 10^6\,\Omega$$

where we used $t = 0.500$ s given (implicitly) in the problem.

53. At $t = 0$ the capacitor is completely uncharged and the current in the capacitor branch is as it would be if the capacitor were replaced by a wire. Let i_1 be the current in R_1 and take it to be positive if it is to the right. Let i_2 be the current in R_2 and take it to be positive if it is downward. Let i_3 be the current in R_3 and take it to be positive if it is downward. The junction rule produces $i_1 = i_2 + i_3$, the loop rule applied to the left-hand loop produces

$$\varepsilon - i_1 R_1 - i_2 R_2 = 0\ ,$$

and the loop rule applied to the right-hand loop produces

$$i_2 R_2 - i_3 R_3 = 0\ .$$

Since the resistances are all the same we can simplify the mathematics by replacing R_1, R_2, and R_3 with R.

(a) Solving the three simultaneous equations, we find

$$i_1 = \frac{2\varepsilon}{3R} = \frac{2\left(1.2 \times 10^3\,\mathrm{V}\right)}{3\left(0.73 \times 10^6\,\Omega\right)} = 1.1 \times 10^{-3}\,\mathrm{A}$$

(b) $i_2 = \dfrac{\varepsilon}{3R} = \dfrac{1.2 \times 10^3\,\mathrm{V}}{3\left(0.73 \times 10^6\,\Omega\right)} = 5.5 \times 10^{-4}\,\mathrm{A}.$

(c) $i_3 = i_2 = 5.5 \times 10^{-4}\,\mathrm{A}.$

At $t = \infty$ the capacitor is fully charged and the current in the capacitor branch is 0. Thus, $i_1 = i_2$, and the loop rule yields

$$\varepsilon - i_1 R_1 - i_1 R_2 = 0\ .$$

(d) The solution is

$$i_1 = \frac{\varepsilon}{2R} = \frac{1.2 \times 10^3 \, \text{V}}{2\left(0.73 \times 10^6 \, \Omega\right)} = 8.2 \times 10^{-4} \, \text{A}.$$

(e) $i_2 = i_1 = 8.2 \times 10^{-4} \, \text{A}.$

(f) As stated before, the current in the capacitor branch is $i_3 = 0$.

We take the upper plate of the capacitor to be positive. This is consistent with current flowing into that plate. The junction equation is $i_1 = i_2 + i_3$, and the loop equations are

$$\varepsilon - i_1 R - i_2 R = 0 \quad \text{and} \quad -\frac{q}{C} - i_3 R + i_2 R = 0.$$

We use the first equation to substitute for i_1 in the second and obtain $\varepsilon - 2i_2 R - i_3 R = 0$. Thus $i_2 = (\varepsilon - i_3 R)/2R$. We substitute this expression into the third equation above to obtain $-(q/C) - (i_3 R) + (\varepsilon/2) - (i_3 R/2) = 0$. Now we replace i_3 with dq/dt to obtain

$$\frac{3R}{2}\frac{dq}{dt} + \frac{q}{C} = \frac{\varepsilon}{2}.$$

This is just like the equation for an *RC* series circuit, except that the time constant is $\tau = 3RC/2$ and the impressed potential difference is $\varepsilon/2$. The solution is

$$q = \frac{C\varepsilon}{2}\left(1 - e^{-2t/3RC}\right).$$

The current in the capacitor branch is

$$i_3(t) = \frac{dq}{dt} = \frac{\varepsilon}{3R}e^{-2t/3RC}.$$

The current in the center branch is

$$i_2(t) = \frac{\varepsilon}{2R} - \frac{i_3}{2} = \frac{\varepsilon}{2R} - \frac{\varepsilon}{6R}e^{-2t/3RC} = \frac{\varepsilon}{6R}\left(3 - e^{-2t/3RC}\right)$$

and the potential difference across R_2 is

$$V_2(t) = i_2 R = \frac{\varepsilon}{6}\left(3 - e^{-2t/3RC}\right).$$

(g) For $t = 0, e^{-2t/3RC}$ is 1 and $V_2 = \varepsilon/3 = (1.2 \times 10^3 \text{ V})/3 = 4.0 \times 10^2 \text{ V}$.

(h) For $t = \infty, e^{-2t/3RC}$ is 0 and $V_2 = \varepsilon/2 = (1.2 \times 20^3 \text{ V})/2 = 6.0 \times 10^2 \text{ V}$.

(i) A plot of V_2 as a function of time is shown in the following graph.

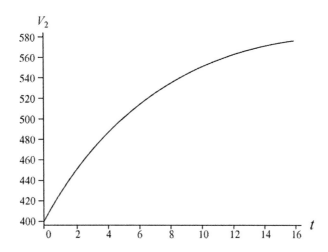

54. In the steady state situation, the capacitor voltage will equal the voltage across $R_2 = 15 \text{ k}\Omega$:

$$V_0 = R_2 \frac{\varepsilon}{R_1 + R_2} = (15.0 \text{k}\Omega)\left(\frac{20.0 \text{V}}{10.0 \text{k}\Omega + 15.0 \text{k}\Omega}\right) = 12.0 \text{V}.$$

Now, multiplying Eq. 27-39 by the capacitance leads to $V = V_0 e^{-t/RC}$ describing the voltage across the capacitor (and across $R_2 = 15.0 \text{ k}\Omega$) after the switch is opened (at $t = 0$). Thus, with $t = 0.00400$ s, we obtain

$$V = (12)e^{-0.004/(15000)(0.4 \times 10^{-6})} = 6.16 \text{V}.$$

Therefore, using Ohm's law, the current through R_2 is $6.16/15000 = 4.11 \times 10^{-4}$ A.

55. (a) The charge on the positive plate of the capacitor is given by

$$q = C\varepsilon(1 - e^{-t/\tau}),$$

where ε is the emf of the battery, C is the capacitance, and τ is the time constant. The value of τ is

$$\tau = RC = (3.00 \times 10^6 \ \Omega)(1.00 \times 10^{-6} \text{ F}) = 3.00 \text{ s}.$$

At $t = 1.00$ s, $t/\tau = (1.00\text{ s})/(3.00\text{ s}) = 0.333$ and the rate at which the charge is increasing is

$$\frac{dq}{dt} = \frac{C\varepsilon}{\tau}e^{-t/\tau} = \frac{(1.00\times10^{-6})(4.00\text{ V})}{3.00\text{ s}}e^{-0.333} = 9.55\times10^{-7}\text{ C/s}.$$

(b) The energy stored in the capacitor is given by $U_C = \dfrac{q^2}{2C}$, and its rate of change is

$$\frac{dU_C}{dt} = \frac{q}{C}\frac{dq}{dt}.$$

Now

$$q = C\varepsilon\left(1-e^{-t/\tau}\right) = (1.00\times10^{-6})(4.00\text{ V})\left(1-e^{-0.333}\right) = 1.13\times10^{-6}\text{ C},$$

so

$$\frac{dU_C}{dt} = \left(\frac{1.13\times10^{-6}\text{ C}}{1.00\times10^{-6}\text{ F}}\right)(9.55\times10^{-7}\text{ C/s}) = 1.08\times10^{-6}\text{ W}.$$

(c) The rate at which energy is being dissipated in the resistor is given by $P = i^2R$. The current is 9.55×10^{-7} A, so

$$P = (9.55\times10^{-7}\text{ A})^2(3.00\times10^{6}\ \Omega) = 2.74\times10^{-6}\text{ W}.$$

(d) The rate at which energy is delivered by the battery is

$$i\varepsilon = (9.55\times10^{-7}\text{ A})(4.00\text{ V}) = 3.82\times10^{-6}\text{ W}.$$

The energy delivered by the battery is either stored in the capacitor or dissipated in the resistor. Conservation of energy requires that $i\varepsilon = (q/C)\,(dq/dt) + i^2R$. Except for some round-off error the numerical results support the conservation principle.

56. We apply Eq. 27-39 to each capacitor, demand their initial charges are in a ratio of 3:2 as described in the problem, and solve for the time: we obtain

$$t = \frac{\ln\!\left(\frac{3}{2}\right)}{\left(\dfrac{1}{R_2C_2} - \dfrac{1}{R_1C_1}\right)} = 162\ \mu\text{s}.$$

57. We use the result of problem 50: $R = t/[C \ln(V_0/V)]$.

(a) Then, for $t_{min} = 10.0\ \mu s$

$$R_{min} = \frac{10.0\ \mu s}{(0.220\ \mu F)\ln(5.00/0.800)} = 24.8\ \Omega.$$

(b) For $t_{max} = 6.00$ ms,

$$R_{max} = \left(\frac{6.00\ ms}{10.0\ \mu s}\right)(24.8\ \Omega) = 1.49 \times 10^4\ \Omega,$$

where in the last equation we used $\tau = RC$.

58. (a) We denote $L = 10$ km and $\alpha = 13\ \Omega/km$. Measured from the east end we have

$$R_1 = 100\ \Omega = 2\alpha(L - x) + R,$$

and measured from the west end $R_2 = 200\ \Omega = 2\alpha x + R$. Thus,

$$x = \frac{R_2 - R_1}{4\alpha} + \frac{L}{2} = \frac{200\Omega - 100\Omega}{4(13\Omega/km)} + \frac{10\ km}{2} = 6.9\ km.$$

(b) Also, we obtain

$$R = \frac{R_1 + R_2}{2} - \alpha L = \frac{100\Omega + 200\Omega}{2} - (13\Omega/km)(10\ km) = 20\Omega.$$

59. (a) From symmetry we see that the current through the top set of batteries (i) is the same as the current through the second set. This implies that the current through the $R = 4.0\ \Omega$ resistor at the bottom is $i_R = 2i$. Thus, with r denoting the internal resistance of each battery (equal to 4.0 Ω) and ε denoting the 20 V emf, we consider one loop equation (the outer loop), proceeding counterclockwise:

$$3(\varepsilon - ir) - (2i)R = 0.$$

This yields $i = 3.0$ A. Consequently, $i_R = 6.0$ A.

(b) The terminal voltage of each battery is $\varepsilon - ir = 8.0$ V.

(c) Using Eq. 27-17, we obtain $P = i\varepsilon = (3)(20) = 60$ W.

(d) Using Eq. 26-27, we have $P = i^2 r = 36$ W.

60. The equivalent resistance in Fig. 27-59 (with n parallel resistors) is

$$R_{eq} = R + \frac{R}{n} = \frac{(n+1)R}{n}.$$

The current in the battery in this case should be $i_n = \dfrac{V_{battery}}{R_{eq}} = \dfrac{n\,V_{battery}}{(n+1)R}$. If there were $n+1$ parallel resistors, then

$$i_{n+1} = \frac{(n+1)V_{battery}}{(n+2)R}.$$

For the relative increase to be 0.0125 ($=$ 1/80), we require

$$\frac{i_{n+1} - i_n}{i_n} = \frac{i_{n+1}}{i_n} - 1 = \frac{\left(\dfrac{n+1}{n+2}\right)}{\left(\dfrac{n}{n+1}\right)} - 1 = \frac{1}{80}.$$

This leads to the second-degree equation $n^2 + 2n - 80 = (n+10)(n-8) = 0$. Clearly the only physically interesting solution to this is $n = 8$. Thus, there are eight resistors in parallel (as well as that resistor in series shown towards the bottom) in Fig. 27-59.

61. (a) The magnitude of the current density vector is

$$J_A = \frac{i}{A} = \frac{V}{(R_1 + R_2)A} = \frac{4V}{(R_1 + R_2)\pi D^2} = \frac{4(60.0V)}{\pi(0.127\Omega + 0.729\Omega)(2.60\times10^{-3}m)^2}$$

$$= 1.32\times10^7 \text{ A/m}^2.$$

(b) $V_A = V R_1/(R_1 + R_2) = (60.0 \text{ V})(0.127 \text{ }\Omega)/(0.127 \text{ }\Omega + 0.729 \text{ }\Omega) = 8.90 \text{ V}$,

(c) The resistivity for A is

$$\rho_A = R_A A/L_A = \pi R_A D^2/4L_A = \pi(0.127\Omega)(2.60\times10^{-3}\text{ m})^2/\left[4(40.0\text{m})\right] = 1.69\times10^{-8}\Omega\cdot\text{m}.$$

So A is made of copper.

(d) $J_B = J_A = 1.32\times10^7 \text{ A/m}^2.$

(e) $V_B = V - V_A = 60.0 \text{ V} - 8.9 \text{ V} = 51.1 \text{ V}$.

(f) The resistivity for B is $\rho_B = 9.68\times10^{-8} \Omega\cdot\text{m}$, so B is made of iron.

62. Line 1 has slope $R_1 = 6$ kΩ. Line 2 has slope $R_2 = 4$ kΩ. Line 3 has slope $R_3 = 2$ kΩ. The parallel pair equivalence is $R_{12} = R_1 R_2/(R_1+R_2) = 2.4$ kΩ. That in series with R_3 gives an equivalence of 4.4 kΩ. The current through the battery is therefore (6 V)/(4.4 kΩ) and the voltage drop across R_3 is (6 V)(2 kΩ)/(4.4 kΩ) = 2.73 V. Subtracting this (because of the loop rule) from the battery voltage leaves us with the voltage across R_2. Then Ohm's law gives the current through R_2: (6 V $-$ 2.73 V)/(4 kΩ) = 0.82 mA

63. (a) Since $R_{tank} = 140\ \Omega, i = 12\,\mathrm{V}/(10\Omega + 140\Omega) = 8.0 \times 10^{-2}$ A.

(b) Now, $R_{tank} = (140\ \Omega + 20\ \Omega)/2 = 80\ \Omega$, so $i = 12$ V/(10 Ω + 80 Ω) = 0.13 A.

(c) When full, $R_{tank} = 20\ \Omega$ so $i = 12$ V/(10 Ω + 20 Ω) = 0.40 A.

64. (a) The loop rule leads to a voltage-drop across resistor 3 equal to 5.0 V (since the total drop along the upper branch must be 12 V). The current there is consequently $i = (5.0$ V$)/(200\ \Omega) = 25$ mA. Then the resistance of resistor 1 must be $(2.0$ V$)/i = 80\ \Omega$.

(b) Resistor 2 has the same voltage-drop as resistor 3; its resistance is 200 Ω, also.

65. (a) Here we denote the battery emf as V. See Fig. 27-4(a): $V_T = V - ir$.

(b) Doing a least squares fit for the V_T versus i values listed, we obtain

$$V_T = 13.61 - 0.0599i$$

which implies $V = 13.6$ volts.

(c) It also implies the internal resistance is 0.060 Ω.

66. (a) The loop rule (proceeding counterclockwise around the right loop) leads to $\varepsilon_2 - i_1 R_1 = 0$ (where i_1 was assumed downward). This yields $i_1 = 0.0600$ A.

(b) The direction of i_1 is downward.

(c) The loop rule (counterclockwise around the left loop) gives

$$\left(+\varepsilon_1\right) + \left(+i_1 R_1\right) + \left(-i_2 R_2\right) = 0$$

where i_2 has been assumed leftward. This yields $i_3 = 0.180$ A.

(d) A positive value of i_3 implies that our assumption on the direction is correct, i.e., it flows leftward.

(e) The junction rule tells us that the current through the 12 V battery is 0.180 + 0.0600 = 0.240 A.

(f) The direction is upward.

67. (a) The charge q on the capacitor as a function of time is $q(t) = (\varepsilon C)(1 - e^{-t/RC})$, so the charging current is $i(t) = dq/dt = (\varepsilon/R)e^{-t/RC}$. The energy supplied by the emf is then

$$U = \int_0^\infty \varepsilon i \, dt = \frac{\varepsilon^2}{R} \int_0^\infty e^{-t/RC} \, dt = C\varepsilon^2 = 2U_C$$

where $U_C = \frac{1}{2} C\varepsilon^2$ is the energy stored in the capacitor.

(b) By directly integrating $i^2 R$ we obtain

$$U_R = \int_0^\infty i^2 R \, dt = \frac{\varepsilon^2}{R} \int_0^\infty e^{-2t/RC} \, dt = \frac{1}{2} C\varepsilon^2.$$

68. (a) Using Eq. 27-4, we take the derivative of the power $P = i^2 R$ with respect to R and set the result equal to zero:

$$\frac{dP}{dR} = \frac{d}{dR} \left(\frac{\varepsilon^2 R}{(R+r)^2} \right) = \frac{\varepsilon^2(r-R)}{(R+r)^3} = 0$$

which clearly has the solution $R = r$.

(b) When $R = r$, the power dissipated in the external resistor equals

$$P_{max} = \left. \frac{\varepsilon^2 R}{(R+r)^2} \right|_{R=r} = \frac{\varepsilon^2}{4r}.$$

69. Here we denote the battery emf as V. Eq. 27-30 leads to

$$i = \frac{V}{R} - \frac{q}{RC} = \frac{12}{4} - \frac{8}{(4)(4)} = 2.5 \text{ A}.$$

70. The equivalent resistance of the series pair of $R_3 = R_4 = 2.0\ \Omega$ is $R_{34} = 4.0\ \Omega$, and the equivalent resistance of the parallel pair of $R_1 = R_2 = 4.0\ \Omega$ is $R_{12} = 2.0\ \Omega$. Since the voltage across R_{34} must equal that across R_{12}:

$$V_{34} = V_{12} \quad \Rightarrow \quad i_{34} R_{34} = i_{12} R_{12} \quad \Rightarrow \quad i_{34} = \frac{1}{2} i_{12}$$

This relation, plus the junction rule condition $I = i_{12} + i_{34} = 6.00$ A leads to the solution $i_{12} = 4.0$ A. It is clear by symmetry that $i_1 = i_{12}/2 = 2.00$ A.

71. (a) The work done by the battery relates to the potential energy change:

$$q\Delta V = eV = e(12.0\,\text{V}) = 12.0 \text{ eV}.$$

(b) $P = iV = neV = (3.40 \times 10^{18}/\text{s})(1.60 \times 10^{-19}\,\text{C})(12.0\text{ V}) = 6.53$ W.

72. (a) Since $P = \varepsilon^2/R_{eq}$, the higher the power rating the smaller the value of R_{eq}. To achieve this, we can let the low position connect to the larger resistance (R_1), middle position connect to the smaller resistance (R_2), and the high position connect to both of them in parallel.

(b) For $P = 300$ W, $R_{eq} = R_1R_2/(R_1 + R_2) = (144\ \Omega)R_2/(144\ \Omega + R_2) = (120\text{ V})^2/300$ W. We obtain $R_2 = 72\ \Omega$.

(c) For $P = 100$ W, $R_{eq} = R_1 = \varepsilon^2/P = (120\text{ V})^2/100$ W $= 144\ \Omega$;

73. The internal resistance of the battery is $r = (12$ V -11.4 V$)/50$ A $= 0.012\ \Omega < 0.020\ \Omega$, so the battery is OK. The resistance of the cable is

$$R = 3.0 \text{ V}/50 \text{ A} = 0.060\ \Omega > 0.040\ \Omega,$$

so the cable is defective.

74. (a) If S_1 is closed, and S_2 and S_3 are open, then $i_a = \varepsilon/2R_1 = 120$ V$/40.0\ \Omega = 3.00$ A.

(b) If S_3 is open while S_1 and S_2 remain closed, then

$$R_{eq} = R_1 + R_1\,(R_1 + R_2)\,/(2R_1 + R_2) = 20.0\ \Omega + (20.0\ \Omega) \times (30.0\ \Omega)/(50.0\ \Omega) = 32.0\ \Omega,$$

so $i_a = \varepsilon/R_{eq} = 120$ V$/32.0\ \Omega = 3.75$ A.

(c) If all three switches S_1, S_2 and S_3 are closed, then $R_{eq} = R_1 + R_1\,R'/(R_1 + R')$ where

$$R' = R_2 + R_1\,(R_1 + R_2)/(2R_1 + R_2) = 22.0\ \Omega,$$

i.e.,

$$R_{eq} = 20.0\ \Omega + (20.0\ \Omega)\,(22.0\ \Omega)/(20.0\ \Omega + 22.0\ \Omega) = 30.5\ \Omega,$$

so $i_a = \varepsilon/R_{eq} = 120$ V$/30.5\ \Omega = 3.94$ A.

75. (a) Reducing the bottom two series resistors to a single $R' = 4.00\ \Omega$ (with current i_1 through it), we see we can make a path (for use with the loop rule) that passes through R, the $\varepsilon_4 = 5.00$ V battery, the $\varepsilon_1 = 20.0$ V battery, and the $\varepsilon_3 = 5.00$ V. This leads to

$$i_1 = \frac{\varepsilon_1 + \varepsilon_3 + \varepsilon_4}{4.0\ \Omega} = \frac{30.0\ \text{V}}{4.0\ \Omega} = 7.50\ \text{A}.$$

(b) The direction of i_1 is leftward.

(c) The voltage across the bottom series pair is $i_1 R' = 30.0$ V. This must be the same as the voltage across the two resistors directly above them, one of which has current i_2 through it and the other (by symmetry) has current $\frac{1}{2}\,i_2$ through it. Therefore,

$$30.0\ \text{V} = i_2\,(2.00\ \Omega) + \frac{1}{2}\,i_2\,(2.00\ \Omega)$$

leads to $i_2 = 10.0$ A.

(d) The direction of i_2 is also leftward.

(e) We use Eq. 27-17: $P_4 = (i_1 + i_2)\varepsilon_4 = 87.5$ W.

(f) The energy is being supplied to the circuit since the current is in the "forward" direction through the battery.

76. The bottom two resistors are in parallel, equivalent to a $2.0R$ resistance. This, then, is in series with resistor R on the right, so that their equivalence is $R' = 3.0R$. Now, near the top left are two resistors ($2.0R$ and $4.0R$) which are in series, equivalent to $R'' = 6.0R$. Finally, R' and R'' are in parallel, so the net equivalence is

$$R_{eq} = \frac{(R')\,(R'')}{R' + R''} = 2.0R = 20\ \Omega$$

where in the final step we use the fact that $R = 10\ \Omega$.

77. (a) The four resistors R_1, R_2, R_3 and R_4 on the left reduce to

$$R_{eq} = R_{12} + R_{34} = \frac{R_1 R_2}{R_1 + R_2} + \frac{R_3 R_4}{R_3 + R_4} = 7.0\Omega + 3.0\Omega = 10\Omega$$

With $\varepsilon = 30$ V across R_{eq} the current there is $i_2 = 3.0$ A.

(b) The three resistors on the right reduce to

$$R'_{eq} = R_{56} + R_7 = \frac{R_5 R_6}{R_5 + R_6} + R_7 = \frac{(6.0\Omega)(2.0\Omega)}{6.0\Omega + 2.0\Omega} + 1.5\Omega = 3.0\Omega.$$

With $\varepsilon = 30$ V across R'_{eq} the current there is $i_4 = 10$ A.

(c) By the junction rule, $i_1 = i_2 + i_4 = 13$ A.

(d) By symmetry, $i_3 = \frac{1}{2} i_2 = 1.5$ A.

(e) By the loop rule (proceeding clockwise),

$$30\text{V} - i_4(1.5 \ \Omega) - i_5(2.0 \ \Omega) = 0$$

readily yields $i_5 = 7.5$ A.

78. (a) We analyze the lower left loop and find $i_1 = \varepsilon_1/R = (12.0 \text{ V})/(4.00 \ \Omega) = 3.00$ A.

(b) The direction of i_1 is downward.

(c) Letting $R = 4.00 \ \Omega$, we apply the loop rule to the tall rectangular loop in the center of the figure (proceeding clockwise):

$$\varepsilon_2 + (+i_1 R) + (- i_2 R) + \left(-\frac{i_2}{2} R\right) + (- i_2 R) = 0 \ .$$

Using the result from part (a), we find $i_2 = 1.60$ A.

(d) The direction of i_2 is downward (as was assumed in writing the equation as we did).

(e) Battery 1 is supplying this power since the current is in the "forward" direction through the battery.

(f) We apply Eq. 27-17: The current through the $\varepsilon_1 = 12.0$ V battery is, by the junction rule, $1.60 + 3.00 = 4.60$ A and $P = (4.60)(12.0 \text{ V}) = 55.2$ W.

(g) Battery 2 is supplying this power since the current is in the "forward" direction through the battery.

(h) $P = i_2(4.00 \text{ V}) = 6.40$ W.

79. (a) We reduce the parallel pair of resistors (at the bottom of the figure) to a single R' $= 1.00 \ \Omega$ resistor and then reduce it with its series 'partner' (at the lower left of the figure) to obtain an equivalence of $R'' = 2.00 \ \Omega + 1.00\Omega = 3.00 \ \Omega$. It is clear that the current through R'' is the i_1 we are solving for. Now, we employ the loop rule, choose a path that

includes R'' and all the batteries (proceeding clockwise). Thus, assuming i_1 goes leftward through R'', we have

$$5.00 \text{ V} + 20.0 \text{ V} - 10.0 \text{ V} - i_1 R'' = 0$$

which yields $i_1 = 5.00$ A.

(b) Since i_1 is positive, our assumption regarding its direction (leftward) was correct.

(c) Since the current through the $\varepsilon_1 = 20.0$ V battery is "forward", battery 1 is supplying energy.

(d) The rate is $P_1 = (5.00 \text{ A})(20.0 \text{ V}) = 100$ W.

(e) Reducing the parallel pair (which are in parallel to the $\varepsilon_2 = 10.0$ V battery) to a single $R' = 1.00 \ \Omega$ resistor (and thus with current $i' = (10.0 \text{ V})/(1.00 \ \Omega) = 10.0$ A downward through it), we see that the current through the battery (by the junction rule) must be $i = i' - i_1 = 5.00$ A *upward* (which is the "forward" direction for that battery). Thus, battery 2 is supplying energy.

(f) Using Eq. 27-17, we obtain $P_2 = 50.0$ W.

(g) The set of resistors that are in parallel with the $\varepsilon_3 = 5$ V battery is reduced to $R''' = 0.800 \ \Omega$ (accounting for the fact that two of those resistors are actually reduced in series, first, before the parallel reduction is made), which has current $i''' = (5.00 \text{ V})/(0.800 \ \Omega) = 6.25$ A downward through it. Thus, the current through the battery (by the junction rule) must be $i = i''' + i_1 = 11.25$ A *upward* (which is the "forward" direction for that battery). Thus, battery 3 is supplying energy.

(h) Eq. 27-17 leads to $P_3 = 56.3$ W.

80. (a) The parallel set of three identical $R_2 = 18 \ \Omega$ resistors reduce to $R = 6.0 \ \Omega$, which is now in series with the $R_1 = 6.0 \ \Omega$ resistor at the top right, so that the total resistive load across the battery is $R' = R_1 + R = 12 \ \Omega$. Thus, the current through R' is $(12\text{V})/R' = 1.0$ A, which is the current through R. By symmetry, we see one-third of that passes through any one of those $18 \ \Omega$ resistors; therefore, $i_1 = 0.333$ A.

(b) The direction of i_1 is clearly rightward.

(c) We use Eq. 26-27: $P = i^2 R' = (1.0)^2(12) = 12$ W. Thus, in 60 s, the energy dissipated is $(12 \text{ J/s})(60 \text{ s}) = 720$ J.

81. We denote silicon with subscript s and iron with i. Let $T_0 = 20°$. If

$$R(T) = R_s(T) + R_i(T) = R_s(T_0)\left[1 + \alpha(T - T_0)\right] + R_i(T_0)\left[1 + \alpha_i(T - T_0)\right]$$
$$= \left(R_s(T)_0 \alpha_s + R_i(T_0)\alpha_i\right) + \left(\text{temperature independent terms}\right)$$

is to be temperature-independent, we must require that $R_s(T_0)\alpha_s + R_i(T_0)\alpha_i = 0$. Also note that $R_s(T_0) + R_i(T_0) = R = 1000\ \Omega$. We solve for $R_s(T_0)$ and $R_i(T_0)$ to obtain

$$R_s(T_0) = \frac{R\alpha_i}{\alpha_i - \alpha_s} = \frac{(1000\Omega)\left(6.5\times10^{-3}\right)}{6.5\times10^{-3} + 70\times10^{-3}} = 85.0\Omega.$$

(b) $R_i(T_0) = 1000\ \Omega - 85.0\ \Omega = 915\ \Omega$.

82. (a) Since $R_{eq} < R$, the two resistors ($R = 12.0\ \Omega$ and R_x) must be connected in parallel:

$$R_{eq} = 3.00\Omega = \frac{R_x R}{R + R_x} = \frac{R_x(12.0\Omega)}{12.0\Omega + R_x}.$$

We solve for R_x: $R_x = R_{eq}R/(R - R_{eq}) = (3.00\ \Omega)(12.0\ \Omega)/(12.0\ \Omega - 3.00\ \Omega) = 4.00\ \Omega$.

(b) As stated above, the resistors must be connected in parallel.

83. Consider the lowest branch with the two resistors $R_4 = 3.00\ \Omega$ and $R_5 = 5.00\ \Omega$. The voltage difference across R_5 is

$$V = i_5 R_5 = \frac{\varepsilon R_5}{R_4 + R_5} = \frac{(120\text{V})(5.00\Omega)}{3.00\Omega + 5.00\Omega} = 7.50\text{V}.$$

84. When connected in series, the rate at which electric energy dissipates is $P_s = \varepsilon^2/(R_1 + R_2)$. When connected in parallel, the corresponding rate is $P_p = \varepsilon^2(R_1 + R_2)/R_1 R_2$. Letting $P_p/P_s = 5$, we get $(R_1 + R_2)^2/R_1 R_2 = 5$, where $R_1 = 100\ \Omega$. We solve for R_2: $R_2 = 38\ \Omega$ or $260\ \Omega$.

(a) Thus, the smaller value of R_2 is $38\ \Omega$.

(b) The larger value of R_2 is $260\ \Omega$.

85. (a) We reduce the parallel pair of identical $2.0\ \Omega$ resistors (on the right side) to $R' = 1.0\ \Omega$, and we reduce the series pair of identical $2.0\ \Omega$ resistors (on the upper left side) to $R'' = 4.0\ \Omega$. With R denoting the $2.0\ \Omega$ resistor at the bottom (between V_2 and V_1), we now have three resistors in series which are equivalent to

$$R + R' + R'' = 7.0\ \Omega$$

across which the voltage is 7.0 V (by the loop rule, this is 12 V – 5.0 V), implying that the current is 1.0 A (clockwise). Thus, the voltage across R' is (1.0 A)(1.0 Ω) = 1.0 V, which means that (examining the right side of the circuit) the voltage difference between *ground* and V_1 is 12 – 1 = 11 V. Noting the orientation of the battery, we conclude $V_1 = -11$ V .

(b) The voltage across R'' is (1.0 A)(4.0 Ω) = 4.0 V, which means that (examining the left side of the circuit) the voltage difference between *ground* and V_2 is 5.0 + 4.0 = 9.0 V. Noting the orientation of the battery, we conclude $V_2 = -9.0$ V. This can be verified by considering the voltage across R and the value we obtained for V_1.

86. (a) From $P = V^2/R$ we find $V = \sqrt{PR} = \sqrt{(10\,\text{W})(0.10\,\Omega)} = 1.0\,\text{V}$.

(b) From $i = V/R = (\varepsilon - V)/r$ we find

$$r = R\left(\frac{\varepsilon - V}{V}\right) = (0.10\,\Omega)\left(\frac{1.5\,\text{V} - 1.0\,\text{V}}{1.0\,\text{V}}\right) = 0.050\,\Omega.$$

87. (a) $R_{eq}(AB) = 20.0\ \Omega/3 = 6.67\ \Omega$ (three 20.0 Ω resistors in parallel).

(b) $R_{eq}(AC) = 20.0\ \Omega/3 = 6.67\ \Omega$ (three 20.0 Ω resistors in parallel).

(c) $R_{eq}(BC) = 0$ (as B and C are connected by a conducting wire).

88. Note that there is no voltage drop across the ammeter. Thus, the currents in the bottom resistors are the same, which we call i (so the current through the battery is $2i$ and the voltage drop across each of the bottom resistors is iR). The resistor network can be reduced to an equivalence of

$$R_{eq} = \frac{(2R)(R)}{2R + R} + \frac{(R)(R)}{R + R} = \frac{7}{6}R$$

which means that we can determine the current through the battery (and also through each of the bottom resistors):

$$2i = \frac{\varepsilon}{R_{eq}} \Rightarrow i = \frac{3\varepsilon}{7R}.$$

By the loop rule (going around the left loop, which includes the battery, resistor $2R$ and one of the bottom resistors), we have

$$\varepsilon - i_{2R}(2R) - iR = 0 \Rightarrow i_{2R} = \frac{\varepsilon - iR}{2R}.$$

Substituting $i = 3\varepsilon/7R$, this gives $i_{2R} = 2\varepsilon/7R$. The difference between i_{2R} and i is the current through the ammeter. Thus,

$$i_{ammeter} = i - i_{2R} = \frac{3\varepsilon}{7R} - \frac{2\varepsilon}{7R} = \frac{\varepsilon}{7R} \quad \Rightarrow \quad \frac{i_{ammeter}}{\varepsilon/R} = \frac{1}{7} = 0.143.$$

89. When S is open for a long time, the charge on C is $q_i = \varepsilon_2 C$. When S is closed for a long time, the current i in R_1 and R_2 is

$$i = (\varepsilon_2 - \varepsilon_1)/(R_1 + R_2) = (3.0 \text{ V} - 1.0 \text{ V})/(0.20 \ \Omega + 0.40 \ \Omega) = 3.33 \text{ A}.$$

The voltage difference V across the capacitor is then

$$V = \varepsilon_2 - iR_2 = 3.0 \text{ V} - (3.33 \text{ A}) (0.40 \ \Omega) = 1.67 \text{ V}.$$

Thus the final charge on C is $q_f = VC$. So the change in the charge on the capacitor is

$$\Delta q = q_f - q_i = (V - \varepsilon_2)C = (1.67 \text{ V} - 3.0 \text{ V}) (10 \ \mu\text{F}) = -13 \ \mu\text{C}.$$

90. Using the junction and the loop rules, we have

$$20.0 - i_1 R_1 - i_3 R_3 = 0$$
$$20.0 - i_1 R_1 - i_2 R_2 - 50 = 0$$
$$i_2 + i_3 = i_1$$

Requiring no current through the battery 1 means that $i_1=0$, or $i_2 = i_3$. Solving the above equations with $R_1 = 10.0\Omega$ and $R_2 = 20.0\Omega$, we obtain

$$i_1 = \frac{40 - 3R_3}{20 + 3R_3} = 0 \quad \Rightarrow \quad R_3 = \frac{40}{3} = 13.3\Omega$$

91. (a) The capacitor is *initially* uncharged, which implies (by the loop rule) that there is zero voltage (at $t = 0$) across the $R_2 = 10 \text{ k}\Omega$ resistor, and that 30 V is across the $R_1 = 20$ kΩ resistor. Therefore, by Ohm;s law, $i_{10} = (30 \text{ V})/(20 \text{ k}\Omega) = 1.5 \times 10^{-3}$ A.

(b) $i_{20} = 0$,

(c) As $t \rightarrow \infty$ the current to the capacitor reduces to zero and the 20 kΩ and 10 kΩ resistors behave more like a series pair (having the same current), equivalent to 30 kΩ. The current through them, then, at long times, is

$$i = (30 \text{ V})/(30 \text{ k}\Omega) = 1.0 \times 10^{-3} \text{ A}.$$

92. (a) The six resistors to the left of $\varepsilon_1 = 16$ V battery can be reduced to a single resistor $R = 8.0\ \Omega$, through which the current must be $i_R = \varepsilon_1/R = 2.0$ A. Now, by the loop rule, the current through the 3.0 Ω and 1.0 Ω resistors at the upper right corner is

$$i' = \frac{16.0\,\text{V} - 8.0\,\text{V}}{3.0\,\Omega + 1.0\,\Omega} = 2.0\,\text{A}$$

in a direction that is "backward" relative to the $\varepsilon_2 = 8.0$ V battery. Thus, by the junction rule,

$$i_1 = i_R + i' = 4.0\,\text{A}\ .$$

(b) The direction of i_1 is upward (that is, in the "forward" direction relative to ε_1).

(c) The current i_2 derives from a succession of symmetric splittings of i_R (reversing the procedure of reducing those six resistors to find R in part (a)). We find

$$i_2 = \frac{1}{2}\left(\frac{1}{2}i_R\right) = 0.50\,\text{A}\ .$$

(d) The direction of i_2 is clearly downward.

(e) Using our conclusion from part (a) in Eq. 27-17, we have $P = i_1\varepsilon_1 = (4.0)(16) = 64$ W.

(f) Using results from part (a) in Eq. 27-17, we obtain $P = i'\varepsilon_2 = (2.0)(8.0) = 16$ W.

(g) Energy is being supplied in battery 1.

(h) Energy is being absorbed in battery 2.

93. With the unit Ω understood, the equivalent resistance for this circuit is

$$R_{eq} = \frac{20R_3 + 100}{R_3 + 10}\ .$$

Therefore, the power supplied by the battery (equal to the power dissipated in the resistors) is

$$P = \frac{V^2}{R_3} = V^2\,\frac{R_3 + 10}{20R_3 + 100}$$

where $V = 12$ V. We attempt to extremize the expression by working through the $dP/dR_3 = 0$ condition and do not find a value of R_3 that satisfies it.

(a) We note, then, that the function is a monotonically decreasing function of R_3, with R_3 = 0 giving the maximum possible value (since $R_3 < 0$ values are not being allowed).

(b) With the value $R_3 = 0$, we obtain $P = 14.4$ W.

94. (a) The symmetry of the problem allows us to use i_2 as the current in *both* of the R_2 resistors and i_1 for the R_1 resistors. We see from the junction rule that $i_3 = i_1 - i_2$. There are only two independent loop rule equations:

$$\varepsilon - i_2 R_2 - i_1 R_1 = 0$$
$$\varepsilon - 2i_1 R_1 - (i_1 - i_2) R_3 = 0$$

where in the latter equation, a zigzag path through the bridge has been taken. Solving, we find $i_1 = 0.002625$ A, $i_2 = 0.00225$ A and $i_3 = i_1 - i_2 = 0.000375$ A. Therefore, $V_A - V_B = i_1 R_1 = 5.25$ V.

(b) It follows also that $V_B - V_C = i_3 R_3 = 1.50$ V.

(c) We find $V_C - V_D = i_1 R_1 = 5.25$ V.

(d) Finally, $V_A - V_C = i_2 R_2 = 6.75$ V.

95. (a) Using the junction rule ($i_1 = i_2 + i_3$) we write two loop rule equations:

$$\varepsilon_1 - i_2 R_2 - (i_2 + i_3) R_1 = 0$$
$$\varepsilon_2 - i_3 R_3 - (i_2 + i_3) R_1 = 0.$$

Solving, we find $i_2 = 0.0109$ A (rightward, as was assumed in writing the equations as we did), $i_3 = 0.0273$ A (leftward), and $i_1 = i_2 + i_3 = 0.0382$ A (downward).

(b) downward. See the results in part (a).

(c) $i_2 = 0.0109$ A . See the results in part (a).

(d) rightward. See the results in part (a).

(e) $i_3 = 0.0273$ A. See the results in part (a).

(f) leftward. See the results in part (a).

(g) The voltage across R_1 equals V_A: (0.0382 A)(100 Ω) = +3.82 V.

96. (a) R_2 and R_3 are in parallel; their equivalence is in series with R_1. Therefore,

$$R_{eq} = R_1 + \frac{R_2 R_3}{R_2 + R_3} = 300\ \Omega.$$

(b) The current through the battery is $\varepsilon/R_{eq} = 0.0200$ A, which is also the current through R_1. Hence, the voltage across R_1 is $V_1 = (0.0200\text{ A})(100\ \Omega) = 2.00$ V.

(c) From the loop rule,

$$\varepsilon - V_1 - i_3 R_3 = 0$$

which yields $i_3 = 6.67 \times 10^{-3}$ A.

97. From $V_a - \varepsilon_1 = V_c - ir_1 - iR$ and $i = (\varepsilon_1 - \varepsilon_2)/(R + r_1 + r_2)$, we get

$$V_a - V_c = \varepsilon_1 - i(r_1 + R) = \varepsilon_1 - \left(\frac{\varepsilon_1 - \varepsilon_2}{R + r_1 + r_2} \right)(r_1 + R)$$

$$= 4.4\,\text{V} - \left(\frac{4.4\,\text{V} - 2.1\,\text{V}}{5.5\,\Omega + 1.8\,\Omega + 2.3\,\Omega} \right)(2.3\,\Omega + 5.5\,\Omega)$$

$$= 2.5\,\text{V}.$$

98. The potential difference across R_2 is

$$V_2 = iR_2 = \frac{\varepsilon\,R_2}{R_1 + R_2 + R_3} = \frac{(12\text{ V})(4.0\,\Omega)}{3.0\,\Omega + 4.0\,\Omega + 5.0\,\Omega} = 4.0\,\text{V}.$$

99. (a) By symmetry, we see that i_3 is half the current that goes through the battery. The battery current is found by dividing ε by the equivalent resistance of the circuit, which is easily found to be $6.00\ \Omega$. Thus,

$$i_3 = \frac{1}{2} i_{bat} = \frac{1}{2}\left(\frac{12\,\text{V}}{6.0\,\Omega} \right) = 1.00\,\text{A}$$

and is clearly downward (in the figure).

(b) We use Eq. 27-17: $P = i_{bat}\varepsilon = 24.0$ W.

100. The current in the ammeter is given by $i_A = \varepsilon/(r + R_1 + R_2 + R_A)$. The current in R_1 and R_2 without the ammeter is $i = \varepsilon/(r + R_1 + R_2)$. The percent error is then

$$\frac{\Delta i}{i} = \frac{i - i_A}{i} = 1 - \frac{r + R_1 + R_2}{r + R_1 + R_2 + R_A} = \frac{R_A}{r + R_1 + R_2 + R_A}$$

$$= \frac{0.10\,\Omega}{2.0\,\Omega + 5.0\,\Omega + 4.0\,\Omega + 0.10\,\Omega} = 0.90\%.$$

101. When all the batteries are connected in parallel, each supplies a current i; thus, $i_R = Ni$. Then from $\varepsilon = ir + i_R R = ir + Nir$, we get $i_R = N\varepsilon/[(N + 1)r]$. When all the batteries are connected in series, $i_r = i_R$ and

$$\varepsilon_{total} = N\varepsilon = Ni_r r + i_R R = Ni_R r + i_R r,$$

so $i_R = N\varepsilon/[(N + 1)r]$.

102. (a) Dividing Eq. 27-39 by capacitance turns it into an equation that describes the dependence of the voltage on time: $V_C = V_0\, e^{-t/\tau}$;

(b) Taking logarithms of this equation produces a form amenable to a least squares fit:

$$\ln(V_C) = -\frac{1}{\tau}\, t + \ln(V_0)$$

$$\ln(V_C) = -1.2994\, t + 2.525$$

Thus, we have the emf equal to $V_0 = e^{2.525} = 12.49\ \text{V} \approx 12\ \text{V}$;

(c) This also tells us that the time constant is $\tau = 1/1.2994 = 0.77$ s.

(d) Since $\tau = RC$ then we find $C = 3.8\ \mu\text{F}$.

103. Here we denote the supply emf as V (understood to be in volts). The situation is much like that shown in Fig. 27-4, with r now interpreted as the resistance of the transmission line and R interpreted as the resistance of the "consumer" (the reason the circuit has been turned on in the first place – to supply power to some resistive load R). From Eq. 27-4 and Eq. 26-27 (remembering that we are asked to find the power dissipated in the *transmission line*) we obtain

$$P_{line} = \left(\frac{V^2}{(R + r)^2}\right) r\,.$$

Now r is considered constant, certainly, but what about R? The load will not be the same in the two cases (where V = 110000 and V' =110) because the problem requires us to consider the *total* power supplied to be constant, so

$$P_{total} = \left(\frac{V^2}{(R + r)^2}\right)(R + r) \;=\; P'_{total} = \left(\frac{V'^2}{(R' + r)^2}\right)(R' + r)$$

which implies (taking ratio of P_{total} to P'_{total})

$$1 = \frac{V^2 (R' + r)}{V'^2 (R + r)}.$$

Now, as the problem directs, we take ratio of P_{line} to P'_{line} and obtain

$$\frac{P_{line}}{P'_{line}} = \frac{V^2 (R' + r)^2}{V'^2 (R + r)^2} = \frac{V'^2}{V^2} = 1.00 \times 10^{-6}.$$

104. The resistor by the letter i is above three other resistors; together, these four resistors are equivalent to a resistor $R = 10 \; \Omega$ (with current i). As if we were presented with a maze, we find a path through R that passes through any number of batteries (10, it turns out) but no other resistors, which — as in any good maze — winds "all over the place." Some of the ten batteries are opposing each other (particularly the ones along the outside), so that their net emf is only $\varepsilon = 40$ V.

(a) The current through R is then $i = \varepsilon/R = 4.0$ A.

(b) The direction is upward in the figure.

105. The maximum power output is (120 V)(15 A) = 1800 W. Since 1800 W/500 W = 3.6, the maximum number of 500 W lamps allowed is 3.

106. The part of R_0 connected in parallel with R is given by $R_1 = R_0 x/L$, where $L = 10$ cm. The voltage difference across R is then $V_R = \varepsilon R'/R_{eq}$, where $R' = RR_1/(R + R_1)$ and $R_{eq} = R_0(1 - x/L) + R'$. Thus

$$P_R = \frac{V_R^2}{R} = \frac{1}{R}\left(\frac{\varepsilon RR_1/(R + R_1)}{R_0(1 - x/L) + RR_1/(R + R_1)}\right)^2 = \frac{100R(\varepsilon x/R_0)^2}{(100\,R/R_0 + 10x - x^2)^2},$$

where x is measured in cm.

107. The power delivered by the motor is $P = (2.00 \text{ V})(0.500 \text{ m/s}) = 1.00$ W. From $P = i^2 R_{motor}$ and $\varepsilon = i(r + R_{motor})$ we then find $i^2 r - i\varepsilon + P = 0$ (which also follows directly from the conservation of energy principle). We solve for i:

$$i = \frac{\varepsilon \pm \sqrt{\varepsilon^2 - 4rP}}{2r} = \frac{2.00 \text{ V} \pm \sqrt{(2.00 \text{ V})^2 - 4(0.500\,\Omega)(1.00 \text{ W})}}{2(0.500\,\Omega)}.$$

The answer is either 3.41 A or 0.586 A.

(a) The larger i is 3.41 A.

(b) We use $V = \varepsilon - ir = 2.00$ V $- i(0.500$ $\Omega)$. We substitute value of i obtained in part (a) into the above formula to get $V = 0.293$ V.

(c) The smaller i is 0.586 A.

(d) The corresponding V is 1.71 V.

108. (a) Placing a wire (of resistance r) with current i running directly from point a to point b in Fig. 27-51 divides the top of the picture into a left and a right triangle. If we label the currents through each resistor with the corresponding subscripts (for instance, i_s goes toward the lower right through R_s and i_x goes toward the upper right through R_x), then the currents must be related as follows:

$$i_0 = i_1 + i_s \quad \text{and} \quad i_1 = i + i_2$$
$$i_s + i = i_x \quad \text{and} \quad i_2 + i_x = i_0$$

where the last relation is not independent of the previous three. The loop equations for the two triangles and also for the bottom loop (containing the battery and point b) lead to

$$i_s R_s - i_1 R_1 - ir = 0$$
$$i_2 R_s - i_x R_x - ir = 0$$
$$\varepsilon - i_0 R_0 - i_s R_s - i_x R_x = 0.$$

We incorporate the current relations from above into these loop equations in order to obtain three well-posed "simultaneous" equations, for three unknown currents (i_s, i_1 and i):

$$i_s R_s - i_1 R_1 - ir = 0$$
$$i_1 R_2 - i_s R_x - i(r + R_x + R_2) = 0$$
$$\varepsilon - i_s(R_0 + R_s + R_x) - i_1 R_0 - iR_x = 0$$

The problem statement further specifies $R_1 = R_2 = R$ and $R_0 = 0$, which causes our solution for i to simplify significantly. It becomes

$$i = \frac{\varepsilon(R_s - R_x)}{2rR_s + 2R_x R_s + R_s R + 2rR_x + R_x R}$$

which is equivalent to the result shown in the problem statement.

(b) Examining the numerator of our final result in part (a), we see that the condition for $i = 0$ is $R_s = R_x$. Since $R_1 = R_2 = R$, this is equivalent to $R_x = R_s R_2 / R_1$, consistent with the result of Problem 43.

109. (a) They are in parallel and the portions of A and B between the load and their respective sliding contacts have the same potential difference. It is clearly important not to "short" the system (particularly if the load turns out to have very little resistance) by having the sliding contacts too close to the load-ends of A and B to start with. Thus, we suggest putting the contacts roughly in the middle of each. Since $R_A > R_B$, larger currents generally go through B (depending on the position of the sliding contact) than through A. Therefore, B is analogous to a "coarse" control, as A is to a "fine control." Hence, we recommend adjusting the current roughly with B, and then making fine adjustments with A.

(b) Relatively large percentage changes in A cause only small percentage charges in the resistance of the parallel combination, thus permitting fine adjustment; any change in A causes half as much change in this combination.

110. (a) In the process described in the problem, no charge is gained or lost. Thus, $q =$ constant. Hence,

$$q = C_1 V_1 = C_2 V_2 \Rightarrow V_2 = V_1 \frac{C_1}{C_2} = (200)\left(\frac{150}{10}\right) = 3.0 \times 10^3 \text{ V}.$$

(b) Eq. 27-39, with $\tau = RC$, describes not only the discharging of q but also of V. Thus,

$$V = V_0 e^{-t/\tau} \Rightarrow t = RC \ln\left(\frac{V_0}{V}\right) = \left(300 \times 10^9 \ \Omega\right)\left(10 \times 10^{-12} \text{ F}\right)\ln\left(\frac{3000}{100}\right)$$

which yields $t = 10$ s. This is a longer time than most people are inclined to wait before going on to their next task (such as handling the sensitive electronic equipment).

(c) We solve $V = V_0 e^{-t/RC}$ for R with the new values $V_0 = 1400$ V and $t = 0.30$ s. Thus,

$$R = \frac{t}{C \ln(V_0/V)} = \frac{0.30 \text{ s}}{\left(10 \times 10^{-12} \text{ F}\right)\ln(1400/100)} = 1.1 \times 10^{10} \ \Omega.$$

111. In the steady state situation, there is no current going to the capacitors, so the resistors all have the same current. By the loop rule,

$$20.0 \text{ V} = (5.00 \ \Omega)i + (10.0 \ \Omega)i + (15.0 \ \Omega)i$$

which yields $i = \frac{2}{3}$ A. Consequently, the voltage across the $R_1 = 5.00 \ \Omega$ resistor is (5.00 Ω)(2/3 A) = 10/3 V, and is equal to the voltage V_1 across the $C_1 = 5.00 \ \mu\text{F}$ capacitor. Using Eq. 26-22, we find the stored energy on that capacitor:

$$U_1 \; = \; \tfrac{1}{2}C_1(V_1)^2 \; = \; \tfrac{1}{2}(5.00 \times 10^{-6})\left(\tfrac{10}{3}\right)^2 \; = \; 2.78 \times 10^{-5} \text{ J} \; .$$

Similarly, the voltage across the $R_2 = 10.0 \; \Omega$ resistor is $(10.0 \; \Omega)(2/3 \text{ A}) = 20/3$ V and is equal to the voltage V_2 across the $C_2 = 10.0 \; \mu$F capacitor. Hence,

$$U_2 \; = \; \tfrac{1}{2}C_2(V_2)^2 \; = \; \tfrac{1}{2}(10 \times 10^{-6})\left(\tfrac{20}{3}\right)^2 \; = \; 2.22 \times 10^{-4} \text{ J} \; .$$

Therefore, the total capacitor energy is $U_1 + U_2 = 2.50 \times 10^{-4}$ J.

112. (a) Applying the junction rule twice and the loop rule three times, we obtain

$$i_1 - i_2 + i_4 + i_5 = 0$$
$$i_3 + i_4 + i_5 = 0,$$
$$-16 \text{ V} + 4 \text{ V} + (7 \; \Omega)i_1 + (5 \; \Omega)i_2 = 0,$$
$$10 \text{ V} - 4 \text{ V} - (5 \; \Omega)i_2 + (8 \; \Omega)i_3 - (9 \; \Omega)i_4 = 0,$$
$$12 \text{ V} + (9 \; \Omega)i_4 - (4 \; \Omega)i_5 = 0;$$

(b) Examining the coefficients of the currents in the above relations, we find

$$[A] = \begin{bmatrix} 1 & -1 & 0 & 1 & 1 \\ 0 & 0 & 1 & 1 & 1 \\ 7 & 5 & 0 & 0 & 0 \\ 0 & -5 & 8 & -9 & 0 \\ 0 & 0 & 0 & 9 & -4 \end{bmatrix} \text{ ohms,} \quad [C] = \begin{bmatrix} 0 \\ 0 \\ 12 \\ -6 \\ -12 \end{bmatrix} \text{ volts;}$$

(c) $i_1 = 306/427 \approx 0.717$ A,
$i_2 = 426/305 \approx 1.40$ A,
$i_3 = -1452/2135 \approx -0.680$ A,
$i_4 = -1524/2135 \approx -0.714$ A,
$i_5 = 2976/2135 \approx 1.39$ A

113. (a) $V_9 = (9.00 \; \Omega)(|i_4|) = 6.43$ V;

(b) $P_7 = (i_1)^2(7.00 \; \Omega) = 3.60$ W;

(c) $(12.0 \text{ V})(i_5) = 16.7$ W;

(d) $-(4.00 \text{ V})(i_2) = -5.60$ W;

(e) a (see the true direction of the current labeled i_4)

114. (a) Next, we graph Eq. 27-4 (scaled by a factor of 100) for $\varepsilon = 2.0$ V and $r = 100 \ \Omega$ over the range $0 \le R \le 500 \ \Omega$. We multiplied the SI output of Eq. 27-4 by 100 so that this graph would not be vanishingly small with the other graph (see part (b)) when they are plotted together.

(b) In the same graph, we show $V_R = iR$ over the same range. The graph of current i is the one that starts at 2 (which corresponds to 0.02 A in SI units) and the graph of voltage V_R is the one that starts at 0 (when $R = 0$). The value of V_R are in SI units (not scaled by any factor).

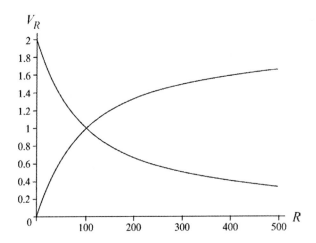

(c) In our final graph, we show the dependence of power $P = iV_R$ (dissipated in resistor R) as a function of R. The units of the vertical axis are Watts. We note that it is maximum when $R = r$.

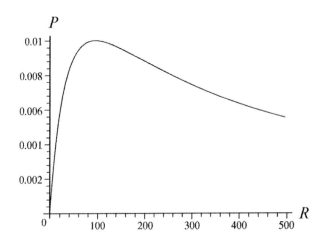

Chapter 28

1. (a) Eq. 28-3 leads to

$$v = \frac{F_B}{eB \sin \phi} = \frac{6.50 \times 10^{-17}\,\text{N}}{(1.60 \times 10^{-19}\,\text{C})(2.60 \times 10^{-3}\,\text{T})\sin 23.0°} = 4.00 \times 10^{5}\,\text{m/s}.$$

(b) The kinetic energy of the proton is

$$K = \frac{1}{2}mv^2 = \frac{1}{2}(1.67 \times 10^{-27}\,\text{kg})(4.00 \times 10^{5}\,\text{m/s})^2 = 1.34 \times 10^{-16}\,\text{J}.$$

This is $(1.34 \times 10^{-16}\,\text{J}) / (1.60 \times 10^{-19}\,\text{J/eV}) = 835$ eV.

2. (a) We use Eq. 28-3:

$$F_B = |q|\,vB \sin \phi = (+3.2 \times 10^{-19}\,\text{C})\,(550\,\text{m/s})\,(0.045\,\text{T})\,(\sin 52°) = 6.2 \times 10^{-18}\,\text{N}.$$

(b) $a = F_B/m = (6.2 \times 10^{-18}\,\text{N}) / (6.6 \times 10^{-27}\,\text{kg}) = 9.5 \times 10^{8}\,\text{m/s}^2$.

(c) Since it is perpendicular to $\vec{v}$, $\vec{F}_B$ does not do any work on the particle. Thus from the work-energy theorem both the kinetic energy and the speed of the particle remain unchanged.

3. (a) The force on the electron is

$$\vec{F}_B = q\vec{v} \times \vec{B} = q\left(v_x\hat{i} + v_y\hat{j}\right) \times \left(B_x\hat{i} + B_y\hat{j}\right) = q\left(v_x B_y - v_y B_x\right)\hat{k}$$
$$= \left(-1.6 \times 10^{-19}\,\text{C}\right)\left[\left(2.0 \times 10^{6}\,\text{m/s}\right)\left(-0.15\,\text{T}\right) - \left(3.0 \times 10^{6}\,\text{m/s}\right)\left(0.030\,\text{T}\right)\right]$$
$$= \left(6.2 \times 10^{-14}\,\text{N}\right)\hat{k}.$$

Thus, the magnitude of $\vec{F}_B$ is 6.2×10^{14} N, and $\vec{F}_B$ points in the positive z direction.

(b) This amounts to repeating the above computation with a change in the sign in the charge. Thus, $\vec{F}_B$ has the same magnitude but points in the negative z direction, namely, $\vec{F}_B = -\left(6.2 \times 10^{-14}\,\text{N}\right)\hat{k}.$

4. The magnetic force on the proton is

$$\vec{F} = q\,\vec{v} \times \vec{B}$$

where $q = +e$. Using Eq. 3-30 this becomes

$$(4 \times 10^{-17})\hat{\mathbf{i}} + (2 \times 10^{-17})\hat{\mathbf{j}} = e[(0.03v_y + 40)\hat{\mathbf{i}} + (20 - 0.03v_x)\hat{\mathbf{j}} - (0.02v_x + 0.01v_y)\hat{\mathbf{k}}]$$

with SI units understood. Equating corresponding components, we find

(a) $v_x = -3.5 \times 10^3$ m/s, and

(b) $v_y = 7.0 \times 10^3$ m/s.

5. Using Eq. 28-2 and Eq. 3-30, we obtain

$$\vec{F} = q\left(v_x B_y - v_y B_x\right)\hat{\mathbf{k}} = q\left(v_x(3B_x) - v_y B_x\right)\hat{\mathbf{k}}$$

where we use the fact that $B_y = 3B_x$. Since the force (at the instant considered) is $F_z\,\hat{\mathbf{k}}$ where $F_z = 6.4 \times 10^{-19}$ N, then we are led to the condition

$$q\left(3v_x - v_y\right)B_x = F_z \Rightarrow B_x = \frac{F_z}{q\left(3v_x - v_y\right)}.$$

Substituting $V_x = 2.0$ m/s, $v_y = 4.0$ m/s and $q = -1.6 \times 10^{-19}$ C, we obtain $B_x = -2.0$ T.

6. Letting $\vec{F} = q\left(\vec{E} + \vec{v} \times \vec{B}\right) = 0$, we get $vB \sin \phi = E$. We note that (for given values of the fields) this gives a minimum value for speed whenever the $\sin \phi$ factor is at its maximum value (which is 1, corresponding to $\phi = 90°$). So

$$v_{min} = E / B = (1.50 \times 10^3 \text{ V / m}) / (0.400 \text{ T}) = 3.75 \times 10^3 \text{ m / s}.$$

7. Straight line motion will result from zero net force acting on the system; we ignore gravity. Thus, $\vec{F} = q\left(\vec{E} + \vec{v} \times \vec{B}\right) = 0$. Note that $\vec{v} \perp \vec{B}$ so $\left|\vec{v} \times \vec{B}\right| = vB$. Thus, obtaining the speed from the formula for kinetic energy, we obtain

$$B = \frac{E}{v} = \frac{E}{\sqrt{2K/m_e}} = \frac{100\,\text{V}/\left(20 \times 10^{-3}\text{m}\right)}{\sqrt{2\left(1.0 \times 10^3\,\text{V}\right)\left(1.60 \times 10^{-19}\text{C}\right)/\left(9.11 \times 10^{-31}\text{kg}\right)}} = 2.67 \times 10^{-4}\,\text{T}.$$

In unit-vector notation, $\vec{B} = -(2.67 \times 10^{-4}\,\text{T})\hat{\mathbf{k}}$.

8. We apply $\vec{F} = q\left(\vec{E} + \vec{v} \times \vec{B}\right) = m_e\vec{a}$ to solve for $\vec{E}$:

$$\vec{E} = \frac{m_e \vec{a}}{q} + \vec{B} \times \vec{v}$$

$$= \frac{(9.11 \times 10^{-31}\,\text{kg})(2.00 \times 10^{12}\,\text{m/s}^2)\hat{i}}{-1.60 \times 10^{-19}\,\text{C}} + (400\,\mu\text{T})\hat{i} \times \left[(12.0\,\text{km/s})\hat{j} + (15.0\,\text{km/s})\hat{k}\right]$$

$$= \left(-11.4\hat{i} - 6.00\hat{j} + 4.80\hat{k}\right)\text{V/m}.$$

9. Since the total force given by $\vec{F} = e(\vec{E} + \vec{v} \times \vec{B})$ vanishes, the electric field $\vec{E}$ must be perpendicular to both the particle velocity $\vec{v}$ and the magnetic field $\vec{B}$. The magnetic field is perpendicular to the velocity, so $\vec{v} \times \vec{B}$ has magnitude vB and the magnitude of the electric field is given by $E = vB$. Since the particle has charge e and is accelerated through a potential difference V, $\frac{1}{2}mv^2 = eV$ and $v = \sqrt{2eV/m}$. Thus,

$$E = B\sqrt{\frac{2eV}{m}} = (1.2\,\text{T})\sqrt{\frac{2(1.60 \times 10^{-19}\,\text{C})(10 \times 10^3\,\text{V})}{(9.99 \times 10^{-27}\,\text{kg})}} = 6.8 \times 10^5\,\text{V/m}.$$

10. (a) The force due to the electric field ($\vec{F} = q\vec{E}$) is distinguished from that associated with the magnetic field ($\vec{F} = q\vec{v} \times \vec{B}$) in that the latter vanishes at the speed is zero and the former is independent of speed. The graph (Fig.28-34) shows that the force (y-component) is negative at $v = 0$ (specifically, its value is -2.0×10^{-19} N there) which (because $q = -e$) implies that the electric field points in the $+y$ direction. Its magnitude is

$$E = (2.0 \times 10^{-19})/(1.60 \times 10^{-19}) = 1.25\,\text{V/m}.$$

(b) We are told that the x and z components of the force remain zero throughout the motion, implying that the electron continues to move along the x axis, even though magnetic forces generally cause the paths of charged particles to curve (Fig. 28-11). The exception to this is discussed in section 28-3, where the forces due to the electric and magnetic fields cancel. This implies (Eq. 28-7) $B = E/v = 2.50 \times 10^{-2}$ T.

For $\vec{F} = q\vec{v} \times \vec{B}$ to be in the opposite direction of $\vec{F} = q\vec{E}$ we must have $\vec{v} \times \vec{B}$ in the opposite direction from $\vec{E}$ which points in the $+y$ direction, as discussed in part (a). Since the velocity is in the $+x$ direction, then (using the right-hand rule) we conclude that the magnetic field must point in the $+z$ direction ($\hat{i} \times \hat{k} = -\hat{j}$). In unit-vector notation, we have $\vec{B} = (2.50 \times 10^{-2}\,\text{T})\hat{k}$.

11. For a free charge q inside the metal strip with velocity $\vec{v}$ we have $\vec{F} = q(\vec{E} + \vec{v} \times \vec{B})$. We set this force equal to zero and use the relation between (uniform) electric field and potential difference. Thus,

$$v = \frac{E}{B} = \frac{|V_x - V_y|/d_{xy}}{B} = \frac{(3.90 \times 10^{-9}\,\mathrm{V})}{(1.20 \times 10^{-3}\,\mathrm{T})(0.850 \times 10^{-2}\,\mathrm{m})} = 0.382\,\mathrm{m/s}.$$

12. We use Eq. 28-12 to solve for V:

$$V = \frac{iB}{nle} = \frac{(23\mathrm{A})(0.65\,\mathrm{T})}{(8.47 \times 10^{28}/\mathrm{m}^3)(150\,\mu\mathrm{m})(1.6 \times 10^{-19}\,\mathrm{C})} = 7.4 \times 10^{-6}\,\mathrm{V}.$$

13. (a) We seek the electrostatic field established by the separation of charges (brought on by the magnetic force). With Eq. 28-10, we define the magnitude of the electric field as $|\vec{E}| = v|\vec{B}| = (20.0\ \mathrm{m/s})(0.030\ \mathrm{T}) = 0.6\,\mathrm{V/m}$. Its direction may be inferred from Figure 28-8; its direction is opposite to that defined by $\vec{v} \times \vec{B}$. In summary,

$$\vec{E} = -(0.600\,\mathrm{V/m})\hat{k}$$

which insures that $\vec{F} = q\left(\vec{E} + \vec{v} \times \vec{B}\right)$ vanishes.

(b) Eq. 28-9 yields $V = (0.600\ \mathrm{V/m})(2.00\ \mathrm{m}) = 1.20$ V.

14. We note that $\vec{B}$ must be along the x axis because when the velocity is along that axis there is no induced voltage. Combining Eq. 28-7 and Eq. 28-9 leads to $d = V/vB$ where one must interpret the symbols carefully to ensure that $\vec{d}$, $\vec{v}$ and $\vec{B}$ are mutually perpendicular. Thus, when the velocity if parallel to the y axis the absolute value of the voltage (which is considered in the same "direction" as $\vec{d}$) is 0.012 V, and

$$d = d_z = (0.012)/(3)(0.02) = 0.20\ \mathrm{m}.$$

And when the velocity if parallel to the z axis the absolute value of the appropriate voltage is 0.018 V, and $d = d_y = (0.018)/(3)(0.02) = 0.30$ m. Thus, our answers are

(a) $d_x = 25$ cm (which we arrive at "by elimination" – since we already have figured out d_y and d_z),

(b) $d_y = 30$ cm, and

(c) $d_z = 20$ cm

15. From Eq. 28-16, we find

$$B = \frac{m_e v}{er} = \frac{(9.11 \times 10^{-31}\,\mathrm{kg})(1.30 \times 10^{6}\ \mathrm{m/s})}{(1.60 \times 10^{-19}\,\mathrm{C})(0.350\ \mathrm{m})} = 2.11 \times 10^{-5}\ \mathrm{T}.$$

16. (a) The accelerating process may be seen as a conversion of potential energy eV into kinetic energy. Since it starts from rest, $\frac{1}{2}m_e v^2 = eV$ and

$$v = \sqrt{\frac{2eV}{m_e}} = \sqrt{\frac{2(1.60 \times 10^{-19}\,\text{C})(350\,\text{V})}{9.11 \times 10^{-31}\,\text{kg}}} = 1.11 \times 10^7 \text{ m/s}.$$

(b) Eq. 28-16 gives

$$r = \frac{m_e v}{eB} = \frac{(9.11 \times 10^{-31}\,\text{kg})(1.11 \times 10^7\,\text{m/s})}{(1.60 \times 10^{-19}\,\text{C})(200 \times 10^{-3}\,\text{T})} = 3.16 \times 10^{-4}\,\text{m}.$$

17. (a) From $K = \frac{1}{2}m_e v^2$ we get

$$v = \sqrt{\frac{2K}{m_e}} = \sqrt{\frac{2(1.20 \times 10^3\,\text{eV})(1.60 \times 10^{-19}\,\text{eV/J})}{9.11 \times 10^{-31}\,\text{kg}}} = 2.05 \times 10^7 \text{ m/s}.$$

(b) From $r = m_e v/qB$ we get

$$B = \frac{m_e v}{qr} = \frac{(9.11 \times 10^{-31}\,\text{kg})(2.05 \times 10^7\,\text{m/s})}{(1.60 \times 10^{-19}\,\text{C})(25.0 \times 10^{-2}\,\text{m})} = 4.67 \times 10^{-4}\text{ T}.$$

(c) The "orbital" frequency is

$$f = \frac{v}{2\pi r} = \frac{2.07 \times 10^7\,\text{m/s}}{2\pi(25.0 \times 10^{-2}\,\text{m})} = 1.31 \times 10^7 \text{ Hz}.$$

(d) $T = 1/f = (1.31 \times 10^7 \text{ Hz})^{-1} = 7.63 \times 10^{-8}$ s.

18. (a) Using Eq. 28-16, we obtain

$$v = \frac{rqB}{m_\alpha} = \frac{2eB}{4.00\,\text{u}} = \frac{2(4.50 \times 10^{-2}\,\text{m})(1.60 \times 10^{-19}\,\text{C})(1.20\,\text{T})}{(4.00\,\text{u})(1.66 \times 10^{-27}\,\text{kg/u})} = 2.60 \times 10^6 \text{ m/s}.$$

(b) $T = 2\pi r/v = 2\pi(4.50 \times 10^{-2} \text{ m})/(2.60 \times 10^6 \text{ m/s}) = 1.09 \times 10^{-7}$ s.

(c) The kinetic energy of the alpha particle is

$$K = \frac{1}{2}m_\alpha v^2 = \frac{(4.00\,\mathrm{u})(1.66 \times 10^{-27}\ \mathrm{kg/u})(2.60 \times 10^6\ \mathrm{m/s})^2}{2(1.60 \times 10^{-19}\ \mathrm{J/eV})} = 1.40 \times 10^5\ \mathrm{eV}\ .$$

(d) $\Delta V = K/q = 1.40 \times 10^5\ \mathrm{eV}/2e = 7.00 \times 10^4\ \mathrm{V}$.

19. (a) The frequency of revolution is

$$f = \frac{Bq}{2\pi m_e} = \frac{(35.0 \times 10^{-6}\ \mathrm{T})(1.60 \times 10^{-19}\ \mathrm{C})}{2\pi(9.11 \times 10^{-31}\ \mathrm{kg})} = 9.78 \times 10^5\ \mathrm{Hz}.$$

(b) Using Eq. 28-16, we obtain

$$r = \frac{m_e v}{qB} = \frac{\sqrt{2m_e K}}{qB} = \frac{\sqrt{2(9.11 \times 10^{-31}\ \mathrm{kg})(100\,\mathrm{eV})(1.60 \times 10^{-19}\ \mathrm{J/eV})}}{(1.60 \times 10^{-19}\ \mathrm{C})(35.0 \times 10^{-6}\ \mathrm{T})} = 0.964\ \mathrm{m}\ .$$

20. Referring to the solution of problem 19 part (b), we see that $r = \sqrt{2mK}/qB$ implies $K = (rqB)^2/2m \propto q^2 m^{-1}$. Thus,

(a) $K_\alpha = \left(q_\alpha/q_p\right)^2 \left(m_p/m_\alpha\right) K_p = (2)^2 (1/4) K_p = K_p = 1.0\,\mathrm{MeV}$;

(b) $K_d = \left(q_d/q_p\right)^2 \left(m_p/m_d\right) K_p = (1)^2 (1/2) K_p = 1.0\ \mathrm{MeV}/2 = 0.50\,\mathrm{MeV}$.

21. Reference to Fig. 28-11 is very useful for interpreting this problem. The distance traveled parallel to $\vec{B}$ is $d_\| = v_\| T = v_\|(2\pi m_e /|q|B)$ using Eq. 28-17. Thus,

$$v_\| = \frac{d_\| e B}{2\pi m_e} = 50.3\ \mathrm{km/s}$$

using the values given in this problem. Also, since the magnetic force is $|q|Bv_\perp$, then we find $v_\perp = 41.7$ km/s. The speed is therefore $v = \sqrt{v_\perp^2 + v_\|^2} = 65.3$ km/s.

22. Using $F = \frac{m v^2}{r}$ (for the centripetal force) and $K = \frac{1}{2}mv^2$, we can easily derive the relation

$$K = \frac{1}{2}Fr\ .$$

With the values given in the problem, we thus obtain $K = 2.09 \times 10^{-22}$ J.

23. (a) If v is the speed of the positron then $v \sin \phi$ is the component of its velocity in the plane that is perpendicular to the magnetic field. Here ϕ is the angle between the velocity and the field (89°). Newton's second law yields $eBv \sin \phi = m_e(v \sin \phi)^2/r$, where r is the radius of the orbit. Thus $r = (m_e v/eB) \sin \phi$. The period is given by

$$T = \frac{2\pi r}{v \sin \phi} = \frac{2\pi m_e}{eB} = \frac{2\pi\left(9.11 \times 10^{-31} \text{ kg}\right)}{\left(1.60 \times 10^{-19} \text{ C}\right)\left(0.100 \text{ T}\right)} = 3.58 \times 10^{-10} \text{ s}.$$

The equation for r is substituted to obtain the second expression for T.

(b) The pitch is the distance traveled along the line of the magnetic field in a time interval of one period. Thus $p = vT \cos \phi$. We use the kinetic energy to find the speed: $K = \frac{1}{2} m_e v^2$ means

$$v = \sqrt{\frac{2K}{m_e}} = \sqrt{\frac{2\left(2.00 \times 10^3 \text{ eV}\right)\left(1.60 \times 10^{-19} \text{ J/eV}\right)}{9.11 \times 10^{-31} \text{ kg}}} = 2.65 \times 10^7 \text{ m/s}.$$

Thus

$$p = \left(2.65 \times 10^7 \text{ m/s}\right)\left(3.58 \times 10^{-10} \text{ s}\right) \cos 89° = 1.66 \times 10^{-4} \text{ m}.$$

(c) The orbit radius is

$$R = \frac{m_e v \sin \phi}{eB} = \frac{\left(9.11 \times 10^{-31} \text{ kg}\right)\left(2.65 \times 10^7 \text{ m/s}\right) \sin 89°}{\left(1.60 \times 10^{-19} \text{ C}\right)\left(0.100 \text{ T}\right)} = 1.51 \times 10^{-3} \text{ m}.$$

24. We consider the point at which it enters the field-filled region, velocity vector pointing downward. The field points out of the page so that $\vec{v} \times \vec{B}$ points leftward, which indeed seems to be the direction it is "pushed"; therefore, $q > 0$ (it is a proton).

(a) Eq. 28-17 becomes $T = 2\pi m_p /e|\vec{B}|$, or

$$2\left(130 \times 10^{-9}\right) = \frac{2\pi\left(1.67 \times 10^{-27}\right)}{\left(1.60 \times 10^{-19}\right)|\vec{B}|}$$

which yields $|\vec{B}| = 0.252 \text{ T}$.

(b) Doubling the kinetic energy implies multiplying the speed by $\sqrt{2}$. Since the period T does not depend on speed, then it remains the same (even though the radius increases by a factor of $\sqrt{2}$). Thus, $t = T/2 = 130$ ns, again.

25. (a) We solve for B from $m = B^2qx^2/8V$ (see Sample Problem 28-3):

$$B = \sqrt{\frac{8Vm}{qx^2}}.$$

We evaluate this expression using $x = 2.00$ m:

$$B = \sqrt{\frac{8(100 \times 10^3 \text{ V})(3.92 \times 10^{-25} \text{ kg})}{(3.20 \times 10^{-19} \text{ C})(2.00 \text{ m})^2}} = 0.495 \text{T}.$$

(b) Let N be the number of ions that are separated by the machine per unit time. The current is $i = qN$ and the mass that is separated per unit time is $M = mN$, where m is the mass of a single ion. M has the value

$$M = \frac{100 \times 10^{-6} \text{ kg}}{3600 \text{ s}} = 2.78 \times 10^{-8} \text{ kg/s}.$$

Since $N = M/m$ we have

$$i = \frac{qM}{m} = \frac{(3.20 \times 10^{-19} \text{ C})(2.78 \times 10^{-8} \text{ kg/s})}{3.92 \times 10^{-25} \text{ kg}} = 2.27 \times 10^{-2} \text{ A}.$$

(c) Each ion deposits energy qV in the cup, so the energy deposited in time Δt is given by

$$E = NqV \, \Delta t = \frac{iqV}{q} \Delta t = iV \, \Delta t.$$

For $\Delta t = 1.0$ h,

$$E = (2.27 \times 10^{-2} \text{ A})(100 \times 10^3 \text{ V})(3600 \text{ s}) = 8.17 \times 10^6 \text{ J}.$$

To obtain the second expression, i/q is substituted for N.

26. Eq. 28-17 gives $T = 2\pi m_e/eB$. Thus, the total time is

$$\left(\frac{T}{2}\right)_1 + t_{gap} + \left(\frac{T}{2}\right)_2 = \frac{\pi m_e}{e}\left(\frac{1}{B_1} + \frac{1}{B_2}\right) + t_{gap}.$$

The time spent in the gap (which is where the electron is accelerating in accordance with Eq. 2-15) requires a few steps to figure out: letting $t = t_{gap}$ then we want to solve

$$d = v_0 t + \frac{1}{2} a t^2$$

$$0.25 \text{ m} = \sqrt{\frac{2K_0}{m_e}} t + \frac{1}{2} \frac{e \, \Delta V}{m_e \, d} t^2$$

for t. We find in this way that the time spent in the gap is $t \approx 6$ ns. Thus, the total time is 8.7 ns.

27. Each of the two particles will move in the same circular path, initially going in the opposite direction. After traveling half of the circular path they will collide. Therefore, using Eq. 28-17, the time is given by

$$t = \frac{T}{2} = \frac{\pi m}{Bq} = \frac{\pi (9.11 \times 10^{-31} \text{kg})}{(3.53 \times 10^{-3} \text{T})(1.60 \times 10^{-19} \text{C})} = 5.07 \times 10^{-9} \text{ s.}$$

28. (a) Using Eq. 28-23 and Eq. 28-18, we find

$$f_{osc} = \frac{qB}{2\pi m_p} = \frac{(1.60 \times 10^{-19} \text{C})(1.20\text{T})}{2\pi (1.67 \times 10^{-27} \text{kg})} = 1.83 \times 10^7 \text{ Hz.}$$

(b) From $r = m_p v / qB = \sqrt{2 m_p k}/qB$ we have

$$K = \frac{(rqB)^2}{2m_p} = \frac{\left[(0.500\text{m})(1.60 \times 10^{-19} \text{C})(1.20\text{T}) \right]^2}{2(1.67 \times 10^{-27} \text{kg})(1.60 \times 10^{-19} \text{J/eV})} = 1.72 \times 10^7 \text{ eV.}$$

29. We approximate the total distance by the number of revolutions times the circumference of the orbit corresponding to the average energy. This should be a good approximation since the deuteron receives the same energy each revolution and its period does not depend on its energy. The deuteron accelerates twice in each cycle, and each time it receives an energy of $qV = 80 \times 10^3$ eV. Since its final energy is 16.6 MeV, the number of revolutions it makes is

$$n = \frac{16.6 \times 10^6 \text{ eV}}{2(80 \times 10^3 \text{eV})} = 104 \; .$$

Its average energy during the accelerating process is 8.3 MeV. The radius of the orbit is given by $r = mv/qB$, where v is the deuteron's speed. Since this is given by $v = \sqrt{2K/m}$, the radius is

$$r = \frac{m}{qB}\sqrt{\frac{2K}{m}} = \frac{1}{qB}\sqrt{2Km}\,.$$

For the average energy

$$r = \frac{\sqrt{2(8.3\times 10^6\ \mathrm{eV})(1.60\times 10^{-19}\ \mathrm{J/eV})(3.34\times 10^{-27}\ \mathrm{kg})}}{(1.60\times 10^{-19}\ \mathrm{C})(1.57\,\mathrm{T})} = 0.375\,\mathrm{m}\,.$$

The total distance traveled is about $n2\pi r = (104)(2\pi)(0.375) = 2.4\times 10^2$ m.

30. (a) The magnitude of the field required to achieve resonance is

$$B = \frac{2\pi f m_p}{q} = \frac{2\pi(12.0\times 10^6\ \mathrm{Hz})(1.67\times 10^{-27}\,\mathrm{kg})}{1.60\times 10^{-19}\,\mathrm{C}} = 0.787\,\mathrm{T}.$$

(b) The kinetic energy is given by

$$K = \tfrac{1}{2}mv^2 = \frac{1}{2}m(2\pi Rf)^2 = \frac{1}{2}(1.67\times 10^{-27}\,\mathrm{kg})4\pi^2(0.530\ \mathrm{m})^2(12.0\times 10^6\ \mathrm{Hz})^2$$
$$= 1.33\times 10^{-12}\,\mathrm{J} = 8.34\times 10^6\,\mathrm{eV}.$$

(c) The required frequency is

$$f = \frac{qB}{2\pi m_p} = \frac{(1.60\times 10^{-19}\,\mathrm{C})(1.57\,\mathrm{T})}{2\pi(1.67\times 10^{-27}\,\mathrm{kg})} = 2.39\times 10^7\ \mathrm{Hz}.$$

(d) The kinetic energy is given by

$$K = \tfrac{1}{2}mv^2 = \frac{1}{2}m(2\pi Rf)^2 = \frac{1}{2}(1.67\times 10^{-27}\,\mathrm{kg})4\pi^2(0.530\ \mathrm{m})^2(2.39\times 10^7\ \mathrm{Hz})^2$$
$$= 5.3069\times 10^{-12}\,\mathrm{J} = 3.32\times 10^7\,\mathrm{eV}.$$

31. (a) By conservation of energy (using qV for the potential energy which is converted into kinetic form) the kinetic energy gained in each pass is 200 eV.

(b) Multiplying the part (a) result by $n = 100$ gives $\Delta K = n(200\ \mathrm{eV}) = 20.0$ keV.

(c) Combining Eq. 28-16 with the kinetic energy relation $(n(200 \text{ eV}) = \frac{1}{2} m_p v^2$ in this particular application) leads to the expression

$$r = \frac{m_p}{e\,B} \sqrt{\frac{2n(200 \text{ eV})}{m_p}}.$$

which shows that r is proportional to $\sqrt{n}$. Thus, the percent increase defined in the problem in going from $n = 100$ to $n = 101$ is $\sqrt{101/100} - 1 = 0.00499$ or 0.499%.

32. The magnetic force on the (straight) wire is

$$F_B = iBL \sin \theta = (13.0\text{A})\,(1.50\text{T})\,(1.80\text{m})\,(\sin 35.0°) = 20.1\text{N}.$$

33. (a) The magnitude of the magnetic force on the wire is given by $F_B = iLB \sin \phi$, where i is the current in the wire, L is the length of the wire, B is the magnitude of the magnetic field, and ϕ is the angle between the current and the field. In this case $\phi = 70°$. Thus,

$$F_B = (5000\,\text{A})(100\,\text{m})(60.0 \times 10^{-6}\,\text{T})\sin 70° = 28.2\,\text{N} .$$

(b) We apply the right-hand rule to the vector product $\vec{F}_B = i\vec{L} \times \vec{B}$ to show that the force is to the west.

34. (a) From symmetry, we conclude that any x-component of force will vanish (evaluated over the entirety of the bent wire as shown). By the right-hand rule, a field in the $\hat{k}$ direction produces on each part of the bent wire a y-component of force pointing in the $-\hat{j}$ direction; each of these components has magnitude

$$|F_y| = i\ell |\vec{B}| \sin 30° = 8\,\text{N}.$$

Therefore, the force (in Newtons) on the wire shown in the figure is $-16\hat{j}$.

(b) The force exerted on the left half of the bent wire points in the $-\hat{k}$ direction, by the right-hand rule, and the force exerted on the right half of the wire points in the $+\hat{k}$ direction. It is clear that the magnitude of each force is equal, so that the force (evaluated over the entirety of the bent wire as shown) must necessarily vanish.

35. (a) The magnetic force on the wire must be upward and have a magnitude equal to the gravitational force mg on the wire. Since the field and the current are perpendicular to each other the magnitude of the magnetic force is given by $F_B = iLB$, where L is the length of the wire. Thus,

$$iLB = mg \Rightarrow i = \frac{mg}{LB} = \frac{(0.0130\,\text{kg})(9.8\,\text{m/s}^2)}{(0.620\,\text{m})(0.440\,\text{T})} = 0.467\,\text{A}.$$

(b) Applying the right-hand rule reveals that the current must be from left to right.

36. The magnetic force on the wire is

$$\vec{F}_B = i\vec{L} \times \vec{B} = iL\hat{i} \times \left(B_y\hat{j} + B_z\hat{k}\right) = iL\left(-B_z\hat{j} + B_y\hat{k}\right)$$

$$= (0.500\,\text{A})\,(0.500\,\text{m})\left[-(0.0100\,\text{T})\hat{j} + (0.00300\,\text{T})\hat{k}\right]$$

$$= \left(-2.50\times10^{-3}\hat{j} + 0.750\times10^{-3}\,\hat{k}\right)\text{N}.$$

37. (a) The magnetic force must push horizontally on the rod to overcome the force of friction, but it can be oriented so that it also pulls up on the rod and thereby reduces both the normal force and the force of friction. The forces acting on the rod are: $\vec{F}$, the force of the magnetic field; mg, the magnitude of the (downward) force of gravity; $\vec{F}_N$, the normal force exerted by the stationary rails upward on the rod; and $\vec{f}$, the (horizontal) force of friction. For definiteness, we assume the rod is on the verge of moving eastward, which means that $\vec{f}$ points westward (and is equal to its maximum possible value $\mu_s F_N$). Thus, $\vec{F}$ has an eastward component F_x and an upward component F_y, which can be related to the components of the magnetic field once we assume a direction for the current in the rod. Thus, again for definiteness, we assume the current flows northward. Then, by the right-hand rule, a downward component (B_d) of $\vec{B}$ will produce the eastward F_x, and a westward component (B_w) will produce the upward F_y. Specifically,

$$F_x = iLB_d \quad \text{and} \quad F_y = iLB_w.$$

Considering forces along a vertical axis, we find

$$F_N = mg - F_y = mg - iLB_w$$

so that

$$f = f_{s,\text{max}} = \mu_s\left(mg - iLB_w\right).$$

It is on the verge of motion, so we set the horizontal acceleration to zero:

$$F_x - f = 0 \Rightarrow iLB_d = \mu_s\left(mg - iLB_w\right).$$

The angle of the field components is adjustable, and we can minimize with respect to it. Defining the angle by $B_w = B \sin\theta$ and $B_d = B \cos\theta$ (which means θ is being measured from a vertical axis) and writing the above expression in these terms, we obtain

$$iLB\cos\theta = \mu_s\left(mg - iLB\sin\theta\right) \Rightarrow B = \frac{\mu_s mg}{iL\left(\cos\theta + \mu_s\sin\theta\right)}$$

which we differentiate (with respect to θ) and set the result equal to zero. This provides a determination of the angle:

$$\theta = \tan^{-1}\left(\mu_s\right) = \tan^{-1}\left(0.60\right) = 31°.$$

Consequently,

$$B_{min} = \frac{0.60\left(1.0\,\text{kg}\right)\left(9.8\,\text{m/s}^2\right)}{\left(50\,\text{A}\right)\left(1.0\,\text{m}\right)\left(\cos 31° + 0.60\sin 31°\right)} = 0.10\,\text{T}.$$

(b) As shown above, the angle is $\theta = \tan^{-1}\left(\mu_s\right) = \tan^{-1}\left(0.60\right) = 31°$.

38. We use $d\vec{F}_B = id\vec{L} \times \vec{B}$, where $d\vec{L} = dx\hat{i}$ and $\vec{B} = B_x\hat{i} + B_y\hat{j}$. Thus,

$$\vec{F}_B = \int id\vec{L} \times \vec{B} = \int_{x_i}^{x_f} idx\hat{i} \times \left(B_x\hat{i} + B_y\hat{j}\right) = i\int_{x_i}^{x_f} B_y\,dx\hat{k}$$

$$= \left(-5.0\,\text{A}\right)\left(\int_{1.0}^{3.0}\left(8.0x^2\,dx\right)\left(\text{m}\cdot\text{mT}\right)\right)\hat{k} = (-0.35\text{N})\hat{k}.$$

39. The applied field has two components: $B_x > 0$ and $B_z > 0$. Considering each straight-segment of the rectangular coil, we note that Eq. 28-26 produces a non-zero force only for the component of $\vec{B}$ which is perpendicular to that segment; we also note that the equation is effectively multiplied by $N = 20$ due to the fact that this is a 20-turn coil. Since we wish to compute the torque about the hinge line, we can ignore the force acting on the straight-segment of the coil which lies along the y axis (forces acting at the axis of rotation produce no torque about that axis). The top and bottom straight-segments experience forces due to Eq. 28-26 (caused by the B_z component), but these forces are (by the right-hand rule) in the $\pm y$ directions and are thus unable to produce a torque about the y axis. Consequently, the torque derives completely from the force exerted on the straight-segment located at $x = 0.050$ m, which has length $L = 0.10$ m and is shown in Figure 28-41 carrying current in the $-y$ direction. Now, the B_z component will produce a force on this straight-segment which points in the $-x$ direction (back towards the hinge) and thus will exert no torque about the hinge. However, the B_x component (which is equal to $B \cos\theta$ where $B = 0.50$ T and $\theta = 30°$) produces a force equal to $NiLB_x$ which points (by the right-hand rule) in the $+z$ direction. Since the action of this force is perpendicular

to the plane of the coil, and is located a distance x away from the hinge, then the torque has magnitude

$$\tau = (NiLB_x)(x) = NiLxB\cos\theta = (20)(0.10)(0.10)(0.050)(0.50)\cos 30° = 0.0043$$

in SI units (N·m). Since $\vec{\tau} = \vec{r} \times \vec{F}$, the direction of the torque is $-y$. In unit-vector notation, the torque is $\vec{\tau} = (-4.3 \times 10^{-3} \text{ N} \cdot \text{m})\hat{j}$

An alternative way to do this problem is through the use of Eq. 28-37. We do not show those details here, but note that the magnetic moment vector (a necessary part of Eq. 28-37) has magnitude

$$|\vec{\mu}| = NiA = (20)(0.10\,\text{A})(0.0050\,\text{m}^2)$$

and points in the $-z$ direction. At this point, Eq. 3-30 may be used to obtain the result for the torque vector.

40. We establish coordinates such that the two sides of the right triangle meet at the origin, and the $\ell_y = 50$ cm side runs along the $+y$ axis, while the $\ell_x = 120$ cm side runs along the $+x$ axis. The angle made by the hypotenuse (of length 130 cm) is $\theta = \tan^{-1}(50/120) = 22.6°$, relative to the 120 cm side. If one measures the angle counterclockwise from the $+x$ direction, then the angle for the hypotenuse is $180° - 22.6° = +157°$. Since we are only asked to find the magnitudes of the forces, we have the freedom to assume the current is flowing, say, counterclockwise in the triangular loop (as viewed by an observer on the $+z$ axis. We take $\vec{B}$ to be in the same direction as that of the current flow in the hypotenuse. Then, with $B = |\vec{B}| = 0.0750\,\text{T}$,

$$B_x = -B\cos\theta = -0.0692\,\text{T} \quad \text{and} \quad B_y = B\sin\theta = 0.0288\,\text{T}.$$

(a) Eq. 28-26 produces zero force when $\vec{L} \| \vec{B}$ so there is no force exerted on the hypotenuse of length 130 cm.

(b) On the 50 cm side, the B_x component produces a force $i\ell_y B_x \hat{k}$, and there is no contribution from the B_y component. Using SI units, the magnitude of the force on the ℓ_y side is therefore

$$(4.00\,\text{A})(0.500\,\text{m})(0.0692\,\text{T}) = 0.138\,\text{N}.$$

(c) On the 120 cm side, the B_y component produces a force $i\ell_x B_y \hat{k}$, and there is no contribution from the B_x component. Using SI units, the magnitude of the force on the ℓ_x side is also

$$(4.00\,\mathrm{A})(1.20\,\mathrm{m})(0.0288\,\mathrm{T}) = 0.138\,\mathrm{N}.$$

(d) The net force is

$$i\ell_y B_x \hat{k} + i\ell_x B_y \hat{k} = 0,$$

keeping in mind that $B_x < 0$ due to our initial assumptions. If we had instead assumed $\vec{B}$ went the opposite direction of the current flow in the hypotenuse, then $B_x > 0$ but $B_y < 0$ and a zero net force would still be the result.

41. Consider an infinitesimal segment of the loop, of length ds. The magnetic field is perpendicular to the segment, so the magnetic force on it has magnitude $dF = iB\,ds$. The horizontal component of the force has magnitude $dF_h = (iB\cos\theta)\,ds$ and points inward toward the center of the loop. The vertical component has magnitude $dF_v = (iB\sin\theta)\,ds$ and points upward. Now, we sum the forces on all the segments of the loop. The horizontal component of the total force vanishes, since each segment of wire can be paired with another, diametrically opposite, segment. The horizontal components of these forces are both toward the center of the loop and thus in opposite directions. The vertical component of the total force is

$$F_v = iB\sin\theta \int ds = 2\pi a i B \sin\theta = 2\pi(0.018\ \mathrm{m})(4.6\times10^{-3}\ \mathrm{A})(3.4\times10^{-3}\ \mathrm{T})\sin20^\circ$$
$$= 6.0\times10^{-7}\ \mathrm{N}.$$

We note that i, B, and θ have the same value for every segment and so can be factored from the integral.

42. We use $\tau_{max} = |\vec{\mu}\times\vec{B}|_{max} = \mu B = i\pi r^2 B$, and note that $i = qf = qv/2\pi r$. So

$$\tau_{max} = \left(\frac{qv}{2\pi r}\right)\pi r^2 B = \frac{1}{2}qvrB = \frac{1}{2}(1.60\times10^{-19}\ \mathrm{C})(2.19\times10^6\ \mathrm{m/s})(5.29\times10^{-11}\ \mathrm{m})(7.10\times10^{-3}\ \mathrm{T})$$
$$= 6.58\times10^{-26}\ \mathrm{N\cdot m}.$$

43. (a) The current in the galvanometer should be 1.62 mA when the potential difference across the resistor-galvanometer combination is 1.00 V. The potential difference across the galvanometer alone is $iR_g = (1.62\times10^{-3}\ \mathrm{A})(75.3\Omega) = 0.122\ \mathrm{V}$, so the resistor must be in series with the galvanometer and the potential difference across it must be 1.00 V – 0.122 V = 0.878V. The resistance should be

$$R = (0.878\ \mathrm{V})/(1.62\times10^{-3}\ \mathrm{A}) = 542\,\Omega.$$

(b) As stated above, the resistor is in series with the galvanometer.

(c) The current in the galvanometer should be 1.62 mA when the total current in the resistor and galvanometer combination is 50.0 mA. The resistor should be in parallel with the galvanometer, and the current through it should be 50.0 mA – 1.62 mA = 48.38 mA. The potential difference across the resistor is the same as that across the galvanometer, 0.122 V, so the resistance should be $R = (0.122 \text{ V})/(48.38 \times 10^{-3} \text{ A}) = 2.52 \, \Omega$.

(d) As stated in (c), the resistor is in parallel with the galvanometer.

44. The insight central to this problem is that for a given length of wire (formed into a rectangle of various possible aspect ratios), the maximum possible area is enclosed when the ratio of height to width is 1 (that is, when it is a square). The maximum possible value for the width, the problem says, is $x = 4$ cm (this is when the height is very close to zero, so the total length of wire is effectively 8 cm). Thus, when it takes the shape of a square the value of x must be ¼ of 8 cm; that is, $x = 2$ cm when it encloses maximum area (which leads to a maximum torque by Eq. 28-35 and Eq. 28-37) of $A = (0.02 \text{ m})^2 = 0.0004 \text{ m}^2$. Since $N = 1$ and the torque in this case is given as $4.8 \times 10^{-4} \text{ N·m}$, then the aforementioned equations lead immediately to $i = 0.0030$ A.

45. We use Eq. 28-37 where $\vec{\mu}$ is the magnetic dipole moment of the wire loop and $\vec{B}$ is the magnetic field, as well as Newton's second law. Since the plane of the loop is parallel to the incline the dipole moment is normal to the incline. The forces acting on the cylinder are the force of gravity mg, acting downward from the center of mass, the normal force of the incline F_N, acting perpendicularly to the incline through the center of mass, and the force of friction f, acting up the incline at the point of contact. We take the x axis to be positive down the incline. Then the x component of Newton's second law for the center of mass yields

$$mg \sin \theta - f = ma.$$

For purposes of calculating the torque, we take the axis of the cylinder to be the axis of rotation. The magnetic field produces a torque with magnitude $\mu B \sin\theta$, and the force of friction produces a torque with magnitude fr, where r is the radius of the cylinder. The first tends to produce an angular acceleration in the counterclockwise direction, and the second tends to produce an angular acceleration in the clockwise direction. Newton's second law for rotation about the center of the cylinder, $\tau = I\alpha$, gives

$$fr - \mu B \sin \theta = I\alpha.$$

Since we want the current that holds the cylinder in place, we set $a = 0$ and $\alpha = 0$, and use one equation to eliminate f from the other. The result is $mgr = \mu B$. The loop is rectangular with two sides of length L and two of length $2r$, so its area is $A = 2rL$ and the dipole moment is $\mu = NiA = 2NirL$. Thus, $mgr = 2NirLB$ and

$$i = \frac{mg}{2NLB} = \frac{(0.250\,\text{kg})(9.8\,\text{m/s}^2)}{2(10.0)(0.100\,\text{m})(0.500\,\text{T})} = 2.45\,\text{A}.$$

46. From $\mu = NiA = i\pi r^2$ we get

$$i = \frac{\mu}{\pi r^2} = \frac{8.00 \times 10^{22}\,\text{J/T}}{\pi(3500 \times 10^3\,\text{m})^2} = 2.08 \times 10^9\,\text{A}.$$

47. (a) The magnitude of the magnetic dipole moment is given by $\mu = NiA$, where N is the number of turns, i is the current in each turn, and A is the area of a loop. In this case the loops are circular, so $A = \pi r^2$, where r is the radius of a turn. Thus

$$i = \frac{\mu}{N\pi r^2} = \frac{2.30\,\text{A}\cdot\text{m}^2}{(160)(\pi)(0.0190\,\text{m})^2} = 12.7\,\text{A}.$$

(b) The maximum torque occurs when the dipole moment is perpendicular to the field (or the plane of the loop is parallel to the field). It is given by

$$\tau_{\text{max}} = \mu B = (2.30\ \text{A}\cdot\text{m}^2)(35.0 \times 10^{-3}\,\text{T}) = 8.05 \times 10^{-2}\,\text{N}\cdot\text{m}.$$

48. (a) $\mu = NAi = \pi r^2 i = \pi(0.150\,\text{m})^2(2.60\,\text{A}) = 0.184\text{A}\cdot\text{m}^2.$

(b) The torque is

$$\tau = |\vec{\mu} \times \vec{B}| = \mu B \sin\theta = (0.184\ \text{A}\cdot\text{m}^2)(12.0\,\text{T})\sin 41.0° = 1.45\,\text{N}\cdot\text{m}.$$

49. (a) The area of the loop is $A = \frac{1}{2}(30\,\text{cm})(40\,\text{cm}) = 6.0 \times 10^2\,\text{cm}^2$, so

$$\mu = iA = (5.0\,\text{A})(6.0 \times 10^{-2}\,\text{m}^2) = 0.30\,\text{A}\cdot\text{m}^2.$$

(b) The torque on the loop is

$$\tau = \mu B \sin\theta = (0.30\,\text{A}\cdot\text{m}^2)(80 \times 10^3\,\text{T})\sin 90° = 2.4 \times 10^{-2}\,\text{N}\cdot\text{m}.$$

50. (a) The kinetic energy gained is due to the potential energy decrease as the dipole swings from a position specified by angle θ to that of being aligned (zero angle) with the field. Thus,

$$K = U_i - U_f = -\mu B \cos\theta - (-\mu B \cos 0°).$$

Therefore, using SI units, the angle is

$$\theta = \cos^{-1}\left(1 - \frac{K}{\mu B}\right) = \cos^{-1}\left(1 - \frac{0.00080}{(0.020)(0.052)}\right) = 77^\circ.$$

(b) Since we are making the assumption that no energy is dissipated in this process, then the dipole will continue its rotation (similar to a pendulum) until it reaches an angle $\theta = 77^\circ$ on the other side of the alignment axis.

51. (a) The magnitude of the magnetic moment vector is

$$\mu = \sum_n i_n A_n = \pi r_1^2 i_1 + \pi r_2^2 i_2 = \pi(7.00\text{A})\left[(0.200\text{m})^2 + (0.300\text{m})^2\right] = 2.86\,\text{A}\cdot\text{m}^2.$$

(b) Now,

$$\mu = \pi r_2^2 i_2 - \pi r_1^2 i_1 = \pi(7.00\text{A})\left[(0.300\text{m})^2 - (0.200\text{m})^2\right] = 1.10\,\text{A}\cdot\text{m}^2.$$

52. Let $a = 30.0$ cm, $b = 20.0$ cm, and $c = 10.0$ cm. From the given hint, we write

$$\vec{\mu} = \vec{\mu}_1 + \vec{\mu}_2 = iab\left(-\hat{k}\right) + iac\left(\hat{j}\right) = ia\left(c\hat{j} - b\hat{k}\right) = (5.00\text{A})(0.300\text{m})\left[(0.100\text{m})\hat{j} - (0.200\text{m})\hat{k}\right]$$
$$= \left(0.150\hat{j} - 0.300\hat{k}\right)\text{A}\cdot\text{m}^2.$$

53. The magnetic dipole moment is $\vec{\mu} = \mu\left(0.60\,\hat{i} - 0.80\,\hat{j}\right)$, where

$$\mu = NiA = Ni\pi r^2 = 1(0.20\text{ A})\pi(0.080\text{ m})^2 = 4.02 \times 10^{-4}\text{ A}\cdot\text{m}^2.$$

Here i is the current in the loop, N is the number of turns, A is the area of the loop, and r is its radius.

(a) The torque is

$$\vec{\tau} = \vec{\mu} \times \vec{B} = \mu\left(0.60\hat{i} - 0.80\hat{j}\right) \times \left(0.25\hat{i} + 0.30\hat{k}\right)$$
$$= \mu\left[(0.60)(0.30)\left(\hat{i} \times \hat{k}\right) - (0.80)(0.25)\left(\hat{j} \times \hat{i}\right) - (0.80)(0.30)\left(\hat{j} \times \hat{k}\right)\right]$$
$$= \mu\left[-0.18\hat{j} + 0.20\hat{k} - 0.24\hat{i}\right].$$

Here $\hat{i} \times \hat{k} = -\hat{j}, \hat{j} \times \hat{i} = -\hat{k}$, and $\hat{j} \times \hat{k} = \hat{i}$ are used. We also use $\hat{i} \times \hat{i} = 0$. Now, we substitute the value for μ to obtain

$$\vec{\tau} = \left(-9.7 \times 10^{-4}\hat{i} - 7.2 \times 10^{-4}\hat{j} + 8.0 \times 10^{-4}\hat{k}\right) N \cdot m.$$

(b) The potential energy of the dipole is given by

$$U = -\vec{\mu} \cdot \vec{B} = -\mu\left(0.60\hat{i} - 0.80\hat{j}\right) \cdot \left(0.25\hat{i} + 0.30\hat{k}\right)$$
$$= -\mu(0.60)(0.25) = -0.15\mu = -6.0 \times 10^{-4} J.$$

Here $\hat{i} \cdot \hat{i} = 1, \hat{i} \cdot \hat{k} = 0, \hat{j} \cdot \hat{i} = 0,$ and $\hat{j} \cdot \hat{k} = 0$ are used.

54. Looking at the point in the graph (Fig. 28-47-2(b)) corresponding to $i_2 = 0$ (which means that coil 2 has no magnetic moment) we are led to conclude that the magnetic moment of coil 1 must be 2.0×10^{-5} A·m². Looking at the point where the line crosses the axis (at $i_2 = 5$ mA) we conclude (since the magnetic moments cancel there) that the magnitude of coil 2's moment must also be 2.0×10^{-5} A·m² when $i_2 = 0.005$ A which means (Eq. 28-35) $N_2A_2 = (2.0 \times 10^{-5})/(0.005) = 0.004$ in SI units. Now the problem has us consider the direction of coil 2's current changed so that the net moment is the sum of two (positive) contributions – from coil 1 and coil 2 – specifically for the case that $i_2 = 0.007$ A. We find that total moment is $(2.0 \times 10^{-5}$ A·m²$) + (N_2A_2 i_2) = 4.8 \times 10^{-5}$ A·m².

55. If N closed loops are formed from the wire of length L, the circumference of each loop is L/N, the radius of each loop is $R = L/2\pi N$, and the area of each loop is $A = \pi R^2 = \pi\left(L/2\pi N\right)^2 = L^2/4\pi N^2$.

(a) For maximum torque, we orient the plane of the loops parallel to the magnetic field, so the dipole moment is perpendicular (i.e., at a 90° angle) to the field.

(b) The magnitude of the torque is then

$$\tau = NiAB = \left(Ni\right)\left(\frac{L^2}{4\pi N^2}\right)B = \frac{iL^2 B}{4\pi N}.$$

To maximize the torque, we take the number of turns N to have the smallest possible value, 1. Then $\tau = iL^2 B/4\pi$.

(c) The magnitude of the maximum torque is

$$\tau = \frac{iL^2 B}{4\pi} = \frac{(4.51 \times 10^{-3} \text{ A})(0.250 \text{ m})^2 (5.71 \times 10^{-3}\text{T})}{4\pi} = 1.28 \times 10^{-7} \text{ N} \cdot \text{m}$$

56. Eq. 28-39 gives $U = -\vec{\mu} \cdot \vec{B} = -\mu B \cos\phi$, so at $\phi = 0$ (corresponding to the lowest point on the graph in Fig. 28-48) the mechanical energy is

$$K + U = K_o + (-\mu B) = 6.7 \times 10^{-4} \, \text{J} + (-5 \times 10^{-4} \, \text{J}) = 1.7 \times 10^{-4} \, \text{J}.$$

The turning point occurs where $K = 0$, which implies $U_{\text{turn}} = 1.7 \times 10^{-4} \, \text{J}$. So the angle where this takes place is given by

$$\phi = -\cos^{-1}\left(\frac{1.7 \times 10^{-4} \, \text{J}}{\mu B}\right) = 110°$$

where we have used the fact (see above) that $\mu B = 5 \times 10^{-4} \, \text{J}$.

57. Let $v_\| = v \cos\theta$. The electron will proceed with a uniform speed $v_\|$ in the direction of $\vec{B}$ while undergoing uniform circular motion with frequency f in the direction perpendicular to B: $f = eB/2\pi m_e$. The distance d is then

$$d = v_\| T = \frac{v_\|}{f} = \frac{(v \cos\theta) 2\pi m_e}{eB} = \frac{2\pi (1.5 \times 10^7 \, \text{m/s})(9.11 \times 10^{-31} \, \text{kg})(\cos 10°)}{(1.60 \times 10^{-19} \, \text{C})(1.0 \times 10^{-3} \, \text{T})} = 0.53 \text{m}.$$

58. Combining Eq. 28-16 with energy conservation ($eV = \frac{1}{2} m_e v^2$ in this particular application) leads to the expression

$$r = \frac{m_e}{e \, B} \sqrt{\frac{2eV}{m_e}}$$

which suggests that the slope of the r versus $\sqrt{V}$ graph should be $\sqrt{2m_e/eB^2}$. From Fig. 28-49, we estimate the slope to be 5×10^{-5} in SI units. Setting this equal to $\sqrt{2m_e/eB^2}$ and solving we find $B = 6.7 \times 10^{-2} \, \text{T}$.

59. The period of revolution for the iodine ion is $T = 2\pi r/v = 2\pi m/Bq$, which gives

$$m = \frac{BqT}{2\pi} = \frac{(45.0 \times 10^{-3} \, \text{T})(1.60 \times 10^{-19} \, \text{C})(1.29 \times 10^{-3} \, \text{s})}{(7)(2\pi)(1.66 \times 10^{-27} \, \text{kg/u})} = 127 \, \text{u}.$$

60. Let ξ stand for the ratio ($m/|q|$) we wish to solve for. Then Eq. 28-17 can be written as $T = 2\pi\xi/B$. Noting that the horizontal axis of the graph (Fig. 28-50) is inverse-field $(1/B)$ then we conclude (from our previous expression) that the slope of the line in the graph must be equal to $2\pi\xi$. We estimate that slope as 7.5×10^{-9} T·s, which implies $\xi = 1.2 \times 10^{-9} \, \text{kg/C}$.

61. The fact that the fields are uniform, with the feature that the charge moves in a straight line, implies the speed is constant (if it were not, then the magnetic *force* would vary while the electric force could not — causing it to deviate from straight-line motion). This is then the situation leading to Eq. 28-7, and we find

$$|\vec{E}| = v|\vec{B}| = 500\,\text{V/m}.$$

Its direction (so that $\vec{F} = q(\vec{E} + \vec{v} \times \vec{B})$ vanishes) is downward, or $-\hat{j}$, in the "page" coordinates. In unit-vector notation, $\vec{E} = (-500\,\text{V/m})\hat{j}$

62. The unit vector associated with the current element (of magnitude $d\ell$) is $-\hat{j}$. The (infinitesimal) force on this element is

$$d\vec{F} = i\,d\ell\left(-\hat{j}\right) \times \left(0.3y\hat{i} + 0.4y\hat{j}\right)$$

with SI units (and 3 significant figures) understood. Since $\hat{j} \times \hat{i} = -\hat{k}$ and $\hat{j} \times \hat{j} = 0$, we obtain

$$d\vec{F} = 0.3iy\,d\ell\,\hat{k} = \left(6.00 \times 10^{-4}\,\text{N/m}^2\right)y\,d\ell\,\hat{k}.$$

We integrate the force element found above, using the symbol ξ to stand for the coefficient 6.00×10^{-4} N/m^2, and obtain

$$\vec{F} = \int d\vec{F} = \xi\hat{k}\int_0^{0.25} y\,dy = \xi\hat{k}\left(\frac{0.25^2}{2}\right) = (1.88 \times 10^{-5}\,\text{N})\hat{k}.$$

63. By the right-hand rule, we see that $\vec{v} \times \vec{B}$ points along $-\hat{k}$. From Eq. 28-2 $\left(\vec{F} = q\vec{v} \times \vec{B}\right)$, we find that for the force to point along $+\hat{k}$, we must have $q < 0$. Now, examining the magnitudes (in SI units) in Eq. 28-3, we find $|\vec{F}| = |q||v||\vec{B}|\sin\phi$, or

$$0.48 = |q|(4000)(0.0050)\sin 35°$$

which yields $|q| = 0.040$ C. In summary, then, $q = -40$ mC.

64. (a) The net force on the proton is given by

$$\vec{F} = \vec{F}_E + \vec{F}_B = q\vec{E} + q\vec{v} \times \vec{B}$$
$$= \left(1.60 \times 10^{-19}\,\text{C}\right)\left[\left(4.00\,\text{V/m}\right)\hat{k} + \left(2000\,\text{m/s}\right)\hat{j} \times \left(-2.50 \times 10^{-3}\,\text{T}\right)\hat{i}\right]$$
$$= \left(1.44 \times 10^{-18}\,\text{N}\right)\hat{k}.$$

(b) In this case

$$\vec{F} = \vec{F}_E + \vec{F}_B = q\vec{E} + q\vec{v} \times \vec{B}$$
$$= \left(1.60 \times 10^{-19}\,\text{C}\right)\left[\left(-4.00\,\text{V/m}\right)\hat{k} + \left(2000\,\text{m/s}\right)\hat{j} \times \left(-2.50\,\text{mT}\right)\hat{i}\right]$$
$$= \left(1.60 \times 10^{-19}\,\text{N}\right)\hat{k}.$$

(c) In the final case,

$$\vec{F} = \vec{F}_E + \vec{F}_B = q\vec{E} + q\vec{v} \times \vec{B}$$
$$= \left(1.60 \times 10^{-19}\,\text{C}\right)\left[\left(4.00\,\text{V/m}\right)\hat{i} + \left(2000\,\text{m/s}\right)\hat{j} \times \left(-2.50\,\text{mT}\right)\hat{i}\right]$$
$$= \left(6.41 \times 10^{-19}\,\text{N}\right)\hat{i} + \left(8.01 \times 10^{-19}\,\text{N}\right)\hat{k}.$$

65. Letting $B_x = B_y = B_1$ and $B_z = B_2$ and using Eq. 28-2 and Eq. 3-30, we obtain (with SI units understood)

$$\vec{F} = q\vec{v} \times \vec{B}$$
$$4\hat{i} - 20\hat{j} + 12\hat{k} = 2\left(\left(4B_2 - 6B_1\right)\hat{i} + \left(6B_1 - 2B_2\right)\hat{j} + \left(2B_1 - 4B_1\right)\hat{k}\right).$$

Equating like components, we find $B_1 = -3$ and $B_2 = -4$. In summary (with the unit Tesla understood), $\vec{B} = -3.0\hat{i} - 3.0\hat{j} - 4.0\hat{k}$.

66. (a) Eq. 3-20 gives $\phi = \cos^{-1}(2/19) = 84°$.

(b) No, the magnetic field can only change the direction of motion of a free (unconstrained) particle, not its speed or its kinetic energy.

(c) No, as reference to to Fig. 28-11 should make clear.

(d) We find $v_\perp = v\,\sin\phi = 61.3$ m/s, so $r = mv_\perp/eB = 5.7$ nm.

67. (a) Using Eq. 28-35 and Figure 28-23, we have

$$\vec{\mu} = (NiA)\,(-\hat{j}) = -0.0240\hat{j}\ \text{A·m}^2\ .$$

Then, Eq. 28-38 gives $U = -\vec{\mu} \cdot \vec{B} = -(-0.0240)\,(-3.00 \times 10^{-3}) = -7.20 \times 10^{-5}$ J .

(b) Using the fact that $\hat{j} \times \hat{j} = 0$, Eq. 28-37 leads to

$$\vec{\tau} = \vec{\mu} \times \vec{B} = (-0.0240\hat{j}) \times (2.00 \times 10^{-3}\,\hat{i}) + (-0.0240\hat{j}) \times (-4.00 \times 10^{-3}\,\hat{k})$$
$$= (4.80 \times 10^{-5}\,\hat{k} + 9.60 \times 10^{-5}\,\hat{i})\ \text{N·m}.$$

68. (a) We use Eq. 28-10: $v_d = E/B = (10 \times 10^{-6}\text{V}/1.0 \times 10^{-2}\,\text{m})/(1.5\,\text{T}) = 6.7 \times 10^{-4}$ m/s.

(b) We rewrite Eq. 28-12 in terms of the electric field:

$$n = \frac{Bi}{V\ell e} = \frac{Bi}{(Ed)\ell e} = \frac{Bi}{EAe}$$

which we use $A = \ell d$. In this experiment, $A = (0.010 \text{ m})(10 \times 10^{-6} \text{ m}) = 1.0 \times 10^{-7} \text{ m}^2$. By Eq. 28-10, v_d equals the ratio of the fields (as noted in part (a)), so we are led to

$$n = \frac{Bi}{E \, Ae} = \frac{i}{v_d \, Ae} = \frac{3.0 \text{ A}}{(6.7 \times 10^{-4} \text{ m/s})(1.0 \times 10^{-7} \text{ m}^2)(1.6 \times 10^{-19} \text{ C})} = 2.8 \times 10^{29}/\text{m}^3.$$

(c) Since a drawing of an inherently 3-D situation can be misleading, we describe it in terms of horizontal *north, south, east, west* and vertical *up* and *down* directions. We assume $\vec{B}$ points up and the conductor's width of 0.010 m is along an east-west line. We take the current going northward. The conduction electrons experience a westward magnetic force (by the right-hand rule), which results in the west side of the conductor being negative and the east side being positive (with reference to the Hall voltage which becomes established).

69. The contribution to the force by the magnetic field $\left(\vec{B} = B_x\hat{i} = -0.020\hat{i} \text{ T}\right)$ is given by Eq. 28-2:

$$\vec{F}_B = q\vec{v} \times \vec{B} = q\left(\left(17000\hat{i} \times B_x\hat{i}\right) + \left(-11000\hat{j} \times B_x\hat{i}\right) + \left(7000\hat{k} \times B_x\hat{i}\right)\right)$$
$$= q\left(-220\hat{k} - 140\hat{j}\right)$$

in SI units. And the contribution to the force by the electric field $\left(\vec{E} = E_y\hat{j} = 300\hat{j} \text{ V/m}\right)$ is given by Eq. 23-1: $\vec{F}_E = qE_y\hat{j}$. Using $q = 5.0 \times 10^{-6}$ C, the net force on the particle is

$$\vec{F} = (0.00080\hat{j} - 0.0011\hat{k}) \text{ N}.$$

70. (a) We use Eq. 28-2 and Eq. 3-30:

$$\vec{F} = q\vec{v} \times \vec{B} = (+e)\left(\left(v_y B_z - v_z B_y\right)\hat{i} + \left(v_z B_x - v_x B_z\right)\hat{j} + \left(v_x B_y - v_y B_x\right)\hat{k}\right)$$
$$= \left(1.60 \times 10^{-19}\right)\left(((4)(0.008) - (-6)(-0.004))\hat{i} + \right.$$
$$\left. ((-6)(0.002) - (-2)(0.008))\hat{j} + ((-2)(-0.004) - (4)(0.002))\hat{k}\right)$$
$$= \left(1.28 \times 10^{-21}\right)\hat{i} + \left(6.41 \times 10^{-22}\right)\hat{j}$$

with SI units understood.

(b) By definition of the cross product, $\vec{v} \perp \vec{F}$. This is easily verified by taking the dot (scalar) product of $\vec{v}$ with the result of part (a), yielding zero, provided care is taken not to introduce any round-off error.

(c) There are several ways to proceed. It may be worthwhile to note, first, that if B_z were 6.00 mT instead of 8.00 mT then the two vectors would be exactly antiparallel. Hence, the angle θ between $\vec{B}$ and $\vec{v}$ is presumably "close" to 180°. Here, we use Eq. 3-20:

$$\theta = \cos^{-1}\left(\frac{\vec{v} \cdot \vec{B}}{|\vec{v}||\vec{B}|}\right) = \cos^{-1}\left(\frac{-68}{\sqrt{56}\sqrt{84}}\right) = 173°$$

71. (a) The magnetic force on the wire is $F_B = idB$, pointing to the left. Thus

$$v = at = \frac{F_B t}{m} = \frac{idBt}{m}$$

$$= \frac{(9.13 \times 10^{-3}\,\text{A})(2.56 \times 10^{-2}\,\text{m})(5.63 \times 10^{-2}\,\text{T})(0.0611\text{s})}{2.41 \times 10^{-5}\,\text{kg}} = 3.34 \times 10^{-2}\,\text{m/s}.$$

(b) The direction is to the left (away from the generator).

72. (a) We are given $\vec{B} = B_x\hat{\imath} = 6 \times 10^{-5}\hat{\imath}\,\text{T}$, so that $\vec{v} \times \vec{B} = -v_y B_x\hat{k}$ where $v_y = 4 \times 10^4$ m/s. We note that the magnetic force on the electron is $(-e)\left(-v_y B_x\hat{k}\right)$ and therefore points in the $+\hat{k}$ direction, at the instant the electron enters the field-filled region. In these terms, Eq. 28-16 becomes

$$r = \frac{m_e v_y}{e\,B_x} = 0.0038\,\text{m}.$$

(b) One revolution takes $T = 2\pi r/v_y = 0.60$ μs, and during that time the "drift" of the electron in the x direction (which is the *pitch* of the helix) is $\Delta x = v_x T = 0.019$ m where $v_x = 32 \times 10^3$ m/s.

(c) Returning to our observation of force direction made in part (a), we consider how this is perceived by an observer at some point on the $-x$ axis. As the electron moves away from him, he sees it enter the region with positive v_y (which he might call "upward") but "pushed" in the $+z$ direction (to his right). Hence, he describes the electron's spiral as clockwise.

73. The force associated with the magnetic field must point in the $\hat{j}$ direction in order to cancel the force of gravity in the $-\hat{j}$ direction. By the right-hand rule, $\vec{B}$ points in the $-\hat{k}$ direction (since $\hat{i} \times (-\hat{k}) = \hat{j}$). Note that the charge is positive; also note that we need to assume $B_y = 0$. The magnitude $|B_z|$ is given by Eq. 28-3 (with $\phi = 90°$). Therefore, with $m = 10 \times 10^{-3}$ kg, $v = 2.0 \times 10^4$ m/s and $q = 80 \times 10^{-6}$ C, we find

$$\vec{B} = B_z\hat{k} = -\left(\frac{mg}{qv}\right)\hat{k} = (-0.061 \text{ T})\hat{k}$$

74. With the $\vec{B}$ pointing "out of the page," we evaluate the force (using the right-hand rule) at, say, the dot shown on the left edge of the particle's path, where its velocity is down. If the particle were positively charged, then the force at the dot would be toward the left, which is at odds with the figure (showing it being bent towards the right). Therefore, the particle is negatively charged; it is an electron.

(a) Using Eq. 28-3 (with angle ϕ equal to 90°), we obtain

$$v = \frac{|\vec{F}|}{e|\vec{B}|} = 4.99 \times 10^6 \text{ m/s}.$$

(b) Using either Eq. 28-14 or Eq. 28-16, we find $r = 0.00710$ m.

(c) Using Eq. 28-17 (in either its first or last form) readily yields $T = 8.93 \times 10^{-9}$ s.

75. The current is in the $+\hat{i}$ direction. Thus, the $\hat{i}$ component of $\vec{B}$ has no effect, and (with x in meters) we evaluate

$$\vec{F} = (3.00\text{A}) \int_0^1 (-0.600\,\text{T/m}^2)x^2 dx (\hat{i} \times \hat{j}) = \left(-1.80\frac{1^3}{3}\,\text{A}\cdot\text{T}\cdot\text{m}\right)\hat{k} = (-0.600\text{N})\hat{k}.$$

76. (a) The largest value of force occurs if the velocity vector is perpendicular to the field. Using Eq. 28-3,

$F_{B,\text{max}} = |q|\, vB \sin (90°) = ev\, B = (1.60 \times 10^{-19} \text{ C}) (7.20 \times 10^6 \text{ m/s}) (83.0 \times 10^{-3} \text{ T})$
$= 9.56 \times 10^{-14}$ N.

(b) The smallest value occurs if they are parallel: $F_{B,\text{min}} = |q|\, vB \sin (0) = 0$.

(c) By Newton's second law, $a = F_B/m_e = |q|\, vB \sin \theta /m_e$, so the angle θ between $\vec{v}$ and $\vec{B}$ is

$$\theta = \sin^{-1}\left(\frac{m_e a}{|q|vB}\right) = \sin^{-1}\left[\frac{\left(9.11 \times 10^{-31}\,\text{kg}\right)\left(4.90 \times 10^{14}\,\text{m/s}^2\right)}{\left(1.60 \times 10^{-16}\,\text{C}\right)\left(7.20 \times 10^6\,\text{m/s}\right)\left(83.0 \times 10^{-3}\,\text{T}\right)}\right] = 0.267°.$$

77. (a) We use $\vec{\tau} = \vec{\mu} \times \vec{B}$, where $\vec{\mu}$ points into the wall (since the current goes clockwise around the clock). Since $\vec{B}$ points towards the one-hour (or "5-minute") mark, and (by the properties of vector cross products) $\vec{\tau}$ must be perpendicular to it, then (using the right-hand rule) we find $\vec{\tau}$ points at the 20-minute mark. So the time interval is 20 min.

(b) The torque is given by

$$\tau = |\vec{\mu} \times \vec{B}| = \mu B \sin 90° = NiAB = \pi N i r^2 B = 6\pi (2.0\,\text{A})(0.15\,\text{m})^2 \left(70 \times 10^{-3}\,\text{T}\right)$$
$$= 5.9 \times 10^{-2}\,\text{N} \cdot \text{m}.$$

78. From $m = B^2 q x^2 / 8V$ we have $\Delta m = (B^2 q/8V)(2x\Delta x)$. Here $x = \sqrt{8Vm/B^2 q}$, which we substitute into the expression for Δm to obtain

$$\Delta m = \left(\frac{B^2 q}{8V}\right) 2\sqrt{\frac{8mV}{B^2 q}}\Delta x = B\sqrt{\frac{mq}{2V}}\Delta x.$$

Thus, the distance between the spots made on the photographic plate is

$$\Delta x = \frac{\Delta m}{B}\sqrt{\frac{2V}{mq}}$$
$$= \frac{(37\,\text{u} - 35\,\text{u})\left(1.66 \times 10^{-27}\,\text{kg/u}\right)}{0.50\,\text{T}}\sqrt{\frac{2\left(7.3 \times 10^3\,\text{V}\right)}{\left(36\,\text{u}\right)\left(1.66 \times 10^{-27}\,\text{kg/u}\right)\left(1.60 \times 10^{-19}\,\text{C}\right)}}$$
$$= 8.2 \times 10^{-3}\,\text{m}.$$

79. (a) Since $K = qV$ we have $K_p = \frac{1}{2}K_\alpha \left(\text{as } q_\alpha = 2K_p\right)$, or $K_p / K_\alpha = 0.50$.

(b) Similarly, $q_\alpha = 2K_d$, $K_d / K_\alpha = 0.50$.

(c) Since $r = \sqrt{2mK}/qB \propto \sqrt{mK}/q$, we have

$$r_d = \sqrt{\frac{m_d K_d}{m_p K_p}}\frac{q_p r_p}{q_d} = \sqrt{\frac{(2.00\,\text{u})K_p}{(1.00\,\text{u})K_p}}r_p = 10\sqrt{2}\,\text{cm} = 14\,\text{cm}.$$

(d) Similarly, for the alpha particle, we have

$$r_\alpha = \sqrt{\frac{m_\alpha K_\alpha}{m_p K_p} \frac{q_p r_p}{q_\alpha}} = \sqrt{\frac{(4.00\,u) K_\alpha}{(1.00\,u)(K_\alpha/2)} \frac{e r_p}{2e}} = 10\sqrt{2}\,\text{cm} = 14\,\text{cm}.$$

80. (a) Equating the magnitude of the electric force ($F_e = eE$) with that of the magnetic force (Eq. 28-3), we obtain $B = E / v \sin \phi$. The field is smallest when the $\sin \phi$ factor is at its largest value; that is, when $\phi = 90°$. Now, we use $K = \frac{1}{2}mv^2$ to find the speed:

$$v = \sqrt{\frac{2K}{m_e}} = \sqrt{\frac{2(2.5\times 10^3\,\text{eV})(1.60\times 10^{-19}\,\text{J/eV})}{9.11\times 10^{-31}\,\text{kg}}} = 2.96\times 10^7\,\text{m/s}.$$

Thus,

$$B = \frac{E}{v} = \frac{10\times 10^3\,\text{V/m}}{2.96\times 10^7\,\text{m/s}} = 3.4\times 10^{-4}\,\text{T}.$$

The direction of the magnetic field must be perpendicular to both the electric field $(-\hat{j})$ and the velocity of the electron $(+\hat{i})$. Since the electric force $\vec{F}_e = (-e)\vec{E}$ points in the $+\hat{j}$ direction, the magnetic force $\vec{F}_B = (-e)\vec{v}\times\vec{B}$ points in the $-\hat{j}$ direction. Hence, the direction of the magnetic field is $-\hat{k}$. In unit-vector notation, $\vec{B} = (-3.4\times 10^{-4}\,\text{T})\hat{k}$.

81. (a) In Chapter 27, the electric field (called E_C in this problem) which "drives" the current through the resistive material is given by Eq. 27-11, which (in magnitude) reads $E_C = \rho J$. Combining this with Eq. 27-7, we obtain

$$E_C = \rho n e v_d.$$

Now, regarding the Hall effect, we use Eq. 28-10 to write $E = v_d B$. Dividing one equation by the other, we get $E/E_c = B/ne\rho$.

(b) Using the value of copper's resistivity given in Chapter 27, we obtain

$$\frac{E}{E_c} = \frac{B}{ne\rho} = \frac{0.65\,\text{T}}{(8.47\times 10^{28}/\text{m}^3)(1.60\times 10^{-19}\,\text{C})(1.69\times 10^{-8}\,\Omega\cdot\text{m})} = 2.84\times 10^{-3}.$$

82. (a) For the magnetic field to have an effect on the moving electrons, we need a non-negligible component of $\vec{B}$ to be perpendicular to $\vec{v}$ (the electron velocity). It is most efficient, therefore, to orient the magnetic field so it is perpendicular to the plane of the page. The magnetic force on an electron has magnitude $F_B = evB$, and the acceleration of the electron has magnitude $a = v^2/r$. Newton's second law yields $evB = m_e v^2/r$, so the

radius of the circle is given by $r = m_e v / eB$ in agreement with Eq. 28-16. The kinetic energy of the electron is $K = \frac{1}{2} m_e v^2$, so $v = \sqrt{2K/m_e}$. Thus,

$$r = \frac{m_e}{eB} \sqrt{\frac{2K}{m_e}} = \sqrt{\frac{2m_e K}{e^2 B^2}} \ .$$

This must be less than d, so $\sqrt{\dfrac{2m_e K}{e^2 B^2}} \leq d$, or $B \geq \sqrt{\dfrac{2m_e K}{e^2 d^2}}$.

(b) If the electrons are to travel as shown in Fig. 28-33, the magnetic field must be out of the page. Then the magnetic force is toward the center of the circular path, as it must be (in order to make the circular motion possible).

83. The equation of motion for the proton is

$$\vec{F} = q\vec{v} \times \vec{B} = q\left(v_x \hat{i} + v_y \hat{j} + v_z \hat{k}\right) \times B\hat{i} = qB\left(v_z \hat{j} - v_y \hat{k}\right)$$

$$= m_p \vec{a} = m_p \left[\left(\frac{dv_x}{dt}\right) \hat{i} + \left(\frac{dv_y}{dt}\right) \hat{j} + \left(\frac{dv_z}{dt}\right) \hat{k} \right].$$

Thus,

$$\frac{dv_x}{dt} = 0, \quad \frac{dv_y}{dt} = \omega v_z, \quad \frac{dv_z}{dt} = -\omega v_y,$$

where $\omega = eB/m$. The solution is $v_x = v_{0x}$, $v_y = v_{0y} \cos \omega t$ and $v_z = -v_{0y} \sin \omega t$. In summary, we have $\vec{v}(t) = v_{0x} \hat{i} + v_{0y} \cos(\omega t) \hat{j} - v_{0y} (\sin \omega t) \hat{k}$.

84. Referring to the solution of problem 19 part (b), we see that $r = \sqrt{2mK}/qB$ implies the proportionality: $r \propto \sqrt{mK}/qB$. Thus,

(a) $\dfrac{r_d}{r_p} = \sqrt{\dfrac{m_d K_d}{m_p K_p}} \dfrac{q_p}{q_d} = \sqrt{\dfrac{2.0u}{1.0u}} \dfrac{e}{e} = \sqrt{2} \approx 1.4$, and

(b) $\dfrac{r_\alpha}{r_p} = \sqrt{\dfrac{m_\alpha K_\alpha}{m_p K_p}} \dfrac{q_p}{q_\alpha} = \sqrt{\dfrac{4.0u}{1.0u}} \dfrac{e}{2e} = 1.0$.

85. (a) The textbook uses "geomagnetic north" to refer to Earth's magnetic pole lying in the northern hemisphere. Thus, the electrons are traveling northward. The vertical component of the magnetic field is downward. The right-hand rule indicates that $\vec{v} \times \vec{B}$ is

to the west, but since the electron is negatively charged (and $\vec{F} = q\vec{v} \times \vec{B}$), the magnetic force on it is to the east.

We combine $F = m_e a$ with $F = evB \sin \phi$. Here, $B \sin \phi$ represents the downward component of Earth's field (given in the problem). Thus, $a = evB / m_e$. Now, the electron speed can be found from its kinetic energy. Since $K = \frac{1}{2}mv^2$,

$$v = \sqrt{\frac{2K}{m_e}} = \sqrt{\frac{2(12.0 \times 10^3 \, \text{eV})(1.60 \times 10^{-19} \, \text{J/eV})}{9.11 \times 10^{-31} \, \text{kg}}} = 6.49 \times 10^7 \, \text{m/s}.$$

Therefore,

$$a = \frac{evB}{m_e} = \frac{(1.60 \times 10^{-19} \, \text{C})(6.49 \times 10^7 \, \text{m/s})(55.0 \times 10^{-6} \, \text{T})}{9.11 \times 10^{-31} \, \text{kg}} = 6.27 \times 10^{14} \, \text{m/s}^2 \approx 6.3 \times 10^{14} \, \text{m/s}^2.$$

(b) We ignore any vertical deflection of the beam which might arise due to the horizontal component of Earth's field. Technically, then, the electron should follow a circular arc. However, the deflection is so small that many of the technicalities of circular geometry may be ignored, and a calculation along the lines of projectile motion analysis (see Chapter 4) provides an adequate approximation:

$$\Delta x = vt \Rightarrow t = \frac{\Delta x}{v} = \frac{0.200 \text{m}}{6.49 \times 10^7 \, \text{m/s}}$$

which yields a time of $t = 3.08 \times 10^{-9}$ s. Then, with our y axis oriented eastward,

$$\Delta y = \frac{1}{2}at^2 = \frac{1}{2}\left(6.27 \times 10^{14}\right)\left(3.08 \times 10^{-9}\right)^2 = 0.00298 \text{m} \approx 0.0030 \text{ m}.$$

86. We replace the current loop of arbitrary shape with an assembly of small adjacent rectangular loops filling the same area which was enclosed by the original loop (as nearly as possible). Each rectangular loop carries a current i flowing in the same sense as the original loop. As the sizes of these rectangles shrink to infinitesimally small values, the assembly gives a current distribution equivalent to that of the original loop. The magnitude of the torque $\Delta\vec{\tau}$ exerted by $\vec{B}$ on the nth rectangular loop of area ΔA_n is given by $\Delta\tau_n = NiB \sin\theta \Delta A_n$. Thus, for the whole assembly

$$\tau = \sum_n \Delta\tau_n = NiB\sum_n \Delta A_n = NiAB \sin\theta.$$

87. The total magnetic force on the loop L is

$$\vec{F}_B = i \oint_L \left(d\vec{L} \times \vec{B} \right) = i \left(\oint_L d\vec{L} \right) \times \vec{B} = 0.$$

We note that $\oint_L d\vec{L} = 0$. If $\vec{B}$ is not a constant, however, then the equality

$$\oint_L \left(d\vec{L} \times \vec{B} \right) = \left(\oint_L d\vec{L} \right) \times \vec{B}$$

is not necessarily valid, so $\vec{F}_B$ is not always zero.

88. (a) Since $\vec{B}$ is uniform,

$$\vec{F}_B = \int_{\text{wire}} i d\vec{L} \times \vec{B} = i \left(\int_{\text{wire}} d\vec{L} \right) \times \vec{B} = i \vec{L}_{ab} \times \vec{B},$$

where we note that $\int_{\text{wire}} d\vec{L} = \vec{L}_{ab}$, with $\vec{L}_{ab}$ being the displacement vector from a to b.

(b) Now $\vec{L}_{ab} = 0$, so $\vec{F}_B = i \vec{L}_{ab} \times \vec{B} = 0$.

89. With $F_z = v_z = B_x = 0$, Eq. 28-2 (and Eq. 3-30) gives

$$F_x \hat{\text{i}} + F_y \hat{\text{j}} = q \left(v_y B_z \hat{\text{i}} - v_x B_z \hat{\text{j}} + v_x B_y \hat{\text{k}} \right)$$

where $q = -e$ for the electron.   The last term immediately implies $B_y = 0$, and either of the other two terms (along with the values stated in the problem, bearing in mind that "fN" means femtonewtons or 10^{-15} N) can be used to solve for B_z.   We therefore find that the magnetic field is given by $\vec{B} = (0.75 \text{ T}) \hat{\text{k}}$.

Chapter 29

1. (a) The magnitude of the magnetic field due to the current in the wire, at a point a distance r from the wire, is given by

$$B = \frac{\mu_0 i}{2\pi r}.$$

With $r = 20$ ft $= 6.10$ m, we have

$$B = \frac{\left(4\pi \times 10^{-7}\, \text{T} \cdot \text{m/A}\right)\left(100\,\text{A}\right)}{2\pi\left(6.10\,\text{m}\right)} = 3.3 \times 10^{-6}\,\text{T} = 3.3\,\mu\text{T}.$$

(b) This is about one-sixth the magnitude of the Earth's field. It will affect the compass reading.

2. The straight segment of the wire produces no magnetic field at C (see the *straight sections* discussion in Sample Problem 29-1). Also, the fields from the two semi-circular loops cancel at C (by symmetry). Therefore, $B_C = 0$.

3. (a) The field due to the wire, at a point 8.0 cm from the wire, must be 39 μT and must be directed due south. Since $B = \mu_0 i / 2\pi r$,

$$i = \frac{2\pi r B}{\mu_0} = \frac{2\pi\left(0.080\,\text{m}\right)\left(39 \times 10^{-6}\,\text{T}\right)}{4\pi \times 10^{-7}\,\text{T} \cdot \text{m/A}} = 16\,\text{A}.$$

(b) The current must be from west to east to produce a field which is directed southward at points below it.

4. (a) Recalling the *straight sections* discussion in Sample Problem 29-1, we see that the current in segments AH and JD do not contribute to the field at point C. Using Eq. 29-9 (with $\phi = \pi$) and the right-hand rule, we find that the current in the semicircular arc HJ contributes $\mu_0 i / 4 R_1$ (into the page) to the field at C. Also, arc DA contributes $\mu_0 i / 4 R_2$ (out of the page) to the field there. Thus, the net field at C is

$$B = \frac{\mu_0 i}{4}\left(\frac{1}{R_1} - \frac{1}{R_2}\right) = \frac{(4\pi \times 10^{-7}\,\text{T} \cdot \text{m/A})(0.281\text{A})}{4}\left(\frac{1}{0.0315\text{m}} - \frac{1}{0.0780\text{m}}\right) = 1.67 \times 10^{-6}\,\text{T}.$$

(b) The direction of the field is into the page.

5. (a) Recalling the *straight sections* discussion in Sample Problem 29-1, we see that the current in the straight segments collinear with P do not contribute to the field at that point. Using Eq. 29-9 (with $\phi = \theta$) and the right-hand rule, we find that the current in the semicircular arc of radius b contributes $\mu_0 i \theta / 4\pi b$ (out of the page) to the field at P. Also, the current in the large radius arc contributes $\mu_0 i \theta / 4\pi a$ (into the page) to the field there. Thus, the net field at P is

$$B = \frac{\mu_0 i \theta}{4}\left(\frac{1}{b}-\frac{1}{a}\right) = \frac{(4\pi\times10^{-7}\,\text{T}\cdot\text{m/A})(0.411\text{A})(74°\cdot\pi/180°)}{4\pi}\left(\frac{1}{0.107\text{m}}-\frac{1}{0.135\text{m}}\right)$$
$$=1.02\times10^{-7}\,\text{T}.$$

(b) The direction is out of the page.

6. (a) Recalling the *straight sections* discussion in Sample Problem 29-1, we see that the current in the straight segments collinear with C do not contribute to the field at that point.

Eq. 29-9 (with $\phi = \pi$) indicates that the current in the semicircular arc contributes $\mu_0 i / 4R$ to the field at C. Thus, the magnitude of the magnetic field is

$$B = \frac{\mu_0 i}{4R} = \frac{(4\pi\times10^{-7}\,\text{T}\cdot\text{m/A})(0.0348\text{A})}{4(0.0926\text{m})} = 1.18\times10^{-7}\,\text{T}.$$

(b) The right-hand rule shows that this field is into the page.

7. (a) The currents must be opposite or antiparallel, so that the resulting fields are in the same direction in the region between the wires. If the currents are parallel, then the two fields are in opposite directions in the region between the wires. Since the currents are the same, the total field is zero along the line that runs halfway between the wires.

(b) At a point halfway between they have the same magnitude, $\mu_0 i / 2\pi r$. Thus the total field at the midpoint has magnitude $B = \mu_0 i / \pi r$ and

$$i = \frac{\pi r B}{\mu_0} = \frac{\pi(0.040\,\text{m})(300\times10^{-6}\,\text{T})}{4\pi\times10^{-7}\,\text{T}\cdot\text{m/A}} = 30\,\text{A}.$$

8. (a) Since they carry current in the same direction, then (by the right-hand rule) the only region in which their fields might cancel is between them. Thus, if the point at which we are evaluating their field is r away from the wire carrying current i and is $d - r$ away from the wire carrying current $3.00i$, then the canceling of their fields leads to

$$\frac{\mu_0 i}{2\pi r} = \frac{\mu_0(3i)}{2\pi(d-r)} \quad\Rightarrow\quad r = \frac{d}{4} = \frac{16.0\,\text{cm}}{4} = 4.0\,\text{cm}.$$

(b) Doubling the currents does not change the location where the magnetic field is zero.

9. (a) $B_{P1} = \mu_0 i_1/2\pi r_1$ where $i_1 = 6.5$ A and $r_1 = d_1 + d_2 = 0.75$ cm + 1.5 cm = 2.25 cm, and $B_{P2} = \mu_0 i_2/2\pi r_2$ where $r_2 = d_2 = 1.5$ cm. From $B_{P1} = B_{P2}$ we get

$$i_2 = i_1\left(\frac{r_2}{r_1}\right) = (6.5\,\mathrm{A})\left(\frac{1.5\,\mathrm{cm}}{2.25\,\mathrm{cm}}\right) = 4.3\,\mathrm{A}.$$

(b) Using the right-hand rule, we see that the current i_2 carried by wire 2 must be out of the page.

10. With the "usual" x and y coordinates used in Fig. 29-40, then the vector $\vec{r}$ pointing from a current element to P is $\vec{r} = -s\,\hat{i} + R\,\hat{j}$. Since $d\vec{s} = ds\,\hat{i}$, then $|d\vec{s} \times \vec{r}| = R\,ds$. Therefore, with $r = \sqrt{s^2 + R^2}$, Eq. 29-3 becomes

$$dB = \frac{\mu_0}{4\pi}\frac{i\,R\,ds}{(s^2 + R^2)^{3/2}}.$$

(a) Clearly, considered as a function of s (but thinking of "ds" as some finite-sized constant value), the above expression is maximum for $s = 0$. Its value in this case is $dB_{max} = \mu_0 i\,ds/4\pi R^2$.

(b) We want to find the s value such that $dB = \frac{1}{10}dB_{max}$. This is a non-trivial algebra exercise, but is nonetheless straightforward. The result is $s = \sqrt{10^{2/3} - 1}\,R$. If we set $R = 2.00$ cm, then we obtain $s = 3.82$ cm.

11. We assume the current flows in the $+x$ direction and the particle is at some distance d in the $+y$ direction (away from the wire). Then, the magnetic field at the location of a proton with charge q is $\vec{B} = \frac{\mu_0 i}{2\pi d}\,\hat{k}$. Thus,

$$\vec{F} = q\vec{v} \times \vec{B} = \frac{\mu_0 iq}{2\pi d}\left(\vec{v} \times \hat{k}\right).$$

In this situation, $\vec{v} = v\left(-\hat{j}\right)$ (where v is the speed and is a positive value), and $q > 0$. Thus,

$$\vec{F} = \frac{\mu_0 iqv}{2\pi d}\left(\left(-\hat{j}\right) \times \hat{k}\right) = -\frac{\mu_0 iqv}{2\pi d}\,\hat{i} = -\frac{(4\pi \times 10^{-7}\,\mathrm{T \cdot m/A})(0.350\mathrm{A})(1.60 \times 10^{-19}\,\mathrm{C})(200\mathrm{m/s})}{2\pi(0.0289\,\mathrm{m})}\,\hat{i}$$

$$= (-7.75 \times 10^{-23}\,\mathrm{N})\hat{i}.$$

12. The fact that $B_y = 0$ at $x = 10$ cm implies the currents are in opposite directions. Thus

$$B_y = \frac{\mu_0 i_1}{2\pi (L + x)} - \frac{\mu_0 i_2}{2 \pi x} = \frac{\mu_0 i_2}{2\pi} \left(\frac{4}{L + x} - \frac{1}{x} \right).$$

using Eq. 29-4 and the fact that $i_1 = 4 i_2$. To get the maximum, we take the derivative with respect to x and set equal to zero. This leads to $3x^2 - 2Lx - L^2 = 0$ which factors and becomes $(3x + L)(x - L) = 0$, which has the physically acceptable solution: $x = L$. This produces the maximum B_y: $\mu_0 i_2 /2\pi L$. To proceed further, we must determine L. Examination of the datum at $x = 10$ cm in Fig. 29-42(b) leads (using our expression above for B_y and setting that to zero) to $L = 30$ cm.

(a) The maximum value of B_y occurs at $x = L = 30$ cm.

(b) With $i_2 = 0.003$ A we find $\mu_0 i_2 /2\pi L = 2.0$ nT.

(c) and (d)   Fig. 29-42(b) shows that as we get very close to wire 2 (where its field strongly dominates over that of the more distant wire 1) B_y points along the $-y$ direction. The right-hand rule leads us to conclude that wire 2's current is consequently is *into the page*. We previously observed that the currents were in opposite directions, so wire 1's current is *out of the page*.

13. Each of the semi-infinite straight wires contributes $\mu_0 i/4\pi R$ (Eq. 29-7) to the field at the center of the circle (both contributions pointing "out of the page"). The current in the arc contributes a term given by Eq. 29-9 pointing into the page, and this is able to produce zero total field at that location if $B_{arc} = 2.00 B_{semi\,infinite}$, or

$$\frac{\mu_0 i \phi}{4\pi R} = 2.00 \left(\frac{\mu_0 i}{4\pi R} \right)$$

which yields $\phi = 2.00$ rad.

14. Initially, $B_{net\,y} = 0$, and $B_{net\,x} = B_2 + B_4 = 2(\mu_0 i /2\pi d)$ using Eq. 29-4, where $d = 0.15$ m. To obtain the $30°$ condition described in the problem, we must have

$$B_{net\,y} = B_{net\,x} \tan(30°)$$

$$B_1' - B_3 = 2(\mu_0 i /2\pi d) \tan(30°)$$

where $B_3 = \mu_0 i /2\pi d$ and $B_1' = \mu_0 i /2\pi d'$. Since $\tan(30°) = 1/\sqrt{3}$, this leads to

$$d' = \frac{\sqrt{3}\, d}{\sqrt{3} + 2}.$$

(a) With $d = 15.0$ cm, this gives $d' = 7.0$ cm. Being very careful about the geometry of the situation, then we conclude that we must move wire 1 to $x = -7.0$ cm.

(b) To restore the initial symmetry, we would have to move wire 3 to $x = +7.0$ cm.

15. Each wire produces a field with magnitude given by $B = \mu_0 i/2\pi r$, where r is the distance from the corner of the square to the center. According to the Pythagorean theorem, the diagonal of the square has length $\sqrt{2}a$, so $r = a/\sqrt{2}$ and $B = \mu_0 i/\sqrt{2}\pi a$. The fields due to the wires at the upper left and lower right corners both point toward the upper right corner of the square. The fields due to the wires at the upper right and lower left corners both point toward the upper left corner. The horizontal components cancel and the vertical components sum to

$$B_{total} = 4\frac{\mu_0 i}{\sqrt{2}\pi a}\cos 45° = \frac{2\mu_0 i}{\pi a} = \frac{2\left(4\pi\times 10^{-7}\ \text{T}\cdot\text{m/A}\right)\left(20\ \text{A}\right)}{\pi\left(0.20\ \text{m}\right)} = 8.0\times 10^{-5}\ \text{T}.$$

In the calculation $\cos 45°$ was replaced with $1/\sqrt{2}$. The total field points upward, or in the $+y$ direction. Thus, $\vec{B}_{total} = (8.0\times 10^{-5}\ \text{T})\hat{j}$.

16. We consider Eq. 29-6 but with a finite upper limit ($L/2$ instead of ∞). This leads to

$$B = \frac{\mu_0\ i}{2\pi R}\frac{L/2}{\sqrt{R^2 + (L/2)^2}}\ .$$

In terms of this expression, the problem asks us to see how large L must be (compared with R) such that the infinite wire expression B_∞ (Eq. 29-4) can be used with no more than a 1% error. Thus we must solve

$$\frac{B_\infty - B}{B} = 0.01\ .$$

This is a non-trivial algebra exercise, but is nonetheless straightforward. The result is

$$L = \frac{200R}{\sqrt{201}} \approx 14.1R \quad\Rightarrow\quad \frac{L}{R} \approx 14.1$$

17. Our x axis is along the wire with the origin at the midpoint. The current flows in the positive x direction. All segments of the wire produce magnetic fields at P_1 that are out of the page. According to the Biot-Savart law, the magnitude of the field any (infinitesimal) segment produces at P_1 is given by

$$dB = \frac{\mu_0 i}{4\pi}\frac{\sin\theta}{r^2}dx$$

where θ (the angle between the segment and a line drawn from the segment to P_1) and r (the length of that line) are functions of x. Replacing r with $\sqrt{x^2 + R^2}$ and $\sin \theta$ with $R/r = R/\sqrt{x^2 + R^2}$, we integrate from $x = -L/2$ to $x = L/2$. The total field is

$$B = \frac{\mu_0 iR}{4\pi} \int_{-L/2}^{L/2} \frac{dx}{\left(x^2 + R^2\right)^{3/2}} = \frac{\mu_0 iR}{4\pi} \frac{1}{R^2} \frac{x}{\left(x^2 + R^2\right)^{1/2}} \bigg|_{-L/2}^{L/2} = \frac{\mu_0 i}{2\pi R} \frac{L}{\sqrt{L^2 + 4R^2}}$$

$$= \frac{\left(4\pi \times 10^{-7}\, \text{T} \cdot \text{m/A}\right)\left(0.0582\, \text{A}\right)}{2\pi\left(0.131\, \text{m}\right)} \frac{0.180\, \text{m}}{\sqrt{\left(0.180\, \text{m}\right)^2 + 4\left(0.131\, \text{m}\right)^2}} = 5.03 \times 10^{-8}\, \text{T}.$$

18. Using the law of cosines and the requirement that $B = 100$ nT, we have

$$\theta = \cos^{-1}\left(\frac{B_1^2 + B_2^2 - B^2}{-2B_1 B_2}\right) = 144°.$$

where Eq. 29-10 has been used to determine B_1 (168 nT) and B_2 (151 nT).

19. Our x axis is along the wire with the origin at the right endpoint, and the current is in the positive x direction. All segments of the wire produce magnetic fields at P_2 that are out of the page. According to the Biot-Savart law, the magnitude of the field any (infinitesimal) segment produces at P_2 is given by

$$dB = \frac{\mu_0 i}{4\pi} \frac{\sin \theta}{r^2} dx$$

where θ (the angle between the segment and a line drawn from the segment to P_2) and r (the length of that line) are functions of x. Replacing r with $\sqrt{x^2 + R^2}$ and $\sin \theta$ with $R/r = R/\sqrt{x^2 + R^2}$, we integrate from $x = -L$ to $x = 0$. The total field is

$$B = \frac{\mu_0 iR}{4\pi} \int_{-L}^{0} \frac{dx}{\left(x^2 + R^2\right)^{3/2}} = \frac{\mu_0 iR}{4\pi} \frac{1}{R^2} \frac{x}{\left(x^2 + R^2\right)^{1/2}} \bigg|_{-L}^{0} = \frac{\mu_0 i}{4\pi R} \frac{L}{\sqrt{L^2 + R^2}}$$

$$= \frac{\left(4\pi \times 10^{-7}\, \text{T} \cdot \text{m/A}\right)\left(0.693\, \text{A}\right)}{4\pi\left(0.251\, \text{m}\right)} \frac{0.136\, \text{m}}{\sqrt{\left(0.136\, \text{m}\right)^2 + \left(0.251\, \text{m}\right)^2}} = 1.32 \times 10^{-7}\, \text{T}.$$

20. In the one case we have $B_{small} + B_{big} = 47.25\ \mu\text{T}$, and the other case gives $B_{small} - B_{big} = 15.75\ \mu\text{T}$ (cautionary note about our notation: B_{small} refers to the field at the center of the small-radius arc, which is actually a bigger field than B_{big}!). Dividing one of these equations by the other and canceling out common factors (see Eq. 29-9) we obtain

$$\frac{\dfrac{1}{r_{\text{small}}}+\dfrac{1}{r_{\text{big}}}}{\dfrac{1}{r_{\text{small}}}-\dfrac{1}{r_{\text{big}}}}=3\,.$$

The solution of this is straightforward: $r_{\text{small}}=\frac{1}{2}\,r_{\text{big}}$. Using the given fact that the big radius 4.00 cm, then we conclude that the small radius is 2.00 cm.

21. (a) The contribution to B_C from the (infinite) straight segment of the wire is

$$B_{C1}=\frac{\mu_0 i}{2\pi R}.$$

The contribution from the circular loop is $B_{C2}=\dfrac{\mu_0 i}{2R}$. Thus,

$$B_C=B_{C1}+B_{C2}=\frac{\mu_0 i}{2R}\left(1+\frac{1}{\pi}\right)=\frac{\left(4\pi\times10^{-7}\,\text{T}\cdot\text{m/A}\right)\left(5.78\times10^{-3}\,\text{A}\right)}{2\left(0.0189\,\text{m}\right)}\left(1+\frac{1}{\pi}\right)=2.53\times10^{-7}\,\text{T}.$$

$\vec{B}_C$ points out of the page, or in the $+z$ direction. In unit-vector notation, $\vec{B}_C=(2.53\times10^{-7}\,\text{T})\hat{k}$

(b) Now $\vec{B}_{C1}\perp\vec{B}_{C2}$ so

$$B_C=\sqrt{B_{C1}^2+B_{C2}^2}=\frac{\mu_0 i}{2R}\sqrt{1+\frac{1}{\pi^2}}=\frac{\left(4\pi\times10^{-7}\,\text{T}\cdot\text{m/A}\right)\left(5.78\times10^{-3}\,\text{A}\right)}{2\left(0.0189\,\text{m}\right)}\sqrt{1+\frac{1}{\pi^2}}=2.02\times10^{-7}\,\text{T}.$$

and $\vec{B}_C$ points at an angle (relative to the plane of the paper) equal to

$$\tan^{-1}\left(\frac{B_{C1}}{B_{C2}}\right)=\tan^{-1}\left(\frac{1}{\pi}\right)=17.66°.$$

In unit-vector notation,

$$\vec{B}_C=2.02\times10^{-7}\,\text{T}\left(\cos17.66°\hat{i}+\sin17.66°\hat{k}\right)=(1.92\times10^{-7}\,\text{T})\hat{i}+(6.12\times10^{-8}\,\text{T})\hat{k}$$

22. Letting "out of the page" in Fig. 29-50(a) be the positive direction, the net field is

$$B=\frac{\mu_0\,i_1\,\phi}{4\pi R}-\frac{\mu_0\,i_2}{2\pi(R/2)}$$

from Eqs. 29-9 and 29-4. Referring to Fig. 29-50, we see that $B = 0$ when $i_2 = 0.5$ A, so (solving the above expression with B set equal to zero) we must have

$$\phi = 4(i_2/i_1) = 4(0.5/2) = 1.00 \text{ rad (or } 57.3°).$$

23. Consider a section of the ribbon of thickness dx located a distance x away from point P. The current it carries is $di = i\, dx/w$, and its contribution to B_P is

$$dB_P = \frac{\mu_0 di}{2\pi x} = \frac{\mu_0 i\, dx}{2\pi x w}.$$

Thus,

$$B_P = \int dB_P = \frac{\mu_0 i}{2\pi w}\int_d^{d+w}\frac{dx}{x} = \frac{\mu_0 i}{2\pi w}\ln\left(1+\frac{w}{d}\right) = \frac{\left(4\pi\times10^{-7}\text{ T·m/A}\right)\left(4.61\times10^{-6}\text{ A}\right)}{2\pi\left(0.0491\text{ m}\right)}\ln\left(1+\frac{0.0491}{0.0216}\right)$$

$$= 2.23\times10^{-11}\text{T}.$$

and $\vec{B}_P$ points upward. In unit-vector notation, $\vec{B}_P = (2.23\times10^{-11}\text{T})\hat{j}$

24. Initially we have

$$B_i = \frac{\mu_0\, i\, \phi}{4\pi R} + \frac{\mu_0\, i\, \phi}{4\pi r}$$

using Eq. 29-9. In the final situation we use Pythagorean theorem and write

$$B_f^{\,2} = B_z^{\,2} + B_y^{\,2} = \left(\frac{\mu_0\, i\, \phi}{4\pi R}\right)^2 + \left(\frac{\mu_0\, i\, \phi}{4\pi r}\right)^2.$$

If we square B_i and divide by $B_f^{\,2}$, we obtain

$$\left(\frac{B_i}{B_f}\right)^2 = \frac{\left(\dfrac{1}{R}+\dfrac{1}{r}\right)^2}{\dfrac{1}{R^2}+\dfrac{1}{r^2}}.$$

From the graph (see Fig. 29-52(c) – note the maximum and minimum values) we estimate $B_i/B_f = 12/10 = 1.2$, and this allows us to solve for r in terms of R:

$$r = R\frac{1\pm1.2\sqrt{2-1.2^2}}{1.2^2-1} = 2.3 \text{ cm} \quad\text{or}\quad 43.1 \text{ cm}.$$

Since we require $r < R$, then the acceptable answer is $r = 2.3$ cm.

25. (a) Recalling the *straight sections* discussion in Sample Problem 29-1, we see that the current in the straight segments collinear with P do not contribute to the field at that point. We use the result of problem 16 to evaluate the contributions to the field at P, noting that the nearest wire-segments (each of length a) produce magnetism into the page at P and the further wire-segments (each of length 2a) produce magnetism pointing out of the page at P. Thus, we find (into the page)

$$B_P = 2\left(\frac{\sqrt{2}\mu_0 i}{8\pi a}\right) - 2\left(\frac{\sqrt{2}\mu_0 i}{8\pi(2a)}\right) = \frac{\sqrt{2}\mu_0 i}{8\pi a} = \frac{\sqrt{2}\left(4\pi \times 10^{-7}\,\text{T} \cdot \text{m/A}\right)(13\,\text{A})}{8\pi(0.047\,\text{m})}$$

$$= 1.96 \times 10^{-5}\,\text{T} \approx 2.0 \times 10^{-5}\,\text{T}.$$

(b) The direction of the field is into the page.

26. By the right-hand rule (which is "built-into" Eq. 29-3) the field caused by wire 1's current, evaluated at the coordinate origin, is along the $+y$ axis. Its magnitude B_1 is given by Eq. 29-4. The field caused by wire 2's current will generally have both an x and a y component which are related to its magnitude B_2 (given by Eq. 29-4) and sines and cosines of some angle. A little trig (and the use of the right-hand rule) leads us to conclude that when wire 2 is at angle θ_2 (shown in Fig. 29-54) then its components are

$$B_{2x} = B_2 \sin\theta_2, \quad B_{2y} = -B_2 \cos\theta_2.$$

The magnitude-squared of their net field is then (by Pythagoras' theorem) the sum of the square of their net x-component and the square of their net y-component:

$$B^2 = (B_2 \sin\theta_2)^2 + (B_1 - B_2 \cos\theta_2)^2 = B_1^2 + B_2^2 - 2B_1 B_2 \cos\theta_2.$$

(since $\sin^2 + \cos^2 = 1$), which we could also have gotten directly by using the law of cosines. We have $B_1 = \mu_0 i_1 / 2\pi R = 60$ nT and $B_2 = \mu_0 i_2 / 2\pi R = 40$ nT, so with the requirement that the net field have magnitude $B = 80$ nT, we find

$$\theta_2 = \cos^{-1}\left(\frac{B_1^2 + B_2^2 - B^2}{2B_1 B_2}\right) = \cos^{-1}(-1/4) = 104°,$$

where the positive value has been chosen.

27. Eq. 29-13 gives the magnitude of the force between the wires, and finding the x-component of it amounts to multiplying that magnitude by $\cos\phi = \dfrac{d_2}{\sqrt{d_1^2 + d_2^2}}$. Therefore, the x-component of the force per unit length is

$$\frac{F_x}{L} = \frac{\mu_0 i_1 i_2 d_2}{2\pi (d_1^2 + d_2^2)} = 8.84 \times 10^{-11}\,\text{N/m}.$$

28. Using a magnifying glass, we see that all but i_2 are directed into the page. Wire 3 is therefore attracted to all but wire 2. Letting $d = 0.500$ m, we find the net force (per meter length) using Eq. 29-13, with positive indicated a rightward force:

$$\frac{|\vec{F}|}{\ell} = \frac{\mu_0 i_3}{2\pi}\left(-\frac{i_1}{2d} + \frac{i_2}{d} + \frac{i_4}{d} + \frac{i_5}{2d}\right)$$

which yields $|\vec{F}|/\ell = 8.00 \times 10^{-7}$ N/m .

29. We label these wires 1 through 5, left to right, and use Eq. 29-13. Then,

(a) The magnetic force on wire 1 is

$$\vec{F}_1 = \frac{\mu_0 i^2 l}{2\pi}\left(\frac{1}{d} + \frac{1}{2d} + \frac{1}{3d} + \frac{1}{4d}\right)\hat{j} = \frac{25\mu_0 i^2 l}{24\pi d}\hat{j} = \frac{25\left(4\pi \times 10^{-7}\,\text{T}\cdot\text{m/A}\right)\left(3.00\text{A}\right)^2\left(10.0\text{m}\right)}{24\pi\left(8.00 \times 10^{-2}\,\text{m}\right)}\hat{j}$$

$$= (4.69 \times 10^{-4}\,\text{N})\hat{j}$$

(b) Similarly, for wire 2, we have

$$\vec{F}_2 = \frac{\mu_0 i^2 l}{2\pi}\left(\frac{1}{2d} + \frac{1}{3d}\right)\hat{j} = \frac{5\mu_0 i^2 l}{12\pi d}\hat{j} = (1.88 \times 10^{-4}\,\text{N})\hat{j}.$$

(c) $F_3 = 0$ (because of symmetry).

(d) $\vec{F}_4 = -\vec{F}_2 = (-1.88 \times 10^{-4}\,\text{N})\hat{j}$, and

(e) $\vec{F}_5 = -\vec{F}_1 = -(4.69 \times 10^{-4}\,\text{N})\hat{j}$.

30. Using Eq. 29-13, the force on, say, wire 1 (the wire at the upper left of the figure) is along the diagonal (pointing towards wire 3 which is at the lower right). Only the forces (or their components) along the diagonal direction contribute. With $\theta = 45°$, we find

$$F_1 = |\vec{F}_{12} + \vec{F}_{13} + \vec{F}_{14}| = 2F_{12}\cos\theta + F_{13} = 2\left(\frac{\mu_0 i^2}{2\pi a}\right)\cos 45° + \frac{\mu_0 i^2}{2\sqrt{2}\pi a} = \frac{3}{2\sqrt{2}\pi}\left(\frac{\mu_0 i^2}{a}\right)$$

$$= \frac{3}{2\sqrt{2}\pi}\frac{\left(4\pi \times 10^{-7}\,\text{T}\cdot\text{m/A}\right)\left(15.0\text{A}\right)^2}{\left(8.50 \times 10^{-2}\,\text{m}\right)} = 1.12 \times 10^{-3}\,\text{N/m}.$$

The direction of $\vec{F}_1$ is along $\hat{r} = (\hat{i} - \hat{j})/\sqrt{2}$. In unit-vector notation, we have

$$\vec{F}_1 = \frac{(1.12 \times 10^{-3} \, \text{N/m})}{\sqrt{2}} (\hat{i} - \hat{j}) = (7.94 \times 10^{-4} \, \text{N/m})\hat{i} + (-7.94 \times 10^{-4} \, \text{N/m})\hat{j}$$

31. We use Eq. 29-13 and the superposition of forces: $\vec{F}_4 = \vec{F}_{14} + \vec{F}_{24} + \vec{F}_{34}$. With $\theta = 45°$, the situation is as shown next:

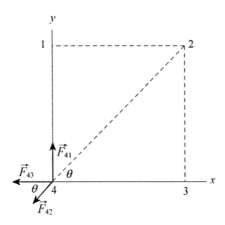

The components of $\vec{F}_4$ are given by

$$F_{4x} = -F_{43} - F_{42}\cos\theta = -\frac{\mu_0 i^2}{2\pi a} - \frac{\mu_0 i^2 \cos 45°}{2\sqrt{2}\pi a} = -\frac{3\mu_0 i^2}{4\pi a}$$

and

$$F_{4y} = F_{41} - F_{42}\sin\theta = \frac{\mu_0 i^2}{2\pi a} - \frac{\mu_0 i^2 \sin 45°}{2\sqrt{2}\pi a} = \frac{\mu_0 i^2}{4\pi a}.$$

Thus,

$$F_4 = \left(F_{4x}^2 + F_{4y}^2\right)^{1/2} = \left[\left(-\frac{3\mu_0 i^2}{4\pi a}\right)^2 + \left(\frac{\mu_0 i^2}{4\pi a}\right)^2\right]^{1/2} = \frac{\sqrt{10}\mu_0 i^2}{4\pi a} = \frac{\sqrt{10}\left(4\pi \times 10^{-7}\,\text{T}\cdot\text{m/A}\right)(7.50\text{A})^2}{4\pi(0.135\text{m})}$$

$$= 1.32 \times 10^{-4} \, \text{N/m}.$$

and $\vec{F}_4$ makes an angle ϕ with the positive x axis, where

$$\phi = \tan^{-1}\left(\frac{F_{4y}}{F_{4x}}\right) = \tan^{-1}\left(-\frac{1}{3}\right) = 162°.$$

In unit-vector notation, we have

$$\vec{F}_1 = (1.32 \times 10^{-4} \, \text{N/m})[\cos 162° \hat{i} + \sin 162° \hat{j}] = (-1.25 \times 10^{-4} \, \text{N/m})\hat{i} + (4.17 \times 10^{-5} \, \text{N/m})\hat{j}$$

32. (a) The fact that the curve in Fig. 29-57(b) passes through zero implies that the currents in wires 1 and 3 exert forces in opposite directions on wire 2. Thus, current i_1 points *out of the page*. When wire 3 is a great distance from wire 2, the only field that affects wire 2 is that caused by the current in wire 1; in this case the force is negative according to Fig. 29-57(b). This means wire 2 is attracted to wire 1, which implies (by the discussion in section 29-2) that wire 2's current is in the same direction as wire 1's current: *out of the page*. With wire 3 infinitely far away, the force per unit length is given (in magnitude) as $6.27 \times 10^{-7} \, \text{N/m}$. We set this equal to $F_{12} = \mu_0 i_1 i_2 / 2\pi d$. When wire 3 is at $x = 0.04$ m the curve passes through the zero point previously mentioned, so the force between 2 and 3 must equal F_{12} there. This allows us to solve for the distance between wire 1 and wire 2:

$$d = (0.04 \text{ m})(0.750 \text{ A})/(0.250 \text{ A}) = 0.12 \text{ m}.$$

Then we solve $6.27 \times 10^{-7} \, \text{N/m} = \mu_0 i_1 i_2 / 2\pi d$ and obtain $i_2 = 0.50$ A.

(b) The direction of i_2 is out of the page.

33. The magnitudes of the forces on the sides of the rectangle which are parallel to the long straight wire (with $i_1 = 30.0$ A) are computed using Eq. 29-13, but the force on each of the sides lying perpendicular to it (along our y axis, with the origin at the top wire and $+y$ downward) would be figured by integrating as follows:

$$F_{\perp \text{ sides}} = \int_a^{a+b} \frac{i_2 \mu_0 i_1}{2\pi y} \, dy.$$

Fortunately, these forces on the two perpendicular sides of length b cancel out. For the remaining two (parallel) sides of length L, we obtain

$$F = \frac{\mu_0 i_1 i_2 L}{2\pi}\left(\frac{1}{a} - \frac{1}{a+d}\right) = \frac{\mu_0 i_1 i_2 b}{2\pi a(a+b)}$$

$$= \frac{(4\pi \times 10^{-7} \, \text{T} \cdot \text{m/A})(30.0\text{A})(20.0\text{A})(8.00\text{cm})(300 \times 10^{-2}\text{ m})}{2\pi(1.00\text{cm} + 8.00\text{cm})} = 3.20 \times 10^{-3} \, \text{N},$$

and $\vec{F}$ points toward the wire, or $+\hat{j}$. In unit-vector notation, we have $\vec{F} = (3.20 \times 10^{-3} \, \text{N})\hat{j}$

34. A close look at the path reveals that only currents 1, 3, 6 and 7 are enclosed. Thus, noting the different current directions described in the problem, we obtain

$$\oint \vec{B} \cdot d\vec{s} = \mu_0 \left(7i - 6i + 3i + i\right) = 5\mu_0 i = 5\left(4\pi \times 10^{-7}\,\text{T} \cdot \text{m/A}\right)\left(4.50 \times 10^{-3}\,\text{A}\right) = 2.83 \times 10^{-8}\,\text{T} \cdot \text{m}.$$

35. (a) Two of the currents are out of the page and one is into the page, so the net current enclosed by the path is 2.0 A, out of the page. Since the path is traversed in the clockwise sense, a current into the page is positive and a current out of the page is negative, as indicated by the right-hand rule associated with Ampere's law. Thus,

$$\oint \vec{B} \cdot d\vec{s} = -\mu_0 i = -(2.0\,\text{A})\left(4\pi \times 10^{-7}\,\text{T} \cdot \text{m/A}\right) = -2.5 \times 10^{-6}\,\text{T} \cdot \text{m}.$$

(b) The net current enclosed by the path is zero (two currents are out of the page and two are into the page), so $\oint \vec{B} \cdot d\vec{s} = \mu_0 i_{\text{enc}} = 0$.

36. We use Ampere's law: $\oint \vec{B} \cdot d\vec{s} = \mu_0 i$, where the integral is around a closed loop and i is the net current through the loop.

(a) For path 1, the result is

$$\oint_1 \vec{B} \cdot d\vec{s} = \mu_0 \left(-5.0\,\text{A} + 3.0\,\text{A}\right) = (-2.0\,\text{A})\left(4\pi \times 10^{-7}\,\text{T} \cdot \text{m/A}\right) = -2.5 \times 10^{-6}\,\text{T} \cdot \text{m}.$$

(b) For path 2, we find

$$\oint_2 \vec{B} \cdot d\vec{s} = \mu_0 \left(-5.0\,\text{A} - 5.0\,\text{A} - 3.0\,\text{A}\right) = (-13.0\,\text{A})\left(4\pi \times 10^{-7}\,\text{T} \cdot \text{m/A}\right) = -1.6 \times 10^{-5}\,\text{T} \cdot \text{m}.$$

37. We use Eq. 29-20 $B = \mu_0 ir / 2\pi a^2$ for the B-field inside the wire ($r < a$) and Eq. 29-17 $B = \mu_0 i / 2\pi r$ for that outside the wire ($r > a$).

(a) At $r = 0$, $B = 0$.

(b) At $r = 0.0100\,\text{m}$, $B = \dfrac{\mu_0 ir}{2\pi a^2} = \dfrac{(4\pi \times 10^{-7}\,\text{T} \cdot \text{m/A})(170\,\text{A})(0.0100\,\text{m})}{2\pi (0.0200\,\text{m})^2} = 8.50 \times 10^{-4}\,\text{T}.$

(c) At $r = a = 0.0200\,\text{m}$, $B = \dfrac{\mu_0 ir}{2\pi a^2} = \dfrac{(4\pi \times 10^{-7}\,\text{T} \cdot \text{m/A})(170\,\text{A})(0.0200\,\text{m})}{2\pi (0.0200\,\text{m})^2} = 1.70 \times 10^{-3}\,\text{T}.$

(d) At $r = 0.0400\,\text{m}$, $B = \dfrac{\mu_0 i}{2\pi r} = \dfrac{(4\pi \times 10^{-7}\,\text{T} \cdot \text{m/A})(170\,\text{A})}{2\pi (0.0400\,\text{m})} = 8.50 \times 10^{-4}\,\text{T}.$

38. (a) The field at the center of the pipe (point C) is due to the wire alone, with a magnitude of

$$B_C = \frac{\mu_0 i_{wire}}{2\pi(3R)} = \frac{\mu_0 i_{wire}}{6\pi R}.$$

For the wire we have $B_{P,\,wire} > B_{C,\,wire}$. Thus, for $B_P = B_C = B_{C,\,wire}$, i_{wire} must be into the page:

$$B_P = B_{P,wire} - B_{P,pipe} = \frac{\mu_0 i_{wire}}{2\pi R} - \frac{\mu_0 i}{2\pi(2R)}.$$

Setting $B_C = -B_P$ we obtain $i_{wire} = 3i/8 = 3(8.00\times10^{-3}\,\text{A})/8 = 3.00\times10^{-3}\,\text{A}$.

(b) The direction is into the page.

39. For $r \le a$,

$$B(r) = \frac{\mu_0 i_{enc}}{2\pi r} = \frac{\mu_0}{2\pi r}\int_0^r J(r)\,2\pi r\,dr = \frac{\mu_0}{2\pi}\int_0^r J_0\left(\frac{r}{a}\right)2\pi r\,dr = \frac{\mu_0 J_0 r^2}{3a}.$$

(a) At $r = 0$, $B = 0$.

(b) At $r = a/2$, we have

$$B(r) = \frac{\mu_0 J_0 r^2}{3a} = \frac{(4\pi\times10^{-7}\,\text{T}\cdot\text{m/A})(310\,\text{A/m}^2)(3.1\times10^{-3}\,\text{m}/2)^2}{3(3.1\times10^{-3}\,\text{m})} = 1.0\times10^{-7}\,\text{T}.$$

(c) At $r = a$,

$$B(r = a) = \frac{\mu_0 J_0 a}{3} = \frac{(4\pi\times10^{-7}\,\text{T}\cdot\text{m/A})(310\,\text{A/m}^2)(3.1\times10^{-3}\,\text{m})}{3} = 4.0\times10^{-7}\,\text{T}.$$

40. It is possible (though tedious) to use Eq. 29-26 and evaluate the contributions (with the intent to sum them) of all 1200 loops to the field at, say, the center of the solenoid. This would make use of all the information given in the problem statement, but this is not the method that the student is expected to use here. Instead, Eq. 29-23 for the ideal solenoid (which does not make use of the coil radius) is the preferred method:

$$B = \mu_0 i n = \mu_0 i \left(\frac{N}{\ell}\right)$$

where $i = 3.60$ A, $\ell = 0.950$ m and $N = 1200$. This yields $B = 0.00571$ T.

41. It is possible (though tedious) to use Eq. 29-26 and evaluate the contributions (with the intent to sum them) of all 200 loops to the field at, say, the center of the solenoid.

This would make use of all the information given in the problem statement, but this is not the method that the student is expected to use here. Instead, Eq. 29-23 for the ideal solenoid (which does not make use of the coil diameter) is the preferred method:

$$B = \mu_0 in = \mu_0 i\left(\frac{N}{\ell}\right)$$

where $i = 0.30$ A, $\ell = 0.25$ m and $N = 200$. This yields $B = 0.0030$ T.

42. We find N, the number of turns of the solenoid, from the magnetic field $B = \mu_0 in = \mu_0 iN / \ell : N = B\ell / \mu_0 i$. Thus, the total length of wire used in making the solenoid is

$$2\pi rN = \frac{2\pi rB\ell}{\mu_0 i} = \frac{2\pi(2.60 \times 10^{-2}\,\text{m})(23.0 \times 10^{-3}\,\text{T})(1.30\,\text{m})}{2(4\pi \times 10^{-7}\,\text{T}\cdot\text{m}/\text{A})(18.0\,\text{A})} = 108\ \text{m}.$$

43. (a) We use Eq. 29-24. The inner radius is $r = 15.0$ cm, so the field there is

$$B = \frac{\mu_0 iN}{2\pi r} = \frac{(4\pi \times 10^{-7}\,\text{T}\cdot\text{m}/\text{A})(0.800\,\text{A})(500)}{2\pi(0.150\,\text{m})} = 5.33 \times 10^{-4}\ \text{T}.$$

(b) The outer radius is $r = 20.0$ cm. The field there is

$$B = \frac{\mu_0 iN}{2\pi r} = \frac{(4\pi \times 10^{-7}\,\text{T}\cdot\text{m}/\text{A})(0.800\,\text{A})(500)}{2\pi(0.200\text{m})} = 4.00 \times 10^{-4}\ \text{T}.$$

44. The orbital radius for the electron is

$$r = \frac{mv}{eB} = \frac{mv}{e\mu_0 ni}$$

which we solve for i:

$$i = \frac{mv}{e\mu_0 nr} = \frac{(9.11 \times 10^{-31}\,\text{kg})(0.0460)(3.00 \times 10^8\ \text{m/s})}{(1.60 \times 10^{-19}\,\text{C})(4\pi \times 10^{-7}\ \text{T}\cdot\text{m/A})(100/0.0100\text{m})(2.30 \times 10^{-2}\,\text{m})}$$

$$= 0.272\,\text{A}.$$

45. (a) We denote the $\vec{B}$-fields at point P on the axis due to the solenoid and the wire as $\vec{B}_s$ and $\vec{B}_w$, respectively. Since $\vec{B}_s$ is along the axis of the solenoid and $\vec{B}_w$ is

perpendicular to it, $\vec{B}_s \perp \vec{B}_w$ respectively. For the net field $\vec{B}$ to be at $45°$ with the axis we then must have $B_s = B_w$. Thus,

$$B_s = \mu_0 i_s n = B_w = \frac{\mu_0 i_w}{2\pi d} ,$$

which gives the separation d to point P on the axis:

$$d = \frac{i_w}{2\pi i_s n} = \frac{6.00\,\text{A}}{2\pi(20.0 \times 10^{-3}\,\text{A})(10\,\text{turns/cm})} = 4.77\,\text{cm} .$$

(b) The magnetic field strength is

$$B = \sqrt{2}B_s = \sqrt{2}\left(4\pi \times 10^{-7}\ \text{T·m/A}\right)\left(20.0 \times 10^{-3}\,\text{A}\right)(10\,\text{turns}/0.0100\,\text{m}) = 3.55 \times 10^{-5}\,\text{T} .$$

46. As the problem states near the end, some idealizations are being made here to keep the calculation straightforward (but are slightly unrealistic). For circular motion (with speed $v_\perp$ which represents the magnitude of the component of the velocity perpendicular to the magnetic field [the field is shown in Fig. 29-19]), the period is (see Eq. 28-17)

$$T = 2\pi r/v_\perp = 2\pi m/eB.$$

Now, the time to travel the length of the solenoid is $t = L/v_\parallel$ where $v_\parallel$ is the component of the velocity in the direction of the field (along the coil axis) and is equal to $v\cos\theta$ where $\theta = 30°$. Using Eq. 29-23 ($B = \mu_0 in$) with $n = N/L$, we find the number of revolutions made is $t/T = 1.6 \times 10^6$.

47. The magnitude of the magnetic dipole moment is given by $\mu = NiA$, where N is the number of turns, i is the current, and A is the area. We use $A = \pi R^2$, where R is the radius. Thus,

$$\mu = (200)(0.30\,\text{A})\pi(0.050\,\text{m})^2 = 0.47\,\text{A·m}^2 .$$

48. (a) We set $z = 0$ in Eq. 29-26 (which is equivalent using to Eq. 29-10 multiplied by the number of loops). Thus, $B(0) \propto i/R$. Since case b has two loops,

$$\frac{B_b}{B_a} = \frac{2i/R_b}{i/R_a} = \frac{2R_a}{R_b} = 4.0 .$$

(b) The ratio of their magnetic dipole moments is

$$\frac{\mu_b}{\mu_a} = \frac{2iA_b}{iA_a} = \frac{2R_b^2}{R_a^2} = 2\left(\frac{1}{2}\right)^2 = \frac{1}{2} = 0.50.$$

49. (a) The magnitude of the magnetic dipole moment is given by $\mu = NiA$, where N is the number of turns, i is the current, and A is the area. We use $A = \pi R^2$, where R is the radius. Thus,

$$\mu = Ni\pi R^2 = (300)(4.0\,\text{A})\pi(0.025\,\text{m})^2 = 2.4\,\text{A}\cdot\text{m}^2\ .$$

(b) The magnetic field on the axis of a magnetic dipole, a distance z away, is given by Eq. 29-27:

$$B = \frac{\mu_0}{2\pi}\frac{\mu}{z^3}\ .$$

We solve for z:

$$z = \left(\frac{\mu_0}{2\pi}\frac{\mu}{B}\right)^{1/3} = \left(\frac{(4\pi\times10^{-7}\,\text{T}\cdot\text{m/A})(2.36\,\text{A}\cdot\text{m}^2)}{2\pi(5.0\times10^{-6}\,\text{T})}\right)^{1/3} = 46\,\text{cm}\ .$$

50. We use Eq. 29-26 and note that the contributions to $\vec{B}_p$ from the two coils are the same. Thus,

$$B_p = \frac{2\mu_0 i R^2 N}{2\left[R^2+(R/2)^2\right]^{3/2}} = \frac{8\mu_0 Ni}{5\sqrt{5}R} = \frac{8(4\pi\times10^{-7}\,\text{T}\cdot\text{m/A})(200)(0.0122\,\text{A})}{5\sqrt{5}(0.25\,\text{m})} = 8.78\times10^{-6}\,\text{T}.$$

$\vec{B}_p$ is in the positive x direction.

51. (a) To find the magnitude of the field, we use Eq. 29-9 for each semicircle ($\phi = \pi$ rad), and use superposition to obtain the result:

$$B = \frac{\mu_0 i\pi}{4\pi a} + \frac{\mu_0 i\pi}{4\pi b} = \frac{\mu_0 i}{4}\left(\frac{1}{a}+\frac{1}{b}\right) = \frac{(4\pi\times10^{-7}\,\text{T}\cdot\text{m/A})(0.0562\,\text{A})}{4}\left(\frac{1}{0.0572\,\text{m}}+\frac{1}{0.0936\,\text{m}}\right)$$
$$= 4.97\times10^{-7}\,\text{T}.$$

(b) By the right-hand rule, $\vec{B}$ points into the paper at P (see Fig. 29-6(c)).

(c) The enclosed area is $A = (\pi a^2 + \pi b^2)/2$ which means the magnetic dipole moment has magnitude

$$|\vec{\mu}| = \frac{\pi i}{2}(a^2 + b^2) = \frac{\pi(0.0562\text{A})}{2}[(0.0572\text{m})^2 + (0.0936\text{m})^2] = 1.06 \times 10^{-3}\,\text{A} \cdot \text{m}^2.$$

(d) The direction of $\vec{\mu}$ is the same as the $\vec{B}$ found in part (a): into the paper.

52. By imagining that each of the segments *bg* and *cf* (which are shown in the figure as having no current) actually has a pair of currents, where both currents are of the same magnitude (*i*) but opposite direction (so that the pair effectively cancels in the final sum), one can justify the superposition.

(a) The dipole moment of path *abcdefgha* is

$$\vec{\mu} = \vec{\mu}_{bcfgb} + \vec{\mu}_{abgha} + \vec{\mu}_{cdefc} = \left(ia^2\right)\left(\hat{j} - \hat{i} + \hat{i}\right) = ia^2\hat{j}$$
$$= (6.0\,\text{A})(0.10\,\text{m})^2\,\hat{j} = (6.0 \times 10^{-2}\,\text{A} \cdot \text{m}^2)\hat{j}.$$

(b) Since both points are far from the cube we can use the dipole approximation. For $(x, y, z) = (0, 5.0\text{ m}, 0)$

$$\vec{B}(0, 5.0\text{ m}, 0) \approx \frac{\mu_0}{2\pi}\frac{\vec{\mu}}{y^3} = \frac{(1.26 \times 10^{-6}\,\text{T} \cdot \text{m/A})(6.0 \times 10^{-2}\text{m}^2 \cdot \text{A})\hat{j}}{2\pi(5.0\text{ m})^3} = (9.6 \times 10^{-11}\,\text{T})\hat{j}.$$

53. (a) We denote the large loop and small coil with subscripts 1 and 2, respectively.

$$B_1 = \frac{\mu_0 i_1}{2R_1} = \frac{(4\pi \times 10^{-7}\,\text{T} \cdot \text{m/A})(15\,\text{A})}{2(0.12\,\text{m})} = 7.9 \times 10^{-5}\,\text{T}.$$

(b) The torque has magnitude equal to

$$\tau = |\vec{\mu}_2 \times \vec{B}_1| = \mu_2 B_1 \sin 90° = N_2 i_2 A_2 B_1 = \pi N_2 i_2 r_2^2 B_1$$
$$= \pi(50)(1.3\text{A})(0.82 \times 10^{-2}\,\text{m})^2(7.9 \times 10^{-5}\,\text{T}) = 1.1 \times 10^{-6}\,\text{N} \cdot \text{m}.$$

54. Using Eq. 29-26, we find that the net *y*-component field is

$$B_y = \frac{\mu_0\,i_1\,R^2}{2(R^2 + z_1^2)^{3/2}} - \frac{\mu_0\,i_2\,R^2}{2(R^2 + z_2^2)^{3/2}},$$

where $z_1^2 = L^2$ (see Fig. 29-68(a)) and $z_2^2 = y^2$ (because the central axis here is denoted *y* instead of *z*). The fact that there is a minus sign between the two terms, above, is due to the observation that the datum in Fig. 29-68(b) corresponding to $B_y = 0$ would be impossible without it (physically, this means that one of the currents is clockwise and the other is counterclockwise).

(a) As $y \to \infty$, only the first term contributes and (with $B_y = 7.2 \times 10^{-6}$ T given in this case) we can solve for i_1. We obtain $i_1 = (45/16\pi)$ A ≈ 0.90 A.

(b) With loop 2 at $y = 0.06$ m (see Fig. 29-68(b)) we are able to determine i_2 from

$$\frac{\mu_0 \, i_1 \, R^2}{2(R^2 + L^2)^{3/2}} = \frac{\mu_0 \, i_2 \, R^2}{2(R^2 + y^2)^{3/2}} .$$

We obtain $i_2 = (117\sqrt{13} / 50\pi)$ A ≈ 2.7 A.

55. (a) We find the field by superposing the results of two semi-infinite wires (Eq. 29-7) and a semicircular arc (Eq. 29-9 with $\phi = \pi$ rad). The direction of $\vec{B}$ is out of the page, as can be checked by referring to Fig. 29-6(c). The magnitude of $\vec{B}$ at point a is therefore

$$B_a = 2\left(\frac{\mu_0 i}{4\pi R}\right) + \frac{\mu_0 i \pi}{4\pi R} = \frac{\mu_0 i}{2R}\left(\frac{1}{\pi} + \frac{1}{2}\right).$$

With $i = 10$ A and $R = 0.0050$ m, we obtain $B_a = 1.0 \times 10^{-3}$ T.

(b) The direction of this field is out of the page, as Fig. 29-6(c) makes clear.

(c) The last remark in the problem statement implies that treating b as a point midway between two infinite wires is a good approximation. Thus, using Eq. 29-4,

$$B_b = 2\left(\frac{\mu_0 i}{2\pi R}\right) = 8.0 \times 10^{-4} \text{ T}.$$

(d) This field, too, points out of the page.

56. Using the Pythagorean theorem, we have

$$B^2 = B_1^2 + B_2^2 = \left(\frac{\mu_0 i_1 \phi}{4\pi R}\right)^2 + \left(\frac{\mu_0 i_2}{2\pi R}\right)^2$$

which, when thought of as the equation for a line in a B^2 versus i_2^2 graph, allows us to identify the first term as the "y-intercept" (1×10^{-10}) and the part of the second term which multiplies i_2^2 as the "slope" (5×10^{-10}). The latter observation leads to the conclusion that $R = 8.9$ mm, and then our observation about the "y-intercept" determines the angle subtended by the arc: $\phi = 1.8$ rad.

57. We refer to the center of the circle (where we are evaluating $\vec{B}$) as C. Recalling the *straight sections* discussion in Sample Problem 29-1, we see that the current in the straight segments which are collinear with C do not contribute to the field there. Eq. 29-9

(with $\phi = \pi/2$ rad) and the right-hand rule indicates that the currents in the two arcs contribute

$$\frac{\mu_0 i(\pi/2)}{4\pi R} - \frac{\mu_0 i(\pi/2)}{4\pi R} = 0$$

to the field at C. Thus, the non-zero contributions come from those straight-segments which are not collinear with C. There are two of these "semi-infinite" segments, one a vertical distance R above C and the other a horizontal distance R to the left of C. Both contribute fields pointing out of the page (see Fig. 29-6(c)). Since the magnitudes of the two contributions (governed by Eq. 29-7) add, then the result is

$$B = 2\left(\frac{\mu_0 i}{4\pi R}\right) = \frac{\mu_0 i}{2\pi R}$$

exactly what one would expect from a single infinite straight wire (see Eq. 29-4). For such a wire to produce such a field (out of the page) with a leftward current requires that the point of evaluating the field be below the wire (again, see Fig. 29-6(c)).

58. We use Eq. 29-4 to relate the magnitudes of the magnetic fields B_1 and B_2 to the currents (i_1 and i_2, respectively) in the two long wires. The angle of their net field is

$$\theta = \tan^{-1}(B_2/B_1) = \tan^{-1}(i_2/i_1) = 53.13°.$$

The accomplish the net field rotation described in the problem, we must achieve a final angle $\theta' = 53.13° - 20° = 33.13°$. Thus, the final value for the current i_1 must be $i_2/\tan\theta'$ = 61.3 mA.

59. Using the right-hand rule (and symmetry), we see that $\vec{B}_{net}$ points along what we will refer to as the y axis (passing through P), consisting of two equal magnetic field y-components. Using Eq. 29-17,

$$|\vec{B}_{net}| = 2\frac{\mu_0 i}{2\pi r}\sin\theta$$

where $i = 4.00$ A, $r = r = \sqrt{d_2^2 + d_1^2/4} = 5.00$ m, and

$$\theta = \tan^{-1}\left(\frac{d_2}{\frac{1}{2}d_1}\right) = 53.1° \quad .$$

Therefore, $|\vec{B}_{net}| = 2.56 \times 10^{-7}$ T.

60. The radial segments do not contribute to $\vec{B}$ (at the center) and the arc-segments contribute according to Eq. 29-9 (with angle in radians). If $\hat{k}$ designates the direction "out of the page" then

$$\vec{B} = \frac{\mu_0 i(\pi \text{ rad})}{4\pi(4.00 \text{ m})}\hat{k} + \frac{\mu_0 i\left(\frac{\pi}{2}\text{ rad}\right)}{4\pi(2.00 \text{ m})}\hat{k} - \frac{\mu_0 i\left(\frac{\pi}{2}\text{ rad}\right)}{4\pi(2.00 \text{ m})}\hat{k}$$

where $i = 2.00$ A. This yields $\vec{B} = (1.57 \times 10^{-7} \text{ T})\hat{k}$, or $|\vec{B}| = 1.57 \times 10^{-7}$ T.

61. (a) The magnetic field at a point within the hole is the sum of the fields due to two current distributions. The first is that of the solid cylinder obtained by filling the hole and has a current density that is the same as that in the original cylinder (with the hole). The second is the solid cylinder that fills the hole. It has a current density with the same magnitude as that of the original cylinder but is in the opposite direction. If these two situations are superposed the total current in the region of the hole is zero. Now, a solid cylinder carrying current i which is uniformly distributed over a cross section, produces a magnetic field with magnitude

$$B = \frac{\mu_0 i r}{2\pi R^2}$$

at a distance r from its axis, inside the cylinder. Here R is the radius of the cylinder. For the cylinder of this problem the current density is

$$J = \frac{i}{A} = \frac{i}{\pi(a^2 - b^2)},$$

where $A = \pi(a^2 - b^2)$ is the cross-sectional area of the cylinder with the hole. The current in the cylinder without the hole is

$$I_1 = JA = \pi J a^2 = \frac{ia^2}{a^2 - b^2}$$

and the magnetic field it produces at a point inside, a distance r_1 from its axis, has magnitude

$$B_1 = \frac{\mu_0 I_1 r_1}{2\pi a^2} = \frac{\mu_0 i r_1 a^2}{2\pi a^2(a^2 - b^2)} = \frac{\mu_0 i r_2}{2\pi(a^2 - b^2)}.$$

The current in the cylinder that fills the hole is

$$I_2 = \pi J b^2 = \frac{ib^2}{a^2 - b^2}$$

and the field it produces at a point inside, a distance r_2 from the its axis, has magnitude

$$B_2 = \frac{\mu_0 I_2 r_2}{2\pi b^2} = \frac{\mu_0 i r_2 b^2}{2\pi b^2 (a^2 - b^2)} = \frac{\mu_0 i r_2}{2\pi (a^2 - b^2)}.$$

At the center of the hole, this field is zero and the field there is exactly the same as it would be if the hole were filled. Place $r_1 = d$ in the expression for B_1 and obtain

$$B = \frac{\mu_0 i d}{2\pi (a^2 - b^2)} = \frac{(4\pi \times 10^{-7}\,\mathrm{T \cdot m/A})(5.25\,\mathrm{A})(0.0200\,\mathrm{m})}{2\pi[(0.0400\,\mathrm{m})^2 - (0.0150\,\mathrm{m})^2]} = 1.53 \times 10^{-5}\,\mathrm{T}$$

for the field at the center of the hole. The field points upward in the diagram if the current is out of the page.

(b) If $b = 0$ the formula for the field becomes

$$B = \frac{\mu_0 i d}{2\pi a^2}.$$

This correctly gives the field of a solid cylinder carrying a uniform current i, at a point inside the cylinder a distance d from the axis. If $d = 0$ the formula gives $B = 0$. This is correct for the field on the axis of a cylindrical shell carrying a uniform current.

(c) Consider a rectangular path with two long sides (side 1 and 2, each with length L) and two short sides (each of length less than b). If side 1 is directly along the axis of the hole, then side 2 would be also parallel to it and also in the hole. To ensure that the short sides do not contribute significantly to the integral in Ampere's law, we might wish to make L *very* long (perhaps longer than the length of the cylinder), or we might appeal to an argument regarding the angle between $\vec{B}$ and the short sides (which is 90° at the axis of the hole). In any case, the integral in Ampere's law reduces to

$$\oint_{\text{rectangle}} \vec{B} \cdot d\vec{s} = \mu_0 i_{\text{enclosed}}$$

$$\int_{\text{side 1}} \vec{B} \cdot d\vec{s} + \int_{\text{side 2}} \vec{B} \cdot d\vec{s} = \mu_0 i_{\text{in hole}}$$

$$\left(B_{\text{side 1}} - B_{\text{side 2}}\right)L = 0$$

where $B_{\text{side 1}}$ is the field along the axis found in part (a). This shows that the field at off-axis points (where $B_{\text{side 2}}$ is evaluated) is the same as the field at the center of the hole; therefore, the field in the hole is uniform.

62. We note that when there is no y-component of magnetic field from wire 1 (which, by the right-hand rule, relates to when wire 1 is at 90° = π/2 rad), the total y-component of magnetic field is zero (see Fig. 29-76(c)). This means wire #2 is either at +π/2 rad or −π/2 rad.

(a) We now make the assumption that wire #2 must be at −π/2 rad (−90°, the bottom of the cylinder) since it would pose an obstacle for the motion of wire #1 (which is needed to make these graphs) if it were anywhere in the top semicircle.

(b) Looking at the θ_1 = 90° datum in Fig. 29-76(b)) – where there is a *maximum* in $B_{net\,x}$ (equal to +6 μT) – we are led to conclude that B_{1x} = 6.0 μT – 2.0 μT = +4.0 μT in that situation. Using Eq. 29-4, we obtain $i_1 = B_{1x}\,2\pi R\,/\mu_o$ = 4.0 A .

(c) The fact that Fig. 29-76(b) increases as θ_1 progresses from 0 to 90° implies that wire 1's current is *out of the page*, and this is consistent with the cancellation of $B_{net\,y}$ at θ_1 = 90°, noted earlier (with regard to Fig. 29-76(c)).

(d) Referring now to Fig. 29-76(b) we note that there is no x-component of magnetic field from wire 1 when θ_1 = 0, so that plot tells us that B_{2x} = +2.0 μT. Using Eq. 29-4, we have $i_2 = B_{2x}\,2\pi R\,/\mu_o$ = 2.0 A for the magnitudes of the currents.

(e) We can conclude (by the right-hand rule) that wire 2's current is *into the page*.

63. Using Eq. 29-20 and Eq. 29-17, we have

$$|\vec{B_1}|=\left(\frac{\mu_0 i}{2\pi R^2}\right)r_1 \quad |\vec{B_2}|=\frac{\mu_0 i}{2\pi r_2}$$

where $r_1 = 0.0040\,\mathrm{m}$, $\left|\vec{B_1}\right| = 2.8\times10^{-4}\,\mathrm{T}$, $r_2 = 0.010\,\mathrm{m}$ and $|\vec{B_2}| = 2.0\times10^{-4}$ T. Point 2 is known to be external to the wire since $|\vec{B_2}|<|\vec{B_1}|$. From the second equation, we find i = 10 A. Plugging this into the first equation yields R = 5.3 × 10^{-3} m.

64. Eq. 29-1 is maximized (with respect to angle) by setting θ = 90° (= π/2 rad). Its value in this case is dB_{max} = $\mu_o\,i\,ds\,/4\pi R^2$. From Fig. 29-77(b), we have B_{max} = 60 × 10^{-12} T. We can relate this B_{max} to our dB_{max} by setting "ds" equal to 1 × 10^{-6} m and R = 0.025 m. This allows us to solve for the current: i = 0.375 A. Plugging this into Eq. 29-4 (for the infinite wire) gives B_∞ = 3.0 μT.

65. Eq. 29-4 gives

$$i = \frac{2\pi R B}{\mu_0} = \frac{2\pi(0.880\,\mathrm{m})(7.30\times10^{-6}\,\mathrm{T})}{4\pi\times10^{-7}\,\mathrm{T\cdot m/A}} = 32.1\,\mathrm{A}.$$

66. (a) By the right-hand rule, the magnetic field $\vec{B}_1$ (evaluated at a) produced by wire 1 (the wire at bottom left) is at $\phi = 150°$ (measured counterclockwise from the $+x$ axis, in the xy plane), and the field produced by wire 2 (the wire at bottom right) is at $\phi = 210°$. By symmetry $(\vec{B}_1 = \vec{B}_2)$ we observe that only the x-components survive, yielding

$$\vec{B}_1 + \vec{B}_2 = \left(2 \frac{\mu_0 i}{2\pi\ell} \cos 150° \right) \hat{i} = (-3.46 \times 10^{-5} \text{T})\hat{i}$$

where $i = 10$ A, $\ell = 0.10$ m, and Eq. 29-4 has been used. To cancel this, wire b must carry current into the page (that is, the $-\hat{k}$ direction) of value

$$i_b = \left(3.46 \times 10^{-5} \right) \frac{2\pi r}{\mu_0} = 15\text{A}$$

where $r = \sqrt{3}\,\ell/2 = 0.087$ m and Eq. 29-4 has again been used.

(b) As stated above, to cancel this, wire b must carry current into the page (that is, the $-z$ direction)

67. (a) The field in this region is entirely due to the long wire (with, presumably, negligible thickness). Using Eq. 29-17,

$$\left| \vec{B} \right| = \frac{\mu_0 i_w}{2\pi r} = 4.8 \times 10^{-3} \text{ T}$$

where $i_w = 24$ A and $r = 0.0010$ m.

(b) Now the field consists of two contributions (which are anti-parallel) — from the wire (Eq. 29-17) and from a portion of the conductor (Eq. 29-20 modified for annular area):

$$|\vec{B}| = \frac{\mu_0 i_w}{2\pi r} - \frac{\mu_0 i_{enc}}{2\pi r} = \frac{\mu_0 i_w}{2\pi r} - \frac{\mu_0 i_c}{2\pi r} \left(\frac{\pi r^2 - \pi R_i^2}{\pi R_0^2 - \pi R_i^2} \right)$$

where $r = 0.0030$ m, $R_i = 0.0020$ m, $R_o = 0.0040$ m and $i_c = 24$ A. Thus, we find $|\vec{B}| = 9.3 \times 10^{-4}$ T.

(c) Now, in the external region, the individual fields from the two conductors cancel completely (since $i_c = i_w$): $\vec{B} = 0$.

68. (a) We designate the wire along $y = r_A = 0.100$ m wire A and the wire along $y = r_B = 0.050$ m wire B. Using Eq. 29-4, we have

$$\vec{B}_{net} = \vec{B}_A + \vec{B}_B = -\frac{\mu_0 i_A}{2\pi r_A}\hat{k} - \frac{\mu_0 i_B}{2\pi r_B}\hat{k} = (-52.0\times10^{-6}\text{ T})\hat{k}.$$

(b) This will occur for some value $r_B < y < r_A$ such that

$$\frac{\mu_0 i_A}{2\pi(r_A - y)} = \frac{\mu_0 i_B}{2\pi(y - r_B)}.$$

Solving, we find $y = 13/160 \approx 0.0813$ m.

(c) We eliminate the $y < r_B$ possibility due to wire B carrying the larger current. We expect a solution in the region $y > r_A$ where

$$\frac{\mu_0 i_A}{2\pi(y - r_A)} = \frac{\mu_0 i_B}{2\pi(y - r_B)}.$$

Solving, we find $y = 7/40 \approx 0.0175$ m.

69. (a) As illustrated in Sample Problem 29-1, the radial segments do not contribute to $\vec{B}_p$ and the arc-segments contribute according to Eq. 29-9 (with angle in radians). If $\hat{k}$ designates the direction "out of the page" then

$$\vec{B} = \frac{\mu_0(0.40\text{A})(\pi\text{ rad})}{4\pi(0.050\text{ m})}\hat{k} - \frac{\mu_0(0.80\text{A})(\frac{2\pi}{3}\text{ rad})}{4\pi(0.040\text{ m})}\hat{k}$$

which yields $\vec{B} = -1.7\times10^{-6}\hat{k}\text{T}$, or $|\vec{B}| = 1.7\times10^{-6}$ T.

(b) The direction is $-\hat{k}$, or into the page.

(c) If the direction of i_1 is reversed, we then have

$$\vec{B} = -\frac{\mu_0(0.40\text{A})(\pi\text{ rad})}{4\pi(0.050\text{ m})}\hat{k} - \frac{\mu_0(0.80\text{A})(\frac{2\pi}{3}\text{ rad})}{4\pi(0.040\text{ m})}\hat{k}$$

which yields $\vec{B} = (-6.7\times10^{-6}\text{ T})\hat{k}$, or $|\vec{B}| = 6.7\times10^{-6}\text{T}$.

(d) The direction is $-\hat{k}$, or into the page.

70. We note that the distance from each wire to P is $r = d/\sqrt{2} = 0.071$m. In both parts, the current is $i = 100$ A.

(a) With the currents parallel, application of the right-hand rule (to determine each of their contributions to the field at P) reveals that the vertical components cancel and the horizontal components add, yielding the result:

$$B = 2\left(\frac{\mu_0 i}{2\pi r}\right)\cos 45.0° = 4.00 \times 10^{-4} \text{ T}$$

and directed in the $-x$ direction. In unit-vector notation, we have $\vec{B} = (-4.00 \times 10^{-4} \text{ T})\hat{i}$.

(b) Now, with the currents anti-parallel, application of the right-hand rule shows that the horizontal components cancel and the vertical components add. Thus,

$$B = 2\left(\frac{\mu_0 i}{2\pi r}\right)\sin 45.0° = 4.00 \times 10^{-4} \text{ T}$$

and directed in the $+y$ direction. In unit-vector notation, we have $\vec{B} = (4.00 \times 10^{-4} \text{ T})\hat{j}$.

71. Since the radius is $R = 0.0013$ m, then the $i = 50$ A produces

$$B = \frac{\mu_0 i}{2\pi R} = 0.0077 \text{ T}$$

at the edge of the wire. The three equations, Eq. 29-4, Eq. 29-17 and Eq. 29-20, agree at this point.

72. The area enclosed by the loop L is $A = \frac{1}{2}(4d)(3d) = 6d^2$. Thus

$$\oint_c \vec{B} \cdot d\vec{s} = \mu_0 i = \mu_0 jA = \left(4\pi \times 10^{-7} \text{ T·m/A}\right)\left(15 \text{ A/m}^2\right)(6)(0.20\text{m})^2 = 4.5 \times 10^{-6} \text{ T·m}.$$

73. (a) With cylindrical symmetry, we have, external to the conductors,

$$\left|\vec{B}\right| = \frac{\mu_0 i_{enc}}{2\pi r}$$

which produces $i_{enc} = 25$ mA from the given information. Therefore, the thin wire must carry 5.0 mA.

(b) The direction is downward, opposite to the 30 mA carried by the thin conducting surface.

74. (a) All wires carry parallel currents and attract each other; thus, the "top" wire is pulled downward by the other two:

$$|\vec{F}| = \frac{\mu_0 \, L(5.0\,\text{A})(3.2\,\text{A})}{2\pi(0.10\,\text{m})} + \frac{\mu_0 \, L(5.0\,\text{A})(5.0\,\text{A})}{2\pi(0.20\,\text{m})}$$

where $L = 3.0$ m. Thus, $|\vec{F}| = 1.7 \times 10^{-4}$ N.

(b) Now, the "top" wire is pushed upward by the center wire and pulled downward by the bottom wire:

$$|\vec{F}| = \frac{\mu_0 L(5.0\,\text{A})(3.2\,\text{A})}{2\pi(0.10\text{m})} - \frac{\mu_0 L(5.0\,\text{A})(5.0\,\text{A})}{2\pi(0.20\text{m})} = 2.1 \times 10^{-5}\,\text{N} .$$

75. We use $B(x,y,z) = (\mu_0/4\pi)i\Delta\vec{s} \times \vec{r}/r^3$, where $\Delta\vec{s} = \Delta s\hat{j}$ and $\vec{r} = x\hat{i} + y\hat{j} + z\hat{k}$. Thus,

$$\vec{B}(x,y,z) = \left(\frac{\mu_0}{4\pi}\right)\frac{i\,\Delta s\hat{j} \times \left(x\vec{i} + y\hat{j} + z\vec{k}\right)}{\left(x^2 + y^2 + z^2\right)^{3/2}} = \frac{\mu_0 i\,\Delta s\left(z\hat{i} - x\hat{k}\right)}{4\pi\left(x^2 + y^s + z^2\right)^{3/2}} .$$

(a) The field on the z axis (at $z = 5.0$ m) is

$$\vec{B}(0,0,5.0\text{m}) = \frac{\left(4\pi \times 10^{-7}\,\text{T} \cdot \text{m/A}\right)(2.0\,\text{A})\left(3.0 \times 10^{-2}\,\text{m}\right)(5.0\text{m})\hat{i}}{4\pi\left(0^2 + 0^2 + (5.0\text{m})^2\right)^{3/2}} = (2.4 \times 10^{-10}\,\text{T})\hat{i}.$$

(b) $\vec{B}$ (0, 6.0 m, 0), since $x = z = 0$.

(c) The field in the xy plane, at $(x, y) = (7,7)$, is

$$\vec{B}(7.0\text{m},7.0\text{m},0) = \frac{\left(4\pi \times 10^{-7}\,\text{T} \cdot \text{m/A}\right)(2.0\,\text{A})(3.0 \times 10^{-2}\,\text{m})(-7.0\text{m})\hat{k}}{4\pi\left((7.0\text{m})^2 + (7.0\text{m})^2 + 0^2\right)^{3/2}} = (-4.3 \times 10^{-11}\,\text{T})\hat{k}.$$

(d) The field in the xy plane, at $(x, y) = (-3, -4)$, is

$$\vec{B}(-3.0\text{m},-4.0\text{m},0) = \frac{\left(4\pi \times 10^{-7}\,\text{T} \cdot \text{m/A}\right)(2.0\,\text{A})(3.0 \times 10^{-2}\,\text{m})(3.0\text{m})\hat{k}}{4\pi\left((-3.0\text{m})^2 + (-4.0\text{m})^2 + 0^2\right)^{3/2}} = (1.4 \times 10^{-10}\,\text{T})\hat{k}.$$

76. (a) The radial segments do not contribute to $\vec{B}_P$ and the arc-segments contribute according to Eq. 29-9 (with angle in radians). If $\hat{k}$ designates the direction "out of the page" then

$$\vec{B}_P = \frac{\mu_0\, i \left(\frac{7\pi}{4}\,\text{rad}\right)}{4\pi(4.00\ \text{m})}\,\hat{k} - \frac{\mu_0\, i \left(\frac{7\pi}{4}\,\text{rad}\right)}{4\pi(2.00\ \text{m})}\,\hat{k}$$

where $i = 0.200$ A. This yields $\vec{B} = -2.75 \times 10^{-8}\,\hat{k}$ T, or $|\vec{B}| = 2.75 \times 10^{-8}$ T.

(b) The direction is $-\hat{k}$, or into the page.

77. The contribution to $\vec{B}_{\text{net}}$ from the first wire is (using Eq. 29-4)

$$\vec{B}_1 = \frac{\mu_0(30\ \text{A})}{2\pi(2.0\ \text{m})}\,\hat{k} = (3.0 \times 10^{-6}\text{T})\,\hat{k}\ .$$

The distance from the second wire to the point where we are evaluating $\vec{B}_{\text{net}}$ is 4 m − 2 m = 2 m. Thus,

$$\vec{B}_2 = \frac{\mu_0(40\ \text{A})}{2\pi(2\ \text{m})}\,\hat{i} = (4.0 \times 10^{-6}\ \text{T})\,\hat{i}$$

and consequently is perpendicular to $\vec{B}_1$. The magnitude of $|\vec{B}_{\text{net}}|$ is therefore $\sqrt{3.0^2 + 4.0^2} = 5.0\ \mu$T.

78. Using Eq. 29-20, $|\vec{B}| = \left(\frac{\mu_0\, i}{2\pi R^2}\right) r,$ we find that $r = 0.00128$ m gives the desired field value.

79. The points must be along a line parallel to the wire and a distance r from it, where r satisfies $B_{\text{wire}} = \dfrac{\mu_0 i}{2\pi r} = B_{\text{ext}}$, or

$$r = \frac{\mu_0 i}{2\pi B_{\text{ext}}} = \frac{\left(1.26 \times 10^{-6}\ \text{T}\cdot\text{m/A}\right)(100\,\text{A})}{2\pi\left(5.0 \times 10^{-3}\ \text{T}\right)} = 4.0 \times 10^{-3}\,\text{m}.$$

80. (a) The magnitude of the magnetic field on the axis of a circular loop, a distance z from the loop center, is given by Eq. 29-26:

$$B = \frac{N\mu_0 i R^2}{2(R^2 + z^2)^{3/2}},$$

where R is the radius of the loop, N is the number of turns, and i is the current. Both of the loops in the problem have the same radius, the same number of turns, and carry the same current. The currents are in the same sense, and the fields they produce are in the same direction in the region between them. We place the origin at the center of the left-hand loop and let x be the coordinate of a point on the axis between the loops. To calculate the field of the left-hand loop, we set $z = x$ in the equation above. The chosen point on the axis is a distance $s - x$ from the center of the right-hand loop. To calculate the field it produces, we put $z = s - x$ in the equation above. The total field at the point is therefore

$$B = \frac{N\mu_0 iR^2}{2}\left[\frac{1}{(R^2+x^2)^{3/2}} + \frac{1}{(R^2+x^2-2sx+s^2)^{3/2}}\right].$$

Its derivative with respect to x is

$$\frac{dB}{dx} = -\frac{N\mu_0 iR^2}{2}\left[\frac{3x}{(R^2+x^2)^{5/2}} + \frac{3(x-s)}{(R^2+x^2-2sx+s^2)^{5/2}}\right].$$

When this is evaluated for $x = s/2$ (the midpoint between the loops) the result is

$$\frac{dB}{dx}\bigg|_{s/2} = -\frac{N\mu_0 iR^2}{2}\left[\frac{3s/2}{(R^2+s^2/4)^{5/2}} - \frac{3s/2}{(R^2+s^2/4-s^2+s^2)^{5/2}}\right] = 0$$

independent of the value of s.

(b) The second derivative is

$$\frac{d^2 B}{dx^2} = \frac{N\mu_0 iR^2}{2}\left[-\frac{3}{(R^2+x^2)^{5/2}} + \frac{15x^2}{(R^2+x^2)^{7/2}}\right.$$
$$\left. -\frac{3}{(R^2+x^2-2sx+s^2)^{5/2}} + \frac{15(x-s)^2}{(R^2+x^2-2sx+s^2)^{7/2}}\right]$$

At $x = s/2$,

$$\frac{d^2 B}{dx^2}\bigg|_{s/2} = \frac{N\mu_0 iR^2}{2}\left[-\frac{6}{(R^2+s^2/4)^{5/2}} + \frac{30s^2/4}{(R^2+s^2/4)^{7/2}}\right]$$
$$= \frac{N\mu_0 R^2}{2}\left[\frac{-6(R^2+s^2/4)+30s^2/4}{(R^2+s^2/4)^{7/2}}\right] = 3N\mu_0 iR^2 \frac{s^2-R^2}{(R^2+s^2/4)^{7/2}}.$$

Clearly, this is zero if $s = R$.

81. The center of a square is a distance $R = a/2$ from the nearest side (each side being of length $L = a$). There are four sides contributing to the field at the center. The result is

$$B_{center} = 4\left(\frac{\mu_0 i}{2\pi(a/2)}\right)\left(\frac{a}{\sqrt{a^2 + 4(a/2)^2}}\right) = \frac{2\sqrt{2}\mu_0 i}{\pi a}.$$

82. We refer to the side of length L as the long side and that of length W as the short side. The center is a distance $W/2$ from the midpoint of each long side, and is a distance $L/2$ from the midpoint of each short side. There are two of each type of side, so the result of problem 11 leads to

$$B = 2\frac{\mu_0 i}{2\pi(W/2)}\frac{L}{\sqrt{L^2 + 4(W/2)^2}} + 2\frac{\mu_0 i}{2\pi(L/2)}\frac{W}{\sqrt{W^2 + 4(L/2)^2}}.$$

The final form of this expression, shown in the problem statement, derives from finding the common denominator of the above result and adding them, while noting that

$$\frac{L^2 + W^2}{\sqrt{W^2 + L^2}} = \sqrt{W^2 + L^2}.$$

83. We imagine the square loop in the yz plane (with its center at the origin) and the evaluation point for the field being along the x axis (as suggested by the notation in the problem). The origin is a distance $a/2$ from each side of the square loop, so the distance from the evaluation point to each side of the square is, by the Pythagorean theorem,

$$R = \sqrt{(a/2)^2 + x^2} = \frac{1}{2}\sqrt{a^2 + 4x^2}.$$

Only the x components of the fields (contributed by each side) will contribute to the final result (other components cancel in pairs), so a trigonometric factor of

$$\frac{a/2}{R} = \frac{a}{\sqrt{a^2 + 4x^2}}$$

multiplies the expression of the field given by the result of problem 11 (for each side of length $L = a$). Since there are four sides, we find

$$B(x) = 4\left(\frac{\mu_0 i}{2\pi R}\right)\left(\frac{a}{\sqrt{a^2 + 4R^2}}\right)\left(\frac{a}{\sqrt{a^2 + 4x^2}}\right) = \frac{4\mu_0 i a^2}{2\pi(\frac{1}{2})\left(\sqrt{a^2 + 4x^2}\right)^2 \sqrt{a^2 + 4(a/2)^2 + 4x^2}}$$

which simplifies to the desired result. It is straightforward to set $x = 0$ and see that this reduces to the expression found in problem 12 (noting that $\frac{4}{\sqrt{2}} = 2\sqrt{2}$).

84. Using the result of problem 12 and Eq. 29-10, we wish to show that

$$\frac{2\sqrt{2}\mu_0 i}{\pi a} > \frac{\mu_0 i}{2R}, \quad \text{or} \quad \frac{4\sqrt{2}}{\pi a} > \frac{1}{R},$$

but to do this we must relate the parameters a and R. If both wires have the same length L then the geometrical relationships $4a = L$ and $2\pi R = L$ provide the necessary connection:

$$4a = 2\pi R \Rightarrow a = \frac{\pi R}{2}.$$

Thus, our proof consists of the observation that

$$\frac{4\sqrt{2}}{\pi a} = \frac{8\sqrt{2}}{\pi^2 R} > \frac{1}{R},$$

as one can check numerically (that $8\sqrt{2}/\pi^2 > 1$).

85. The two small wire-segments, each of length $a/4$, shown in Fig. 29-83 nearest to point P, are labeled 1 and 8 in the figure below.

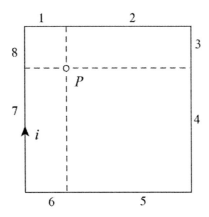

Let $-\hat{k}$ be a unit vector pointing into the page. We use the results of problem 19 to calculate B_{P1} through B_{P8}:

$$B_{P1} = B_{P8} = \frac{\sqrt{2}\mu_0 i}{8\pi(a/4)} = \frac{\sqrt{2}\mu_0 i}{2\pi a},$$

$$B_{P4} = B_{P5} = \frac{\sqrt{2}\mu_0 i}{8\pi(3a/4)} = \frac{\sqrt{2}\mu_0 i}{6\pi a},$$

$$B_{P2} = B_{P7} = \frac{\mu_0 i}{4\pi(a/4)} \cdot \frac{3a/4}{\left[(3a/4)^2 + (a/4)^2\right]^{1/2}} = \frac{3\mu_0 i}{\sqrt{10}\pi a},$$

and

$$B_{P3} = B_{P6} = \frac{\mu_0 i}{4\pi(3a/4)} \cdot \frac{a/4}{\left[(a/4)^2 + (3a/4)^2\right]^{1/2}} = \frac{\mu_0 i}{3\sqrt{10}\pi a}.$$

Finally,

$$\vec{B}_P = \sum_{n=1}^{8} B_{Pn}(-\hat{k}) = 2\frac{\mu_0 i}{\pi a}\left(\frac{\sqrt{2}}{2} + \frac{\sqrt{2}}{6} + \frac{3}{\sqrt{10}} + \frac{1}{3\sqrt{10}}\right)(-\hat{k})$$

$$= \frac{2\left(4\pi\times10^{-7}\,\text{T}\cdot\text{m/A}\right)(10\text{A})}{\pi\left(8.0\times10^{-2}\,\text{m}\right)}\left(\frac{\sqrt{2}}{2} + \frac{\sqrt{2}}{6} + \frac{3}{\sqrt{10}} + \frac{1}{3\sqrt{10}}\right)(-\hat{k})$$

$$= \left(2.0\times10^{-4}\,\text{T}\right)(-\hat{k}).$$

86. (a) Consider a segment of the projectile between y and $y + dy$. We use Eq. 29-12 to find the magnetic force on the segment, and Eq. 29-7 for the magnetic field of each semi-infinite wire (the top rail referred to as wire 1 and the bottom as wire 2). The current in rail 1 is in the $+\hat{i}$ direction, and the current in rail 2 is in the $-\hat{i}$ direction. The field (in the region between the wires) set up by wire 1 is into the paper (the $-\hat{k}$ direction) and that set up by wire 2 is also into the paper. The force element (a function of y) acting on the segment of the projectile (in which the current flows in the $-\hat{j}$ direction) is given below. The coordinate origin is at the bottom of the projectile.

$$d\vec{F} = d\vec{F}_1 + d\vec{F}_2 = i\,dy\left(-\hat{j}\right)\times\vec{B}_1 + dy\left(-\hat{j}\right)\times\vec{B}_2 = i\left[B_1 + B_2\right]\hat{i}\,dy = i\left[\frac{\mu_0 i}{4\pi(2R+w-y)} + \frac{\mu_0 i}{4\pi y}\right]\hat{i}\,dy.$$

Thus, the force on the projectile is

$$\vec{F} = \int d\vec{F} = \frac{i^2\mu_0}{4\pi}\int_R^{R+w}\left(\frac{1}{2R+w-y} + \frac{1}{y}\right)dy\,\hat{i} = \frac{\mu_0 i^2}{2\pi}\ln\left(1 + \frac{w}{R}\right)\hat{i}.$$

(b) Using the work-energy theorem, we have

$$\Delta K = \tfrac{1}{2}mv_f^2 = W_{\text{ext}} = \int\vec{F}\cdot d\vec{s} = FL.$$

Thus, the final speed of the projectile is

$$v_f = \left(\frac{2W_{ext}}{m}\right)^{1/2} = \left[\frac{2}{m}\frac{\mu_0 i^2}{2\pi}\ln\left(1+\frac{w}{R}\right)L\right]^{1/2}$$

$$= \left[\frac{2(4\pi\times10^{-7}\,\text{T}\cdot\text{m}/\text{A})(450\times10^3\,\text{A})^2\,\ln(1+1.2\,\text{cm}/6.7\,\text{cm})(4.0\,\text{m})}{2\pi(10\times10^{-3}\,\text{kg})}\right]^{1/2}$$

$$= 2.3\times10^3\,\text{m}/\text{s}.$$

87. We take the current ($i = 50$ A) to flow in the $+x$ direction, and the electron to be at a point P which is $r = 0.050$ m above the wire (where "up" is the $+y$ direction). Thus, the field produced by the current points in the $+z$ direction at P. Then, combining Eq. 29-4 with Eq. 28-2, we obtain $\vec{F}_e = (-e\mu_0 i/2\pi r)(\vec{v}\times\hat{k})$.

(a) The electron is moving down: $\vec{v} = -v\hat{j}$ (where $v = 1.0\times10^7$ m/s is the speed) so

$$\vec{F}_e = \frac{-e\mu_0 iv}{2\pi r}(-\hat{i}) = (3.2\times10^{-16}\,\text{N})\hat{i},$$

or $|\vec{F}_e| = 3.2\times10^{-16}$ N.

(b) In this case, the electron is in the same direction as the current: $\vec{v} = v\hat{i}$ so

$$\vec{F}_e = \frac{-e\mu_0 iv}{2\pi r}(-\hat{j}) = (3.2\times10^{-16}\,\text{N})\hat{j},$$

or $|\vec{F}_e| = 3.2\times10^{-16}$ N.

(c) Now, $\vec{v} = \pm v\hat{k}$ so $\vec{F}_e \propto \hat{k}\times\hat{k} = 0$.

88. Eq. 29-17 applies for each wire, with $r = \sqrt{R^2 + (d/2)^2}$ (by the Pythagorean theorem). The vertical components of the fields cancel, and the two (identical) horizontal components add to yield the final result

$$B = 2\left(\frac{\mu_0 i}{2\pi r}\right)\left(\frac{d/2}{r}\right) = \frac{\mu_0 id}{2\pi\left(R^2 + (d/2)^2\right)} = 1.25\times10^{-6}\,\text{T},$$

where $(d/2)/r$ is a trigonometric factor to select the horizontal component. It is clear that this is equivalent to the expression in the problem statement. Using the right-hand rule, we find both horizontal components point in the $+x$ direction. Thus, in unit-vector notation, we have $\vec{B} = (1.25\times10^{-6}\,\text{T})\hat{i}$.

89. The "current per unit x-length" may be viewed as current density multiplied by the thickness Δy of the sheet; thus, $\lambda = J\Delta y$. Ampere's law may be (and often is) expressed in terms of the current density vector as follows

$$\oint \vec{B} \cdot d\vec{s} = \mu_0 \int \vec{J} \cdot d\vec{A}$$

where the area integral is over the region enclosed by the path relevant to the line integral (and $\vec{J}$ is in the $+z$ direction, out of the paper). With J uniform throughout the sheet, then it is clear that the right-hand side of this version of Ampere's law should reduce, in this problem, to $\mu_0 JA = \mu_0 J\Delta y\Delta x = \mu_0 \lambda\Delta x$.

(a) Figure 29-86 certainly has the horizontal components of $\vec{B}$ drawn correctly at points P and P' (as reference to Fig. 29-4 will confirm [consider the current elements nearest each of those points]), so the question becomes: is it possible for $\vec{B}$ to have vertical components in the figure? Our focus is on point P. Fig. 29-4 suggests that the current element just to the right of the nearest one (the one directly under point P) will contribute a downward component, but by the same reasoning the current element just to the left of the nearest one should contribute an upward component to the field at P. The current elements are all equivalent, as is reflected in the horizontal-translational symmetry built into this problem; therefore, all vertical components should cancel in pairs. The field at P must be purely horizontal, as drawn.

(b) The path used in evaluating $\oint \vec{B} \cdot d\vec{s}$ is rectangular, of horizontal length Δx (the horizontal sides passing through points P and P' respectively) and vertical size $\delta y > \Delta y$. The vertical sides have no contribution to the integral since $\vec{B}$ is purely horizontal (so the scalar dot product produces zero for those sides), and the horizontal sides contribute two equal terms, as shown next. Ampere's law yields

$$2B\Delta x = \mu_0 \lambda\Delta x \Rightarrow B = \frac{1}{2}\mu_0 \lambda.$$

90. In this case $L = 2\pi r$ is roughly the length of the toroid so

$$B = \mu_0 i_0 \left(\frac{N}{2\pi r}\right) = \mu_0 n i_0$$

This result is expected, since from the perspective of a point inside the toroid the portion of the toroid in the vicinity of the point resembles part of a long solenoid.

91. (a) For the circular path L of radius r concentric with the conductor

$$\oint_L \vec{B} \cdot d\vec{s} = 2\pi r B = \mu_0 i_{enc} = \mu_0 i \frac{\pi(r^2 - b^2)}{\pi(a^2 - b^2)}.$$

Thus, $B = \dfrac{\mu_0 i}{2\pi(a^2 - b^2)} \left(\dfrac{r^2 - b^2}{r} \right).$

(b) At $r = a$, the magnetic field strength is

$$\frac{\mu_0 i}{2\pi(a^2 - b^2)} \left(\frac{a^2 - b^2}{a} \right) = \frac{\mu_0 i}{2\pi a}.$$

At $r = b, B \propto r^2 - b^2 = 0$. Finally, for $b = 0$

$$B = \frac{\mu_0 i}{2\pi a^2} \frac{r^2}{r} = \frac{\mu_0 i r}{2\pi a^2}$$

which agrees with Eq. 29-20.

(c) The field is zero for $r < b$ and is equal to Eq. 29-17 for $r > a$, so this along with the result of part (a) provides a determination of B over the full range of values. The graph (with SI units understood) is shown below.

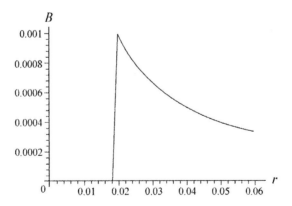

92. (a) Eq. 29-20 applies for $r < c$. Our sign choice is such that i is positive in the smaller cylinder and negative in the larger one.

$$B = \frac{\mu_0 i r}{2\pi c^2}, \quad r \le c.$$

(b) Eq. 29-17 applies in the region between the conductors.

$$B = \frac{\mu_0 i}{2\pi r}, \quad c \le r \le b.$$

(c) Within the larger conductor we have a superposition of the field due to the current in the inner conductor (still obeying Eq. 29-17) plus the field due to the (negative) current in that part of the outer conductor at radius less than r. The result is

$$B = \frac{\mu_0 i}{2\pi r} - \frac{\mu_0 i}{2\pi r}\left(\frac{r^2 - b^2}{a^2 - b^2}\right), \quad b < r \le a.$$

If desired, this expression can be simplified to read

$$B = \frac{\mu_0 i}{2\pi r}\left(\frac{a^2 - r^2}{a^2 - b^2}\right).$$

(d) Outside the coaxial cable, the net current enclosed is zero. So $B = 0$ for $r \ge a$.

(e) We test these expressions for one case. If $a \to \infty$ and $b \to \infty$ (such that $a > b$) then we have the situation described on page 696 of the textbook.

(f) Using SI units, the graph of the field is shown below:

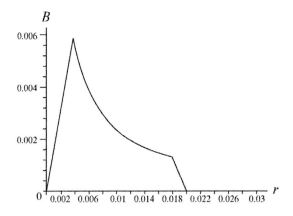

93. We use Ampere's law. For the dotted loop shown on the diagram $i = 0$. The integral $\int \vec{B} \cdot d\vec{s}$ is zero along the bottom, right, and top sides of the loop. Along the right side the field is zero, along the top and bottom sides the field is perpendicular to $d\vec{s}$. If ℓ is the length of the left edge, then direct integration yields $\oint \vec{B} \cdot d\vec{s} = B\ell$, where B is the magnitude of the field at the left side of the loop. Since neither B nor ℓ is zero, Ampere's law is contradicted. We conclude that the geometry shown for the magnetic field lines is in error. The lines actually bulge outward and their density decreases gradually, not discontinuously as suggested by the figure.

Chapter 30

1. The amplitude of the induced emf in the loop is

$$\varepsilon_m = A\mu_0 n i_0 \omega = (6.8 \times 10^{-6} \, \text{m}^2)(4\pi \times 10^{-7} \, \text{T} \cdot \text{m/A})(85400 / \text{m})(1.28 \, \text{A})(212 \, \text{rad/s})$$
$$= 1.98 \times 10^{-4} \, \text{V}.$$

2. (a) $|\varepsilon| = \left| \dfrac{d\Phi_B}{dt} \right| = \dfrac{d}{dt}(6.0t^2 + 7.0t) = 12t + 7.0 = 12(2.0) + 7.0 = 31 \, \text{mV}.$

(b) Appealing to Lenz's law (especially Fig. 30-5(a)) we see that the current flow in the loop is clockwise. Thus, the current is to left through R.

3. (a) We use $\varepsilon = -d\Phi_B/dt = -\pi r^2 dB/dt$. For $0 < t < 2.0$ s:

$$\varepsilon = -\pi r^2 \frac{dB}{dt} = -\pi(0.12\text{m})^2 \left(\frac{0.5\,\text{T}}{2.0\,\text{s}} \right) = -1.1 \times 10^{-2} \, \text{V}.$$

(b) 2.0 s $< t <$ 4.0 s: $\varepsilon \propto dB/dt = 0$.

(c) 4.0 s $< t <$ 6.0 s:

$$\varepsilon = -\pi r^2 \frac{dB}{dt} = -\pi(0.12\text{m})^2 \left(\frac{-0.5\,\text{T}}{6.0\,\text{s} - 4.0\,\text{s}} \right) = 1.1 \times 10^{-2} \, \text{V}.$$

4. The resistance of the loop is

$$R = \rho \frac{L}{A} = (1.69 \times 10^{-8} \, \Omega \cdot \text{m}) \left[\frac{\pi(0.10\,\text{m})}{\pi(2.5 \times 10^{-3})^2 / 4} \right] = 1.1 \times 10^{-3} \, \Omega.$$

We use $i = |\varepsilon|/R = |d\Phi_B/dt|/R = (\pi r^2/R)|dB/dt|$. Thus

$$\left| \frac{dB}{dt} \right| = \frac{iR}{\pi r^2} = \frac{(10\,\text{A})(1.1 \times 10^{-3}\,\Omega)}{\pi(0.05\,\text{m})^2} = 1.4 \ \text{T/s}.$$

5. The total induced emf is given by

$$\varepsilon = -N\frac{d\Phi_B}{dt} = -NA\left(\frac{dB}{dt}\right) = -NA\frac{d}{dt}(\mu_0 ni) = -N\mu_0 nA\frac{di}{dt} = -N\mu_0 n(\pi r^2)\frac{di}{dt}$$

$$= -(120)(4\pi \times 10^{-7}\,\text{T}\cdot\text{m/A})(22000/\text{m})\,\pi\,(0.016\text{m})^2 \left(\frac{1.5\text{ A}}{0.025\text{ s}}\right)$$

$$= 0.16\,\text{V}.$$

Ohm's law then yields $i = |\varepsilon|/R = 0.016\text{ V}/5.3\Omega = 0.030\text{ A}$.

6. Using Faraday's law, the induced emf is

$$\varepsilon = -\frac{d\Phi_B}{dt} = -\frac{d(BA)}{dt} = -B\frac{dA}{dt} = -B\frac{d(\pi r^2)}{dt} = -2\pi r B\frac{dr}{dt}$$

$$= -2\pi(0.12\text{m})(0.800\text{T})(-0.750\text{m/s})$$

$$= 0.452\,\text{V}.$$

7. The flux $\Phi_B = BA\cos\theta$ does not change as the loop is rotated. Faraday's law only leads to a nonzero induced emf when the flux is changing, so the result in this instance is 0.

8. The field (due to the current in the straight wire) is out-of-the-page in the upper half of the circle and is into the page in the lower half of the circle, producing zero net flux, at any time. There is no induced current in the circle.

9. (a) Let L be the length of a side of the square circuit. Then the magnetic flux through the circuit is $\Phi_B = L^2 B/2$, and the induced emf is

$$\varepsilon_i = -\frac{d\Phi_B}{dt} = -\frac{L^2}{2}\frac{dB}{dt}.$$

Now $B = 0.042 - 0.870t$ and $dB/dt = -0.870$ T/s. Thus,

$$\varepsilon_i = \frac{(2.00\text{ m})^2}{2}(0.870\text{ T}/\text{s}) = 1.74\text{ V}.$$

The magnetic field is out of the page and decreasing so the induced emf is counterclockwise around the circuit, in the same direction as the emf of the battery. The total emf is

$$\varepsilon + \varepsilon_i = 20.0\text{ V} + 1.74\text{ V} = 21.7\text{ V}.$$

(b) The current is in the sense of the total emf (counterclockwise).

10. Fig. 30-41(b) demonstrates that $\frac{dB}{dt}$ (the slope of that line) is 0.003 T/s. Thus, in absolute value, Faraday's law becomes

$$\varepsilon = -\frac{d\Phi_B}{dt} = -A\frac{dB}{dt}$$

where $A = 8 \times 10^{-4}$ m^2. We related the induced emf to resistance and current using Ohm's law. The current is estimated from Fig. 30-41(c) to be $i = \frac{dq}{dt} = 0.002$ A (the slope of that line). Therefore, the resistance of the loop is

$$R = |\varepsilon| / i = \frac{(8 \times 10^{-4})(0.003)}{0.002} = 0.0012 \ \Omega .$$

11. (a) It should be emphasized that the result, given in terms of $\sin(2\pi ft)$, could as easily be given in terms of $\cos(2\pi ft)$ or even $\cos(2\pi ft + \phi)$ where ϕ is a phase constant as discussed in Chapter 15. The angular position θ of the rotating coil is measured from some reference line (or plane), and which line one chooses will affect whether the magnetic flux should be written as $BA\cos\theta$, $BA\sin\theta$ or $BA\cos(\theta + \phi)$. Here our choice is such that $\Phi_B = BA\cos\theta$. Since the coil is rotating steadily, θ increases linearly with time. Thus, $\theta = \omega t$ (equivalent to $\theta = 2\pi ft$) if θ is understood to be in radians (and ω would be the angular velocity). Since the area of the rectangular coil is $A=ab$, Faraday's law leads to

$$\varepsilon = -N\frac{d(BA\cos\theta)}{dt} = -NBA\frac{d\cos(2\pi ft)}{dt} = N\,Bab2\pi\, f\sin(2\pi\, ft)$$

which is the desired result, shown in the problem statement. The second way this is written ($\varepsilon_0 \sin(2\pi ft)$) is meant to emphasize that the voltage output is sinusoidal (in its time dependence) and has an amplitude of $\varepsilon_0 = 2\pi f N\, abB$.

(b) We solve $\varepsilon_0 = 150$ V $= 2\pi f N\, abB$ when $f = 60.0$ rev/s and $B = 0.500$ T. The three unknowns are N, a, and b which occur in a product; thus, we obtain $N\,ab = 0.796$ m^2.

12. (a) Since the flux arises from a dot product of vectors, the result of one sign for B_1 and B_2 and of the opposite sign for B_3 (we choose the minus sign for the flux from B_1 and B_2, and therefore a plus sign for the flux from B_3). The induced emf is

$$\varepsilon = -\Sigma\frac{d\Phi_B}{dt} = A\left(\frac{dB_1}{dt} + \frac{dB_2}{dt} - \frac{dB_3}{dt}\right)$$
$$= (0.10 \text{ m})(0.20 \text{ m})(2.0 \times 10^{-6} \text{ T/s} + 1.0 \times 10^{-6} \text{ T/s} - 5.0 \times 10^{-6} \text{ T/s})$$
$$= -4.0 \times 10^{-8} \text{ V}.$$

The minus sign meaning that the effect is dominated by the changes in B_3. Its magnitude (using Ohm's law) is $|\varepsilon| / R = 8.0$ μA.

(b) Consideration of Lenz's law leads to the conclusion that the induced current is therefore counterclockwise.

13. The amount of charge is

$$q(t) = \frac{1}{R}[\Phi_B(0) - \Phi_B(t)] = \frac{A}{R}[B(0) - B(t)] = \frac{1.20 \times 10^{-3}\,\text{m}^2}{13.0\,\Omega}[1.60\,\text{T} - (-1.60\,\text{T})]$$

$$= 2.95 \times 10^{-2}\,\text{C}.$$

14. We note that 1 gauss $= 10^{-4}$ T. The amount of charge is

$$q(t) = \frac{N}{R}[BA\cos 20° - (-BA\cos 20°)] = \frac{2NBA\cos 20°}{R}$$

$$= \frac{2(1000)(0.590 \times 10^{-4}\,\text{T})\pi(0.100\,\text{m})^2(\cos 20°)}{85.0\,\Omega + 140\,\Omega} = 1.55 \times 10^{-5}\,\text{C}.$$

Note that the axis of the coil is at 20°, not 70°, from the magnetic field of the Earth.

15. (a) The frequency is

$$f = \frac{\omega}{2\pi} = \frac{(40\text{ rev/s})(2\pi\text{ rad/rev})}{2\pi} = 40\text{ Hz}.$$

(b) First, we define angle relative to the plane of Fig. 30-44, such that the semicircular wire is in the $\theta = 0$ position and a quarter of a period (of revolution) later it will be in the $\theta = \pi/2$ position (where its midpoint will reach a distance of a above the plane of the figure). At the moment it is in the $\theta = \pi/2$ position, the area enclosed by the "circuit" will appear to us (as we look down at the figure) to that of a simple rectangle (call this area A_0 which is the area it will again appear to enclose when the wire is in the $\theta = 3\pi/2$ position). Since the area of the semicircle is $\pi a^2/2$ then the area (as it appears to us) enclosed by the circuit, as a function of our angle θ, is

$$A = A_0 + \frac{\pi a^2}{2}\cos\theta$$

where (since θ is increasing at a steady rate) the angle depends linearly on time, which we can write either as $\theta = \omega t$ or $\theta = 2\pi f t$ if we take $t = 0$ to be a moment when the arc is in the $\theta = 0$ position. Since $\vec{B}$ is uniform (in space) and constant (in time), Faraday's law leads to

$$\varepsilon = -\frac{d\Phi_B}{dt} = -B\frac{dA}{dt} = -B\frac{d\left(A_0 + \frac{\pi a^2}{2}\cos\theta\right)}{dt} = -B\frac{\pi a^2}{2}\frac{d\cos(2\pi f t)}{dt}$$

which yields $\varepsilon = B\pi^2 a^2 f \sin(2\pi ft)$. This (due to the sinusoidal dependence) reinforces the conclusion in part (a) and also (due to the factors in front of the sine) provides the voltage amplitude:

$$\varepsilon_m = B\pi^2 a^2 f = (0.020 \text{ T})\pi^2 (0.020 \text{ m})^2 (40/\text{s}) = 3.2 \times 10^{-3} \text{V}.$$

16. To have an induced emf, the magnetic field must be perpendicular (or have a nonzero component perpendicular) to the coil, and must be changing with time.

(a) For $\vec{B} = (4.00 \times 10^{-2} \text{T/m}) y\hat{k}$, $dB/dt = 0$ and hence $\varepsilon = 0$.

(b) None.

(c) For $\vec{B} = (6.00 \times 10^{-2} \text{T/s}) t\hat{k}$,

$$\varepsilon = -\frac{d\Phi_B}{dt} = -A\frac{dB}{dt} = -(0.400 \text{ m} \times 0.250 \text{ m})(0.0600 \text{ T/s}) = -6.00 \text{ mV},$$

or $|\varepsilon| = 6.00$ mV.

(d) Clockwise;

(e) For $\vec{B} = (8.00 \times 10^{-2} \text{T/m·s}) y t\hat{k}$,

$$\Phi_B = (0.400)(0.0800t) \int y \, dy = 1.00 \times 10^{-3} t,$$

in SI units. The induced emf is $\varepsilon = -d\Phi B/dt = -1.00$ mV, or $|\varepsilon| = 1.00$ mV.

(f) Clockwise.

(g) $\Phi_B = 0 \Rightarrow \varepsilon = 0$.

(h) None.

(i) $\Phi_B = 0 \Rightarrow \varepsilon = 0$

(j) None.

17. First we write $\Phi_B = BA \cos \theta$. We note that the angular position θ of the rotating coil is measured from some reference line or plane, and we are implicitly making such a choice by writing the magnetic flux as $BA \cos \theta$ (as opposed to, say, $BA \sin \theta$). Since the coil is rotating steadily, θ increases linearly with time. Thus, $\theta = \omega t$ if θ is understood to

be in radians (here, $\omega = 2\pi f$ is the angular velocity of the coil in radians per second, and f = 1000 rev/min $\approx$ 16.7 rev/s is the frequency). Since the area of the rectangular coil is A = $0.500 \times 0.300 = 0.150$ m^2, Faraday's law leads to

$$\varepsilon = -N\frac{d(BA\cos\theta)}{dt} = -NBA\frac{d\cos(2\pi ft)}{dt} = NBA2\pi f \sin(2\pi ft)$$

which means it has a voltage amplitude of

$$\varepsilon_{max} = 2\pi f NAB = 2\pi(16.7\,\text{rev/s})(100\,\text{turns})(0.15\,\text{m}^2)(3.5\,\text{T}) = 5.50\times10^3\,\text{V}\ .$$

18. (a) Since $\vec{B} = B\hat{i}$ uniformly, then only the area "projected" onto the yz plane will contribute to the flux (due to the scalar [dot] product). This "projected" area corresponds to one-fourth of a circle. Thus, the magnetic flux Φ_B through the loop is

$$\Phi_B = \int \vec{B}\cdot d\vec{A} = \frac{1}{4}\pi r^2 B\ .$$

Thus,

$$|\varepsilon| = \left|\frac{d\Phi_B}{dt}\right| = \left|\frac{d}{dt}\left(\frac{1}{4}\pi r^2 B\right)\right| = \frac{\pi r^2}{4}\left|\frac{dB}{dt}\right|$$
$$= \frac{1}{4}\pi(0.10\,\text{m})^2(3.0\times10^{-3}\,\text{T/s}) = 2.4\times10^{-5}\,\text{V}\ .$$

(b) We have a situation analogous to that shown in Fig. 30-5(a). Thus, the current in segment bc flows from c to b (following Lenz's law).

19. (a) In the region of the smaller loop the magnetic field produced by the larger loop may be taken to be uniform and equal to its value at the center of the smaller loop, on the axis. Eq. 29-27, with $z = x$ (taken to be much greater than R), gives

$$\vec{B} = \frac{\mu_0 i R^2}{2x^3}\hat{i}$$

where the $+x$ direction is upward in Fig. 30-47. The magnetic flux through the smaller loop is, to a good approximation, the product of this field and the area (πr^2) of the smaller loop:

$$\Phi_B = \frac{\pi\mu_0 i r^2 R^2}{2x^3}.$$

(b) The emf is given by Faraday's law:

$$\varepsilon = -\frac{d\Phi_B}{dt} = -\left(\frac{\pi\mu_0 ir^2 R^2}{2}\right)\frac{d}{dt}\left(\frac{1}{x^3}\right) = -\left(\frac{\pi\mu_0 ir^2 R^2}{2}\right)\left(-\frac{3}{x^4}\frac{dx}{dt}\right) = \frac{3\pi\mu_0 ir^2 R^2 v}{2x^4}.$$

(c) As the smaller loop moves upward, the flux through it decreases, and we have a situation like that shown in Fig. 30-5(b). The induced current will be directed so as to produce a magnetic field that is upward through the smaller loop, in the same direction as the field of the larger loop. It will be counterclockwise as viewed from above, in the same direction as the current in the larger loop.

20. Since $\dfrac{d\cos\phi}{dt} = -\dfrac{d\phi}{dt}\sin\phi$, Faraday's law (with $N = 1$) becomes (in absolute value)

$$\varepsilon = -\frac{d\Phi_B}{dt} = -B A \frac{d\phi}{dt}\sin\phi$$

which yields $|\varepsilon| = 0.018$ V.

21. (a) Eq. 29-10 gives the field at the center of the large loop with $R = 1.00$ m and current $i(t)$. This is approximately the field throughout the area ($A = 2.00 \times 10^{-4}$ m^2) enclosed by the small loop. Thus, with $B = \mu_0 i/2R$ and $i(t) = i_0 + kt$, where $i_0 = 200$ A and

$$k = (-200 \text{ A} - 200 \text{ A})/1.00 \text{ s} = -400 \text{ A/s},$$

we find

(a) $B(t=0) = \dfrac{\mu_0 i_0}{2R} = \dfrac{\left(4\pi\times10^{-7}\text{ H/m}\right)(200\text{A})}{2(1.00\text{m})} = 1.26\times10^{-4}$ T,

(b) $B(t=0.500\text{s}) = \dfrac{\left(4\pi\times10^{-7}\text{ H/m}\right)\left[200\text{A}-(400\text{A/s})(0.500\text{s})\right]}{2(1.00\text{m})} = 0.$

(c) $B(t=1.00\text{s}) = \dfrac{\left(4\pi\times10^{-7}\text{ H/m}\right)\left[200\text{A}-(400\text{A/s})(1.00\text{s})\right]}{2(1.00\text{m})} = -1.26\times10^{-4}$ T,

or $|B(t=1.00\text{s})| = 1.26\times10^{-4}$ T.

(d) yes, as indicated by the flip of sign of $B(t)$ in (c).

(e) Let the area of the small loop be a. Then $\Phi_B = Ba$, and Faraday's law yields

$$\varepsilon = -\frac{d\Phi_B}{dt} = -\frac{d(Ba)}{dt} = -a\frac{dB}{dt} = -a\left(\frac{\Delta B}{\Delta t}\right)$$

$$= -(2.00\times10^{-4}\,\text{m}^2)\left(\frac{-1.26\times10^{-4}\,\text{T} - 1.26\times10^{-4}\,\text{T}}{1.00\,\text{s}}\right)$$

$$= 5.04\times10^{-8}\,\text{V}.$$

22. (a) First, we observe that a large portion of the figure contributes flux which "cancels out." The field (due to the current in the long straight wire) through the part of the rectangle above the wire is out of the page (by the right-hand rule) and below the wire it is into the page. Thus, since the height of the part above the wire is $b - a$, then a strip below the wire (where the strip borders the long wire, and extends a distance $b - a$ away from it) has exactly the equal-but-opposite flux which cancels the contribution from the part above the wire. Thus, we obtain the non-zero contributions to the flux:

$$\Phi_B = \int B dA = \int_{b-a}^{a}\left(\frac{\mu_0 i}{2\pi r}\right)(b\,dr) = \frac{\mu_0 i b}{2\pi}\ln\left(\frac{a}{b-a}\right).$$

Faraday's law, then, (with SI units and 3 significant figures understood) leads to

$$\varepsilon = -\frac{d\Phi_B}{dt} = -\frac{d}{dt}\left[\frac{\mu_0 i b}{2\pi}\ln\left(\frac{a}{b-a}\right)\right] = -\frac{\mu_0 b}{2\pi}\ln\left(\frac{a}{b-a}\right)\frac{di}{dt}$$

$$= -\frac{\mu_0 b}{2\pi}\ln\left(\frac{a}{b-a}\right)\frac{d}{dt}\left(\frac{9}{2}t^2 - 10t\right)$$

$$= \frac{-\mu_0 b(9t - 10)}{2\pi}\ln\left(\frac{a}{b-a}\right).$$

With $a = 0.120$ m and $b = 0.160$ m, then, at $t = 3.00$ s, the magnitude of the emf induced in the rectangular loop is

$$|\varepsilon| = \frac{(4\pi\times10^{-7})(0.16)(9(3) - 10)}{2\pi}\ln\left(\frac{0.12}{0.16 - 0.12}\right) = 5.98\times10^{-7}\,\text{V}.$$

(b) We note that $di/dt > 0$ at $t = 3$ s. The situation is roughly analogous to that shown in Fig. 30-5(c). From Lenz's law, then, the induced emf (hence, the induced current) in the loop is counterclockwise.

23. (a) Consider a (thin) strip of area of height dy and width $\ell = 0.020\,\text{m}$. The strip is located at some $0 < y < \ell$. The element of flux through the strip is

$$d\Phi_B = B dA = (4t^2 y)(\ell\,dy)$$

where SI units (and 2 significant figures) are understood. To find the total flux through the square loop, we integrate:

$$\Phi_B = \int d\Phi_B = \int_0^\ell \left(4t^2 y\ell\right)dy = 2t^2\ell^3 .$$

Thus, Faraday's law yields

$$|\varepsilon| = \left|\frac{d\Phi_B}{dt}\right| = 4t\ell^3 .$$

At $t = 2.5$ s, we find the magnitude of the induced emf is 8.0×10^{-5} V.

(b) Its "direction" (or "sense") is clockwise, by Lenz's law.

24. (a) We assume the flux is entirely due to the field generated by the long straight wire (which is given by Eq. 29-17). We integrate according to Eq. 30-1, not worrying about the possibility of an overall minus sign since we are asked to find the absolute value of the flux.

$$|\Phi_B| = \int_{r-b/2}^{r+b/2} \left(\frac{\mu_0 i}{2\pi r}\right)(a\,dr) = \frac{\mu_0 i a}{2\pi} \ln\left(\frac{r+\frac{b}{2}}{r-\frac{b}{2}}\right).$$

When $r = 1.5b$, we have

$$|\Phi_B| = \frac{(4\pi\times10^{-7}\,\text{T}\cdot\text{m/A})(4.7\text{A})(0.022\text{m})}{2\pi}\ln(2.0) = 1.4\times10^{-8}\,\text{Wb}.$$

(b) Implementing Faraday's law involves taking a derivative of the flux in part (a), and recognizing that $\frac{dr}{dt} = v$. The magnitude of the induced emf divided by the loop resistance then gives the induced current:

$$i_{\text{loop}} = \left|\frac{\varepsilon}{R}\right| = -\frac{\mu_0 i a}{2\pi R}\left|\frac{d}{dt}\ln\left(\frac{r+\frac{b}{2}}{r-\frac{b}{2}}\right)\right| = \frac{\mu_0 i a b v}{2\pi R[r^2 - (b/2)^2]}$$

$$= \frac{(4\pi\times10^{-7}\,\text{T}\cdot\text{m/A})(4.7\text{A})(0.022\text{m})(0.0080\text{m})(3.2\times10^{-3}\,\text{m/s})}{2\pi(4.0\times10^{-4}\,\Omega)[2(0.0080\text{m})^2]}$$

$$= 1.0\times10^{-5}\,\text{A}.$$

25. (a) We refer to the (very large) wire length as L and seek to compute the flux per meter: Φ_B/L. Using the right-hand rule discussed in Chapter 29, we see that the net field in the region between the axes of anti-parallel currents is the addition of the magnitudes of their individual fields, as given by Eq. 29-17 and Eq. 29-20. There is an evident

reflection symmetry in the problem, where the plane of symmetry is midway between the two wires (at what we will call $x = \ell/2$, where $\ell = 20\,\text{mm} = 0.020\,\text{m}$); the net field at any point $0 < x < \ell/2$ is the same at its "mirror image" point $\ell - x$. The central axis of one of the wires passes through the origin, and that of the other passes through $x = \ell$. We make use of the symmetry by integrating over $0 < x < \ell/2$ and then multiplying by 2:

$$\Phi_B = 2\int_0^{\ell/2} B\, dA = 2\int_0^{d/2} B(L\,dx) + 2\int_{d/2}^{\ell/2} B(L\,dx)$$

where $d = 0.0025$ m is the diameter of each wire. We will use $R = d/2$, and r instead of x in the following steps. Thus, using the equations from Ch. 29 referred to above, we find

$$\frac{\Phi_B}{L} = 2\int_0^R \left(\frac{\mu_0 i}{2\pi R^2} r + \frac{\mu_0 i}{2\pi(\ell - r)}\right) dr + 2\int_R^{\ell/2} \left(\frac{\mu_0 i}{2\pi r} + \frac{\mu_0 i}{2\pi(\ell - r)}\right) dr$$

$$= \frac{\mu_0 i}{2\pi}\left(1 - 2\ln\left(\frac{\ell - R}{\ell}\right)\right) + \frac{\mu_0 i}{\pi}\ln\left(\frac{\ell - R}{R}\right)$$

$$= 0.23 \times 10^{-5}\,\text{T}\cdot\text{m} + 1.08 \times 10^{-5}\,\text{T}\cdot\text{m}$$

which yields $\Phi_B/L = 1.3 \times 10^{-5}$ T·m or 1.3×10^{-5} Wb/m.

(b) The flux (per meter) existing within the regions of space occupied by one or the other wires was computed above to be 0.23×10^{-5} T·m. Thus,

$$\frac{0.23 \times 10^{-5}\,\text{T}\cdot\text{m}}{1.3 \times 10^{-5}\,\text{T}\cdot\text{m}} = 0.17 = 17\%\ .$$

(c) What was described in part (a) as a symmetry plane at $x = \ell/2$ is now (in the case of parallel currents) a plane of vanishing field (the fields subtract from each other in the region between them, as the right-hand rule shows). The flux in the $0 < x < \ell/2$ region is now of opposite sign of the flux in the $\ell/2 < x < \ell$ region which causes the total flux (or, in this case, flux per meter) to be zero.

26. Noting that $|\Delta B| = B$, we find the thermal energy is

$$P_{\text{thermal}}\Delta t = \frac{\varepsilon^2 \Delta t}{R} = \frac{1}{R}\left(-\frac{d\Phi_B}{dt}\right)^2 \Delta t = \frac{1}{R}\left(-A\frac{\Delta B}{\Delta t}\right)^2 \Delta t = \frac{A^2 B^2}{R\Delta t}$$

$$= \frac{(2.00 \times 10^{-4}\,\text{m}^2)^2(17.0 \times 10^{-6}\,\text{T})^2}{(5.21 \times 10^{-6}\,\Omega)(2.96 \times 10^{-3}\,\text{s})}$$

$$= 7.50 \times 10^{-10}\,\text{J}.$$

27. Thermal energy is generated at the rate $P = \varepsilon^2/R$ (see Eq. 27-23). Using Eq. 27-16, the resistance is given by $R = \rho L/A$, where the resistivity is 1.69×10^{-8} Ω·m (by Table 27-1)

and $A = \pi d^2/4$ is the cross-sectional area of the wire ($d = 0.00100$ m is the wire thickness). The area *enclosed* by the loop is

$$A_{\text{loop}} = \pi r_{\text{loop}}^2 = \pi \left(\frac{L}{2\pi} \right)^2$$

since the length of the wire ($L = 0.500$ m) is the circumference of the loop. This enclosed area is used in Faraday's law (where we ignore minus signs in the interest of finding the magnitudes of the quantities):

$$\varepsilon = \frac{d\Phi_B}{dt} = A_{\text{loop}} \frac{dB}{dt} = \frac{L^2}{4\pi} \frac{dB}{dt}$$

where the rate of change of the field is $dB/dt = 0.0100$ T/s. Consequently, we obtain

$$P = \frac{\left(\frac{L^2}{4\pi} \frac{dB^2}{dt} \right)}{4\rho L / \pi d^2} = \frac{d^2 L^3}{64\pi\rho} \left(\frac{dB}{dt} \right)^2 = 3.68 \times 10^{-6} \, \text{W} \, .$$

28. Eq. 27-23 gives ε^2/R as the rate of energy transfer into thermal forms (dE_{th}/dt, which, from Fig. 30-51(c), is roughly 40 nJ/s). Interpreting ε as the induced emf (in absolute value) in the single-turn loop ($N = 1$) from Faraday's law, we have

$$\varepsilon = \frac{d\Phi_B}{dt} = A \frac{dB}{dt} \, .$$

Eq. 29-23 gives $B = \mu_0 n i$ for the solenoid (and note that the field is zero outside of the solenoid – which implies that $A = A_{\text{coil}}$), so our expression for the magnitude of the induced emf becomes

$$\varepsilon = \mu_0 n A_{\text{coil}} \frac{di_{\text{coil}}}{dt}$$

where Fig. 30-51(b) suggests that $di_{\text{coil}}/dt = 0.5$ A/s. With $n = 8000$ (in SI units) and $A_{\text{coil}} = \pi(0.02)^2$ (note that the loop radius does not come into the computations of this problem, just the coil's), we find V = 6.3 microvolts. Returning to our earlier observations, we can now solve for the resistance: $R = \varepsilon^2/(dE_{\text{th}}/dt) = 1.0$ mΩ.

29. (a) Eq. 30-8 leads to

$$\varepsilon = BLv = (0.350 \text{ T})(0.250 \text{ m})(0.55 \text{ m}/\text{s}) = 0.0481 \text{ V} \, .$$

(b) By Ohm's law, the induced current is $i = 0.0481$ V/18.0 Ω = 0.00267 A. By Lenz's law, the current is clockwise in Fig. 30-52.

(c) Eq. 26-22 leads to $P = i^2 R = 0.000129$ W.

30. Noting that $F_{net} = BiL - mg = 0$, we solve for the current:

$$i = \frac{mg}{BL} = \frac{|\varepsilon|}{R} = \frac{1}{R}\left|\frac{d\Phi_B}{dt}\right| = \frac{B}{R}\left|\frac{dA}{dt}\right| = \frac{Bv_t L}{R},$$

which yields $v_t = mgR/B^2L^2$.

31. (a) Eq. 30-8 leads to

$$\varepsilon = BLv = (1.2\,\text{T})(0.10\text{ m})(5.0\text{ m/s}) = 0.60\text{ V}.$$

(b) By Lenz's law, the induced emf is clockwise. In the rod itself, we would say the emf is directed up the page.

(c) By Ohm's law, the induced current is $i = 0.60\text{ V}/0.40\,\Omega = 1.5\text{ A}.$

(d) The direction is clockwise.

(e) Eq. 27-22 leads to $P = i^2R = 0.90\text{ W}.$

(f) From Eq. 29-2, we find that the force on the rod associated with the uniform magnetic field is directed rightward and has magnitude

$$F = iLB = (1.5\text{ A})(0.10\text{ m})(1.2\text{ T}) = 0.18\text{ N}.$$

To keep the rod moving at constant velocity, therefore, a leftward force (due to some external agent) having that same magnitude must be continuously supplied to the rod.

(g) Using Eq. 7-48, we find the power associated with the force being exerted by the external agent: $P = Fv = (0.18\text{ N})(5.0\text{ m/s}) = 0.90\text{ W}$, which is the same as our result from part (e).

32. (a) The "height" of the triangular area enclosed by the rails and bar is the same as the distance traveled in time v: $d = vt$, where $v = 5.20$ m/s. We also note that the "base" of that triangle (the distance between the intersection points of the bar with the rails) is $2d$. Thus, the area of the triangle is

$$A = \frac{1}{2}(\text{base})(\text{height}) = \frac{1}{2}(2vt)(vt) = v^2t^2.$$

Since the field is a uniform $B = 0.350$ T, then the magnitude of the flux (in SI units) is

$$\Phi_B = BA = (0.350)(5.20)^2t^2 = 9.46t^2.$$

At $t = 3.00$ s, we obtain $\Phi_B = 85.2$ Wb.

(b) The magnitude of the emf is the (absolute value of) Faraday's law:

$$\varepsilon = \frac{d\Phi_B}{dt} = 9.46\frac{dt^2}{dt} = 18.9t$$

in SI units. At $t = 3.00$ s, this yields $\varepsilon = 56.8$ V.

(c) Our calculation in part (b) shows that $n = 1$.

33. (a) Letting x be the distance from the right end of the rails to the rod, we find an expression for the magnetic flux through the area enclosed by the rod and rails. By Eq. 29-17, the field is $B = \mu_0 i / 2\pi r$, where r is the distance from the long straight wire. We consider an infinitesimal horizontal strip of length x and width dr, parallel to the wire and a distance r from it; it has area $A = x\, dr$ and the flux $d\Phi_B = (\mu_0 i x / 2\pi r)dr$. By Eq. 30-1, the total flux through the area enclosed by the rod and rails is

$$\Phi_B = \frac{\mu_0 i x}{2\pi} \int_a^{a+L} \frac{dr}{r} = \frac{\mu_0 i x}{2\pi} \ln\left(\frac{a+L}{a}\right).$$

According to Faraday's law the emf induced in the loop is

$$\varepsilon = \frac{d\Phi_B}{dt} = \frac{\mu_0 i}{2\pi} \frac{dx}{dt} \ln\left(\frac{a+L}{a}\right) = \frac{\mu_0 i v}{2\pi} \ln\left(\frac{a+L}{a}\right)$$

$$= \frac{\left(4\pi \times 10^{-7}\,\text{T}\cdot\text{m/A}\right)(100\,\text{A})(5.00\,\text{m/s})}{2\pi} \ln\left(\frac{1.00\text{cm}+10.0\text{cm}}{1.00\text{cm}}\right) = 2.40 \times 10^{-4}\,\text{V}.$$

(b) By Ohm's law, the induced current is

$$i_\ell = \varepsilon / R = \left(2.40 \times 10^{-4}\,\text{V}\right)/\left(0.400\,\Omega\right) = 6.00 \times 10^{-4}\,\text{A}.$$

Since the flux is increasing the magnetic field produced by the induced current must be into the page in the region enclosed by the rod and rails. This means the current is clockwise.

(c) Thermal energy is being generated at the rate

$$P = i_\ell^2 R = \left(6.00 \times 10^{-4}\,\text{A}\right)^2 \left(0.400\,\Omega\right) = 1.44 \times 10^{-7}\,\text{W}.$$

(d) Since the rod moves with constant velocity, the net force on it is zero. The force of the external agent must have the same magnitude as the magnetic force and must be in the opposite direction. The magnitude of the magnetic force on an infinitesimal segment of the rod, with length dr at a distance r from the long straight wire, is

$$dF_B = i_\ell B \, dr = \left(\mu_0 i_\ell i / 2\pi r\right) dr.$$

We integrate to find the magnitude of the total magnetic force on the rod:

$$
\begin{aligned}
F_B &= \frac{\mu_0 i_\ell i}{2\pi} \int_a^{a+L} \frac{dr}{r} = \frac{\mu_0 i_\ell i}{2\pi} \ln\left(\frac{a+L}{a}\right) \\
&= \frac{\left(4\pi \times 10^{-7}\,\text{T·m/A}\right)\left(6.00 \times 10^{-4}\,\text{A}\right)\left(100\,\text{A}\right)}{2\pi} \ln\left(\frac{1.00\,\text{cm} + 10.0\,\text{cm}}{1.00\,\text{cm}}\right) \\
&= 2.87 \times 10^{-8}\,\text{N}.
\end{aligned}
$$

Since the field is out of the page and the current in the rod is upward in the diagram, the force associated with the magnetic field is toward the right. The external agent must therefore apply a force of 2.87×10^{-8} N, to the left.

(e) By Eq. 7-48, the external agent does work at the rate

$$P = Fv = (2.87 \times 10^{-8}\,\text{N})(5.00\,\text{m/s}) = 1.44 \times 10^{-7}\,\text{W}.$$

This is the same as the rate at which thermal energy is generated in the rod. All the energy supplied by the agent is converted to thermal energy.

34. (a) For path 1, we have

$$
\begin{aligned}
\oint_1 \vec{E} \cdot d\vec{s} &= -\frac{d\Phi_{B1}}{dt} = -\frac{d}{dt}(B_1 A_1) = A_1 \frac{dB_1}{dt} = \pi r_1^2 \frac{dB_1}{dt} = \pi(0.200\,\text{m})^2\left(-8.50 \times 10^{-3}\,\text{T/s}\right) \\
&= -1.07 \times 10^{-3}\,\text{V}
\end{aligned}
$$

(b) For path 2, the result is

$$\oint_2 \vec{E} \cdot d\vec{s} = -\frac{d\Phi_{B2}}{dt} = \pi r_2^2 \frac{dB_2}{dt} = \pi(0.300\,\text{m})^2\left(-8.50 \times 10^{-3}\,\text{T/s}\right) = -2.40 \times 10^{-3}\,\text{V}$$

(c) For path 3, we have

$$\oint_3 \vec{E} \cdot d\vec{s} = \oint_1 \vec{E} \cdot d\vec{s} - \oint_2 \vec{E} \cdot d\vec{s} = -1.07 \times 10^{-3}\,\text{V} - \left(-2.4 \times 10^{-3}\,\text{V}\right) = 1.33 \times 10^{-3}\,\text{V}$$

35. (a) The point at which we are evaluating the field is inside the solenoid, so Eq. 30-25 applies. The magnitude of the induced electric field is

$$E = \frac{1}{2}\frac{dB}{dt}r = \frac{1}{2}(6.5 \times 10^{-3} \text{ T/s})(0.0220 \text{ m}) = 7.15 \times 10^{-5} \text{ V/m}.$$

(b) Now the point at which we are evaluating the field is outside the solenoid and Eq. 30-27 applies. The magnitude of the induced field is

$$E = \frac{1}{2}\frac{dB}{dt}\frac{R^2}{r} = \frac{1}{2}(6.5 \times 10^{-3} \text{ T/s})\frac{(0.0600 \text{ m})^2}{(0.0820 \text{ m})} = 1.43 \times 10^{-4} \text{ V/m}.$$

36. From the "kink" in the graph of Fig. 30-57, we conclude that the radius of the circular region is 2.0 cm. For values of r less than that, we have (from the absolute value of Eq. 30-20)

$$E(2\pi r) = \frac{d\Phi_B}{dt} = A\frac{dB}{dt} = \pi r^2 \frac{dB}{dt} = \pi r^2 a\, t$$

which means that $E/r = a/2$. This corresponds to the slope of that graph (the linear portion for small values of r) which we estimate to be 0.015 (in SI units). Thus, $a = 0.030$ T/s.

37. The magnetic field B can be expressed as

$$B(t) = B_0 + B_1 \sin(\omega t + \phi_0),$$

where $B_0 = (30.0 \text{ T} + 29.6 \text{ T})/2 = 29.8$ T and $B_1 = (30.0 \text{ T} - 29.6 \text{ T})/2 = 0.200$ T. Then from Eq. 30-25

$$E = \frac{1}{2}\left(\frac{dB}{dt}\right)r = \frac{r}{2}\frac{d}{dt}[B_0 + B_1 \sin(\omega t + \phi_0)] = \frac{1}{2}B_1 \omega r \cos(\omega t + \phi_0).$$

We note that $\omega = 2\pi f$ and that the factor in front of the cosine is the maximum value of the field. Consequently,

$$E_{max} = \frac{1}{2}B_1(2\pi f)r = \frac{1}{2}(0.200 \text{ T})(2\pi)(15 \text{ Hz})(1.6 \times 10^{-2} \text{ m}) = 0.15 \text{ V/m}.$$

38. (a) We interpret the question as asking for N multiplied by the flux through one turn:

$$\Phi_{turns} = N\Phi_B = NBA = NB(\pi r^2) = (30.0)(2.60 \times 10^{-3} \text{ T})(\pi)(0.100 \text{ m})^2 = 2.45 \times 10^{-3} \text{ Wb}.$$

(b) Eq. 30-33 leads to

$$L = \frac{N\Phi_B}{i} = \frac{2.45 \times 10^{-3} \text{ Wb}}{3.80 \text{ A}} = 6.45 \times 10^{-4} \text{ H.}$$

39. Since $N\Phi_B = Li$, we obtain

$$\Phi_B = \frac{Li}{N} = \frac{(8.0 \times 10^{-3} \text{ H})(5.0 \times 10^{-3} \text{ A})}{400} = 1.0 \times 10^{-7} \text{ Wb.}$$

40. (a) We imagine dividing the one-turn solenoid into N small circular loops placed along the width W of the copper strip. Each loop carries a current $\Delta i = i/N$. Then the magnetic field inside the solenoid is

$$B = \mu_0 n \Delta i = \mu_0 (N/W)(i/N) = \mu_0 i / W = \frac{(4\pi \times 10^{-7} \text{ T} \cdot \text{m/A})(0.035\text{A})}{0.16\text{m}} = 2.7 \times 10^{-7} \text{ T.}$$

(b) Eq. 30-33 leads to

$$L = \frac{\Phi_B}{i} = \frac{\pi R^2 B}{i} = \frac{\pi R^2 (\mu_0 i / W)}{i} = \frac{\pi \mu_0 R^2}{W} = \frac{\pi (4\pi \times 10^{-7} \text{ T} \cdot \text{m/A})(0.018\text{m})^2}{0.16\text{m}} = 8.0 \times 10^{-9} \text{ H.}$$

41. We refer to the (very large) wire length as ℓ and seek to compute the flux per meter: Φ_B / ℓ. Using the right-hand rule discussed in Chapter 29, we see that the net field in the region between the axes of antiparallel currents is the addition of the magnitudes of their individual fields, as given by Eq. 29-17 and Eq. 29-20. There is an evident reflection symmetry in the problem, where the plane of symmetry is midway between the two wires (at $x = d/2$); the net field at any point $0 < x < d/2$ is the same at its "mirror image" point $d - x$. The central axis of one of the wires passes through the origin, and that of the other passes through $x = d$. We make use of the symmetry by integrating over $0 < x < d/2$ and then multiplying by 2:

$$\Phi_B = 2\int_0^{d/2} B \, dA = 2\int_0^a B(\ell \, dx) + 2\int_a^{d/2} B(\ell \, dx)$$

where $d = 0.0025$ m is the diameter of each wire. We will use r instead of x in the following steps. Thus, using the equations from Ch. 29 referred to above, we find

$$\frac{\Phi_B}{\ell} = 2\int_0^a \left(\frac{\mu_0 i}{2\pi a^2} r + \frac{\mu_0 i}{2\pi(d-r)} \right) dr + 2\int_a^{d/2} \left(\frac{\mu_0 i}{2\pi r} + \frac{\mu_0 i}{2\pi(d-r)} \right) dr$$

$$= \frac{\mu_0 i}{2\pi} \left(1 - 2\ln\left(\frac{d-a}{d}\right) \right) + \frac{\mu_0 i}{\pi} \ln\left(\frac{d-a}{a}\right)$$

where the first term is the flux within the wires and will be neglected (as the problem suggests). Thus, the flux is approximately $\Phi_B \approx \mu_0 i\ell / \pi \, \ln((d-a)/a)$. Now, we use Eq. 30-33 (with $N = 1$) to obtain the inductance per unit length:

$$\frac{L}{\ell} = \frac{\Phi_B}{\ell i} = \frac{\mu_0}{\pi} \ln\left(\frac{d-a}{a}\right) = \frac{(4\pi \times 10^{-7}\,\text{T·m/A})}{\pi} \ln\left(\frac{142-1.53}{1.53}\right) = 1.81 \times 10^{-6}\,\text{H/m}.$$

42. (a) Speaking anthropomorphically, the coil wants to fight the changes—so if it wants to push current rightward (when the current is already going rightward) then i must be in the process of decreasing.

(b) From Eq. 30-35 (in absolute value) we get

$$L = \left|\frac{\varepsilon}{di/dt}\right| = \frac{17\,\text{V}}{2.5\,\text{kA/s}} = 6.8 \times 10^{-4}\,\text{H}.$$

43. Since $\varepsilon = -L(di/dt)$, we may obtain the desired induced emf by setting

$$\frac{di}{dt} = -\frac{\varepsilon}{L} = -\frac{60\,\text{V}}{12\,\text{H}} = -5.0\,\text{A/s},$$

or $|di/dt| = 5.0\,\text{A/s}$. We might, for example, uniformly reduce the current from 2.0 A to zero in 40 ms.

44. During periods of time when the current is varying linearly with time, Eq. 30-35 (in absolute values) becomes $|\varepsilon| = L|\Delta i/\Delta t|$. For simplicity, we omit the absolute value signs in the following.

(a) For $0 < t < 2$ ms

$$\varepsilon = L\frac{\Delta i}{\Delta t} = \frac{(4.6\,\text{H})(7.0\,\text{A} - 0)}{2.0 \times 10^{-3}\,\text{s}} = 1.6 \times 10^4\,\text{V}.$$

(b) For 2 ms $< t < 5$ ms

$$\varepsilon = L\frac{\Delta i}{\Delta t} = \frac{(4.6\,\text{H})(5.0\,\text{A} - 7.0\,\text{A})}{(5.0 - 2.0)10^{-3}\,\text{s}} = 3.1 \times 10^3\,\text{V}.$$

(c) For 5 ms $< t < 6$ ms

$$\varepsilon = L\frac{\Delta i}{\Delta t} = \frac{(4.6\,\text{H})(0 - 5.0\,\text{A})}{(6.0 - 5.0)10^{-3}\,\text{s}} = 2.3 \times 10^4\,\text{V}.$$

45. (a) Voltage is proportional to inductance (by Eq. 30-35) just as, for resistors, it is proportional to resistance. Since the (independent) voltages for series elements add ($V_1 + V_2$), then inductances in series must *add*, $L_{eq} = L_1 + L_2$, just as was the case for resistances.

Note that to ensure the independence of the voltage values, it is important that the inductors not be too close together (the related topic of mutual inductance is treated in §30-12). The requirement is that magnetic field lines from one inductor should not have significant presence in any other.

(b) Just as with resistors, $L_{eq} = \sum_{n=1}^{N} L_n$.

46. (a) Voltage is proportional to inductance (by Eq. 30-35) just as, for resistors, it is proportional to resistance. Now, the (independent) voltages for parallel elements are equal ($V_1 = V_2$), and the currents (which are generally functions of time) add ($i_1 (t) + i_2 (t) = i(t)$). This leads to the Eq. 27-21 for resistors. We note that this condition on the currents implies

$$\frac{di_1(t)}{dt} + \frac{di_2(t)}{dt} = \frac{di(t)}{dt}.$$

Thus, although the inductance equation Eq. 30-35 involves the rate of change of current, as opposed to current itself, the conditions that led to the parallel resistor formula also applies to inductors. Therefore,

$$\frac{1}{L_{eq}} = \frac{1}{L_1} + \frac{1}{L_2}.$$

Note that to ensure the independence of the voltage values, it is important that the inductors not be too close together (the related topic of mutual inductance is treated in §30-12). The requirement is that the field of one inductor not have significant influence (or "coupling") in the next.

(b) Just as with resistors, $\frac{1}{L_{eq}} = \sum_{n=1}^{N} \frac{1}{L_n}$.

47. Using the results from Problems 45 and 46, the equivalent resistance is

$$L_{eq} = L_1 + L_4 + L_{23} = L_1 + L_4 + \frac{L_2 L_3}{L_2 + L_3}$$

$$= 30.0 \text{mH} + 15.0 \text{mH} + \frac{(50.0 \text{mH})(20.0 \text{mH})}{50.0 \text{mH} + 20.0 \text{mH}}$$

$$= 59.3 \text{ mH}.$$

48. The steady state value of the current is also its maximum value, ε/R, which we denote as i_m. We are told that $i = i_m/3$ at $t_0 = 5.00$ s. Eq. 30-41 becomes $i = i_m \left(1 - e^{-t_0/\tau_L}\right)$ which leads to

$$\tau_L = -\frac{t_0}{\ln(1 - i/i_m)} = -\frac{5.00\,\text{s}}{\ln(1 - 1/3)} = 12.3\,\text{s}.$$

49. Starting with zero current at $t = 0$ (the moment the switch is closed) the current in the circuit increases according to

$$i = \frac{\varepsilon}{R}\left(1 - e^{-t/\tau_L}\right),$$

where $\tau_L = L/R$ is the inductive time constant and ε is the battery emf. To calculate the time at which $i = 0.9990\varepsilon/R$, we solve for t:

$$0.990\frac{\varepsilon}{R} = \frac{\varepsilon}{R}\left(1 - e^{-t/\tau_L}\right) \Rightarrow \ln(0.0010) = -(t/\tau) \Rightarrow t/\tau_L = 6.91.$$

50. (a) Immediately after the switch is closed $\varepsilon - \varepsilon_L = iR$. But $i = 0$ at this instant, so $\varepsilon_L = \varepsilon$, or $\varepsilon_L/\varepsilon = 1.00$

(b) $\varepsilon_L(t) = \varepsilon e^{-t/\tau_L} = \varepsilon e^{-2.0\tau_L/\tau_L} = \varepsilon e^{-2.0} = 0.135\varepsilon$, or $\varepsilon_L/\varepsilon = 0.135$.

(c) From $\varepsilon_L(t) = \varepsilon e^{-t/\tau_L}$ we obtain

$$\frac{t}{\tau_L} = \ln\left(\frac{\varepsilon}{\varepsilon_L}\right) = \ln 2 \Rightarrow t = \tau_L \ln 2 = 0.693\tau_L \Rightarrow t/\tau_L = 0.693.$$

51. The current in the circuit is given by $i = i_0 e^{-t/\tau_L}$, where i_0 is the current at time $t = 0$ and τ_L is the inductive time constant (L/R). We solve for τ_L. Dividing by i_0 and taking the natural logarithm of both sides, we obtain

$$\ln\left(\frac{i}{i_0}\right) = -\frac{t}{\tau_L}.$$

This yields

$$\tau_L = -\frac{t}{\ln(i/i_0)} = -\frac{1.0\,\text{s}}{\ln\left((10\times10^{-3}\,\text{A})/(1.0\,\text{A})\right)} = 0.217\,\text{s}.$$

Therefore, $R = L/\tau_L = 10 \text{ H}/0.217 \text{ s} = 46 \ \Omega$.

52. (a) The inductor prevents a fast build-up of the current through it, so immediately after the switch is closed, the current in the inductor is zero. It follows that

$$i_1 = \frac{\varepsilon}{R_1 + R_2} = \frac{100 \text{ V}}{10.0\Omega + 20.0\Omega} = 3.33 \text{ A}.$$

(b) $i_2 = i_1 = 3.33 \text{ A}$.

(c) After a suitably long time, the current reaches steady state. Then, the emf across the inductor is zero, and we may imagine it replaced by a wire. The current in R_3 is $i_1 - i_2$. Kirchhoff's loop rule gives

$$\varepsilon - i_1 R_1 - i_2 R_2 = 0 \qquad \text{and} \qquad \varepsilon - i_1 R_1 - (i_1 - i_2) R_3 = 0.$$

We solve these simultaneously for i_1 and i_2, and find

$$i_1 = \frac{\varepsilon(R_2 + R_3)}{R_1 R_2 + R_1 R_3 + R_2 R_3} = \frac{(100 \text{ V})(20.0\Omega + 30.0\Omega)}{(10.0\Omega)(20.0\Omega) + (10.0\Omega)(30.0\Omega) + (20.0\Omega)(30.0\Omega)}$$
$$= 4.55 \text{ A},$$

(d) and

$$i_2 = \frac{\varepsilon R_3}{R_1 R_2 + R_1 R_3 + R_2 R_3} = \frac{(100 \text{ V})(30.0\Omega)}{(10.0\Omega)(20.0\Omega) + (10.0\Omega)(30.0\Omega) + (20.0\Omega)(30.0\Omega)}$$
$$= 2.73 \text{ A}.$$

(e) The left-hand branch is now broken. We take the current (immediately) as zero in that branch when the switch is opened (that is, $i_1 = 0$).

(f) The current in R_3 changes less rapidly because there is an inductor in its branch. In fact, immediately after the switch is opened it has the same value that it had before the switch was opened. That value is $4.55 \text{ A} - 2.73 \text{ A} = 1.82 \text{ A}$. The current in R_2 is the same but in the opposite direction as that in R_3, i.e., $i_2 = -1.82 \text{ A}$.

A long time later after the switch is reopened, there are no longer any sources of emf in the circuit, so all currents eventually drop to zero. Thus,

(g) $i_1 = 0$, and

(h) $i_2 = 0$.

53. (a) If the battery is switched into the circuit at $t = 0$, then the current at a later time t is given by

$$i = \frac{\varepsilon}{R}\left(1 - e^{-t/\tau_L}\right),$$

where $\tau_L = L/R$. Our goal is to find the time at which $i = 0.800\varepsilon/R$. This means

$$0.800 = 1 - e^{-t/\tau_L} \Rightarrow e^{-t/\tau_L} = 0.200.$$

Taking the natural logarithm of both sides, we obtain $-(t/\tau_L) = \ln(0.200) = -1.609$. Thus

$$t = 1.609\tau_L = \frac{1.609\,L}{R} = \frac{1.609(6.30\times10^{-6}\,\text{H})}{1.20\times10^{3}\,\Omega} = 8.45\times10^{-9}\,\text{s}.$$

(b) At $t = 1.0\,\tau_L$ the current in the circuit is

$$i = \frac{\varepsilon}{R}\left(1 - e^{-1.0}\right) = \left(\frac{14.0\,\text{V}}{1.20\times10^{3}\,\Omega}\right)(1 - e^{-1.0}) = 7.37\times10^{-3}\,\text{A}.$$

54. From the graph we get $\Phi/i = 2\times10^{-4}$ in SI units. Therefore, with $N = 25$, we find the self-inductance is $L = N\Phi/i = 5\times10^{-3}\,\text{H}$. From the derivative of Eq. 30-41 (or a combination of that equation and Eq. 30-39) we find (using the symbol V to stand for the battery emf)

$$\frac{d\,i}{d\,t} = \frac{V}{R}\frac{R}{L}e^{-t/\tau_L} = \frac{V}{L}e^{-t/\tau_L} = 7.1\times10^{2}\,\text{A/s}.$$

55. Applying the loop theorem

$$\varepsilon - L\left(\frac{di}{dt}\right) = iR,$$

we solve for the (time-dependent) emf, with SI units understood:

$$\varepsilon = L\frac{di}{dt} + iR = L\frac{d}{dt}(3.0 + 5.0t) + (3.0 + 5.0t)R = (6.0)(5.0) + (3.0 + 5.0t)(4.0)$$
$$= (42 + 20t).$$

56. (a) Our notation is as follows: h is the height of the toroid, a its inner radius, and b its outer radius. Since it has a square cross section, $h = b - a = 0.12\,\text{m} - 0.10\,\text{m} = 0.02\,\text{m}$. We derive the flux using Eq. 29-24 and the self-inductance using Eq. 30-33:

$$\Phi_B = \int_a^b B\,dA = \int_a^b \left(\frac{\mu_0 Ni}{2\pi r}\right) h\,dr = \frac{\mu_0 Nih}{2\pi}\ln\left(\frac{b}{a}\right)$$

and

$$L = N\Phi_B/i = (\mu_0 N^2 h/2\pi)\ln(b/a).$$

Now, since the inner circumference of the toroid is $l = 2\pi a = 2\pi(10 \text{ cm}) \approx 62.8$ cm, the number of turns of the toroid is roughly $N \approx 62.8 \text{ cm}/1.0 \text{ mm} = 628$. Thus

$$L = \frac{\mu_0 N^2 h}{2\pi}\ln\left(\frac{b}{a}\right) \approx \frac{\left(4\pi\times10^{-7} \text{ H/m}\right)(628)^2 (0.02\text{m})}{2\pi}\ln\left(\frac{12}{10}\right) = 2.9\times10^{-4}\text{H}.$$

(b) Noting that the perimeter of a square is four times its sides, the total length ℓ of the wire is $\ell = (628)4(2.0\text{cm}) = 50\,\text{m}$, the resistance of the wire is

$$R = (50 \text{ m})(0.02 \ \Omega/\text{m}) = 1.0 \ \Omega.$$

Thus

$$\tau_L = \frac{L}{R} = \frac{2.9\times10^{-4}\text{H}}{1.0\Omega} = 2.9\times10^{-4}\text{s}.$$

57. (a) We assume i is from left to right through the closed switch. We let i_1 be the current in the resistor and take it to be downward. Let i_2 be the current in the inductor, also assumed downward. The junction rule gives $i = i_1 + i_2$ and the loop rule gives $i_1 R - L(di_2/dt) = 0$. According to the junction rule, $(di_1/dt) = -(di_2/dt)$. We substitute into the loop equation to obtain

$$L\frac{di_1}{dt} + i_1 R = 0.$$

This equation is similar to Eq. 30-46, and its solution is the function given as Eq. 30-47:

$$i_1 = i_0 e^{-Rt/L},$$

where i_0 is the current through the resistor at $t = 0$, just after the switch is closed. Now just after the switch is closed, the inductor prevents the rapid build-up of current in its branch, so at that moment $i_2 = 0$ and $i_1 = i$. Thus $i_0 = i$, so

$$i_1 = ie^{-Rt/L} \qquad \text{and} \qquad i_2 = i - i_1 = i\left(1 - e^{-Rt/L}\right).$$

(b) When $i_2 = i_1$,

$$e^{-Rt/L} = 1 - e^{-Rt/L} \Rightarrow e^{-Rt/L} = \frac{1}{2}.$$

Taking the natural logarithm of both sides (and using $\ln(1/2) = -\ln 2$) we obtain

$$\left(\frac{Rt}{L}\right) = \ln 2 \Rightarrow t = \frac{L}{R}\ln 2.$$

58. Let $U_B(t) = \frac{1}{2}Li^2(t)$. We require the energy at time t to be half of its final value: $U(t) = \frac{1}{2}U_B(t \to \infty) = \frac{1}{4}Li_f^2$. This gives $i(t) = i_f/\sqrt{2}$. But $i(t) = i_f(1 - e^{-t/\tau_L})$, so

$$1 - e^{-t/\tau_L} = \frac{1}{\sqrt{2}} \Rightarrow \frac{t}{\tau_L} = -\ln\left(1 - \frac{1}{\sqrt{2}}\right) = 1.23.$$

59. From Eq. 30-49 and Eq. 30-41, the rate at which the energy is being stored in the inductor is

$$\frac{dU_B}{dt} = \frac{d\left(\frac{1}{2}Li^2\right)}{dt} = Li\frac{di}{dt} = L\left(\frac{\varepsilon}{R}\left(1 - e^{-t/\tau_L}\right)\right)\left(\frac{\varepsilon}{R}\frac{1}{\tau_L}e^{-t/\tau_L}\right) = \frac{\varepsilon^2}{R}\left(1 - e^{-t/\tau_L}\right)e^{-t/\tau_L}$$

where $\tau_L = L/R$ has been used. From Eq. 26-22 and Eq. 30-41, the rate at which the resistor is generating thermal energy is

$$P_{\text{thermal}} = i^2 R = \frac{\varepsilon^2}{R^2}\left(1 - e^{-t/\tau_L}\right)^2 R = \frac{\varepsilon^2}{R}\left(1 - e^{-t/\tau_L}\right)^2.$$

We equate this to dU_B/dt, and solve for the time:

$$\frac{\varepsilon^2}{R}\left(1 - e^{-t/\tau_L}\right)^2 = \frac{\varepsilon^2}{R}\left(1 - e^{-t/\tau_L}\right)e^{-t/\tau_L} \Rightarrow t = \tau_L \ln 2 = (37.0\,\text{ms})\ln 2 = 25.6\,\text{ms}.$$

60. (a) From Eq. 30-49 and Eq. 30-41, the rate at which the energy is being stored in the inductor is

$$\frac{dU_B}{dt} = \frac{d\left(\frac{1}{2}Li^2\right)}{dt} = Li\frac{di}{dt} = L\left(\frac{\varepsilon}{R}\left(1 - e^{-t/\tau_L}\right)\right)\left(\frac{\varepsilon}{R}\frac{1}{\tau_L}e^{-t/\tau_L}\right) = \frac{\varepsilon^2}{R}\left(1 - e^{-t/\tau_L}\right)e^{-t/\tau_L}.$$

Now,

$$\tau_L = L/R = 2.0\,\text{H}/10\,\Omega = 0.20\,\text{s}$$

and $\varepsilon = 100$ V, so the above expression yields $dU_B/dt = 2.4 \times 10^2$ W when $t = 0.10$ s.

(b) From Eq. 26-22 and Eq. 30-41, the rate at which the resistor is generating thermal energy is

$$P_{thermal} = i^2 R = \frac{\varepsilon^2}{R^2}\left(1 - e^{-t/\tau_L}\right)^2 R = \frac{\varepsilon^2}{R}\left(1 - e^{-t/\tau_L}\right)^2.$$

At $t = 0.10$ s, this yields $P_{thermal} = 1.5 \times 10^2$ W.

(c) By energy conservation, the rate of energy being supplied to the circuit by the battery is

$$P_{battery} = P_{thermal} + \frac{dU_B}{dt} = 3.9 \times 10^2 \text{ W}.$$

We note that this result could alternatively have been found from Eq. 28-14 (with Eq. 30-41).

61. (a) If the battery is applied at time $t = 0$ the current is given by

$$i = \frac{\varepsilon}{R}\left(1 - e^{-t/\tau_L}\right),$$

where ε is the emf of the battery, R is the resistance, and τ_L is the inductive time constant (L/R). This leads to

$$e^{-t/\tau_L} = 1 - \frac{iR}{\varepsilon} \Rightarrow -\frac{t}{\tau_L} = \ln\left(1 - \frac{iR}{\varepsilon}\right).$$

Since

$$\ln\left(1 - \frac{iR}{\varepsilon}\right) = \ln\left[1 - \frac{\left(2.00 \times 10^{-3} \text{A}\right)\left(10.0 \times 10^3 \Omega\right)}{50.0 \text{ V}}\right] = -0.5108,$$

the inductive time constant is

$$\tau_L = t/0.5108 = (5.00 \times 10^{-3} \text{ s})/0.5108 = 9.79 \times 10^{-3} \text{ s}$$

and the inductance is

$$L = \tau_L R = \left(9.79 \times 10^{-3} \text{s}\right)\left(10.0 \times 10^3 \Omega\right) = 97.9 \text{ H}.$$

(b) The energy stored in the coil is

$$U_B = \frac{1}{2}Li^2 = \frac{1}{2}(97.9\,\text{H})(2.00 \times 10^{-3}\,\text{A})^2 = 1.96 \times 10^{-4}\,\text{J}.$$

62. (a) The energy delivered by the battery is the integral of Eq. 28-14 (where we use Eq. 30-41 for the current):

$$\int_0^{t'} P_{\text{battery}}\,dt = \int_0^{t'} \frac{\varepsilon^2}{R}\left(1 - e^{-Rt/L}\right)dt = \frac{\varepsilon^2}{R}\left[t + \frac{L}{R}\left(e^{-Rt/L} - 1\right)\right]$$

$$= \frac{(10.0\,\text{V})^2}{6.70\,\Omega}\left[2.00\,\text{s} + \frac{(5.50\,\text{H})\left(e^{-(6.70\,\Omega)(2.00\,\text{s})/5.50\,\text{H}} - 1\right)}{6.70\,\Omega}\right]$$

$$= 18.7\,\text{J}.$$

(b) The energy stored in the magnetic field is given by Eq. 30-49:

$$U_B = \frac{1}{2}Li^2(t) = \frac{1}{2}L\left(\frac{\varepsilon}{R}\right)^2\left(1 - e^{-Rt/L}\right)^2$$

$$= \frac{1}{2}(5.50\,\text{H})\left(\frac{10.0\,\text{V}}{6.70\,\Omega}\right)^2\left[1 - e^{-(6.70\,\Omega)(2.00\,\text{s})/5.50\,\text{H}}\right]^2$$

$$= 5.10\,\text{J}.$$

(c) The difference of the previous two results gives the amount "lost" in the resistor: $18.7\,\text{J} - 5.10\,\text{J} = 13.6\,\text{J}.$

63. (a) At any point the magnetic energy density is given by $u_B = B^2/2\mu_0$, where B is the magnitude of the magnetic field at that point. Inside a solenoid $B = \mu_0 ni$, where n, for the solenoid of this problem, is $(950\ \text{turns})/(0.850\ \text{m}) = 1.118 \times 10^3\ \text{m}^{-1}$. The magnetic energy density is

$$u_B = \frac{1}{2}\mu_0 n^2 i^2 = \frac{1}{2}(4\pi \times 10^{-7}\,\text{T}\cdot\text{m/A})(1.118 \times 10^3\,\text{m}^{-1})^2(6.60\,\text{A})^2 = 34.2\,\text{J/m}^3.$$

(b) Since the magnetic field is uniform inside an ideal solenoid, the total energy stored in the field is $U_B = u_B V$, where V is the volume of the solenoid. V is calculated as the product of the cross-sectional area and the length. Thus

$$U_B = (34.2\,\text{J/m}^3)(17.0 \times 10^{-4}\,\text{m}^2)(0.850\,\text{m}) = 4.94 \times 10^{-2}\,\text{J}.$$

64. The magnetic energy stored in the toroid is given by $U_B = \frac{1}{2} L i^2$, where L is its inductance and i is the current. By Eq. 30-54, the energy is also given by $U_B = u_B V$, where u_B is the average energy density and V is the volume. Thus

$$i = \sqrt{\frac{2u_B V}{L}} = \sqrt{\frac{2(70.0\,\text{J/m}^3)(0.0200\,\text{m}^3)}{90.0 \times 10^{-3}\,\text{H}}} = 5.58\,\text{A} \;.$$

65. We set $u_E = \frac{1}{2}\varepsilon_0 E^2 = u_B = \frac{1}{2} B^2/\mu_0$ and solve for the magnitude of the electric field:

$$E = \frac{B}{\sqrt{\varepsilon_0 \mu_0}} = \frac{0.50\,\text{T}}{\sqrt{(8.85 \times 10^{-12}\,\text{F/m})(4\pi \times 10^{-7}\,\text{H/m})}} = 1.5 \times 10^8 \,\text{V/m} \;.$$

66. (a) The magnitude of the magnetic field at the center of the loop, using Eq. 29-9, is

$$B = \frac{\mu_0 i}{2R} = \frac{(4\pi \times 10^{-7}\,\text{H/m})(100\,\text{A})}{2(50 \times 10^{-3}\,\text{m})} = 1.3 \times 10^{-3}\,\text{T} \;.$$

(b) The energy per unit volume in the immediate vicinity of the center of the loop is

$$u_B = \frac{B^2}{2\mu_0} = \frac{(1.3 \times 10^{-3}\,\text{T})^2}{2(4\pi \times 10^{-7}\,\text{H/m})} = 0.63\,\text{J/m}^3 \;.$$

67. (a) The energy per unit volume associated with the magnetic field is

$$u_B = \frac{B^2}{2\mu_0} = \frac{1}{2\mu_0}\left(\frac{\mu_0 i}{2R}\right)^2 = \frac{\mu_0 i^2}{8R^2} = \frac{(4\pi \times 10^{-7}\,\text{H/m})(10\,\text{A})^2}{8(2.5 \times 10^{-3}\,\text{m/2})^2} = 1.0\,\text{J/m}^3 \;.$$

(b) The electric energy density is

$$u_E = \frac{1}{2}\varepsilon_0 E^2 = \frac{\varepsilon_0}{2}(\rho J)^2 = \frac{\varepsilon_0}{2}\left(\frac{iR}{\ell}\right)^2 = \frac{1}{2}(8.85 \times 10^{-12}\,\text{F/m})\left[(10\,\text{A})(3.3\Omega/10^3\,\text{m})\right]^2$$

$$= 4.8 \times 10^{-15}\,\text{J/m}^3 \;.$$

Here we used $J = i/A$ and $R = \rho\ell/A$ to obtain $\rho J = iR/\ell$.

68. We use $\varepsilon_2 = -M\,di_1/dt \approx M|\Delta i/\Delta t|$ to find M:

$$M = \left| \frac{\varepsilon}{\Delta i_1 / \Delta t} \right| = \frac{30 \times 10^3 \, \text{V}}{6.0 \, \text{A} / (2.5 \times 10^{-3} \, \text{s})} = 13 \, \text{H} \,.$$

69. (a) Eq. 30-65 yields

$$M = \frac{\varepsilon_1}{|di_2 / dt|} = \frac{25.0 \, \text{mV}}{15.0 \, \text{A/s}} = 1.67 \, \text{mH} \,.$$

(b) Eq. 30-60 leads to

$$N_2 \Phi_{21} = M i_1 = (1.67 \, \text{mH})(3.60 \, \text{A}) = 6.00 \, \text{mWb} \,.$$

70. (a) The flux in coil 1 is

$$\frac{L_1 i_1}{N_1} = \frac{(25 \text{mH})(6.0 \text{mA})}{100} = 1.5 \, \mu \text{Wb}.$$

(b) The magnitude of the self-induced emf is

$$L_1 \frac{di_1}{dt} = (25 \text{mH})(4.0 \, \text{A/s}) = 1.0 \times 10^2 \, \text{mV}.$$

(c) In coil 2, we find

$$\Phi_{21} = \frac{M i_1}{N_2} = \frac{(3.0 \text{mH})(6.0 \text{mA})}{200} = 90 \text{nWb} \,.$$

(d) The mutually induced emf is

$$\varepsilon_{21} = M \frac{di_1}{dt} = (3.0 \text{mH})(4.0 \, \text{A/s}) = 12 \text{mV}.$$

71. (a) We assume the current is changing at (nonzero) rate di/dt and calculate the total emf across both coils. First consider the coil 1. The magnetic field due to the current in that coil points to the right. The magnetic field due to the current in coil 2 also points to the right. When the current increases, both fields increase and both changes in flux contribute emf's in the same direction. Thus, the induced emf's are

$$\varepsilon_1 = -(L_1 + M) \frac{di}{dt} \quad \text{and} \quad \varepsilon_2 = -(L_2 + M) \frac{di}{dt} \,.$$

Therefore, the total emf across both coils is

$$\varepsilon = \varepsilon_1 + \varepsilon_2 = -\left(L_1 + L_2 + 2M\right)\frac{di}{dt}$$

which is exactly the emf that would be produced if the coils were replaced by a single coil with inductance $L_{eq} = L_1 + L_2 + 2M$.

(b) We imagine reversing the leads of coil 2 so the current enters at the back of coil rather than the front (as pictured in the diagram). Then the field produced by coil 2 at the site of coil 1 is opposite to the field produced by coil 1 itself. The fluxes have opposite signs. An increasing current in coil 1 tends to increase the flux in that coil, but an increasing current in coil 2 tends to decrease it. The emf across coil 1 is

$$\varepsilon_1 = -\left(L_1 - M\right)\frac{di}{dt}.$$

Similarly, the emf across coil 2 is

$$\varepsilon_2 = -\left(L_2 - M\right)\frac{di}{dt}.$$

The total emf across both coils is

$$\varepsilon = -\left(L_1 + L_2 - 2M\right)\frac{di}{dt}.$$

This the same as the emf that would be produced by a single coil with inductance

$$L_{eq} = L_1 + L_2 - 2M.$$

72. The coil-solenoid mutual inductance is

$$M = M_{cs} = \frac{N\Phi_{cs}}{i_s} = \frac{N\left(\mu_0 i_s n\pi R^2\right)}{i_s} = \mu_0 \pi R^2 nN .$$

As long as the magnetic field of the solenoid is entirely contained within the cross-section of the coil we have $\Phi_{sc} = B_s A_s = B_s \pi R^2$, regardless of the shape, size, or possible lack of close-packing of the coil.

73. The flux over the loop cross section due to the current i in the wire is given by

$$\Phi = \int_a^{a+b} B_{wire} l\, dr = \int_a^{a+b} \frac{\mu_0 i l}{2\pi r}\, dr = \frac{\mu_0 i l}{2\pi} \ln\left(1 + \frac{b}{a}\right).$$

Thus,

$$M = \frac{N\Phi}{i} = \frac{N\mu_0 l}{2\pi}\ln\left(1+\frac{b}{a}\right).$$

From the formula for M obtained above, we have

$$M = \frac{(100)(4\pi \times 10^{-7}\,\text{H/m})(0.30\,\text{m})}{2\pi}\ln\left(1+\frac{8.0}{1.0}\right) = 1.3 \times 10^{-5}\,\text{H}.$$

74. (a) The current is given by Eq. 30-41

$$i = (\varepsilon/R)(1 - e^{-t/\tau_L}) = 2.00\,\text{A},$$

where $L = 0.018$ H and $\varepsilon = 12$ V. If $R = 1.00\,\Omega$ (so $\tau_L = L/R = 0.018$ s), we obtain $t = 0.00328$ s when we solve this equation.

(b) For $R = 5.00\,\Omega$ we find $t = 0.00645$ s.

(c) If we set $R = 6.00\,\Omega$ then $\varepsilon/R = 2.00$ A so $e^{-t/\tau_L} = 0$, which means $t = \infty$.

(d) The trend in our answers to parts (a), (b) and (c) lead us to expect the smaller the resistance then the smaller to value of t. If we consider what happens to Eq. 30-39 in the extreme case where $R \to 0$, we find that the time-derivative of the current becomes equal to the emf divided by the self-inductance, which leads to a linear dependence of current on time: $i = (\varepsilon/L)t$. In fact, this is what one have obtained starting from Eq. 30-41 and considering its $R \to 0$ limit. Thus, this case seems self-consistent, so we conclude that it is meaningful and that $R = 0$ is actually a valid answer here.

(e) Thus $t = Li/\varepsilon = 0.00300$ s in this "least-time" scenario.

75. Faraday's law (for a single turn, with B changing in time) gives

$$\varepsilon = -\frac{d\Phi_B}{dt} = -A\frac{dB}{dt} = -\pi r^2\frac{dB}{dt}.$$

In this problem, we find $\frac{dB}{dt} = -\frac{B_0}{\tau}e^{-t/\tau}$. Thus, $\varepsilon = \pi r^2 \frac{B_0}{\tau}e^{-t/\tau}$.

76. From the datum at $t = 0$ in Fig. 30-69(b) we see 0.0015 A $= V_{battery}/R$, which implies that the resistance is $R = (6\,\mu\text{V})/(0.0015\,\text{A}) = 0.004\,\Omega$. Now, the value of the current during 10 s $< t <$ 20 s leads us to equate $(V_{battery} + \varepsilon_{nduced})/R = 0.0005$ A. This shows that the induced emf is $\varepsilon_{induced} = -4\,\mu\text{V}$. Now we use Faraday's law:

$$\varepsilon = -\frac{d\Phi_B}{dt} = -A\frac{dB}{dt} = -A\,a.$$

Plugging in $\varepsilon = -4 \times 10^{-6}$ V and $A = 5 \times 10^{-4}\,\text{m}^2$, we obtain $a = 0.0080$ T/s.

77. Using Ohm's law, we relate the induced current to the emf and (the absolute value of) Faraday's law:

$$i = \varepsilon /R = \frac{1}{R}\frac{d\Phi_B}{dt}.$$

As the loop is crossing the boundary between regions 1 and 2 (so that "x" amount of its length is in region 2 while "$D - x$" amount of its length remains in region 1) the flux is

$$\Phi_B = xHB_2 + (D - x)HB_1 = DHB_1 + xH(B_2 - B_1)$$

which means

$$\frac{d\Phi_B}{dt} = \frac{dx}{dt}H(B_2 - B_1) = vH(B_2 - B_1) \quad\Rightarrow\quad i = vH(B_2 - B_1)/R.$$

Similar considerations hold (replacing "B_1" with 0 and "B_2" with B_1) for the loop crossing initially from the zero-field region (to the left of Fig. 30-70(a)) into region 1.

(a) In this latter case, appeal to Fig. 30-70(b) leads to

$$3.0 \times 10^{-6}\,\text{A} = (0.40\ \text{m/s})(0.015\ \text{m})\,B_1 /(0.020\ \Omega)$$

which yields $B_1 = 10\ \mu\text{T}$.

(b) Lenz's law considerations lead us to conclude that the direction of the region 1 field is *out of the page*.

(c) Similarly, $i = vH(B_2 - B_1)/R$ leads to $\vec{B_2} = 3.3\ \mu\text{T}$,

(d) The direction of $\vec{B_2}$ is out of the page.

78. The energy stored when the current is i is

$$U_B = \tfrac{1}{2} L\, i^2$$

where L is the self-inductance. The rate at which this is developed is

$$\frac{d\,U_B}{d\,t} = L\,i\,\frac{d\,i}{d\,t}$$

where i is given by Eq. 30-41 and di/dt is gotten by taking the derivative of that equation (or by using Eq. 30-37). Thus, using the symbol V to stand for the battery voltage (12.0 volts) and R for the resistance (20.0 Ω), we have

$$\frac{dU_B}{dt} = \frac{V^2}{R}(1 - e^{-t/\tau_L})e^{-t/\tau_L} = 1.15 \text{ W} .$$

79. (a) Before the fuse blows, the current through the resistor remains zero. We apply the loop theorem to the battery-fuse-inductor loop: $\varepsilon - L\, di/dt = 0$. So $i = \varepsilon t/L$. As the fuse blows at $t = t_0$, $i = i_0 = 3.0$ A. Thus,

$$t_0 = \frac{i_0 L}{\varepsilon} = \frac{(3.0\,\text{A})(5.0\,\text{H})}{10\,\text{V}} = 1.5 \text{ s}.$$

(b) We do not show the graph here; qualitatively, it would be similar to Fig. 30-15.

80. Since $A = \ell^2$, we have $dA/dt = 2\ell\, d\ell/dt$. Thus, Faraday's law, with $N = 1$, becomes (in absolute value)

$$\varepsilon = \frac{d\Phi_B}{dt} = B\frac{dA}{dt} = 2\ell\, B\, \frac{d\ell}{dt}$$

which yields $\varepsilon = 0.0029$ V.

81. We write $i = i_0 e^{-t/\tau_L}$ and note that $i = 10\%\ i_0$. We solve for t:

$$t = \tau_L \ln\left(\frac{i_0}{i}\right) = \frac{L}{R}\ln\left(\frac{i_0}{i}\right) = \frac{2.00\,\text{H}}{3.00\,\Omega}\ln\left(\frac{i_0}{0.100\,i_0}\right) = 1.54 \text{ s} .$$

82. It is important to note that the x that is used in the graph of Fig. 30-72(b) is not the x at which the energy density is being evaluated. The x in Fig. 30-72(b) is the location of wire 2. The energy density (Eq. 30-54) is being evaluated at the coordinate origin throughout this problem. We note the curve in Fig. 30-72(b) has a zero; this implies that the magnetic fields (caused by the individual currents) are in opposite directions (at the origin), which further implies that the currents have the same direction. Since the magnitudes of the fields must be equal (for them to cancel) when the x of Fig. 30-72(b) is equal to 0.20 m, then we have (using Eq. 29-4) $B_1 = B_2$, or

$$\frac{\mu_0\, i_1}{2\pi d} = \frac{\mu_0\, i_2}{2\pi(0.20\text{ m})}$$

which leads to $d = \frac{1}{3}(0.20 \text{ m})$ once we substitute $i_1 = \frac{1}{3}i_2$ and simplify. We can also use the given fact that when the energy density is completely caused by B_1 (this occurs when x becomes infinitely large because then $B_2 = 0$) its value is $u_B = 1.96 \times 10^{-9}$ (in SI units) in order to solve for B_1:

$$B_1 = \sqrt{2\mu_0 u_B} .$$

(a) This combined with $B_1 = \frac{\mu_0 i_1}{2\pi d}$ allows us to find wire 1's current: $i_1 \approx 23$ mA.

(b) Since $i_2 = 3i_1$ then $i_2 = 70$ mA (approximately).

83. (a) As the switch closes at $t = 0$, the current being zero in the inductor serves as an initial condition for the building-up of current in the circuit. Thus, at $t = 0$ any current through the battery is also that through the 20 Ω and 10 Ω resistors. Hence,

$$i = \frac{\varepsilon}{30\Omega} = 0.40 \text{ A}$$

which results in a voltage drop across the 10 Ω resistor equal to $(0.40)(10) = 4.0$ V. The inductor must have this same voltage across it $|\varepsilon_L|$, and we use (the absolute value of) Eq. 30-35:

$$\frac{di}{dt} = \frac{|\varepsilon_L|}{L} = \frac{4.0}{0.010} = 400 \text{ A/s}.$$

(b) Applying the loop rule to the outer loop, we have

$$\varepsilon - (0.50 \text{ A})(20\Omega) - |\varepsilon_L| = 0.$$

Therefore, $|\varepsilon_L| = 2.0$ V, and Eq. 30-35 leads to

$$\frac{di}{dt} = \frac{|\varepsilon_L|}{L} = \frac{2.0}{0.010} = 200 \text{ A/s}.$$

(c) As $t \to \infty$, the inductor has $\varepsilon_L = 0$ (since the current is no longer changing). Thus, the loop rule (for the outer loop) leads to

$$\varepsilon - i(20\Omega) - |\varepsilon_L| = 0 \Rightarrow i = 0.60 \text{ A}.$$

84. (a) From Eq. 30-35, we find $L = (3.00 \text{ mV})/(5.00 \text{ A/s}) = 0.600$ mH.

(b) Since $N\Phi = iL$ (where $\Phi = 40.0$ μWb and $i = 8.00$ A), we obtain $N = 120$.

85. (a) The magnitude of the average induced emf is

$$\varepsilon_{avg} = \left| \frac{-d\Phi_B}{dt} \right| = \left| \frac{\Delta\Phi_B}{\Delta t} \right| = \frac{BA_i}{t} = \frac{(2.0\,\text{T})(0.20\,\text{m})^2}{0.20\,\text{s}} = 0.40\,\text{V}.$$

(b) The average induced current is

$$i_{avg} = \frac{\varepsilon_{avg}}{R} = \frac{0.40\,\text{V}}{20\times10^{-3}\,\Omega} = 20\,\text{A}.$$

86. In absolute value, Faraday's law (for a single turn, with B changing in time) gives

$$\frac{d\Phi_B}{dt} = A\frac{dB}{dt} = \pi R^2 \frac{dB}{dt}$$

for the magnitude of the induced emf. Dividing it by R^2 then allows us to relate this to the slope of the graph in Fig. 30-74(b) [particularly the first part of the graph], which we estimate to be 80 μV/m^2.

(a) Thus, $\dfrac{dB_1}{dt} = (80\ \mu\text{V/m}^2)/\pi \approx 25\ \mu\text{T/s}$.

(b) Similar reasoning for region 2 (corresponding to the slope of the second part of the graph in Fig. 30-74(b)) leads to an emf equal to

$$\pi r_1^2 \left(\frac{dB_1}{dt} - \frac{dB_2}{dt} \right) + \pi R^2 \frac{dB_2}{dt} \ .$$

which means the second slope (which we estimate to be 40 μV/m^2) is equal to $\pi \dfrac{dB_2}{dt}$.

Therefore, $\dfrac{dB_2}{dt} = (40\ \mu\text{V/m}^2)/\pi \approx 13\ \mu\text{T/s}$.

(c) Considerations of Lenz's law leads to the conclusion that B_2 is increasing.

87. The induced electric field E as a function of r is given by $E(r) = (r/2)(dB/dt)$.

(a) The acceleration of the electron released at point a is

$$\vec{a}_a = \frac{eE}{m}\hat{i} = \frac{er}{2m}\left(\frac{dB}{dt} \right)\hat{i} = \frac{(1.60\times10^{-19}\,\text{C})(5.0\times10^{-2}\,\text{m})(10\times10^{-3}\,\text{T/s})}{2(9.11\times10^{-27}\,\text{kg})}\hat{i} = (4.4\times10^{7}\ \text{m/s}^2)\hat{i}.$$

(b) At point b we have $a_b \propto r_b = 0$.

(c) The acceleration of the electron released at point c is

$$\vec{a}_c = -\vec{a}_a = -(4.4 \times 10^7 \text{ m/s}^2)\hat{\text{i}}.$$

88. Because of the decay of current (Eq. 30-45) that occurs after the switches are closed on B, the flux will decay according to

$$\Phi_1 = \Phi_{10}e^{-t/\tau_{L1}}, \quad \Phi_2 = \Phi_{20}e^{-t/\tau_{L2}} \quad .$$

where each time-constant is given by Eq. 30-42. Setting the fluxes equal to each other and solving for time leads to

$$t = \frac{\ln\left(\dfrac{\Phi_{20}}{\Phi_{10}}\right)}{\dfrac{R_2}{L_2} - \dfrac{R_1}{L_1}} = \frac{\ln(1.5)}{\dfrac{30}{0.003} - \dfrac{25}{0.005}} \quad .$$

Thus, $t = 81.1$ μs.

89. (a) When switch S is just closed, $V_1 = \varepsilon$ and $i_1 = \varepsilon/R_1 = 10$ V/5.0 Ω = 2.0 A.

(b) Since now $\varepsilon_L = \varepsilon$, we have $i_2 = 0$.

(c) $i_s = i_1 + i_2 = 2.0$ A + 0 = 2.0 A.

(d) Since $V_L = \varepsilon$, $V_2 = \varepsilon - \varepsilon_L = 0$.

(e) $V_L = \varepsilon = 10$ V.

(f) $di_2/dt = V_L/L = \varepsilon/L = 10$ V / 5.0 H = 2.0 A/s.

(g) After a long time, we still have $V_1 = \varepsilon$, so $i_1 = 2.0$ A.

(h) Since now $V_L = 0$, $i_2 = \varepsilon/R_2 = 10$ V/10 Ω = 1.0 A.

(i) $i_s = i_1 + i_2 = 2.0$ A + 1.0 A = 3.0 A.

(j) Since $V_L = 0$, $V_2 = \varepsilon - V_L = \varepsilon = 10$ V.

(k) $V_L = 0$.

(l) $di_2/dt = V_L/L = 0$.

90. Eq. 30-41 applies, and the problem requires

$$iR = L\frac{di}{dt} = \varepsilon - iR$$

at some time t (where Eq. 30-39 has been used in that last step). Thus, we have $2iR = \varepsilon$, or

$$2[(\varepsilon / R)(1 - e^{-t/\tau_L})]R = \varepsilon$$

where Eq. 30-42 gives the inductive time constant as $\tau_L = L/R$. We note that the emf ε cancels out of that final equation, and we are able to rearrange (and take natural log) and solve. We obtain $t = 0.520$ ms.

91. Taking the derivative of Eq. 30-41, we have

$$\frac{di}{dt} = (\varepsilon / R\tau_L)\, e^{-t/\tau_L} = (\varepsilon / L)\, e^{-t/\tau_L} \ .$$

With $\tau_L = L/R$ (Eq. 30-42), $L = 0.023$ H and $\varepsilon = 12$ V, $t = 0.00015$ s, and $di/dt = 280$ A/s, we obtain $e^{-t/\tau_L} = 0.537$. Taking the natural log and rearranging leads to $R = 95.4\ \Omega$.

92. We use the expression for the flux Φ_B over the toroid cross-section derived in our solution to problem 52 to obtain the coil-toroid mutual inductance:

$$M_{ct} = \frac{N_c \Phi_{ct}}{i_t} = \frac{N_c}{i_t}\frac{\mu_0 i_t N_t h}{2\pi}\ln\!\left(\frac{b}{a}\right) = \frac{\mu_0 N_1 N_2 h}{2\pi}\ln\!\left(\frac{b}{a}\right)$$

where $N_t = N_1$ and $N_c = N_2$.

93. From the given information, we find

$$\frac{dB}{dt} = \frac{0.030\,\mathrm{T}}{0.015\,\mathrm{s}} = 2.0\,\mathrm{T/s}.$$

Thus, with $N = 1$ and $\cos 30° = \sqrt{3}/2$, and using Faraday's law with Ohm's law, we have

$$i = \frac{|\varepsilon|}{R} = \frac{N\pi r^2}{R}\frac{\sqrt{3}}{2}\frac{dB}{dt} = 0.021\,\mathrm{A}.$$

94. The self-inductance and resistance of the coil may be treated as a "pure" inductor in series with a "pure" resistor, in which case the situation described in the problem may be addressed by using Eq. 30-41. The derivative of that solution is

$$\frac{di}{dt} = (\varepsilon / R\tau_L)\, e^{-t/\tau_L} = (\varepsilon / L)\, e^{-t/\tau_L} \ .$$

With $\tau_L = 0.28$ ms (by Eq. 30-42), $L = 0.050$ H and $\varepsilon = 45$ V, we obtain $di/dt = 12$ A/s when $t = 1.2$ ms.

95. (a) The energy density is

$$u_B = \frac{B_e^2}{2\mu_0} = \frac{\left(50 \times 10^{-6}\ \text{T}\right)^2}{2\left(4\pi \times 10^{-7}\ \text{H/m}\right)} = 1.0 \times 10^{-3}\ \text{J/m}^3 \ .$$

(b) The volume of the shell of thickness h is $V \approx 4\pi R_e^2 h$, where R_e is the radius of the Earth. So

$$U_B \approx V u_B \approx 4\pi \left(6.4 \times 10^6\ \text{m}\right)^2 \left(16 \times 10^3\ \text{m}\right)\left(1.0 \times 10^{-3}\ \text{J/m}^3\right) = 8.4 \times 10^{15}\ \text{J}.$$

96. (a) From Eq. 30-28, we have

$$L = N\Phi/i = (150)(50 \times 10^{-9})/(0.002) = 3.75\ \text{mH}.$$

(b) The answer for L (which should be considered the *constant* of proportionality in Eq. 30-35) does not change; it is still 3.75 mH.

(c) The equations of Chapter 28 display a simple proportionality between magnetic field and the current that creates it. Thus, if the current has doubled, so has the field (and consequently the flux). The answer is $2(50) = 100$ nWb.

(d) The magnitude of the induced emf is (from Eq. 30-35)

$$L\frac{di}{dt}\Big|_{\text{max}} = (0.00375\ \text{H})(0.003\ \text{A})(377\ \text{rad/s}) = 0.00424\ \text{V}.$$

97. (a) At $t = 0.50$, the magnetic field is decreasing at a rate of 3/2 mT/s, leading to

$$i = \frac{|\varepsilon|}{R} = \frac{A|dB/dt|}{R} = \frac{(3.0)(3/2)}{9.0} = 0.50\ \text{mA} \ .$$

(b) By Lenz's law, the current is counterclockwise.

(c) At $t = 1.5$ s, the magnetic field is decreasing at a rate of 3/2 mT/s, same as that in (a). Thus, $i = 0.50$ mA.

(d) By Lenz's law, the current is counterclockwise.

(e) For $t = 3.0$ s, there is no change in flux and therefore no induced current.

(f) None.

98. For $t < 0$, no current goes through L_2, so $i_2 = 0$ and $i_1 = \varepsilon/R$. As the switch is opened there will be a very brief sparking across the gap. i_1 drops while i_2 increases, both very quickly. The loop rule can be written as

$$\varepsilon - i_1 R - L_1 \frac{di_1}{dt} - i_2 R - L_2 \frac{di_2}{dt} = 0 \,,$$

where the initial value of i_1 at $t = 0$ is given by ε/R and that of i_2 at $t = 0$ is 0. We consider the situation shortly after $t = 0$. Since the sparking is very brief, we can reasonably assume that both i_1 and i_2 get equalized quickly, before they can change appreciably from their respective initial values. Here, the loop rule requires that $L_1(di_1/dt)$, which is large and negative, must roughly cancel $L_2(di_2/dt)$, which is large and positive:

$$L_1 \frac{di_1}{dt} \approx -L_2 \frac{di_2}{dt} \,.$$

Let the common value reached by i_1 and i_2 be i, then

$$\frac{di_1}{dt} \approx \frac{\Delta i_1}{\Delta t} = \frac{i - \varepsilon/R}{\Delta t}$$

and

$$\frac{di_2}{dt} \approx \frac{\Delta i_2}{\Delta t} = \frac{i - 0}{\Delta t} \,.$$

The equations above yield

$$L_1\left(i - \frac{\varepsilon}{R}\right) = -L_2\left(i - 0\right) \Rightarrow i = \frac{\varepsilon L_1}{L_2 R_1 + L_1 R_2} = \frac{L_1}{L_1 + L_2}\frac{\varepsilon}{R} \,.$$

99. (a) As the switch closes at $t = 0$, the current being zero in the inductor serves as an initial condition for the building-up of current in the circuit. Thus, at $t = 0$ the current through the battery is also zero.

(b) With no current anywhere in the circuit at $t = 0$, the loop rule requires the emf of the inductor ε_L to cancel that of the battery ($\varepsilon = 40$ V). Thus, the absolute value of Eq. 30-35 yields

$$\frac{di_{bat}}{dt} = \frac{|\varepsilon_L|}{L} = \frac{40}{0.050} = 8.0 \times 10^2 \text{ A/s} \,.$$

(c) This circuit becomes equivalent to that analyzed in §30-9 when we replace the parallel set of 20000 Ω resistors with $R = 10000$ Ω. Now, with $\tau_L = L/R = 5 \times 10^{-6}$ s, we have $t/\tau_L = 3/5$, and we apply Eq. 30-41:

$$i_{bat} = \frac{\varepsilon}{R}\left(1 - e^{-3/5}\right) \approx 1.8 \times 10^{-3} \text{ A.}$$

(d) The rate of change of the current is figured from the loop rule (and Eq. 30-35):

$$\varepsilon - i_{bat}R - |\varepsilon_L| = 0.$$

Using the values from part (c), we obtain $|\varepsilon_L| \approx 22$ V. Then,

$$\frac{di_{bat}}{dt} = \frac{|\varepsilon_L|}{L} = \frac{22}{0.050} \approx 4.4 \times 10^2 \text{ A/s .}$$

(e) As $t \to \infty$, the circuit reaches a steady state condition, so that $di_{bat}/dt = 0$ and $\varepsilon_L = 0$. The loop rule then leads to

$$\varepsilon - i_{bat}R - |\varepsilon_L| = 0 \implies i_{bat} = \frac{40}{10000} = 4.0 \times 10^{-3} \text{ A.}$$

(f) As $t \to \infty$, the circuit reaches a steady state condition, $di_{bat}/dt = 0$.

100. (a) $i_0 = \varepsilon/R = 100$ V/10 Ω = 10 A.

(b) $U_B = \frac{1}{2}Li_0^2 = \frac{1}{2}(2.0\text{H})(10\text{A})^2 = 1.0 \times 10^2 \text{ J .}$

101. (a) The magnetic flux Φ_B through the loop is given by

$$\Phi_B = 2B\left(\pi r^2/2\right)\left(\cos 45°\right) = \pi r^2 B/\sqrt{2} .$$

Thus

$$\varepsilon = -\frac{d\Phi_B}{dt} = -\frac{d}{dt}\left(\frac{\pi r^2 B}{\sqrt{2}}\right) = -\frac{\pi r^2}{\sqrt{2}}\left(\frac{\Delta B}{\Delta t}\right) = -\frac{\pi\left(3.7 \times 10^{-2}\text{ m}\right)^2}{\sqrt{2}}\left(\frac{0 - 76 \times 10^{-3}\text{ T}}{4.5 \times 10^{-3}\text{ s}}\right)$$
$$= 5.1 \times 10^{-2} \text{ V .}$$

(a) The direction of the induced current is clockwise when viewed along the direction of $\vec{B}$.

102. Using Eq. 30-41

$$i = \frac{\varepsilon}{R}\left(1 - e^{-t/\tau_L}\right)$$

where $\tau_L = 2.0$ ns, we find

$$t = \tau_L \ln\left(\frac{1}{1 - iR/\varepsilon}\right) \approx 1.0\,\text{ns}.$$

103. The area enclosed by any turn of the coil is πr^2 where $r = 0.15$ m, and the coil has $N = 50$ turns. Thus, the magnitude of the induced emf, using Eq. 30-5, is

$$|\varepsilon| = N\pi r^2 \left|\frac{dB}{dt}\right| = (3.53\,\text{m}^2)\left|\frac{dB}{dt}\right|$$

where $\left|\frac{dB}{dt}\right| = (0.0126\,\text{T/s})|\cos\omega t|$. Thus, using Ohm's law, we have

$$i = \frac{|\varepsilon|}{R} = \frac{(3.53)(0.0126)}{4.0}|\cos\omega t|.$$

When $t = 0.020$ s, this yields $i = 0.011$ A.

104. (a) $L = \Phi/i = 26 \times 10^{-3}$ Wb/5.5 A $= 4.7 \times 10^{-3}$ H.

(b) We use Eq. 30-41 to solve for t:

$$t = -\tau_L \ln\left(1 - \frac{iR}{\varepsilon}\right) = -\frac{L}{R}\ln\left(1 - \frac{iR}{\varepsilon}\right) = -\frac{4.7 \times 10^{-3}\,\text{H}}{0.75\,\Omega}\ln\left[1 - \frac{(2.5\,\text{A})(0.75\,\Omega)}{6.0\,\text{V}}\right]$$

$$= 2.4 \times 10^{-3}\,\text{s}.$$

105. (a) We use $U_B = \frac{1}{2}Li^2$ to solve for the self-inductance:

$$L = \frac{2U_B}{i^2} = \frac{2(25.0 \times 10^{-3}\,\text{J})}{(60.0 \times 10^{-3}\,\text{A})^2} = 13.9\,\text{H}.$$

(b) Since $U_B \propto i^2$, for U_B to increase by a factor of 4, i must increase by a factor of 2. Therefore, i should be increased to $2(60.0\,\text{mA}) = 120$ mA.

106. (a) The self-inductance per meter is

$$\frac{L}{\ell} = \mu_0 n^2 A = \left(4\pi \times 10^{-7}\,\mathrm{H/m}\right)\left(100\,\mathrm{turns/cm}\right)^2 (\pi)\left(1.6\,\mathrm{cm}\right)^2 = 0.10\,\mathrm{H/m}.$$

(b) The induced emf per meter is

$$\frac{\varepsilon}{\ell} = \frac{L}{\ell}\frac{di}{dt} = \left(0.10\,\mathrm{H/m}\right)\left(13\,\mathrm{A/s}\right) = 1.3\,\mathrm{V/m}.$$

107. Using Eq. 30-41, we find

$$i = \frac{\varepsilon}{R}\left(1 - e^{-t/\tau_L}\right) \Rightarrow \tau_L = \frac{t}{\ln\left(\dfrac{1}{1 - iR/\varepsilon}\right)} = 22.4\,\mathrm{s}.$$

Thus, from Eq. 30-42 (the definition of the time constant), we obtain

$$L = (22.4\ \mathrm{s})(2.0\ \Omega) = 45\ \mathrm{H}.$$

108. (a) As the switch closes at $t = 0$, the current being zero in the inductors serves as an initial condition for the building-up of current in the circuit. Thus, the current through any element of this circuit is also zero at that instant. Consequently, the loop rule requires the emf (ε_{L1}) of the $L_1 = 0.30$ H inductor to cancel that of the battery. We now apply (the absolute value of) Eq. 30-35

$$\frac{di}{dt} = \frac{|\varepsilon_{L1}|}{L_1} = \frac{6.0}{0.30} = 20\,\mathrm{A/s}.$$

(b) What is being asked for is essentially the current in the battery when the emf's of the inductors vanish (as $t \to \infty$). Applying the loop rule to the outer loop, with $R_1 = 8.0\ \Omega$, we have

$$\varepsilon - i\,R_1 - |\varepsilon_{L1}| - |\varepsilon_{L2}| = 0 \Rightarrow i = \frac{6.0\,\mathrm{V}}{R_1} = 0.75\,\mathrm{A}.$$

Chapter 31

1. (a) All the energy in the circuit resides in the capacitor when it has its maximum charge. The current is then zero. If Q is the maximum charge on the capacitor, then the total energy is

$$U = \frac{Q^2}{2C} = \frac{(2.90 \times 10^{-6} \, \text{C})^2}{2(3.60 \times 10^{-6} \, \text{F})} = 1.17 \times 10^{-6} \, \text{J}.$$

(b) When the capacitor is fully discharged, the current is a maximum and all the energy resides in the inductor. If I is the maximum current, then $U = LI^2/2$ leads to

$$I = \sqrt{\frac{2U}{L}} = \sqrt{\frac{2(1.168 \times 10^{-6} \, \text{J})}{75 \times 10^{-3} \, \text{H}}} = 5.58 \times 10^{-3} \, \text{A}.$$

2. According to $U = \frac{1}{2} LI^2 = \frac{1}{2} Q^2/C$, the current amplitude is

$$I = \frac{Q}{\sqrt{LC}} = \frac{3.00 \times 10^{-6} \, \text{C}}{\sqrt{(1.10 \times 10^{-3} \, \text{H})(4.00 \times 10^{-6} \, \text{F})}} = 4.52 \times 10^{-2} \, \text{A}.$$

3. We find the capacitance from $U = \frac{1}{2} Q^2/C$:

$$C = \frac{Q^2}{2U} = \frac{(1.60 \times 10^{-6} \, \text{C})^2}{2(140 \times 10^{-6} \, \text{J})} = 9.14 \times 10^{-9} \, \text{F}.$$

4. (a) The period is $T = 4(1.50 \, \mu s) = 6.00 \, \mu s$.

(b) The frequency is the reciprocal of the period:

$$f = \frac{1}{T} = \frac{1}{6.00 \mu s} = 1.67 \times 10^5 \, \text{Hz}.$$

(c) The magnetic energy does not depend on the direction of the current (since $U_B \propto i^2$), so this will occur after one-half of a period, or $3.00 \, \mu s$.

5. (a) We recall the fact that the period is the reciprocal of the frequency. It is helpful to refer also to Fig. 31-1. The values of t when plate A will again have maximum positive charge are multiples of the period:

$$t_A = nT = \frac{n}{f} = \frac{n}{2.00 \times 10^3 \, \text{Hz}} = n(5.00 \, \mu s),$$

where $n = 1, 2, 3, 4, \ldots$. The earliest time is ($n=1$) $t_A = 5.00 \, \mu s$.

(b) We note that it takes $t = \frac{1}{2}T$ for the charge on the other plate to reach its maximum positive value for the first time (compare steps a and e in Fig. 31-1). This is when plate A acquires its most negative charge. From that time onward, this situation will repeat once every period. Consequently,

$$t = \frac{1}{2}T + (n-1)T = \frac{1}{2}(2n-1)T = \frac{(2n-1)}{2f} = \frac{(2n-1)}{2(2\times10^3 \, \text{Hz})} = (2n-1)(2.50 \, \mu s),$$

where $n = 1, 2, 3, 4, \ldots$. The earliest time is ($n=1$) $t = 2.50 \, \mu s$.

(c) At $t = \frac{1}{4}T$, the current and the magnetic field in the inductor reach maximum values for the first time (compare steps a and c in Fig. 31-1). Later this will repeat every half-period (compare steps c and g in Fig. 31-1). Therefore,

$$t_L = \frac{T}{4} + \frac{(n-1)T}{2} = (2n-1)\frac{T}{4} = (2n-1)(1.25 \, \mu s),$$

where $n = 1, 2, 3, 4, \ldots$. The earliest time is ($n=1$) $t = 1.25 \, \mu s$.

6. (a) The angular frequency is

$$\omega = \sqrt{\frac{k}{m}} = \sqrt{\frac{F/x}{m}} = \sqrt{\frac{8.0 \, \text{N}}{(2.0\times10^{-13} \, \text{m})(0.50 \, \text{kg})}} = 89 \, \text{rad/s}.$$

(b) The period is $1/f$ and $f = \omega/2\pi$. Therefore,

$$T = \frac{2\pi}{\omega} = \frac{2\pi}{89 \, \text{rad/s}} = 7.0 \times 10^{-2} \, \text{s}.$$

(c) From $\omega = (LC)^{-1/2}$, we obtain

$$C = \frac{1}{\omega^2 L} = \frac{1}{(89 \, \text{rad/s})^2 (5.0 \, \text{H})} = 2.5 \times 10^{-5} \, \text{F}.$$

7. (a) The mass m corresponds to the inductance, so $m = 1.25$ kg.

(b) The spring constant k corresponds to the reciprocal of the capacitance. Since the total energy is given by $U = Q^2/2C$, where Q is the maximum charge on the capacitor and C is the capacitance,

$$C = \frac{Q^2}{2U} = \frac{\left(175 \times 10^{-6} \text{ C}\right)^2}{2\left(5.70 \times 10^{-6} \text{ J}\right)} = 2.69 \times 10^{-3} \text{ F}$$

and

$$k = \frac{1}{2.69 \times 10^{-3} \text{ m / N}} = 372 \text{ N / m}.$$

(c) The maximum displacement corresponds to the maximum charge, so $x_{max} = 1.75 \times 10^{-4}$ m.

(d) The maximum speed v_{max} corresponds to the maximum current. The maximum current is

$$I = Q\omega = \frac{Q}{\sqrt{LC}} = \frac{175 \times 10^{-6} \text{ C}}{\sqrt{(1.25 \text{ H})(2.69 \times 10^{-3} \text{ F})}} = 3.02 \times 10^{-3} \text{ A}.$$

Consequently, $v_{max} = 3.02 \times 10^{-3}$ m/s.

8. We apply the loop rule to the entire circuit:

$$\varepsilon_{total} = \varepsilon_{L_1} + \varepsilon_{C_1} + \varepsilon_{R_1} + \cdots = \sum_j \left(\varepsilon_{L_j} + \varepsilon_{C_j} + \varepsilon_{R_j}\right) = \sum_j \left(L_j \frac{di}{dt} + \frac{q}{C_j} + iR_j\right)$$

$$= L\frac{di}{dt} + \frac{q}{C} + iR \qquad \text{with } L = \sum_j L_j, \ \frac{1}{C} = \sum_j \frac{1}{C_j}, R = \sum_j R_j$$

where we require $\varepsilon_{total} = 0$. This is equivalent to the simple LRC circuit shown in Fig. 31-24(b).

9. The time required is $t = T/4$, where the period is given by $T = 2\pi/\omega = 2\pi\sqrt{LC}$. Consequently,

$$t = \frac{T}{4} = \frac{2\pi\sqrt{LC}}{4} = \frac{2\pi\sqrt{(0.050\,\text{H})(4.0\times10^{-6}\,\text{F})}}{4} = 7.0\times10^{-4}\,\text{s}.$$

10. We find the inductance from $f = \omega/2\pi = \left(2\pi\sqrt{LC}\right)^{-1}$.

$$L = \frac{1}{4\pi^2 f^2 C} = \frac{1}{4\pi^2\left(10\times10^3\,\text{Hz}\right)^2\left(6.7\times10^{-6}\,\text{F}\right)} = 3.8\times10^{-5}\,\text{H}.$$

11. (a) $Q = CV_{max} = (1.0\times10^{-9}\,\text{F})(3.0\,\text{V}) = 3.0\times10^{-9}\,\text{C}.$

(b) From $U = \frac{1}{2}LI^2 = \frac{1}{2}Q^2/C$ we get

$$I = \frac{Q}{\sqrt{LC}} = \frac{3.0\times10^{-9}\,\text{C}}{\sqrt{\left(3.0\times10^{-3}\,\text{H}\right)\left(1.0\times10^{-9}\,\text{F}\right)}} = 1.7\times10^{-3}\,\text{A}.$$

(c) When the current is at a maximum, the magnetic field is at maximum:

$$U_{B,max} = \frac{1}{2}LI^2 = \frac{1}{2}\left(3.0\times10^{-3}\,\text{H}\right)\left(1.7\times10^{-3}\,\text{A}\right)^2 = 4.5\times10^{-9}\,\text{J}.$$

12. The capacitors C_1 and C_2 can be used in four different ways: (1) C_1 only; (2) C_2 only; (3) C_1 and C_2 in parallel; and (4) C_1 and C_2 in series.

(a) The smallest oscillation frequency is

$$f_3 = \frac{1}{2\pi\sqrt{L(C_1+C_2)}} = \frac{1}{2\pi\sqrt{\left(1.0\times10^{-2}\,\text{H}\right)\left(2.0\times10^{-6}\,\text{F}+5.0\times10^{-6}\,\text{F}\right)}} = 6.0\times10^2\,\text{Hz}$$

(b) The second smallest oscillation frequency is

$$f_1 = \frac{1}{2\pi\sqrt{LC_1}} = \frac{1}{2\pi\sqrt{\left(1.0\times10^{-2}\,\text{H}\right)\left(5.0\times10^{-6}\,\text{F}\right)}} = 7.1\times10^2\,\text{Hz}$$

(c) The second largest oscillation frequency is

$$f_2 = \frac{1}{2\pi\sqrt{LC_2}} = \frac{1}{2\pi\sqrt{\left(1.0\times10^{-2}\,\text{H}\right)\left(2.0\times10^{-6}\,\text{F}\right)}} = 1.1\times10^3\,\text{Hz}$$

(d) The largest oscillation frequency is

$$f_4 = \frac{1}{2\pi\sqrt{LC_1C_2/(C_1+C_2)}} = \frac{1}{2\pi}\sqrt{\frac{2.0\times10^{-6}\,\text{F}+5.0\times10^{-6}\,\text{F}}{(1.0\times10^{-2}\,\text{H})(2.0\times10^{-6}\,\text{F})(5.0\times10^{-6}\,\text{F})}} = 1.3\times10^3\,\text{Hz}$$

13. (a) After the switch is thrown to position b the circuit is an LC circuit. The angular frequency of oscillation is $\omega = 1/\sqrt{LC}$. Consequently,

$$f = \frac{\omega}{2\pi} = \frac{1}{2\pi\sqrt{LC}} = \frac{1}{2\pi\sqrt{(54.0\times10^{-3}\,\text{H})(6.20\times10^{-6}\,\text{F})}} = 275\,\text{Hz.}$$

(b) When the switch is thrown, the capacitor is charged to $V = 34.0$ V and the current is zero. Thus, the maximum charge on the capacitor is $Q = VC = (34.0\text{ V})(6.20 \times 10^{-6}\text{ F}) = 2.11 \times 10^{-4}$ C. The current amplitude is

$$I = \omega Q = 2\pi f Q = 2\pi(275\,\text{Hz})(2.11\times10^{-4}\,\text{C}) = 0.365\,\text{A.}$$

14. For the first circuit $\omega = (L_1C_1)^{-1/2}$, and for the second one $\omega = (L_2C_2)^{-1/2}$. When the two circuits are connected in series, the new frequency is

$$\omega' = \frac{1}{\sqrt{L_{eq}C_{eq}}} = \frac{1}{\sqrt{(L_1+L_2)C_1C_2/(C_1+C_2)}} = \frac{1}{\sqrt{(L_1C_1C_2+L_2C_2C_1)/(C_1+C_2)}}$$

$$= \frac{1}{\sqrt{L_1C_1}}\frac{1}{\sqrt{(C_1+C_2)/(C_1+C_2)}} = \omega,$$

where we use $\omega^{-1} = \sqrt{L_1C_1} = \sqrt{L_2C_2}$.

15. (a) Since the frequency of oscillation f is related to the inductance L and capacitance C by $f = 1/2\pi\sqrt{LC}$, the smaller value of C gives the larger value of f. Consequently, $f_{max} = 1/2\pi\sqrt{LC_{min}}$, $f_{min} = 1/2\pi\sqrt{LC_{max}}$, and

$$\frac{f_{max}}{f_{min}} = \frac{\sqrt{C_{max}}}{\sqrt{C_{min}}} = \frac{\sqrt{365\,\text{pF}}}{\sqrt{10\,\text{pF}}} = 6.0.$$

(b) An additional capacitance C is chosen so the ratio of the frequencies is

$$r = \frac{1.60\,\text{MHz}}{0.54\,\text{MHz}} = 2.96.$$

Since the additional capacitor is in parallel with the tuning capacitor, its capacitance adds to that of the tuning capacitor. If C is in picofarads, then

$$\frac{\sqrt{C+365\,\text{pF}}}{\sqrt{C+10\,\text{pF}}} = 2.96.$$

The solution for C is

$$C = \frac{(365\,\text{pF}) - (2.96)^2(10\,\text{pF})}{(2.96)^2 - 1} = 36\,\text{pF}.$$

(c) We solve $f = 1/2\pi\sqrt{LC}$ for L. For the minimum frequency C = 365 pF + 36 pF = 401 pF and f = 0.54 MHz. Thus

$$L = \frac{1}{(2\pi)^2\,Cf^2} = \frac{1}{(2\pi)^2(401\times10^{-12}\,\text{F})(0.54\times10^6\,\text{Hz})^2} = 2.2\times10^{-4}\,\text{H}.$$

16. The linear relationship between θ (the knob angle in degrees) and frequency f is

$$f = f_0\left(1 + \frac{\theta}{180°}\right) \Rightarrow \theta = 180°\left(\frac{f}{f_0} - 1\right)$$

where $f_0 = 2 \times 10^5$ Hz. Since $f = \omega/2\pi = 1/2\pi\sqrt{LC}$, we are able to solve for C in terms of θ:

$$C = \frac{1}{4\pi^2\,Lf_0^2\left(1 + \frac{\theta}{180°}\right)^2} = \frac{81}{400000\pi^2(180° + \theta)^2}$$

with SI units understood. After multiplying by 10^{12} (to convert to picofarads), this is plotted, below.

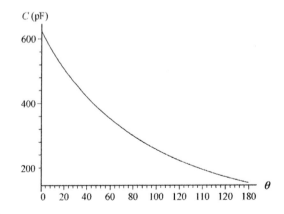

17. (a) The total energy U is the sum of the energies in the inductor and capacitor:

$$U = U_E + U_B = \frac{q^2}{2C} + \frac{i^2 L}{2} = \frac{\left(3.80 \times 10^{-6}\,\text{C}\right)^2}{2\left(7.80 \times 10^{-6}\,\text{F}\right)} + \frac{\left(9.20 \times 10^{-3}\,\text{A}\right)^2 \left(25.0 \times 10^{-3}\,\text{H}\right)}{2} = 1.98 \times 10^{-6}\,\text{J}.$$

(b) We solve $U = Q^2/2C$ for the maximum charge:

$$Q = \sqrt{2CU} = \sqrt{2\left(7.80 \times 10^{-6}\,\text{F}\right)\left(1.98 \times 10^{-6}\,\text{J}\right)} = 5.56 \times 10^{-6}\,\text{C}.$$

(c) From $U = I^2 L/2$, we find the maximum current:

$$I = \sqrt{\frac{2U}{L}} = \sqrt{\frac{2\left(1.98 \times 10^{-6}\,\text{J}\right)}{25.0 \times 10^{-3}\,\text{H}}} = 1.26 \times 10^{-2}\,\text{A}.$$

(d) If q_0 is the charge on the capacitor at time $t = 0$, then $q_0 = Q \cos \phi$ and

$$\phi = \cos^{-1}\left(\frac{q}{Q}\right) = \cos^{-1}\left(\frac{3.80 \times 10^{-6}\,\text{C}}{5.56 \times 10^{-6}\,\text{C}}\right) = \pm 46.9°.$$

For $\phi = +46.9°$ the charge on the capacitor is decreasing, for $\phi = -46.9°$ it is increasing. To check this, we calculate the derivative of q with respect to time, evaluated for $t = 0$. We obtain $-\omega Q \sin \phi$, which we wish to be positive. Since $\sin(+46.9°)$ is positive and $\sin(-46.9°)$ is negative, the correct value for increasing charge is $\phi = -46.9°$.

(e) Now we want the derivative to be negative and $\sin \phi$ to be positive. Thus, we take $\phi = +46.9°$.

18. (a) Since the percentage of energy stored in the electric field of the capacitor is $(1 - 75.0\%) = 25.0\%$, then

$$\frac{U_E}{U} = \frac{q^2/2C}{Q^2/2C} = 25.0\%$$

which leads to $q/Q = \sqrt{0.250} = 0.500$.

(b) From

$$\frac{U_B}{U} = \frac{Li^2/2}{LI^2/2} = 75.0\%,$$

we find $i/I = \sqrt{0.750} = 0.866$.

19. (a) The charge (as a function of time) is given by $q = Q \sin \omega t$, where Q is the maximum charge on the capacitor and ω is the angular frequency of oscillation. A sine function was chosen so that $q = 0$ at time $t = 0$. The current (as a function of time) is

$$i = \frac{dq}{dt} = \omega Q \cos \omega t,$$

and at $t = 0$, it is $I = \omega Q$. Since $\omega = 1/\sqrt{LC}$,

$$Q = I\sqrt{LC} = (2.00 \text{ A})\sqrt{(3.00 \times 10^{-3} \text{ H})(2.70 \times 10^{-6} \text{ F})} = 1.80 \times 10^{-4} \text{ C}.$$

(b) The energy stored in the capacitor is given by

$$U_E = \frac{q^2}{2C} = \frac{Q^2 \sin^2 \omega t}{2C}$$

and its rate of change is

$$\frac{dU_E}{dt} = \frac{Q^2 \omega \sin \omega t \cos \omega t}{C}$$

We use the trigonometric identity $\cos \omega t \sin \omega t = \frac{1}{2} \sin(2\omega t)$ to write this as

$$\frac{dU_E}{dt} = \frac{\omega Q^2}{2C} \sin(2\omega t).$$

The greatest rate of change occurs when $\sin(2\omega t) = 1$ or $2\omega t = \pi/2$ rad. This means

$$t = \frac{\pi}{4\omega} = \frac{\pi}{4}\sqrt{LC} = \frac{\pi}{4}\sqrt{(3.00 \times 10^{-3} \text{ H})(2.70 \times 10^{-6} \text{ F})} = 7.07 \times 10^{-5} \text{ s}.$$

(c) Substituting $\omega = 2\pi/T$ and $\sin(2\omega t) = 1$ into $dU_E/dt = (\omega Q^2/2C) \sin(2\omega t)$, we obtain

$$\left(\frac{dU_E}{dt}\right)_{\text{max}} = \frac{2\pi Q^2}{2TC} = \frac{\pi Q^2}{TC}.$$

Now $T = 2\pi\sqrt{LC} = 2\pi\sqrt{(3.00 \times 10^{-3} \text{ H})(2.70 \times 10^{-6} \text{ F})} = 5.655 \times 10^{-4} \text{ s}$, so

$$\left(\frac{dU_E}{dt}\right)_{max} = \frac{\pi\left(1.80\times10^{-4}\,\text{C}\right)^2}{\left(5.655\times10^{-4}\,\text{s}\right)\left(2.70\times10^{-6}\,\text{F}\right)} = 66.7\,\text{W}.$$

We note that this is a positive result, indicating that the energy in the capacitor is indeed increasing at $t = T/8$.

20. (a) We use $U = \frac{1}{2}LI^2 = \frac{1}{2}Q^2/C$ to solve for L:

$$L = \frac{1}{C}\left(\frac{Q}{I}\right)^2 = \frac{1}{C}\left(\frac{CV_{max}}{I}\right)^2 = C\left(\frac{V_{max}}{I}\right)^2 = \left(4.00\times10^{-6}\,\text{F}\right)\left(\frac{1.50\,\text{V}}{50.0\times10^{-3}\,\text{A}}\right)^2 = 3.60\times10^{-3}\,\text{H}.$$

(b) Since $f = \omega/2\pi$, the frequency is

$$f = \frac{1}{2\pi\sqrt{LC}} = \frac{1}{2\pi\sqrt{\left(3.60\times10^{-3}\,\text{H}\right)\left(4.00\times10^{-6}\,\text{F}\right)}} = 1.33\times10^3\,\text{Hz}.$$

(c) Referring to Fig. 31-1, we see that the required time is one-fourth of a period (where the period is the reciprocal of the frequency). Consequently,

$$t = \frac{1}{4}T = \frac{1}{4f} = \frac{1}{4\left(1.33\times10^3\,\text{Hz}\right)} = 1.88\times10^{-4}\,\text{s}.$$

21. (a) We compare this expression for the current with $i = I\sin(\omega t + \phi_0)$. Setting $(\omega t + \phi) = 2500t + 0.680 = \pi/2$, we obtain $t = 3.56\times10^{-4}$ s.

(b) Since $\omega = 2500$ rad/s $= (LC)^{-1/2}$,

$$L = \frac{1}{\omega^2 C} = \frac{1}{\left(2500\,\text{rad}/\text{s}\right)^2\left(64.0\times10^{-6}\,\text{F}\right)} = 2.50\times10^{-3}\,\text{H}.$$

(c) The energy is

$$U = \frac{1}{2}LI^2 = \frac{1}{2}\left(2.50\times10^{-3}\,\text{H}\right)\left(1.60\,\text{A}\right)^2 = 3.20\times10^{-3}\,\text{J}.$$

22. (a) From $V = IX_C$ we find $\omega = I/CV$. The period is then $T = 2\pi/\omega = 2\pi CV/I = 46.1\,\mu\text{s}$.

(b) $\frac{1}{2}CV^2 = 6.88$ nJ.

(c) The answer is again 6.88 nJ (see Fig. 31-4).

(d) We apply Eq. 30-35 as $V = L(di/dt)_{max}$. We can substitute $L = CV^2/I^2$ (combining what we found in part (a) with Eq. 31-4) into Eq. 30-35 (as written above) and solve for $(di/dt)_{max}$. Our result is 1.02×10^3 A/s.

(e) The derivative of $U = \frac{1}{2}Li^2$ leads to $dU/dt = LI^2\omega \sin(\omega t)\cos(\omega t) = \frac{1}{2}LI^2\omega \sin(2\omega t)$. Therefore, $(dU/dt)_{max} = \frac{1}{2}LI^2\omega = \frac{1}{2}IV = 0.938$ mW.

23. The loop rule, for just two devices in the loop, reduces to the statement that the magnitude of the voltage across one of them must equal the magnitude of the voltage across the other. Consider that the capacitor has charge q and a voltage (which we'll consider positive in this discussion) $V = q/C$. Consider at this moment that the current in the inductor at this moment is directed in such a way that the capacitor charge is increasing (so $i = +dq/dt$). Eq. 30-35 then produces a positive result equal to the V across the capacitor: $V = -L(di/dt)$, and we interpret the fact that $-di/dt > 0$ in this discussion to mean that $d(dq/dt)/dt = d^2q/dt^2 < 0$ represents a "deceleration" of the charge-buildup process on the capacitor (since it is approaching its maximum value of charge). In this way we can "check" the signs in Eq. 31-11 (which states $q/C = -L\,d^2q/dt^2$) to make sure we have implemented the loop rule correctly.

24. The charge q after N cycles is obtained by substituting $t = NT = 2\pi N/\omega'$ into Eq. 31-25:

$$q = Qe^{-Rt/2L}\cos(\omega't + \phi) = Qe^{-RNT/2L}\cos\left[\omega'\left(2\pi N/\omega'\right) + \phi\right]$$
$$= Qe^{-RN\left(2\pi\sqrt{L/C}\right)/2L}\cos(2\pi N + \phi)$$
$$= Qe^{-N\pi R\sqrt{C/L}}\cos\phi.$$

We note that the initial charge (setting $N = 0$ in the above expression) is $q_0 = Q\cos\phi$, where $q_0 = 6.2$ μC is given (with 3 significant figures understood). Consequently, we write the above result as $q_N = q_0e^{-N\pi R\sqrt{C/L}}$.

(a) For $N = 5$,

$$q_5 = (6.2\,\mu C)e^{-5\pi(7.2\Omega)\sqrt{0.0000032\ F/12H}} = 5.85\,\mu C.$$

(b) For $N = 10$,

$$q_{10} = (6.2\,\mu C)e^{-10\pi(7.2\Omega)\sqrt{0.0000032\ F/12H}} = 5.52\,\mu C.$$

(c) For $N = 100$,

$$q_{100} = (6.2\,\mu\text{C})e^{-100\pi(7.2\Omega)\sqrt{0.0000032\,\text{F}/12\text{H}}} = 1.93\,\mu\text{C}.$$

25. Since $\omega \approx \omega'$, we may write $T = 2\pi/\omega$ as the period and $\omega = 1/\sqrt{LC}$ as the angular frequency. The time required for 50 cycles (with 3 significant figures understood) is

$$t = 50T = 50\left(\frac{2\pi}{\omega}\right) = 50\left(2\pi\sqrt{LC}\right) = 50\left(2\pi\sqrt{(220\times10^{-3}\,\text{H})(12.0\times10^{-6}\,\text{F})}\right)$$

$$= 0.5104\,\text{s}.$$

The maximum charge on the capacitor decays according to $q_{max} = Qe^{-Rt/2L}$ (this is called the *exponentially decaying amplitude* in §31-5), where Q is the charge at time $t = 0$ (if we take $\phi = 0$ in Eq. 31-25). Dividing by Q and taking the natural logarithm of both sides, we obtain

$$\ln\left(\frac{q_{max}}{Q}\right) = -\frac{Rt}{2L}$$

which leads to

$$R = -\frac{2L}{t}\ln\left(\frac{q_{max}}{Q}\right) = -\frac{2(220\times10^{-3}\,\text{H})}{0.5104\,\text{s}}\ln(0.99) = 8.66\times10^{-3}\ \Omega.$$

26. The assumption stated at the end of the problem is equivalent to setting $\phi = 0$ in Eq. 31-25. Since the maximum energy in the capacitor (each cycle) is given by $q_{max}^2/2C$, where q_{max} is the maximum charge (during a given cycle), then we seek the time for which

$$\frac{q_{max}^2}{2C} = \frac{1}{2}\frac{Q^2}{2C} \Rightarrow q_{max} = \frac{Q}{\sqrt{2}}.$$

Now q_{max} (referred to as the *exponentially decaying amplitude* in §31-5) is related to Q (and the other parameters of the circuit) by

$$q_{max} = Qe^{-Rt/2L} \Rightarrow \ln\left(\frac{q_{max}}{Q}\right) = -\frac{Rt}{2L}.$$

Setting $q_{max} = Q/\sqrt{2}$, we solve for t:

$$t = -\frac{2L}{R}\ln\left(\frac{q_{max}}{Q}\right) = -\frac{2L}{R}\ln\left(\frac{1}{\sqrt{2}}\right) = \frac{L}{R}\ln 2.$$

The identities $\ln(1/\sqrt{2}) = -\ln\sqrt{2} = -\frac{1}{2}\ln 2$ were used to obtain the final form of the result.

27. Let t be a time at which the capacitor is fully charged in some cycle and let $q_{max\,1}$ be the charge on the capacitor then. The energy in the capacitor at that time is

$$U(t) = \frac{q_{max1}^2}{2C} = \frac{Q^2}{2C}e^{-Rt/L}$$

where

$$q_{max1} = Qe^{-Rt/2L}$$

(see the discussion of the *exponentially decaying amplitude* in §31-5). One period later the charge on the fully charged capacitor is

$$q_{max2} = Qe^{-R(t+T)2/L} \qquad \text{where } T = \frac{2\pi}{\omega'},$$

and the energy is

$$U(t+T) = \frac{q_{max2}^2}{2C} = \frac{Q^2}{2C}e^{-R(t+T)/L} \ .$$

The fractional loss in energy is

$$\frac{|\Delta U|}{U} = \frac{U(t) - U(t+T)}{U(t)} = \frac{e^{-Rt/L} - e^{-R(t+T)/L}}{e^{-Rt/L}} = 1 - e^{-RT/L} \ .$$

Assuming that RT/L is very small compared to 1 (which would be the case if the resistance is small), we expand the exponential (see Appendix E). The first few terms are:

$$e^{-RT/L} \approx 1 - \frac{RT}{L} + \frac{R^2T^2}{2L^2} + \cdots .$$

If we approximate $\omega \approx \omega'$, then we can write T as $2\pi/\omega$. As a result, we obtain

$$\frac{|\Delta U|}{U} \approx 1 - \left(1 - \frac{RT}{L} + \cdots\right) \approx \frac{RT}{L} = \frac{2\pi R}{\omega L}.$$

28. (a) We use $I = \varepsilon/X_c = \omega_d C\varepsilon$.

$$I = \omega_d C \varepsilon_m = 2\pi f_d C \varepsilon_m = 2\pi(1.00 \times 10^3 \, \text{Hz})(1.50 \times 10^{-6} \, \text{F})(30.0 \, \text{V}) = 0.283 \, \text{A} \, .$$

(b) $I = 2\pi(8.00 \times 10^3 \, \text{Hz})(1.50 \times 10^{-6} \, \text{F})(30.0 \, \text{V}) = 2.26 \, \text{A}.$

29. (a) The current amplitude I is given by $I = V_L/X_L$, where $X_L = \omega_d L = 2\pi f_d L$. Since the circuit contains only the inductor and a sinusoidal generator, $V_L = \varepsilon_m$. Therefore,

$$I = \frac{V_L}{X_L} = \frac{\varepsilon_m}{2\pi f_d L} = \frac{30.0 \, \text{V}}{2\pi(1.00 \times 10^3 \, \text{Hz})(50.0 \times 10^{-3} \, \text{H})} = 0.0955 \, \text{A} = 95.5 \, \text{mA}.$$

(b) The frequency is now eight times larger than in part (a), so the inductive reactance X_L is eight times larger and the current is one-eighth as much. The current is now

$$I = (0.0955 \, \text{A})/8 = 0.0119 \, \text{A} = 11.9 \, \text{mA}.$$

30. (a) The current through the resistor is

$$I = \frac{\varepsilon_m}{R} = \frac{30.0 \, \text{V}}{50.0 \, \Omega} = 0.600 \, \text{A} \, .$$

(b) Regardless of the frequency of the generator, the current is the same, $I = 0.600 \, \text{A} \, .$

31. (a) The inductive reactance for angular frequency ω_d is given by $X_L = \omega_d L$, and the capacitive reactance is given by $X_C = 1/\omega_d C$. The two reactances are equal if $\omega_d L = 1/\omega_d C$, or $\omega_d = 1/\sqrt{LC}$. The frequency is

$$f_d = \frac{\omega_d}{2\pi} = \frac{1}{2\pi\sqrt{LC}} = \frac{1}{2\pi\sqrt{(6.0 \times 10^{-3} \, \text{H})(10 \times 10^{-6} \, \text{F})}} = 6.5 \times 10^2 \, \text{Hz}.$$

(b) The inductive reactance is

$$X_L = \omega_d L = 2\pi f_d L = 2\pi(650 \, \text{Hz})(6.0 \times 10^{-3} \, \text{H}) = 24 \, \Omega.$$

The capacitive reactance has the same value at this frequency.

(c) The natural frequency for free LC oscillations is $f = \omega/2\pi = 1/2\pi\sqrt{LC}$, the same as we found in part (a).

32. (a) The circuit consists of one generator across one inductor; therefore, $\varepsilon_m = V_L$. The current amplitude is

$$I = \frac{\varepsilon_m}{X_L} = \frac{\varepsilon_m}{\omega_d L} = \frac{25.0 \, \text{V}}{(377 \, \text{rad/s})(12.7 \, \text{H})} = 5.22 \times 10^{-3} \, \text{A} \, .$$

(b) When the current is at a maximum, its derivative is zero. Thus, Eq. 30-35 gives $\varepsilon_L = 0$ at that instant. Stated another way, since $\varepsilon(t)$ and $i(t)$ have a 90° phase difference, then $\varepsilon(t)$ must be zero when $i(t) = I$. The fact that $\phi = 90° = \pi/2$ rad is used in part (c).

(c) Consider Eq. 32-28 with $\varepsilon = -\frac{1}{2}\varepsilon_m$. In order to satisfy this equation, we require $\sin(\omega_d t) = -1/2$. Now we note that the problem states that ε is increasing *in magnitude*, which (since it is already negative) means that it is becoming more negative. Thus, differentiating Eq. 32-28 with respect to time (and demanding the result be negative) we must also require $\cos(\omega_d t) < 0$. These conditions imply that ωt must equal $(2n\pi - 5\pi/6)$ [*n* = integer]. Consequently, Eq. 31-29 yields (for all values of *n*)

$$i = I\sin\left(2n\pi - \frac{5\pi}{6} - \frac{\pi}{2}\right) = (5.22 \times 10^{-3}\,\text{A})\left(\frac{\sqrt{3}}{2}\right) = 4.51 \times 10^{-3}\,\text{A} \; .$$

33. (a) The generator emf is a maximum when $\sin(\omega_d t - \pi/4) = 1$ or $\omega_d t - \pi/4 = (\pi/2) \pm 2n\pi$ [*n* = integer]. The first time this occurs after $t = 0$ is when $\omega_d t - \pi/4 = \pi/2$ (that is, *n* = 0). Therefore,

$$t = \frac{3\pi}{4\omega_d} = \frac{3\pi}{4(350\,\text{rad}/\text{s})} = 6.73 \times 10^{-3}\,\text{s} \; .$$

(b) The current is a maximum when $\sin(\omega_d t - 3\pi/4) = 1$, or $\omega_d t - 3\pi/4 = (\pi/2) \pm 2n\pi$ [*n* = integer]. The first time this occurs after $t = 0$ is when $\omega_d t - 3\pi/4 = \pi/2$ (as in part (a), *n* = 0). Therefore,

$$t = \frac{5\pi}{4\omega_d} = \frac{5\pi}{4(350\,\text{rad}/\text{s})} = 1.12 \times 10^{-2}\,\text{s} \; .$$

(c) The current lags the emf by $+\pi/2$ rad, so the circuit element must be an inductor.

(d) The current amplitude I is related to the voltage amplitude V_L by $V_L = IX_L$, where X_L is the inductive reactance, given by $X_L = \omega_d L$. Furthermore, since there is only one element in the circuit, the amplitude of the potential difference across the element must be the same as the amplitude of the generator emf: $V_L = \varepsilon_m$. Thus, $\varepsilon_m = I\omega_d L$ and

$$L = \frac{\varepsilon_m}{I\omega_d} = \frac{30.0\,\text{V}}{(620 \times 10^{-3}\,\text{A})(350\,\text{rad}/\text{s})} = 0.138\,\text{H}.$$

34. (a) The circuit consists of one generator across one capacitor; therefore, $\varepsilon_m = V_C$. Consequently, the current amplitude is

$$I = \frac{\varepsilon_m}{X_C} = \omega C \varepsilon_m = (377 \text{ rad/s})(4.15 \times 10^{-6} \text{F})(25.0 \text{ V}) = 3.91 \times 10^{-2} \text{A} .$$

(b) When the current is at a maximum, the charge on the capacitor is changing at its largest rate. This happens not when it is fully charged ($\pm q_{max}$), but rather as it passes through the (momentary) states of being uncharged ($q = 0$). Since $q = CV$, then the voltage across the capacitor (and at the generator, by the loop rule) is zero when the current is at a maximum. Stated more precisely, the time-dependent emf $\varepsilon(t)$ and current $i(t)$ have a $\phi = -90°$ phase relation, implying $\varepsilon(t) = 0$ when $i(t) = I$. The fact that $\phi = -90°$ $= -\pi/2$ rad is used in part (c).

(c) Consider Eq. 32-28 with $\varepsilon = -\frac{1}{2}\varepsilon_m$. In order to satisfy this equation, we require $\sin(\omega_d t) = -1/2$. Now we note that the problem states that ε is increasing in magnitude, which (since it is already negative) means that it is becoming more negative. Thus, differentiating Eq. 32-28 with respect to time (and demanding the result be negative) we must also require $\cos(\omega_d t) < 0$. These conditions imply that ωt must equal $(2n\pi - 5\pi/6)$ [n = integer]. Consequently, Eq. 31-29 yields (for all values of n)

$$i = I\sin\left(2n\pi - \frac{5\pi}{6} + \frac{\pi}{2}\right) = (3.91 \times 10^{-3} \text{ A})\left(-\frac{\sqrt{3}}{2}\right) = -3.38 \times 10^{-2} \text{ A},$$

or $|i| = 3.38 \times 10^{-2}$ A.

35. (a) Now $X_C = 0$, while $R = 200 \ \Omega$ and $X_L = \omega L = 2\pi f_d L = 86.7 \ \Omega$ remain unchanged. Therefore, the impedance is

$$Z = \sqrt{R^2 + X_L^2} = \sqrt{(200 \ \Omega)^2 + (86.7 \ \Omega)^2} = 218 \ \Omega .$$

(b) The phase angle is, from Eq. 31-65,

$$\phi = \tan^{-1}\left(\frac{X_L - X_C}{R}\right) = \tan^{-1}\left(\frac{86.7\Omega - 0}{200 \ \Omega}\right) = 23.4° .$$

(c) The current amplitude is now found to be

$$I = \frac{\varepsilon_m}{Z} = \frac{36.0 \text{ V}}{218 \ \Omega} = 0.165 \text{ A} .$$

(d) We first find the voltage amplitudes across the circuit elements:

$$V_R = IR = (0.165 \text{ A})(200\Omega) \approx 33 \text{ V}$$
$$V_L = IX_L = (0.165\text{A})(86.7\Omega) \approx 14.3 \text{ V}$$

This is an inductive circuit, so ε_m leads I. The phasor diagram is drawn to scale below.

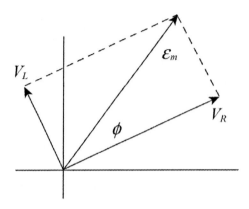

36. (a) The graph shows that the resonance angular frequency is 25000 rad/s, which means (using Eq. 31-4)

$$C = (\omega^2 L)^{-1} = [(25000)^2 \times 200 \times 10^{-6}]^{-1} = 8.0 \text{ }\mu\text{F.}$$

(b) The graph also shows that the current amplitude at resonance is 4.0 A, but at resonance the impedance Z becomes purely resistive ($Z = R$) so that we can divide the emf amplitude by the current amplitude at resonance to find R: 8.0/4.0 = 2.0 Ω.

37. (a) Now $X_L = 0$, while $R = 200$ Ω and $X_C = 1/2\pi f_d C = 177$ Ω. Therefore, the impedance is

$$Z = \sqrt{R^2 + X_C^2} = \sqrt{(200\Omega)^2 + (177\Omega)^2} = 267\Omega.$$

(b) The phase angle is

$$\phi = \tan^{-1}\left(\frac{X_L - X_C}{R}\right) = \tan^{-1}\left(\frac{0 - 177\Omega}{200\Omega}\right) = -41.5°$$

(c) The current amplitude is

$$I = \frac{\varepsilon_m}{Z} = \frac{36.0 \text{ V}}{267\Omega} = 0.135 \text{ A .}$$

(d) We first find the voltage amplitudes across the circuit elements:

$$V_R = IR = (0.135\,\text{A})(200\,\Omega) \approx 27.0\,\text{V}$$
$$V_C = IX_C = (0.135\,\text{A})(177\,\Omega) \approx 23.9\,\text{V}$$

The circuit is capacitive, so I leads ε_m. The phasor diagram is drawn to scale next.

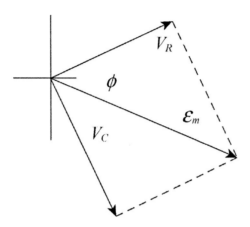

38. (a) The circuit has a resistor and a capacitor (but no inductor). Since the capacitive reactance decreases with frequency, then the asymptotic value of Z must be the resistance: $R = 500\ \Omega$.

(b) We describe three methods here (each using information from different points on the graph):

<u>method 1</u>: At $\omega_d = 50$ rad/s, we have $Z \approx 700\ \Omega$ which gives $C = (\omega_d\sqrt{Z^2 - R^2}\,)^{-1} = 41\ \mu\text{F}$.

<u>method 2</u>: At $\omega_d = 50$ rad/s, we have $X_C \approx 500\ \Omega$ which gives $C = (\omega_d X_C)^{-1} = 40\ \mu\text{F}$.

<u>method 3</u>: At $\omega_d = 250$ rad/s, we have $X_C \approx 100\ \Omega$ which gives $C = (\omega_d X_C)^{-1} = 40\ \mu\text{F}$.

39. (a) The capacitive reactance is

$$X_C = \frac{1}{\omega_d C} = \frac{1}{2\pi f_d C} = \frac{1}{2\pi(60.0\,\text{Hz})(70.0\times10^{-6}\,\text{F})} = 37.9\,\Omega\,.$$

The inductive reactance 86.7 Ω is unchanged. The new impedance is

$$Z = \sqrt{R^2 + (X_L - X_C)^2} = \sqrt{(200\,\Omega)^2 + (37.9\,\Omega - 86.7\,\Omega)^2} = 206\,\Omega.$$

(b) The phase angle is

$$\phi = \tan^{-1}\left(\frac{X_L - X_C}{R}\right) = \tan^{-1}\left(\frac{86.7\,\Omega - 37.9\,\Omega}{200\,\Omega}\right) = 13.7°\,.$$

(c) The current amplitude is

$$I = \frac{\varepsilon_m}{Z} = \frac{36.0 \text{ V}}{206 \,\Omega} = 0.175 \,\text{A}.$$

(d) We first find the voltage amplitudes across the circuit elements:

$$V_R = IR = (0.175 \text{ A})(200 \,\Omega) = 35.0 \text{ V}$$
$$V_L = IX_L = (0.175 \text{ A})(86.7 \,\Omega) = 15.2 \text{ V}$$
$$V_C = IX_C = (0.175 \text{ A})(37.9 \,\Omega) = 6.62 \text{ V}$$

Note that $X_L > X_C$, so that ε_m leads I. The phasor diagram is drawn to scale below.

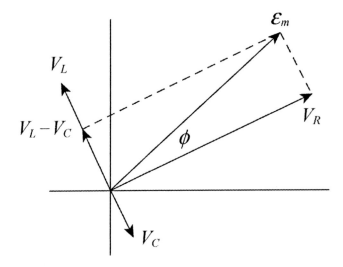

40. (a) Since $Z = \sqrt{R^2 + X_L^2}$ and $X_L = \omega_d L$, then as $\omega_d \to 0$ we find $Z \to R = 40 \,\Omega$.

(b) $L = X_L / \omega_d = slope = 60$ mH.

41. The resistance of the coil is related to the reactances and the phase constant by Eq. 31-65. Thus,

$$\frac{X_L - X_C}{R} = \frac{\omega_d L - 1/\omega_d C}{R} = \tan\phi \,,$$

which we solve for R:

$$R = \frac{1}{\tan\phi}\left(\omega_d L - \frac{1}{\omega_d C}\right) = \frac{1}{\tan 75°}\left[(2\pi)(930 \,\text{Hz}(8.8\times10^{-2}\,\text{H}) - \frac{1}{(2\pi)(930 \,\text{Hz})(0.94\times10^{-6}\,\text{F}}\right]$$
$$= 89\,\Omega.$$

42. A phasor diagram very much like Fig. 31-11(c) leads to the condition:

$$V_L - V_C = (6.00 \text{ V})\sin(30°) = 3.00 \text{ V}.$$

With the magnitude of the capacitor voltage at 5.00 V, this gives a inductor voltage magnitude equal to 8.00 V.

43. (a) Yes, the voltage amplitude across the inductor can be much larger than the amplitude of the generator emf.

(b) The amplitude of the voltage across the inductor in an RLC series circuit is given by $V_L = IX_L = I\omega_d L$. At resonance, the driving angular frequency equals the natural angular frequency: $\omega_d = \omega = 1/\sqrt{LC}$. For the given circuit

$$X_L = \frac{L}{\sqrt{LC}} = \frac{1.0 \text{ H}}{\sqrt{(1.0 \text{ H})(1.0 \times 10^{-6} \text{F})}} = 1000 \ \Omega.$$

At resonance the capacitive reactance has this same value, and the impedance reduces simply: $Z = R$. Consequently,

$$I = \left.\frac{\varepsilon_m}{Z}\right|_{\text{resonance}} = \frac{\varepsilon_m}{R} = \frac{10 \text{ V}}{10 \ \Omega} = 1.0 \text{ A}.$$

The voltage amplitude across the inductor is therefore

$$V_L = IX_L = (1.0 \text{ A})(1000 \ \Omega) = 1.0 \times 10^3 \text{ V}$$

which is much larger than the amplitude of the generator emf.

44. (a) With both switches closed (which effectively removes the resistor from the circuit), the impedance is just equal to the (net) reactance and is equal to (12 V)/(0.447 A) = 26.85 Ω. With switch 1 closed but switch 2 open, we have the same (net) reactance as just discussed, but now the resistor is part of the circuit; using Eq. 31-65 we find

$$R = X_{\text{net}}/\tan\phi = 26.85/\tan(15°) = 100 \ \Omega.$$

(b) For the first situation described in the problem (both switches open) we can reverse our reasoning of part (a) and find $X_{\text{net first}} = R \tan(-30.9°) = -59.96 \ \Omega$. We observe that the effect of switch 1 implies

$$X_C = X_{\text{net}} - X_{\text{net firsrt}} = 26.85 - (-59.96) = 86.81 \ \Omega.$$

Then Eq. 31-39 leads to $C = 1/\omega X_C = 30.6 \ \mu\text{F}$.

(c) Since $X_{net} = X_L - X_C$, then we find $L = X_L/\omega = 301$ mH

45. (a) For a given amplitude ε_m of the generator emf, the current amplitude is given by

$$I = \frac{\varepsilon_m}{Z} = \frac{\varepsilon_m}{\sqrt{R^2 + (\omega_d L - 1/\omega_d C)^2}}.$$

We find the maximum by setting the derivative with respect to ω_d equal to zero:

$$\frac{dI}{d\omega_d} = -(E)_m [R^2 + (\omega_d L - 1/\omega_d C)^2]^{-3/2} \left[\omega_d L - \frac{1}{\omega_d C}\right]\left[L + \frac{1}{\omega_d^2 C}\right].$$

The only factor that can equal zero is $\omega_d L - (1/\omega_d C)$; it does so for $\omega_d = 1/\sqrt{LC} = \omega$. For this

$$\omega_d = \frac{1}{\sqrt{LC}} = \frac{1}{\sqrt{(1.00 \text{ H})(20.0 \times 10^{-6} \text{ F})}} = 224 \text{ rad/s}.$$

(b) When $\omega_d = \omega$, the impedance is $Z = R$, and the current amplitude is

$$I = \frac{\varepsilon_m}{R} = \frac{30.0 \text{ V}}{5.00 \text{ }\Omega} = 6.00 \text{ A}.$$

(c) We want to find the (positive) values of ω_d for which $I = \varepsilon_m/2R$:

$$\frac{\varepsilon_m}{\sqrt{R^2 + (\omega_d L - 1/\omega_d C)^2}} = \frac{\varepsilon_m}{2R}.$$

This may be rearranged to yield

$$\left(\omega_d L - \frac{1}{\omega_d C}\right)^2 = 3R^2.$$

Taking the square root of both sides (acknowledging the two ± roots) and multiplying by $\omega_d C$, we obtain

$$\omega_d^2 (LC) \pm \omega_d \left(\sqrt{3}CR\right) - 1 = 0.$$

Using the quadratic formula, we find the smallest positive solution

$$\omega_2 = \frac{-\sqrt{3}CR + \sqrt{3C^2R^2 + 4LC}}{2LC} = \frac{-\sqrt{3}(20.0\times10^{-6}\ \text{F})(5.00\ \Omega)}{2(1.00\ \text{H})(20.0\times10^{-6}\text{F})}$$

$$+\frac{\sqrt{3(20.0\times10^{-6}\ \text{F})^2(5.00\ \Omega)^2 + 4(1.00\ \text{H})(20.0\times10^{-6}\text{F})}}{2(1.00\ \text{H})(20.0\times10^{-6}\text{F})}$$

$$= 219\ \text{rad/s} ,$$

(d) and the largest positive solution

$$\omega_1 = \frac{+\sqrt{3}CR + \sqrt{3C^2R^2 + 4LC}}{2LC} = \frac{+\sqrt{3}(20.0\times10^{-6}\ \text{F})(5.00\Omega)}{2(1.00\ \text{H})(20.0\times10^{-6}\text{F})}$$

$$+\frac{\sqrt{3(20.0\times10^{-6}\ \text{F})^2(5.00\ \Omega)^2 + 4(1.00\ \text{H})(20.0\times10^{-6}\text{F})}}{2(1.00\ \text{H})(20.0\times10^{-6}\text{F})}$$

$$= 228\ \text{rad/s} .$$

(e) The fractional width is

$$\frac{\omega_1 - \omega_2}{\omega_0} = \frac{228\,\text{rad/s} - 219\,\text{rad/s}}{224\ \text{rad/s}} = 0.040 .$$

46. (a) The capacitive reactance is

$$X_C = \frac{1}{2\pi fC} = \frac{1}{2\pi(400\ \text{Hz})(24.0\times10^{-6}\text{F})} = 16.6\ \Omega .$$

(b) The impedance is

$$Z = \sqrt{R^2 + (X_L - X_C)^2} = \sqrt{R^2 + (2\pi fL - X_C)^2}$$

$$= \sqrt{(220\,\Omega)^2 + [2\pi(400\ \text{Hz})(150\times10^{-3}\ \text{H}) - 16.6\ \Omega]^2} = 422\ \Omega .$$

(c) The current amplitude is

$$I = \frac{\mathcal{E}_m}{Z} = \frac{220\,\text{V}}{422\,\Omega} = 0.521\ \text{A} .$$

(d) Now $X_C \propto C_{eq}^{-1}$. Thus, X_C increases as C_{eq} decreases.

(e) Now $C_{eq} = C/2$, and the new impedance is

$$Z = \sqrt{(220\ \Omega)^2 + [2\pi(400\ \text{Hz})(150 \times 10^{-3}\ \text{H}) - 2(16.6\ \Omega)]^2} = 408\ \Omega < 422\ \Omega.$$

Therefore, the impedance decreases.

(f) Since $I \propto Z^{-1}$, it increases.

47. We use the expressions found in Problem 45:

$$\omega_1 = \frac{+\sqrt{3}CR + \sqrt{3C^2R^2 + 4LC}}{2LC}$$

$$\omega_2 = \frac{-\sqrt{3}CR + \sqrt{3C^2R^2 + 4LC}}{2LC}$$

We also use Eq. 31-4. Thus,

$$\frac{\Delta\omega_d}{\omega} = \frac{\omega_1 - \omega_2}{\omega} = \frac{2\sqrt{3}CR\sqrt{LC}}{2LC} = R\sqrt{\frac{3C}{L}}.$$

For the data of Problem 45,

$$\frac{\Delta\omega_d}{\omega} = (5.00\ \Omega)\sqrt{\frac{3(20.0 \times 10^{-6}\ \text{F})}{1.00\ \text{H}}} = 3.87 \times 10^{-2}.$$

This is in agreement with the result of Problem 45. The method of Problem 45, however, gives only one significant figure since two numbers close in value are subtracted ($\omega_1 - \omega_2$). Here the subtraction is done algebraically, and three significant figures are obtained.

48. (a) A sketch of the phasors would be very much like Fig. 31-10(c) but with the label "I_L" on the green arrow replaced with "V_R."

(b) We have $V_R = V_L$, which implies

$$I R = I X_L \quad \rightarrow \quad R = \omega_d L$$

which yields $f = \omega_d/2\pi = R/2\pi L = 318$ Hz.

(c) $\phi = \tan^{-1}(V_L/V_R) = +45°$.

(d) $\omega_d = R/L = 2.00 \times 10^3$ rad/s.

(e) $I = (6\ \text{V})/\sqrt{R^2 + X_L^2} = 3/(40\sqrt{2}) \approx 53.0$ mA.

49. (a) Since $L_{eq} = L_1 + L_2$ and $C_{eq} = C_1 + C_2 + C_3$ for the circuit, the resonant frequency is

$$\omega = \frac{1}{2\pi\sqrt{L_{eq}C_{eq}}} = \frac{1}{2\pi\sqrt{(L_1 + L_2)(C_1 + C_2 + C_3)}}$$

$$= \frac{1}{2\pi\sqrt{(1.70 \times 10^{-3}\,\text{H} + 2.30 \times 10^{-3}\,\text{H})(4.00 \times 10^{-6}\,\text{F} + 2.50 \times 10^{-6}\,\text{F} + 3.50 \times 10^{-6}\,\text{F})}}$$

$$= 796\,\text{Hz}.$$

(b) The resonant frequency does not depend on R so it will not change as R increases.

(c) Since $\omega \propto (L_1 + L_2)^{-1/2}$, it will decrease as L_1 increases.

(d) Since $\omega \propto C_{eq}^{-1/2}$ and C_{eq} decreases as C_3 is removed, ω will increase.

50. Since the impedance of the voltmeter is large, it will not affect the impedance of the circuit when connected in parallel with the circuit. So the reading will be 100 V in all three cases.

51. The average power dissipated in resistance R when the current is alternating is given by $P_{avg} = I_{rms}^2 R$, where I_{rms} is the root-mean-square current. Since $I_{rms} = I/\sqrt{2}$, where I is the current amplitude, this can be written $P_{avg} = I^2 R/2$. The power dissipated in the same resistor when the current i_d is direct is given by $P = i_d^2 R$. Setting the two powers equal to each other and solving, we obtain

$$i_d = \frac{I}{\sqrt{2}} = \frac{2.60\,\text{A}}{\sqrt{2}} = 1.84\,\text{A}.$$

52. The amplitude (peak) value is

$$V_{max} = \sqrt{2}V_{rms} = \sqrt{2}(100\,\text{V}) = 141\,\text{V}.$$

53. (a) Using Eq. 31-61, the impedance is

$$Z = \sqrt{(12.0\,\Omega)^2 + (1.30\,\Omega - 0)^2} = 12.1\,\Omega.$$

(b) We use the result of problem 54:

$$P_{avg} = \frac{\varepsilon_{rms}^2 R}{Z^2} = \frac{(120\,\text{V})^2(12.0\,\Omega)}{(12.07\,\Omega)^2} = 1.186 \times 10^3\,\text{W} \approx 1.19 \times 10^3\,\text{W}.$$

54. This circuit contains no reactances, so $\varepsilon_{rms} = I_{rms}R_{total}$. Using Eq. 31-71, we find the average dissipated power in resistor R is

$$P_R = I_{rms}^2 R = \left(\frac{\varepsilon_m}{r + R}\right)^2 R.$$

In order to maximize P_R we set the derivative equal to zero:

$$\frac{dP_R}{dR} = \frac{\varepsilon_m^2\left[(r + R)^2 - 2(r + R)R\right]}{(r + R)^4} = \frac{\varepsilon_m^2(r - R)}{(r + R)^3} = 0 \Rightarrow R = r$$

55. (a) The power factor is $\cos \phi$, where ϕ is the phase constant defined by the expression $i = I \sin(\omega t - \phi)$. Thus, $\phi = -42.0°$ and $\cos \phi = \cos(-42.0°) = 0.743$.

(b) Since $\phi < 0$, $\omega t - \phi > \omega t$. The current leads the emf.

(c) The phase constant is related to the reactance difference by $\tan \phi = (X_L - X_C)/R$. We have $\tan \phi = \tan(-42.0°) = -0.900$, a negative number. Therefore, $X_L - X_C$ is negative, which leads to $X_C > X_L$. The circuit in the box is predominantly capacitive.

(d) If the circuit were in resonance X_L would be the same as X_C, $\tan \phi$ would be zero, and ϕ would be zero. Since ϕ is not zero, we conclude the circuit is not in resonance.

(e) Since $\tan \phi$ is negative and finite, neither the capacitive reactance nor the resistance are zero. This means the box must contain a capacitor and a resistor.

(f) The inductive reactance may be zero, so there need not be an inductor.

(g) Yes, there is a resistor.

(h) The average power is

$$P_{avg} = \frac{1}{2}\varepsilon_m I \cos \phi = \frac{1}{2}(75.0\,\text{V})(1.20\,\text{A})(0.743) = 33.4\,\text{W}.$$

(i) The answers above depend on the frequency only through the phase constant ϕ, which is given. If values were given for R, L and C then the value of the frequency would also be needed to compute the power factor.

56. (a) The power consumed by the light bulb is $P = I^2R/2$. So we must let $P_{max}/P_{min} = (I/I_{min})^2 = 5$, or

$$\left(\frac{I}{I_{min}}\right)^2 = \left(\frac{\varepsilon_m / Z_{min}}{\varepsilon_m / Z_{max}}\right)^2 = \left(\frac{Z_{max}}{Z_{min}}\right)^2 = \left(\frac{\sqrt{R^2 + (\omega L_{max})^2}}{R}\right)^2 = 5.$$

We solve for L_{max}:

$$L_{max} = \frac{2R}{\omega} = \frac{2(120\,\text{V})^2 / 1000\,\text{W}}{2\pi(60.0\,\text{Hz})} = 7.64 \times 10^{-2}\,\text{H}.$$

(b) Yes, one could use a variable resistor.

(c) Now we must let

$$\left(\frac{R_{max} + R_{bulb}}{R_{bulb}}\right)^2 = 5,$$

or

$$R_{max} = \left(\sqrt{5} - 1\right)R_{bulb} = \left(\sqrt{5} - 1\right)\frac{(120\,\text{V})^2}{1000\,\text{W}} = 17.8\,\Omega.$$

(d) This is not done because the resistors would consume, rather than temporarily store, electromagnetic energy.

57. We shall use

$$P_{avg} = \frac{\varepsilon_m^2 R}{2Z^2} = \frac{\varepsilon_m^2 R}{2\left[R^2 + (\omega_d L - 1/\omega_d C)^2\right]}.$$

where $Z = \sqrt{R^2 + (\omega_d L - 1/\omega_d C)^2}$ is the impedance.

(a) Considered as a function of C, P_{avg} has its largest value when the factor $R^2 + (\omega_d L - 1/\omega_d C)^2$ has the smallest possible value. This occurs for $\omega_d L = 1/\omega_d C$, or

$$C = \frac{1}{\omega_d^2 L} = \frac{1}{(2\pi)^2 (60.0\,\text{Hz})^2 (60.0 \times 10^{-3}\,\text{H})} = 1.17 \times 10^{-4}\,\text{F}.$$

The circuit is then at resonance.

(b) In this case, we want Z^2 to be as large as possible. The impedance becomes large without bound as C becomes very small. Thus, the smallest average power occurs for $C = 0$ (which is not very different from a simple open switch).

(c) When $\omega_d L = 1/\omega_d C$, the expression for the average power becomes

$$P_{avg} = \frac{\varepsilon_m^2}{2R},$$

so the maximum average power is in the resonant case and is equal to

$$P_{avg} = \frac{(30.0\,\text{V})^2}{2(5.00\,\Omega)} = 90.0\,\text{W}.$$

(d) At maximum power, the reactances are equal: $X_L = X_C$. The phase angle ϕ in this case may be found from

$$\tan\phi = \frac{X_L - X_C}{R} = 0,$$

which implies $\phi = 0°$.

(e) At maximum power, the power factor is $\cos\phi = \cos 0° = 1$,

(f) The minimum average power is $P_{avg} = 0$ (as it would be for an open switch).

(g) On the other hand, at minimum power $X_C \propto 1/C$ is infinite, which leads us to set $\tan\phi = -\infty$. In this case, we conclude that $\phi = -90°$.

(h) At minimum power, the power factor is $\cos\phi = \cos(-90°) = 0$.

58. The current in the circuit satisfies $i(t) = I\sin(\omega_d t - \phi)$, where

$$I = \frac{\varepsilon_m}{Z} = \frac{\varepsilon_m}{\sqrt{R^2 + (\omega_d L - 1/\omega_d C)^2}}$$

$$= \frac{45.0\,\text{V}}{\sqrt{(16.0\,\Omega)^2 + \{(3000\,\text{rad/s})(9.20\,\text{mH}) - 1/[(3000\,\text{rad/s})(31.2\,\mu\text{F})]\}^2}}$$

$$= 1.93\,\text{A}$$

and

$$\phi = \tan^{-1}\left(\frac{X_L - X_C}{R}\right) = \tan^{-1}\left(\frac{\omega_d L - 1/\omega_d C}{R}\right)$$

$$= \tan^{-1}\left[\frac{(3000\,\text{rad/s})(9.20\,\text{mH})}{16.0\,\Omega} - \frac{1}{(3000\,\text{rad/s})(16.0\,\Omega)(31.2\,\mu\text{F})}\right]$$

$$= 46.5°.$$

(a) The power supplied by the generator is

$$P_g = i(t)\varepsilon(t) = I\sin(\omega_d t - \phi)\varepsilon_m \sin\omega_d t$$
$$= (1.93\,\text{A})(45.0\,\text{V})\sin\left[(3000\,\text{rad/s})(0.442\text{ ms})\right]\sin\left[(3000\,\text{rad/s})(0.442\text{ ms}) - 46.5°\right]$$
$$= 41.4\text{ W}.$$

(b) The rate at which the energy in the capacitor changes is

$$P_c = -\frac{d}{dt}\left(\frac{q^2}{2C}\right) = -i\frac{q}{C} = -iV_c$$

$$= -I\sin(\omega_d t - \phi)\left(\frac{I}{\omega_d C}\right)\cos(\omega_d t - \phi) = -\frac{I^2}{2\omega_d C}\sin\left[2(\omega_d t - \phi)\right]$$

$$= -\frac{(1.93\,\text{A})^2}{2(3000\,\text{rad/s})(31.2\times10^{-6}\text{F})}\sin\left[2(3000\,\text{rad/s})(0.442\text{ ms}) - 2(46.5°)\right]$$

$$= -17.0\text{ W}.$$

(c) The rate at which the energy in the inductor changes is

$$P_L = \frac{d}{dt}\left(\frac{1}{2}Li^2\right) = Li\frac{di}{dt} = LI\sin(\omega_d t - \phi)\frac{d}{dt}\left[I\sin(\omega_d t - \phi)\right] = \frac{1}{2}\omega_d LI^2\sin\left[2(\omega_d t - \phi)\right]$$

$$= \frac{1}{2}(3000\,\text{rad/s})(1.93\,\text{A})^2(9.20\,\text{mH})\sin\left[2(3000\,\text{rad/s})(0.442\text{ ms}) - 2(46.5°)\right]$$

$$= 44.1\text{ W}.$$

(d) The rate at which energy is being dissipated by the resistor is

$$P_R = i^2 R = I^2 R\sin^2(\omega_d t - \phi) = (1.93\,\text{A})^2(16.0\,\Omega)\sin^2\left[(3000\,\text{rad/s})(0.442\text{ ms}) - 46.5°\right]$$
$$= 14.4\text{ W}.$$

(e) Equal. $P_L + P_R + P_c = 44.1\text{ W} - 17.0\text{ W} + 14.4\text{ W} = 41.5\text{ W} = P_g$.

59. (a) The rms current is

$$I_{rms} = \frac{\varepsilon_{rms}}{Z} = \frac{\varepsilon_{rms}}{\sqrt{R^2 + \left(2\pi fL - 1/2\pi fC\right)^2}}$$

$$= \frac{75.0\,\text{V}}{\sqrt{(15.0\,\Omega)^2 + \left\{2\pi(550\text{Hz})(25.0\text{mH}) - 1/\left[2\pi(550\text{Hz})(4.70\,\mu\text{F})\right]\right\}^2}}$$

$$= 2.59\,\text{A}.$$

(b) The rms voltage across R is

$$V_{ab} = I_{rms}R = (2.59\,\text{A})(15.0\,\Omega) = 38.8\,\text{V}.$$

(c) The rms voltage across C is

$$V_{bc} = I_{rms}X_C = \frac{I_{rms}}{2\pi fC} = \frac{2.59\text{A}}{2\pi(550\,\text{Hz})(4.70\,\mu\text{F})} = 159\,\text{V}.$$

(d) The rms voltage across L is

$$V_{cd} = I_{rms}X_L = 2\pi I_{rms}fL = 2\pi(2.59\,\text{A})(550\,\text{Hz})(25.0\,\text{mH}) = 224\,\text{V}.$$

(e) The rms voltage across C and L together is

$$V_{bd} = |V_{bc} - V_{cd}| = |159.5\,\text{V} - 223.7\,\text{V}| = 64.2\text{V}$$

(f) The rms voltage across R, C and L together is

$$V_{ad} = \sqrt{V_{ab}^2 + V_{bd}^2} = \sqrt{(38.8\,\text{V})^2 + (64.2\,\text{V})^2} = 75.0\,\text{V}$$

(g) For R,

$$P_R = \frac{V_{ab}^2}{R} = \frac{(38.8\,\text{V})^2}{15.0\,\Omega} = 100\,\text{W}.$$

(h) No energy dissipation in C.

(i) No energy dissipation in L.

60. We use Eq. 31-79 to find

$$V_s = V_p \left(\frac{N_s}{N_p} \right) = (100 \text{ V}) \left(\frac{500}{50} \right) = 1.00 \times 10^3 \text{ V}.$$

61. (a) The stepped-down voltage is

$$V_s = V_p \left(\frac{N_s}{N_p} \right) = (120 \text{ V}) \left(\frac{10}{500} \right) = 2.4 \text{ V}.$$

(b) By Ohm's law, the current in the secondary is

$$I_s = \frac{V_s}{R_s} = \frac{2.4 \text{ V}}{15 \Omega} = 0.16 \text{ A}.$$

We find the primary current from Eq. 31-80:

$$I_p = I_s \left(\frac{N_s}{N_p} \right) = (0.16 \text{ A}) \left(\frac{10}{500} \right) = 3.2 \times 10^{-3} \text{ A}.$$

(c) As shown above, the current in the secondary is $I_s = 0.16 \text{ A}$.

62. For step-up trasnformer:

(a) The smallest value of the ratio V_s / V_p is achieved by using $T_2 T_3$ as primary and $T_1 T_3$ as secondary coil: $V_{13}/V_{23} = (800 + 200)/800 = 1.25$.

(b) The second smallest value of the ratio V_s / V_p is achieved by using $T_1 T_2$ as primary and $T_2 T_3$ as secondary coil: $V_{23}/V_{13} = 800/200 = 4.00$.

(c) The largest value of the ratio V_s / V_p is achieved by using $T_1 T_2$ as primary and $T_1 T_3$ as secondary coil: $V_{13}/V_{12} = (800 + 200)/200 = 5.00$.

For the step-down transformer, we simply exchange the primary and secondary coils in each of the three cases above.

(d) The smallest value of the ratio V_s / V_p is $1/5.00 = 0.200$.

(e) The second smallest value of the ratio V_s / V_p is $1/4.00 = 0.250$.

(f) The largest value of the ratio V_s / V_p is $1/1.25 = 0.800$.

63. (a) The rms current in the cable is $I_{rms} = P/V_t = 250 \times 10^3 \, W/(80 \times 10^3 \, V) = 3.125 \, A$. The rms voltage drop is then $\Delta V = I_{rms} R = (3.125 \, A)(2)(0.30 \, \Omega) = 1.9 \, V$.

(b) The rate of energy dissipation is $P_d = I_{rms}^2 R = (3.125 \, A)^2 (0.60 \, \Omega) = 5.9 \, W$.

(c) Now $I_{rms} = 250 \times 10^3 \, W/(8.0 \times 10^3 \, V) = 31.25 \, A$, so $\Delta V = (31.25 \, A)(0.60 \, \Omega) = 19 \, V$.

(d) $P_d = (3.125 \, A)^2 (0.60 \, \Omega) = 5.9 \times 10^2 \, W$.

(e) $I_{rms} = 250 \times 10^3 \, W/(0.80 \times 10^3 \, V) = 312.5 \, A$, so $\Delta V = (312.5 \, A)(0.60 \, \Omega) = 1.9 \times 10^2 \, V$.

(f) $P_d = (312.5 \, A)^2 (0.60 \, \Omega) = 5.9 \times 10^4 \, W$.

64. (a) The amplifier is connected across the primary windings of a transformer and the resistor R is connected across the secondary windings.

(b) If I_s is the rms current in the secondary coil then the average power delivered to R is $P_{avg} = I_s^2 R$. Using $I_s = (N_p/N_s)I_p$, we obtain

$$P_{avg} = \left(\frac{I_p N_p}{N_s} \right)^2 R.$$

Next, we find the current in the primary circuit. This is effectively a circuit consisting of a generator and two resistors in series. One resistance is that of the amplifier (r), and the other is the equivalent resistance R_{eq} of the secondary circuit. Therefore,

$$I_p = \frac{\mathcal{E}_{rms}}{r + R_{eq}} = \frac{\mathcal{E}_{rms}}{r + (N_p/N_s)^2 R}$$

where Eq. 31-82 is used for R_{eq}. Consequently,

$$P_{avg} = \frac{\varepsilon^2 (N_p/N_s)^2 R}{\left[r + (N_p/N_s)^2 R \right]^2}.$$

Now, we wish to find the value of N_p/N_s such that P_{avg} is a maximum. For brevity, let $x = (N_p/N_s)^2$. Then

$$P_{\text{avg}} = \frac{\varepsilon^2 Rx}{(r + xR)^2},$$

and the derivative with respect to x is

$$\frac{dP_{\text{avg}}}{dx} = \frac{\varepsilon^2 R(r - xR)}{(r + xR)^3}.$$

This is zero for $x = r / R = (1000\,\Omega)/(10\,\Omega) = 100$. We note that for small x, P_{avg} increases linearly with x, and for large x it decreases in proportion to $1/x$. Thus $x = r/R$ is indeed a maximum, not a minimum. Recalling $x = (N_p/N_s)^2$, we conclude that the maximum power is achieved for

$$N_p / N_s = \sqrt{x} = 10.$$

The diagram that follows is a schematic of a transformer with a ten to one turns ratio. An actual transformer would have many more turns in both the primary and secondary coils.

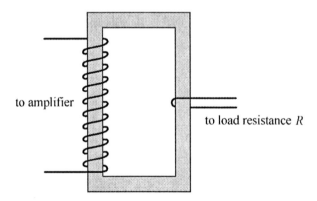

65. (a) We consider the following combinations: $\Delta V_{12} = V_1 - V_2$, $\Delta V_{13} = V_1 - V_3$, and $\Delta V_{23} = V_2 - V_3$. For ΔV_{12},

$$\Delta V_{12} = A \sin(\omega_d t) - A \sin(\omega_d t - 120°) = 2A \sin\left(\frac{120°}{2}\right)\cos\left(\frac{2\omega_d t - 120°}{2}\right) = \sqrt{3} A \cos(\omega_d t - 60°)$$

where we use $\sin \alpha - \sin \beta = 2 \sin[(\alpha - \beta)/2]\cos[(\alpha + \beta)/2]$ and $\sin 60° = \sqrt{3}/2$. Similarly,

$$\Delta V_{13} = A \sin(\omega_d t) - A \sin(\omega_d t - 240°) = 2A \sin\left(\frac{240°}{2}\right)\cos\left(\frac{2\omega_d t - 240°}{2}\right)$$

$$= \sqrt{3} A \cos(\omega_d t - 120°)$$

and

$$\Delta V_{23} = A\sin(\omega_d t - 120°) - A\sin(\omega_d t - 240°) = 2A\sin\left(\frac{120°}{2}\right)\cos\left(\frac{2\omega_d t - 360°}{2}\right)$$

$$= \sqrt{3}A\cos(\omega_d t - 180°)$$

All three expressions are sinusoidal functions of t with angular frequency ω_d.

(b) We note that each of the above expressions has an amplitude of $\sqrt{3}A$.

66. We start with Eq. 31-76:

$$P_{avg} = \varepsilon_{rms} I_{rms} \cos\phi = \varepsilon_{rms}\left(\frac{\varepsilon_{rms}}{Z}\right)\left(\frac{R}{Z}\right) = \frac{\varepsilon_{rms}^2 R}{Z^2}.$$

For a purely resistive circuit, $Z = R$, and this result reduces to Eq. 27-23 (with V replaced with ε_{rms}). This is also the case for a series RLC circuit at resonance. The average rate for dissipating energy is, of course, zero if $R = 0$, as would be the case for a purely inductive circuit.

67. (a) The effective resistance R_{eff} satisfies $I_{rms}^2 R_{eff} = P_{mechanical}$, or

$$R_{eff} = \frac{P_{mechanical}}{I_{rms}^2} = \frac{(0.100\,\text{hp})(746\,\text{W}/\text{hp})}{(0.650\,\text{A})^2} = 177\,\Omega.$$

(b) This is not the same as the resistance R of its coils, but just the effective resistance for power transfer from electrical to mechanical form. In fact $I_{rms}^2 R$ would not give $P_{mechanical}$ but rather the rate of energy loss due to thermal dissipation.

68. From Eq. 31-4 we get $f = 1/2\pi\sqrt{LC} = 1.84$ kHz

69. The rms current in the motor is

$$I_{rms} = \frac{\varepsilon_{rms}}{Z} = \frac{\varepsilon_{rms}}{\sqrt{R^2 + X_L^2}} = \frac{420\,\text{V}}{\sqrt{(45.0\,\Omega)^2 + (32.0\,\Omega)^2}} = 7.61\,\text{A}.$$

70. (a) A sketch of the phasors would be very much like Fig. 31-9(c) but with the label "I_C" on the green arrow replaced with "V_R."

(b) We have $IR = IX_C$, or

$$IR = IX_C \rightarrow R = \frac{1}{\omega_d C}$$

which yields $f = \omega_d/2\pi = 1/2\pi RC = 159$ Hz.

(c) $\phi = \tan^{-1}(-V_C/V_R) = -45°$.

(d) $\omega_d = 1/RC = 1.00 \times 10^3$ rad/s.

(e) $I = (12\text{ V})/\sqrt{R^2 + X_C^2} = 6/(25\sqrt{2}) \approx 170$ mA.

71. (a) The energy stored in the capacitor is given by $U_E = q^2/2C$. Since q is a periodic function of t with period T, so must be U_E. Consequently, U_E will not be changed over one complete cycle. Actually, U_E has period $T/2$, which does not alter our conclusion.

(b) Similarly, the energy stored in the inductor is $U_B = \frac{1}{2}i^2 L$. Since i is a periodic function of t with period T, so must be U_B.

(c) The energy supplied by the generator is

$$P_{avg}T = (I_{rms}\varepsilon_{rms}\cos\phi)T = \left(\frac{1}{2}T\right)\varepsilon_m I \cos\phi$$

where we substitute $I_{rms} = I/\sqrt{2}$ and $\varepsilon_{rms} = \varepsilon_m/\sqrt{2}$.

(d) The energy dissipated by the resistor is

$$P_{avg,resistor}T = (I_{rms}V_R)T = I_{rms}(I_{rms}R)T = \left(\frac{1}{2}T\right)I^2 R.$$

(e) Since $\varepsilon_m I \cos\phi = \varepsilon_m I(V_R/\varepsilon_m) = \varepsilon_m I(IR/\varepsilon_m) = I^2 R$, the two quantities are indeed the same.

72. (a) Eq. 31-39 gives $f = \omega/2\pi = (2\pi C X_C)^{-1} = 8.84$ kHz.

(b) Because of its inverse relationship with frequency, then the reactance will go down by a factor of 2 when f increases by a factor of 2. The answer is $X_C = 6.00$ Ω.

73. (a) The impedance is $Z = \dfrac{\varepsilon_m}{I} = \dfrac{125\text{ V}}{3.20\text{ A}} = 39.1$ Ω.

(b) From $V_R = IR = \varepsilon_m \cos\phi$, we get

$$R = \frac{\varepsilon_m \cos\phi}{I} = \frac{(125\text{ V})\cos(0.982\text{ rad})}{3.20\text{ A}} = 21.7\,Ω.$$

(c) Since $X_L - X_C \propto \sin\phi = \sin(-0.982\,\text{rad})$, we conclude that $X_L < X_C$. The circuit is predominantly capacitive.

74. (a) Eq. 31-4 directly gives $1/\sqrt{LC} \approx 5.77\times10^3$ rad/s.

(b) Eq. 16-5 then yields $T = 2\pi/\omega = 1.09$ ms.

(c) Although we do not show the graph here, we describe it: it is a cosine curve with amplitude 200 μC and period given in part (b).

75. (a) The phase constant is given by

$$\phi = \tan^{-1}\left(\frac{V_L - V_C}{R}\right) = \tan^{-1}\left(\frac{V_L - V_L/2.00}{V_L/2.00}\right) = \tan^{-1}(1.00) = 45.0°.$$

(b) We solve R from $\varepsilon_m \cos\phi = IR$:

$$R = \frac{\varepsilon_m \cos\phi}{I} = \frac{(30.0\,\text{V})(\cos45°)}{300\times10^{-3}\,\text{A}} = 70.7\ \Omega.$$

76. From Eq. 31-4, we have $C = (\omega^2 L)^{-1} = ((2\pi f)^2 L)^{-1} = 1.59$ μF.

77. (a) We solve L from Eq. 31-4, using the fact that $\omega = 2\pi f$:

$$L = \frac{1}{4\pi^2 f^2 C} = \frac{1}{4\pi^2 (10.4\times10^3\,\text{Hz})^2 (340\times10^{-6}\,\text{F})} = 6.89\times10^{-7}\,\text{H}.$$

(b) The total energy may be figured from the inductor (when the current is at maximum):

$$U = \frac{1}{2}LI^2 = \frac{1}{2}(6.89\times10^{-7}\,\text{H})(7.20\times10^{-3}\,\text{A})^2 = 1.79\times10^{-11}\,\text{J}.$$

(c) We solve for Q from $U = \frac{1}{2}Q^2/C$:

$$Q = \sqrt{2CU} = \sqrt{2(340\times10^{-6}\,\text{F})(1.79\times10^{-11}\,\text{J})} = 1.10\times10^{-7}\,\text{C}.$$

78. (a) With a phase constant of 45° the (net) reactance must equal the resistance in the circuit, which means the circuit impedance becomes $Z = R\sqrt{2} \Rightarrow R = Z/\sqrt{2} = 707\ \Omega$.

(b) Since $f = 8000$ Hz then $\omega_d = 2\pi(8000)$ rad/s. The net reactance (which, as observed, must equal the resistance) is therefore $X_L - X_C = \omega_d L - (\omega_d C)^{-1} = 707\ \Omega$. We are also told that the resonance frequency is 6000 Hz, which (by Eq. 31-4) means $C = (\omega^2 L)^{-1} = ((2\pi(6000))^2 L)^{-1}$. Substituting this in for C in our previous expression (for the net reactance) we obtain an equation that can be solved for the self-inductance. Our result is $L = 32.2$ mH.

(c) $C = ((2\pi(6000))^2 L)^{-1} = 21.9$ nF.

79. (a) Let $\omega t - \pi/4 = \pi/2$ to obtain $t = 3\pi/4\omega = 3\pi/[4(350\,\text{rad}/\text{s})] = 6.73 \times 10^{-3}$ s.

(b) Let $\omega t + \pi/4 = \pi/2$ to obtain $t = \pi/4\omega = \pi/[4(350\,\text{rad}/\text{s})] = 2.24 \times 10^{-3}$ s.

(c) Since i leads ε in phase by $\pi/2$, the element must be a capacitor.

(d) We solve C from $X_C = (\omega C)^{-1} = \varepsilon_m / I$:

$$C = \frac{I}{\varepsilon_m \omega} = \frac{6.20 \times 10^{-3}\,\text{A}}{(30.0\,\text{V})(350\,\text{rad/s})} = 5.90 \times 10^{-5}\,\text{F}.$$

80. Resonance occurs when the inductive reactance equals the capacitive reactance. Reactances of a certain type add (in series) just like resistances did in Chapter 28. Thus, since the resonance ω values are the same for both circuits, we have for each circuit:

$$\omega L_1 = \frac{1}{\omega C_1}, \quad \omega L_2 = \frac{1}{\omega C_2}$$

and adding these equations we find

$$\omega(L_1 + L_2) = \frac{1}{\omega}\left(\frac{1}{C_1} + \frac{1}{C_2}\right)$$

$$\omega L_{eq} = \frac{1}{\omega C_{eq}} \Rightarrow \text{resonance in the combined circuit.}$$

81. (a) From Eq. 31-4, we have $L = (\omega^2 C)^{-1} = ((2\pi f)^2 C)^{-1} = 2.41\,\mu\text{H}$.

(b) The total energy is the maximum energy on either device (see Fig. 31-4). Thus, we have $U_{max} = \frac{1}{2}LI^2 = 21.4$ pJ.

(c) Of several methods available to do this part, probably the one most "in the spirit" of this problem (considering the energy that was calculated in part (b)) is to appeal to $U_{max} = \frac{1}{2}Q^2/C$ (from Chapter 26) to find the maximum charge: $Q = \sqrt{2CU_{max}} = 82.2$ nC.

82. (a) From Eq. 31-65, we have

$$\phi = \tan^{-1}\left(\frac{V_L - V_C}{V_R}\right) = \tan^{-1}\left(\frac{V_L - (V_L / 1.50)}{(V_L / 2.00)}\right)$$

which becomes $\tan^{-1} 2/3 = 33.7°$ or 0.588 rad.

(b) Since $\phi > 0$, it is inductive ($X_L > X_C$).

(c) We have $V_R = IR = 9.98$ V, so that $V_L = 2.00V_R = 20.0$ V and $V_C = V_L/1.50 = 13.3$ V. Therefore, from Eq. 31-60.

$$\varepsilon_m = \sqrt{V_R^2 + (V_L - V_C)^2}$$

we find $\varepsilon_m = 12.0$ V .

83. When switch S_1 is closed and the others are open, the inductor is essentially out of the circuit and what remains is an *RC* circuit. The time constant is $\tau_C = RC$. When switch S_2 is closed and the others are open, the capacitor is essentially out of the circuit. In this case, what we have is an *LR* circuit with time constant $\tau_L = L/R$. Finally, when switch S_3 is closed and the others are open, the resistor is essentially out of the circuit and what remains is an *LC* circuit that oscillates with period $T = 2\pi\sqrt{LC}$. Substituting $L = R\tau_L$ and $C = \tau_C/R$, we obtain $T = 2\pi\sqrt{\tau_C\tau_L}$.

84. (a) The impedance is $Z = (80.0$ V)/(1.25 A) = 64.0 Ω.

(b) We can write $\cos\phi = R/Z \Rightarrow R = (64.0$ $\Omega)\cos(0.650$ rad) = 50.9 Ω.

(c) Since the "current leads the emf" the circuit is capacitive.

85. (a) We find L from $X_L = \omega L = 2\pi fL$:

$$f = \frac{X_L}{2\pi L} = \frac{1.30 \times 10^3 \Omega}{2\pi(45.0 \times 10^{-3} \text{H})} = 4.60 \times 10^3 \text{Hz}.$$

(b) The capacitance is found from $X_C = (\omega C)^{-1} = (2\pi fC)^{-1}$:

$$C = \frac{1}{2\pi fX_C} = \frac{1}{2\pi(4.60 \times 10^3 \text{Hz})(1.30 \times 10^3 \Omega)} = 2.66 \times 10^{-8} \text{F}.$$

(c) Noting that $X_L \propto f$ and $X_C \propto f^{-1}$, we conclude that when f is doubled, X_L doubles and X_C reduces by half. Thus, $X_L = 2(1.30 \times 10^3$ $\Omega) = 2.60 \times 10^3 \Omega$.

(d) $X_C = 1.30 \times 10^3 \, \Omega/2 = 6.50 \times 10^2 \, \Omega$.

86. (a) Using $\omega = 2\pi f$, $X_L = \omega L$, $X_C = 1/\omega C$ and $\tan(\phi) = (X_L - X_C)/R$, we find

$$\phi = \tan^{-1}[(16.022 - 33.157)/40.0] = -0.40473 \approx -0.405 \text{ rad}.$$

(b) Eq. 31-63 gives $I = 120/\sqrt{40^2 + (16-33)^2} = 2.7576 \approx 2.76$ A.

(c) $X_C > X_L \implies$ capacitive.

87. When the switch is open, we have a series LRC circuit involving just the one capacitor near the upper right corner. Eq. 31-65 leads to

$$\frac{\omega_d L - \dfrac{1}{\omega_d C}}{R} = \tan \phi_o = \tan(-20°) = -\tan 20°.$$

Now, when the switch is in position 1, the equivalent capacitance in the circuit is $2C$. In this case, we have

$$\frac{\omega_d L - \dfrac{1}{2\omega_d C}}{R} = \tan \phi_1 = \tan 10.0°.$$

Finally, with the switch in position 2, the circuit is simply an LC circuit with current amplitude

$$I_2 = \frac{\varepsilon_m}{Z_{LC}} = \frac{\varepsilon_m}{\sqrt{\left(\omega_d L - \dfrac{1}{\omega_d C}\right)^2}} = \frac{\varepsilon_m}{\dfrac{1}{\omega_d C} - \omega_d L}$$

where we use the fact that $(\omega_d C)^{-1} > \omega_d L$ in simplifying the square root (this fact is evident from the description of the first situation, when the switch was open). We solve for L, R and C from the three equations above, and the results are

(a) $R = \dfrac{-\varepsilon_m}{I_2 \tan \phi_o} = \dfrac{120\text{V}}{(2.00 \text{ A}) \tan 20.0°} = 165 \, \Omega.$

(b) $L = \dfrac{\varepsilon_m}{\omega_d I_2}\left(1 - 2\dfrac{\tan \phi_1}{\tan \phi_o}\right) = \dfrac{120\text{ V}}{2\pi(60.0 \text{ Hz})(2.00 \text{ A})}\left(1 + 2\dfrac{\tan 10.0°}{\tan 20.0°}\right) = 0.313 \text{ H}.$

(c) $C = \dfrac{I_2}{2\omega_d \varepsilon_m \left(1 - \frac{\tan\phi_1}{\tan\phi_o}\right)} = \dfrac{2.00\ \text{A}}{2(2\pi)(60.0\ \text{Hz})(120\ \text{V})\left(1 + \frac{\tan 10.0°}{\tan 20.0°}\right)} = 1.49 \times 10^{-5}\ \text{F}$

88. From $U_{max} = \frac{1}{2} L I^2$ we get $I = 0.115$ A.

89. (a) At any time, the total energy U in the circuit is the sum of the energy U_E in the capacitor and the energy U_B in the inductor. When $U_E = 0.500 U_B$ (at time t), then $U_B = 2.00 U_E$ and $U = U_E + U_B = 3.00 U_E$. Now, U_E is given by $q^2/2C$, where q is the charge on the capacitor at time t. The total energy U is given by $Q^2/2C$, where Q is the maximum charge on the capacitor. Thus, $Q^2/2C = 3.00 q^2/2C$ or $q = Q/\sqrt{3.00} = 0.577 Q$.

(b) If the capacitor is fully charged at time $t = 0$, then the time-dependent charge on the capacitor is given by $q = Q\cos\omega t$. This implies that the condition $q = 0.577 Q$ is satisfied when $\cos\omega t = 0.557$, or $\omega t = 0.955$ rad. Since $\omega = 2\pi/T$ (where T is the period of oscillation), $t = 0.955 T/2\pi = 0.152 T$, or $t/T = 0.152$.

90. (a) The computations are as follows:

$$X_L = 2\pi f_d L = 60.82\ \Omega$$

$$X_C = (2\pi f_d C)^{-1} = 32.88\ \Omega$$

$$Z = \sqrt{20^2 + (61\text{-}33)^2} = 34.36\ \Omega$$

$$I = \varepsilon/Z = 2.62\ \text{A} \quad \Rightarrow \quad I_{rms} = I/\sqrt{2} = 1.85\ \text{A}.$$

Therefore, $V_{R\,rms} = I_{rms}\, R = 37.0$ V.

(b) $V_{C\,rms} = I_{rms}\, X_C = 60.9$ V.

(c) $V_{L\,rms} = I_{rms}\, X_L = 113$ V.

(d) $P_{avg} = (I_{rms})^2 R = 68.6$ W.

91. (a) Eqs. 31-4 and 31-14 lead to

$$Q = \frac{1}{\omega} = I\sqrt{LC} = 1.27 \times 10^{-6}\,\text{C}.$$

(b) We choose the phase constant in Eq. 31-12 to be $\phi = -\pi/2$, so that $i_0 = I$ in Eq. 31-15). Thus, the energy in the capacitor is

$$U_E = \frac{q^2}{2C} = \frac{Q^2}{2C}(\sin \omega t)^2 .$$

Differentiating and using the fact that $2 \sin \theta \cos \theta = \sin 2\theta$, we obtain

$$\frac{dU_E}{dt} = \frac{Q^2}{2C} \omega \sin 2\omega t .$$

We find the maximum value occurs whenever $\sin 2\omega t = 1$, which leads (with $n = odd$ integer) to

$$t = \frac{1}{2\omega}\frac{n\pi}{2} = \frac{n\pi}{4\omega} = \frac{n\pi}{4}\sqrt{LC} = 8.31 \times 10^{-5}\,\text{s}, \ 2.49 \times 10^{-4}\,\text{s}, \dots .$$

The earliest time is $t = 8.31 \times 10^{-5}\,\text{s}$.

(c) Returning to the above expression for dU_E/dt with the requirement that $\sin 2\omega t = 1$, we obtain

$$\left(\frac{dU_E}{dt}\right)_{max} = \frac{Q^2}{2C}\omega = \frac{\left(I\sqrt{LC}\right)^2}{2C}\frac{I}{\sqrt{LC}} = \frac{I^2}{2}\sqrt{\frac{L}{C}} = 5.44 \times 10^{-3}\,\text{J/s} .$$

92. (a) We observe that $\omega = 6597$ rad/s, and, consequently, $X_L = 594\ \Omega$ and $X_C = 303\ \Omega$. Since $X_L > X_C$, the phase angle is positive: $\phi = +60.0°$.

From Eq. 31-65, we obtain $R = \dfrac{X_L - X_C}{\tan \phi} = 168\ \Omega$.

(b) Since we are already on the "high side" of resonance, increasing f will only decrease the current further, but *decreasing* f brings us closer to resonance and, consequently, large values of I.

(c) Increasing L increases X_L, but we already have $X_L > X_C$. Thus, if we wish to move closer to resonance (where X_L must equal X_C), we need to *decrease* the value of L.

(d) To change the present condition of $X_C < X_L$ to something closer to $X_C = X_L$ (resonance, large current), we can increase X_C. Since X_C depends inversely on C, this means *decreasing C*.

93. (a) We observe that $\omega_d = 12566$ rad/s. Consequently, $X_L = 754\ \Omega$ and $X_C = 199\ \Omega$. Hence, Eq. 31-65 gives

$$\phi = \tan^{-1}\left(\frac{X_L - X_C}{R}\right) = 1.22 \text{ rad} .$$

(b) We find the current amplitude from Eq. 31-60:

$$I = \frac{\varepsilon_m}{\sqrt{R^2 + (X_L - X_C)^2}} = 0.288\,\text{A} .$$

94. From Eq. 31-60, we have

$$\left(\frac{220\,\text{V}}{3.00\,\text{A}}\right)^2 = R^2 + X_L^2 \Rightarrow X_L = 69.3\,\Omega .$$

95. From Eq. 31-4, with $\omega = 2\pi f = 4.49 \times 10^3$ rad / s, we obtain

$$L = \frac{1}{\omega^2 C} = 7.08 \times 10^{-3}\,\text{H}.$$

96. (a) From Eq. 31-4, with $\omega = 2\pi f$, $C = 2.00$ nF and $L = 2.00$ mH, we have

$$f = \frac{1}{2\pi\sqrt{LC}} = 7.96 \times 10^4\,\text{Hz}.$$

(b) The maximum current in the oscillator is

$$i_{max} = I_C = \frac{V_C}{X_C} = \omega C v_{max} = 4.00 \times 10^{-3}\,\text{A}.$$

(c) Using Eq. 30-49, we find the maximum magnetic energy:

$$U_{B,max} = \frac{1}{2} L i_{max}^2 = 1.60 \times 10^{-8}\,\text{J}.$$

(d) Adapting Eq. 30-35 to the notation of this chapter,

$$v_{max} = L\left|\frac{di}{dt}\right|_{max}$$

which yields a (maximum) time rate of change (for i) equal to 2.00×10^3 A/s.

97. Reading carefully, we note that the driving frequency of the source is permanently set at the resonance frequency of the *initial* circuit (with switches open); it is set at $\omega_d =$

$1/\sqrt{LC} = 1.58 \times 10^4$ rad/s and does not correspond to the resonance frequency once the switches are closed. In our table, below, C_{eq} is in μF, f is in kHz, and R_{eq} and Z are in Ω. Steady state conditions are assumed in calculating the current amplitude (which is in amperes); this I is the current through the source (or through the inductor), as opposed to the (generally smaller) current in one of the resistors. Resonant frequencies f are computed with $\omega = 2\pi f$. Reducing capacitor and resistor combinations is explained in chapters 26 and 28, respectively.

switch	(a) $C_{eq}(\mu F)$	(b) $f(kHz)$	(c) $R_{eq}(\Omega)$	(d) $Z(\Omega)$	(e) $I(A)$
S_1	4.00	1.78	12.0	19.8	0.605
S_2	5.00	1.59	12.0	22.4	0.535
S_3	5.00	1.59	6.0	19.9	0.603
S_4	5.00	1.59	4.0	19.4	0.619

98. (a) We note that we obtain the maximum value in Eq. 31-28 when we set

$$t = \frac{\pi}{2\omega_d} = \frac{1}{4f} = \frac{1}{4(60)} = 0.00417 \text{ s}$$

or 4.17 ms. The result is $\varepsilon_m \sin(\pi/2) = \varepsilon_m \sin(90°) = 36.0$ V.

(b) At $t = 4.17$ ms, the current is

$i = I\sin(\omega_d t - \phi) = I\sin(90° - (-24.3°)) = (0.164\,\text{A})\cos(24.3°) = 0.1495\,\text{A} \approx 0.150$ A.

using Eq. 31-29 and the results of the Sample Problem. Ohm's law directly gives

$$v_R = iR = (0.1495\,\text{A})(200\,\Omega) = 29.9\,\text{V}.$$

(c) The capacitor voltage phasor is 90° less than that of the current. Thus, at $t = 4.17$ ms, we obtain

$v_C = I\sin(90° - (-24.3°) - 90°)X_C = IX_C\sin(24.3°) = (0.164\,\text{A})(177\,\Omega)\sin(24.3°) = 11.9\,\text{V}.$

(d) The inductor voltage phasor is 90° more than that of the current. Therefore, at $t = 4.17$ ms, we find

$v_L = I\sin(90° - (-24.3°) + 90°)X_L = -IX_L\sin(24.3°) = -(0.164\,\text{A})(86.7\,\Omega)\sin(24.3°)$
$= -5.85\,\text{V}.$

(e) Our results for parts (b), (c) and (d) add to give 36.0 V, the same as the answer for part (a).

99. (a) Since $T = 2\pi / \omega = 2\pi\sqrt{LC}$, we may rewrite the power on the exponential factor as

$$-\pi R\sqrt{\frac{C}{L}}\frac{t}{T} = -\pi R\sqrt{\frac{C}{L}}\frac{t}{2\pi\sqrt{LC}} = -\frac{Rt}{2L}.$$

Thus $e^{-Rt/2L} = e^{-\pi R\sqrt{C/L}(t/T)}$.

(b) Since $-\pi R\sqrt{C/L}(t/T)$ must be dimensionless (as is t/T), $R\sqrt{C/L}$ must also be dimensionless. Thus, the SI unit of $\sqrt{C/L}$ must be Ω^{-1}. In other words, the SI unit for $\sqrt{L/C}$ is Ω.

(c) Since the amplitude of oscillation reduces by a factor of $e^{-\pi R\sqrt{C/L}(T/T)} = e^{-\pi R\sqrt{C/L}}$ after each cycle, the condition is equivalent to $\pi R\sqrt{C/L} \ll 1$, or $R \ll \sqrt{L/C}$.

100. (a) The curves are shown in the graph below. We have also included here the impedance curve (which is asked for later in the problem statement). The curve sloping towards zero at high frequencies is X_C, and the linearly rising line is X_L. The vertical axis is in ohms. For simplicity of notation, we have omitted the "d" subscript from f.

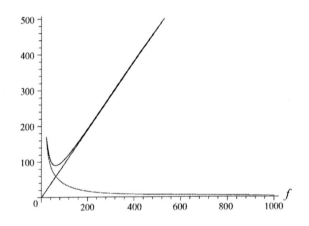

(b) The reactance curves cross each other (to the extent that we can estimate from our graph) at a value near 60 Hz. A more careful calculation (setting the reactances equal to each other) leads to the resonance value: $f = 61.26 \approx 61$ Hz.

(c) Z is at its lowest value at resonance: $Z_{resonance} = R = 90\ \Omega$.

(d) As noted in our solution of part (b), the resonance value is $f = 61.26 \approx 61$ Hz.

Chapter 32

1. (a)   The flux through the top is $+(0.30 \text{ T})\pi r^2$ where $r = 0.020$ m. The flux through the bottom is $+0.70$ mWb as given in the problem statement. Since the *net* flux must be zero then the flux through the sides must be negative and exactly cancel the total of the previously mentioned fluxes. Thus (in magnitude) the flux though the sides is 1.1 mWb.

(b) The fact that it is negative means it is inward.

2. We use $\sum_{n=1}^{6} \Phi_{Bn} = 0$ to obtain

$$\Phi_{B6} = -\sum_{n=1}^{5} \Phi_{Bn} = -(-1\,\text{Wb} + 2\,\text{Wb} - 3\,\text{Wb} + 4\,\text{Wb} - 5\,\text{Wb}) = +3\,\text{Wb} .$$

3. (a) We use Gauss' law for magnetism: $\oint \vec{B} \cdot d\vec{A} = 0$. Now, $\oint \vec{B} \cdot d\vec{A} = \Phi_1 + \Phi_2 + \Phi_C$, where Φ_1 is the magnetic flux through the first end mentioned, Φ_2 is the magnetic flux through the second end mentioned, and Φ_C is the magnetic flux through the curved surface. Over the first end the magnetic field is inward, so the flux is $\Phi_1 = -25.0\ \mu\text{Wb}$. Over the second end the magnetic field is uniform, normal to the surface, and outward, so the flux is $\Phi_2 = AB = \pi r^2 B$, where A is the area of the end and r is the radius of the cylinder. It value is

$$\Phi_2 = \pi(0.120\,\text{m})^2 (1.60 \times 10^{-3}\,\text{T}) = +7.24 \times 10^{-5}\,\text{Wb} = +72.4\ \mu\text{Wb} .$$

Since the three fluxes must sum to zero,

$$\Phi_C = -\Phi_1 - \Phi_2 = 25.0\ \mu\text{Wb} - 72.4\ \mu\text{Wb} = -47.4\ \mu\text{Wb} .$$

Thus, the magnitude is $|\Phi_C| = 47.4\ \mu\text{Wb}$.

(b) The minus sign in Φ_c indicates that the flux is inward through the curved surface.

4. From Gauss' law for magnetism, the flux through S_1 is equal to that through S_2, the portion of the *xz* plane that lies within the cylinder. Here the normal direction of S_2 is *+y*. Therefore,

$$\Phi_B\left(S_1\right) = \Phi_B\left(S_2\right) = \int_{-r}^{r} B(x)L\,dx = 2\int_{-r}^{r} B_{\text{left}}\left(x\right)L\,dx = 2\int_{-r}^{r} \frac{\mu_0 i}{2\pi}\frac{1}{2r-x}L\,dx$$

$$= \frac{\mu_0 iL}{\pi}\ln 3\,.$$

5. We use the result of part (b) in Sample Problem 32-1:

$$B = \frac{\mu_0 \varepsilon_0 R^2}{2r}\frac{dE}{dt}\qquad\left(\text{for } r \geq R\right)$$

to solve for dE/dt:

$$\frac{dE}{dt} = \frac{2Br}{\mu_0 \varepsilon_0 R^2} = \frac{2\left(2.0\times10^{-7}\,\text{T}\right)\left(6.0\times10^{-3}\,\text{m}\right)}{\left(4\pi\times10^{-7}\,\text{T}\cdot\text{m/A}\right)\left(8.85\times10^{-12}\,\frac{\text{C}^2}{\text{N}\cdot\text{m}^2}\right)\left(3.0\times10^{-3}\,\text{m}\right)^2} = 2.4\times10^{13}\,\frac{\text{V}}{\text{m}\cdot\text{s}}.$$

6. From Sample Problem 32-1 we know that $B \propto r$ for $r \leq R$ and $B \propto r^{-1}$ for $r \geq R$. So the maximum value of B occurs at $r = R$, and there are two possible values of r at which the magnetic field is 75% of B_{max}. We denote these two values as r_1 and r_2, where $r_1 < R$ and $r_2 > R$.

(a) Inside the capacitor, $0.75\,B_{\text{max}}/B_{\text{max}} = r_1/R$, or $r_1 = 0.75\,R = 0.75\,(40\text{ mm}) = 30$ mm.

(b) Outside the capacitor, $0.75\,B_{\text{max}}/B_{\text{max}} = (r_2/R)^{-1}$, or $r_2 = R/0.75 = 4R/3 = (4/3)(40\text{ mm}) = 53$ mm..

(c) From Eqs. 32-15 and 32-17,

$$B_{\text{max}} = \frac{\mu_0 i_d}{2\pi R} = \frac{\mu_0 i}{2\pi R} = \frac{\left(4\pi\times10^{-7}\,\text{T}\cdot\text{m/A}\right)\left(6.0\,\text{A}\right)}{2\pi\left(0.040\,\text{m}\right)} = 3.0\times10^{-5}\,\text{T}.$$

7. (a) Noting that the magnitude of the electric field (assumed uniform) is given by $E = V/d$ (where $d = 5.0$ mm), we use the result of part (a) in Sample Problem 32-1

$$B = \frac{\mu_0 \varepsilon_0 r}{2}\frac{dE}{dt} = \frac{\mu_0 \varepsilon_0 r}{2d}\frac{dV}{dt}\qquad\left(\text{for } r \leq R\right).$$

We also use the fact that the time derivative of $\sin\left(\omega t\right)$ (where $\omega = 2\pi f = 2\pi(60) \approx 377/\text{s}$ in this problem) is $\omega\cos(\omega t)$. Thus, we find the magnetic field as a function of r (for $r \leq R$; note that this neglects "fringing" and related effects at the edges):

$$B = \frac{\mu_0 \varepsilon_0 r}{2d}V_{\text{max}}\omega\cos\left(\omega t\right) \Rightarrow B_{\text{max}} = \frac{\mu_0 \varepsilon_0 r V_{\text{max}}\omega}{2d}$$

where $V_{max} = 150$ V. This grows with r until reaching its highest value at $r = R = 30$ mm:

$$B_{max}\Big|_{r=R} = \frac{\mu_0\varepsilon_0 R V_{max}\omega}{2d} = \frac{\left(4\pi\times10^{-7}\,\text{H/m}\right)\left(8.85\times10^{-12}\,\text{F/m}\right)\left(30\times10^{-3}\,\text{m}\right)\left(150\,\text{V}\right)\left(377/\text{s}\right)}{2\left(5.0\times10^{-3}\,\text{m}\right)}$$

$$= 1.9\times10^{-12}\,\text{T}.$$

(b) For $r \le 0.03$ m, we use the $B_{max} = \frac{\mu_0\varepsilon_0 r V_{max}\omega}{2d}$ expression found in part (a) (note the

$B \propto r$ dependence), and for $r \ge 0.03$ m we perform a similar calculation starting with the result of part (b) in Sample Problem 32-1:

$$B_{max} = \left(\frac{\mu_0\varepsilon_0 R^2}{2r}\frac{dE}{dt}\right)_{max} = \left(\frac{\mu_0\varepsilon_0 R^2}{2rd}\frac{dV}{dt}\right)_{max} = \left(\frac{\mu_0\varepsilon_0 R^2}{2rd}V_{max}\omega\cos(\omega t)\right)_{max}$$

$$= \frac{\mu_0\varepsilon_0 R^2 V_{max}\omega}{2rd} \quad (\text{for}\,r \ge R)$$

(note the $B \propto r^{-1}$ dependence — See also Eqs. 32-16 and 32-17). The plot (with SI units understood) is shown below.

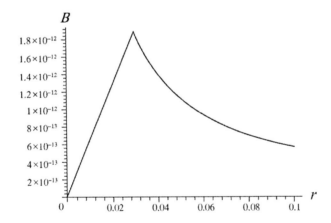

8. (a) Inside we have (by Eq. 32-16) $B = \mu_0 i_d\, r_1/2\pi R^2$, where $r_1 = 0.0200$, $R = 0.0300$, and the displacement current is given by Eq. 32-38: $i_d = \varepsilon_0\, d\Phi_E/dt = \varepsilon_0(0.00300)$, in SI units. Thus we find $B = 1.18 \times 10^{-19}$ T.

(b) Outside we have (by Eq. 32-17) $B = \mu_0 i_d/2\pi r_2$ where $r_2 = 0.0500$ in SI units. Here we obtain $B = 1.06 \times 10^{-19}$ T.

9. (a) Application of Eq. 32-3 along the circle referred to in the second sentence of the problem statement (and taking the derivative of the flux expression given in that sentence) leads to

$$B (2\pi r) = \varepsilon_0 \mu_0 (0.60 \text{ V·m/s}) \frac{r}{R} .$$

Using $r = 0.0200$ (which, in any case, cancels out) and $R = 0.0500$ (SI units understood) leads to $B = 3.54 \times 10^{-17}$ T.

(b) For a value of r larger than R, we must note that the flux enclosed has already reached its full amount (when $r = R$ in the given flux expression). Referring to the equation we wrote in our solution of part (a), this means that the final fraction ($\frac{r}{R}$) should be replaced with unity. On the left hand side of that equation, we set $r = 0.0500$ m and solve. We now find $B = 2.13 \times 10^{-17}$ T.

10. (a) Application of Eq. 32-7 with $A = \pi r^2$ (and taking the derivative of the field expression given in the problem) leads to

$$B (2\pi r) = \varepsilon_0 \mu_0 \pi r^2 (0.00450 \text{ V/m·s}) .$$

With $r = 0.0200$ m, this gives $B = 5.01 \times 10^{-22}$ T.

(b) With $r > R$, the expression above must replaced by

$$B (2\pi r) = \varepsilon_0 \mu_0 \pi R^2 (0.00450 \text{ V/m·s}) .$$

Substituting $r = 0.050$ m and $R = 0.030$ m, we obtain $B = 4.51 \times 10^{-22}$ T.

11. (a) Here, the enclosed electric flux is found by integrating

$$\Phi_E = \int_0^r E \, 2\pi r \, dr = t(0.500 \text{ V/m·s})(2\pi) \int_0^r \left(1 - \frac{r}{R}\right) r \, dr = t \, \pi \left(\frac{1}{2} r^2 - \frac{r^3}{3R}\right)$$

with SI units understood. Then (after taking the derivative with respect to time) Eq. 32-3 leads to

$$B (2\pi r) = \varepsilon_0 \mu_0 \pi \left(\frac{1}{2} r^2 - \frac{r^3}{3R}\right) .$$

With $r = 0.0200$ m and $R = 0.0300$ m, this gives $B = 3.09 \times 10^{-20}$ T.

(b) The integral shown above will no longer (since now $r > R$) have r as the upper limit; the upper limit is now R. Thus, $\Phi_E = t \, \pi \left(\frac{1}{2} R^2 - \frac{R^3}{3R}\right) = \frac{1}{6} t \, \pi R^2$. Consequently, Eq. 32-3 becomes

$$B (2\pi r) = \frac{1}{6} \varepsilon_0 \mu_0 \pi R^2$$

which yields (for $r = 0.0500$ m) $B = 1.67 \times 10^{-20}$ T.

12. Let the area plate be A and the plate separation be d. We use Eq. 32-10:

$$i_d = \varepsilon_0 \frac{d\Phi_E}{dt} = \varepsilon_0 \frac{d}{dt}(AE) = \varepsilon_0 A \frac{d}{dt}\left(\frac{V}{d}\right) = \frac{\varepsilon_0 A}{d}\left(\frac{dV}{dt}\right),$$

or

$$\frac{dV}{dt} = \frac{i_d d}{\varepsilon_0 A} = \frac{i_d}{C} = \frac{1.5\,\text{A}}{2.0 \times 10^{-6}\,\text{F}} = 7.5 \times 10^5\,\text{V/s}.$$

Therefore, we need to change the voltage difference across the capacitor at the rate of 7.5×10^5 V/s.

13. The displacement current is given by $i_d = \varepsilon_0 A(dE/dt)$, where A is the area of a plate and E is the magnitude of the electric field between the plates. The field between the plates is uniform, so $E = V/d$, where V is the potential difference across the plates and d is the plate separation. Thus

$$i_d = \frac{\varepsilon_0 A}{d}\frac{dV}{dt}.$$

Now, $\varepsilon_0 A/d$ is the capacitance C of a parallel-plate capacitor (not filled with a dielectric), so

$$i_d = C\frac{dV}{dt}.$$

14. We use Eq. 32-14: $i_d = \varepsilon_0 A(dE/dt)$. Note that, in this situation, A is the area over which a changing electric field is present. In this case $r > R$, so $A = \pi R^2$. Thus,

$$\frac{dE}{dt} = \frac{i_d}{\varepsilon_0 A} = \frac{i_d}{\varepsilon_0 \pi R^2} = \frac{2.0\,\text{A}}{\pi\left(8.85 \times 10^{-12}\,\frac{\text{C}^2}{\text{N·m}^2}\right)(0.10\,\text{m})^2} = 7.2 \times 10^{12}\,\frac{\text{V}}{\text{m·s}}.$$

15. Consider an area A, normal to a uniform electric field $\vec{E}$. The displacement current density is uniform and normal to the area. Its magnitude is given by $J_d = i_d/A$. For this situation, $i_d = \varepsilon_0 A(dE/dt)$, so

$$J_d = \frac{1}{A}\varepsilon_0 A\frac{dE}{dt} = \varepsilon_0 \frac{dE}{dt}.$$

16. (a) From Eq. 32-10,

$$i_d = \varepsilon_0 \frac{d\Phi_E}{dt} = \varepsilon_0 A \frac{dE}{dt} \varepsilon_0 A \frac{d}{dt}\left[\left(4.0\times10^5\right) - \left(6.0\times10^4 t\right)\right] = -\varepsilon_0 A\left(6.0\times10^4 \text{ V/m}\cdot\text{s}\right)$$

$$= -\left(8.85\times10^{-12}\,\frac{\text{C}^2}{\text{N}\cdot\text{m}^2}\right)\left(4.0\times10^{-2}\,\text{m}^2\right)\left(6.0\times10^4 \text{ V/m}\cdot\text{s}\right)$$

$$= -2.1\times10^{-8}\,\text{A}.$$

Thus, the magnitude of the displacement current is $|i_d| = 2.1\times10^{-8}\,\text{A}$.

(b) The negative sign in i_d implies that the direction is downward.

(c) If one draws a counterclockwise circular loop s around the plates, then according to Eq. 32-18

$$\oint_s \vec{B}\cdot d\vec{s} = \mu_0 i_d < 0,$$

which means that $\vec{B}\cdot d\vec{s} < 0$. Thus $\vec{B}$ must be clockwise.

17. (a) We use $\oint \vec{B}\cdot d\vec{s} = \mu_0 I_{\text{enclosed}}$ to find

$$B = \frac{\mu_0 I_{\text{enclosed}}}{2\pi r} = \frac{\mu_0\left(J_d \pi r^2\right)}{2\pi r} = \frac{1}{2}\mu_0 J_d r = \frac{1}{2}\left(1.26\times10^{-6}\,\text{H/m}\right)\left(20\,\text{A/m}^2\right)\left(50\times10^{-3}\,\text{m}\right)$$

$$= 6.3\times10^{-7}\,\text{T}.$$

(b) From $i_d = J_d \pi r^2 = \varepsilon_0 \frac{d\Phi_E}{dt} = \varepsilon_0 \pi r^2 \frac{dE}{dt}$, we get

$$\frac{dE}{dt} = \frac{J_d}{\varepsilon_0} = \frac{20\,\text{A/m}^2}{8.85\times10^{-12}\,\text{F/m}} = 2.3\times10^{12}\,\frac{\text{V}}{\text{m}\cdot\text{s}}.$$

18. (a) Since $i = i_d$ (Eq. 32-15) then the portion of displacement current enclosed is

$$i_{d,\text{enc}} = i\,\frac{\pi\left(\frac{R}{3}\right)^2}{\pi R^2} = i\frac{1}{9} = 1.33\,\text{A}.$$

(b) We see from Sample Problems 32-1 and 32-2 that the maximum field is at $r = R$ and that (in the interior) the field is simply proportional to r. Therefore,

$$\frac{B}{B_{\text{max}}} = \frac{3.00\,\text{mT}}{12.0\,\text{mT}} = \frac{r}{R}$$

which yields $r = R/4 = (1.20 \text{ cm})/4 = 0.300 \text{ cm}$.

(c) We now look for a solution in the exterior region, where the field is inversely proportional to r (by Eq. 32-17):

$$\frac{B}{B_{\max}} = \frac{3.00 \text{ mT}}{12.0 \text{ mT}} = \frac{R}{r}$$

which yields $r = 4R = 4(1.20 \text{ cm}) = 4.80 \text{ cm}$.

19. (a) In region a of the graph,

$$|i_d| = \varepsilon_0 \left| \frac{d\Phi_E}{dt} \right| = \varepsilon_0 A \left| \frac{dE}{dt} \right|$$

$$= \left(8.85 \times 10^{-12} \text{ F/m} \right) \left(1.6 \text{ m}^2 \right) \left| \frac{4.5 \times 10^5 \text{ N/C} - 6.0 \times 10^5 \text{ N/C}}{4.0 \times 10^{-6} \text{ s}} \right| = 0.71 \text{ A}.$$

(b) $i_d \propto dE/dt = 0$.

(c) In region c of the graph,

$$|i_d| = \varepsilon_0 A \left| \frac{dE}{dt} \right| = \left(8.85 \times 10^{-12} \text{ F/m} \right) \left(1.6 \text{ m}^2 \right) \left| \frac{-4.0 \times 10^5 \text{ N/C}}{2.0 \times 10^{-6} \text{ s}} \right| = 2.8 \text{ A}.$$

20. From Eq. 28-11, we have $i = (\varepsilon / R) \, e^{-t/\tau}$ since we are ignoring the self-inductance of the capacitor. Eq. 32-16 gives

$$B = \frac{\mu_0 i_d \, r}{2\pi R^2} \, .$$

Furthermore, Eq. 25-9 yields the capacitance

$$C = \frac{\varepsilon_0 \pi (0.05 \text{ m})^2}{0.003 \text{ m}} = 2.318 \times 10^{-11} \text{ F},$$

so that the capacitive time constant is $\tau = (20.0 \times 10^6 \, \Omega)(2.318 \times 10^{-11} \text{ F}) = 4.636 \times 10^{-4} \text{ s}$.

At $t = 250 \times 10^{-6} \text{ s}$, the current is

$$i = \frac{12.0 \text{ V}}{20.0 \times 10^6 \, \Omega} \, e^{-t/\tau} = 3.50 \times 10^{-7} \text{ A}.$$

Since $i = i_d$ (see Eq. 32-15) and $r = 0.0300$ m, then (with plate radius $R = 0.0500$ m) we find

$$B = \frac{\mu_0 i_d\, r}{2\pi R^2} = \frac{\mu_0 (3.50 \times 10^{-7})(0.03)}{2\pi (0.05)^2} = 8.40 \times 10^{-13}\ \text{T}.$$

21. (a) At any instant the displacement current i_d in the gap between the plates equals the conduction current i in the wires. Thus $i_d = i = 2.0$ A.

(b) The rate of change of the electric field is

$$\frac{dE}{dt} = \frac{1}{\varepsilon_0 A}\left(\varepsilon_0 \frac{d\Phi_E}{dt}\right) = \frac{i_d}{\varepsilon_0 A} = \frac{2.0\,\text{A}}{(8.85 \times 10^{-12}\ \text{F/m})(1.0\,\text{m})^2} = 2.3 \times 10^{11}\ \frac{\text{V}}{\text{m·s}}.$$

(c) The displacement current through the indicated path is

$$i'_d = i_d \times \left(\frac{d^2}{L^2}\right) = (2.0\,\text{A})\left(\frac{0.50\text{m}}{1.0\text{m}}\right)^2 = 0.50\,\text{A}.$$

(d) The integral of the field around the indicated path is

$$\oint \vec{B} \cdot d\vec{s} = \mu_0 i'_d = (1.26 \times 10^{-16}\ \text{H/m})(0.50\,\text{A}) = 6.3 \times 10^{-7}\ \text{T·m}.$$

22. (a) Fig. 32-34 indicates that $i = 4.0$ A when $t = 20$ ms. Thus, $B_i = \mu_0 i/2\pi r = 0.89$ mT.

(b) Fig. 32-34 indicates that $i = 8.0$ A when $t = 40$ ms. Thus, $B_i \approx 0.18$ mT.

(c) Fig. 32-34 indicates that $i = 10$ A when $t > 50$ ms. Thus, $B_i \approx 0.220$ mT.

(d) Eq. 32-4 gives the displacement current in terms of the time-derivative of the electric field: $i_d = \varepsilon_0 A(dE/dt)$, but using Eq. 26-5 and Eq. 26-10 we have $E = \rho i/A$ (in terms of the real current); therefore, $i_d = \varepsilon_0 \rho(di/dt)$. For $0 < t < 50$ ms, Fig. 32-34 indicates that $di/dt = 200$ A/s. Thus, $B_{id} = \mu_0 i_d/2\pi r = 6.4 \times 10^{-22}$ T.

(e) As in (d), $B_{id} = \mu_0 i_d/2\pi r = 6.4 \times 10^{-22}$ T.

(f) Here $di/dt = 0$, so (by the reasoning in the previous step) $B = 0$.
(g) By the right-hand rule, the direction of $\vec{B}_i$ at $t = 20$ s is out of page.

(h) By the right-hand rule, the direction of $\vec{B}_{id}$ at $t = 20$ s is out of page.

23. (a) Eq. 32-16 (with Eq. 26-5) gives

$$B = \frac{\mu_0 i_d\, r}{2\pi R^2} = \frac{\mu_0 J_d A\, r}{2\pi R^2} = 75.4 \text{ nT}$$

where we set $A = \pi R^2$ (which led to several cancellations).

(b) Similarly, Eq. 32-17 gives $B = \dfrac{\mu_0 i_d}{2\pi r} = \dfrac{\mu_0 J_d\, \pi R^2}{2\pi r} = 67.9 \text{ nT}$.

24. (a) Eq. 32-16 gives $B = \dfrac{\mu_0 i_d\, r}{2\pi R^2} = 2.22 \ \mu\text{T}$.

(b) Eq. 32-17 gives $B = \dfrac{\mu_0 i_d}{2\pi r} = 2.00 \ \mu\text{T}$.

25. (a) Eq. 32-11 applies (though the last term is zero) but we must be careful with $i_{d,enc}$. It is the enclosed portion of the displacement current, and if we related this to the displacement current density J_d, then

$$i_{d,enc} = \int_0^r J_d\, 2\pi r\ dr = (4.00 \text{ A/m·s})(2\pi) \int_0^r \left(1 - \tfrac{r}{R}\right) r\ dr = 8\pi \left(\tfrac{1}{2} r^2 - \tfrac{r^3}{3R}\right)$$

with SI units understood. Now, we apply Eq. 32-17 (with i_d replaced with $i_{d,enc}$) or start from scratch with Eq. 32-11, to get $B = \dfrac{\mu_0\, i_{d\,enc}}{2\pi r} = 27.9 \text{ nT}$.

(b) The integral shown above will no longer (since now $r > R$) have r as the upper limit; the upper limit is now R. Thus,

$$i_{d,enc} = i_d = 8\pi \left(\tfrac{1}{2} R^2 - \tfrac{R^3}{3R}\right) = \tfrac{4}{3}\pi R^2.$$

Now Eq. 32-17 gives $B = \dfrac{\mu_0 i_d}{2\pi r} = 15.1 \text{ nT}$.

26. (a) Eq. 32-11 applies (though the last term is zero) but we must be careful with $i_{d,enc}$. It is the enclosed portion of the displacement current. Thus Eq. 32-17 (which derives from Eq. 32-11) becomes, with i_d replaced with $i_{d,enc}$,

$$B = \frac{\mu_0\, i_{d\,enc}}{2\pi r} = \frac{\mu_0\, (3.00 \text{ A})\, r}{2\pi r\, R}$$

which yields (after canceling r, and setting $R = 0.0300$ m) $B = 20.0 \ \mu\text{T}$.

(b) Here $i_d = 3.00$ A, and we get $B = \dfrac{\mu_0 i_d}{2\pi r} = 12.0 \ \mu\text{T}$.

27. The horizontal component of the Earth's magnetic field is given by $B_h = B\cos\phi_i$, where B is the magnitude of the field and ϕ_i is the inclination angle. Thus

$$B = \frac{B_h}{\cos\phi_i} = \frac{16\,\mu\text{T}}{\cos 73°} = 55\,\mu\text{T} .$$

28. (a) The flux through Arizona is

$$\Phi = -B_r A = -(43\times 10^{-6}\,\text{T})(295{,}000\,\text{km}^2)(10^3\,\text{m/km})^2 = -1.3\times 10^7\ \text{Wb} ,$$

inward. By Gauss' law this is equal to the negative value of the flux Φ' through the rest of the surface of the Earth. So $\Phi' = 1.3 \times 10^7$ Wb.

(b) The direction is outward.

29. We use Eq. 32-31: $\mu_{\text{orb},z} = -m_\ell \mu_B$.

(a) For $m_\ell = 1$, $\mu_{\text{orb},z} = -(1)\,(9.3\times 10^{-24}\ \text{J/T}) = -9.3\times 10^{-24}$ J/T.

(b) For $m_\ell = -2$, $\mu_{\text{orb},z} = -(-2)\,(9.3\times 10^{-24}\ \text{J/T}) = 1.9\times 10^{-23}$ J/T.

30. We use Eq. 32-27 to obtain $\Delta U = -\Delta(\mu_{s,z}B) = -B\Delta\mu_{s,z}$, where $\mu_{s,z} = \pm eh/4\pi m_e = \pm\mu_B$ (see Eqs. 32-24 and 32-25). Thus,

$$\Delta U = -B\left[\mu_B - (-\mu_B)\right] = 2\mu_B B = 2(9.27\times 10^{-24}\ \text{J/T})(0.25\,\text{T}) = 4.6\times 10^{-24}\ \text{J} .$$

31. (a) Since $m_\ell = 0$, $L_{\text{orb},z} = m_\ell h/2\pi = 0$.

(b) Since $m_\ell = 0$, $\mu_{\text{orb},z} = -m_\ell \mu_B = 0$.

(c) Since $m_\ell = 0$, then from Eq. 32-32, $U = -\mu_{\text{orb},z}B_{\text{ext}} = -m_\ell \mu_B B_{\text{ext}} = 0$.

(d) Regardless of the value of m_ℓ, we find for the spin part

$$U = -\mu_{s,z}B = \pm\mu_B B = \pm(9.27\times 10^{-24}\ \text{J/T})(35\,\text{mT}) = \pm 3.2\times 10^{-25}\ \text{J} .$$

(e) Now $m_\ell = -3$, so

$$L_{orb,z} = \frac{m_\ell h}{2\pi} = \frac{(-3)\,(6.63\times10^{-27}\,\text{J}\cdot\text{s})}{2\pi} = -3.16\times10^{-34}\,\text{J}\cdot\text{s} \approx -3.2\times10^{-34}\,\text{J}\cdot\text{s}$$

(f) and

$$\mu_{orb,z} = -m_\ell \mu_B = -(-3)\,(9.27\times10^{-24}\,\text{J/T}) = 2.78\times10^{-23}\,\text{J/T} \approx 2.8\times10^{-23}\,\text{J/T}.$$

(g) The potential energy associated with the electron's orbital magnetic moment is now

$$U = -\mu_{orb,z}B_{ext} = -(2.78\times10^{-23}\,\text{J/T})(35\times10^{-3}\,\text{T}) = -9.7\times10^{-25}\,\text{J}.$$

(h) On the other hand, the potential energy associated with the electron spin, being independent of m_ℓ, remains the same: $\pm 3.2 \times 10^{-25}$ J.

32. Combining Eq. 32-27 with Eqs. 32-22 and 32-23, we see that the energy difference is

$$\Delta U = 2\,\mu_B\,B$$

where μ_B is the Bohr magneton (given in Eq. 32-25). With $\Delta U = 6.00 \times 10^{-25}$ J, we obtain $B = 32.3$ mT.

33. (a) The potential energy of the atom in association with the presence of an external magnetic field $\vec{B}_{ext}$ is given by Eqs. 32-31 and 32-32:

$$U = -\vec{\mu}_{orb}\cdot\vec{B}_{ext} = -\mu_{orb,z}B_{ext} = -m_\ell \mu_B B_{ext}.$$

For level E_1 there is no change in energy as a result of the introduction of $\vec{B}_{ext}$, so $U \propto m_\ell = 0$, meaning that $m_\ell = 0$ for this level.

(b) For level E_2 the single level splits into a triplet (i.e., three separate ones) in the presence of $\vec{B}_{ext}$, meaning that there are three different values of m_ℓ. The middle one in the triplet is unshifted from the original value of E_2 so its m_ℓ must be equal to 0. The other two in the triplet then correspond to $m_\ell = -1$ and $m_\ell = +1$, respectively.

(c) For any pair of adjacent levels in the triplet $|\Delta m_\ell| = 1$. Thus, the spacing is given by

$$\Delta U = |\Delta(-m_\ell \mu_B B)| = |\Delta m_\ell|\,\mu_B B = \mu_B B = (9.27\times10^{-24}\,\text{J/T})(0.50\text{T}) = 4.64\times10^{-24}\,\text{J}.$$

34. (a) A sketch of the field lines (due to the presence of the bar magnet) in the vicinity of the loop is shown below:

(b) The primary conclusion of §32-9 is two-fold: $\vec{u}$ is opposite to $\vec{B}$, and the effect of $\vec{F}$ is to move the material towards regions of smaller $|\vec{B}|$ values. The direction of the magnetic moment vector (of our loop) is toward the right in our sketch, or in the $+x$ direction.

(c) The direction of the current is clockwise (from the perspective of the bar magnet.)

(d) Since the size of $|\vec{B}|$ relates to the "crowdedness" of the field lines, we see that $\vec{F}$ is towards the right in our sketch, or in the $+x$ direction.

35. An electric field with circular field lines is induced as the magnetic field is turned on. Suppose the magnetic field increases linearly from zero to B in time t. According to Eq. 31-27, the magnitude of the electric field at the orbit is given by

$$E = \left(\frac{r}{2}\right)\frac{dB}{dt} = \left(\frac{r}{2}\right)\frac{B}{t} \, ,$$

where r is the radius of the orbit. The induced electric field is tangent to the orbit and changes the speed of the electron, the change in speed being given by

$$\Delta v = at = \frac{eE}{m_e}t = \left(\frac{e}{m_e}\right)\left(\frac{r}{2}\right)\left(\frac{B}{t}\right)t = \frac{erB}{2m_e} \, .$$

The average current associated with the circulating electron is $i = ev/2\pi r$ and the dipole moment is

$$\mu = Ai = \left(\pi r^2\right)\left(\frac{ev}{2\pi r}\right) = \frac{1}{2}evr \, .$$

The change in the dipole moment is

$$\Delta\mu = \frac{1}{2}er\Delta v = \frac{1}{2}er\left(\frac{erB}{2m_e}\right) = \frac{e^2 r^2 B}{4m_e} \, .$$

36. Reviewing Sample Problem 32-3 before doing this exercise is helpful. Let

$$K = \frac{3}{2}kT = \left|\vec{\mu}\cdot\vec{B} - \left(-\vec{\mu}\cdot\vec{B}\right)\right| = 2\mu B$$

which leads to

$$T = \frac{4\mu B}{3k} = \frac{4(1.0\times10^{-23}\,\text{J/T})(0.50\,\text{T})}{3(1.38\times10^{-23}\,\text{J/K})} = 0.48\,\text{K}\ .$$

37. The magnetization is the dipole moment per unit volume, so the dipole moment is given by $\mu = MV$, where M is the magnetization and V is the volume of the cylinder ($V = \pi r^2 L$, where r is the radius of the cylinder and L is its length). Thus,

$$\mu = M\pi r^2 L = (5.30\times10^3\,\text{A/m})\pi(0.500\times10^{-2}\,\text{m})^2(5.00\times10^{-2}\,\text{m}) = 2.08\times10^{-2}\,\text{J/T}\ .$$

38. (a) From Fig. 32-14 we estimate a slope of $B/T = 0.50$ T/K when $M/M_{max} = 50\%$. So

$$B = 0.50\ \text{T} = (0.50\ \text{T/K})(300\ \text{K}) = 1.5\times10^2\ \text{T}.$$

(b) Similarly, now $B/T \approx 2$ so $B = (2)(300) = 6.0\times10^2$ T.

(c) Except for very short times and in very small volumes, these values are not attainable in the lab.

39. For the measurements carried out, the largest ratio of the magnetic field to the temperature is $(0.50\ \text{T})/(10\ \text{K}) = 0.050$ T/K. Look at Fig. 32-14 to see if this is in the region where the magnetization is a linear function of the ratio. It is quite close to the origin, so we conclude that the magnetization obeys Curie's law.

40. Section 32-10 explains the terms used in this problem and the connection between M and μ. The graph in Fig. 32-37 gives a slope of

$$\frac{M/M_{max}}{B_{ext}/T} = \frac{0.15}{0.20} = 3/4$$

in Kelvins per Tesla. Thus we can write

$$\frac{\mu}{\mu_{max}} = ¾\ (0.800\ \text{T})/(2.00\ \text{K}) = 0.30\ .$$

41. (a) A charge e traveling with uniform speed v around a circular path of radius r takes time $T = 2\pi r/v$ to complete one orbit, so the average current is

$$i = \frac{e}{T} = \frac{ev}{2\pi r}.$$

The magnitude of the dipole moment is this multiplied by the area of the orbit:

$$\mu = \frac{ev}{2\pi r}\pi r^2 = \frac{evr}{2}.$$

Since the magnetic force with magnitude evB is centripetal, Newton's law yields $evB = m_e v^2/r$, so $r = m_e v / eB$. Thus,

$$\mu = \frac{1}{2}(ev)\left(\frac{m_e v}{eB}\right) = \left(\frac{1}{B}\right)\left(\frac{1}{2}m_e v^2\right) = \frac{K_e}{B}.$$

The magnetic force $-e\vec{v} \times \vec{B}$ must point toward the center of the circular path. If the magnetic field is directed out of the page (defined to be $+z$ direction), the electron will travel counterclockwise around the circle. Since the electron is negative, the current is in the opposite direction, clockwise and, by the right-hand rule for dipole moments, the dipole moment is into the page, or in the $-z$ direction. That is, the dipole moment is directed opposite to the magnetic field vector.

(b) We note that the charge canceled in the derivation of $\mu = K_e/B$. Thus, the relation $\mu = K_i/B$ holds for a positive ion.

(c) The direction of the dipole moment is $-z$, opposite to the magnetic field.

(d) The magnetization is given by $M = \mu_e n_e + \mu_i n_i$, where μ_e is the dipole moment of an electron, n_e is the electron concentration, μ_i is the dipole moment of an ion, and n_i is the ion concentration. Since $n_e = n_i$, we may write n for both concentrations. We substitute $\mu_e = K_e/B$ and $\mu_i = K_i/B$ to obtain

$$M = \frac{n}{B}(K_e + K_i) = \frac{5.3 \times 10^{21}\,\mathrm{m}^{-3}}{1.2\,\mathrm{T}}\left(6.2 \times 10^{-20}\,\mathrm{J} + 7.6 \times 10^{-21}\,\mathrm{J}\right) = 3.1 \times 10^2\,\mathrm{A/m}.$$

42. The Curie temperature for iron is 770°C. If x is the depth at which the temperature has this value, then $10°C + (30°C/km)x = 770°C$. Therefore,

$$x = \frac{770°\,C - 10°\,C}{30°\,C/km} = 25\,km.$$

43. (a) The field of a dipole along its axis is given by Eq. 30-29:

$$B = \frac{\mu_0}{2\pi} \frac{\mu}{z^3},$$

where μ is the dipole moment and z is the distance from the dipole. Thus,

$$B = \frac{\left(4\pi \times 10^{-7}\,\text{T}\cdot\text{m}/A\right)\left(1.5 \times 10^{-23}\,\text{J}/\text{T}\right)}{2\pi\left(10 \times 10^{-9}\,\text{m}\right)} = 3.0 \times 10^{-6}\,\text{T}.$$

(b) The energy of a magnetic dipole $\vec{\mu}$ in a magnetic field $\vec{B}$ is given by $U = -\vec{\mu}\cdot\vec{B} = -\mu B\cos\phi$, where ϕ is the angle between the dipole moment and the field. The energy required to turn it end-for-end (from $\phi = 0°$ to $\phi = 180°$) is

$$\Delta U = 2\mu B = 2\left(1.5 \times 10^{-23}\,\text{J}/\text{T}\right)\left(3.0 \times 10^{-6}\,\text{T}\right) = 9.0 \times 10^{-29}\,\text{J} = 5.6 \times 10^{-10}\,\text{eV}.$$

The mean kinetic energy of translation at room temperature is about 0.04 eV. Thus, if dipole-dipole interactions were responsible for aligning dipoles, collisions would easily randomize the directions of the moments and they would not remain aligned.

44. (a) The number of iron atoms in the iron bar is

$$N = \frac{\left(7.9\,\text{g}/\text{cm}^3\right)\left(5.0\,\text{cm}\right)\left(1.0\,\text{cm}^2\right)}{\left(55.847\,\text{g}/\text{mol}\right)/\left(6.022 \times 10^{23}/\text{mol}\right)} = 4.3 \times 10^{23}.$$

Thus the dipole moment of the iron bar is

$$\mu = \left(2.1 \times 10^{-23}\,\text{J}/\text{T}\right)\left(4.3 \times 10^{23}\right) = 8.9\,\text{A}\cdot\text{m}^2.$$

(b) $\tau = \mu B \sin 90° = (8.9\,\text{A}\cdot\text{m}^2)(1.57\,\text{T}) = 13\,\text{N}\cdot\text{m}.$

45. The saturation magnetization corresponds to complete alignment of all atomic dipoles and is given by $M_{\text{sat}} = \mu n$, where n is the number of atoms per unit volume and μ is the magnetic dipole moment of an atom. The number of nickel atoms per unit volume is $n = \rho/m$, where ρ is the density of nickel. The mass of a single nickel atom is calculated using $m = M/N_A$, where M is the atomic mass of nickel and N_A is Avogadro's constant. Thus,

$$n = \frac{\rho N_A}{M} = \frac{\left(8.90\,\text{g}/\text{cm}^3\right)\left(6.02 \times 10^{23}\,\text{atoms}/\text{mol}\right)}{58.71\,\text{g}/\text{mol}} = 9.126 \times 10^{22}\,\text{atoms}/\text{cm}^3$$
$$= 9.126 \times 10^{28}\,\text{atoms}/\text{m}^3.$$

The dipole moment of a single atom of nickel is

$$\mu = \frac{M_{sat}}{n} = \frac{4.70 \times 10^5 \text{ A/m}}{9.126 \times 10^{28} \text{ m}^3} = 5.15 \times 10^{-24} \text{ A} \cdot \text{m}^2.$$

46. (a) Eq. 29-36 gives

$$\tau = \mu_{rod} B \sin\theta = (2700 \text{ A/m})(0.06 \text{ m})\pi(0.003 \text{ m})^2(0.035 \text{ T})\sin(68°) = 1.49 \times 10^{-4} \text{ N·m}.$$

We have used the fact that the volume of a cylinder is its length times its (circular) cross sectional area.

(b) Using Eq. 29-38, we have

$$\begin{aligned}
\Delta U &= -\mu_{rod} B(\cos\theta_f - \cos\theta_i) \\
&= -(2700 \text{ A/m})(0.06 \text{ m})\pi(0.003\text{m})^2(0.035\text{T})[\cos(34°) - \cos(68°)] \\
&= -72.9 \text{ }\mu\text{J}.
\end{aligned}$$

47. (a) The magnitude of the toroidal field is given by $B_0 = \mu_0 n i_p$, where n is the number of turns per unit length of toroid and i_p is the current required to produce the field (in the absence of the ferromagnetic material). We use the average radius ($r_{avg} = 5.5$ cm) to calculate n:

$$n = \frac{N}{2\pi r_{avg}} = \frac{400 \text{ turns}}{2\pi(5.5 \times 10^{-2}\text{m})} = 1.16 \times 10^3 \text{ turns/m} .$$

Thus,

$$i_p = \frac{B_0}{\mu_0 n} = \frac{0.20 \times 10^{-3} \text{ T}}{(4\pi \times 10^{-7}\text{T} \cdot \text{m/A})(1.16 \times 10^3 \text{ /m})} = 0.14 \text{ A} .$$

(b) If Φ is the magnetic flux through the secondary coil, then the magnitude of the emf induced in that coil is $\varepsilon = N(d\Phi/dt)$ and the current in the secondary is $i_s = \varepsilon/R$, where R is the resistance of the coil. Thus,

$$i_s = \left(\frac{N}{R}\right)\frac{d\Phi}{dt}.$$

The charge that passes through the secondary when the primary current is turned on is

$$q = \int i_s \, dt = \frac{N}{R}\int\frac{d\Phi}{dt}dt = \frac{N}{R}\int_0^\Phi d\Phi = \frac{N\Phi}{R} .$$

The magnetic field through the secondary coil has magnitude $B = B_0 + B_M = 801B_0$, where B_M is the field of the magnetic dipoles in the magnetic material. The total field is perpendicular to the plane of the secondary coil, so the magnetic flux is $\Phi = AB$, where A

is the area of the Rowland ring (the field is inside the ring, not in the region between the ring and coil). If r is the radius of the ring's cross section, then $A = \pi r^2$. Thus,

$$\Phi = 801\pi r^2 B_0 .$$

The radius r is (6.0 cm – 5.0 cm)/2 = 0.50 cm and

$$\Phi = 801\pi (0.50 \times 10^{-2} \text{ m})^2 (0.20 \times 10^{-3} \text{T}) = 1.26 \times 10^{-5} \text{ Wb} .$$

Consequently,

$$q = \frac{50(1.26 \times 10^{-5} \text{ Wb})}{8.0 \ \Omega} = 7.9 \times 10^{-5} \text{ C} .$$

48. From Eq. 29-37 (see also Eq. 29-36) we write the torque as $\tau = -\mu B_h \sin\theta$ where the minus indicates that the torque opposes the angular displacement θ (which we will assume is small and in radians). The small angle approximation leads to $\tau \approx -\mu B_h \theta$, which is an indicator for simple harmonic motion (see section 16-5, especially Eq. 16-22). Comparing with Eq. 16-23, we then find the period of oscillation is

$$T = 2\pi \sqrt{\frac{I}{\mu B_h}}$$

where I is the rotational inertial that we asked to solve for. Since the frequency is given as 0.312 Hz, then the period is $T = 1/f = 1/0.312$ in SI units. Similarly, $B_h = 18.0 \times 10^{-6}$ and $\mu = 6.80 \times 10^{-4}$. The above relation then yields $I = 3.19 \times 10^{-9} \text{ kg·m}^2$.

49. (a) If the magnetization of the sphere is saturated, the total dipole moment is $\mu_{\text{total}} = N\mu$, where N is the number of iron atoms in the sphere and μ is the dipole moment of an iron atom. We wish to find the radius of an iron sphere with N iron atoms. The mass of such a sphere is Nm, where m is the mass of an iron atom. It is also given by $4\pi\rho R^3/3$, where ρ is the density of iron and R is the radius of the sphere. Thus $Nm = 4\pi\rho R^3/3$ and

$$N = \frac{4\pi\rho R^3}{3m} .$$

We substitute this into $\mu_{\text{total}} = N\mu$ to obtain

$$\mu_{\text{total}} = \frac{4\pi\rho R^3 \mu}{3m} .$$

We solve for R and obtain

$$R = \left(\frac{3m\mu_{total}}{4\pi\rho\mu} \right)^{1/3}.$$

The mass of an iron atom is $m = 56\,u = (56\,u)(1.66 \times 10^{-27}\ \text{kg/u}) = 9.30 \times 10^{-26}\ \text{kg}$. Therefore,

$$R = \left[\frac{3(9.30 \times 10^{-26}\ \text{kg})(8.0 \times 10^{22}\ \text{J/T})}{4\pi(14 \times 10^{3}\ \text{kg/m}^{3})(2.1 \times 10^{-23}\ \text{J/T})} \right]^{1/3} = 1.8 \times 10^{5}\ \text{m}.$$

(b) The volume of the sphere is $V_s = \dfrac{4\pi}{3} R^3 = \dfrac{4\pi}{3}(1.82 \times 10^5\ \text{m})^3 = 2.53 \times 10^{16}\ \text{m}^3$ and the volume of the Earth is

$$V_e = \frac{4\pi}{3}(6.37 \times 10^6\ \text{m})^3 = 1.08 \times 10^{21}\ \text{m}^3,$$

so the fraction of the Earth's volume that is occupied by the sphere is

$$\frac{2.53 \times 10^{16}\ \text{m}^3}{1.08 \times 10^{21}\ \text{m}^3} = 2.3 \times 10^{-5}.$$

50. The integral of the field along the indicated path is, by Eq. 32-18 and Eq. 32-19, equal to

$\mu_0 i_d$(enclosed area)/(total area) = μ_0(0.75 A)(4 cm ×2 cm)/(12 cm)^2 = 52 nT·m.

51. (a) Inside the gap of the capacitor, $B_1 = \mu_0 i_d\, r_1/2\pi R^2$ (Eq. 32-16); outside the gap the magnetic field is $B_2 = \mu_0 i_d/2\pi r_2$ (Eq. 32-17). Consequently, $B_2 = B_1 R^2/r_1 r_2 = 16.7$ nT.

(b) The displacement current is $i_d = 2\pi B_1 R^2/\mu_0 r_1 = 5.00$ mA.

52. (a) The period of rotation is $T = 2\pi/\omega$ and in this time all the charge passes any fixed point near the ring. The average current is $i = q/T = q\omega/2\pi$ and the magnitude of the magnetic dipole moment is

$$\mu = iA = \frac{q\omega}{2\pi}\pi r^2 = \frac{1}{2} q\omega r^2.$$

(b) We curl the fingers of our right hand in the direction of rotation. Since the charge is positive, the thumb points in the direction of the dipole moment. It is the same as the direction of the angular momentum vector of the ring.

53. (a) We use the result of part (a) in Sample Problem 32-1:

$$B = \frac{\mu_0 \varepsilon_0 r}{2} \frac{dE}{dt} \quad (\text{for } r \le R),$$

where $r = 0.80R$ and

$$\frac{dE}{dt} = \frac{d}{dt}\left(\frac{V}{d}\right) = \frac{1}{d}\frac{d}{dt}\left(V_0 e^{-t/\tau}\right) = -\frac{V_0}{\tau d} e^{-t/\tau}.$$

Here $V_0 = 100$ V. Thus

$$B(t) = \left(\frac{\mu_0 \varepsilon_0 r}{2}\right)\left(-\frac{V_0}{\tau d} e^{-t/\tau}\right) = -\frac{\mu_0 \varepsilon_0 V_0 r}{2\tau d} e^{-t/\tau}$$

$$= -\frac{\left(4\pi \times 10^{-7} \text{ T} \cdot \text{m/A}\right)\left(8.85 \times 10^{-12} \frac{\text{C}^2}{\text{N} \cdot \text{m}^2}\right)(100 \text{ V})(0.80)(16 \text{ mm})}{2\left(12 \times 10^{-3} \text{ s}\right)(5.0 \text{ mm})} e^{-t/12 \text{ ms}}$$

$$= -\left(1.2 \times 10^{-13} \text{ T}\right) e^{-t/12 \text{ ms}}.$$

The magnitude is $\left|B(t)\right| = \left(1.2 \times 10^{-13} \text{ T}\right) e^{-t/12 \text{ ms}}$.

(b) At time $t = 3\tau$, $B(t) = -(1.2 \times 10^{-13} \text{ T})e^{-3\tau/\tau} = -5.9 \times 10^{-15}$ T, with a magnitude $|B(t)| = 5.9 \times 10^{-15}$ T.

54. (a) Eq. 30-22 gives $B = \dfrac{\mu_0 i r}{2\pi R^2} = 222$ μT.

(b) Eq. 30-19 (or Eq. 30-6) gives $B = \dfrac{\mu_0 i}{2\pi r} = 167$ μT.

(c) As in part (b), we obtain a field of $\dfrac{\mu_0 i}{2\pi r} = 22.7$ μT.

(d) Eq. 32-16 (with Eq. 32-15) gives $B = \dfrac{\mu_0 i_d r}{2\pi R^2} = 1.25$ μT.

(e) As in part (d), we get $\dfrac{\mu_0 i_d r}{2\pi R^2} = 3.75$ μT.

(f) Eq. 32-17 yields $B = 22.7$ μT.

(g) Because the displacement current in the gap is spread over a larger cross-sectional area, values of B within that area are relatively small. Outside that cross-sectional area, the two values of B are identical. See Fig. 32-23b.

55. (a) Again from Fig. 32-14, for $M/M_{\max} = 50\%$ we have $B/T = 0.50$. So $T = B/0.50 = 2/0.50 = 4$ K.

(b) Now $B/T = 2.0$, so $T = 2/2.0 = 1$ K.

56. (a) The complete set of values are $\{-4, -3, -2, -1, 0, +1, +2, +3, +4\} \Rightarrow$ nine values in all.

(b) The maximum value is $4\mu_B = 3.71 \times 10^{-23}$ J/T.

(c) Multiplying our result for part (b) by 0.250 T gives $U = +9.27 \times 10^{-24}$ J.

(d) Similarly, for the lower limit, $U = -9.27 \times 10^{-24}$ J.

57. (a) Using Eq. 27-10, we find $E = \rho J = \dfrac{\rho i}{A} = \dfrac{\left(1.62 \times 10^{-8}\,\Omega \cdot \text{m}\right)(100\,\text{A})}{5.00 \times 10^{-6}\,\text{m}^2} = 0.324$ V/m.

(b) The displacement current is

$$i_d = \varepsilon_0 \frac{d\Phi_E}{dt} = \varepsilon_0 A \frac{dE}{dt} = \varepsilon_0 A \frac{d}{dt}\left(\frac{\rho i}{A}\right) = \varepsilon_0 \rho \frac{di}{dt} = \left(8.85 \times 10^{-12}\,\text{F}\right)\left(1.62 \times 10^{-8}\,\Omega\right)(2000\,\text{A/s})$$

$$= 2.87 \times 10^{-16}\,\text{A}.$$

(c) The ratio of fields is $\dfrac{B(\text{due to } i_d)}{B(\text{due to } i)} = \dfrac{\mu_0 i_d / 2\pi r}{\mu_0 i / 2\pi r} = \dfrac{i_d}{i} = \dfrac{2.87 \times 10^{-16}\,\text{A}}{100\,\text{A}} = 2.87 \times 10^{-18}.$

58. (a) Using Eq. 32-31, we find

$$\mu_{\text{orb},z} = -3\mu_B = -2.78 \times 10^{-23} \text{ J/T}.$$

(That these are acceptable units for magnetic moment is seen from Eq. 32-32 or Eq. 32-27; they are equivalent to A·m^2).

(b) Similarly, for $m_\ell = -4$ we obtain $\mu_{\text{orb},z} = 3.71 \times 10^{-23}$ J/T.

59. Let the area of each circular plate be A and that of the central circular section be a, then

$$\frac{A}{a} = \frac{\pi R^2}{\pi(R/2)^2} = 4 \ .$$

Thus, from Eqs. 32-14 and 32-15 the total discharge current is given by $i = i_d = 4(2.0\text{ A}) = 8.0\text{ A}$.

60. The interacting potential energy between the magnetic dipole of the compass and the Earth's magnetic field is $U = -\vec{\mu} \cdot \vec{B}_e = -\mu B_e \cos\theta$, where θ is the angle between $\vec{\mu}$ and $\vec{B}_e$. For small angle θ

$$U(\theta) = -\mu B_e \cos\theta \approx -\mu B_e \left(1 - \frac{\theta^2}{2}\right) = \frac{1}{2}\kappa\theta^2 - \mu B_e$$

where $\kappa = \mu B_e$. Conservation of energy for the compass then gives

$$\frac{1}{2}I\left(\frac{d\theta}{dt}\right)^2 + \frac{1}{2}\kappa\theta^2 = \text{const.}$$

This is to be compared with the following expression for the mechanical energy of a spring-mass system:

$$\frac{1}{2}m\left(\frac{dx}{dt}\right)^2 + \frac{1}{2}kx^2 = \text{const.},$$

which yields $\omega = \sqrt{k/m}$. So by analogy, in our case

$$\omega = \sqrt{\frac{\kappa}{I}} = \sqrt{\frac{\mu B_e}{I}} = \sqrt{\frac{\mu B_e}{ml^2/12}},$$

which leads to

$$\mu = \frac{ml^2\omega^2}{12B_e} = \frac{(0.050\,\text{kg})(4.0\times 10^{-2}\,\text{m})^2(45\,\text{rad/s})^2}{12(16\times 10^{-6}\,\text{T})} = 8.4\times 10^2\,\text{J/T}.$$

61. (a) At any instant the displacement current i_d in the gap between the plates equals the conduction current i in the wires. Thus $i_{max} = i_{d\,max} = 7.60\ \mu\text{A}$.

(b) Since $i_d = \varepsilon_0\,(d\Phi_E/dt)$,

$$\left(\frac{d\Phi_E}{dt}\right)_{max} = \frac{i_{d\,max}}{\varepsilon_0} = \frac{7.60\times 10^{-6}\,\text{A}}{8.85\times 10^{-12}\,\text{F/m}} = 8.59\times 10^5\,\text{V}\cdot\text{m/s}.$$

(c) According to problem 13,

$$i_d = C \frac{dV}{dt} = \frac{\varepsilon_0 A}{d} \frac{dV}{dt}.$$

Now the potential difference across the capacitor is the same in magnitude as the emf of the generator, so $V = \varepsilon_m \sin \omega t$ and $dV/dt = \omega \varepsilon_m \cos \omega t$. Thus, $i_d = (\varepsilon_0 A \omega \varepsilon_m / d) \cos \omega t$ and $i_{d\,max} = \varepsilon_0 A \omega \varepsilon_m / d$. This means

$$d = \frac{\varepsilon_0 A \omega \varepsilon_m}{i_{d\,max}} = \frac{\left(8.85 \times 10^{-12}\ \text{F/m}\right) \pi \left(0.180\ \text{m}\right)^2 \left(130\ \text{rad/s}\right) \left(220\ \text{V}\right)}{7.60 \times 10^{-6}\ \text{A}} = 3.39 \times 10^{-3}\ \text{m},$$

where $A = \pi R^2$ was used.

(d) We use the Ampere-Maxwell law in the form $\oint \vec{B} \cdot d\vec{s} = \mu_0 I_d$, where the path of integration is a circle of radius r between the plates and parallel to them. I_d is the displacement current through the area bounded by the path of integration. Since the displacement current density is uniform between the plates, $I_d = (r^2/R^2) i_d$, where i_d is the total displacement current between the plates and R is the plate radius. The field lines are circles centered on the axis of the plates, so $\vec{B}$ is parallel to $d\vec{s}$. The field has constant magnitude around the circular path, so $\oint \vec{B} \cdot d\vec{s} = 2\pi r B$. Thus,

$$2\pi r B = \mu_0 \left(\frac{r^2}{R^2} \right) i_d \quad \Rightarrow \quad B = \frac{\mu_0 i_d r}{2\pi R^2}.$$

The maximum magnetic field is given by

$$B_{max} = \frac{\mu_0 i_{d\,max} r}{2\pi R^2} = \frac{\left(4\pi \times 10^{-7}\ \text{T}\cdot\text{m/A}\right)\left(7.6 \times 10^{-6}\ \text{A}\right)\left(0.110\,\text{m}\right)}{2\pi \left(0.180\,\text{m}\right)^2} = 5.16 \times 10^{-12}\ \text{T}.$$

62. The definition of displacement current is Eq. 32-10, and the formula of greatest convenience here is Eq. 32-17:

$$i_d = \frac{2\pi r\ B}{\mu_0} = \frac{2\pi \left(0.0300\,\text{m}\right)\left(2.00 \times 10^{-6}\ \text{T}\right)}{4\pi \times 10^{-7}\ \text{T}\cdot\text{m/A}} = 0.300\ \text{A}.$$

63. (a) For a given value of ℓ, m_ℓ varies from $-\ell$ to $+\ell$. Thus, in our case $\ell = 3$, and the number of different m_ℓ's is $2\ell + 1 = 2(3) + 1 = 7$. Thus, since $L_{orb,z} \propto m_\ell$, there are a total of seven different values of $L_{orb,z}$.

(b) Similarly, since $\mu_{orb,z} \propto m_\ell$, there are also a total of seven different values of $\mu_{orb,z}$.

(c) Since $L_{orb,z} = m_\ell h/2\pi$, the greatest allowed value of $L_{orb,z}$ is given by $|m_\ell|_{max} h/2\pi = 3h/2\pi$.

(d) Similar to part (c), since $\mu_{orb,z} = -m_\ell \mu_B$, the greatest allowed value of $\mu_{orb,z}$ is given by $|m_\ell|_{max}\mu_B = 3eh/4\pi m_e$.

(e) From Eqs. 32-23 and 32-29 the z component of the net angular momentum of the electron is given by

$$L_{net,z} = L_{orb,z} + L_{s,z} = \frac{m_\ell h}{2\pi} + \frac{m_s h}{2\pi}.$$

For the maximum value of $L_{net,z}$ let $m_\ell = [m_\ell]_{max} = 3$ and $m_s = \frac{1}{2}$. Thus

$$[L_{net,z}]_{max} = \left(3 + \frac{1}{2}\right)\frac{h}{2\pi} = \frac{3.5h}{2\pi}.$$

(f) Since the maximum value of $L_{net,z}$ is given by $[m_J]_{max}h/2\pi$ with $[m_J]_{max} = 3.5$ (see the last part above), the number of allowed values for the z component of $L_{net,z}$ is given by $2[m_J]_{max} + 1 = 2(3.5) + 1 = 8$.

64. Ignoring points where the determination of the slope is problematic, we find the interval of largest $\Delta|\vec{E}|/\Delta t$ is 6 $\mu s < t < 7$ μs. During that time, we have, from Eq. 32-14,

$$i_d = \varepsilon_0 A \frac{\Delta|\vec{E}|}{\Delta t} = \varepsilon_0 (2.0\,\text{m}^2)(2.0 \times 10^6 \text{ V/m})$$

which yields $i_d = 3.5 \times 10^{-5}$ A.

65. (a) A sketch of the field lines (due to the presence of the bar magnet) in the vicinity of the loop is shown below:

(b) For paramagnetic materials, the dipole moment $\vec{\mu}$ is in the same direction as $\vec{B}$. From the above figure, $\vec{\mu}$ points in the $-x$ direction.

(c) Form the right-hand rule, since $\vec{\mu}$ points in the $-x$ direction, the current flows counterclockwise, from the perspective of the bar magnet.

(d) The effect of $\vec{F}$ is to move the material towards regions of larger $\left|\vec{B}\right|$ values. Since the size of $\left|\vec{B}\right|$ relates to the "crowdedness" of the field lines, we see that $\vec{F}$ is towards the left, or $-x$.

66. (a) From Eq. 21-3,

$$E = \frac{e}{4\pi\varepsilon_0 r^2} = \frac{\left(1.60 \times 10^{-19}\ \text{C}\right)\left(8.99 \times 10^9\ \text{N} \cdot \text{m}^2/\text{C}^2\right)}{\left(5.2 \times 10^{-11}\ \text{m}\right)^2} = 5.3 \times 10^{11}\ \text{N/C} .$$

(b) We use Eq. 29-28:

$$B = \frac{\mu_0}{2\pi}\frac{\mu_p}{r^3} = \frac{\left(4\pi \times 10^{-7}\ \text{T} \cdot \text{m/A}\right)\left(1.4 \times 10^{-26}\ \text{J/T}\right)}{2\pi\left(5.2 \times 10^{-11}\ \text{m}\right)^3} = 2.0 \times 10^{-2}\ \text{T} .$$

(c) From Eq. 32-30,

$$\frac{\mu_{\text{orb}}}{\mu_p} = \frac{eh/4\pi m_e}{\mu_p} = \frac{\mu_B}{\mu_p} = \frac{9.27 \times 10^{-24}\ \text{J/T}}{1.4 \times 10^{-26}\ \text{J/T}} = 6.6 \times 10^2 .$$

67. (a) From $\mu = iA = i\pi R_e^2$ we get

$$i = \frac{\mu}{\pi R_e^2} = \frac{8.0 \times 10^{22}\ \text{J}/\text{T}}{\pi(6.37 \times 10^6\ \text{m})^2} = 6.3 \times 10^8\ \text{A} .$$

(b) Yes, because far away from the Earth the fields of both the Earth itself and the current loop are dipole fields. If these two dipoles cancel each other out, then the net field will be zero.

(c) No, because the field of the current loop is not that of a magnetic dipole in the region close to the loop.

68. (a) Using Eq. 32-13 but noting that the capacitor is being *discharged*, we have

$$\frac{d\left|\vec{E}\right|}{dt} = -\frac{i}{\varepsilon_0 A} = -8.8 \times 10^{15}$$

where $A = (0.0080)^2$ and SI units are understood.

(b) Assuming a perfectly uniform field, even so near to an edge (which is consistent with the fact that fringing is neglected in §32-4), we follow part (a) of Sample Problem 32-2 and relate the (absolute value of the) line integral to the portion of displacement current enclosed.

$$\left| \oint \vec{B} \cdot d\vec{s} \right| = \mu_0 i_{d,\text{enc}} = \mu_0 \left(\frac{WH}{L^2} i \right) = 5.9 \times 10^{-7} \text{ Wb/m}.$$

69. (a) We use the notation $P(\mu)$ for the probability of a dipole being parallel to $\vec{B}$, and $P(-\mu)$ for the probability of a dipole being antiparallel to the field. The magnetization may be thought of as a "weighted average" in terms of these probabilities:

$$M = \frac{N\mu P(\mu) - N\mu P(-\mu)}{P(\mu) + P(-\mu)} = \frac{N\mu \left(e^{\mu B/KT} - e^{-\mu B/KT} \right)}{e^{\mu B/KT} + e^{-\mu B/KT}} = N\mu \tanh\left(\frac{\mu B}{kT} \right).$$

(b) For $\mu B \ll kT$ (that is, $\mu B / kT \ll 1$) we have $e^{\pm \mu B/kT} \approx 1 \pm \mu B/kT$, so

$$M = N\mu \tanh\left(\frac{\mu B}{kT} \right) \approx \frac{N\mu \left[(1 + \mu B/kT) - (1 - \mu B/kT) \right]}{(1 + \mu B/kT) + (1 - \mu B/kT)} = \frac{N\mu^2 B}{kT}.$$

(c) For $\mu B \gg kT$ we have $\tanh(\mu B/kT) \approx 1$, so $M = N\mu \tanh\left(\frac{\mu B}{kT} \right) \approx N\mu$.

(d) One can easily plot the tanh function using, for instance, a graphical calculator. One can then note the resemblance between such a plot and Fig. 32-14. By adjusting the parameters used in one's plot, the curve in Fig. 32-14 can reliably be fit with a tanh function.

70. (a) From Eq. 32-1, we have

$$\left(\Phi_B \right)_{\text{in}} = \left(\Phi_B \right)_{\text{out}} = 0.0070 \text{ Wb} + (0.40 \text{ T})(\pi r^2) = 9.2 \times 10^{-3} \text{ Wb}.$$

Thus, the magnetic of the magnetic flux is 9.2 mWb.

(b) The flux is inward.

71. (a) The Pythagorean theorem leads to

$$B = \sqrt{B_h^2 + B_v^2} = \sqrt{\left(\frac{\mu_0 \mu}{4\pi r^3} \cos\lambda_m\right)^2 + \left(\frac{\mu_0 \mu}{2\pi r^3} \sin\lambda_m\right)^2} = \frac{\mu_0 \mu}{4\pi r^3}\sqrt{\cos^2\lambda_m + 4\sin^2\lambda_m}$$

$$= \frac{\mu_0 \mu}{4\pi r^3}\sqrt{1 + 3\sin^2\lambda_m},$$

where $\cos^2\lambda_m + \sin^2\lambda_m = 1$ was used.

(b) We use Eq. 3-6:

$$\tan\phi_i = \frac{Bv}{Bh} = \frac{\left(\mu_0\mu/2\pi r^3\right)\sin\lambda_m}{\left(\mu_0\mu/4\pi r^3\right)\cos\lambda_m} = 2\tan\lambda_m .$$

72. (a) At the magnetic equator ($\lambda_m = 0$), the field is

$$B = \frac{\mu_0 \mu}{4\pi r^3} = \frac{\left(4\pi\times10^{-7}\text{ T·m/A}\right)\left(8.00\times10^{22}\text{ A·m}^2\right)}{4\pi\left(6.37\times10^6\text{ m}\right)^3} = 3.10\times10^{-5}\text{ T.}$$

(b) $\phi_i = \tan^{-1}(2\tan\lambda_m) = \tan^{-1}(0) = 0°$.

(c) At $\lambda_m = 60.0°$, we find

$$B = \frac{\mu_0\mu}{4\pi r^3}\sqrt{1 + 3\sin^2\lambda_m} = \left(3.10\times10^{-5}\right)\sqrt{1 + 3\sin^2 60.0°} = 5.59\times10^{-5}\text{ T.}$$

(d) $\phi_i = \tan^{-1}(2\tan 60.0°) = 73.9°$.

(e) At the north magnetic pole ($\lambda_m = 90.0°$), we obtain

$$B = \frac{\mu_0\mu}{4\pi r^3}\sqrt{1 + 3\sin^2\lambda_m} = \left(3.10\times10^{-5}\right)\sqrt{1 + 3(1.00)^2} = 6.20\times10^{-5}\text{ T.}$$

(f) $\phi_i = \tan^{-1}(2\tan 90.0°) = 90.0°$.

73. (a) At a distance r from the center of the Earth, the magnitude of the magnetic field is given by

$$B = \frac{\mu_0\mu}{4\pi r^3}\sqrt{1 + 3\sin^2\lambda_m} ,$$

where μ is the Earth's dipole moment and λ_m is the magnetic latitude. The ratio of the field magnitudes for two different distances at the same latitude is

$$\frac{B_2}{B_1} = \frac{r_1^3}{r_2^3} .$$

With B_1 being the value at the surface and B_2 being half of B_1, we set r_1 equal to the radius R_e of the Earth and r_2 equal to $R_e + h$, where h is altitude at which B is half its value at the surface. Thus,

$$\frac{1}{2} = \frac{R_e^3}{\left(R_e + h\right)^3} .$$

Taking the cube root of both sides and solving for h, we get

$$h = \left(2^{1/3} - 1\right) R_e = \left(2^{1/3} - 1\right)(6370\,\text{km}) = 1.66 \times 10^3\,\text{km}.$$

(b) We use the expression for B obtained in problem 6, part (a). For maximum B, we set $\sin \lambda_m = 1.00$. Also, $r = 6370\,\text{km} - 2900\,\text{km} = 3470\,\text{km}$. Thus,

$$B_{\text{max}} = \frac{\mu_0 \mu}{4\pi r^3}\sqrt{1 + 3\sin^2 \lambda_m} = \frac{\left(4\pi \times 10^{-7}\,\text{T} \cdot \text{m/A}\right)\left(8.00 \times 10^{22}\,\text{A} \cdot \text{m}^2\right)}{4\pi\left(3.47 \times 10^6\,\text{m}\right)^3}\sqrt{1 + 3(1.00)^2}$$

$$= 3.83 \times 10^{-4}\,\text{T}.$$

(c) The angle between the magnetic axis and the rotational axis of the Earth is $11.5°$, so $\lambda_m = 90.0° - 11.5° = 78.5°$ at Earth's geographic north pole. Also $r = R_e = 6370\,\text{km}$. Thus,

$$B = \frac{\mu_0 \mu}{4\pi R_E^3}\sqrt{1 + 3\sin^2 \lambda_m} = \frac{\left(4\pi \times 10^{-7}\,\text{T} \cdot \text{m/A}\right)\left(8.0 \times 10^{22}\,\text{J/T}\right)\sqrt{1 + 3\sin^2 78.5°}}{4\pi\left(6.37 \times 10^6\,\text{m}\right)^3}$$

$$= 6.11 \times 10^{-5}\,\text{T}.$$

(d) $\phi_i = \tan^{-1}\left(2 \tan 78.5°\right) = 84.2°$.

(e) A plausible explanation to the discrepancy between the calculated and measured values of the Earth's magnetic field is that the formulas we obtained in problem 6 are based on dipole approximation, which does not accurately represent the Earth's actual magnetic field distribution on or near its surface. (Incidentally, the dipole approximation becomes more reliable when we calculate the Earth's magnetic field far from its center.)

74. Let R be the radius of a capacitor plate and r be the distance from axis of the capacitor. For points with $r \leq R$, the magnitude of the magnetic field is given by

$$B = \frac{\mu_0 \varepsilon_0 r}{2} \frac{dE}{dt},$$

and for $r \geq R$, it is

$$B = \frac{\mu_0 \varepsilon_0 R^2}{2r} \frac{dE}{dt}.$$

The maximum magnetic field occurs at points for which $r = R$, and its value is given by either of the formulas above:

$$B_{max} = \frac{\mu_0 \varepsilon_0 R}{2} \frac{dE}{dt}.$$

There are two values of r for which $B = B_{max}/2$: one less than R and one greater.

(a) To find the one that is less than R, we solve

$$\frac{\mu_0 \varepsilon_0 r}{2} \frac{dE}{dt} = \frac{\mu_0 \varepsilon_0 R}{4} \frac{dE}{dt}$$

for r. The result is $r = R/2 = (55.0 \text{ mm})/2 = 27.5$ mm.

(b) To find the one that is greater than R, we solve

$$\frac{\mu_0 \varepsilon_0 R^2}{2r} \frac{dE}{dt} = \frac{\mu_0 \varepsilon_0 R}{4} \frac{dE}{dt}$$

for r. The result is $r = 2R = 2(55.0 \text{ mm}) = 110$ mm.

75. (a) Since the field lines of a bar magnet point towards its South pole, then the $\vec{B}$ arrows in one's sketch should point generally towards the left and also towards the central axis.

(b) The sign of $\vec{B} \cdot d\vec{A}$ for every $d\vec{A}$ on the side of the paper cylinder is negative.

(c) No, because Gauss' law for magnetism applies to an *enclosed* surface only. In fact, if we include the top and bottom of the cylinder to form an enclosed surface S then $\oint_S \vec{B} \cdot d\vec{A} = 0$ will be valid, as the flux through the open end of the cylinder near the magnet is positive.

Chapter 33

1. In air, light travels at roughly $c = 3.0 \times 10^8$ m/s. Therefore, for $t = 1.0$ ns, we have a distance of

$$d = ct = (3.0 \times 10^8 \text{ m/s})(1.0 \times 10^{-9} \text{ s}) = 0.30 \text{ m.}$$

2. (a) From Fig. 33-2 we find the smaller wavelength in question to be about 515 nm,

(b) and the larger wavelength to be approximately 610 nm.

(c) From Fig. 33-2 the wavelength at which the eye is most sensitive is about 555 nm.

(d) Using the result in (c), we have

$$f = \frac{c}{\lambda} = \frac{3.00 \times 10^8 \text{ m/s}}{555 \text{ nm}} = 5.41 \times 10^{14} \text{ Hz.}$$

(e) The period is $(5.41 \times 10^{14} \text{ Hz})^{-1} = 1.85 \times 10^{-15}$ s.

3. (a) The frequency of the radiation is

$$f = \frac{c}{\lambda} = \frac{3.0 \times 10^8 \text{ m/s}}{(1.0 \times 10^5)(6.4 \times 10^6 \text{ m})} = 4.7 \times 10^{-3} \text{ Hz.}$$

(b) The period of the radiation is

$$T = \frac{1}{f} = \frac{1}{4.7 \times 10^{-3} \text{ Hz}} = 212 \text{ s} = 3 \min 32 \text{ s.}$$

4. Since $\Delta\lambda \ll \lambda$, we find Δf is equal to

$$\left| \Delta\left(\frac{c}{\lambda}\right) \right| \approx \frac{c\Delta\lambda}{\lambda^2} = \frac{(3.0 \times 10^8 \text{ m/s})(0.0100 \times 10^{-9} \text{ m})}{(632.8 \times 10^{-9} \text{ m})^2} = 7.49 \times 10^9 \text{ Hz.}$$

5. If f is the frequency and λ is the wavelength of an electromagnetic wave, then $f\lambda = c$. The frequency is the same as the frequency of oscillation of the current in the LC circuit of the generator. That is, $f = 1/2\pi\sqrt{LC}$, where C is the capacitance and L is the inductance. Thus

$$\frac{\lambda}{2\pi\sqrt{LC}} = c.$$

The solution for L is

$$L = \frac{\lambda^2}{4\pi^2 Cc^2} = \frac{\left(550 \times 10^{-9}\ \text{m}\right)^2}{4\pi^2 \left(17 \times 10^{-12}\ \text{F}\right)\left(2.998 \times 10^8\ \text{m}/\text{s}\right)^2} = 5.00 \times 10^{-21}\ \text{H}.$$

This is exceedingly small.

6. The emitted wavelength is

$$\lambda = \frac{c}{f} = 2\pi c\sqrt{LC} = 2\pi\left(2.998 \times 10^8\ \text{m/s}\right)\sqrt{\left(0.253 \times 10^{-6}\ \text{H}\right)\left(25.0 \times 10^{-12}\ \text{F}\right)} = 4.74\,\text{m}.$$

7. The amplitude of the magnetic field in the wave is

$$B_m = \frac{E_m}{c} = \frac{3.20 \times 10^{-4}\ \text{V}/\text{m}}{2.998 \times 10^8\ \text{m}/\text{s}} = 1.07 \times 10^{-12}\ \text{T}.$$

8. (a) The amplitude of the magnetic field is

$$B_m = \frac{E_m}{c} = \frac{2.0\,\text{V/m}}{2.998 \times 10^8\ \text{m/s}} = 6.67 \times 10^{-9}\,\text{T} \approx 6.7 \times 10^{-9}\,\text{T}..$$

(b) Since the $\vec{E}$-wave oscillates in the z direction and travels in the x direction, we have $B_x = B_z = 0$. So, the oscillation of the magnetic field is parallel to the y axis.

(c) The direction $(+x)$ of the electromagnetic wave propagation is determined by $\vec{E} \times \vec{B}$. If the electric field points in $+z$, then the magnetic field must point in the $-y$ direction.

With SI units understood, we may write

$$B_y = B_m \cos\left[\pi \times 10^{15}\left(t - \frac{x}{c}\right)\right] = \frac{2.0\cos\left[10^{15}\pi\left(t - x/c\right)\right]}{3.0 \times 10^8}$$

$$= \left(6.7 \times 10^{-9}\right)\cos\left[10^{15}\pi\left(t - \frac{x}{c}\right)\right]$$

9. If P is the power and Δt is the time interval of one pulse, then the energy in a pulse is

$$E = P\Delta t = \left(100 \times 10^{12}\ \text{W}\right)\left(1.0 \times 10^{-9}\ \text{s}\right) = 1.0 \times 10^5\ \text{J}.$$

10. The intensity of the signal at Proxima Centauri is

$$I = \frac{P}{4\pi r^2} = \frac{1.0 \times 10^6 \text{ W}}{4\pi\left[(4.3 \text{ly})(9.46 \times 10^{15} \text{ m/ly})\right]^2} = 4.8 \times 10^{-29} \text{ W/m}^2.$$

11. The intensity is the average of the Poynting vector:

$$I = S_{\text{avg}} = \frac{cB_m^2}{2\mu_0} = \frac{(3.0 \times 10^8 \text{ m/s})(1.0 \times 10^{-4} \text{ T})^2}{2(1.26 \times 10^{-6} \text{ H/m})^2} = 1.2 \times 10^6 \text{ W/m}^2.$$

12. (a) The amplitude of the magnetic field in the wave is

$$B_m = \frac{E_m}{c} = \frac{5.00 \text{ V/m}}{2.998 \times 10^8 \text{ m/s}} = 1.67 \times 10^{-8} \text{ T}.$$

(b) The intensity is the average of the Poynting vector:

$$I = S_{\text{avg}} = \frac{E_m^2}{2\mu_0 c} = \frac{(5.00 \text{ V/m})^2}{2(4\pi \times 10^{-7} \text{ T·m/A})(2.998 \times 10^8 \text{ m/s})} = 3.31 \times 10^{-2} \text{ W/m}^2.$$

13. (a) We use $I = E_m^2/2\mu_0 c$ to calculate E_m:

$$E_m = \sqrt{2\mu_0 I_c} = \sqrt{2(4\pi \times 10^{-7} \text{ T·m/A})(1.40 \times 10^3 \text{ W/m}^2)(2.998 \times 10^8 \text{ m/s})}$$
$$= 1.03 \times 10^3 \text{ V/m}.$$

(b) The magnetic field amplitude is therefore

$$B_m = \frac{E_m}{c} = \frac{1.03 \times 10^4 \text{ V/m}}{2.998 \times 10^8 \text{ m/s}} = 3.43 \times 10^{-6} \text{ T}.$$

14. (a) The power received is

$$P_r = (1.0 \times 10^{-12} \text{ W})\frac{\pi(300 \text{ m})^2/4}{4\pi(6.37 \times 10^6 \text{ m})^2} = 1.4 \times 10^{-22} \text{ W}.$$

(b) The power of the source would be

$$P = 4\pi r^2 I = 4\pi \left[\left(2.2 \times 10^4 \text{ ly} \right) \left(9.46 \times 10^{15} \text{ m/ly} \right) \right]^2 \left[\frac{1.0 \times 10^{-12} \text{ W}}{4\pi \left(6.37 \times 10^6 \text{ m} \right)^2} \right] = 1.1 \times 10^{15} \text{ W}.$$

15. (a) The magnetic field amplitude of the wave is

$$B_m = \frac{E_m}{c} = \frac{2.0 \text{ V/m}}{2.998 \times 10^8 \text{ m/s}} = 6.7 \times 10^{-9} \text{ T}.$$

(b) The intensity is

$$I = \frac{E_m^2}{2\mu_0 c} = \frac{\left(2.0 \text{ V/m} \right)^2}{2 \left(4\pi \times 10^{-7} \text{ T} \cdot \text{m/A} \right) \left(2.998 \times 10^8 \text{ m/s} \right)} = 5.3 \times 10^{-3} \text{ W/m}^2.$$

(c) The power of the source is

$$P = 4\pi r^2 I_{\text{avg}} = 4\pi \left(10 \text{ m} \right)^2 \left(5.3 \times 10^{-3} \text{ W/m}^2 \right) = 6.7 \text{ W}.$$

16. From the equation immediately preceding Eq. 33-12, we see that the maximum value of $\partial B / \partial t$ is ωB_m. We can relate B_m to the intensity: $B_m = E_m / c = \sqrt{2 c\mu_0 I} / c$, and relate the intensity to the power P (and distance r) using Eq. 33-27.   Finally, we relate ω to wavelength λ using $\omega = kc = 2\pi c / \lambda$. Putting all this together, we obtain

$$\left(\frac{\partial B}{\partial t} \right)_{\text{max}} = \sqrt{\frac{2 \mu_0 P}{4 \pi c}} \frac{2 \pi c}{\lambda r} = 3.44 \times 10^6 \text{ T/s}.$$

17. (a) The average rate of energy flow per unit area, or intensity, is related to the electric field amplitude E_m by $I = E_m^2 / 2\mu_0 c$, so

$$E_m = \sqrt{2\mu_0 cI} = \sqrt{2 \left(4\pi \times 10^{-7} \text{ H/m} \right) \left(2.998 \times 10^8 \text{ m/s} \right) \left(10 \times 10^{-6} \text{ W/m}^2 \right)}$$
$$= 8.7 \times 10^{-2} \text{ V/m}.$$

(b) The amplitude of the magnetic field is given by

$$B_m = \frac{E_m}{c} = \frac{8.7 \times 10^{-2} \text{ V/m}}{2.998 \times 10^8 \text{ m/s}} = 2.9 \times 10^{-10} \text{ T}.$$

(c) At a distance r from the transmitter, the intensity is $I = P / 2\pi r^2$, where P is the power of the transmitter over the hemisphere having a surface area $2\pi r^2$. Thus

$$P = 2\pi r^2 I = 2\pi \left(10 \times 10^3 \text{ m}\right)^2 \left(10 \times 10^{-6} \text{ W/m}^2\right) = 6.3 \times 10^3 \text{ W}.$$

18. (a) The expression $E_y = E_m \sin(kx - \omega t)$ it fits the requirement "at point P ... [it] is decreasing with time" if we imagine P is just to the right ($x > 0$) of the coordinate origin (but at a value of x less than $\pi/2k = \lambda/4$ which is where there would be a maximum, at $t = 0$). It is important to bear in mind, in this description, that the wave is moving to the right. Specifically, $x_P = \frac{1}{k} \sin^{-1}(1/4)$ so that $E_y = (1/4) E_m$ at $t = 0$, there. Also, $E_y = 0$ with our choice of expression for E_y. Therefore, part (a) is answered simply by solving for x_P. Since $k = 2\pi f/c$ we find

$$x_P = \frac{c}{2\pi f} \sin^{-1}(1/4) = 30.1 \text{ nm}.$$

(b) If we proceed to the right on the x axis (still studying this "snapshot" of the wave at $t = 0$) we find another point where $E_y = 0$ at a distance of one-half wavelength from the previous point where $E_y = 0$. Thus (since $\lambda = c/f$) the next point is at $x = \frac{1}{2}\lambda = \frac{1}{2}c/f$ and is consequently a distance $c/2f - x_P = 345$ nm to the right of P.

19. The plasma completely reflects all the energy incident on it, so the radiation pressure is given by $p_r = 2I/c$, where I is the intensity. The intensity is $I = P/A$, where P is the power and A is the area intercepted by the radiation. Thus

$$p_r = \frac{2P}{Ac} = \frac{2\left(1.5 \times 10^9 \text{ W}\right)}{\left(1.00 \times 10^{-6} \text{ m}^2\right)\left(2.998 \times 10^8 \text{ m/s}\right)} = 1.0 \times 10^7 \text{ Pa}.$$

20. The radiation pressure is

$$p_r = \frac{I}{c} = \frac{10 \text{ W/m}^2}{2.998 \times 10^8 \text{ m/s}} = 3.3 \times 10^{-8} \text{ Pa}.$$

21. Since the surface is perfectly absorbing, the radiation pressure is given by $p_r = I/c$, where I is the intensity. Since the bulb radiates uniformly in all directions, the intensity a distance r from it is given by $I = P/4\pi r^2$, where P is the power of the bulb. Thus

$$p_r = \frac{P}{4\pi r^2 c} = \frac{500 \text{ W}}{4\pi \left(1.5 \text{ m}\right)^2 \left(2.998 \times 10^8 \text{ m/s}\right)} = 5.9 \times 10^{-8} \text{ Pa}.$$

22. (a) The radiation pressure produces a force equal to

$$F_r = p_r \left(\pi R_e^2\right) = \left(\frac{I}{c}\right)\left(\pi R_e^2\right) = \frac{\pi \left(1.4 \times 10^3 \text{ W/m}^2\right)\left(6.37 \times 10^6 \text{ m}\right)^2}{2.998 \times 10^8 \text{ m/s}} = 6.0 \times 10^8 \text{ N}.$$

(b) The gravitational pull of the Sun on Earth is

$$F_{grav} = \frac{GM_s M_e}{d_{es}^2} = \frac{\left(6.67 \times 10^{-11} \, \text{N} \cdot \text{m}^2 / \text{kg}^2\right)\left(2.0 \times 10^{30} \, \text{kg}\right)\left(5.98 \times 10^{24} \, \text{kg}\right)}{\left(1.5 \times 10^{11} \, \text{m}\right)^2}$$

$$= 3.6 \times 10^{22} \, \text{N},$$

which is much greater than F_r.

23. (a) Since $c = \lambda f$, where λ is the wavelength and f is the frequency of the wave,

$$f = \frac{c}{\lambda} = \frac{2.998 \times 10^8 \, \text{m/s}}{3.0 \, \text{m}} = 1.0 \times 10^8 \, \text{Hz}.$$

(b) The angular frequency is

$$\omega = 2\pi f = 2\pi (1.0 \times 10^8 \, \text{Hz}) = 6.3 \times 10^8 \, \text{rad/s}.$$

(c) The angular wave number is

$$k = \frac{2\pi}{\lambda} = \frac{2\pi}{3.0 \, \text{m}} = 2.1 \, \text{rad/m}.$$

(d) The magnetic field amplitude is

$$B_m = \frac{E_m}{c} = \frac{300 \, \text{V/m}}{2.998 \times 10^8 \, \text{m/s}} = 1.0 \times 10^{-6} \, \text{T}.$$

(e) $\vec{B}$ must be in the positive z direction when $\vec{E}$ is in the positive y direction in order for $\vec{E} \times \vec{B}$ to be in the positive x direction (the direction of propagation).

(f) The intensity of the wave is

$$I = \frac{E_m^2}{2\mu_0 c} = \frac{(300 \, \text{V/m})^2}{2(4\pi \times 10^{-7} \, \text{H/m})(2.998 \times 10^8 \, \text{m/s})} = 119 \, \text{W/m}^2 \approx 1.2 \times 10^2 \, \text{W/m}^2.$$

(g) Since the sheet is perfectly absorbing, the rate per unit area with which momentum is delivered to it is I/c, so

$$\frac{dp}{dt} = \frac{IA}{c} = \frac{(119 \, \text{W/m}^2)(2.0 \, \text{m}^2)}{2.998 \times 10^8 \, \text{m/s}} = 8.0 \times 10^{-7} \, \text{N}.$$

(h) The radiation pressure is

$$p_r = \frac{dp/dt}{A} = \frac{8.0 \times 10^{-7} \text{ N}}{2.0 \text{ m}^2} = 4.0 \times 10^{-7} \text{ Pa.}$$

24. (a) We note that the cross section area of the beam is $\pi d^2/4$, where d is the diameter of the spot ($d = 2.00\lambda$). The beam intensity is

$$I = \frac{P}{\pi d^2/4} = \frac{5.00 \times 10^{-3} \text{ W}}{\pi \left[(2.00)(633 \times 10^{-9} \text{ m})\right]^2 / 4} = 3.97 \times 10^9 \text{ W/m}^2.$$

(b) The radiation pressure is

$$p_r = \frac{I}{c} = \frac{3.97 \times 10^9 \text{ W/m}^2}{2.998 \times 10^8 \text{ m/s}} = 13.2 \text{ Pa.}$$

(c) In computing the corresponding force, we can use the power and intensity to eliminate the area (mentioned in part (a)). We obtain

$$F_r = \left(\frac{\pi d^2}{4}\right) p_r = \left(\frac{P}{I}\right) p_r = \frac{(5.00 \times 10^{-3} \text{ W})(13.2 \text{ Pa})}{3.97 \times 10^9 \text{ W/m}^2} = 1.67 \times 10^{-11} \text{ N.}$$

(d) The acceleration of the sphere is

$$a = \frac{F_r}{m} = \frac{F_r}{\rho(\pi d^3/6)} = \frac{6(1.67 \times 10^{-11} \text{ N})}{\pi(5.00 \times 10^3 \text{ kg/m}^3)[(2.00)(633 \times 10^{-9} \text{ m})]^3}$$
$$= 3.14 \times 10^3 \text{ m/s}^2.$$

25. Let f be the fraction of the incident beam intensity that is reflected. The fraction absorbed is $1 - f$. The reflected portion exerts a radiation pressure of

$$p_r = \frac{2f I_0}{c}$$

and the absorbed portion exerts a radiation pressure of

$$p_a = \frac{(1-f)I_0}{c},$$

where I_0 is the incident intensity. The factor 2 enters the first expression because the momentum of the reflected portion is reversed. The total radiation pressure is the sum of the two contributions:

$$P_{total} = p_r + p_a = \frac{2fI_0 + (1-f)I_0}{c} = \frac{(1+f)I_0}{c}.$$

To relate the intensity and energy density, we consider a tube with length ℓ and cross-sectional area A, lying with its axis along the propagation direction of an electromagnetic wave. The electromagnetic energy inside is $U = uA\ell$, where u is the energy density. All this energy passes through the end in time $t = \ell / c$, so the intensity is

$$I = \frac{U}{At} = \frac{uA\ell c}{A\ell} = uc.$$

Thus $u = I/c$. The intensity and energy density are positive, regardless of the propagation direction. For the partially reflected and partially absorbed wave, the intensity just outside the surface is $I = I_0 + fI_0 = (1 + f)I_0$, where the first term is associated with the incident beam and the second is associated with the reflected beam. Consequently, the energy density is

$$u = \frac{I}{c} = \frac{(1+f)I_0}{c},$$

the same as radiation pressure.

26. The mass of the cylinder is $m = \rho(\pi D^2 / 4)H$, where D is the diameter of the cylinder. Since it is in equilibrium

$$F_{net} = mg - F_r = \frac{\pi H D^2 g \rho}{4} - \left(\frac{\pi D^2}{4}\right)\left(\frac{2I}{c}\right) = 0.$$

We solve for H:

$$H = \frac{2I}{gc\rho} = \left(\frac{2P}{\pi D^2 / 4}\right)\frac{1}{gc\rho}$$

$$= \frac{2(4.60\,\text{W})}{[\pi(2.60\times10^{-3}\,\text{m})^2 / 4](9.8\,\text{m/s}^2)(3.0\times10^8\,\text{m/s})(1.20\times10^3\,\text{kg/m}^3)}$$

$$= 4.91\times10^{-7}\,\text{m}.$$

27. If the beam carries energy U away from the spaceship, then it also carries momentum $p = U/c$ away. Since the total momentum of the spaceship and light is conserved, this is the magnitude of the momentum acquired by the spaceship. If P is the power of the laser,

then the energy carried away in time t is $U = Pt$. We note that there are 86400 seconds in a day. Thus, $p = Pt/c$ and, if m is mass of the spaceship, its speed is

$$v = \frac{p}{m} = \frac{Pt}{mc} = \frac{(10 \times 10^3 \text{ W})(86400 \text{ s})}{(1.5 \times 10^3 \text{ kg})(2.998 \times 10^8 \text{ m/s})} = 1.9 \times 10^{-3} \text{ m/s}.$$

28. We require $F_{grav} = F_r$ or

$$G\frac{mM_s}{d_{es}^2} = \frac{2IA}{c},$$

and solve for the area A:

$$A = \frac{cGmM_s}{2Id_{es}^2} = \frac{(6.67 \times 10^{-11} \text{ N} \cdot \text{m}^2/\text{kg}^2)(1500 \text{ kg})(1.99 \times 10^{30} \text{ kg})(2.998 \times 10^8 \text{ m/s})}{2(1.40 \times 10^3 \text{ W/m}^2)(1.50 \times 10^{11} \text{ m})^2}$$

$$= 9.5 \times 10^5 \text{ m}^2 = 0.95 \text{ km}^2.$$

29. Eq. 33-27 suggests that the slope in an intensity versus inverse-square-distance graph (I plotted versus r^{-2}) is $P/4\pi$. We estimate the slope to be about 20 (in SI units) which means the power is $P = 4\pi(30) \approx 2.5 \times 10^2$ W.

30. (a) The upward force supplied by radiation pressure in this case (Eq. 33-32) must be equal to the magnitude of the pull of gravity (mg). For a sphere, the "projected" area (which is a factor in Eq. 33-32) is that of a circle $A = \pi r^2$ (not the entire surface area of the sphere) and the volume (needed because the mass is given by the density multiplied by the volume: $m = \rho V$) is $V = \frac{4}{3}\pi r^3$. Finally, the intensity is related to the power P of the light source and another area factor $4\pi R^2$, given by Eq. 33-27. In this way, with $\rho = 19000$ in SI units, equating the forces leads to

$$P = 4\pi R^2 c \rho \frac{4}{3}\pi r^3 g / \pi r^2 = 4.68 \times 10^{11} \text{ W}.$$

(b) Any chance disturbance could move the sphere from being directly above the source, and then the two force vectors would no longer be along the same axis.

31. The angle between the direction of polarization of the light incident on the first polarizing sheet and the polarizing direction of that sheet is $\theta_1 = 70°$. If I_0 is the intensity of the incident light, then the intensity of the light transmitted through the first sheet is

$$I_1 = I_0 \cos^2 \theta_1 = (43 \text{ W/m}^2) \cos^2 70° = 5.03 \text{ W/m}^2.$$

The direction of polarization of the transmitted light makes an angle of 70° with the vertical and an angle of $\theta_2 = 20°$ with the horizontal. θ_2 is the angle it makes with the polarizing direction of the second polarizing sheet. Consequently, the transmitted intensity is

$$I_2 = I_1 \cos^2 \theta_2 = (5.03 \text{ W} / \text{m}^2) \cos^2 20° = 4.4 \text{ W} / \text{m}^2.$$

32. In this case, we replace $I_0 \cos^2 70°$ by $\frac{1}{2} I_0$ as the intensity of the light after passing through the first polarizer. Therefore,

$$I_f = \frac{1}{2} I_0 \cos^2 (90°-70°) = \frac{1}{2}(43 \text{ W} / \text{m}^2)(\cos^2 20°) = 19 \text{ W} / \text{m}^2.$$

33. Let I_0 be the intensity of the unpolarized light that is incident on the first polarizing sheet. The transmitted intensity is $I_1 = \frac{1}{2} I_0$, and the direction of polarization of the transmitted light is $\theta_1 = 40°$ counterclockwise from the y axis in the diagram. The polarizing direction of the second sheet is $\theta_2 = 20°$ clockwise from the y axis, so the angle between the direction of polarization that is incident on that sheet and the polarizing direction of the sheet is $40° + 20° = 60°$. The transmitted intensity is

$$I_2 = I_1 \cos^2 60° = \frac{1}{2} I_0 \cos^2 60°,$$

and the direction of polarization of the transmitted light is 20° clockwise from the y axis. The polarizing direction of the third sheet is $\theta_3 = 40°$ counterclockwise from the y axis. Consequently, the angle between the direction of polarization of the light incident on that sheet and the polarizing direction of the sheet is $20° + 40° = 60°$. The transmitted intensity is

$$I_3 = I_2 \cos^2 60° = \frac{1}{2} I_0 \cos^4 60° = 3.1 \times 10^{-2}.$$

Thus, 3.1% of the light's initial intensity is transmitted.

34. After passing through the first polarizer the initial intensity I_0 reduces by a factor of 1/2. After passing through the second one it is further reduced by a factor of $\cos^2 (\pi - \theta_1 - \theta_2) = \cos^2 (\theta_1 + \theta_2)$. Finally, after passing through the third one it is again reduced by a factor of $\cos^2 (\pi - \theta_2 - \theta_3) = \cos^2 (\theta_2 + \theta_3)$. Therefore,

$$\frac{I_f}{I_0} = \frac{1}{2} \cos^2 (\theta_1 + \theta_2) \cos^2 (\theta_2 + \theta_3) = \frac{1}{2} \cos^2 (50° + 50°) \cos^2 (50° + 50°)$$

$$= 4.5 \times 10^{-4}.$$

Thus, 0.045% of the light's initial intensity is transmitted.

35. (a) Since the incident light is unpolarized, half the intensity is transmitted and half is absorbed. Thus the transmitted intensity is $I = 5.0$ mW/m^2. The intensity and the electric field amplitude are related by $I = E_m^2 / 2\mu_0 c$, so

$$E_m = \sqrt{2\mu_0 cI} = \sqrt{2(4\pi \times 10^{-7} \text{ H/m})(3.00 \times 10^8 \text{ m/s})(5.0 \times 10^{-3} \text{ W/m}^2)}$$
$$= 1.9 \text{ V/m}.$$

(b) The radiation pressure is $p_r = I_a/c$, where I_a is the absorbed intensity. Thus

$$p_r = \frac{5.0 \times 10^{-3} \text{ W/m}^2}{3.00 \times 10^8 \text{ m/s}} = 1.7 \times 10^{-11} \text{ Pa}.$$

36. We examine the point where the graph reaches zero: $\theta_2 = 160°$. Since the polarizers must be "crossed" for the intensity to vanish, then $\theta_1 = 160° - 90° = 70°$. Now we consider the case $\theta_2 = 90°$ (which is hard to judge from the graph). Since θ_1 is still equal to 70°, then the angle between the polarizers is now $\Delta\theta = 20°$. Accounting for the "automatic" reduction (by a factor of one-half) whenever unpolarized light passes through any polarizing sheet, then our result is $\frac{1}{2}\cos^2(\Delta\theta) = 0.442 \approx 44\%$.

37. As the polarized beam of intensity I_0 passes the first polarizer, its intensity is reduced to $I_0 \cos^2\theta$. After passing through the second polarizer which makes a 90° angle with the first filter, the intensity is $I = (I_0\cos^2\theta)\sin^2\theta = I_0/10$ which implies $\sin^2\theta\cos^2\theta = 1/10$, or $\sin\theta\cos\theta = \sin 2\theta/2 = 1/\sqrt{10}$. This leads to $\theta = 70°$ or $20°$.

38. We note the points at which the curve is zero ($\theta_2 = 0°$ and $90°$) in Fig. 33-44(b). We infer that sheet 2 is perpendicular to one of the other sheets at $\theta_2 = 0°$, and that it is perpendicular to the *other* of the other sheets when $\theta_2 = 90°$. Without loss of generality, we choose $\theta_1 = 0°$, $\theta_3 = 90°$. Now, when $\theta_2 = 30°$, it will be $\Delta\theta = 30°$ relative to sheet 1 and $\Delta\theta' = 60°$ relative to sheet 3. Therefore,

$$\frac{I_f}{I_i} = \frac{1}{2}\cos^2(\Delta\theta)\cos^2(\Delta\theta') = 9.4\%.$$

39. Let I_0 be the intensity of the incident beam and f be the fraction that is polarized. Thus, the intensity of the polarized portion is fI_0. After transmission, this portion contributes $fI_0 \cos^2\theta$ to the intensity of the transmitted beam. Here θ is the angle between the direction of polarization of the radiation and the polarizing direction of the filter. The intensity of the unpolarized portion of the incident beam is $(1-f)I_0$ and after transmission, this portion contributes $(1-f)I_0/2$ to the transmitted intensity. Consequently, the transmitted intensity is

$$I = fI_0 \cos^2 \theta + \frac{1}{2}(1-f)I_0.$$

As the filter is rotated, $\cos^2 \theta$ varies from a minimum of 0 to a maximum of 1, so the transmitted intensity varies from a minimum of

$$I_{min} = \frac{1}{2}(1-f)I_0$$

to a maximum of

$$I_{max} = fI_0 + \frac{1}{2}(1-f)I_0 = \frac{1}{2}(1+f)I_0.$$

The ratio of I_{max} to I_{min} is

$$\frac{I_{max}}{I_{min}} = \frac{1+f}{1-f}.$$

Setting the ratio equal to 5.0 and solving for f, we get $f = 0.67$.

40. We apply Eq. 33-40 (once) and Eq. 33-42 (twice) to obtain

$$I = \frac{1}{2} I_0 \cos^2 \theta_2 \cos^2 (90° - \theta_2).$$

Using trig identities, we rewrite this as

$$\frac{I}{I_0} = \frac{1}{8} \sin^2 (2\theta_2).$$

(a) Therefore we find $\theta_2 = \frac{1}{2}\sin^{-1}\sqrt{0.40} = 19.6°$.

(b) Since the first expression we wrote is symmetric under the exchange: $\theta_2 \leftrightarrow 90° - \theta_2$, then we see that the angle's complement, $70.4°$, is also a solution.

41. (a) The fraction of light which is transmitted by the glasses is

$$\frac{I_f}{I_0} = \frac{E_f^2}{E_0^2} = \frac{E_v^2}{E_v^2 + E_h^2} = \frac{E_v^2}{E_v^2 + (2.3E_v)^2} = 0.16.$$

(b) Since now the horizontal component of $\vec{E}$ will pass through the glasses,

$$\frac{I_f}{I_0} = \frac{E_h^2}{E_v^2 + E_h^2} = \frac{(2.3E_v)^2}{E_v^2 + (2.3E_v)^2} = 0.84.$$

42. We note the points at which the curve is zero ($\theta_2 = 60°$ and $140°$) in Fig. 33-44(b). We infer that sheet 2 is perpendicular to one of the other sheets at $\theta_2 = 60°$, and that it is perpendicular to the *other* of the other sheets when $\theta_2 = 140°$. Without loss of generality, we choose $\theta_1 = 150°$, $\theta_3 = 50°$. Now, when $\theta_2 = 90°$, it will be $|\Delta\theta| = 60°$ relative to sheet 1 and $|\Delta\theta'| = 40°$ relative to sheet 3. Therefore,

$$\frac{I_f}{I_i} = \frac{1}{2}(\cos(\Delta\theta))^2 (\cos(\Delta\theta'))^2 = 7.3\% .$$

43. (a) The rotation cannot be done with a single sheet. If a sheet is placed with its polarizing direction at an angle of $90°$ to the direction of polarization of the incident radiation, no radiation is transmitted. It can be done with two sheets. We place the first sheet with its polarizing direction at some angle θ, between 0 and $90°$, to the direction of polarization of the incident radiation. Place the second sheet with its polarizing direction at $90°$ to the polarization direction of the incident radiation. The transmitted radiation is then polarized at $90°$ to the incident polarization direction. The intensity is $I_0 \cos^2 \theta \cos^2 (90° - \theta) = I_0 \cos^2 \theta \sin^2 \theta$, where I_0 is the incident radiation. If θ is not 0 or $90°$, the transmitted intensity is not zero.

(b) Consider n sheets, with the polarizing direction of the first sheet making an angle of $\theta = 90°/n$ relative to the direction of polarization of the incident radiation. The polarizing direction of each successive sheet is rotated $90°/n$ in the same sense from the polarizing direction of the previous sheet. The transmitted radiation is polarized, with its direction of polarization making an angle of $90°$ with the direction of polarization of the incident radiation. The intensity is

$$I = I_0 \cos^{2n} (90°/n).$$

We want the smallest integer value of n for which this is greater than $0.60I_0$. We start with $n = 2$ and calculate $\cos^{2n} (90°/n)$. If the result is greater than 0.60, we have obtained the solution. If it is less, increase n by 1 and try again. We repeat this process, increasing n by 1 each time, until we have a value for which $\cos^{2n} (90°/n)$ is greater than 0.60. The first one will be $n = 5$.

44. The angle of incidence for the light ray on mirror B is $90° - \theta$. So the outgoing ray r' makes an angle $90° - (90° - \theta) = \theta$ with the vertical direction, and is antiparallel to the incoming one. The angle between i and r' is therefore $180°$.

45. The law of refraction states

$$n_1 \sin\theta_1 = n_2 \sin\theta_2.$$

We take medium 1 to be the vacuum, with $n_1 = 1$ and $\theta_1 = 32.0°$. Medium 2 is the glass, with $\theta_2 = 21.0°$. We solve for n_2:

$$n_2 = n_1 \frac{\sin \theta_1}{\sin \theta_2} = (1.00)\left(\frac{\sin 32.0°}{\sin 21.0°}\right) = 1.48.$$

46. (a) For the angles of incidence and refraction to be equal, the graph in Fig. 33-48(b) would consist of a "$y = x$" line at 45° in the plot. Instead, the curve for material 1 falls under such a "$y = x$" line, which tells us that all refraction angles are less than incident ones. With $\theta_2 < \theta_1$ Snell's law implies $n_2 > n_1$.

(b) Using the same argument as in (a), the value of n_2 for material 2 is also greater than that of water (n_1).

(c) It's easiest to examine the right end-point of each curve. With $\theta_1 = 90°$ and $\theta_2 = \frac{3}{4}(90°)$, and with $n_1 = 1.33$ (Table 33-1) we find, from Snell's law, $n_2 = 1.4$ for material 1.

(d) Similarly, with $\theta_1 = 90°$ and $\theta_2 = \frac{1}{2}(90°)$, we obtain $n_2 = 1.9$.

47. Note that the normal to the refracting surface is vertical in the diagram. The angle of refraction is $\theta_2 = 90°$ and the angle of incidence is given by $\tan \theta_1 = L/D$, where D is the height of the tank and L is its width. Thus

$$\theta_1 = \tan^{-1}\left(\frac{L}{D}\right) = \tan^{-1}\left(\frac{1.10 \text{ m}}{0.850 \text{ m}}\right) = 52.31°.$$

The law of refraction yields

$$n_1 = n_2 \frac{\sin \theta_2}{\sin \theta_1} = (1.00)\left(\frac{\sin 90°}{\sin 52.31°}\right) = 1.26,$$

where the index of refraction of air was taken to be unity.

48. (a) For the angles of incidence and refraction to be equal, the graph in Fig. 33-48(b) would consist of a "$y = x$" line at 45° in the plot. Instead, the curve for material 1 falls under such a "$y = x$" line, which tells us that all refraction angles are less than incident ones. With $\theta_2 < \theta_1$ Snell's law implies $n_2 > n_1$.

(b) Using the same argument as in (a), the value of n_2 for material 2 is also greater than that of water (n_1).

(c) It's easiest to examine the topmost point of each curve. With $\theta_2 = 90°$ and $\theta_1 = \frac{1}{2}(90°)$, and with $n_2 = 1.33$ (Table 33-1) we find $n_1 = 1.9$ from Snell's law.

(d) Similarly, with $\theta_2 = 90°$ and $\theta_1 = \frac{3}{4}(90°)$, we obtain $n_1 = 1.4$.

49. Consider a ray that grazes the top of the pole, as shown in the diagram that follows.

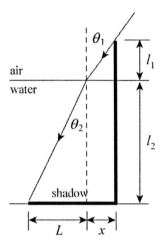

Here $\theta_1 = 90° - \theta = 35°$, $\ell_1 = 0.50$ m, and $\ell_2 = 1.50$ m. The length of the shadow is $x + L$. x is given by

$$x = \ell_1 \tan\theta_1 = (0.50 \text{ m}) \tan 35° = 0.35 \text{ m}.$$

According to the law of refraction, $n_2 \sin\theta_2 = n_1 \sin\theta_1$. We take $n_1 = 1$ and $n_2 = 1.33$ (from Table 33-1). Then,

$$\theta_2 = \sin^{-1}\left(\frac{\sin\theta_1}{n_2}\right) = \sin^{-1}\left(\frac{\sin 35.0°}{1.33}\right) = 25.55°.$$

L is given by

$$L = \ell_2 \tan\theta_2 = (1.50 \text{ m}) \tan 25.55° = 0.72 \text{ m}.$$

The length of the shadow is 0.35 m + 0.72 m = 1.07 m.

50. (a) A simple implication of Snell's law is that $\theta_2 = \theta_1$ when $n_1 = n_2$. Since the angle of incidence is shown in Fig. 33-52(a) to be 30°, then we look for a point in Fig. 33-52(b) where $\theta_2 = 30°$. This seems to occur when $n_2 = 1.7$. By inference, then, $n_1 = 1.7$.

(b) From $1.7\sin(60°) = 2.4\sin(\theta_2)$ we get $\theta_2 = 38°$.

51. (a) Approximating $n = 1$ for air, we have

$$n_1 \sin\theta_1 = (1)\sin\theta_5 \Rightarrow 56.9° = \theta_5$$

and with the more accurate value for n_{air} in Table 33-1, we obtain 56.8°.

(b) Eq. 33-44 leads to

$$n_1 \sin\theta_1 = n_2 \sin\theta_2 = n_3 \sin\theta_3 = n_4 \sin\theta_4$$

so that

$$\theta_4 = \sin^{-1}\left(\frac{n_1}{n_4}\sin\theta_1\right) = 35.3°.$$

52. (a) We use subscripts b and r for the blue and red light rays. Snell's law gives

$$\theta_{2b} = \sin^{-1}\left(\frac{1}{1.343}\sin(70°)\right) = 44.403°$$

$$\theta_{2r} = \sin^{-1}\left(\frac{1}{1.331}\sin(70°)\right) = 44.911°$$

for the refraction angles at the first surface (where the normal axis is vertical). These rays strike the second surface (where A is) at complementary angles to those just calculated (since the normal axis is horizontal for the second surface). Taking this into consideration, we again use Snell's law to calculate the second refractions (with which the light re-enters the air):

$$\theta_{3b} = \sin^{-1}[1.343\sin(90°-\theta_{2b})] = 73.636°$$
$$\theta_{3r} = \sin^{-1}[1.331\sin(90°-\theta_{2r})] = 70.497°$$

which differ by 3.1° (thus giving a rainbow of angular width 3.1°).

(b) Both of the refracted rays emerges from the bottom side with the same angle (70°) with which they were incident on the topside (the occurrence of an intermediate reflection [from side 2] does not alter this overall fact: light comes into the block at the same angle that it emerges with from the opposite parallel side). There is thus no difference (the difference is 0°) and thus there is no rainbow in this case.

53. We label the light ray's point of entry A, the vertex of the prism B, and the light ray's exit point C. Also, the point in Fig. 33-55 where ψ is defined (at the point of intersection of the extrapolations of the incident and emergent rays) is denoted D. The angle indicated by ADC is the supplement of ψ, so we denote it $\psi_s = 180° - \psi$. The angle of refraction in the glass is $\theta_2 = \frac{1}{n}\sin\theta$. The angles between the interior ray and the nearby surfaces is the complement of θ_2, so we denote it $\theta_{2c} = 90° - \theta_2$. Now, the angles in the triangle ABC must add to 180°:

$$180° = 2\theta_{2c} + \phi \Rightarrow \theta_2 = \frac{\phi}{2}.$$

Also, the angles in the triangle ADC must add to $180°$:

$$180° = 2(\theta - \theta_2) + \psi_s \Rightarrow \theta = 90° + \theta_2 - \frac{1}{2}\psi_s$$

which simplifies to $\theta = \theta_2 + \frac{1}{2}\psi$. Combining this with our previous result, we find $\theta = \frac{1}{2}(\phi + \psi)$. Thus, the law of refraction yields

$$n = \frac{\sin(\theta)}{\sin(\theta_2)} = \frac{\sin\left(\frac{1}{2}(\phi + \psi)\right)}{\sin\left(\frac{1}{2}\phi\right)}.$$

54. The critical angle is $\theta_c = \sin^{-1}\left(\frac{1}{n}\right) = \sin^{-1}\left(\frac{1}{1.8}\right) = 34°$.

55. Reference to Fig. 33-24 may help in the visualization of why there appears to be a "circle of light" (consider revolving that picture about a vertical axis). The depth and the radius of that circle (which is from point a to point f in that figure) is related to the tangent of the angle of incidence. Thus, the diameter D of the circle in question is

$$D = 2h \tan \theta_c = 2h \tan\left[\sin^{-1}\left(\frac{1}{n_w}\right)\right] = 2(80.0\,\text{cm}) \tan\left[\sin^{-1}\left(\frac{1}{1.33}\right)\right] = 182 \text{ cm.}$$

56. (a) We note that the complement of the angle of refraction (in material 2) is the critical angle. Thus,

$$n_1 \sin \theta = n_2 \cos \theta_c = n_2 \sqrt{1 - \left(\frac{n_3}{n_2}\right)^2} = \sqrt{(n_2)^2 - (n_3)^2}$$

leads to $\theta = 26.8°$.

(b) Increasing θ leads to a decrease of the angle with which the light strikes the interface between materials 2 and 3, so it becomes greater than the critical angle; therefore, there will be some transmission of light into material 3.

57. (a) In the notation of this problem, Eq. 33-47 becomes

$$\theta_c = \sin^{-1} \frac{n_3}{n_2}$$

which yields $n_3 = 1.39$ for $\theta_c = \phi = 60°$.

(b) Applying Eq. 33-44 law to the interface between material 1 and material 2, we have

$$n_2 \sin 30° = n_1 \sin \theta$$

which yields $\theta = 28.1°$.

(c) Decreasing θ will increase ϕ and thus cause the ray to strike the interface (between materials 2 and 3) at an angle larger than θ_c. Therefore, no transmission of light into material 3 can occur.

58. (a) The angle of incidence $\theta_{B,1}$ at B is the complement of the critical angle at A; its sine is

$$\sin \theta_{B,1} = \cos \theta_c = \sqrt{1 - \left(\frac{n_3}{n_2}\right)^2}$$

so that the angle of refraction $\theta_{B,2}$ at B becomes

$$\theta_{B,2} = \sin^{-1}\left(\frac{n_2}{n_3}\sqrt{1 - \left(\frac{n_3}{n_2}\right)^2}\right) = \sin^{-1}\sqrt{\left(\frac{n_2}{n_3}\right)^2 - 1} = 35.1° .$$

(b) From $n_1 \sin \theta = n_2 \sin \theta_c = n_2(n_3/n_2)$, we find

$$\theta = \sin^{-1}\frac{n_3}{n_1} = 49.9° .$$

(c) The angle of incidence $\theta_{A,1}$ at A is the complement of the critical angle at B; its sine is

$$\sin \theta_{A,1} = \cos \theta_c = \sqrt{1 - \left(\frac{n_3}{n_2}\right)^2}$$

so that the angle of refraction $\theta_{A,2}$ at A becomes

$$\theta_{A,2} = \sin^{-1}\left(\frac{n_2}{n_3}\sqrt{1 - \left(\frac{n_3}{n_2}\right)^2}\right) = \sin^{-1}\sqrt{\left(\frac{n_2}{n_3}\right)^2 - 1} = 35.1° .$$

(d) From

$$n_1 \sin \theta = n_2 \sin \theta_{A,1} = n_2\sqrt{1 - \left(\frac{n_3}{n_2}\right)^2} = \sqrt{(n_2)^2 - (n_3)^2}$$

we find

$$\theta = \sin^{-1} \frac{\sqrt{(n_2)^2 - (n_3)^2}}{n_1} = 26.1° \quad .$$

(e) The angle of incidence $\theta_{B,1}$ at B is the complement of the Brewster angle at A; its sine is

$$\sin \theta_{B,1} = \frac{n_2}{\sqrt{(n_2)^2 + (n_3)^2}}$$

so that the angle of refraction $\theta_{B,2}$ at B becomes

$$\theta_{B,2} = \sin^{-1} \left(\frac{(n_2)^2}{n_3 \sqrt{(n_2)^2 + (n_3)^2}} \right) = 60.7° \quad .$$

(f) From

$$n_1 \sin \theta = n_2 \sin \theta_{Brewster} = n_2 \frac{n_3}{\sqrt{(n_2)^2 + (n_3)^2}}$$

we find

$$\theta = \sin^{-1} \frac{n_2 n_3}{n_1 \sqrt{(n_2)^2 + (n_3)^2}} = 35.3° \quad .$$

59. When examining Fig. 33-59, it is important to note that the angle (measured from the central axis) for the light ray in air, θ, is not the angle for the ray in the glass core, which we denote θ'. The law of refraction leads to

$$\sin \theta' = \frac{1}{n_1} \sin \theta$$

assuming $n_{air} = 1$. The angle of incidence for the light ray striking the coating is the complement of θ', which we denote as θ'_{comp} and recall that

$$\sin \theta'_{comp} = \cos \theta' = \sqrt{1 - \sin^2 \theta'}.$$

In the critical case, θ'_{comp} must equal θ_c specified by Eq. 33-47. Therefore,

$$\frac{n_2}{n_1} = \sin \theta'_{comp} = \sqrt{1 - \sin^2 \theta'} = \sqrt{1 - \left(\frac{1}{n_1} \sin \theta \right)^2}$$

which leads to the result: $\sin \theta = \sqrt{n_1^2 - n_2^2}$. With $n_1 = 1.58$ and $n_2 = 1.53$, we obtain

$$\theta = \sin^{-1}\left(1.58^2 - 1.53^2\right) = 23.2°.$$

60. (a) We note that the upper-right corner is at an angle (measured from the point where the light enters, and measured relative to a normal axis established at that point [the normal at that point would be horizontal in Fig. 33-60) is at $\tan^{-1}(2/3) = 33.7°$. The angle of refraction is given by

$$n_{air} \sin 40° = 1.56 \sin \theta_2$$

which yields $\theta_2 = 24.33°$ if we use the common approximation $n_{air} = 1.0$, and yields $\theta_2 = 24.34°$ if we use the more accurate value for n_{air} found in Table 33-1. The value is less than $33.7°$ which means that the light goes to side 3.

(b) The ray strikes a point on side 3 which is 0.643 cm below that upper-right corner, and then (using the fact that the angle is symmetrical upon reflection) strikes the top surface (side 2) at a point 1.42 cm to the left of that corner. Since 1.42 cm is certainly less than 3 cm we have a self-consistency check to the effect that the ray does indeed strike side 2 as its second reflection (if we had gotten 3.42 cm instead of 1.42 cm, then the situation would be quite different).

(c) The normal axes for sides 1 and 3 are both horizontal, so the angle of incidence (in the plastic) at side 3 is the same as the angle of refraction was at side 1. Thus,

$$1.56 \sin 24.3° = n_{air} \sin \theta_{air} \quad \Rightarrow \quad \theta_{air} = 40°.$$

(d) It strikes the top surface (side 2) at an angle (measured from the normal axis there, which in this case would be a vertical axis) of $90° - \theta_2 = 66°$ which is much greater than the critical angle for total internal reflection ($\sin^{-1}(n_{air}/1.56) = 39.9°$). Therefore, no refraction occurs when the light strikes side 2.

(e) In this case, we have $n_{air} \sin 70° = 1.56 \sin \theta_2$ which yields $\theta_2 = 37.04°$ if we use the common approximation $n_{air} = 1.0$, and yields $\theta_2 = 37.05°$ if we use the more accurate value for n_{air} found in Table 33-1. This is greater than the $33.7°$ mentioned above (regarding the upper-right corner), so the ray strikes side 2 instead of side 3.

(f) After bouncing from side 2 (at a point fairly close to that corner) to goes to side 3.

(g) When it bounced from side 2, its angle of incidence (because the normal axis for side 2 is orthogonal to that for side 1) is $90° - \theta_2 = 53°$ which is much greater than the critical angle for total internal reflection (which, again, is $\sin^{-1}(n_{air}/1.56) = 39.9°$). Therefore, no refraction occurs when the light strikes side 2.

(h) For the same reasons implicit in the calculation of part (c), the refracted ray emerges from side 3 with the same angle (70°) that it entered side 1 at (we see that the occurrence of an intermediate reflection [from side 2] does not alter this overall fact: light comes into the block at the same angle that it emerges with from the opposite parallel side.

61. (a) No refraction occurs at the surface ab, so the angle of incidence at surface ac is $90° - \phi$. For total internal reflection at the second surface, $n_g \sin (90° - \phi)$ must be greater than n_a. Here n_g is the index of refraction for the glass and n_a is the index of refraction for air. Since $\sin (90° - \phi) = \cos \phi$, we want the largest value of ϕ for which $n_g \cos \phi \geq n_a$. Recall that $\cos \phi$ decreases as ϕ increases from zero. When ϕ has the largest value for which total internal reflection occurs, then $n_g \cos \phi = n_a$, or

$$\phi = \cos^{-1}\left(\frac{n_a}{n_g}\right) = \cos^{-1}\left(\frac{1}{1.52}\right) = 48.9°.$$

The index of refraction for air is taken to be unity.

(b) We now replace the air with water. If $n_w = 1.33$ is the index of refraction for water, then the largest value of ϕ for which total internal reflection occurs is

$$\phi = \cos^{-1}\left(\frac{n_w}{n_g}\right) = \cos^{-1}\left(\frac{1.33}{1.52}\right) = 29.0°.$$

62. (a) We refer to the entry point for the original incident ray as point A (which we take to be on the left side of the prism, as in Fig. 33-55), the prism vertex as point B, and the point where the interior ray strikes the right surface of the prism as point C. The angle between line AB and the interior ray is β (the complement of the angle of refraction at the first surface), and the angle between the line BC and the interior ray is α (the complement of its angle of incidence when it strikes the second surface). When the incident ray is at the minimum angle for which light is able to exit the prism, the light exits along the second face. That is, the angle of refraction at the second face is 90°, and the angle of incidence there for the interior ray is the critical angle for total internal reflection. Let θ_1 be the angle of incidence for the original incident ray and θ_2 be the angle of refraction at the first face, and let θ_3 be the angle of incidence at the second face. The law of refraction, applied to point C, yields $n \sin \theta_3 = 1$, so

$$\sin \theta_3 = 1/n = 1/1.60 = 0.625 \Rightarrow \theta_3 = 38.68°.$$

The interior angles of the triangle ABC must sum to 180°, so $\alpha + \beta = 120°$. Now, $\alpha = 90° - \theta_3 = 51.32°$, so $\beta = 120° - 51.32° = 69.68°$. Thus, $\theta_2 = 90° - \beta = 21.32°$. The law of refraction, applied to point A, yields

$$\sin \theta_1 = n \sin \theta_2 = 1.60 \sin 21.32° = 0.5817.$$

Thus $\theta_1 = 35.6°$.

(b) We apply the law of refraction to point C. Since the angle of refraction there is the same as the angle of incidence at A, $n \sin \theta_3 = \sin \theta_1$. Now, $\alpha + \beta = 120°$, $\alpha = 90° - \theta_3$, and $\beta = 90° - \theta_2$, as before. This means $\theta_2 + \theta_3 = 60°$. Thus, the law of refraction leads to

$$\sin \theta_1 = n \sin(60° - \theta_2) \Rightarrow \sin \theta_1 = n \sin 60° \cos \theta_2 - n \cos 60° \sin \theta_2$$

where the trigonometric identity $\sin(A - B) = \sin A \cos B - \cos A \sin B$ is used. Next, we apply the law of refraction to point A:

$$\sin \theta_1 = n \sin \theta_2 \Rightarrow \sin \theta_2 = (1/n)\sin \theta_1$$

which yields $\cos \theta_2 = \sqrt{1 - \sin^2 \theta_2} = \sqrt{1 - (1/n^2)\sin^2 \theta_1}$. Thus,

$$\sin \theta_1 = n \sin 60° \sqrt{1 - (1/n)^2 \sin^2 \theta_1} - \cos 60° \sin \theta_1$$

or

$$(1 + \cos 60°)\sin \theta_1 = \sin 60° \sqrt{n^2 - \sin^2 \theta_1}.$$

Squaring both sides and solving for $\sin \theta_1$, we obtain

$$\sin \theta_1 = \frac{n \sin 60°}{\sqrt{(1 + \cos 60°)^2 + \sin^2 60°}} = \frac{1.60 \sin 60°}{\sqrt{(1 + \cos 60°)^2 + \sin^2 60°}} = 0.80$$

and $\theta_1 = 53.1°$.

63. (a) A ray diagram is shown below.

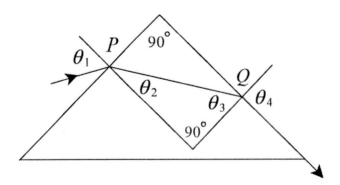

Let θ_1 be the angle of incidence and θ_2 be the angle of refraction at the first surface. Let θ_3 be the angle of incidence at the second surface. The angle of refraction there is $\theta_4 = 90°$. The law of refraction, applied to the second surface, yields $n \sin \theta_3 = \sin \theta_4 = 1$. As shown in the diagram, the normals to the surfaces at P and Q are perpendicular to each other. The interior angles of the triangle formed by the ray and the two normals must sum to 180°, so $\theta_3 = 90° - \theta_2$ and

$$\sin\theta_3 = \sin(90° - \theta_2) = \cos\theta_2 = \sqrt{1 - \sin^2\theta_2}.$$

According to the law of refraction, applied at Q, $n\sqrt{1 - \sin^2\theta_2} = 1$. The law of refraction, applied to point P, yields $\sin\theta_1 = n\sin\theta_2$, so $\sin\theta_2 = (\sin\theta_1)/n$ and

$$n\sqrt{1 - \frac{\sin^2\theta_1}{n^2}} = 1.$$

Squaring both sides and solving for n, we get

$$n = \sqrt{1 + \sin^2\theta_1}.$$

(b) The greatest possible value of $\sin^2\theta_1$ is 1, so the greatest possible value of n is $n_{max} = \sqrt{2} = 1.41$.

(c) For a given value of n, if the angle of incidence at the first surface is greater than θ_1, the angle of refraction there is greater than θ_2 and the angle of incidence at the second face is less than θ_3 (= 90° − θ_2). That is, it is less than the critical angle for total internal reflection, so light leaves the second surface and emerges into the air.

(d) If the angle of incidence at the first surface is less than θ_1, the angle of refraction there is less than θ_2 and the angle of incidence at the second surface is greater than θ_3. This is greater than the critical angle for total internal reflection, so all the light is reflected at Q.

64. (a) We use Eq. 33-49: $\theta_B = \tan^{-1}n_w = \tan^{-1}(1.33) = 53.1°$.

(b) Yes, since n_w depends on the wavelength of the light.

65. The angle of incidence θ_B for which reflected light is fully polarized is given by Eq. 33-48 of the text. If n_1 is the index of refraction for the medium of incidence and n_2 is the index of refraction for the second medium, then

$$\theta_B = \tan^{-1}(n_2/n_1) = \tan^{-1}(1.53/1.33) = 49.0°.$$

66. Since the layers are parallel, the angle of refraction regarding the first surface is the same as the angle of incidence regarding the second surface (as is suggested by the notation in Fig. 33-63). We recall that as part of the derivation of Eq. 33-49 (Brewster's angle), the refracted angle is the complement of the incident angle:

$$\theta_2 = (\theta_1)_c = 90°-\theta_1.$$

We apply Eq. 33-49 to both refractions, setting up a product:

$$\left(\frac{n_2}{n_1}\right)\left(\frac{n_3}{n_2}\right) = (\tan\theta_{B1\to2})(\tan\theta_{B2\to3}) \quad \Rightarrow \quad \frac{n_3}{n_1} = (\tan\theta_1)(\tan\theta_2).$$

Now, since θ_2 is the complement of θ_1 we have

$$\tan\theta_2 = \tan(\theta_1)_c = \frac{1}{\tan\theta_1}.$$

Therefore, the product of tangents cancel and we obtain $n_3/n_1 = 1$. Consequently, the third medium is air: $n_3 = 1.0$.

67. Since some of the angles in Fig. 33-64 are measured from vertical axes and some are measured from horizontal axes, we must be very careful in taking differences. For instance, the angle difference between the first polarizer struck by the light and the second is 110° (or 70° depending on how we measure it; it does not matter in the final result whether we put $\Delta\theta_1 = 70°$ or put $\Delta\theta_1 = 110°$). Similarly, the angle difference between the second and the third is $\Delta\theta_2 = 40°$, and between the third and the fourth is $\Delta\theta_3 = 40°$, also. Accounting for the "automatic" reduction (by a factor of one-half) whenever unpolarized light passes through any polarizing sheet, then our result is the incident intensity multiplied by

$$\tfrac{1}{2}\cos^2(\Delta\theta_1)\,\cos^2(\Delta\theta_2)\,\cos^2(\Delta\theta_3).$$

Thus, the light that emerges from the system has intensity equal to 0.50 W/m².

68. (a) Suppose there are a total of N transparent layers ($N = 5$ in our case). We label these layers from left to right with indices 1, 2, ..., N. Let the index of refraction of the air be n_0. We denote the initial angle of incidence of the light ray upon the air-layer boundary as θ_i and the angle of the emerging light ray as θ_f. We note that, since all the boundaries are parallel to each other, the angle of incidence θ_j at the boundary between the j-th and the $(j + 1)$-th layers is the same as the angle between the transmitted light ray and the normal in the j-th layer. Thus, for the first boundary (the one between the air and the first layer)

$$\frac{n_1}{n_0} = \frac{\sin\theta_i}{\sin\theta_1},$$

for the second boundary

$$\frac{n_2}{n_1} = \frac{\sin\theta_1}{\sin\theta_2},$$

and so on. Finally, for the last boundary

$$\frac{n_0}{n_N} = \frac{\sin\theta_N}{\sin\theta_f},$$

Multiplying these equations, we obtain

$$\left(\frac{n_1}{n_0}\right)\left(\frac{n_2}{n_1}\right)\left(\frac{n_3}{n_2}\right)\cdots\left(\frac{n_0}{n_N}\right) = \left(\frac{\sin\theta_i}{\sin\theta_1}\right)\left(\frac{\sin\theta_1}{\sin\theta_2}\right)\left(\frac{\sin\theta_2}{\sin\theta_3}\right)\cdots\left(\frac{\sin\theta_N}{\sin\theta_f}\right).$$

We see that the L.H.S. of the equation above can be reduced to n_0/n_0 while the R.H.S. is equal to $\sin\theta_i/\sin\theta_f$. Equating these two expressions, we find

$$\sin\theta_f = \left(\frac{n_0}{n_0}\right)\sin\theta_i = \sin\theta_i,$$

which gives $\theta_i = \theta_f$. So for the two light rays in the problem statement, the angle of the emerging light rays are both the same as their respective incident angles. Thus, $\theta_f = 0$ for ray a,

(b) and $\theta_f = 20°$ for ray b.

(c) In this case, all we need to do is to change the value of n_0 from 1.0 (for air) to 1.5 (for glass). This does not change the result above. That is, we still have $\theta_f = 0$ for ray a,

(d) and $\theta_f = 20°$ for ray b.

Note that the result of this problem is fairly general. It is independent of the number of layers and the thickness and index of refraction of each layer.

69. (a) The Sun is far enough away that we approximate its rays as "parallel" in this Figure. That is, if the sunray makes angle θ from horizontal when the bird is in one position, then it makes the same angle θ when the bird is any other position. Therefore, its shadow on the ground moves as the bird moves: at 15 m/s.

(b) If the bird is in a position, a distance $x > 0$ from the wall, such that its shadow is on the wall at a distance $0 \geq y \geq h$ from the top of the wall, then it is clear from the Figure that $\tan\theta = y/x$. Thus,

$$\frac{dy}{dt} = \frac{dx}{dt}\tan\theta = (-15 \text{ m/s})\tan 30° = -8.7 \text{ m/s},$$

which means that the distance y (which was measured as a positive number downward from the top of the wall) is shrinking at the rate of 8.7 m/s.

(c) Since $\tan\theta$ grows as $0 \le \theta < 90°$ increases, then a larger value of $|dy/dt|$ implies a larger value of θ. The Sun is higher in the sky when the hawk glides by.

(d) With $|dy/dt| = 45$ m/s, we find

$$v_{hawk} = \left|\frac{dx}{dt}\right| = \frac{\left|\frac{dy}{dt}\right|}{\tan\theta}$$

so that we obtain $\theta = 72°$ if we assume $v_{hawk} = 15$ m/s.

70. (a) From $n_1\sin\theta_1 = n_2\sin\theta_2$ and $n_2\sin\theta_2 = n_3\sin\theta_3$, we find $n_1\sin\theta_1 = n_3\sin\theta_3$. This has a simple implication: that $\theta_1 = \theta_3$ when $n_1 = n_3$. Since we are given $\theta_1 = 40°$ in Fig. 33-67(a) then we look for a point in Fig. 33-67(b) where $\theta_3 = 40°$. This seems to occur at $n_3 = 1.6$, so we infer that $n_1 = 1.6$.

(b) Our first step in our solution to part (a) shows that information concerning n_2 disappears (cancels) in the manipulation. Thus, we cannot tell; we need more information.

(c) From $1.6\sin70° = 2.4\sin\theta_3$ we obtain $\theta_3 = 39°$.

71. (a) Reference to Fig. 33-24 may help in the visualization of why there appears to be a "circle of light" (consider revolving that picture about a vertical axis). The depth and the radius of that circle (which is from point a to point f in that figure) is related to the tangent of the angle of incidence. The diameter of the circle in question is given by $d = 2h \tan\theta_c$. For water $n = 1.33$, so Eq. 33-47 gives $\sin\theta_c = 1/1.33$, or $\theta_c = 48.75°$. Thus,

$$d = 2h\tan\theta_c = 2(2.00 \text{ m})(\tan 48.75°) = 4.56 \text{ m}.$$

(b) The diameter d of the circle will increase if the fish descends (increasing h).

72. (a) Snell's law gives $n_{air}\sin(50°) = n_{2b}\sin\theta_{2b}$ and $n_{air}\sin(50°) = n_{2r}\sin\theta_{2r}$ where we use subscripts b and r for the blue and red light rays. Using the common approximation for air's index ($n_{air} = 1.0$) we find the two angles of refraction to be 30.176° and 30.507°. Therefore, $\Delta\theta = 0.33°$.

(b) Both of the refracted rays emerges from the other side with the same angle (50°) with which they were incident on the first side (generally speaking, light comes into a block at the same angle that it emerges with from the opposite parallel side). There is thus no difference (the difference is 0°) and thus there is no dispersion in this case.

73. (a) The wave is traveling in the $-y$ direction (see §16-5 for the significance of the relative sign between the spatial and temporal arguments of the wave function).

(b) Figure 33-5 may help in visualizing this. The direction of propagation (along the y axis) is perpendicular to $\vec{B}$ (presumably along the x axis, since the problem gives B_x and no other component) and both are perpendicular to $\vec{E}$ (which determines the axis of polarization). Thus, the wave is z-polarized.

(c) Since the magnetic field amplitude is $B_m = 4.00 \ \mu\text{T}$, then (by Eq. 33-5) $E_m = 1199$ V/m $\approx 1.20 \times 10^3$ V/m. Dividing by $\sqrt{2}$ yields $E_{rms} = 848$ V/m. Then, Eq. 33-26 gives

$$I = \frac{I}{c\mu_0} E_{rms}^2 = 1.91 \times 10^3 \ \text{W}/\text{m}^2.$$

(d) Since $kc = \omega$ (equivalent to $c = f\lambda$), we have

$$k = \frac{2.00 \times 10^{15}}{c} = 6.67 \times 10^6 \ \text{m}^{-1}.$$

Summarizing the information gathered so far, we have (with SI units understood)

$$E_z = (1.2 \times 10^3)\sin\left(\left(6.67 \times 10^6\right)y + \left(2.00 \times 10^{15}\right)t\right).$$

(e) $\lambda = 2\pi/k = 942$ nm.

(f) This is an infrared light.

74. (a) The condition (in Eq. 33-44) required in the critical angle calculation is $\theta_3 = 90°$. Thus (with $\theta_2 = \theta_c$, which we don't compute here),

$$n_1 \sin\theta_1 = n_2 \sin\theta_2 = n_3 \sin\theta_3$$

leads to $\theta_1 = \theta = \sin^{-1} n_3/n_1 = 54.3°$.

(b) Yes. Reducing θ leads to a reduction of θ_2 so that it becomes less than the critical angle; therefore, there will be some transmission of light into material 3.

(c) We note that the complement of the angle of refraction (in material 2) is the critical angle. Thus,

$$n_1 \sin\theta = n_2 \cos\theta_c = n_2 \sqrt{1 - \left(\frac{n_3}{n_2}\right)^2} = \sqrt{n_2^2 - n_3^2}$$

leads to $\theta = 51.1°$.

(d) No. Reducing θ leads to an increase of the angle with which the light strikes the interface between materials 2 and 3, so it becomes greater than the critical angle. Therefore, there will be no transmission of light into material 3.

75. Let $\theta_1 = 45°$ be the angle of incidence at the first surface and θ_2 be the angle of refraction there. Let θ_3 be the angle of incidence at the second surface. The condition for total internal reflection at the second surface is $n \sin \theta_3 \geq 1$. We want to find the smallest value of the index of refraction n for which this inequality holds. The law of refraction, applied to the first surface, yields $n \sin \theta_2 = \sin \theta_1$. Consideration of the triangle formed by the surface of the slab and the ray in the slab tells us that $\theta_3 = 90° - \theta_2$. Thus, the condition for total internal reflection becomes $1 \leq n \sin(90° - \theta_2) = n \cos \theta_2$. Squaring this equation and using $\sin^2 \theta_2 + \cos^2 \theta_2 = 1$, we obtain $1 \leq n^2 (1 - \sin^2 \theta_2)$. Substituting $\sin \theta_2 = (1/n) \sin \theta_1$ now leads to

$$1 \leq n^2 \left(1 - \frac{\sin^2 \theta_1}{n^2} \right) = n^2 - \sin^2 \theta_1.$$

The largest value of n for which this equation is true is given by $1 = n^2 - \sin^2 \theta_1$. We solve for n:

$$n = \sqrt{1 + \sin^2 \theta_1} = \sqrt{1 + \sin^2 45°} = 1.22.$$

76. We write $m = \rho V$ where $V = 4\pi R^3/3$ is the volume. Plugging this into $F = ma$ and then into Eq. 33-32 (with $A = \pi R^2$, assuming the light is in the form of plane waves), we find

$$\rho \frac{4\pi R^3}{3} a = \frac{I \pi R^2}{c}.$$

This simplifies to

$$a = \frac{3I}{4\rho c R}$$

which yields $a = 1.5 \times 10^{-9}$ m/s^2.

77. (a) The first contribution to the overall deviation is at the first refraction: $\delta\theta_1 = \theta_i - \theta_r$. The next contribution to the overall deviation is the reflection. Noting that the angle between the ray right before reflection and the axis normal to the back surface of the sphere is equal to θ_r, and recalling the law of reflection, we conclude that the angle

by which the ray turns (comparing the direction of propagation before and after the reflection) is $\delta\theta_2 = 180° - 2\theta_r$. The final contribution is the refraction suffered by the ray upon leaving the sphere: $\delta\theta_3 = \theta_i - \theta_r$ again. Therefore,

$$\theta_{\text{dev}} = \delta\theta_1 + \delta\theta_2 + \delta\theta_3 = 180° + 2\theta_i - 4\theta_r.$$

(b) We substitute $\theta_r = \sin^{-1}\left(\frac{1}{n}\sin\theta_i\right)$ into the expression derived in part (a), using the two given values for n. The higher curve is for the blue light.

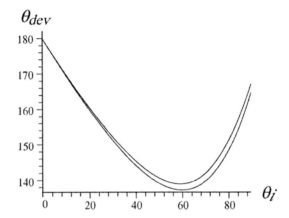

(c) We can expand the graph and try to estimate the minimum, or search for it with a more sophisticated numerical procedure. We find that the θ_{dev} minimum for red light is $137.63° \approx 137.6°$, and this occurs at $\theta_i = 59.52°$.

(d) For blue light, we find that the θ_{dev} minimum is $139.35° \approx 139.4°$, and this occurs at $\theta_i = 59.52°$.

(e) The difference in θ_{dev} in the previous two parts is $1.72°$.

78. (a) The first contribution to the overall deviation is at the first refraction: $\delta\theta_1 = \theta_i - \theta_r$. The next contribution(s) to the overall deviation is (are) the reflection(s). Noting that the angle between the ray right before reflection and the axis normal to the back surface of the sphere is equal to θ_r, and recalling the law of reflection, we conclude that the angle by which the ray turns (comparing the direction of propagation before and after [each] reflection) is $\delta\theta_r = 180° - 2\theta_r$. Thus, for k reflections, we have $\delta\theta_2 = k\theta_r$ to account for these contributions. The final contribution is the refraction suffered by the ray upon leaving the sphere: $\delta\theta_3 = \theta_i - \theta_r$ again. Therefore,

$$\theta_{\text{dev}} = \delta\theta_1 + \delta\theta_2 + \delta\theta_3 = 2(\theta_i - \theta_r) + k(180° - 2\theta_r) = k(180°) + 2\theta_i - 2(k+1)\theta_r.$$

(b) For $k = 2$ and $n = 1.331$ (given in problem 67), we search for the second-order rainbow angle numerically. We find that the θ_{dev} minimum for red light is $230.37° \approx 230.4°$, and this occurs at $\theta_i = 71.90°$.

(c) Similarly, we find that the second-order θ_{dev} minimum for blue light (for which $n = 1.343$) is $233.48° \approx 233.5°$, and this occurs at $\theta_i = 71.52°$.

(d) The difference in θ_{dev} in the previous two parts is approximately $3.1°$.

(e) Setting $k = 3$, we search for the third-order rainbow angle numerically. We find that the θ_{dev} minimum for red light is $317.5°$, and this occurs at $\theta_i = 76.88°$.

(f) Similarly, we find that the third-order θ_{dev} minimum for blue light is $321.9°$, and this occurs at $\theta_i = 76.62°$.

(g) The difference in θ_{dev} in the previous two parts is $4.4°$.

79. (a) and (b) At the Brewster angle, $\theta_{incident} + \theta_{refracted} = \theta_B + 32.0° = 90.0°$, so $\theta_B = 58.0°$ and $n_{glass} = \tan \theta_B = \tan 58.0° = 1.60$.

80. We take the derivative with respect to x of both sides of Eq. 33-11:

$$\frac{\partial}{\partial x}\left(\frac{\partial E}{\partial x}\right) = \frac{\partial^2 E}{\partial x^2} = \frac{\partial}{\partial x}\left(-\frac{\partial B}{\partial t}\right) = -\frac{\partial^2 B}{\partial x \partial t}.$$

Now we differentiate both sides of Eq. 33-18 with respect to t:

$$\frac{\partial}{\partial t}\left(-\frac{\partial B}{\partial x}\right) = -\frac{\partial^2 B}{\partial x \partial t} = \frac{\partial}{\partial t}\left(\varepsilon_0 \mu_0 \frac{\partial E}{\partial t}\right) = \varepsilon_0 \mu_0 \frac{\partial^2 E}{\partial t^2}.$$

Substituting $\partial^2 E/\partial x^2 = -\partial^2 B/\partial x \partial t$ from the first equation above into the second one, we get

$$\varepsilon_0 \mu_0 \frac{\partial^2 E}{\partial t^2} = \frac{\partial^2 E}{\partial x^2} \quad \Rightarrow \quad \frac{\partial^2 E}{\partial t^2} = \frac{1}{\varepsilon_0 \mu_0}\frac{\partial^2 E}{\partial x^2} = c^2 \frac{\partial^2 E}{\partial x^2}.$$

Similarly, we differentiate both sides of Eq. 33-11 with respect to t

$$\frac{\partial^2 E}{\partial x \partial t} = -\frac{\partial^2 B}{\partial t^2},$$

and differentiate both sides of Eq. 33-18 with respect to x

$$-\frac{\partial^2 B}{\partial x^2} = \varepsilon_0 \mu_0 -\frac{\partial^2 E}{\partial x \partial t}.$$

Combining these two equations, we get

$$\frac{\partial^2 B}{\partial t^2} = \frac{1}{\varepsilon_0 \mu_0} \frac{\partial^2 B}{\partial x^2} = c^2 \frac{\partial^2 B}{\partial x^2}.$$

81. We apply Eq. 33-40 (once) and Eq. 33-42 (twice) to obtain

$$I = \frac{1}{2} I_0 \cos^2 \theta_1' \cos^2 \theta_2'$$

where $\theta_1' = 90° - \theta_1 = 60°$ and $\theta_2' = 90° - \theta_2 = 60°$. This yields $I/I_0 = 0.031$.

82. (a) An incident ray which is normal to the water surface is not refracted, so the angle at which it strikes the first mirror is $\theta_1 = 45°$. According to the law of reflection, the angle of reflection is also $45°$. This means the ray is horizontal as it leaves the first mirror, and the angle of incidence at the second mirror is $\theta_2 = 45°$. Since the angle of reflection at the second mirror is also $45°$ the ray leaves that mirror normal again to the water surface. There is no refraction at the water surface, and the emerging ray is parallel to the incident ray.

(b) We imagine that the incident ray makes an angle θ_1 with the normal to the water surface. The angle of refraction θ_2 is found from $\sin \theta_1 = n \sin \theta_2$, where n is the index of refraction of the water. The normal to the water surface and the normal to the first mirror make an angle of $45°$. If the normal to the water surface is continued downward until it meets the normal to the first mirror, the triangle formed has an interior angle of $180° - 45° = 135°$ at the vertex formed by the normal. Since the interior angles of a triangle must sum to $180°$, the angle of incidence at the first mirror satisfies $\theta_3 + \theta_2 + 135° = 180°$, so $\theta_3 = 45° - \theta_2$. Using the law of reflection, the angle of reflection at the first mirror is also $45° - \theta_2$. We note that the triangle formed by the ray and the normals to the two mirrors is a right triangle. Consequently,

$$\theta_3 + \theta_4 + 90° = 180° \implies \theta_4 = 90° - \theta_3 = 90° - 45° + \theta_2 = 45° + \theta_2.$$

The angle of reflection at the second mirror is also $45° + \theta_2$. Now, we continue the normal to the water surface downward from the exit point of the ray to the second mirror. It makes an angle of $45°$ with the mirror. Consider the triangle formed by the second mirror, the ray, and the normal to the water surface. The angle at the intersection of the normal and the mirror is $180° - 45° = 135°$. The angle at the intersection of the ray and the mirror is

$$90° - \theta_4 = 90° - (45° + \theta_2) = 45° - \theta_2.$$

The angle at the intersection of the ray and the water surface is θ_5. These three angles must sum to $180°$, so $135° + 45° - \theta_2 + \theta_5 = 180°$. This means $\theta_5 = \theta_2$. Finally, we use the law of refraction to find θ_6:

$$\sin\theta_6 = n\sin\theta_5 \Rightarrow \sin\theta_6 = n\,\sin\theta_2,$$

since $\theta_5 = \theta_2$. Finally, since $\sin\theta_1 = n\sin\theta_2$, we conclude that $\sin\theta_6 = \sin\theta_1$ and $\theta_6 = \theta_1$. The exiting ray is parallel to the incident ray.

83. We use the result of problem 33-53 to solve for ψ. Note that $\phi = 60.0°$ in our case. Thus, from

$$n = \frac{\sin\frac{1}{2}(\psi + \phi)}{\sin\frac{1}{2}\phi},$$

we get

$$\sin\frac{1}{2}(\psi + \phi) = n\sin\frac{1}{2}\phi = (1.31)\sin\left(\frac{60.0°}{2}\right) = 0.655,$$

which gives $\frac{1}{2}(\psi + \phi) = \sin^{-1}(0.655) = 40.9°$. Thus,

$$\psi = 2(40.9°) - \phi\,2(40.9°) - 60.0° = 21.8°.$$

84. The law of refraction requires that $\sin\theta_1 / \sin\theta_2 = n_{\text{water}} = $ const. We can check that this is indeed valid for any given pair of θ_1 and θ_2. For example $\sin 10° / \sin 8° = 1.3$, and $\sin 20° / \sin 15°30' = 1.3$, etc. Therefore, the index of refraction of water is $n_{\text{water}} = 1.3$.

85. Let θ be the angle of incidence and θ_2 be the angle of refraction at the left face of the plate. Let n be the index of refraction of the glass. Then, the law of refraction yields $\sin\theta = n\sin\theta_2$. The angle of incidence at the right face is also θ_2. If θ_3 is the angle of emergence there, then $n\sin\theta_2 = \sin\theta_3$. Thus $\sin\theta_3 = \sin\theta$ and $\theta_3 = \theta$.

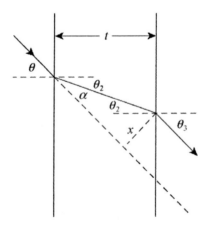

The emerging ray is parallel to the incident ray. We wish to derive an expression for x in terms of θ. If D is the length of the ray in the glass, then $D\cos\theta_2 = t$ and $D = t/\cos\theta_2$. The angle α in the diagram equals $\theta - \theta_2$ and $x = D\sin\alpha = D\sin(\theta - \theta_2)$. Thus

$$x = \frac{t \sin(\theta - \theta_2)}{\cos\theta_2}.$$

If all the angles θ, θ_2, θ_3, and $\theta - \theta_2$ are small and measured in radians, then $\sin \theta \approx \theta$, $\sin \theta_2 \approx \theta_2$, $\sin(\theta - \theta_2) \approx \theta - \theta_2$, and $\cos \theta_2 \approx 1$. Thus $x \approx t(\theta - \theta_2)$. The law of refraction applied to the point of incidence at the left face of the plate is now $\theta \approx n\theta_2$, so $\theta_2 \approx \theta/n$ and

$$x \approx t\left(\theta - \frac{\theta}{n}\right) = \frac{(n-1)t\theta}{n}.$$

86. (a) Setting v = c in the wave relation $kv = \omega = 2\pi f$, we find $f = 1.91 \times 10^8$ Hz.

(b) $E_{rms} = E_m/\sqrt{2} = B_m c/\sqrt{2} = 18.2$ V/m.

(c) $I = (E_{rms})^2/c\mu_0 = 0.878$ W/m^2.

87. From Fig. 33-19 we find $n_{max} = 1.470$ for $\lambda = 400$ nm and $n_{min} = 1.456$ for $\lambda = 700$ nm.
(a) The corresponding Brewster's angles are

$$\theta_{B,max} = \tan^{-1} n_{max} = \tan^{-1} (1.470) = 55.8°,$$

(b) and $\theta_{B,min} = \tan^{-1} (1.456) = 55.5°$.

88. We apply Eq. 33-40 (once) and Eq. 33-42 (twice) to obtain

$$I = \frac{1}{2}I_0 \cos^2 \theta_1' \cos^2 \theta_2'$$

where $\theta_1' = (90° - \theta_1) + \theta_2 = 110°$ is the relative angle between the first and the second polarizing sheets, and $\theta_2' = 90° - \theta_2 = 50°$ is the relative angle between the second and the third polarizing sheets. Thus, we have $I/I_0 = 0.024$.

89. The time for light to travel a distance d in free space is $t = d/c$, where c is the speed of light (3.00×10^8 m/s).

(a) We take d to be 150 km $= 150 \times 10^3$ m. Then,

$$t = \frac{d}{c} = \frac{150 \times 10^3 \text{ m}}{3.00 \times 10^8 \text{ m/s}} = 5.00 \times 10^{-4} \text{ s.}$$

(b) At full moon, the Moon and Sun are on opposite sides of Earth, so the distance traveled by the light is

$$d = (1.5 \times 10^8 \text{ km}) + 2 \ (3.8 \times 10^5 \text{ km}) = 1.51 \times 10^8 \text{ km} = 1.51 \times 10^{11} \text{ m}.$$

The time taken by light to travel this distance is

$$t = \frac{d}{c} = \frac{1.51 \times 10^{11} \text{ m}}{3.00 \times 10^8 \text{ m/s}} = 500 \text{ s} = 8.4 \text{ min}.$$

(c) We take d to be $2(1.3 \times 10^9 \text{ km}) = 2.6 \times 10^{12} \text{ m}$. Then,

$$t = \frac{d}{c} = \frac{2.6 \times 10^{12} \text{ m}}{3.00 \times 10^8 \text{ m/s}} = 8.7 \times 10^3 \text{ s} = 2.4 \text{ h}.$$

(d) We take d to be 6500 ly and the speed of light to be 1.00 ly/y. Then,

$$t = \frac{d}{c} = \frac{6500 \text{ ly}}{1.00 \text{ ly/y}} = 6500 \text{ y}.$$

The explosion took place in the year $1054 - 6500 = -5446$ or 5446 b.c.

90. (a) At $r = 40$ m, the intensity is

$$I = \frac{P}{\pi d^2 / 4} = \frac{P}{\pi (\theta r)^2 / 4} = \frac{4(3.0 \times 10^{-3} \text{ W})}{\pi \left[(0.17 \times 10^{-3} \text{ rad})(40 \text{ m}) \right]^2} = 83 \text{ W/m}^2.$$

(b) $P' = 4\pi r^2 I = 4\pi (40 \text{ m})^2 (83 \text{ W/m}^2) = 1.7 \times 10^6 \text{ W}.$

91. Since intensity is power divided by area (and the area is spherical in the isotropic case), then the intensity at a distance of $r = 20$ m from the source is

$$I = \frac{P}{4\pi r^2} = 0.040 \text{ W/m}^2.$$

as illustrated in Sample Problem 33-2. Now, in Eq. 33-32 for a totally absorbing area A, we note that the exposed area of the small sphere is that on a flat circle $A = \pi (0.020 \text{ m})^2 = 0.0013 \text{ m}^2$. Therefore,

$$F = \frac{IA}{c} = \frac{(0.040)(0.0013)}{3 \times 10^8} = 1.7 \times 10^{-13} \text{ N}.$$

92. (a) Assuming complete absorption, the radiation pressure is

$$p_r = \frac{I}{c} = \frac{1.4 \times 10^3 \text{ W/m}^2}{3.0 \times 10^8 \text{ m/s}} = 4.7 \times 10^{-6} \text{ N/m}^2.$$

(b) We compare values by setting up a ratio:

$$\frac{p_r}{p_0} = \frac{4.7 \times 10^{-6} \text{ N/m}^2}{1.0 \times 10^5 \text{ N/m}^2} = 4.7 \times 10^{-11}.$$

93. (a) From $kc = \omega$ where $k = 1.00 \times 10^6 \text{ m}^{-1}$, we obtain $\omega = 3.00 \times 10^{14}$ rad/s. The magnetic field amplitude is, from Eq. 33-5, $B = (5.00 \text{ V/m})/c = 1.67 \times 10^{-8}$ T. From the fact that $-\hat{k}$ (the direction of propagation), $\vec{E} = E_y\hat{j}$, and $\vec{B}$ are mutually perpendicular, we conclude that the only non-zero component of $\vec{B}$ is B_x, so that we have (in SI units)

$$B_x = 1.67 \times 10^{-8} \sin\left((1.00 \times 10^6)z + (3.00 \times 10^{14})t\right).$$

(b) The wavelength is $\lambda = 2\pi/k = 6.28 \times 10^{-6}$ m.

(c) The period is $T = 2\pi/\omega = 2.09 \times 10^{-14}$ s.

(d) The intensity is

$$I = \frac{1}{c\mu_0}\left(\frac{5.00 \text{ V/m}}{\sqrt{2}}\right)^2 = 0.0332 \text{ W/m}^2.$$

(e) As noted in part (a), the only nonzero component of $\vec{B}$ is B_x. The magnetic field oscillates along the x axis.

(f) The wavelength found in part (b) places this in the infrared portion of the spectrum.

94. It's useful to look back at the beginning of section 20-4 (particularly the steps leading up to Eq. 20-18) when considering "pressure due to collisions" (although using that term with light-interactions might be considered a little misleading). The v_x that occurs in that discussion in section 19-4 would correspond to the component $v\cos\theta$ in this problem because the angle is here being measured from the "normal axis" (instead of from the surface). Since it is the square of v_x that occurs in the section 20-4 discussion, we see therefore how the $\cos^2\theta$ factor comes about in this final result: $p_r(\theta) = p_{r\perp}\cos^2\theta$.

95. (a) The area of a hemisphere is $A = 2\pi r^2$, and we get $I = P/A = 3.5 \text{ }\mu\text{W/m}^2$.

(b) Our part (a) result multiplied by 0.22 m^2 gives 0.78 μW.

(c) The part (b) answer divided by the A of part (a) leads to 1.5×10^{-17} W/m^2.

(d) Then Eq. 33-26 gives $E_{rms} = 76$ nV/m $\Rightarrow E_{max} = \sqrt{2} E_{rms} = 1.1 \times 10^{-7}$ nV/m.

(e) $B_{rms} = E_{rms}/c = 2.5 \times 10^{-16}$ T $= 0.25$ fT.

96. (a) The electric field amplitude is $E_m = \sqrt{2} E_{rms} = 70.7$ V/m, so that the magnetic field amplitude is $B_m = 2.36 \times 10^{-7}$ T by Eq. 33-5. Since the direction of propagation, $\vec{E}$, and $\vec{B}$ are mutually perpendicular, we infer that the only non-zero component of $\vec{B}$ is B_x, and note that the direction of propagation being along the $-z$ axis means the spatial and temporal parts of the wave function argument are of like sign (see §16-5). Also, from $\lambda = 250$ nm, we find that $f = c/\lambda = 1.20 \times 10^{15}$ Hz, which leads to $\omega = 2\pi f = 7.53 \times 10^{15}$ rad/s. Also, we note that $k = 2\pi/\lambda = 2.51 \times 10^{7}$ m^{-1}. Thus, assuming some "initial condition" (that, say the field is zero, with its derivative positive, at $z = 0$ when $t = 0$), we have

$$B_x = 2.36 \times 10^{-7} \sin[(2.51 \times 10^{7})z + (7.53 \times 10^{15})t]$$

in SI units.

(b) The exposed area of the triangular chip is $A = \sqrt{3}\ell^2/8$, where $\ell = 2.00 \times 10^{-6}$ m. The intensity of the wave is

$$I = \frac{1}{c\mu_0} E_{rms}^2 = 6.64 \text{ W/m}^2.$$

Thus, Eq. 33-33 leads to

$$F = \frac{2IA}{c} = 3.83 \times 10^{-20} \text{ N}.$$

97. Accounting for the "automatic" reduction (by a factor of one-half) whenever unpolarized light passes through any polarizing sheet, then our result is $\frac{1}{2}(\cos^2(30°))^3 = 0.21$.

98. The result is

$$B_z = (2.50 \times 10^{-14} \text{ T}) \sin[(1.40 \times 10^{7} \text{ m}^{-1})y + (4.19 \times 10^{15} \text{ s}^{-1})t],$$

and we briefly indicate our reasoning as follows: the amplitude B_m is equal to $E_m/c = \sqrt{2} E_{rms}/c$. The wavenumber k is $2\pi/\lambda = 2\pi (450 \times 10^{-9}$ m$)^{-1}$. The fact that it travels in the negative x direction accounts for the + sign between terms in the sine argument. Finally, $\omega = kc$ gives the angular frequency.

99. We apply Eq. 33-40 (once) and Eq. 33-42 (twice) to obtain

$$I = \frac{1}{2} I_0 \cos^2 \theta' \cos^2 \theta''.$$

With $\theta' = \theta_2 - \theta_1 = 60° - 20° = 40°$ and $\theta'' = \theta_3 + (\pi/2 - \theta_2) = 40° + 30° = 70°$, this yields $I/I_0 = 0.034$.

100. We remind ourselves that when the unpolarized light passes through the first sheet, its intensity is reduced by a factor of 2. Thus, to end up with an overall reduction of one-third, the second sheet must cause a further decrease by a factor of two-thirds (since $(1/2)(2/3) = 1/3$). Thus, $\cos^2 \theta = 2/3 \Rightarrow \theta = 35°$.

101. (a) The magnitude of the magnetic field is

$$B = \frac{E}{c} = \frac{100 \text{ V/m}}{3.0 \times 10^8 \text{ m/s}} = 3.3 \times 10^{-7} \text{ T}.$$

(b) With $\vec{E} \times \vec{B} = \mu_0 \vec{S}$, where $\vec{E} = E\hat{k}$ and $\vec{S} = S(-\hat{j})$, one can verify easily that since $\hat{k} \times (-\hat{i}) = -\hat{j}$, $\vec{B}$ has to be in the negative x direction.

102. We use Eq. 33-33 for the force, where A is the area of the reflecting surface (4.0 m^2). The intensity is gotten from Eq. 33-27 where $P = P_S$ is in Appendix C (see also Sample Problem 33-2) and $r = 3.0 \times 10^{11}$ m (given in the problem statement). Our result for the force is 9.2 μN.

103. From Eq. 33-26, we have $E_{\text{rms}} = \sqrt{\mu_0 c I} = 1941 \text{ V}/\text{m}$, which implies (using Eq. 33-5) that $B_{\text{rms}} = 1941/c = 6.47 \times 10^{-6}$ T. Multiplying by $\sqrt{2}$ yields the magnetic field amplitude $B_m = 9.16 \times 10^{-6}$ T.

104. Eq. 33-5 gives $B = E/c$, which relates the field values at any instant — and so relates rms values to rms values, and amplitude values to amplitude values, as the case may be. Thus, the rms value of the magnetic field is $0.2/3 \times 10^8 = 6.67 \times 10^{-10}$ T, which (upon multiplication by $\sqrt{2}$) yields an amplitude value of magnetic field equal to 9.43×10^{-10} T.

105. (a) From Eq. 33-1,

$$\frac{\partial^2 E}{\partial t^2} = \frac{\partial^2}{\partial t^2} [E_m \sin(kx - \omega t)] = -\omega^2 E_m \sin(kx - \omega t),$$

and

$$c^2 \frac{\partial^2 E}{\partial x^2} = c^2 \frac{\partial^2}{\partial x^2} [E_m \sin(kx - \omega t)] = -k^2 c^2 \sin(kx - \omega t) = -\omega^2 E_m \sin(kx - \omega t).$$

Consequently,

$$\frac{\partial^2 E}{\partial t^2} = c^2 \frac{\partial^2 E}{\partial x^2}$$

is satisfied. Analogously, one can show that Eq. 33-2 satisfies

$$\frac{\partial^2 B}{\partial t^2} = c^2 \frac{\partial^2 B}{\partial x^2}.$$

(b) From $E = E_m f(kx \pm \omega t)$,

$$\frac{\partial^2 E}{\partial t^2} = E_m \frac{\partial^2 f(kx \pm \omega t)}{\partial t^2} = \omega^2 E_m \frac{d^2 f}{du^2}\bigg|_{u=kx\pm\omega t}$$

and

$$c^2 \frac{\partial^2 E}{\partial x^2} = c^2 E_m \frac{\partial^2 f(kx \pm \omega t)}{\partial t^2} = c^2 E_m k^2 \frac{d^2 f}{du^2}\bigg|_{u=kx\pm\omega t}$$

Since $\omega = ck$ the right-hand sides of these two equations are equal. Therefore,

$$\frac{\partial^2 E}{\partial t^2} = c^2 \frac{\partial^2 E}{\partial x^2}.$$

Changing E to B and repeating the derivation above shows that $B = B_m f(kx \pm \omega t)$ satisfies

$$\frac{\partial^2 B}{\partial t^2} = c^2 \frac{\partial^2 B}{\partial x^2}.$$

106. (a) Let r be the radius and ρ be the density of the particle. Since its volume is $(4\pi/3)r^3$, its mass is $m = (4\pi/3)\rho r^3$. Let R be the distance from the Sun to the particle and let M be the mass of the Sun. Then, the gravitational force of attraction of the Sun on the particle has magnitude

$$F_g = \frac{GMm}{R^2} = \frac{4\pi GM\rho r^3}{3R^2}.$$

If P is the power output of the Sun, then at the position of the particle, the radiation intensity is $I = P/4\pi R^2$, and since the particle is perfectly absorbing, the radiation pressure on it is

$$p_r = \frac{I}{c} = \frac{P}{4\pi R^2 c}.$$

All of the radiation that passes through a circle of radius r and area $A = \pi r^2$, perpendicular to the direction of propagation, is absorbed by the particle, so the force of the radiation on the particle has magnitude

$$F_r = p_r A = \frac{\pi P r^2}{4\pi R^2 c} = \frac{P r^2}{4 R^2 c}.$$

The force is radially outward from the Sun. Notice that both the force of gravity and the force of the radiation are inversely proportional to R^2. If one of these forces is larger than the other at some distance from the Sun, then that force is larger at all distances. The two forces depend on the particle radius r differently: F_g is proportional to r^3 and F_r is proportional to r^2. We expect a small radius particle to be blown away by the radiation pressure and a large radius particle with the same density to be pulled inward toward the Sun. The critical value for the radius is the value for which the two forces are equal. Equating the expressions for F_g and F_r, we solve for r:

$$r = \frac{3P}{16\pi GM\rho c}.$$

(b) According to Appendix C, $M = 1.99 \times 10^{30}$ kg and $P = 3.90 \times 10^{26}$ W. Thus,

$$r = \frac{3(3.90 \times 10^{26} \text{ W})}{16\pi(6.67 \times 10^{-11} \text{ N} \cdot \text{m}^2 / \text{kg}^2)(1.99 \times 10^{30} \text{ kg})(1.0 \times 10^3 \text{ kg} / \text{m}^3)(3.00 \times 10^8 \text{ m} / \text{s})}$$

$$= 5.8 \times 10^{-7} \text{ m}.$$

107. (a) The polarization direction is defined by the electric field (which is perpendicular to the magnetic field in the wave, and also perpendicular to the direction of wave travel). The given function indicates the magnetic field is along the x axis (by the subscript on B) and the wave motion is along $-y$ axis (see the argument of the sine function). Thus, the electric field direction must be parallel to the z axis.

(b) Since k is given as 1.57×10^7/m, then $\lambda = 2\pi/k = 4.0 \times 10^{-7}$ m, which means $f = c/\lambda = 7.5 \times 10^{14}$ Hz.

(c) The magnetic field amplitude is given as $B_m = 4.0 \times 10^{-6}$ T. The electric field amplitude E_m is equal to B_m divided by the speed of light c. The rms value of the electric field is then E_m divided by $\sqrt{2}$. Eq. 33-26 then gives $I = 1.9$ kW/m^2.

108. Using Eqs. 33-40 and 33-42, we obtain

$$\frac{I_{final}}{I_0} = \frac{\left(\frac{1}{2}I_0\right)\left(\cos^2 45°\right)\left(\cos^2 45°\right)}{I_0} = \frac{1}{8} = 0.125.$$

109. With the index of refraction $n = 1.456$ at the red end, since $\sin \theta_c = 1/n$, the critical angle is $\theta_c = 43.38°$ for red.

(a) At an angle of incidence of $\theta_1 = 42.00° < \theta_c$, the refracted light is white.

(b) At an angle of incidence of $\theta_1 = 43.10°$ which is slightly less than θ_c, the refracted light is white but dominated by red end.

(c) At an angle of incidence of $\theta_1 = 44.00° > \theta_c$, there is no refracted light.

110. (a) The diagram below shows a cross section, through the center of the cube and parallel to a face. L is the length of a cube edge and S labels the spot. A portion of a ray from the source to a cube face is also shown.

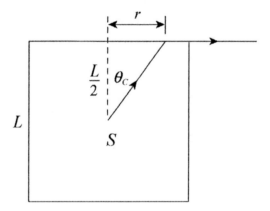

Light leaving the source at a small angle θ is refracted at the face and leaves the cube; light leaving at a sufficiently large angle is totally reflected. The light that passes through the cube face forms a circle, the radius r being associated with the critical angle for total internal reflection. If θ_c is that angle, then

$$\sin \theta_c = \frac{1}{n}$$

where n is the index of refraction for the glass. As the diagram shows, the radius of the circle is given by $r = (L/2) \tan \theta_c$. Now,

$$\tan \theta_c = \frac{\sin \theta_c}{\cos \theta_c} = \frac{\sin \theta_c}{\sqrt{1 - \sin^2 \theta_c}} = \frac{1/n}{\sqrt{1 - (1/n)^2}} = \frac{1}{\sqrt{n^2 - 1}}$$

and the radius of the circle is

$$r = \frac{L}{2\sqrt{n^2 - 1}} = \frac{10\,\text{mm}}{2\sqrt{(1.5)^2 - 1}} = 4.47\,\text{mm}.$$

If an opaque circular disk with this radius is pasted at the center of each cube face, the spot will not be seen (provided internally reflected light can be ignored).

(b) There must be six opaque disks, one for each face. The total area covered by disks is $6\pi r^2$ and the total surface area of the cube is $6L^2$. The fraction of the surface area that must be covered by disks is

$$f = \frac{6\pi r^2}{6L^2} = \frac{\pi r^2}{L^2} = \frac{\pi(4.47\,\text{mm})^2}{(10\,\text{mm})^2} = 0.63.$$

111. (a) Suppose that at time t_1, the moon is starting a revolution (on the verge of going behind Jupiter, say) and that at this instant, the distance between Jupiter and Earth is ℓ_1. The time of the start of the revolution as seen on Earth is $t_1^* = t_1 + \ell_1/c$. Suppose the moon starts the next revolution at time t_2 and at that instant, the Earth-Jupiter distance is ℓ_2. The start of the revolution as seen on Earth is $t_2^* = t_2 + \ell_2/c$. Now, the actual period of the moon is given by $T = t_2 - t_1$ and the period as measured on Earth is

$$T^* = t_2^* - t_1^* = t_2 - t_1 + \frac{\ell_2}{c} - \frac{\ell_1}{c} = T + \frac{\ell_2 - \ell_1}{c}.$$

The period as measured on Earth is longer than the actual period. This is due to the fact that Earth moves during a revolution, and light takes a finite time to travel from Jupiter to Earth. For the situation depicted in Fig. 33-80, light emitted at the end of a revolution travels a longer distance to get to Earth than light emitted at the beginning. Suppose the position of Earth is given by the angle θ, measured from x. Let R be the radius of Earth's orbit and d be the distance from the Sun to Jupiter. The law of cosines, applied to the triangle with the Sun, Earth, and Jupiter at the vertices, yields $\ell^2 = d^2 + R^2 - 2dR\cos\theta$. This expression can be used to calculate ℓ_1 and ℓ_2. Since Earth does not move very far during one revolution of the moon, we may approximate $\ell_2 - \ell_1$ by $(d\ell/dt)T$ and T^* by $T + (d\ell/dt)(T/c)$. Now

$$\frac{d\ell}{dt} = \frac{2Rd\,\sin\theta}{\sqrt{d^2 + R^2 - 2dR\cos\theta}}\frac{d\theta}{dt} = \frac{2vd\,\sin\theta}{\sqrt{d^2 + R^2 - 2dR\cos\theta}},$$

where $v = R\,(d\theta/dt)$ is the speed of Earth in its orbit. For $\theta = 0$, $(d\ell/dt) = 0$ and $T^* = T$. Since Earth is then moving perpendicularly to the line from the Sun to Jupiter, its distance from the planet does not change appreciably during one revolution of the moon. On the other hand, when $\theta = 90°$, $d\ell / dt = vd / \sqrt{d^2 + R^2}$ and

$$T^* = T\left(1 + \frac{vd}{c\sqrt{d^2 + R^2}}\right).$$

The Earth is now moving parallel to the line from the Sun to Jupiter, and its distance from the planet changes during a revolution of the moon.

(b) Our notation is as follows: t is the actual time for the moon to make N revolutions, and t^* is the time for N revolutions to be observed on Earth. Then,

$$t^* = t + \frac{\ell_2 - \ell_1}{c},$$

where ℓ_1 is the Earth-Jupiter distance at the beginning of the interval and ℓ_2 is the Earth-Jupiter distance at the end. Suppose Earth is at position x at the beginning of the interval, and at y at the end. Then, $\ell_1 = d - R$ and $\ell_2 = \sqrt{d^2 + R^2}$. Thus,

$$t^* = t + \frac{\sqrt{d^2 + R^2} - (d - R)}{c}.$$

A value can be found for t by measuring the observed period of revolution when Earth is at x and multiplying by N. We note that the observed period is the true period when Earth is at x. The time interval as Earth moves from x to y is t^*. The difference is

$$t^* - t = \frac{\sqrt{d^2 + R^2} - (d - R)}{c}.$$

If the radii of the orbits of Jupiter and Earth are known, the value for $t^* - t$ can be used to compute c. Since Jupiter is much further from the Sun than Earth, $\sqrt{d^2 + R^2}$ may be approximated by d and $t^* - t$ may be approximated by R/c. In this approximation, only the radius of Earth's orbit need be known.

Chapter 34

1. The image is 10 cm behind the mirror and you are 30 cm in front of the mirror. You must focus your eyes for a distance of 10 cm + 30 cm = 40 cm.

2. The bird is a distance d_2 in front of the mirror; the plane of its image is that same distance d_2 behind the mirror. The lateral distance between you and the bird is $d_3 = 5.00$ m. We denote the distance from the camera to the mirror as d_1, and we construct a right triangle out of d_3 and the distance between the camera and the image plane $(d_1 + d_2)$. Thus, the focus distance is

$$d = \sqrt{(d_1 + d_2)^2 + d_3^2} = \sqrt{(4.30\text{m}+3.30\text{m})^2 + (5.00\text{m})^2} = 9.10\text{m}.$$

3. The intensity of light from a point source varies as the inverse of the square of the distance from the source. Before the mirror is in place, the intensity at the center of the screen is given by $I_P = A/d^2$, where A is a constant of proportionality. After the mirror is in place, the light that goes directly to the screen contributes intensity I_P, as before. Reflected light also reaches the screen. This light appears to come from the image of the source, a distance d behind the mirror and a distance $3d$ from the screen. Its contribution to the intensity at the center of the screen is

$$I_r = \frac{A}{(3d)^2} = \frac{A}{9d^2} = \frac{I_P}{9}.$$

The total intensity at the center of the screen is

$$I = I_p + I_r = I_p + \frac{I_P}{9} = \frac{10}{9}I_P.$$

The ratio of the new intensity to the original intensity is $I/I_P = 10/9 = 1.11$.

4. When S is barely able to see B the light rays from B must reflect to S off the edge of the mirror. The angle of reflection in this case is 45°, since a line drawn from S to the mirror's edge makes a 45° angle relative to the wall. By the law of reflection, we find

$$\frac{x}{d/2} = \tan 45° \Rightarrow x = \frac{d}{2} = \frac{3.0\,\text{m}}{2} = 1.5\,\text{m}.$$

5. We apply the law of refraction, assuming all angles are in radians:

$$\frac{\sin \theta}{\sin \theta'} = \frac{n_w}{n_{air}},$$

which in our case reduces to $\theta' \approx \theta/n_w$ (since both θ and θ' are small, and $n_{air} \approx 1$). We refer to our figure below.

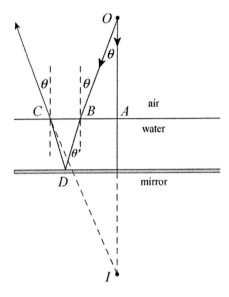

The object O is a vertical distance d_1 above the water, and the water surface is a vertical distance d_2 above the mirror. We are looking for a distance d (treated as a positive number) below the mirror where the image I of the object is formed. In the triangle $O\,AB$

$$|AB| = d_1 \tan \theta \approx d_1 \theta,$$

and in the triangle CBD

$$|BC| = 2d_2 \tan \theta' \approx 2d_2\theta' \approx \frac{2d_2\theta}{n_w}.$$

Finally, in the triangle ACI, we have $|AI| = d + d_2$. Therefore,

$$d = |AI| - d_2 = \frac{|AC|}{\tan \theta} - d_2 \approx \frac{|AB| + |BC|}{\theta} - d_2 = \left(\frac{d_1}{\theta} + \frac{2d_2\theta}{n_w} \right) \frac{1}{\theta} - d_2 = d_1 + \frac{2d_2}{n_w} - d_2$$

$$= 250 \text{cm} + \frac{2(200 \text{cm})}{1.33} - 200 \text{cm} = 351 \text{cm}.$$

6. We note from Fig. 34-32 that $m = \frac{1}{2}$ when $p = 5$ cm. Thus Eq. 34-7 (the magnification equation) gives us $i = -10$ cm in that case. Then, by Eq. 34-9 (which applies to mirrors and thin-lenses) we find the focal length of the mirror is $f = 10$ cm. Next, the problem

asks us to consider $p = 14$ cm. With the focal length value already determined, then Eq. 34-9 yields $i = 35$ cm for this new value of object distance. Then, using Eq. 34-7 again, we find $m = i/p = -2.5$.

7. We use Eqs. 34-3 and 34-4, and note that $m = -i/p$. Thus,

$$\frac{1}{p} - \frac{1}{pm} = \frac{1}{f} = \frac{2}{r}.$$

We solve for p:

$$p = \frac{r}{2}\left(1 - \frac{1}{m}\right) = \frac{35.0\,\text{cm}}{2}\left(1 - \frac{1}{2.50}\right) = 10.5 \text{ cm.}$$

8. The graph in Fig. 34-33 implies that $f = 20$ cm, which we can plug into Eq. 34-9 (with $p = 70$ cm) to obtain $i = +28$ cm.

9. We recall that a concave mirror has a positive value of focal length.

(a) Then (with $f = +12$ cm and $p = +18$ cm), the radius of curvature is $r = 2f = +24$ cm .

(b) Eq. 34-9 yields $i = pf/(p-f) = +36$ cm.

(c)Then, by Eq. 34-7, the lateral magnification is $m = -i/p = -2.0$.

(d) Since the image distance computation produced a positive value, the image is real (R).

(e) The magnification computation produced a negative value, so it is inverted (I).

(f) For a mirror, the side where a real image forms is the <u>same</u> as the side where the object is.

10. A concave mirror has a positive value of focal length.

(a) Then (with $f = +10$ cm and $p = +15$ cm), the radius of curvature is $r = 2f = +20$ cm .

(b) Eq. 34-9 yields $i = pf/(p-f) = +30$ cm.

(c)Then, by Eq. 34-7, $m = -i/p = -2.0$.

(d) Since the image distance computation produced a positive value, the image is real (R).

(e) The magnification computation produced a negative value, so it is inverted (I).

(f) For a mirror, the side where a real image forms is <u>the same</u> as the side where the object is.

11. A concave mirror has a positive value of focal length.

(a) Then (with $f = +18$ cm and $p = +12$ cm) , the radius of curvature is $r = 2f = +36$ cm.

(b) Eq. 34-9 yields $i = pf/(p-f) = -36$ cm.

(c) Then, by Eq. 34-7, $m = -i/p = +3.0$.

(d) Since the image distance is negative, the image is virtual (V).

(e) The magnification computation produced a positive value, so it is upright [not inverted] (NI).

(f) For a mirror, the side where a virtual image forms is <u>opposite</u> from the side where the object is.

12. A concave mirror has a positive value of focal length.

(a) Then (with $f = +36$ cm and $p = +24$ cm), the radius of curvature is $r = 2f = +72$ cm.

(b) Eq. 34-9 yields $i = pf/(p-f) = -72$ cm.

(c) Then, by Eq. 34-7, $m = -i/p = +3.0$.

(d) Since the image distance is negative, the image is virtual (V).

(e) The magnification computation produced a positive value, so it is upright [not inverted] (NI).

(f) For a mirror, the side where a virtual image forms is <u>opposite</u> from the side where the object is.

13. A convex mirror has a negative value of focal length.

(a) Then (with $f = -10$ cm and $p = +8$ cm), the radius of curvature is $r = 2f = -20$ cm.

(b) Eq. 34-9 yields $i = pf/(p-f) = -4.4$ cm.

(c) Then, by Eq. 34-7, $m = -i/p = +0.56$.

(d) Since the image distance is negative, the image is virtual (V).

(e) The magnification computation produced a positive value, so it is upright [not inverted] (NI).

(f) For a mirror, the side where a virtual image forms is opposite from the side where the object is.

14. A convex mirror has a negative value of focal length.

(a) Then (with $f = -35$ cm and $p = +22$ cm), the radius of curvature is $r = 2f = -70$ cm.

(b) Eq. 34-9 yields $i = pf/(p-f) = -14$ cm.

(c) Then, by Eq. 34-7, $m = -i/p = +0.61$.

(d) Since the image distance is negative, the image is virtual (V).

(e) The magnification computation produced a positive value, so it is upright [not inverted] (NI).

(f) For a mirror, the side where a virtual image forms is opposite from the side where the object is.

15. A convex mirror has a negative value of focal length.

(a) Then (with $f = -8$ cm and $p = +10$ cm), the radius of curvature is $r = 2f = -16$ cm.

(b) Eq. 34-9 yields $i = pf/(p-f) = -4.4$ cm.

(c) Then, by Eq. 34-7, $m = -i/p = +0.44$.

(d) Since the image distance is negative, the image is virtual (V).

(e) The magnification computation produced a positive value, so it is upright [not inverted] (NI).

(f) For a mirror, the side where a virtual image forms is opposite from the side where the object is.

16. A convex mirror has a negative value of focal length.

(a) Then (with $f = -14$ cm and $p = +17$ cm), the radius of curvature is $r = 2f = -28$ cm.

(b) Eq. 34-9 yields $i = pf/(p-f) = -7.7$ cm.

(c) Then, by Eq. 34-7, $m = -i/p = +0.45$.

(d) Since the image distance is negative, the image is virtual (V).

(e) The magnification computation produced a positive value, so it is upright [not inverted] (NI).

(f) For a mirror, the side where a virtual image forms is <u>opposite</u> from the side where the object is.

17. (a) From Eqs. 34-3 and 34-4, we obtain

$$i = pf/(p - f) = pr/(2p - r).$$

Differentiating both sides with respect to time and using $v_O = -dp/dt$, we find

$$v_I = \frac{di}{dt} = \frac{d}{dt}\left(\frac{pr}{2p-r}\right) = \frac{-rv_O(2p-r) + 2v_O pr}{(2p-r)^2} = \left(\frac{r}{2p-r}\right)^2 v_O.$$

(b) If $p = 30$ cm, we obtain

$$v_I = \left[\frac{15 \text{ cm}}{2(30 \text{ cm}) - 15 \text{ cm}}\right]^2 (5.0 \text{ cm/s}) = 0.56 \text{ cm/s}.$$

(c) If $p = 8.0$ cm, we obtain

$$v_I = \left[\frac{15 \text{ cm}}{2(8.0 \text{ cm}) - 15 \text{ cm}}\right]^2 (5.0 \text{ cm/s}) = 1.1 \times 10^3 \text{ cm/s}.$$

(d) If $p = 1.0$ cm, we obtain

$$v_I = \left[\frac{15 \text{cm}}{2(1.0 \text{cm}) - 15 \text{cm}}\right]^2 (5.0 \text{cm/s}) = 6.7 \text{cm/s}.$$

18. We note that there is "singularity" in this graph (Fig. 34-34) like there was in Fig. 34-33), which tells us that there is no point where $p = f$ (which causes Eq. 34-9 to "blow up"). Since $p > 0$, as usual, then this means that the focal length is <u>not</u> positive. We know it is not a flat mirror since the curve shown does decrease with p, so we conclude it is a convex mirror. We examine the point where $m = 0.50$ and $p = 10$ cm. Combining Eq. 34-7 and Eq. 34-9 we obtain $m = -i/p = -f/(p - f)$. This yields $f = -10$ cm (verifying our expectation that the mirror is convex). Now, for $p = 21$ cm, we find $m = -f/(p - f) = +0.32$.

19. (a) The mirror is concave.

(b) $f = +20$ cm (positive, because the mirror is concave).

(c) $r = 2f = 2(+20$ cm$) = +40$ cm.

(d) The object distance $p = +10$ cm, as given in the Table.

(e) The image distance is $i = (1/f - 1/p)^{-1} = (1/20$ cm $- 1/10$ cm$)^{-1} = -20$ cm.

(f) $m = -i/p = -(-20$ cm$/10$ cm$) = +2.0$.

(g) The image is virtual (V).

(h) The image is upright or not inverted (NI).

(i) For a mirror, the side where a virtual image forms is <u>opposite</u> from the side where the object is.

20. (a) The fact that the magnification is 1 means that the mirror is flat (plane).

(b) Flat mirrors (and flat "lenses" such as a window pane) have $f = \infty$ (or $f = -\infty$ since the sign does not matter in this extreme case).

(c) The radius of curvature is $r = 2f = \infty$ (or $r = -\infty$) by Eq. 34-3.

(d) $p = +10$ cm, as given in the Table.

(e) Eq. 34-4 readily yields $i = pf/(p-f) = -10$ cm.

(f) The magnification is $m = -i/p = +1.0$.

(g) The image is virtual since $i < 0$.

(h) The image is upright, or not inverted (NI).

(i) For a mirror, the side where a virtual image forms is <u>opposite</u> from the side where the object is.

21. (a) Since $f > 0$, the mirror is concave.

(b) $f = +20$ cm, as given in the Table.

(c) Using Eq. 34-3, we obtain $r = 2f = +40$ cm.

(d) $p = +10$ cm, as given in the Table.

(e) Eq. 34-4 readily yields $i = pf/(p-f) = +60$ cm.

(f) Eq. 34-6 gives $m = -i/p = -2.0$.

(g) Since $i > 0$, the image is real (R).

(h) Since $m < 0$, the image is inverted (I).

(i) For a mirror, the side where a real image forms is the <u>same</u> as the side where the object is.

22. (a) Since $m = -1/2 < 0$, the image is inverted. With that in mind, we examine the various possibilities in Figs. 34-7, 34-9 and 34-10, and note that an inverted image (for reflections from a single mirror) can only occur if the mirror is concave (and if $p > f$).

(b) Next, we find i from Eq. 34-6 (which yields $i = mp = 30$ cm) and then use this value (and Eq. 34-4) to compute the focal length; we obtain $f = +20$ cm.

(c) Then, Eq. 34-3 gives $r = 2f = +40$ cm.

(d) $p = 60$ cm, as given in the Table.

(e) As already noted, $i = +30$ cm.

(f) $m = -1/2$, as given.

(g) Since $i > 0$, the image is real (R).

(h) As already noted, the image is inverted (I).

(i) For a mirror, the side where a real image forms is the <u>same</u> as the side where the object is.

23. (a) Since $r < 0$ then (by Eq. 34-3) $f < 0$, which means the mirror is convex.

(b) The focal length is $f = r/2 = -20$ cm.

(c) $r = -40$ cm, as given in the Table.

(d) Eq. 34-4 leads to $p = +20$ cm.

(e) $i = -10$ cm, as given in the Table.

(f) Eq. 34-6 gives $m = +0.50$.

(g) The image is virtual (V).

(h) The image is upright, or not inverted (NI).

(i) For a mirror, the side where a virtual image forms is <u>opposite</u> from the side where the object is.

24. (a) Since $0 < m < 1$, the image is upright but smaller than the object. With that in mind, we examine the various possibilities in Figs. 34-7, 34-9 and 34-10, and note that such an image (for reflections from a single mirror) can only occur if the mirror is convex.

(b) Thus, we must put a minus sign in front of the "20" value given for f, i.e., $f = -20$ cm.

(c) Eq. 34-3 then gives $r = 2f = -40$ cm.

(d) To solve for i and p we must set up Eq. 34-4 and Eq. 34-6 as a simultaneous set and solve for the two unknowns. The results are $p = +180$ cm $= +1.8$ m, and

(e) $i = -18$ cm.

(f) $m = 0.10$, as given in the Table.

(g) The image is virtual (V) since $i < 0$.

(h) The image is upright, or not inverted (NI), as already noted.

(i) For a mirror, the side where a virtual image forms is <u>opposite</u> from the side where the object is.

25. (a) The mirror is convex, as given.

(b) Knowing the mirror is convex means we must put a minus sign in front of the "40" value given for r. Then, Eq. 34-3 yields $f = r/2 = -20$ cm.

(c) $r = -40$ cm.

(d) The fact that the mirror is convex also means that we need to insert a minus sign in front of the "4.0" value given for i, since the image in this case must be virtual (see Figs. 34-7, 34-9 and 34-10). Eq. 34-4 leads to $p = +5.0$ cm.

(e) As noted above, $i = -4.0$.

(f) Eq. 34-6 gives $m = +0.8$.

(g) The image is virtual (V) since $i < 0$.

(h) The image is upright, or not inverted (NI).

(i) For a mirror, the side where a virtual image forms is <u>opposite</u> from the side where the object is.

26. (a) Since the image is inverted, we can scan Figs. 34-7, 34-9 and 34-10 in the textbook and find that the mirror must be concave.

(b) This also implies that we must put a minus sign in front of the "0.50" value given for m. To solve for f, we first find $i = -pm = +12$ cm from Eq. 34-6 and plug into Eq. 34-4; the result is $f = +8$ cm.

(c) Thus, $r = 2f = +16$ cm.

(d) $p = +24$ cm, as given in the Table.

(e) As shown above, $i = -pm = +12$ cm.

(f) $m = -0.50$, with a minus sign.

(g) The image is real (R) since $i > 0$.

(h) The image is inverted (I), as noted above.

(i) For a mirror, the side where a real image forms is <u>the same</u> as the side where the object is.

27. (a) The fact that the focal length is given as a negative value means the mirror is convex.

(b) $f = -30$ cm, as given in the Table.

(c) The radius of curvature is $r = 2f = -60$ cm.

(d) Eq. 34-9 gives $p = if/(i - f) = +30$ cm.

(e) $i = -15$, as given in the Table.

(f) From Eq. 34-7, we get $m = +1/2 = 0.50$.

(g) The image distance is given as a negative value (as it would have to be, since the mirror is convex), which means the image is virtual (V).

(h) Since $m > 0$, the image is upright (not inverted: NI).

(i) The image is on the side of the mirror <u>opposite</u> to the object.

28. (a) We are told that the image is on the same side as the object; this means the image is real (R) and further implies that the mirror is concave.

(b) The focal distance is $f = +20$ cm.

(c) The radius of curvature is $r = 2f = +40$ cm.

(d) $p = +60$ cm, as given in the Table.

(e) Eq. 34-9 gives $i = pf/(p - f) = +30$ cm.

(f) Eq. 34-7 gives $m = -i/p = -0.50$.

(g) As noted above, the image is real (R).

(h) The image is inverted (I) since $m < 0$.

(i) For a mirror, the side where a real image forms is the <u>same</u> as the side where the object is.

29. (a) As stated in the problem, the image is inverted (I) which implies that it is real (R). It also (more directly) tells us that the magnification is equal to a negative value: $m = -0.40$. By Eq. 34-7, the image distance is consequently found to be $i = +12$ cm. Real images don't arise (under normal circumstances) from convex mirrors, so we conclude that this mirror is concave.

(b) The focal length is $f = +8.6$ cm, using Eq. 34-9 $f = +8.6$ cm.

(c) The radius of curvature is $r = 2f = +17.2$ cm ≈ 17 cm.

(d) $p = +30$ cm, as given in the Table.

(e) As noted above, $i = +12$ cm.

(f) Similarly, $m = -0.40$, with a minus sign.

(g) The image is real (R).

(h) The image is inverted (I).

(i) For a mirror, the side where a real image forms is the <u>same</u> as the side where the object is.

30. (a) From Eq. 34-7, we get $i = -mp = +28$ cm, which implies the image is real (R) and on the same side as the object. Since $m < 0$, we know it was inverted (I). From Eq. 34-9,

we obtain $f = ip/(i + p) = +16$ cm, which tells us (among other things) that the mirror is concave.

(b) $f = ip/(i + p) = +16$ cm.

(c) $r = 2f = +32$ cm.

(d) $p = +40$ cm, as given in the Table.

(e) $i = -mp = +28$ cm.

(f) $m = -0.70$, as given in the Table.

(g) The image is real (R).

(h) The image is inverted (I).

(i) For a mirror, the side where a real image forms is the same as the side where the object is.

31. (a) The fact that the magnification is equal to a positive value means that the image is upright (not inverted: NI), and further implies (by Eq. 34-7) that the image distance (i) is equal to a negative value $\Rightarrow$ the image is virtual (V). Looking at the discussion of mirrors in sections 34-3 and 34-4, we see that a positive magnification of magnitude less than unity is only possible for convex mirrors.

(b) For $0 < m < 1$ this will only give a positive value for $p = f/(1 - 1/m)$ if $f < 0$. Thus, with a minus sign, we have $f = -30$ cm.

(c) $r = 2f = -60$ cm.

(d) $p = f/(1 - 1/m) = + 120$ cm $= 1.2$ m.

(e) $i = -mp = -24$ cm.

(f) $m = +0.20$, as given in the Table.

(g) The image is virtual (V).

(h) The image is upright, or not inverted (NI).

(i) For a mirror, the side where a virtual image forms is opposite from the side where the object is.

32. (a) We use Eq. 34-8 and note that $n_1 = n_{air} = 1.00$, $n_2 = n$, $p = \infty$, and $i = 2r$:

$$\frac{1.00}{\infty} + \frac{n}{2r} = \frac{n-1}{r}.$$

We solve for the unknown index: $n = 2.00$.

(b) Now $i = r$ so Eq. 34-8 becomes

$$\frac{n}{r} = \frac{n-1}{r},$$

which is not valid unless $n \to \infty$ or $r \to \infty$. It is impossible to focus at the center of the sphere.

33. We use Eq. 34-8 (and Fig. 34-10(d) is useful), with $n_1 = 1.6$ and $n_2 = 1$ (using the rounded-off value for air).

$$\frac{1.6}{p} + \frac{1}{i} = \frac{1-1.6}{r}$$

Using the sign convention for r stated in the paragraph following Eq. 34-8 (so that $r = -5.0$ cm), we obtain $i = -2.4$ cm for objects at $p = 3.0$ cm. Returning to Fig. 34-36 (and noting the location of the observer), we conclude that the tabletop seems 7.4 cm away.

34. In addition to $n_1 = 1.0$, we are given (a) $n_2 = 1.5$, (b) $p = +10$ and (c) $r = +30$.

(d) Eq. 34-8 yields

$$i = n_2 \left(\frac{n_2 - n_1}{r} - \frac{n_1}{p} \right)^{-1} = 1.5 \left(\frac{1.5-1.0}{30 \text{ cm}} - \frac{1.0}{10 \text{ cm}} \right) = -18 \text{ cm.}$$

(e) The image is virtual (V) and upright since $i < 0$.

(f) The object and its image are in the same side. The ray diagram would be similar to Fig. 34-11(c) in the textbook.

35. In addition to $n_1 = 1.0$, we are given (a) $n_2 = 1.5$, (b) $p = +10$ and (d) $i = -13$.

(c) Eq. 34-8 yields

$$r = (n_2 - n_1) \left(\frac{n_1}{p} + \frac{n_2}{i} \right)^{-1} = (1.5-1.0) \left(\frac{1.0}{10} + \frac{1.5}{-13} \right)^{-1} = -32.5 \text{cm} \approx -33 \text{ cm.}$$

(e) The image is virtual (V) and upright.

(f) The object and its image are in the same side. The ray diagram would be similar to Fig. 34-11(e).

36. In addition to $n_1 = 1.0$, we are given (a) $n_2 = 1.5$, (c) $r = +30$ and (d) $i = +600$.

(b) Eq. 34-8 gives

$$p = \frac{n_1}{\dfrac{n_2 - n_1}{r} - \dfrac{n_2}{i}} = \frac{1.0}{\dfrac{1.5 - 1.0}{30} - \dfrac{1.5}{600}} = 71 \text{cm}.$$

(d) With $i > 0$, the image is real (R) and inverted.

(e) The object and its image are in the opposite side. The ray diagram would be similar to Fig. 34-11(a) in the textbook.

37. In addition to $n_1 = 1.5$, we are given (a) $n_2 = 1.0$, (b) $p = +10$ and (d) $i = -6.0$.

(c) We manipulate Eq. 34-8 to find r:

$$r = (n_2 - n_1)\left(\frac{n_1}{p} + \frac{n_2}{i}\right)^{-1} = (1.0 - 1.5)\left(\frac{1.5}{10} + \frac{1.0}{-6.0}\right)^{-1} = 30 \text{ cm}.$$

(e) The image is virtual (V) and upright.

(f) The object and its image are in the same side. The ray diagram would be similar to Fig. 34-11(f) in the textbook, but with the object and the image located closer to the surface.

38. In addition to $n_1 = 1.5$, we are given (a) $n_2 = 1.0$, (c) $r = -30$ and (d) $i = -7.5$.

(b)We manipulate Eq. 34-8 to find p:

$$p = \frac{n_1}{\dfrac{n_2 - n_1}{r} - \dfrac{n_2}{i}} = \frac{1.5}{\dfrac{1.0 - 1.5}{-30} - \dfrac{1.0}{-7.5}} = 10 \text{cm}.$$

(e) The image is virtual (V) and upright.

(f) The object and its image are in the same side. The ray diagram would be similar to Fig. 34-11(d) in the textbook.

39. In addition to $n_1 = 1.5$, we are given (a) $n_2 = 1.0$, (b) $p = +70$ and (c) $r = +30$.

(d) We manipulate Eq. 34-8 to find the image distance:

$$i = n_2 \left(\frac{n_2 - n_1}{r} - \frac{n_1}{p} \right)^{-1} = 1.0 \left(\frac{1.0 - 1.5}{30 \text{ cm}} - \frac{1.5}{70 \text{ cm}} \right)^{-1} = -26 \text{ cm}.$$

(e) The image is virtual (V) and upright.

(f) The object and its image are in the same side. The ray diagram would be similar to Fig. 34-11(f) in the textbook.

40. In addition to $n_1 = 1.5$, we are given (b) $p = +100$, (c) $r = -30$ and (d) $i = +600$.

(a) We manipulate Eq. 34-8 to separate the indices:

$$n_2 \left(\frac{1}{r} - \frac{1}{i} \right) = \left(\frac{n_1}{p} + \frac{n_1}{r} \right) \quad \Rightarrow \quad n_2 \left(\frac{1}{-30} - \frac{1}{600} \right) = \left(\frac{1.5}{100} + \frac{1.5}{-30} \right) \quad \Rightarrow \quad n_2 \left(-0.035 \right) = -0.035$$

which implies $n_2 = 1.0$.

(e) The image is real (R) and inverted.

(f) The object and its image are in the opposite side. The ray diagram would be similar to Fig. 34-11(b) in the textbook.

41. Let the diameter of the Sun be d_s and that of the image be d_i. Then, Eq. 34-5 leads to

$$d_i = |m| d_s = \left(\frac{i}{p} \right) d_s \approx \left(\frac{f}{p} \right) d_s = \frac{\left(20.0 \times 10^{-2} \text{ m} \right) \left(2 \right) \left(6.96 \times 10^8 \text{ m} \right)}{1.50 \times 10^{11} \text{ m}} = 1.86 \times 10^{-3} \text{ m}$$

$$= 1.86 \text{ mm}.$$

42. The singularity the graph (where the curve goes to $\pm\infty$) is at $p = 30$ cm, which implies (by Eq. 34-9) that $f = 30$ cm > 0 (converging type lens). For $p = 100$ cm, Eq. 34-9 leads to $i = +43$ cm.

43. We use the lens maker's equation, Eq. 34-10:

$$\frac{1}{f} = (n-1) \left(\frac{1}{r_1} - \frac{1}{r_2} \right)$$

where f is the focal length, n is the index of refraction, r_1 is the radius of curvature of the first surface encountered by the light and r_2 is the radius of curvature of the second surface. Since one surface has twice the radius of the other and since one surface is convex to the incoming light while the other is concave, set $r_2 = -2r_1$ to obtain

$$\frac{1}{f} = (n-1)\left(\frac{1}{r_1} + \frac{1}{2r_1}\right) = \frac{3(n-1)}{2r_1}.$$

(a) We solve for the smaller radius r_1:

$$r_1 = \frac{3(n-1)f}{2} = \frac{3(1.5-1)(60 \text{ mm})}{2} = 45 \text{ mm}.$$

(b) The magnitude of the larger radius is $|r_2| = 2r_1 = 90 \text{ mm}$.

44. Since the focal length is a constant for the whole graph, then $1/p + 1/i = $ constant. Consider the value of the graph at $p = 20$ cm; we estimate its value there to be -10 cm. Therefore, $1/20 + 1/(-10) = 1/70 + 1/i_{\text{new}}$. Thus, $i_{\text{new}} = -16$ cm.

45. (a) We use Eq. 34-10:

$$f = \left[(n-1)\left(\frac{1}{r_1} - \frac{1}{r_2}\right)\right]^{-1} = \left[(1.5-1)\left(\frac{1}{\infty} - \frac{1}{-20 \text{ cm}}\right)\right]^{-1} = +40 \text{ cm}.$$

(b) From Eq. 34-9,

$$i = \left(\frac{1}{f} - \frac{1}{p}\right)^{-1} = \left(\frac{1}{40 \text{ cm}} - \frac{1}{40 \text{ cm}}\right)^{-1} = \infty.$$

46. Combining Eq. 34-7 and Eq. 34-9, we have $m(p-f) = -f$. The graph in Fig. 34-39 indicates that $m = 2$ where $p = 5$ cm, so our expression yields $f = 10$ cm. Plugging this back into our expression and evaluating at $p = 14$ cm yields $m = -2.5$.

47. We solve Eq. 34-9 for the image distance:

$$i = \left(\frac{1}{f} - \frac{1}{p}\right)^{-1} = \frac{fp}{p-f}.$$

The height of the image is thus

$$h_i = mh_p = \left(\frac{i}{p}\right)h_p = \frac{fh_p}{p-f} = \frac{(75 \text{ mm})(1.80 \text{ m})}{27 \text{ m} - 0.075 \text{ m}} = 5.0 \text{ mm}.$$

48. Combining Eq. 34-7 and Eq. 34-9, we have $m(p-f) = -f$. The graph in Fig. 34-40 indicates that $m = 0.5$ where $p = 15$ cm, so our expression yields $f = -15$ cm. Plugging this back into our expression and evaluating at $p = 35$ cm yields $m = +0.30$.

49. Using Eq. 34-9 and noting that $p + i = d = 44$ cm, we obtain $p^2 - dp + df = 0$. Therefore,

$$p = \frac{1}{2}(d \pm \sqrt{d^2 - 4df}) = 22 \text{ cm} \pm \frac{1}{2}\sqrt{(44 \text{ cm})^2 - 4(44 \text{ cm})(11 \text{ cm})} = 22 \text{ cm}.$$

50. We recall that for a converging (C) lens, the focal length value should be positive ($f = +4$ cm).

(a) Eq. 34-9 gives $i = pf/(p-f) = +5.3$ cm.

(b) Eq. 34-7 give $m = -i/p = -0.33$.

(c) The fact that the image distance i is a positive value means the image is real (R).

(d) The fact that the magnification is a negative value means the image is inverted (I).

(e) The image is on the side opposite from the object (see Fig. 34-14).

51. We recall that for a converging (C) lens, the focal length value should be positive ($f = +16$ cm).

(a) Eq. 34-9 gives $i = pf/(p-f) = -48$ cm.

(b) Eq. 34-7 give $m = -i/p = +4.0$.

(c) The fact that the image distance is a negative value means the image is virtual (V).

(d) A positive value of magnification means the image is not inverted (NI).

(e) The image is on the same side as the object (see Fig. 34-14).

52. We recall that for a diverging (D) lens, the focal length value should be negative ($f = -6$ cm).

(a) Eq. 34-9 gives $i = pf/(p-f) = -3.8$ cm.

(b) Eq. 34-7 give $m = -i/p = +0.38$.

(c) The fact that the image distance is a negative value means the image is virtual (V).

(d) A positive value of magnification means the image is not inverted (NI).

(e) The image is on the same side as the object (see Fig. 34-14).

53. We recall that for a diverging (D) lens, the focal length value should be negative ($f = -12$ cm).

(a) Eq. 34-9 gives $i = pf/(p-f) = -4.8$ cm.

(b) Eq. 34-7 give $m = -i/p = +0.60$.

(c) The fact that the image distance is a negative value means the image is virtual (V).

(d) A positive value of magnification means the image is not inverted (NI).

(e) The image is on the same side as the object (see Fig. 34-14).

54. We recall that for a converging (C) lens, the focal length value should be positive ($f = +35$ cm).

(a) Eq. 34-9 gives $i = pf/(p-f) = -88$ cm.

(b) Eq. 34-7 give $m = -i/p = +3.5$.

(c) The fact that the image distance is a negative value means the image is virtual (V).

(d) A positive value of magnification means the image is not inverted (NI).

(e) The image is on the same side as the object (see Fig. 34-14).

55. We recall that for a diverging (D) lens, the focal length value should be negative ($f = -14$ cm).

(a) Eq. 34-9 gives $i = pf/(p-f) = -8.6$ cm.

(b) Eq. 34-7 give $m = -i/p = +0.39$.

(c) The fact that the image distance is a negative value means the image is virtual (V).

(d) A positive value of magnification means the image is not inverted (NI).

(e) The image is on the same side as the object (see Fig. 34-14).

56. We recall that for a diverging (D) lens, the focal length value should be negative ($f = -31$ cm).

(a) Eq. 34-9 gives $i = pf/(p-f) = -8.7$ cm.

(b) Eq. 34-7 give $m = -i/p = +0.72$.

(c) The fact that the image distance is a negative value means the image is virtual (V).

(d) A positive value of magnification means the image is not inverted (NI).

(e) The image is on the same side as the object (see Fig. 34-14).

57. We recall that for a converging (C) lens, the focal length value should be positive ($f =$ +20 cm).

(a) Eq. 34-9 gives $i = pf/(p-f) = +36$ cm.

(b) Eq. 34-7 give $m = -i/p = -0.80$.

(c) The fact that the image distance is a positive value means the image is real (R).

(d) A negative value of magnification means the image is inverted (I).

(e) The image is on the opposite side of the object (see Fig. 34-14).

58. (a) A convex (converging) lens, since a real image is formed.

(b) Since $i = d - p$ and $i/p = 1/2$,

$$p = \frac{2d}{3} = \frac{2(40.0\text{ cm})}{3} = 26.7\text{ cm.}$$

(c) The focal length is

$$f = \left(\frac{1}{i} + \frac{1}{p}\right)^{-1} = \left(\frac{1}{d/3} + \frac{1}{2d/3}\right)^{-1} = \frac{2d}{9} = \frac{2(40.0\text{ cm})}{9} = 8.89\text{ cm.}$$

59. (a) Combining Eq. 34-9 and Eq. 34-10 gives $i = +84$ cm.

(b) Eq. 34-7 give $m = -i/p = -1.4$.

(c) The fact that the image distance is a positive value means the image is real (R).

(d) The fact that the magnification is a negative value means the image is inverted (I).

(e) The image is on the side opposite from the object (see Fig. 34-14).

60. (a) Combining Eq. 34-9 and Eq. 34-10 gives $i = -26$ cm.

(b) Eq. 34-7 give $m = -i/p = +4.3$.

(c) The fact that the image distance is a negative value means the image is virtual (V).

(d) A positive value of magnification means the image is not inverted (NI).

(e) The image it is on the same side as the object (see Fig. 34-14).

61. (a) Combining Eq. 34-9 and Eq. 34-10 gives $i = -18$ cm.

(b) Eq. 34-7 give $m = -i/p = +0.76$.

(c) The fact that the image distance is a negative value means the image is virtual (V).

(d) A positive value of magnification means the image is not inverted (NI).

(e) The image it is on the same side as the object (see Fig. 34-14).

62. (a) Combining Eq. 34-9 and Eq. 34-10 gives $i = -9.7$ cm.

(b) Eq. 34-7 give $m = -i/p = +0.54$.

(c) The fact that the image distance is a negative value means the image is virtual (V).

(d) A positive value of magnification means the image is not inverted (NI).

(e) The image it is on the same side as the object (see Fig. 34-14).

63. (a) Combining Eq. 34-9 and Eq. 34-10 gives $i = -30$ cm.

(b) Eq. 34-7 give $m = -i/p = +0.86$.

(c) The fact that the image distance is a negative value means the image is virtual (V).

(d) A positive value of magnification means the image is not inverted (NI).

(e) The image it is on the same side as the object (see Fig. 34-14).

64. (a) Combining Eq. 34-9 and Eq. 34-10 gives $i = -63$ cm.

(b) Eq. 34-7 give $m = -i/p = +2.2$.

(c) The fact that the image distance is a negative value means the image is virtual (V).

(d) A positive value of magnification means the image is not inverted (NI).

(e) The image it is on the same side as the object (see Fig. 34-14).

65. (a) Combining Eq. 34-9 and Eq. 34-10 gives $i = +55$ cm.

(b) Eq. 34-7 give $m = -i/p = -0.74$.

(c) The fact that the image distance is a positive value means the image is real (R).

(d) The fact that the magnification is a negative value means the image is inverted (I).

(e) The image is on the side opposite from the object (see Fig. 34-14).

66. (a) Eq. 34-10 yields $f = \frac{1}{n-1}(1/r_1 - 1/r_2)^{-1} = +30$ cm. Since $f > 0$, this must be a converging ("C") lens. From Eq. 34-9, we obtain

$$i = \frac{1}{\frac{1}{f} - \frac{1}{p}} = \frac{1}{\frac{1}{30} - \frac{1}{10}} = -15 \text{cm}.$$

(b) Eq. 34-6 yields $m = -i/p = -(-15)/10 = +1.5$.

(c) Since $i < 0$, the image is virtual (V).

(d) Since $m > 0$, the image is upright, or not inverted (NI).

(e) The image it is on the same side as the object. The ray diagram would be similar to Fig. 34-15(b) in the textbook.

67. (a) Eq. 34-10 yields $f = \frac{1}{n-1}(1/r_1 - 1/r_2)^{-1} = -30$ cm. Since $f < 0$, this must be a diverging ("D") lens. From Eq. 34-9, we obtain

$$i = \frac{1}{\frac{1}{f} - \frac{1}{p}} = \frac{1}{\frac{1}{-30} - \frac{1}{10}} = -7.5 \text{cm}.$$

(b) Eq. 34-6 yields $m = -i/p = -(-7.5)/10 = +0.75$.

(c) Since $i < 0$, the image is virtual (V).

(d) Since $m > 0$, the image is upright, or not inverted (NI).

(e) The image it is on the same side as the object. The ray diagram would be similar to Fig. 34-15(c) in the textbook.

68. (a) Eq. 34-10 yields $f = \frac{1}{n-1}(1/r_1 - 1/r_2)^{-1} = -120$ cm. Since $f < 0$, this must be a diverging ("D") lens. From Eq. 34-9, we obtain

$$i = \frac{1}{\dfrac{1}{f} - \dfrac{1}{p}} = \frac{1}{\dfrac{1}{-120} - \dfrac{1}{10}} = -9.2\,\text{cm}.$$

(b) Eq. 34-6 yields $m = -i/p = -(-9.2)/10 = +0.92$.

(c) Since $i < 0$, the image is virtual (V).

(d) Since $m > 0$, the image is upright, or not inverted (NI).

(e) The image it is on the same side as the object. The ray diagram would be similar to Fig. 34-15(c) in the textbook.

69. (a) The fact that $m > 1$ means the lens is of the converging type (C) (it may help to look at Fig. 34-14 to illustrate this).

(b) A converging lens implies $f = +20$ cm, with a plus sign.

(d) Eq. 34-9 then gives $i = -13$ cm.

(e) Eq. 34-7 gives $m = -i/p = +1.7$.

(f) The fact that the image distance i is a negative value means the image is virtual (V).

(g) A positive value of magnification means the image is not inverted (NI).

(h) The image it is on the same side as the object (see Fig. 34-14).

70. (a) The fact that $m < 1$ and that the image is upright (not inverted: NI) means the lens is of the diverging type (D) (it may help to look at Fig. 34-14 to illustrate this).

(b) A diverging lens implies that $f = -20$ cm, with a minus sign.

(d) Eq. 34-9 gives $i = -5.7$ cm.

(e) Eq. 34-7 gives $m = -i/p = +0.71$.

(f) The fact that the image distance i is a negative value means the image is virtual (V).

(h) The image it is on the same side as the object (see Fig. 34-14).

71. (a) Eq. 34-7 yields $i = -mp = -(0.25)(16) = -4.0$ cm. Eq. 34-9 gives $f = -5.3$ cm, which implies the lens is of the diverging type (D).

(b) From (a), we have $f = -5.3$ cm.

(d) Similarly, $i = -4.0$ cm.

(f) The fact that the image distance i is a negative value means the image is virtual (V).

(g) A positive value of magnification means the image is not inverted (NI).

(h) The image it is on the same side as the object (see Fig. 34-14).

72. (a) Eq. 34-7 readily yields $i = +4.0$ cm. Then Eq. 34-9 gives $f = +3.2$ cm, which implies the lens is of the converging type (C).

(b) From (a), we have $f = +3.2$ cm.

(d) Similarly, $i = +4.0$ cm.

(f) The fact that the image distance is a positive value means the image is real (R).

(g) The fact that the magnification is a negative value means the image is inverted (I).

(h) The image is on the side opposite from the object.

73. (a) Eq. 34-7 readily yields $i = -20$ cm. Then Eq. 34-9 gives $f = +80$ cm, which implies the lens is of the converging type (C).

(b) From (a), we have $f = +80$ cm.

(d) Similarly, $i = -20$ cm.

(f) The fact that the image distance i is a negative value means the image is virtual (V).

(g) A positive value of magnification means the image is not inverted (NI).

(h) The image it is on the same side as the object (see Fig. 34-14).

74. (b) Since this is a converging lens ("C") then $f > 0$, so we should put a plus sign in front of the "10" value given for the focal length.

(d) Eq. 34-9 gives

$$i = \frac{1}{\dfrac{1}{f} - \dfrac{1}{p}} = \frac{1}{\dfrac{1}{10} - \dfrac{1}{20}} = +20\,\text{cm}.$$

(e) From Eq. 34-6, $m = -20/20 = -1.0$.

(f) The fact that the image distance is a positive value means the image is real (R).

(g) The fact that the magnification is a negative value means the image is inverted (I).

(h) The image is on the side opposite from the object.

75. (a) Since $f > 0$, this is a converging lens ("C").

(d) Eq. 34-9 gives

$$i = \frac{1}{\dfrac{1}{f} - \dfrac{1}{p}} = \frac{1}{\dfrac{1}{10} - \dfrac{1}{5}} = -10\,\text{cm.}$$

(e) From Eq. 34-6, $m = -(-10)/5 = +2.0$.

(f) The fact that the image distance i is a negative value means the image is virtual (V).

(g) A positive value of magnification means the image is not inverted (NI).

(h) The image it is on the same side as the object (see Fig. 34-14).

76. (a) We are told the magnification is positive and greater than 1. Scanning the single-lens-image figures in the textbook (Figs. 34-14, 34-15 and 34-17), we see that such a magnification (which implies an upright image larger than the object) is only possible if the lens is of the converging ("C") type (and if $p < f$).

(b) We should put a plus sign in front of the "10" value given for the focal length.

(d) Eq. 34-9 gives

$$i = \frac{1}{\dfrac{1}{f} - \dfrac{1}{p}} = \frac{1}{\dfrac{1}{10} - \dfrac{1}{5}} = -10\,\text{cm.}$$

(e) $m = -i/p = +2.0$.

(f) The fact that the image distance i is a negative value means the image is virtual (V).

(g) A positive value of magnification means the image is not inverted (NI).

(h) The image it is on the same side as the object (see Fig. 34-14).

77. (a) We are told the magnification is less than 1, and we note that $p < |f|$). Scanning Figs. 34-14, 34-15 and 34-17, we see that such a magnification (which implies an image

smaller than the object) and object position (being fairly close to the lens) are simultaneously possible only if the lens is of the diverging ("D") type.

(b) Thus, we should put a minus sign in front of the "10" value given for the focal length.

(d) Eq. 34-9 gives

$$i = \frac{1}{\frac{1}{f} - \frac{1}{p}} = \frac{1}{\frac{1}{-10} - \frac{1}{5}} = -3.3\,\text{cm}.$$

(e) $m = -i/p = +0.67$.

(f) The fact that the image distance i is a negative value means the image is virtual (V).

(g) A positive value of magnification means the image is not inverted (NI).

78. (a) We are told the absolute value of the magnification is 0.5 and that the image was upright (NI). Thus, $m = +0.5$. Using Eq. 34-6 and the given value of p, we find $i = -5.0$ cm; it is a virtual image. Eq. 34-9 then yields the focal length: $f = -10$ cm. Therefore, the lens is of the diverging ("D") type.

(b) From (a), we have $f = -10$ cm.

(d) Similarly, $i = -5.0$ cm.

(e) $m = +0.5$, with a plus sign

(f) The fact that the image distance i is a negative value means the image is virtual (V).

(h) The image it is on the same side as the object (see Fig. 34-14).

79. (a) Using Eq. 34-6 (which implies the image is inverted) and the given value of p, we find $i = -mp = +5.0$ cm; it is a real image. Eq. 34-9 then yields the focal length: $f = +3.3$ cm. Therefore, the lens is of the converging ("C") type.

(b) From (a), we have $f = +3.3$ cm.

(d) Similarly, $i = -mp = +5.0$ cm.

(f) The fact that the image distance is a positive value means the image is real (R).

(g) The fact that the magnification is a negative value means the image is inverted (I).

(h) The image is on the side opposite from the object. The ray diagram would be similar to Fig. 34-15(a) in the textbook.

80. (a) The image from lens 1 (which has $f_1 = +15$ cm) is at $i_1 = -30$ cm (by Eq. 34-9). This serves as an "object" for lens 2 (which has $f_2 = +8$ cm) with $p_2 = d - i_1 = 40$ cm. Then Eq. 34-9 (applied to lens 2) yields $i_2 = +10$ cm.

(b) Eq. 34-11 yields $M = m_1 m_2 = (-i_1 / p_1)(-i_2 / p_2) = i_1 i_2 / p_1 p_2 = -0.75$.

(c) The fact that the (final) image distance is a positive value means the image is real (R).

(d) The fact that the magnification is a negative value means the image is inverted (I).

(e) The image it is on the side opposite from the object (relative to lens 2).

81. (a) The image from lens 1 (which has $f_1 = +8$ cm) is at $i_1 = 24$ cm (by Eq. 34-9). This serves as an "object" for lens 2 (which has $f_2 = +6$ cm) with $p_2 = d - i_1 = 8$ cm. Then Eq. 34-9 (applied to lens 2) yields $i_2 = +24$ cm.

(b) Eq. 34-11 yields $M = m_1 m_2 = (-i_1 / p_1)(-i_2 / p_2) = i_1 i_2 / p_1 p_2 = +6.0$.

(c)The fact that the (final) image distance is a positive value means the image is real (R).

(d) The fact that the magnification is positive means the image is not inverted (NI).

(e) The image it is on the side opposite from the object (relative to lens 2).

82. (a) The image from lens 1 (which has $f_1 = +12$ cm) is at $i_1 = +60$ cm (by Eq. 34-9). This serves as an "object" for lens 2 (which has $f_2 = +10$ cm) with $p_2 = d - i_1 = 7$ cm. Then Eq. 34-9 (applied to lens 2) yields $i_2 = -23$ cm.

(b) Eq. 34-11 yields $M = m_1 m_2 = (-i_1 / p_1)(-i_2 / p_2) = i_1 i_2 / p_1 p_2 = -13$.

(c) The fact that the (final) image distance is negative means the image is virtual (V).

(d) The fact that the magnification is a negative value means the image is inverted (I).

(e) The image it is on the same side as the object (relative to lens 2).

83. (a) The image from lens 1 (which has $f_1 = +9$ cm) is at $i_1 = 16.4$ cm (by Eq. 34-9). This serves as an "object" for lens 2 (which has $f_2 = +5$ cm) with $p_2 = d - i_1 = -8.4$ cm. Then Eq. 34-9 (applied to lens 2) yields $i_2 = +3.1$ cm.

(b) Eq. 34-11 yields $M = m_1 m_2 = (-i_1 / p_1)(-i_2 / p_2) = i_1 i_2 / p_1 p_2 = -0.31$.

(c) The fact that the (final) image distance is a positive value means the image is real (R).

(d) The fact that the magnification is a negative value means the image is inverted (I).

(e) The image it is on the side opposite from the object (relative to lens 2). Since this result involves a negative value for p_2 (and perhaps other "non-intuitive" features), we offer a few words of explanation: lens 1 is converging the rays towards an image (that never gets a chance to form due to the intervening presence of lens 2) that would be real and inverted (and 8.4 cm beyond lens 2's location). Lens 2, in a sense, just causes these rays to converge a little more rapidly, and causes the image to form a little closer (to the lens system) than if lens 2 were not present.

84. (a) The image from lens 1 (which has $f_1 = -6$ cm) is at $i_1 = -3.4$ cm (by Eq. 34-9). This serves as an "object" for lens 2 (which has $f_2 = +6$ cm) with $p_2 = d - i_1 = 15.4$ cm. Then Eq. 34-9 (applied to lens 2) yields $i_2 = +9.8$ cm.

(b) Eq. 34-11 yields $M = -0.27$.

(c) The fact that the (final) image distance is a positive value means the image is real (R).

(d) The fact that the magnification is a negative value means the image is inverted (I).

(e) The image it is on the side opposite from the object (relative to lens 2).

85. (a) The image from lens 1 (which has $f_1 = +6$ cm) is at $i_1 = -12$ cm (by Eq. 34-9). This serves as an "object" for lens 2 (which has $f_2 = -6$ cm) with $p_2 = d - i_1 = 20$ cm. Then Eq. 34-9 (applied to lens 2) yields $i_2 = -4.6$ cm.

(b) Eq. 34-11 yields $M = +0.69$.

(c) The fact that the (final) image distance is negative means the image is virtual (V).

(d) The fact that the magnification is positive means the image is not inverted (NI).

(e) The image it is on the same side as the object (relative to lens 2).

86. (a) The image from lens 1 (which has $f_1 = +8$ cm) is at $i_1 = +24$ cm (by Eq. 34-9). This serves as an "object" for lens 2 (which has $f_2 = -8$ cm) with $p_2 = d - i_1 = 6$ cm. Then Eq. 34-9 (applied to lens 2) yields $i_2 = -3.4$ cm.

(b) Eq. 34-11 yields $M = -1.1$.

(c) The fact that the (final) image distance is negative means the image is virtual (V).

(d) The fact that the magnification is a negative value means the image is inverted (I).

(e) The image it is on the same side as the object (relative to lens 2).

87. (a) The image from lens 1 (which has $f_1 = -12$ cm) is at $i_1 = -7.5$ cm (by Eq. 34-9). This serves as an "object" for lens 2 (which has $f_2 = -8$ cm) with $p_2 = d - i_1 = 17.5$ cm. Then Eq. 34-9 (applied to lens 2) yields $i_2 = -5.5$ cm.

(b) Eq. 34-11 yields $M = +0.12$.

(c) The fact that the (final) image distance is negative means the image is virtual (V).

(d) The fact that the magnification is positive means the image is not inverted (NI).

(e) The image it is on the same side as the object (relative to lens 2).

88. The minimum diameter of the eyepiece is given by

$$d_{ey} = \frac{d_{ob}}{m_\theta} = \frac{75 \text{ mm}}{36} = 2.1 \text{ mm.}$$

89. (a) If L is the distance between the lenses, then according to Fig. 34-18, the tube length is

$$s = L - f_{ob} - f_{ey} = 25.0 \text{ cm} - 4.00 \text{ cm} - 8.00 \text{ cm} = 13.0 \text{ cm.}$$

(b) We solve $(1/p) + (1/i) = (1/f_{ob})$ for p. The image distance is

$$i = f_{ob} + s = 4.00 \text{ cm} + 13.0 \text{ cm} = 17.0 \text{ cm,}$$

so

$$p = \frac{i f_{ob}}{i - f_{ob}} = \frac{(17.0 \text{ cm})(4.00 \text{ cm})}{17.0 \text{ cm} - 4.00 \text{ cm}} = 5.23 \text{ cm.}$$

(c) The magnification of the objective is

$$m = -\frac{i}{p} = -\frac{17.0 \text{ cm}}{5.23 \text{ cm}} = -3.25.$$

(d) The angular magnification of the eyepiece is

$$m_\theta = \frac{25 \text{ cm}}{f_{ey}} = \frac{25 \text{ cm}}{8.00 \text{ cm}} = 3.13.$$

(e) The overall magnification of the microscope is

$$M = m m_\theta = (-3.25)(3.13) = -10.2.$$

90. (a) Without the magnifier, $\theta = h/P_n$ (see Fig. 34-17). With the magnifier, letting $i = -|i| = -P_n$, we obtain

$$\frac{1}{p} = \frac{1}{f} - \frac{1}{i} = \frac{1}{f} + \frac{1}{|i|} = \frac{1}{f} + \frac{1}{P_n}.$$

Consequently,

$$m_\theta = \frac{\theta'}{\theta} = \frac{h/p}{h/P_n} = \frac{1/f + 1/P_n}{1/P_n} = 1 + \frac{P_n}{f} = 1 + \frac{25\,\text{cm}}{f}.$$

With $f = 10$ cm, $m_\theta = 1 + \dfrac{25\,\text{cm}}{10\,\text{cm}} = 3.5$.

(b) In the case where the image appears at infinity, let $i = -|i| \to -\infty$, so that $1/p + 1/i = 1/p = 1/f$, we have

$$m_\theta = \frac{\theta'}{\theta} = \frac{h/p}{h/P_n} = \frac{1/f}{1/P_n} = \frac{P_n}{f} = \frac{25\,\text{cm}}{f}.$$

With $f = 10$ cm,

$$m_\theta = \frac{25\,\text{cm}}{10\,\text{cm}} = 2.5.$$

91. (a) When the eye is relaxed, its lens focuses far-away objects on the retina, a distance i behind the lens. We set $p = \infty$ in the thin lens equation to obtain $1/i = 1/f$, where f is the focal length of the relaxed effective lens. Thus, $i = f = 2.50$ cm. When the eye focuses on closer objects, the image distance i remains the same but the object distance and focal length change. If p is the new object distance and f' is the new focal length, then

$$\frac{1}{p} + \frac{1}{i} = \frac{1}{f'}.$$

We substitute $i = f$ and solve for f':

$$f' = \frac{pf}{f + p} = \frac{(40.0\,\text{cm})(2.50\,\text{cm})}{40.0\,\text{cm} + 2.50\,\text{cm}} = 2.35\,\text{cm}.$$

(b) Consider the lens maker's equation

$$\frac{1}{f} = (n-1)\left(\frac{1}{r_1} - \frac{1}{r_2}\right)$$

where r_1 and r_2 are the radii of curvature of the two surfaces of the lens and n is the index of refraction of the lens material. For the lens pictured in Fig. 34-43, r_1 and r_2 have about the same magnitude, r_1 is positive, and r_2 is negative. Since the focal length decreases, the combination $(1/r_1) - (1/r_2)$ must increase. This can be accomplished by decreasing the magnitudes of both radii.

92. We refer to Fig. 34-18. For the intermediate image $p = 10$ mm and

$$i = (f_{ob} + s + f_{ey}) - f_{ey} = 300 \text{ m} - 50 \text{ mm} = 250 \text{ mm},$$

so

$$\frac{1}{f_{ob}} = \frac{1}{i} + \frac{1}{p} = \frac{1}{250 \text{ mm}} + \frac{1}{10 \text{ mm}} \Rightarrow f_{ob} = 9.62 \text{ mm},$$

and

$$s = (f_{ob} + s + f_{ey}) - f_{ob} - f_{ey} = 300 \text{ mm} - 9.62 \text{ mm} - 50 \text{ mm} = 240 \text{ mm}.$$

Then from Eq. 34-14,

$$M = -\frac{s}{f_{ob}} \frac{25 \text{ cm}}{f_{ey}} = -\left(\frac{240 \text{ mm}}{9.62 \text{ mm}}\right)\left(\frac{150 \text{ mm}}{50 \text{ mm}}\right) = -125.$$

93. (a) Now, the lens-film distance is

$$i = \left(\frac{1}{f} - \frac{1}{p}\right)^{-1} = \left(\frac{1}{5.0 \text{ cm}} - \frac{1}{100 \text{ cm}}\right)^{-1} = 5.3 \text{ cm}.$$

(b) The change in the lens-film distance is 5.3 cm – 5.0 cm = 0.30 cm.

94. (a) In the closest mirror M_1, the "first" image I_1 is 10 cm behind M_1 and therefore 20 cm from the object O. This is the smallest distance between the object and an image of the object.

(b) There are images from both O and I_1 in the more distant mirror, M_2: an image I_2 located at 30 cm behind M_2. Since O is 30 cm in front of it, I_2 is 60 cm from O. This is the second smallest distance between the object and an image of the object.

(c) There is also an image I_3 which is 50 cm behind M_2 (since I_1 is 50 cm in front of it). Thus, I_3 is 80 cm from O. In addition, we have another image I_4 which is 70 cm behind M_1 (since I_2 is 70 cm in front of it). The distance from I_4 to O for is 80 cm.

(d) Returning to the closer mirror M_1, there is an image I_5 which is 90 cm behind the mirror (since I_3 is 90 cm in front of it). The distances (measured from O) for I_5 is 100 cm = 1.0 m.

95. (a) Parallel rays are bent by positive-f lenses to their focal points F_1, and rays that come from the focal point positions F_2 in front of positive-f lenses are made to emerge parallel. The key, then, to this type of beam expander is to have the rear focal point F_1 of the first lens coincide with the front focal point F_2 of the second lens. Since the triangles that meet at the coincident focal point are similar (they share the same angle; they are vertex angles), then $W_f/f_2 = W_i/f_1$ follows immediately. Substituting the values given, we have

$$W_f = \frac{f_2}{f_1}W_i = \frac{30.0 \text{ cm}}{12.5 \text{ cm}}(2.5 \text{ mm}) = 6.0 \text{ mm}.$$

(b) The area is proportional to W^2. Since intensity is defined as power P divided by area, we have

$$\frac{I_f}{I_i} = \frac{P/W_f^2}{P/W_i^2} = \frac{W_i^2}{W_f^2} = \frac{f_1^2}{f_2^2} \Rightarrow I_f = \left(\frac{f_1}{f_2}\right)^2 I_i = 1.6 \text{ kW/m}^2.$$

(c) The previous argument can be adapted to the first lens in the expanding pair being of the diverging type, by ensuring that the front focal point of the first lens coincides with the front focal point of the second lens. The distance between the lenses in this case is

$$f_2 - |f_1| = 30.0 \text{ cm} - 26.0 \text{ cm} = 4.0 \text{ cm}.$$

96. By Eq. 34-9, $1/i + 1/p$ is equal to constant $(1/f)$. Thus,

$$1/(-10) + 1/(15) = 1/i_{new} + 1/(70).$$

This leads to $i_{new} = -21$ cm.

97. (a) The "object" for the mirror which results in that box-image is equally in front of the mirror (4 cm). This object is actually the first image formed by the system (produced by the first transmission through the lens); in those terms, it corresponds to $i_1 = 10 - 4 = 6$ cm. Thus, with $f_1 = 2$ cm, Eq. 34-9 leads to

$$\frac{1}{p_1} + \frac{1}{i_1} = \frac{1}{f_1} \Rightarrow p_1 = 3.00 \text{ cm}.$$

(b) The previously mentioned box-image (4 cm behind the mirror) serves as an "object" (at $p_3 = 14$ cm) for the return trip of light through the lens ($f_3 = f_1 = 2$ cm). This time, Eq. 34-9 leads to

$$\frac{1}{p_3} + \frac{1}{i_3} = \frac{1}{f_3} \Rightarrow i_3 = 2.33 \text{ cm.}$$

98. (a) First, the lens forms a real image of the object located at a distance

$$i_1 = \left(\frac{1}{f_1} - \frac{1}{p_1}\right)^{-1} = \left(\frac{1}{f_1} - \frac{1}{2f_1}\right)^{-1} = 2f_1$$

to the right of the lens, or at

$$p_2 = 2(f_1 + f_2) - 2f_1 = 2f_2$$

in front of the mirror. The subsequent image formed by the mirror is located at a distance

$$i_2 = \left(\frac{1}{f_2} - \frac{1}{p_2}\right)^{-1} = \left(\frac{1}{f_2} - \frac{1}{2f_2}\right)^{-1} = 2f_2$$

to the left of the mirror, or at

$$p'_1 = 2(f_1 + f_2) - 2f_2 = 2f_1$$

to the right of the lens. The final image formed by the lens is at a distance i'_1 to the left of the lens, where

$$i'_1 = \left(\frac{1}{f_1} - \frac{1}{p'_1}\right)^{-1} = \left(\frac{1}{f_1} - \frac{1}{2f_1}\right)^{-1} = 2f_1.$$

This turns out to be the same as the location of the original object.

(b) The lateral magnification is

$$m = \left(-\frac{i_1}{p_1}\right)\left(-\frac{i_2}{p_2}\right)\left(-\frac{i'_1}{p'_1}\right) = \left(-\frac{2f_1}{2f_1}\right)\left(-\frac{2f_2}{2f_2}\right)\left(-\frac{2f_1}{2f_1}\right) = -1.0.$$

(c) The final image is real (R).

(d) It is at a distance i'_1 to the left of the lens,

(e) and inverted (I), as shown in the figure below.

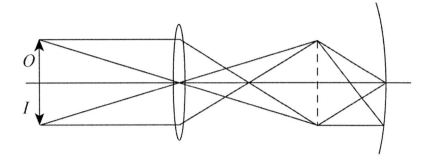

99. We refer to Fig. 34-2 in the textbook. Consider the two light rays, r and r', which are closest to and on either side of the normal ray (the ray that reverses when it reflects). Each of these rays has an angle of incidence equal to θ when they reach the mirror. Consider that these two rays reach the top and bottom edges of the pupil after they have reflected. If ray r strikes the mirror at point A and ray r' strikes the mirror at B, the distance between A and B (call it x) is

$$x = 2d_o \tan\theta$$

where d_o is the distance from the mirror to the object. We can construct a right triangle starting with the image point of the object (a distance d_o behind the mirror; see I in Fig. 34-2). One side of the triangle follows the extended normal axis (which would reach from I to the middle of the pupil), and the hypotenuse is along the extension of ray r (after reflection). The distance from the pupil to I is $d_{ey} + d_o$, and the small angle in this triangle is again θ. Thus,

$$\tan\theta = \frac{R}{d_{ey} + d_o}$$

where R is the pupil radius (2.5 mm). Combining these relations, we find

$$x = 2d_o \frac{R}{d_{ey} + d_o} = 2(100\,\text{mm}) \frac{2.5\,\text{mm}}{300\,\text{mm} + 100\,\text{mm}}$$

which yields $x = 1.67$ mm. Now, x serves as the diameter of a circular area A on the mirror, in which all rays that reflect will reach the eye. Therefore,

$$A = \frac{1}{4}\pi x^2 = \frac{\pi}{4}(1.67\,\text{mm})^2 = 2.2\,\text{mm}^2 \ .$$

100. We use Eq. 34-10, with the conventions for signs discussed in §34-6 and §34-7.

(a) For lens 1, the bi-convex (or double convex) case, we have

$$f = \left[(n-1)\left(\frac{1}{r_1} - \frac{1}{r_2} \right) \right]^{-1} = \left[(1.5-1)\left(\frac{1}{40\,\text{cm}} - \frac{1}{-40\,\text{cm}} \right) \right]^{-1} = 40 \text{ cm}.$$

(b) Since $f > 0$ the lens forms a real image of the Sun.

(c) For lens 2, of the planar convex type, we find

$$f = \left[(1.5-1)\left(\frac{1}{\infty} - \frac{1}{-40\,\text{cm}} \right) \right]^{-1} = 80\,\text{cm}.$$

(d) The image formed is real (since $f > 0$).

(e) Now for lens 3, of the meniscus convex type, we have

$$f = \left[(1.5-1)\left(\frac{1}{40\,\text{cm}} - \frac{1}{60\,\text{cm}} \right) \right]^{-1} = 240\,\text{cm} = 2.4 \text{ m}.$$

(f) The image formed is real (since $f > 0$).

(g) For lens 4, of the bi-concave type, the focal length is

$$f = \left[(1.5-1)\left(\frac{1}{-40\,\text{cm}} - \frac{1}{40\,\text{cm}} \right) \right]^{-1} = -40\,\text{cm}.$$

(h) The image formed is virtual (since $f < 0$).

(i) For lens 5 (plane-concave), we have

$$f = \left[(1.5-1)\left(\frac{1}{\infty} - \frac{1}{40\,\text{cm}} \right) \right]^{-1} = -80\,\text{cm}.$$

(j) The image formed is virtual (since $f < 0$).

(k) For lens 6 (meniscus concave),

$$f = \left[(1.5-1)\left(\frac{1}{60\,\text{cm}} - \frac{1}{40\,\text{cm}} \right) \right]^{-1} = -240\,\text{cm} = -2.4 \text{ m}.$$

(l) The image formed is virtual (since $f < 0$).

101. (a) The first image is figured using Eq. 34-8, with $n_1 = 1$ (using the rounded-off value for air) and $n_2 = 8/5$.

$$\frac{1}{p} + \frac{8}{5i} = \frac{1.6 - 1}{r}$$

For a "flat lens" $r = \infty$, so we obtain

$$i = -8p/5 = -64/5$$

(with the unit cm understood) for that object at $p = 10$ cm. Relative to the second surface, this image is at a distance of $3 + 64/5 = 79/5$. This serves as an object in order to find the final image, using Eq. 34-8 again (and $r = \infty$) but with $n_1 = 8/5$ and $n_2 = 4/3$.

$$\frac{8}{5p'} + \frac{4}{3i'} = 0$$

which produces (for $p' = 79/5$)

$$i' = -5p/6 = -79/6 \approx -13.2.$$

This means the observer appears $13.2 + 6.8 = 20$ cm from the fish.

(b) It is straightforward to "reverse" the above reasoning, the result being that the final fish-image is 7.0 cm to the right of the air-wall interface, and thus 15 cm from the observer.

102. (a) There are three images. Two are formed by single reflections from each of the mirrors and the third is formed by successive reflections from both mirrors. The positions of the images are shown on the two diagrams that follow.

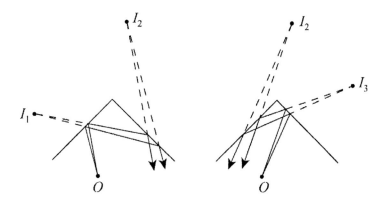

The diagram on the left shows the image I_1, formed by reflections from the left-hand mirror. It is the same distance behind the mirror as the object O is in front, and lies on the line perpendicular to the mirror and through the object. Image I_2 is formed by light that is reflected from both mirrors. We may consider I_2 to be the image of I_1 formed by the right-hand mirror, extended. I_2 is the same distance behind the line of the right-hand mirror as I_1 is in front and it is on the line that is perpendicular to the line of the mirror. The diagram on the right shows image I_3, formed by reflections from the right-hand mirror. It is the same distance behind the mirror as the object is in front, and lies on the line perpendicular to the mirror and through the object. As the diagram shows, light that is first reflected from the right-hand mirror and then from the left-hand mirror forms an image at I_2.

(b) For $\theta = 45°$, we have two images in the second mirror caused by the object and its "first" image, and from these one can construct two new images I and I' behind the first mirror plane. Extending the second mirror plane, we can find two further images of I and I' which are on equal sides of the extension of the first mirror plane. This circumstance implies there are no further images, since these final images are each other's "twins." We show this construction in the figure below. Summarizing, we find $1 + 2 + 2 + 2 = 7$ images in this case.

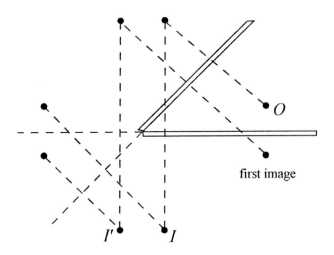

(c) For $\theta = 60°$, we have two images in the second mirror caused by the object and its "first" image, and from these one can construct two new images I and I' behind the first mirror plane. The images I and I' are each other's "twins" in the sense that they are each other's reflections about the extension of the second mirror plane; there are no further images. Summarizing, we find $1 + 2 + 2 = 5$ images in this case.

For $\theta = 120°$, we have two images I'_1 and I_2 behind the extension of the second mirror plane, caused by the object and its "first" image (which we refer to here as I_1). No further images can be constructed from I'_1 and I_2, since the method indicated above would place any further possibilities in front of the mirrors. This construction has the disadvantage of deemphasizing the actual ray-tracing, and thus any dependence on where the observer of these images is actually placing his or her eyes. It turns out in this case that the number of images that can be seen ranges from 1 to 3, depending on the locations of both the object and the observer.

(d) Thus, the smallest number of images that can be seen is 1. For example, if the observer's eye is collinear with I_1 and I'_1, then the observer can only see one image (I_1 and not the one behind it). Note that an observer who stands close to the second mirror would probably be able to see two images, I_1 and I_2.

(e) Similarly, the largest number would be 3. This happens if the observer moves further back from the vertex of the two mirrors. He or she should also be able to see the third image, I'_1, which is essentially the "twin" image formed from I_1 relative to the extension of the second mirror plane.

103. For a thin lens, $(1/p) + (1/i) = (1/f)$, where p is the object distance, i is the image distance, and f is the focal length. We solve for i:

$$i = \frac{fp}{p-f}.$$

Let $p = f + x$, where x is positive if the object is outside the focal point and negative if it is inside. Then,

$$i = \frac{f(f+x)}{x}.$$

Now let $i = f + x'$, where x' is positive if the image is outside the focal point and negative if it is inside. Then,

$$x' = i - f = \frac{f(f+x)}{x} - f = \frac{f^2}{x}$$

and $xx' = f^2$.

104. For an object in front of a thin lens, the object distance p and the image distance i are related by $(1/p) + (1/i) = (1/f)$, where f is the focal length of the lens. For the situation described by the problem, all quantities are positive, so the distance x between the object and image is $x = p + i$. We substitute $i = x - p$ into the thin lens equation and solve for x:

$$x = \frac{p^2}{p-f}.$$

To find the minimum value of x, we set $dx/dp = 0$ and solve for p. Since

$$\frac{dx}{dp} = \frac{p(p-2f)}{(p-f)^2},$$

the result is $p = 2f$. The minimum distance is

$$x_{min} = \frac{p^2}{p-f} = \frac{(2f)^2}{2f-f} = 4f.$$

This is a minimum, rather than a maximum, since the image distance i becomes large without bound as the object approaches the focal point.

105. We place an object far away from the composite lens and find the image distance i. Since the image is at a focal point, $i = f$, where f equals the effective focal length of the composite. The final image is produced by two lenses, with the image of the first lens being the object for the second. For the first lens, $(1/p_1) + (1/i_1) = (1/f_1)$, where f_1 is the focal length of this lens and i_1 is the image distance for the image it forms. Since $p_1 = \infty$, $i_1 = f_1$. The thin lens equation, applied to the second lens, is $(1/p_2) + (1/i_2) = (1/f_2)$, where p_2 is the object distance, i_2 is the image distance, and f_2 is the focal length. If the thicknesses of the lenses can be ignored, the object distance for the second lens is $p_2 = -i_1$. The negative sign must be used since the image formed by the first lens is beyond the second lens if i_1 is positive. This means the object for the second lens is virtual and the object distance is negative. If i_1 is negative, the image formed by the first lens is in front of the second lens and p_2 is positive. In the thin lens equation, we replace p_2 with $-f_1$ and i_2 with f to obtain

$$-\frac{1}{f_1} + \frac{1}{f} = \frac{1}{f_2}$$

or

$$\frac{1}{f} = \frac{1}{f_1} + \frac{1}{f_2} = \frac{f_1 + f_2}{f_1 f_2}.$$

Thus,

$$f = \frac{f_1 f_2}{f_1 + f_2}.$$

106. (a) If the object distance is x, then the image distance is $D - x$ and the thin lens equation becomes

$$\frac{1}{x} + \frac{1}{D-x} = \frac{1}{f}.$$

We multiply each term in the equation by $fx(D - x)$ and obtain $x^2 - Dx + Df = 0$. Solving for x, we find that the two object distances for which images are formed on the screen are

$$x_1 = \frac{D - \sqrt{D(D-4f)}}{2} \quad \text{and} \quad x_2 = \frac{D + \sqrt{D(D-4f)}}{2}.$$

The distance between the two object positions is

$$d = x_2 - x_1 = \sqrt{D(D-4f)}.$$

(b) The ratio of the image sizes is the same as the ratio of the lateral magnifications. If the object is at $p = x_1$, the magnitude of the lateral magnification is

$$|m_1| = \frac{i_1}{p_1} = \frac{D - x_1}{x_1}.$$

Now $x_1 = \frac{1}{2}(D - d)$, where $d = \sqrt{D(D-f)}$, so

$$|m_1| = \frac{D - (D-d)/2}{(D-d)/2} = \frac{D+d}{D-d}.$$

Similarly, when the object is at x_2, the magnitude of the lateral magnification is

$$|m_2| = \frac{I_2}{p_2} = \frac{D - x_2}{x_2} = \frac{D - (D+d)/2}{(D+d)/2} = \frac{D-d}{D+d}.$$

The ratio of the magnifications is

$$\frac{m_2}{m_1} = \frac{(D-d)/(D+d)}{(D+d)/(D-d)} = \left(\frac{D-d}{D+d}\right)^2.$$

107. The sphere (of radius 0.35 m) is a convex mirror with focal length $f = -0.175$ m. We adopt the approximation that the rays are close enough to the central axis for Eq. 34-4 to be applicable.

(a) With $p = 1.0$ m, the equation $1/p + 1/i = 1/f$ yields $i = -0.15$ m, which means the image is 0.15 m from the front surface, appearing to be *inside* the sphere.

(b) The lateral magnification is $m = -i/p$ which yields $m = 0.15$. Therefore, the image distance is $(0.15)(2.0 \text{ m}) = 0.30$ m.

(c) Since $m > 0$, the image is upright, or not inverted (NI).

108. (a) We use Eq. 34-8 (and Fig. 34-11(b) is useful), with $n_1 = 1$ (using the rounded-off value for air) and $n_2 = 1.5$.

$$\frac{1}{p} + \frac{1.5}{i} = \frac{1.5-1}{r}$$

Using the sign convention for r stated in the paragraph following Eq. 34-8 (so that $r =$ +6.0 cm), we obtain $i = -90$ cm for objects at $p = 10$ cm. Thus, the object and image are 80 cm apart.

(b) The image distance i is negative with increasing magnitude as p increases from very small values to some value p_0 at which point $i \to -\infty$. Since $1/(-\infty) = 0$, the above equation yields

$$\frac{1}{p_0} = \frac{1.5-1}{r} \Rightarrow p_0 = 2r.$$

Thus, the range for producing virtual images is $0 < p \le 12$ cm.

109. (a) In this case $i < 0$ so $i = -|i|$, and Eq. 34-9 becomes $1/f = 1/p - 1/|i|$. We differentiate this with respect to time (t) to obtain

$$\frac{d|i|}{dt} = \left(\frac{|i|}{p}\right)^2 \frac{dp}{dt}.$$

As the object is moved toward the lens, p is decreasing, so $dp/dt < 0$. Consequently, the above expression shows that $d|i|/dt < 0$; that is, the image moves in from infinity. The angular magnification $m_\theta = \theta'/\theta$ also increases as the following graph shows ("read" the graph from left to right since we are considering decreasing p from near the focal length to near 0). To obtain this graph of m_θ, we chose $f = 30$ cm and $h = 2$ cm.

(b) When the image appears to be at the near point (that is, $|i| = P_n$), m_θ is at its maximum usable value. The textbook states in section 34-8 that it generally takes P_n to be equal to 25 cm (this value, too, was used in making the above graph).

(c) In this case,

$$p = if/(i - f) = |i| f/(|i| + f) = P_n f/(P_n + f).$$

If we use the small angle approximation, we have $\theta' \approx h'/|i|$ and $\theta \approx h/P_n$ (note: this approximation was not used in obtaining the graph, above). We therefore find $m_\theta \approx (h'/|i|)/(h/P_n)$ which (using Eq. 34-7 relating the ratio of heights to the ratio of distances) becomes

$$m_\theta \approx (h'/h)(P_n/|i|) = (|i|/p)(P_n/|i|) = (P_n/p) = [P_n/(P_n f/(P_n + f))] = \frac{P_n + f}{f}$$

which readily simplifies to the desired result.

(d) The linear magnification (Eq. 34-7) is given by $(h'/h) \approx m_\theta (|i|/P_n)$ (see the first in the chain of equalities, above). Once we set $|i| = P_n$ (see part (b)) then this shows the equality in the magnifications.

110. (a) Suppose one end of the object is a distance p from the mirror and the other end is a distance $p + L$. The position i_1 of the image of the first end is given by

$$\frac{1}{p} + \frac{1}{i_1} = \frac{1}{f}$$

where f is the focal length of the mirror. Thus,

$$i_1 = \frac{fp}{p - f}.$$

The image of the other end is located at

$$i_2 = \frac{f(p + L)}{p + L - f},$$

so the length of the image is

$$L' = i_1 - i_2 = \frac{fp}{p - f} - \frac{f(p + L)}{p + L - f} = \frac{f^2 L}{(p - f)(p + L - f)}.$$

Since the object is short compared to $p - f$, we may neglect the L in the denominator and write

$$L' = L\left(\frac{f}{p - f}\right)^2.$$

(b) The lateral magnification is $m = -i/p$ and since $i = fp/(p - f)$, this can be written $m = -f/(p - f)$. The longitudinal magnification is

$$m' = \frac{L'}{L} = \left(\frac{f}{p-f}\right)^2 = m^2.$$

111. (a) In this case $m > +1$ and we know that lens 1 is converging (producing a virtual image), so that our result for focal length should be positive. Since $|P + i_1| = 20$ cm and $i_1 = -2p_1$, we find $p_1 = 20$ cm and $i_1 = -40$ cm. Substituting these into Eq. 34-9,

$$\frac{1}{p_1} + \frac{1}{i_1} = \frac{1}{f_1}$$

leads to $f_1 = +40$ cm, which is positive as we expected.

(b) The object distance is $p_1 = 20$ cm, as shown in part (a).

(c) In this case $0 < m < 1$ and we know that lens 2 is diverging (producing a virtual image), so that our result for focal length should be negative. Since $|p + i_2| = 20$ cm and $i_2 = -p_2/2$, we find $p_2 = 40$ cm and $i_2 = -20$ cm. Substituting these into Eq. 34-9 leads to $f_2 = -40$ cm, which is negative as we expected.

(d) $p_2 = 40$ cm, as shown in part (c).

112. The water is medium 1, so $n_1 = n_w$ which we simply write as n. The air is medium 2, for which $n_2 \approx 1$. We refer to points where the light rays strike the water surface as A (on the left side of Fig. 34-52) and B (on the right side of the picture). The point midway between A and B (the center point in the picture) is C. The penny P is directly below C, and the location of the "apparent" or Virtual penny is V. We note that the angle $\angle CVB$ (the same as $\angle CVA$) is equal to θ_2, and the angle $\angle CPB$ (the same as $\angle CPA$) is equal to θ_1. The triangles CVB and CPB share a common side, the horizontal distance from C to B (which we refer to as x). Therefore,

$$\tan\theta_2 = \frac{x}{d_a} \qquad \text{and} \qquad \tan\theta_1 = \frac{x}{d}.$$

Using the small angle approximation (so a ratio of tangents is nearly equal to a ratio of sines) and the law of refraction, we obtain

$$\frac{\tan\theta_2}{\tan\theta_1} \approx \frac{\sin\theta_2}{\sin\theta_1} \quad \Rightarrow \quad \frac{\dfrac{x}{d_a}}{\dfrac{x}{d}} \approx \frac{n_1}{n_2} \quad \Rightarrow \quad \frac{d}{d_a} \approx n$$

which yields the desired relation: $d_a = d/n$.

113. A converging lens has a positive-valued focal length, so $f_1 = +6$ cm, $f_2 = +3$ cm, and $f_3 = +3$ cm. We use Eq. 34-9 for each lens separately, "bridging the gap" between the results of one calculation and the next with $p_2 = d_{12} - i_1$ and $p_3 = d_{23} - i_2$. We also use Eq. 34-7 for each magnification (m_1 etc), and $m = m_1 m_2 m_3$ (a generalized version of Eq. 34-11) for the net magnification of the system. Our intermediate results for image distances are $i_1 = 9$ cm and $i_2 = 6$ cm. Our final results are as follows:

(a) $i_3 = +7.5$ cm.

(b) $m = -0.75$.

(c) The image is real (R).

(d) The image is inverted (I).

(e) It is on the opposite side of lens 3 from the object (which is expected for a real final image).

114. A converging lens has a positive-valued focal length, so $f_1 = +6$ cm, $f_2 = +6$ cm, and $f_3 = +5$ cm. We use Eq. 34-9 for each lens separately, "bridging the gap" between the results of one calculation and the next with $p_2 = d_{12} - i_1$ and $p_3 = d_{23} - i_2$. We also use Eq. 34-7 for each magnification (m_1 etc), and $m = m_1 m_2 m_3$ (a generalized version of Eq. 34-11) for the net magnification of the system. Our intermediate results for image distances are $i_1 = -3$ cm and $i_2 = 9$ cm. Our final results are as follows:

(a) $i_3 = +10$ cm.

(b) $m = +0.75$.

(c) The image is real (R).

(d) The image is not inverted (NI).

(e) It is on the opposite side of lens 3 from the object (which is expected for a real final image).

115. A converging lens has a positive-valued focal length, so $f_1 = +8$ cm, $f_2 = +6$ cm, and $f_3 = +6$ cm. We use Eq. 34-9 for each lens separately, "bridging the gap" between the results of one calculation and the next with $p_2 = d_{12} - i_1$ and $p_3 = d_{23} - i_2$. We also use Eq. 34-7 for each magnification (m_1 etc), and $m = m_1 m_2 m_3$ (a generalized version of Eq. 34-11) for the net magnification of the system. Our intermediate results for image distances are $i_1 = 24$ cm and $i_2 = -12$ cm. Our final results are as follows:

(a) $i_3 = +8.6$ cm.

(b) $m = +2.6$.

(c) The image is real (R).

(d) The image is not inverted (NI)

(e) It is on the opposite side of lens 3 from the object (which is expected for a real final image).

116. A converging lens has a positive-valued focal length, and a diverging lens has a negative-valued focal length. Therefore, $f_1 = -6$ cm, $f_2 = +6$ cm, and $f_3 = +4$ cm. We use Eq. 34-9 for each lens separately, "bridging the gap" between the results of one calculation and the next with $p_2 = d_{12} - i_1$ and $p_3 = d_{23} - i_2$. We also use Eq. 34-7 for each magnification (m_1 etc), and $m = m_1 m_2 m_3$ (a generalized version of Eq. 34-11) for the net magnification of the system. Our intermediate results for image distances are $i_1 = -2.4$ cm and $i_2 = 12$ cm. Our final results are as follows:

(a) $i_3 = -4.0$ cm.

(b) $m = -1.2$.

(c) The image is virtual (V).

(d) The image is inverted (I).

(e) It is on the same side as the object (relative to lens 3) as expected for a virtual image.

117. A converging lens has a positive-valued focal length, and a diverging lens has a negative-valued focal length. Therefore, $f_1 = -8.0$ cm, $f_2 = -16$ cm, and $f_3 = +8.0$ cm. We use Eq. 34-9 for each lens separately, "bridging the gap" between the results of one calculation and the next with $p_2 = d_{12} - i_1$ and $p_3 = d_{23} - i_2$. We also use Eq. 34-7 for each magnification (m_1 etc), and $m = m_1 m_2 m_3$ (a generalized version of Eq. 34-11) for the net magnification of the system. Our intermediate results for image distances are $i_1 = -4.0$ cm and $i_2 = -6.86$ cm. Our final results are as follows:

(a) $i_3 = +24.2$ cm.

(b) $m = -0.58$.

(c) The image is real (R).

(d) The image is inverted (I).

(e) It is on the opposite side of lens 3 from the object (as expected for a real image).

118. A converging lens has a positive-valued focal length, and a diverging lens has a negative-valued focal length. Therefore, $f_1 = +6$ cm, $f_2 = -4$ cm, and $f_3 = -12$ cm. We use Eq. 34-9 for each lens separately, "bridging the gap" between the results of one calculation and the next with $p_2 = d_{12} - i_1$ and $p_3 = d_{23} - i_2$. We also use Eq. 34-7 for each magnification (m_1 etc), and $m = m_1 m_2 m_3$ (a generalized version of Eq. 34-11) for the net magnification of the system. Our intermediate results for image distances are $i_1 = -12$ cm and $i_2 = -3.33$ cm. Our final results are as follows:

(a) $i_3 = -5.15$ cm ≈ -5.2 cm.

(b) $m = +0.285 \approx +0.29$.

(c) The image is virtual (V).

(d) The image is not inverted (NI).

(e) It is on the same side as the object (relative to lens 3) as expected for a virtual image.

119. (a) The discussion in the textbook of the refracting telescope (a subsection of §34-8) applies to the Newtonian arrangement if we replace the objective lens of Fig. 34-19 with an objective mirror (with the light incident on it from the right). This might suggest that the incident light would be blocked by the person's head in Fig. 34-19, which is why Newton added the mirror M' in his design (to move the head and eyepiece out of the way of the incoming light). The beauty of the idea of characterizing both lenses and mirrors by focal lengths is that it is easy, in a case like this, to simply carry over the results of the objective-lens telescope to the objective-mirror telescope, so long as we replace a positive f device with another positive f device. Thus, the converging lens serving as the objective of Fig. 34-19 must be replaced (as Newton has done in Fig. 34-54) with a concave mirror. With this change of language, the discussion in the textbook leading up to Eq. 34-15 applies equally as well to the Newtonian telescope: $m_\theta = -f_{ob}/f_{ey}$.

(b) A meter stick (held perpendicular to the line of sight) at a distance of 2000 m subtends an angle of

$$\theta_{stick} \approx \frac{1 \text{ m}}{2000 \text{ m}} = 0.0005 \text{ rad.}$$

multiplying this by the mirror focal length gives $(16.8 \text{ m})(0.0005) = 8.4$ mm for the size of the image.

(c) With $r = 10$ m, Eq. 34-3 gives $f_{ob} = 5$ m. Plugging this into (the absolute value of) Eq. 34-15 leads to $f_{ey} = 5/200 = 2.5$ cm.

120. Consider a single ray from the source to the mirror and let θ be the angle of incidence. The angle of reflection is also θ and the reflected ray makes an angle of 2θ with the incident ray.

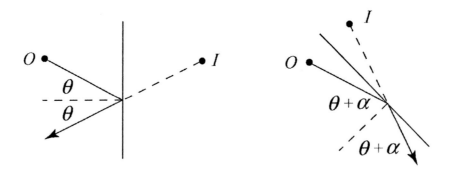

Now we rotate the mirror through the angle α so that the angle of incidence increases to $\theta + \alpha$. The reflected ray now makes an angle of $2(\theta + \alpha)$ with the incident ray. The reflected ray has been rotated through an angle of 2α. If the mirror is rotated so the angle of incidence is decreased by α, then the reflected ray makes an angle of $2(\theta - \alpha)$ with the incident ray. Again it has been rotated through 2α. The diagrams below show the situation for $\alpha = 45°$. The ray from the object to the mirror is the same in both cases and the reflected rays are 90° apart.

121. (a) If we let $p \to \infty$ in Eq. 34-8, we get $i = n_2 r /(n_2 - n_1)$. If we set $n_1 = 1$ (for air) and restrict n_2 so that $1 < n_2 < 2$, then this suggests that $i > 2r$ (so this image does form before the rays strike the opposite side of the sphere). We can still consider this as a sort of "virtual" object for the second imaging event, where this "virtual" object distance is $2r - i = (n - 2) r /(n - 1)$, where we have simplified the notation by writing $n_2 = n$. Putting this in for p in Eq. 34-8 and being careful with the sign convention for r in that equation, we arrive at the final image location: $i' = (0.5)(2 - n)r/(n - 1)$.

(b) The image is to the right of the right side of the sphere.

122. Setting $n_{air} = 1$, $n_{water} = n$, and $p = r/2$ in Eq. 34-8 (and being careful with the sign convention for r in that equation), we obtain $i = -r/(1 + n)$, or $|i| = r/(1 + n)$. Then we use similar triangles (where h is the size of the fish and h' is that of the "virtual fish") to set up the ratio

$$\frac{h'}{r - |i|} = \frac{h}{r/2} \ .$$

Using our previous result for $|i|$, this gives $h'/h = 2(1 - 1/(1 + n)) = 1.14$.

123. (a) Our first step is to form the image from the first lens. With $p_1 = 10$ cm and $f_1 = -15$ cm, Eq. 34-9 leads to

$$\frac{1}{p_1} + \frac{1}{i_1} = \frac{1}{f_1} \Rightarrow i_1 = -6.0\,\text{cm}.$$

The corresponding magnification is $m_1 = -i_1/p_1 = 0.60$. This image serves the role of "object" for the second lens, with $p_2 = 12 + 6.0 = 18$ cm, and $f_2 = 12$ cm. Now, Eq. 34-9 leads to

$$\frac{1}{p_2} + \frac{1}{i_2} = \frac{1}{f_2} \Rightarrow i_2 = 36 \text{ cm}.$$

(b) The corresponding magnification is $m_2 = -i_2/p_2 = -2.0$, which results in a net magnification of $m = m_1 m_2 = -1.2$. The height of the final image is (in absolute value) $(1.2)(1.0 \text{ cm}) = 1.2$ cm.

(c) The fact that i_2 is positive means that the final image is real.

(d) The fact that m is negative means that the orientation of the final image is inverted with respect to the (original) object.

124. (a) Without the diverging lens (lens 2), the real image formed by the converging lens (lens 1) is located at a distance

$$i_1 = \left(\frac{1}{f_1} - \frac{1}{p_1} \right)^{-1} = \left(\frac{1}{20 \text{ cm}} - \frac{1}{40 \text{ cm}} \right)^{-1} = 40 \text{ cm}$$

to the right of lens 1. This image now serves as an object for lens 2, with $p_2 = -(40 \text{ cm} - 10 \text{ cm}) = -30$ cm. So

$$i_2 = \left(\frac{1}{f_2} - \frac{1}{p_2} \right)^{-1} = \left(\frac{1}{-15\,\text{cm}} - \frac{1}{-30\,\text{cm}} \right)^{-1} = -30\,\text{cm}.$$

Thus, the image formed by lens 2 is located 30 cm to the left of lens 2.

(b) The magnification is $m = (-i_1/p_1) \times (-i_2/p_2) = +1.0 > 0$, so the image is not inverted.

(c) The image is virtual since $i_2 < 0$.

(d) The magnification is $m = (-i_1/p_1) \times (-i_2/p_2) = +1.0$, so the image has the same size as the object.

125. (a) For the image formed by the first lens

$$i_1 = \left(\frac{1}{f_1} - \frac{1}{p_1} \right)^{-1} = \left(\frac{1}{10\,\text{cm}} - \frac{1}{20\,\text{cm}} \right)^{-1} = 20\,\text{cm}.$$

For the subsequent image formed by the second lens $p_2 = 30$ cm $- 20$ cm $= 10$ cm, so

$$i_2 = \left(\frac{1}{f_2} - \frac{1}{p_2} \right)^{-1} = \left(\frac{1}{12.5\,\text{cm}} - \frac{1}{10\,\text{cm}} \right)^{-1} = -50\,\text{cm}.$$

Thus, the final image is 50 cm to the left of the second lens, which means that it coincides with the object.

(b) The magnification is

$$m = \left(\frac{i_1}{p_1} \right) \left(\frac{i_2}{p_2} \right) = \left(\frac{20\,\text{cm}}{20\,\text{cm}} \right) \left(\frac{-50\,\text{cm}}{10\,\text{cm}} \right) = -5.0,$$

which means that the final image is five times larger than the original object.

(c) The image is virtual since $i_2 < 0$.

(d) The image is inverted since $m < 0$.

126. (a) We solve Eq. 34-9 for the image distance i: $i = pf/(p - f)$. The lens is diverging, so its focal length is $f = -30$ cm. The object distance is $p = 20$ cm. Thus,

$$i = \frac{(20\,\text{cm})(-30\,\text{cm})}{(20\,\text{cm}) - (-30\,\text{cm})} = -12 \text{ cm}.$$

The negative sign indicates that the image is virtual and is on the same side of the lens as the object.

(b) The ray diagram, drawn to scale, is shown below.

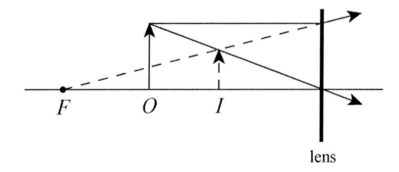

127. We set up an *xyz* coordinate system where the individual planes (*xy*, *yz*, *xz*) serve as the mirror surfaces. Suppose an incident ray of light *A* first strikes the mirror in the *xy* plane. If the unit vector denoting the direction of *A* is given by

$$\cos(\alpha)\hat{i} + \cos(\beta)\hat{j} + \cos(\gamma)\hat{k}$$

where α, β, γ are the angles A makes with the axes, then after reflection off the xy plane the unit vector becomes $\cos(\alpha)\hat{i} + \cos(\beta)\hat{j} - \cos(\gamma)\hat{k}$ (one way to rationalize this is to think of the reflection as causing the angle γ to become $\pi - \gamma$). Next suppose it strikes the mirror in the xz plane. The unit vector of the reflected ray is now $\cos(\alpha)\hat{i} - \cos(\beta)\hat{j} - \cos(\gamma)\hat{k}$. Finally as it reflects off the mirror in the yz plane α becomes $\pi - \alpha$, so the unit vector in the direction of the reflected ray is given by $- \cos(\alpha)\hat{i} - \cos(\beta)\hat{j} - \cos(\gamma)\hat{k}$, exactly reversed from A's original direction. A further observation may be made: this argument would fail if the ray could strike any given surface twice and some consideration (perhaps an illustration) should convince the student that such an occurrence is not possible.

128. The fact that it is inverted implies $m < 0$. Therefore, with $m = -1/2$, we have $i = p/2$, which we substitute into Eq. 34-4:

$$\frac{1}{p} + \frac{1}{i} = \frac{1}{f}$$

$$\frac{1}{p} + \frac{2}{p} = \frac{1}{f}$$

$$\frac{3}{30.0} = \frac{1}{f}$$

with the unit cm understood. Consequently, we find $f = 30/3 = 10.0$ cm. The fact that $f > 0$ implies the mirror is concave.

129. Since $m = -2$ and $p = 4.00$ cm, then $i = 8.00$ cm (and is real). Eq. 34-9 is

$$\frac{1}{p} + \frac{1}{i} = \frac{1}{f}$$

and leads to $f = 2.67$ cm (which is positive, as it must be for a converging lens).

130. (a) The mirror has focal length $f = 12.0$ cm. With $m = +3$, we have $i = -3p$. We substitute this into Eq. 34-4:

$$\frac{1}{p} + \frac{1}{i} = \frac{1}{f}$$

$$\frac{1}{p} + \frac{1}{-3p} = \frac{1}{12}$$

$$\frac{2}{3p} = \frac{1}{12}$$

with the unit cm understood. Consequently, we find $p = 2(12)/3 = 8.0$ cm.

(b) With $m = -3$, we have $i = +3p$, which we substitute into Eq. 34-4:

$$\frac{1}{p} + \frac{1}{i} = \frac{1}{f}$$

$$\frac{1}{p} + \frac{1}{3p} = \frac{1}{12}$$

$$\frac{4}{3p} = \frac{1}{12}$$

with the unit cm understood. Consequently, we find $p = 4(12)/3 = 16$ cm.

(c) With $m = -1/3$, we have $i = p/3$. Thus, Eq. 34-4 leads to

$$\frac{1}{p} + \frac{1}{i} = \frac{1}{f}$$

$$\frac{1}{p} + \frac{3}{p} = \frac{1}{12}$$

$$\frac{4}{p} = \frac{1}{12}.$$

with the unit cm understood. Consequently, we find $p = 4(12) = 48$ cm.

131. (a) Since $m = +0.200$, we have $i = -0.2p$ which indicates that the image is virtual (as well as being diminished in size). We conclude from this that the mirror is convex (and that $f = -40.0$ cm).

(b) Substituting $i = -p/5$ into Eq. 34-4 produces

$$\frac{1}{p} - \frac{5}{p} = -\frac{4}{p} = \frac{1}{f}.$$

Therefore, we find $p = 160$ cm.

132. Since $0 < m < 1$, we conclude the lens is of the diverging type (so $f = -40$ cm). Thus, substituting $i = -3p/10$ into Eq. 34-9 produces

$$\frac{1}{p} - \frac{10}{3p} = -\frac{7}{3p} = \frac{1}{f}.$$

Therefore, we find $p = 93.3$ cm and $i = -28.0$ cm, or $|i| = 28.0$ cm.

133. (a) Our first step is to form the image from the first lens. With $p_1 = 3.00$ cm and $f_1 = +4.00$ cm, Eq. 34-9 leads to

$$\frac{1}{p_1} + \frac{1}{i_1} = \frac{1}{f_1} \Rightarrow i_1 = -12.0 \text{cm}.$$

The corresponding magnification is $m_1 = -i_1/p_1 = 4$. This image serves the role of "object" for the second lens, with $p_2 = 8.00 + 12.0 = 20.0$ cm, and $f_2 = -4.00$ cm. Now, Eq. 34-9 leads to

$$\frac{1}{p_2} + \frac{1}{i_2} = \frac{1}{f_2} \Rightarrow i_2 = -3.33 \text{ cm}.$$

(b) The fact that i_2 is negative means that the final image is virtual (and therefore to the left of the second lens).

(c) The image is virtual.

(d) With $m_2 = -i_2/p_2 = 1/6$, the net magnification is $m = m_1 m_2 = 2/3 > 0$. The fact that m is positive means that the orientation of the final image is the same as the (original) object. Therefore, the image is not inverted.

134. (a) Our first step is to form the image from the first lens. With $p_1 = 4.00$ cm and $f_1 = -4.00$ cm, Eq. 34-9 leads to

$$\frac{1}{p_1} + \frac{1}{i_1} = \frac{1}{f_1} \Rightarrow i_1 = -2.00 \text{cm}.$$

The corresponding magnification is $m_1 = -i_1/p_1 = 1/2$. This image serves the role of "object" for the second lens, with $p_2 = 10.0 + 2.00 = 12.0$ cm, and $f_2 = -4.00$ cm. Now, Eq. 34-9 leads to

$$\frac{1}{p_2} + \frac{1}{i_2} = \frac{1}{f_2} \Rightarrow i_2 = -3.00 \text{ cm},$$

or $|i_2| = 3.00$ cm.

(b) The fact that i_2 is negative means that the final image is virtual (and therefore to the left of the second lens).

(c) The image is virtual.

(d) With $m_2 = -i_2/p_2 = 1/4$, the net magnification is $m = m_1 m_2 = 1/8 > 0$. The fact that m is positive means that the orientation of the final image is the same as the (original) object. Therefore, the image is not inverted.

135. Of course, the shortest possible path between A and B is the straight line path which does not go to the mirror at all. In this problem, we are concerned with only those paths which do strike the mirror. The problem statement suggests that we turn our attention to the mirror-image point of A (call it A') and requests that we construct a proof without calculus. We can see that the length of any line segment AP drawn from A to the mirror (at point P on the mirror surface) is the same as the length of its "mirror segment" $A'P$ drawn from A' to that point P. Thus, the total length of the light path from A to P to B is the same as the total length of segments drawn from A' to P to B. Now, we dismissed (in the first sentence of this solution) the possibility of a straight line path directly from A to B because it does not strike the mirror. However, we *can* construct a straight line path from A' to B which does intersect the mirror surface! Any other pair of segments ($A'P$ and PB) would give greater total length than the straight path (with $A'P$ and PB collinear), so if the straight path $A'B$ obeys the law of reflection, then we have our proof. Now, since $A'P$ is the mirror-twin of AP, then they both approach the mirror surface with the same angle α (one from the front side and the other from the back side). And since $A'P$ is collinear with PB, then PB also makes the same angle α with respect to the mirror surface (by vertex angles). If AP and PB are each α degrees away from the front of the mirror, then they are each θ degrees (where θ is the complement of α) measured from the normal axis. Thus, the law of reflection is consistent with the concept of the shortest light path.

136. (a) Since a beam of parallel light will be focused at a distance f from the (converging) lens, then the shorter the focal length f the greater the ability for the lens to bend the light. A window pane is an example of a "lens" with $f = \infty$, yet it has essentially zero bending ability. Therefore, $P = 1/f$ is a reasonable definition.

(b) First we must consider the two-lens situation in the limit that d (their separation) becomes vanishingly small. We place an object far away from the composite lens and find the image distance i. Since the image is at a focal point $i = f$, the effective focal length of the composite can be determined in this way. The final image is produced by two lenses, with the image of the first lens being the object for the second. For the first lens we have $1/p_1 + 1/i_1 = 1/f_1$, where f_1 is the focal length of the first lens. Since $p_1 = \infty$, we find $i_1 = f_1$. The thin lens equation, applied to the second lens, gives

$$i_2 = i = p_2 f_2 / (p_2 - f_2),$$

where $p_2 = d - i_1 = -f_1$ in this situation. Therefore, i (thought of as f for the equivalent single lens [equivalent to the 2 lens system] as explained above) is equal to

$$-f_1 f_2 / (-f_1 - f_2) \quad \text{or} \quad \frac{1}{f} = \frac{1}{f_1} + \frac{1}{f_2} \ .$$

Next, using the definition for P, we readily get the desired result.

137. (a) Suppose that the lens is placed to the left of the mirror. The image formed by the converging lens is located at a distance

$$i = \left(\frac{1}{f} - \frac{1}{p}\right)^{-1} = \left(\frac{1}{0.50\,\text{m}} - \frac{1}{1.0\,\text{m}}\right)^{-1} = 1.0\,\text{m}$$

to the right of the lens, or 2.0 m − 1.0 m = 1.0 m in front of the mirror. The image formed by the mirror for this real image is then at 1.0 m to the right of the the mirror, or 2.0 m + 1.0 m = 3.0 m to the right of the lens. This image then results in another image formed by the lens, located at a distance

$$i' = \left(\frac{1}{f} - \frac{1}{p'}\right)^{-1} = \left(\frac{1}{0.50\text{m}} - \frac{1}{3.0\text{m}}\right)^{-1} = 0.60\text{m}$$

to the left of the lens (that is, 2.6 cm from the mirror).

(b) The lateral magnification is

$$m = \left(-\frac{i}{p}\right)\left(-\frac{i'}{p'}\right) = \left(-\frac{1.0\,\text{m}}{1.0\,\text{m}}\right)\left(-\frac{0.60\,\text{m}}{3.0\,\text{m}}\right) = +0.20 \,.$$

(c) The final image is real since $i' > 0$.

(d) The image is to the left of the lens.

(e) It also has the same orientation as the object since $m > 0$. Therefore, the image is not inverted.

138. (a) Since $m = +0.250$, we have $i = -0.25p$ which indicates that the image is virtual (as well as being diminished in size). We conclude from this that the mirror is convex and that $f < 0$; in fact, $f = -2.00$ cm. Substituting $i = -p/4$ into Eq. 34-4 produces

$$\frac{1}{p} - \frac{4}{p} = -\frac{3}{p} = \frac{1}{f}$$

Therefore, we find $p = 6.00$ cm and $i = -1.50$ cm, or $|i| = 1.50$ cm.

(b) The focal length is negative.

(c) As shown in (a), the image is virtual.

139. First, we note that — *relative to the water* — the index of refraction of the carbon tetrachloride should be thought of as $n = 1.46/1.33 = 1.1$ (this notation is chosen to be consistent with problem 15). Now, if the observer were in the water, directly above the 40 mm deep carbon tetrachloride layer, then the apparent depth of the penny as measured below the surface of the carbon tetrachloride is $d_a = 40$ mm/1.1 = 36.4 mm. This "apparent penny" serves as an "object" for the rays propagating upward through the 20 mm layer of water, where this "object" should be thought of as being 20 mm + 36.4 mm = 56.4 mm from the top surface. Using the result of problem 15 again, we find the perceived location of the penny, for a person at the normal viewing position above the water, to be 56.4 mm/1.33 = 42 mm below the water surface.

140. (a) We show the $\alpha = 0.500$ rad, $r = 12$ cm, $p = 20$ cm calculation in detail. The understood length unit is the centimeter:

The distance from the object to point x: $d = p - r + x = 8 + x$
$$y = d \tan \alpha = 4.3704 + 0.54630x$$

From the solution of $x^2 + y^2 = r^2$ we get $x = 8.1398$.

$$\beta = \tan^{-1}(y/x) = 0.8253 \text{ rad}$$
$$\gamma = 2\beta - \alpha = 1.151 \text{ rad}$$

From the solution of $\tan(\gamma) = y/(x + i - r)$ we get $i = 7.799$. The other results are shown without the intermediate steps:

For $\alpha = 0.100$ rad, we get $i = 8.544$ cm; for $\alpha = 0.0100$ rad, we get $i = 8.571$ cm. Eq. 34-3 and Eq. 34-4 (the mirror equation) yield $i = 8.571$ cm.

(b) Here the results are: ($\alpha = 0.500$ rad, $i = -13.56$ cm), ($\alpha = 0.100$ rad, $i = -12.05$ cm), ($\alpha = 0.0100$ rad, $i = -12.00$ cm). The mirror equation gives $i = -12.00$ cm.

Chapter 35

1. Comparing the light speeds in sapphire and diamond, we obtain

$$\Delta v = v_s - v_d = c\left(\frac{1}{n_s} - \frac{1}{n_d}\right)$$

$$= \left(2.998 \times 10^8 \text{ m/s}\right)\left(\frac{1}{1.77} - \frac{1}{2.42}\right) = 4.55 \times 10^7 \text{ m/s}.$$

2. (a) The frequency of yellow sodium light is

$$f = \frac{c}{\lambda} = \frac{2.998 \times 10^8 \text{ m/s}}{589 \times 10^{-9} \text{ m}} = 5.09 \times 10^{14} \text{ Hz}.$$

(b) When traveling through the glass, its wavelength is

$$\lambda_n = \frac{\lambda}{n} = \frac{589 \text{ nm}}{1.52} = 388 \text{ nm}.$$

(c) The light speed when traveling through the glass is

$$v = f \lambda_n = \left(5.09 \times 10^{14} \text{ Hz}\right)\left(388 \times 10^{-9} \text{ m}\right) = 1.97 \times 10^8 \text{ m/s}.$$

3. The index of refraction is found from Eq. 35-3:

$$n = \frac{c}{v} = \frac{2.998 \times 10^8 \text{ m/s}}{1.92 \times 10^8 \text{ m/s}} = 1.56.$$

4. Note that Snell's Law (the law of refraction) leads to $\theta_1 = \theta_2$ when $n_1 = n_2$. The graph indicates that $\theta_2 = 30°$ (which is what the problem gives as the value of θ_1) occurs at $n_2 = 1.5$. Thus, $n_1 = 1.5$, and the speed with which light propagates in that medium is

$$v = \frac{c}{1.5} = 2.0 \times 10^8 \text{ m/s}.$$

5. The fact that wave W_2 reflects two additional times has no substantive effect on the calculations, since two reflections amount to a $2(\lambda/2) = \lambda$ phase difference, which is effectively not a phase difference at all. The substantive difference between W_2 and W_1 is the extra distance $2L$ traveled by W_2.

(a) For wave W_2 to be a half-wavelength "behind" wave W_1, we require $2L = \lambda/2$, or $L = \lambda/4 = 155$ nm using the wavelength value given in the problem.

(b) Destructive interference will again appear if W_2 is $\frac{3}{2}\lambda$ "behind" the other wave. In this case, $2L' = 3\lambda/2$, and the difference is

$$L' - L = \frac{3\lambda}{4} - \frac{\lambda}{4} = \frac{\lambda}{2} = 310\,\text{nm} \ .$$

6. In contrast to the initial conditions of problem 30, we now consider waves W_2 and W_1 with an initial effective phase difference (in wavelengths) equal to $\frac{1}{2}$, and seek positions of the sliver which cause the wave to constructively interfere (which corresponds to an integer-valued phase difference in wavelengths). Thus, the extra distance $2L$ traveled by W_2 must amount to $\frac{1}{2}\lambda, \frac{3}{2}\lambda$, and so on. We may write this requirement succinctly as

$$L = \frac{2m+1}{4}\lambda \quad \text{where } m = 0, 1, 2, \dots .$$

(a) Thus, the smallest value of L/λ that results in the final waves being exactly in phase is when $m = 0$, which gives $L/\lambda = 1/4 = 0.25$.

(b) The second smallest value of L/λ that results in the final waves being exactly in phase is when $m = 1$, which gives $L/\lambda = 3/4 = 0.75$.

(c) The third smallest value of L/λ that results in the final waves being exactly in phase is when $m = 2$, which gives $L/\lambda = 5/4 = 1.25$.

7. (a) We take the phases of both waves to be zero at the front surfaces of the layers. The phase of the first wave at the back surface of the glass is given by $\phi_1 = k_1 L - \omega t$, where k_1 ($= 2\pi/\lambda_1$) is the angular wave number and λ_1 is the wavelength in glass. Similarly, the phase of the second wave at the back surface of the plastic is given by $\phi_2 = k_2 L - \omega t$, where k_2 ($= 2\pi/\lambda_2$) is the angular wave number and λ_2 is the wavelength in plastic. The angular frequencies are the same since the waves have the same wavelength in air and the frequency of a wave does not change when the wave enters another medium. The phase difference is

$$\phi_1 - \phi_2 = \left(k_1 - k_2\right)L = 2\pi\left(\frac{1}{\lambda_1} - \frac{1}{\lambda_2}\right)L.$$

Now, $\lambda_1 = \lambda_{\text{air}}/n_1$, where λ_{air} is the wavelength in air and n_1 is the index of refraction of the glass. Similarly, $\lambda_2 = \lambda_{\text{air}}/n_2$, where n_2 is the index of refraction of the plastic. This means that the phase difference is

$$\phi_1 - \phi_2 = (2\pi/\lambda_{air})\,(n_1 - n_2)L.$$

The value of L that makes this 5.65 rad is

$$L = \frac{(\phi_1 - \phi_2)\lambda_{air}}{2\pi(n_1 - n_2)} = \frac{5.65(400 \times 10^{-9}\,\text{m})}{2\pi(1.60 - 1.50)} = 3.60 \times 10^{-6}\,\text{m}.$$

(b) 5.65 rad is less than 2π rad $= 6.28$ rad, the phase difference for completely constructive interference, and greater than π rad $(= 3.14$ rad), the phase difference for completely destructive interference. The interference is, therefore, intermediate, neither completely constructive nor completely destructive. It is, however, closer to completely constructive than to completely destructive.

8. (a) The time t_2 it takes for pulse 2 to travel through the plastic is

$$t_2 = \frac{L}{c/1.55} + \frac{L}{c/1.70} + \frac{L}{c/1.60} + \frac{L}{c/1.45} = \frac{6.30\,L}{c}.$$

Similarly for pulse 1:

$$t_1 = \frac{2L}{c/1.59} + \frac{L}{c/1.65} + \frac{L}{c/1.50} = \frac{6.33\,L}{c}.$$

Thus, pulse 2 travels through the plastic in less time.

(b) The time difference (as a multiple of L/c) is

$$\Delta t = t_2 - t_1 = \frac{6.33\,L}{c} - \frac{6.30\,L}{c} = \frac{0.03\,L}{c}.$$

Thus, the multiple is 0.03.

9. (a) We wish to set Eq. 35-11 equal to $1/2$, since a half-wavelength phase difference is equivalent to a π radians difference. Thus,

$$L_{min} = \frac{\lambda}{2(n_2 - n_1)} = \frac{620\,\text{nm}}{2(1.65 - 1.45)} = 1550\,\text{nm} = 1.55\ \mu\text{m}.$$

(b) Since a phase difference of $\dfrac{3}{2}$ (wavelengths) is effectively the same as what we required in part (a), then

$$L = \frac{3\lambda}{2(n_2 - n_1)} = 3L_{min} = 3(1.55 \mu m) = 4.65 \mu m.$$

10. (a) The exiting angle is 50°, the same as the incident angle, due to what one might call the "transitive" nature of Snell's law: $n_1 \sin\theta_1 = n_2 \sin\theta_2 = n_3 \sin\theta_3 = ...$

(b) Due to the fact that the speed (in a certain medium) is c/n (where n is that medium's index of refraction) and that speed is distance divided by time (while it's constant), we find

$$t = nL/c = (1.45)(25 \times 10^{-19} \, m)/(3.0 \times 10^8 \, m/s) = 1.4 \times 10^{-13} \, s = 0.14 \text{ ps}.$$

11. (a) Eq. 35-11 (in absolute value) yields

$$\frac{L}{\lambda}|n_2 - n_1| = \frac{(8.50 \times 10^{-6} \, m)}{500 \times 10^{-9} \, m}(1.60 - 1.50) = 1.70.$$

(b) Similarly,

$$\frac{L}{\lambda}|n_2 - n_1| = \frac{(8.50 \times 10^{-6} \, m)}{500 \times 10^{-9} \, m}(1.72 - 1.62) = 1.70.$$

(c) In this case, we obtain

$$\frac{L}{\lambda}|n_2 - n_1| = \frac{(3.25 \times 10^{-6} \, m)}{500 \times 10^{-9} \, m}(1.79 - 1.59) = 1.30.$$

(d) Since their phase differences were identical, the brightness should be the same for (a) and (b). Now, the phase difference in (c) differs from an integer by 0.30, which is also true for (a) and (b). Thus, their effective phase differences are equal, and the brightness in case (c) should be the same as that in (a) and (b).

12. (a) We note that ray 1 travels an extra distance $4L$ more than ray 2. To get the least possible L which will result in destructive interference, we set this extra distance equal to half of a wavelength:

$$4L = \tfrac{1}{2}\lambda \quad \Rightarrow \quad L = \frac{\lambda}{8} = 52.50 \text{ nm}.$$

(b) The next case occurs when that extra distance is set equal to $\frac{3}{2}\lambda$. The result is

$$L = \frac{3\lambda}{8} = 157.5 \text{ nm}.$$

13. (a) We choose a horizontal x axis with its origin at the left edge of the plastic. Between $x = 0$ and $x = L_2$ the phase difference is that given by Eq. 35-11 (with L in that equation replaced with L_2). Between $x = L_2$ and $x = L_1$ the phase difference is given by an expression similar to Eq. 35-11 but with L replaced with $L_1 - L_2$ and n_2 replaced with 1 (since the top ray in Fig. 35-36 is now traveling through air, which has index of refraction approximately equal to 1). Thus, combining these phase differences and letting all lengths be in μm (so $\lambda = 0.600$), we have

$$\frac{L_2}{\lambda}(n_2 - n_1) + \frac{L_1 - L_2}{\lambda}(1 - n_1) = \frac{3.50}{0.600}(1.60 - 1.40) + \frac{4.00 - 3.50}{0.600}(1 - 1.40) = 0.833.$$

(b) Since the answer in part (a) is closer to an integer than to a half-integer, the interference is more nearly constructive than destructive.

14. (a) For the maximum adjacent to the central one, we set $m = 1$ in Eq. 35-14 and obtain

$$\theta_1 = \sin^{-1}\left(\frac{m\lambda}{d}\right)\bigg|_{m=1} = \sin^{-1}\left[\frac{(1)(\lambda)}{100\lambda}\right] = 0.010\,\text{rad}.$$

(b) Since $y_1 = D \tan \theta_1$ (see Fig. 35-10(a)), we obtain

$$y_1 = (500\,\text{mm}) \tan (0.010\,\text{rad}) = 5.0\,\text{mm}.$$

The separation is $\Delta y = y_1 - y_0 = y_1 - 0 = 5.0$ mm.

15. The angular positions of the maxima of a two-slit interference pattern are given by $d \sin \theta = m\lambda$, where d is the slit separation, λ is the wavelength, and m is an integer. If θ is small, $\sin \theta$ may be approximated by θ in radians. Then, $\theta = m\lambda/d$ to good approximation. The angular separation of two adjacent maxima is $\Delta\theta = \lambda/d$. Let λ' be the wavelength for which the angular separation is greater by 10.0%. Then, $1.10\lambda/d = \lambda'/d$. or

$$\lambda' = 1.10\lambda = 1.10(589\,\text{nm}) = 648\,\text{nm}.$$

16. (a) We use Eq. 35-14 with $m = 3$:

$$\theta = \sin^{-1}\left(\frac{m\lambda}{d}\right) = \sin^{-1}\left[\frac{2(550 \times 10^{-9}\,\text{m})}{7.70 \times 10^{-6}\,\text{m}}\right] = 0.216\,\text{rad}.$$

(b) $\theta = (0.216)(180°/\pi) = 12.4°$.

17. Interference maxima occur at angles θ such that $d \sin \theta = m\lambda$, where m is an integer. Since $d = 2.0$ m and $\lambda = 0.50$ m, this means that $\sin \theta = 0.25m$. We want all values of m

(positive and negative) for which $|0.25m| \leq 1$. These are –4, –3, –2, –1, 0, +1, +2, +3, and +4. For each of these except –4 and +4, there are two different values for θ. A single value of $\theta (-90°)$ is associated with $m = -4$ and a single value (+90°) is associated with $m = +4$. There are sixteen different angles in all and, therefore, sixteen maxima.

18. (a) The phase difference (in wavelengths) is

$$\phi = d \sin\theta/\lambda = (4.24 \text{ μm})\sin(20°)/(0.500 \text{ μm}) = 2.90 \ .$$

(b) Multiplying this by 2π gives $\phi = 18.2$ rad.

(c) The result from part (a) is greater than $\frac{5}{2}$ (which would indicate the third minimum) and is less than 3 (which would correspond to the third side maximum).

19. The condition for a maximum in the two-slit interference pattern is $d \sin \theta = m\lambda$, where d is the slit separation, λ is the wavelength, m is an integer, and θ is the angle made by the interfering rays with the forward direction. If θ is small, $\sin \theta$ may be approximated by θ in radians. Then, $\theta = m\lambda/d$, and the angular separation of adjacent maxima, one associated with the integer m and the other associated with the integer $m + 1$, is given by $\Delta\theta = \lambda/d$. The separation on a screen a distance D away is given by $\Delta y = D \Delta\theta = \lambda D/d$. Thus,

$$\Delta y = \frac{(500 \times 10^{-9} \text{ m})(5.40 \text{ m})}{1.20 \times 10^{-3} \text{ m}} = 2.25 \times 10^{-3} \text{ m} = 2.25 \text{ mm}.$$

20. In Sample Problem 35-2, an experimentally useful relation is derived: $\Delta y = \lambda D/d$. Dividing both sides by D, this becomes $\Delta\theta = \lambda/d$ with θ in radians. In the steps that follow, however, we will end up with an expression where degrees may be directly used. Thus, in the present case,

$$\Delta\theta_n = \frac{\lambda_n}{d} = \frac{\lambda}{nd} = \frac{\Delta\theta}{n} = \frac{0.20°}{1.33} = 0.15° .$$

21. The maxima of a two-slit interference pattern are at angles θ given by $d \sin \theta = m\lambda$, where d is the slit separation, λ is the wavelength, and m is an integer. If θ is small, $\sin \theta$ may be replaced by θ in radians. Then, $d\theta = m\lambda$. The angular separation of two maxima associated with different wavelengths but the same value of m is $\Delta\theta = (m/d)(\lambda_2 - \lambda_1)$, and their separation on a screen a distance D away is

$$\Delta y = D \tan\Delta\theta \approx D\Delta\theta = \left[\frac{mD}{d}\right](\lambda_2 - \lambda_1)$$

$$= \left[\frac{3(1.0 \text{ m})}{5.0 \times 10^{-3} \text{ m}}\right](600 \times 10^{-9} \text{ m} - 480 \times 10^{-9} \text{ m}) = 7.2 \times 10^{-5} \text{ m}.$$

The small angle approximation $\tan \Delta\theta \approx \Delta\theta$ (in radians) is made.

22. (a) We use Eq. 35-14 to find d:

$$d\sin\theta = m\lambda \quad \Rightarrow \quad d = (4)(450 \text{ nm})/\sin(90°) = 1800 \text{ nm} .$$

For the third order spectrum, the wavelength that corresponds to $\theta = 90°$ is

$$\lambda = d\sin(90°)/3 = 600 \text{ nm} .$$

Any wavelength greater than this will not be seen. Thus, 600 nm < θ ≤ 700 nm are absent.

(b) The slit separation d needs to be decreased.

(c) In this case, the 400 nm wavelength in the $m = 4$ diffraction is to occur at 90°. Thus

$$d_{new}\sin\theta = m\lambda \quad \Rightarrow \quad d_{new} = (4)(400 \text{ nm})/\sin(90°) = 1600 \text{ nm} .$$

This represents a change of $|\Delta d| = d - d_{new} = 200 \text{ nm} = 0.20 \text{ μm}$.

23. Initially, source A leads source B by 90°, which is equivalent to 1/4 wavelength. However, source A also lags behind source B since r_A is longer than r_B by 100 m, which is $100\,\text{m}/400\,\text{m} = 1/4$ wavelength. So the net phase difference between A and B at the detector is zero.

24. Imagine a y axis midway between the two sources in the figure. Thirty points of destructive interference (to be considered in the xy plane of the figure) implies there are $7+1+7=15$ on each side of the y axis. There is no point of destructive interference on the y axis itself since the sources are in phase and any point on the y axis must therefore correspond to a zero phase difference (and corresponds to $\theta = 0$ in Eq. 35-14). In other words, there are 7 "dark" points in the first quadrant, one along the +x axis, and 7 in the fourth quadrant, constituting the 15 dark points on the right-hand side of the y axis. Since the y axis corresponds to a minimum phase difference, we can count (say, in the first quadrant) the m values for the destructive interference (in the sense of Eq. 35-16) beginning with the one closest to the y axis and going clockwise until we reach the x axis (at any point beyond S_2). This leads us to assign $m = 7$ (in the sense of Eq. 35-16) to the point on the x axis itself (where the path difference for waves coming from the sources is simply equal to the separation of the sources, d); this would correspond to $\theta = 90°$ in Eq. 35-16. Thus,

$$d = (7 + \tfrac{1}{2})\lambda = 7.5\lambda \quad \Rightarrow \quad \frac{d}{\lambda} = 7.5 .$$

25. Let the distance in question be x. The path difference (between rays originating from S_1 and S_2 and arriving at points on the $x > 0$ axis) is

$$\sqrt{d^2 + x^2} - x = \left(m + \frac{1}{2}\right)\lambda,$$

where we are requiring destructive interference (half-integer wavelength phase differences) and $m = 0, 1, 2, \cdots$. After some algebraic steps, we solve for the distance in terms of m:

$$x = \frac{d^2}{(2m+1)\lambda} - \frac{(2m+1)\lambda}{4}.$$

To obtain the largest value of x, we set $m = 0$:

$$x_0 = \frac{d^2}{\lambda} - \frac{\lambda}{4} = \frac{(3.00\lambda)^2}{\lambda} - \frac{\lambda}{4} = 8.75\lambda = 8.75(900 \text{ nm}) = 7.88 \times 10^3 \text{ nm}$$
$$= 7.88 \mu\text{m}.$$

26. (a) We note that, just as in the usual discussion of the double slit pattern, the $x = 0$ point on the screen (where that vertical line of length D in the picture intersects the screen) is a bright spot with phase difference equal to zero (it would be the middle fringe in the usual double slit pattern). We are not considering $x < 0$ values here, so that negative phase differences are not relevant (and if we did wish to consider $x < 0$ values, we could limit our discussion to absolute values of the phase difference, so that – again – negative phase differences do not enter it). Thus, the $x = 0$ point is the one with the minimum phase difference.

(b) As noted in part (a), the phase difference $\phi = 0$ at $x = 0$.

(c) The path length difference is greatest at the rightmost "edge" of the screen (which is assumed to go on forever), so ϕ is maximum at $x = \infty$.

(d) In considering $x = \infty$, we can treat the rays from the sources as if they are essentially horizontal. In this way, we see that the difference between the path lengths is simply the distance ($2d$) between the sources. The problem specifies $2d = 6.00\,\lambda$, or $2d/\lambda = 6.00$.

(e) Using the Pythagorean theorem, we have

$$\phi = \frac{\sqrt{D^2 + (x + d)^2}}{\lambda} - \frac{\sqrt{D^2 + (x - d)^2}}{\lambda} = 1.71$$

where we have plugged in $D = 20\lambda$, $d = 3\lambda$ and $x = 6\lambda$. Thus, the phase difference at that point is 1.71 wavelengths.

(f) We note that the answer to part (e) is closer to $\frac{3}{2}$ (destructive interference) than to 2 (constructive interference), so that the point is "intermediate" but closer to a minimum than to a maximum.

27. Consider the two waves, one from each slit, that produce the seventh bright fringe in the absence of the mica. They are in phase at the slits and travel different distances to the seventh bright fringe, where they have a phase difference of $2\pi m = 14\pi$. Now a piece of mica with thickness x is placed in front of one of the slits, and an additional phase difference between the waves develops. Specifically, their phases at the slits differ by

$$\frac{2\pi x}{\lambda_m} - \frac{2\pi x}{\lambda} = \frac{2\pi x}{\lambda}(n-1)$$

where λ_m is the wavelength in the mica and n is the index of refraction of the mica. The relationship $\lambda_m = \lambda/n$ is used to substitute for λ_m. Since the waves are now in phase at the screen,

$$\frac{2\pi x}{\lambda}(n-1) = 14\pi$$

or

$$x = \frac{7\lambda}{n-1} = \frac{7(550 \times 10^{-9}\,\text{m})}{1.58 - 1} = 6.64 \times 10^{-6}\,\text{m}.$$

28. The problem asks for "the greatest value of x... exactly out of phase" which is to be interpreted as the value of x where the curve shown in the figure passes through a phase value of π radians. This happens as some point P on the x axis, which is, of course, a distance x from the top source and (using Pythagoras' theorem) a distance $\sqrt{d^2 + x^2}$ from the bottom source. The difference (in normal length units) is therefore $\sqrt{d^2 + x^2} - x$, or (expressed in radians) is $\frac{2\pi}{\lambda}(\sqrt{d^2 + x^2} - x)$. We note (looking at the leftmost point in the graph) that at $x = 0$, this latter quantity equals 6π, which means $d = 3\lambda$. Using this value for d, we now must solve the condition

$$\frac{2\pi}{\lambda}(\sqrt{d^2 + x^2} - x) = \pi.$$

Straightforward algebra then lead to $x = (35/4)\lambda$, and using $\lambda = 400$ nm we find $x = 3500$ nm, or 3.5 μm.

29. The phasor diagram is shown below.

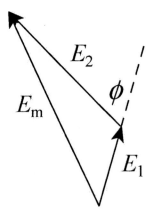

Here $E_1 = 1.00$, $E_2 = 2.00$, and $\phi = 60°$. The resultant amplitude E_m is given by the trigonometric law of cosines:

$$E_m^2 = E_1^2 + E_2^2 - 2E_1E_2 \cos(180° - \phi).$$

Thus,

$$E_m = \sqrt{(1.00)^2 + (2.00)^2 - 2(1.00)(2.00)\cos 120°} = 2.65.$$

30. In adding these with the phasor method (as opposed to, say, trig identities), we may set $t = 0$ (see Sample Problem 35-4) and add them as vectors:

$$y_h = 10\cos 0° + 8.0\cos 30° = 16.9$$
$$y_v = 10\sin 0° + 8.0\sin 30° = 4.0$$

so that

$$y_R = \sqrt{y_h^2 + y_v^2} = 17.4$$

$$\beta = \tan^{-1}\left(\frac{y_v}{y_h}\right) = 13.3°.$$

Thus, $y = y_1 + y_2 = y_R \sin(\omega t + \beta) = 17.4 \sin(\omega t + 13.3°)$. Quoting the answer to two significant figures, we have $y \approx 17\sin(\omega t + 13°)$.

31. In adding these with the phasor method (as opposed to, say, trig identities), we may set $t = 0$ (see Sample Problem 35-4) and add them as vectors:

$$y_h = 10\cos 0° + 15\cos 30° + 5.0\cos(-45°) = 26.5$$
$$y_v = 10\sin 0° + 15\sin 30° + 5.0\sin(-45°) = 4.0$$

so that

$$y_R = \sqrt{y_h^2 + y_v^2} = 26.8 \approx 27$$

$$\beta = \tan^{-1}\left(\frac{y_v}{y_h}\right) = 8.5°.$$

Thus, $y = y_1 + y_2 + y_3 = y_R \sin(\omega t + \beta) = 27 \sin(\omega t + 8.5°)$.

32. (a) Referring to Figure 35-10(a) makes clear that

$$\theta = \tan^{-1}(y/D) = \tan^{-1}(0.205/4) = 2.93°.$$

Thus, the phase difference at point P is $\phi = d\sin\theta/\lambda = 0.397$ wavelengths, which means it is between the central maximum (zero wavelength difference) and the first minimum ($\frac{1}{2}$ wavelength difference). Note that the above computation could have been simplified somewhat by avoiding the explicit use of the tangent and sine functions and making use of the small-angle approximation ($\tan\theta \approx \sin\theta$).

(b) From Eq. 35-22, we get (with $\phi = (0.397)(2\pi) = 2.495$ rad)

$$I = 4I_0(\cos(\phi/2))^2 = 0.404\ I_0$$

at point P and

$$I_{cen} = 4I_0(\cos(0))^2 = 4\ I_0$$

at the center . Thus, $\dfrac{I}{I_{cen}} = \dfrac{0.404}{4} = 0.101$.

33. With phasor techniques, this amounts to a vector addition problem $\vec{R} = \vec{A} + \vec{B} + \vec{C}$ where (in magnitude-angle notation) $\vec{A} = (10\angle 0°)$, $\vec{B} = (5\angle 45°)$, and $\vec{C} = (5\angle -45°)$, where the magnitudes are understood to be in μV/m. We obtain the resultant (especially efficient on a vector-capable calculator in polar mode):

$$\vec{R} = (10\angle 0°) + (5\angle 45°) + (5\angle -45°) = (17.1\angle 0°)$$

which leads to

$$E_R = (17.1\,\mu\,\text{V/m})\sin(\omega t)$$

where $\omega = 2.0 \times 10^{14}$ rad/s.

34. (a) We can use phasor techniques or use trig identities. Here we show the latter approach. Since $\sin a + \sin(a+b) = 2\cos(b/2)\sin(a + b/2)$, we find

$$E_1 + E_2 = 2E_o \cos(\tfrac{\phi}{2}) \sin(\omega t + \tfrac{\phi}{2})$$

where $E_o = 2.00\ \mu V/m$, $\omega = 1.26 \times 10^{15}$ rad/s, and $\phi = 39.6$ rad. This shows that the electric field amplitude of the resultant wave is

$$E = 2E_o \cos(\phi/2) = 2.33\ \mu V/m\ .$$

(b) Eq. 35-22 leads to

$$I = 4I_0(\cos(\phi/2))^2 = 1.35\ I_0$$

at point P, and

$$I_{cen} = 4I_0(\cos(0))^2 = 4\ I_0$$

at the center . Thus, $\dfrac{I}{I_{cen}} = \dfrac{1.35}{4} = 0.338$.

(c) The phase difference ϕ (in wavelengths) is gotten from ϕ in radians by dividing by 2π. Thus, $\phi = 39.6/2\pi = 6.3$ wavelengths. Thus, point P is between the sixth side maximum (at which $\phi = 6$ wavelengths) and the seventh minimum (at which $\phi = 6\tfrac{1}{2}$ wavelengths).

(d) The rate is given by $\omega = 1.26 \times 10^{15}$ rad/s.

(e) The angle between the phasors is $\phi = 39.6$ rad $= 2270°$ (which would look like about $110°$ when drawn in the usual way).

35. For constructive interference, we use Eq. 35-36: $2n_2 L = (m+1/2)\lambda$. For the smallest value of L, let $m = 0$:

$$L_0 = \frac{\lambda/2}{2n_2} = \frac{624\,\text{nm}}{4(1.33)} = 117\,\text{nm} = 0.117\,\mu\text{m}.$$

(b) For the second smallest value, we set $m = 1$ and obtain

$$L_1 = \frac{(1+1/2)\lambda}{2n_2} = \frac{3\lambda}{2n_2} = 3L_0 = 3(0.1173\,\mu\text{m}) = 0.352\,\mu\text{m}.$$

36. (a) On both sides of the soap is a medium with lower index (air) and we are examining the reflected light, so the condition for strong reflection is Eq. 35-36. With lengths in nm,

$$\lambda = \frac{2n_2L}{m+\frac{1}{2}} = \begin{cases} 3360 & \text{for } m=0 \\ 1120 & \text{for } m=1 \\ 672 & \text{for } m=2 \\ 480 & \text{for } m=3 \\ 373 & \text{for } m=4 \\ 305 & \text{for } m=5 \end{cases}$$

from which we see the latter *four* values are in the given range.

(b) We now turn to Eq. 35-37 and obtain

$$\lambda = \frac{2n_2L}{m} = \begin{cases} 1680 & \text{for } m=1 \\ 840 & \text{for } m=2 \\ 560 & \text{for } m=3 \\ 420 & \text{for } m=4 \\ 336 & \text{for } m=5 \end{cases}$$

from which we see the latter *three* values are in the given range.

37. Light reflected from the front surface of the coating suffers a phase change of π rad while light reflected from the back surface does not change phase. If L is the thickness of the coating, light reflected from the back surface travels a distance $2L$ farther than light reflected from the front surface. The difference in phase of the two waves is $2L(2\pi/\lambda_c)$ − π, where λ_c is the wavelength in the coating. If λ is the wavelength in vacuum, then $\lambda_c = \lambda/n$, where n is the index of refraction of the coating. Thus, the phase difference is $2nL(2\pi/\lambda) - \pi$. For fully constructive interference, this should be a multiple of 2π. We solve

$$2nL\left(\frac{2\pi}{\lambda}\right) - \pi = 2m\pi$$

for L. Here m is an integer. The solution is

$$L = \frac{(2m+1)\lambda}{4n}.$$

To find the smallest coating thickness, we take $m = 0$. Then,

$$L = \frac{\lambda}{4n} = \frac{560 \times 10^{-9} \text{ m}}{4(2.00)} = 7.00 \times 10^{-8} \text{ m}.$$

38. (a) We are dealing with a thin film (material 2) in a situation where $n_1 > n_2 > n_3$, looking for strong *reflections*; the appropriate condition is the one expressed by Eq. 35-37. Therefore, with lengths in nm and $L = 500$ and $n_2 = 1.7$, we have

$$\lambda = \frac{2n_2L}{m} = \begin{cases} 1700 & \text{for } m = 1 \\ 850 & \text{for } m = 2 \\ 567 & \text{for } m = 3 \\ 425 & \text{for } m = 4 \end{cases}$$

from which we see the latter *two* values are in the given range. The longer wavelength (m=3) is $\lambda = 567$ nm.

(b) The shorter wavelength (m=4) is $\lambda = 425$ nm.

(c) We assume the temperature dependence of the refraction index is negligible. From the proportionality evident in the part (a) equation, longer L means longer λ.

39. For complete destructive interference, we want the waves reflected from the front and back of the coating to differ in phase by an odd multiple of π rad. Each wave is incident on a medium of higher index of refraction from a medium of lower index, so both suffer phase changes of π rad on reflection. If L is the thickness of the coating, the wave reflected from the back surface travels a distance $2L$ farther than the wave reflected from the front. The phase difference is $2L(2\pi/\lambda_c)$, where λ_c is the wavelength in the coating. If n is the index of refraction of the coating, $\lambda_c = \lambda/n$, where λ is the wavelength in vacuum, and the phase difference is $2nL(2\pi/\lambda)$. We solve

$$2nL\left(\frac{2\pi}{\lambda}\right) = (2m+1)\pi$$

for L. Here m is an integer. The result is

$$L = \frac{(2m+1)\lambda}{4n}.$$

To find the least thickness for which destructive interference occurs, we take $m = 0$. Then,

$$L = \frac{\lambda}{4n} = \frac{600 \times 10^{-9}\,\text{m}}{4(1.25)} = 1.20 \times 10^{-7}\,\text{m}.$$

40. The situation is analogous to that treated in Sample Problem 35-6, in the sense that the incident light is in a low index medium, the thin film of acetone has somewhat higher $n = n_2$, and the last layer (the glass plate) has the highest refractive index. To see very little or no reflection, according to the Sample Problem, the condition

$$2L = \left(m + \tfrac{1}{2}\right)\frac{\lambda}{n_2} \quad \text{where} \quad m = 0, 1, 2, \ldots$$

must hold. This is the same as Eq. 35-36 which was developed for the opposite situation (constructive interference) regarding a thin film surrounded on both sides by air (a very different context than the one in this problem). By analogy, we expect Eq. 35-37 to apply in this problem to reflection *maxima*. A more careful analysis such as that given in §35-7 bears this out. Thus, using Eq. 35-37 with $n_2 = 1.25$ and $\lambda = 700$ nm yields

$$L = 0, 280\,\text{nm}, 560\,\text{nm}, 840\,\text{nm}, 1120\,\text{nm}, \ldots$$

for the first several *m* values. And the equation shown above (equivalent to Eq. 35-36) gives, with $\lambda = 600$ nm,

$$L = 120\,\text{nm}, 360\,\text{nm}, 600\,\text{nm}, 840\,\text{nm}, 1080\,\text{nm}, \ldots$$

for the first several *m* values. The lowest number these lists have in common is $L = 840$ nm.

41. When a thin film of thickness L and index of refraction n_2 is placed between materials 1 and 3 such that $n_1 > n_2$ and $n_3 > n_2$ where n_1 and n_3 are the indexes of refraction of the materials, the general condition for destructive interference for a thin film is

$$2L = m\frac{\lambda}{n_2} \quad \Rightarrow \quad \lambda = \frac{2Ln_2}{m}, \quad m = 0, 1, 2, \ldots$$

where λ is the wavelength of light as measured in air. Thus, we have, for $m = 1$

$$\lambda = 2Ln_2 = 2(200 \text{ nm})(1.40) = 560 \text{ nm}.$$

42. In this setup, we have $n_2 > n_1$ and $n_2 > n_3$, and the condition for constructive interference is

$$2L = \left(m + \frac{1}{2}\right)\frac{\lambda}{n_2} \quad \Rightarrow \quad \lambda = \frac{4Ln_2}{2m+1}, \quad m = 0, 1, 2, \ldots$$

Thus, we have,

$$\lambda = \begin{cases} 4Ln_2 = 4(285 \text{ nm})(1.60) = 1824 \text{ nm} \ (m = 0) \\ 4Ln_2/3 = 4(285 \text{ nm})(1.60)/3 = 608 \text{ nm} \ (m = 1) \end{cases}.$$

For the wavelength to be in the visible range, we choose $m=1$ with $\lambda = 608$ nm.

43. In this setup, we have $n_2 > n_1$ and $n_2 < n_3$, and the condition for destructive interference is

$$2L = \left(m + \frac{1}{2}\right)\frac{\lambda}{n_2} \quad \Rightarrow \quad \lambda = \frac{4Ln_2}{2m+1}, \quad m = 0,1,2,...$$

Thus, we have,

$$\lambda = \begin{cases} 4Ln_2 = 4(210\ \text{nm})(1.46) = 1226\ \text{nm} \ (m = 0) \\ 4Ln_2/3 = 4(210\ \text{nm})(1.46)/3 = 409\ \text{nm} \ (m = 1) \end{cases}.$$

For the wavelength to be in the visible range, we choose $m=1$ with $\lambda = 409$ nm.

44. In this setup, we have $n_2 > n_1$ and $n_2 > n_3$, and the condition for constructive interference is

$$2L = \left(m + \frac{1}{2}\right)\frac{\lambda}{n_2} \quad \Rightarrow \quad \lambda = \frac{4Ln_2}{2m+1}, \quad m = 0,1,2,...$$

Thus, we have,

$$\lambda = \begin{cases} 4Ln_2 = 4(325\ \text{nm})(1.75) = 2275\ \text{nm} \ (m = 0) \\ 4Ln_2/3 = 4(325\ \text{nm})(1.75)/3 = 758\ \text{nm} \ (m = 1) \\ 4Ln_2/5 = 4(325\ \text{nm})(1.75)/5 = 455\ \text{nm} \ (m = 2) \end{cases}.$$

For the wavelength to be in the visible range, we choose $m=2$ with $\lambda = 455$ nm.

45. In this setup, we have $n_2 < n_1$ and $n_2 < n_3$, and the condition for destructive interference is

$$2L = m\frac{\lambda}{n_2} \quad \Rightarrow \quad \lambda = \frac{2Ln_2}{m}, \quad m = 0,1,2,...$$

Thus, we have,

$$\lambda = \begin{cases} 2Ln_2 = 2(380\ \text{nm})(1.34) = 1018\ \text{nm} \ (m = 1) \\ Ln_2 = (380\ \text{nm})(1.34) = 509\ \text{nm} \ (m = 2) \end{cases}.$$

For the wavelength to be in the visible range, we choose $m=2$ with $\lambda = 509$ nm.

46. In this setup, we have $n_2 < n_1$ and $n_2 > n_3$, and the condition for destructive interference is

$$2L = \left(m + \frac{1}{2} \right) \frac{\lambda}{n_2} \quad \Rightarrow \quad \lambda = \frac{4Ln_2}{2m+1}, \quad m = 0,1,2,...$$

Thus, we have,

$$\lambda = \begin{cases} 4Ln_2 = 4(415 \text{ nm})(1.59) = 2639 \text{ nm} \quad (m=0) \\ 4Ln_2/3 = 4(415 \text{ nm})(1.59)/3 = 880 \text{ nm} \quad (m=1) \\ 4Ln_2/5 = 4(415 \text{ nm})(1.59)/5 = 528 \text{ nm} \quad (m=2) \end{cases}.$$

For the wavelength to be in the visible range, we choose $m=3$ with $\lambda = 528$ nm.

47. In this setup, we have $n_2 > n_1$ and $n_2 > n_3$, and the condition for constructive interference is

$$2L = \left(m + \frac{1}{2} \right) \frac{\lambda}{n_2} \quad \Rightarrow \quad L = \left(m + \frac{1}{2} \right) \frac{\lambda}{2n_2}, \quad m = 0,1,2,...$$

The third least thickness is ($m=2$)

$$L = \left(2 + \frac{1}{2} \right) \frac{612 \text{ nm}}{2(1.60)} = 478 \text{ nm}.$$

48. In this setup, we have $n_2 < n_1$ and $n_2 < n_3$, and the condition for constructive interference is

$$2L = \left(m + \frac{1}{2} \right) \frac{\lambda}{n_2} \quad \Rightarrow \quad L = \left(m + \frac{1}{2} \right) \frac{\lambda}{2n_2}, \quad m = 0,1,2,...$$

The second least thickness is ($m=1$)

$$L = \left(1 + \frac{1}{2} \right) \frac{632 \text{ nm}}{2(1.40)} = 339 \text{ nm}.$$

49. In this setup, we have $n_2 > n_1$ and $n_2 > n_3$, and the condition for constructive interference is

$$2L = \left(m + \frac{1}{2} \right) \frac{\lambda}{n_2} \quad \Rightarrow \quad L = \left(m + \frac{1}{2} \right) \frac{\lambda}{2n_2}, \quad m = 0,1,2,...$$

The third least thickness is ($m=2$)

$$L = \left(2 + \frac{1}{2}\right) \frac{382 \text{ nm}}{2(1.75)} = 273 \text{ nm}.$$

50. In this setup, we have $n_2 > n_1$ and $n_2 < n_3$, and the condition for destructive interference is

$$2L = \left(m + \frac{1}{2}\right) \frac{\lambda}{n_2} \quad \Rightarrow \quad L = \left(m + \frac{1}{2}\right) \frac{\lambda}{2n_2}, \quad m = 0, 1, 2, \ldots$$

The second least thickness is (m=1)

$$L = \left(1 + \frac{1}{2}\right) \frac{482 \text{ nm}}{2(1.46)} = 248 \text{ nm}.$$

51. In this setup, we have $n_2 < n_1$ and $n_2 > n_3$, and the condition for destructive interference is

$$2L = \left(m + \frac{1}{2}\right) \frac{\lambda}{n_2} \quad \Rightarrow \quad L = \left(m + \frac{1}{2}\right) \frac{\lambda}{2n_2}, \quad m = 0, 1, 2, \ldots$$

The second least thickness is (m=1)

$$L = \left(1 + \frac{1}{2}\right) \frac{342 \text{ nm}}{2(1.59)} = 161 \text{ nm}.$$

52. In this setup, we have $n_2 < n_1$ and $n_2 < n_3$, and the condition for constructive interference is

$$2L = \left(m + \frac{1}{2}\right) \frac{\lambda}{n_2} \quad \Rightarrow \quad L = \left(m + \frac{1}{2}\right) \frac{\lambda}{2n_2}, \quad m = 0, 1, 2, \ldots$$

The second least thickness is (m=1)

$$L = \left(1 + \frac{1}{2}\right) \frac{587 \text{ nm}}{2(1.34)} = 329 \text{ nm}.$$

53. We solve Eq. 35-36 with $n_2 = 1.33$ and $\lambda = 600$ nm for $m = 1, 2, 3, \ldots$:

$$L = 113 \text{ nm}, 338 \text{ nm}, 564 \text{ nm}, 789 \text{ nm}, \ldots$$

And, we similarly solve Eq. 35-37 with the same n_2 and $\lambda = 450$ nm:

$$L = 0, 169\,\text{nm}, 338\,\text{nm}, 508\,\text{nm}, 677\,\text{nm}, \ldots$$

The lowest number these lists have in common is $L = 338$ nm.

54. The situation is analogous to that treated in Sample Problem 35-6, in the sense that the incident light is in a low index medium, the thin film of oil has somewhat higher $n = n_2$, and the last layer (the glass plate) has the highest refractive index. To see very little or no reflection, according to the Sample Problem, the condition

$$2L = \left(m + \frac{1}{2}\right)\frac{\lambda}{n_2} \quad \text{where} \quad m = 0, 1, 2, \ldots$$

must hold. With $\lambda = 500$ nm and $n_2 = 1.30$, the possible answers for L are

$$L = 96\,\text{nm}, 288\,\text{nm}, 481\,\text{nm}, 673\,\text{nm}, 865\,\text{nm}, \ldots$$

And, with $\lambda = 700$ nm and the same value of n_2, the possible answers for L are

$$L = 135\,\text{nm}, 404\,\text{nm}, 673\,\text{nm}, 942\,\text{nm}, \ldots$$

The lowest number these lists have in common is $L = 673$ nm.

55. The situation is analogous to that treated in Sample Problem 35-6, in the sense that the incident light is in a low index medium, the thin film has somewhat higher $n = n_2$, and the last layer has the highest refractive index. To see very little or no reflection, according to the Sample Problem, the condition

$$2L = \left(m + \frac{1}{2}\right)\frac{\lambda}{n_2} \quad \text{where} \quad m = 0, 1, 2, \ldots$$

must hold. The value of L which corresponds to no reflection corresponds, reasonably enough, to the value which gives maximum transmission of light (into the highest index medium — which in this problem is the water).

(a) If $2L = \left(m + \frac{1}{2}\right)\frac{\lambda}{n_2}$ (Eq. 35-36) gives zero reflection in this type of system, then we might reasonably expect that its counterpart, Eq. 35-37, gives maximum reflection here. A more careful analysis such as that given in §35-7 bears this out. We disregard the $m = 0$ value (corresponding to $L = 0$) since there is *some* oil on the water. Thus, for $m = 1, 2, \ldots$, maximum reflection occurs for wavelengths

$$\lambda = \frac{2n_2 L}{m} = \frac{2(1.20)(460\,\text{nm})}{m} = 1104\,\text{nm}, 552\,\text{nm}, 368\,\text{nm} \ldots$$

We note that only the 552 nm wavelength falls within the visible light range.

(b) As remarked above, maximum transmission into the water occurs for wavelengths given by

$$2L = \left(m + \frac{1}{2}\right)\frac{\lambda}{n_2} \Rightarrow \lambda = \frac{4n_2 L}{2m+1}$$

which yields $\lambda = 2208$ nm, 736 nm, 442 nm … for the different values of m. We note that only the 442 nm wavelength (blue) is in the visible range, though we might expect some red contribution since the 736 nm is very close to the visible range.

56. For constructive interference (which is obtained for $\lambda = 600$ nm) in this circumstance, we require

$$2L = \frac{k}{2}\lambda_n = \frac{k\lambda}{2n}$$

where k = some positive <u>odd</u> integer and n is the index of refraction of the thin film. Rearranging and plugging in $L = 272.7$ nm and the wavelength value, this gives

$$\frac{k}{1.818} = n .$$

Since we expect $n > 1$, then $k = 1$ is ruled out. However, $k = 3$ seems reasonable, since it leads to $n = 1.65$, which is close to the "typical" values found in Table 34-1. Taking this to be the correct index of refraction for the thin film, we now consider the destructive interference part of the question. Now we have $2L = $ (integer)λ_{dest}/n. Thus, $\lambda_{dest} = $ (900 nm)/(integer). We note that setting the integer equal to 1 yields a λ_{dest} value outside the range of the visible spectrum. A similar remark holds for setting the integer equal to 3. Thus, we set it equal to 2 and obtain $\lambda_{dest} = 450$ nm.

57. In this setup, we have $n_2 > n_1$ and $n_2 > n_3$, and the condition for minimum transmission (maximum reflection) or destructive interference is

$$2L = \left(m + \frac{1}{2}\right)\frac{\lambda}{n_2} \Rightarrow \lambda = \frac{4Ln_2}{2m+1}, \quad m = 0,1,2,...$$

Thus, we have,

$$\lambda = \begin{cases} 4Ln_2 = 4(285 \text{ nm})(1.60) = 1824 \text{ nm} & (m = 0) \\ 4Ln_2/3 = 4(415 \text{ nm})(1.59)/3 = 608 \text{ nm} & (m = 1) \end{cases}.$$

For the wavelength to be in the visible range, we choose $m=1$ with $\lambda = 608$ nm.

58. In this setup, we have $n_2 < n_1$ and $n_2 < n_3$, and the condition for maximum transmission (minimum reflection) or constructive interference is

$$2L = m\frac{\lambda}{n_2} \quad \Rightarrow \quad \lambda = \frac{2Ln_2}{m}, \quad m = 0,1,2,...$$

Thus, we have (with $m = 1$),

$$\lambda = 2Ln_2 = 2(200 \text{ nm})(1.40) = 560 \text{ nm}.$$

59. In this setup, we have $n_2 > n_1$ and $n_2 > n_3$, and the condition for minimum transmission (maximum reflection) or destructive interference is

$$2L = \left(m + \frac{1}{2}\right)\frac{\lambda}{n_2} \quad \Rightarrow \quad \lambda = \frac{4Ln_2}{2m+1}, \quad m = 0,1,2,...$$

Thus, we have,

$$\lambda = \begin{cases} 4Ln_2 = 4(325 \text{ nm})(1.75) = 2275 \text{ nm} \ (m = 0) \\ 4Ln_2/3 = 4(415 \text{ nm})(1.59)/3 = 758 \text{ nm} \ (m = 1) \\ 4Ln_2/5 = 4(415 \text{ nm})(1.59)/5 = 455 \text{ nm} \ (m = 2) \end{cases}.$$

For the wavelength to be in the visible range, we choose $m=2$ with $\lambda = 455$ nm.

60. In this setup, we have $n_2 > n_1$ and $n_2 < n_3$, and the condition for maximum transmission (minimum reflection) or constructive interference is

$$2L = \left(m + \frac{1}{2}\right)\frac{\lambda}{n_2} \quad \Rightarrow \quad \lambda = \frac{4Ln_2}{2m+1}, \quad m = 0,1,2,...$$

Thus, we have,

$$\lambda = \begin{cases} 4Ln_2 = 4(210 \text{ nm})(1.46) = 1226 \text{ nm} \ (m = 0) \\ 4Ln_2/3 = 4(210 \text{ nm})(1.46)/3 = 409 \text{ nm} \ (m = 1) \end{cases}.$$

For the wavelength to be in the visible range, we choose $m=1$ with $\lambda = 409$ nm.

61. In this setup, we have $n_2 < n_1$ and $n_2 > n_3$, and the condition for maximum transmission (minimum reflection) or constructive interference is

$$2L = \left(m + \frac{1}{2}\right)\frac{\lambda}{n_2} \quad \Rightarrow \quad \lambda = \frac{4Ln_2}{2m+1}, \quad m = 0,1,2,...$$

Thus, we have,

$$\lambda = \begin{cases} 4Ln_2 = 4(415 \text{ nm})(1.59) = 2639 \text{ nm} \quad (m = 0) \\ 4Ln_2/3 = 4(415 \text{ nm})(1.59)/3 = 880 \text{ nm} \quad (m = 1) \\ 4Ln_2/5 = 4(415 \text{ nm})(1.59)/5 = 528 \text{ nm} \quad (m = 2) \end{cases}.$$

For the wavelength to be in the visible range, we choose $m=3$ with $\lambda = 528$ nm.

62. In this setup, we have $n_2 < n_1$ and $n_2 < n_3$, and the condition for maximum transmission (minimum reflection) or constructive interference is

$$2L = m\frac{\lambda}{n_2} \quad \Rightarrow \quad \lambda = \frac{2Ln_2}{m}, \quad m = 0,1,2,...$$

Thus, we have,

$$\lambda = \begin{cases} 2Ln_2 = 2(380 \text{ nm})(1.34) = 1018 \text{ nm} \quad (m = 1) \\ Ln_2 = (380 \text{ nm})(1.34) = 509 \text{ nm} \quad (m = 2) \end{cases}.$$

For the wavelength to be in the visible range, we choose $m=2$ with $\lambda = 509$ nm.

63. In this setup, we have $n_2 < n_1$ and $n_2 < n_3$, and the condition for minimum transmission (maximum reflection) or destructive interference is

$$2L = \left(m + \frac{1}{2}\right)\frac{\lambda}{n_2} \quad \Rightarrow \quad L = \left(m + \frac{1}{2}\right)\frac{\lambda}{2n_2}, \quad m = 0,1,2,...$$

The second least thickness is ($m=1$)

$$L = \left(1 + \frac{1}{2}\right)\frac{632 \text{ nm}}{2(1.40)} = 339 \text{ nm}.$$

64. In this setup, we have $n_2 > n_1$ and $n_2 > n_3$, and the condition for minimum transmission (maximum reflection) or destructive interference is

$$2L = \left(m + \frac{1}{2}\right)\frac{\lambda}{n_2} \quad \Rightarrow \quad L = \left(m + \frac{1}{2}\right)\frac{\lambda}{2n_2}, \quad m = 0,1,2,...$$

The third least thickness is ($m=2$)

$$L = \left(2 + \frac{1}{2}\right)\frac{612 \text{ nm}}{2(1.60)} = 478 \text{ nm}.$$

65. In this setup, we have $n_2 > n_1$ and $n_2 < n_3$, and the condition for maximum transmission (minimum reflection) or constructive interference is

$$2L = \left(m + \frac{1}{2}\right)\frac{\lambda}{n_2} \quad \Rightarrow L = \left(m + \frac{1}{2}\right)\frac{\lambda}{2n_2}, \quad m = 0, 1, 2, \ldots$$

The second least thickness is ($m=1$)

$$L = \left(1 + \frac{1}{2}\right)\frac{482 \text{ nm}}{2(1.46)} = 248 \text{ nm}.$$

66. In this setup, we have $n_2 > n_1$ and $n_2 > n_3$, and the condition for minimum transmission (maximum reflection) or destructive interference is

$$2L = \left(m + \frac{1}{2}\right)\frac{\lambda}{n_2} \quad \Rightarrow L = \left(m + \frac{1}{2}\right)\frac{\lambda}{2n_2}, \quad m = 0, 1, 2, \ldots$$

The third least thickness is ($m=2$)

$$L = \left(2 + \frac{1}{2}\right)\frac{382 \text{ nm}}{2(1.75)} = 273 \text{ nm}.$$

67. In this setup, we have $n_2 < n_1$ and $n_2 < n_3$, and the condition for minimum transmission (maximum reflection) or destructive interference is

$$2L = \left(m + \frac{1}{2}\right)\frac{\lambda}{n_2} \quad \Rightarrow L = \left(m + \frac{1}{2}\right)\frac{\lambda}{2n_2}, \quad m = 0, 1, 2, \ldots$$

The second least thickness is ($m=1$)

$$L = \left(1 + \frac{1}{2}\right)\frac{587 \text{ nm}}{2(1.34)} = 329 \text{ nm}.$$

68. In this setup, we have $n_2 < n_1$ and $n_2 > n_3$, and the condition for maximum transmission (minimum reflection) or constructive interference is

$$2L = \left(m + \frac{1}{2}\right)\frac{\lambda}{n_2} \quad \Rightarrow L = \left(m + \frac{1}{2}\right)\frac{\lambda}{2n_2}, \quad m = 0, 1, 2, \ldots$$

The second least thickness is ($m=1$)

$$L = \left(1 + \frac{1}{2}\right)\frac{342 \text{ nm}}{2(1.59)} = 161 \text{ nm} .$$

69. Consider the interference of waves reflected from the top and bottom surfaces of the air film. The wave reflected from the upper surface does not change phase on reflection but the wave reflected from the bottom surface changes phase by π rad. At a place where the thickness of the air film is L, the condition for fully constructive interference is $2L = \left(m + \frac{1}{2}\right)\lambda$ where λ (= 683 nm) is the wavelength and m is an integer. This is satisfied for $m = 140$:

$$L = \frac{\left(m + \frac{1}{2}\right)\lambda}{2} = \frac{(140.5)(683 \times 10^{-9} \text{ m})}{2} = 4.80 \times 10^{-5} \text{ m} = 0.048 \text{ mm}.$$

At the thin end of the air film, there is a bright fringe. It is associated with $m = 0$. There are, therefore, 140 bright fringes in all.

70. By the condition $m\lambda = 2y$ where y is the thickness of the air-film between the plates directly underneath the middle of a dark band), the edge of the plates (the edge where they are not touching) are $y = 8\lambda/2 = 2400$ nm apart (where we have assumed that the *middle* of the ninth dark band is at the edge). Increasing that to $y' = 3000$ nm would correspond to $m' = 2y'/\lambda = 10$ (counted as the eleventh dark band, since the first one corresponds to $m = 0$). There are thus 11 dark fringes along the top plate.

71. Assume the wedge-shaped film is in air, so the wave reflected from one surface undergoes a phase change of π rad while the wave reflected from the other surface does not. At a place where the film thickness is L, the condition for fully constructive interference is $2nL = \left(m + \frac{1}{2}\right)\lambda$, where n is the index of refraction of the film, λ is the wavelength in vacuum, and m is an integer. The ends of the film are bright. Suppose the end where the film is narrow has thickness L_1 and the bright fringe there corresponds to $m = m_1$. Suppose the end where the film is thick has thickness L_2 and the bright fringe there corresponds to $m = m_2$. Since there are ten bright fringes, $m_2 = m_1 + 9$. Subtract $2nL_1 = \left(m_1 + \frac{1}{2}\right)\lambda$ from $2nL_2 = \left(m_1 + 9 + \frac{1}{2}\right)\lambda$ to obtain $2n\,\Delta L = 9\lambda$, where $\Delta L = L_2 - L_1$ is the change in the film thickness over its length. Thus,

$$\Delta L = \frac{9\lambda}{2n} = \frac{9(630 \times 10^{-9} \text{ m})}{2(1.50)} = 1.89 \times 10^{-6} \text{ m}.$$

72. We apply Eq. 35-27 to both scenarios: $m = 4001$ and $n_2 = n_{\text{air}}$, and $m = 4000$ and $n_2 = n_{\text{vacuum}} = 1.00000$:

$$2L = (4001)\frac{\lambda}{n_{\text{air}}} \quad \text{and} \quad 2L = (4000)\frac{\lambda}{1.00000} .$$

Since the $2L$ factor is the same in both cases, we set the right hand sides of these expressions equal to each other and cancel the wavelength. Finally, we obtain

$$n_{air} = (1.00000)\frac{4001}{4000} = 1.00025.$$

We remark that this same result can be obtained starting with Eq. 35-43 (which is developed in the textbook for a somewhat different situation) and using Eq. 35-42 to eliminate the $2L/\lambda$ term.

73. Using the relations of §35-7, we find that the (vertical) change between the center of one dark band and the next is

$$\Delta y = \lambda 2 = 2.5 \times 10^{-4} \text{mm}.$$

Thus, with the (horizontal) separation of dark bands given by $\Delta x = 1.2$ mm, we have

$$\theta \approx \tan\theta = \frac{\Delta y}{\Delta x} = 2.08 \times 10^{-4} \text{rad}.$$

Converting this angle into degrees, we arrive at $\theta = 0.012°$.

74. (a) The third sentence of the problem implies $m_o = 9.5$ in $2 d_o = m_o\lambda$ initially. Then, $\Delta t = 15$ s later, we have $m' = 9.0$ in $2d' = m'\lambda$. This means

$$|\Delta d| = d_o - d' = \tfrac{1}{2}(m_o\lambda - m'\lambda) = 155 \text{ nm}.$$

Thus, $|\Delta d|$ divided by Δt gives 10.3 nm/s.

(b) In this case, $m_f = 6$ so that $d_o - d_f = \tfrac{1}{2}(m_o\lambda - m_f\lambda) = \tfrac{7}{4}\lambda = 1085$ nm $= 1.09$ μm.

75. Consider the interference pattern formed by waves reflected from the upper and lower surfaces of the air wedge. The wave reflected from the lower surface undergoes a π rad phase change while the wave reflected from the upper surface does not. At a place where the thickness of the wedge is d, the condition for a maximum in intensity is $2d = (m+\tfrac{1}{2})\lambda$, where λ is the wavelength in air and m is an integer. Thus, $d = (2m + 1)\lambda/4$. As the geometry of Fig. 35-46 shows, $d = R - \sqrt{R^2 - r^2}$, where R is the radius of curvature of the lens and r is the radius of a Newton's ring. Thus, $(2m+1)\lambda/4 = R - \sqrt{R^2 - r^2}$. First, we rearrange the terms so the equation becomes

$$\sqrt{R^2 - r^2} = R - \frac{(2m+1)\lambda}{4}.$$

Next, we square both sides, rearrange to solve for r^2, then take the square root. We get

$$r = \sqrt{\frac{(2m+1)R\lambda}{2} - \frac{(2m+1)^2\lambda^2}{16}}.$$

If R is much larger than a wavelength, the first term dominates the second and

$$r = \sqrt{\frac{(2m+1)R\lambda}{2}}.$$

76. (a) We find m from the last formula obtained in problem 75:

$$m = \frac{r^2}{R\lambda} - \frac{1}{2} = \frac{(10\times10^{-3}\,\text{m})^2}{(5.0\,\text{m})(589\times10^{-9}\,\text{m})} - \frac{1}{2}$$

which (rounding down) yields $m = 33$. Since the first bright fringe corresponds to $m = 0$, $m = 33$ corresponds to the thirty-fourth bright fringe.

(b) We now replace λ by $\lambda_n = \lambda/n_w$. Thus,

$$m_n = \frac{r^2}{R\lambda_n} - \frac{1}{2} = \frac{n_w r^2}{R\lambda} - \frac{1}{2} = \frac{(1.33)(10\times10^{-3}\,\text{m})^2}{(5.0\,\text{m})(589\times10^{-9}\,\text{m})} - \frac{1}{2} = 45.$$

This corresponds to the forty-sixth bright fringe (see remark at the end of our solution in part (a)).

77. We solve for m using the formula $r = \sqrt{(2m+1)R\lambda/2}$ obtained in problem 49 and find $m = r^2/R\lambda - 1/2$. Now, when m is changed to $m + 20$, r becomes r', so

$$m + 20 = r'^2/R\lambda - 1/2.$$

Taking the difference between the two equations above, we eliminate m and find

$$R = \frac{r'^2 - r^2}{20\lambda} = \frac{(0.368\,\text{cm})^2 - (0.162\,\text{cm})^2}{20(546\times10^{-7}\,\text{cm})} = 100\,\text{cm}.$$

78. The time to change from one minimum to the next is $\Delta t = 12$ s. This involves a change in thickness $\Delta L = \lambda/2n_2$ (see Eq. 35-37), and thus a change of volume

$$\Delta V = \pi r^2 \Delta L = \frac{\pi r^2 \lambda}{2n_2} \qquad \Rightarrow \qquad \frac{dV}{dt} = \frac{\pi r^2 \lambda}{2n_2 \, \Delta t} = \frac{\pi (0.0180)^2 \, (550 \times 10^{-9})}{2(1.40) \, (12)}$$

using SI units. Thus, the rate of change of volume is 1.67×10^{-11} m^3/s.

79. A shift of one fringe corresponds to a change in the optical path length of one wavelength. When the mirror moves a distance d the path length changes by $2d$ since the light traverses the mirror arm twice. Let N be the number of fringes shifted. Then, $2d = N\lambda$ and

$$\lambda = \frac{2d}{N} = \frac{2(0.233 \times 10^{-3} \text{ m})}{792} = 5.88 \times 10^{-7} \text{ m} = 588 \text{ nm} \ .$$

80. According to Eq. 35-43, the number of fringes shifted (ΔN) due to the insertion of the film of thickness L is $\Delta N = (2L/\lambda)(n-1)$. Therefore,

$$L = \frac{\lambda \Delta N}{2(n-1)} = \frac{(589 \text{ nm})(7.0)}{2(1.40-1)} = 5.2 \ \mu\text{m} \ .$$

81. Let ϕ_1 be the phase difference of the waves in the two arms when the tube has air in it, and let ϕ_2 be the phase difference when the tube is evacuated. These are different because the wavelength in air is different from the wavelength in vacuum. If λ is the wavelength in vacuum, then the wavelength in air is λ/n, where n is the index of refraction of air. This means

$$\phi_1 - \phi_2 = 2L\left[\frac{2\pi n}{\lambda} - \frac{2\pi}{\lambda}\right] = \frac{4\pi(n-1)L}{\lambda}$$

where L is the length of the tube. The factor 2 arises because the light traverses the tube twice, once on the way to a mirror and once after reflection from the mirror. Each shift by one fringe corresponds to a change in phase of 2π rad, so if the interference pattern shifts by N fringes as the tube is evacuated,

$$\frac{4\pi(n-1)L}{\lambda} = 2N\pi$$

and

$$n = 1 + \frac{N\lambda}{2L} = 1 + \frac{60(500 \times 10^{-9} \text{ m})}{2(5.0 \times 10^{-2} \text{ m})} = 1.00030 .$$

82. We denote the two wavelengths as λ and λ', respectively. We apply Eq. 35-42 to both wavelengths and take the difference:

$$N' - N = \frac{2L}{\lambda'} - \frac{2L}{\lambda} = 2L \left(\frac{1}{\lambda'} - \frac{1}{\lambda} \right) .$$

We now require $N' - N = 1$ and solve for L:

$$L = \frac{1}{2} \left(\frac{1}{\lambda} - \frac{1}{\lambda'} \right)^{-1} = \frac{1}{2} \left(\frac{1}{589.10 \text{nm}} - \frac{1}{589.59 \text{nm}} \right)^{-1} = 3.54 \times 10^5 \text{ nm} = 354 \,\mu\text{m}.$$

83. (a) The path length difference between Rays 1 and 2 is $7d - 2d = 5d$. For this to correspond to a half-wavelength requires $5d = \lambda/2$, so that $d = 50.0$ nm.

(b) The above requirement becomes $5d = \lambda/2n$ in the presence of the solution, with $n = 1.38$. Therefore, $d = 36.2$ nm.

84. (a) Since P_1 is equidistant from S_1 and S_2 we conclude the sources are not in phase with each other. Their phase difference is $\Delta\phi_{\text{source}} = 0.60\,\pi$ rad, which may be expressed in terms of "wavelengths" (thinking of the $\lambda \Leftrightarrow 2\pi$ correspondence in discussing a full cycle) as $\Delta\phi_{\text{source}} = (0.60\,\pi / 2\pi)\,\lambda = 0.3\,\lambda$ (with S_2 "leading" as the problem states). Now S_1 is closer to P_2 than S_2 is. Source S_1 is 80 nm ($\Leftrightarrow 80/400\,\lambda = 0.2\,\lambda$) from P_2 while source S_2 is 1360 nm ($\Leftrightarrow 1360/400\,\lambda = 3.4\,\lambda$) from P_2. Here we find a difference of $\Delta\phi_{\text{path}} = 3.2\,\lambda$ (with S_1 "leading" since it is closer). Thus, the net difference is

$$\Delta\phi_{\text{net}} = \Delta\phi_{\text{path}} - \Delta\phi_{\text{source}} = 2.90\,\lambda,$$

or 2.90 wavelengths.

(b) A whole number (like 3 wavelengths) would mean fully constructive, so our result is of the following nature: intermediate, but close to fully constructive.

85. (a) Applying the law of refraction, we obtain $\sin\theta_2 / \sin\theta_1 = \sin\theta_2 / \sin 30° = v_s/v_d$. Consequently,

$$\theta_2 = \sin^{-1} \left(\frac{v_s \sin 30°}{v_d} \right) = \sin^{-1} \left[\frac{(3.0 \text{ m/s}) \sin 30°}{4.0 \text{ m/s}} \right] = 22°.$$

(b) The angle of incidence is gradually reduced due to refraction, such as shown in the calculation above (from 30° to 22°). Eventually after many refractions, θ_2 will be virtually zero. This is why most waves come in normal to a shore.

86. When the depth of the liquid (L_{liq}) is zero, the phase difference ϕ is 60 wavelengths; this must equal the difference between the number of wavelengths in length $L = 40$ μm (since the liquid initially fills the hole) of the plastic (for ray r_1) and the number in that same length of the air (for ray r_2). That is,

$$\frac{L\, n_{plastic}}{\lambda} - \frac{L\, n_{air}}{\lambda} = 60 .$$

(a) Since $\lambda = 400 \times 10^{-9}$ m and $n_{air} = 1$ (to good approximation), we find $n_{plastic} = 1.6$.

(b) The slope of the graph can be used to determine n_{liq}, but we show an approach more closely based on the above equation:

$$\frac{L\, n_{plastic}}{\lambda} - \frac{L\, n_{liq}}{\lambda} = 20$$

which makes use of the leftmost point of the graph. This readily yields $n_{liq} = 1.4$.

87. Let the $m = 10$ bright fringe on the screen be a distance y from the central maximum. Then from Fig. 35-10(a)

$$r_1 - r_2 = \sqrt{(y+d/2)^2 + D^2} - \sqrt{(y-d/2)^2 + D^2} = 10\lambda,$$

from which we may solve for y. To the order of $(d/D)^2$ we find

$$y = y_0 + \frac{y(y^2 + d^2/4)}{2D^2},$$

where $y_0 = 10D\lambda/d$. Thus, we find the percent error as follows:

$$\frac{y_0(y_0^2 + d^2/4)}{2y_0 D^2} = \frac{1}{2}\left(\frac{10\lambda}{D}\right)^2 + \frac{1}{8}\left(\frac{d}{D}\right)^2 = \frac{1}{2}\left(\frac{5.89\,\mu m}{2000\,\mu m}\right)^2 + \frac{1}{8}\left(\frac{2.0\,mm}{40\,mm}\right)^2$$

which yields 0.032%.

88. (a) The minimum path length difference occurs when both rays are nearly vertical. This would correspond to a point as far up in the picture as possible. Treating the screen as if it extended forever, then the point is at $y = \infty$.

(b) When both rays are nearly vertical, there is no path length difference between them. Thus at $y = \infty$, the phase difference is $\phi = 0$.

(c) At $y = 0$ (where the screen crosses the x axis) both rays are horizontal, with the ray from S_1 being longer than the one from S_2 by distance d.

(d) Since the problem specifies $d = 6.00\lambda$, then the phase difference here is $\phi = 6.00$ wavelengths and is at its maximum value.

(e) With $D = 20\lambda$, use of the Pythagorean theorem leads to

$$\phi = \frac{L_1 - L_2}{\lambda} = \frac{\sqrt{d^2 + (d + D)^2} - \sqrt{d^2 + D^2}}{\lambda} = 5.80$$

which means the rays reaching the point $y = d$ have a phase difference of roughly 5.8 wavelengths.

(f) The result of the previous part is "intermediate" – closer to 6 (constructive interference) than to $5\frac{1}{2}$ (destructive interference).

89. (a) In our solution here, we assume the reader has looked at our solution for problem 98. A light ray traveling directly along the central axis reaches the end in time

$$t_{direct} = \frac{L}{v_1} = \frac{n_1 L}{c}.$$

For the ray taking the critical zig-zag path, only its velocity component along the core axis direction contributes to reaching the other end of the fiber. That component is $v_1 \cos \theta'$, so the time of travel for this ray is

$$t_{zig\ zag} = \frac{L}{v_1 \cos \theta'} = \frac{n_1 L}{c\sqrt{1 - \left(\frac{1}{n_1} \sin \theta\right)^2}}$$

using results from the previous solution. Plugging in $\sin \theta = \sqrt{n_1^2 - n_2^2}$ and simplifying, we obtain

$$t_{zig\ zag} = \frac{n_1 L}{c(n_2 / n_1)} = \frac{n_1^2 L}{n_2 c}.$$

The difference $t_{zig\ zag} - t_{direct}$ readily yields the result shown in the problem statement.

(b) With $n_1 = 1.58$, $n_2 = 1.53$ and $L = 300$ m, we obtain $\Delta t = 51.6$ ns.

90. (a) The graph shows part of a periodic pattern of half-cycle "length" $\Delta n = 0.4$. Thus if we set $n = 1.0 + 2\,\Delta n = 1.8$ then the maximum at $n = 1.0$ should repeat itself there.

(b) Continuing the reasoning of part (a), adding another half-cycle "length" we get $1.8 + \Delta n = 2.2$ for the answer.

(c) Since $\Delta n = 0.4$ represents a half-cycle, then $\Delta n/2$ represents a quarter-cycle. To accumulate a total change of $2.0 - 1.0 = 1.0$ (see problem statement), then we need $2\Delta n + \Delta n/2 = 5/4^{th}$ of a cycle, which corresponds to 1.25 wavelengths.

91. The wave that goes directly to the receiver travels a distance L_1 and the reflected wave travels a distance L_2. Since the index of refraction of water is greater than that of air this last wave suffers a phase change on reflection of half a wavelength. To obtain constructive interference at the receiver, the difference $L_2 - L_1$ must be an odd multiple of a half wavelength. Consider the diagram below. The right triangle on the left, formed by the vertical line from the water to the transmitter T, the ray incident on the water, and the water line, gives $D_a = a/\tan\theta$. The right triangle on the right, formed by the vertical line from the water to the receiver R, the reflected ray, and the water line leads to $D_b = x/\tan\theta$. Since $D_a + D_b = D$,

$$\tan\theta = \frac{a+x}{D}.$$

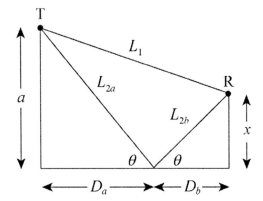

We use the identity $\sin^2\theta = \tan^2\theta\,/\,(1 + \tan^2\theta)$ to show that

$$\sin\theta = (a+x)/\sqrt{D^2 + (a+x)^2}\,.$$

This means

$$L_{2a} = \frac{a}{\sin\theta} = \frac{a\sqrt{D^2 + (a+x)^2}}{a+x}$$

and

$$L_{2b} = \frac{x}{\sin\theta} = \frac{x\sqrt{D^2 + (a+x)^2}}{a+x}.$$

Therefore,

$$L_2 = L_{2a} + L_{2b} = \frac{(a+x)\sqrt{D^2 + (a+x)^2}}{a+x} = \sqrt{D^2 + (a+x)^2}.$$

Using the binomial theorem, with D^2 large and $a^2 + x^2$ small, we approximate this expression: $L_2 \approx D + (a + x)^2 / 2D$. The distance traveled by the direct wave is $L_1 = \sqrt{D^2 + (a-x)^2}$. Using the binomial theorem, we approximate this expression: $L_1 \approx D + (a - x)^2 / 2D$. Thus,

$$L_2 - L_1 \approx D + \frac{a^2 + 2ax + x^2}{2D} - D - \frac{a^2 - 2ax + x^2}{2D} = \frac{2ax}{D}.$$

Setting this equal to $(m + \frac{1}{2})\lambda$, where m is zero or a positive integer, we find $x = (m + \frac{1}{2})(D/2a)\lambda$.

92. (a) Looking at the figure (where a portion of a periodic pattern is shown) we see that half of the periodic pattern is of length $\Delta L = 750$ nm (judging from the maximum at $x = 0$ to the minimum at $x = 750$ nm); this suggests that the wavelength (the full length of the periodic pattern) is $\lambda = 2\Delta L = 1500$ nm. A maximum should be reached again at $x = 1500$ nm (and at $x = 3000$ nm, $x = 4500$ nm, …).

(b) From our discussion in part (b), we expect a minimum to be reached at each value $x = 750$ nm $+ n(1500$ nm), where $n = 1, 2, 3 …$. For instance, for $n = 1$ we would find the minimum at $x = 2250$ nm.

(c) With $\lambda = 1500$ nm (found in part (a)), we can express $x = 1200$ nm as $x = 1200/1500 = 0.80$ wavelength.

93. $v_{min} = c/n = (2.998 \times 10^8$ m/s$)/1.54 = 1.95 \times 10^8$ m/s.

94. We note that $\Delta\phi = 60° = \frac{\pi}{3}$ rad. The phasors rotate with constant angular velocity

$$\omega = \frac{\Delta\phi}{\Delta t} = \frac{\pi/3}{2.5 \times 10^{-16}} = 4.19 \times 10^{15} \text{ rad/s}.$$

Since we are working with light waves traveling in a medium (presumably air) where the wave speed is approximately c, then $kc = \omega$ (where $k = 2\pi/\lambda$), which leads to

$$\lambda = \frac{2\pi c}{\omega} = 450 \text{ nm} .$$

95. We infer from Sample Problem 35-2, that (with angle in radians)

$$\Delta\theta = \frac{\lambda}{d}$$

for adjacent fringes. With the wavelength change ($\lambda' = \lambda/n$ by Eq. 35-8), this equation becomes

$$\Delta\theta' = \frac{\lambda'}{d} .$$

Dividing one equation by the other, the requirement of *radians* can now be relaxed and we obtain

$$\frac{\Delta\theta'}{\Delta\theta} = \frac{\lambda'}{\lambda} = \frac{1}{n} .$$

Therefore, with $n = 1.33$ and $\Delta\theta = 0.30°$, we find $\Delta\theta' = 0.23°$.

96. We note that ray 1 travels an extra distance $4L$ more than ray 2. For constructive interference (which is obtained for $\lambda = 620$ nm) we require

$$4L = m\lambda \quad \text{where } m = \text{some positive integer} .$$

For destructive interference (which is obtained for $\lambda' = 496$ nm) we require

$$4L = \frac{k}{2}\lambda' \quad \text{where } k = \text{some positive } \underline{\text{odd}} \text{ integer} .$$

Equating these two equations (since their left-hand sides are equal) and rearranging, we obtain

$$k = 2 m \frac{\lambda}{\lambda'} = 2 m \frac{620}{496} = 2.5 m .$$

We note that this condition is satisfied for $k = 5$ and $m = 2$. It is satisfied for some larger values, too, but – recalling that we want the least possible value for L – we choose the solution set $(k, m) = (5, 2)$. Plugging back into either of the equations above, we obtain the distance L:

$$4L = 2\lambda \quad \Rightarrow \quad L = \frac{\lambda}{2} = 310.0 \text{ nm} .$$

97. (a) The path length difference is 0.5 μm = 500 nm, which is represents 500/400 = 1.25 wavelengths — that is, a meaningful difference of 0.25 wavelengths. In angular measure, this corresponds to a phase difference of $(0.25)2\pi = \pi/2$ radians ≈ 1.6 rad.

(b) When a difference of index of refraction is involved, the approach used in Eq. 35-9 is quite useful. In this approach, we count the wavelengths between S_1 and the origin

$$N_1 = \frac{Ln}{\lambda} + \frac{L'n'}{\lambda}$$

where $n = 1$ (rounding off the index of air), $L = 5.0$ μm, $n' = 1.5$ and $L' = 1.5$ μm. This yields $N_1 = 18.125$ wavelengths. The number of wavelengths between S_2 and the origin is (with $L_2 = 6.0$ μm) given by

$$N_2 = \frac{L_2 n}{\lambda} = 15.000.$$

Thus, $N_1 - N_2 = 3.125$ wavelengths, which gives us a meaningful difference of 0.125 wavelength and which "converts" to a phase of $\pi/4$ radian ≈ 0.79 rad.

98. (a) The difference in wavelengths, with and without the $n = 1.4$ material, is found using Eq. 35-9:

$$\Delta N = \frac{L(n - 1)}{\lambda} = 1.143.$$

The result is equal to a phase shift of $(1.143)(360°) = 411.4°$, or

(b) more meaningfully -- a shift of $411.4° - 360° = 51.4°$.

99. Using Eq. 35-16 with the small-angle approximation (illustrated in Sample Problem 35-2), we arrive at

$$y = \frac{\left(m + \frac{1}{2}\right)\lambda D}{d}$$

for the position of the $(m + 1)^{th}$ dark band (a simple way to get this is by averaging the expressions in Eq. 35-17 and Eq. 35-18). Thus, with $m = 1$, $y = 0.012$ m and $d = 800\lambda$, we find $D = 6.4$ m.

100. (a) We are dealing with a symmetric situation (with the film index $n_2 = 1.5$ being less than that of the materials bounding it), and with reflected light, so Eqs. 35-36 and -37 apply *with* their stated applicability. Both can be written in the form

$$\frac{2n_2L}{\lambda} = \begin{cases} \text{half-integer for bright} \\ \text{integer for dark} \end{cases}$$

Thus, we find $2n_2L/\lambda = 3$, so that we find the middle of a dark band at the left edge of the figure. Since there is nothing beyond this "middle" then a more appropriate phrasing is that there is half of a dark band next to the left edge, being darkest precisely at the edge.

(b) The right edge, where they touch, satisfies the dark reflection condition for $L = 0$ (where $m = 0$), so there is (essentially half of) a dark band at the right end.

(c) Counting half-bands and whole bands alike, we find four dark bands: $(m = 0, 1, 2, 3)$.

101. (a) In this case, the film has a smaller index material on one side (air) and a larger index material on the other (glass), and we are dealing (in part (a)) with strongly transmitted light, so the condition is given by Eq. 35-37 (which would give dark *reflection* in this scenario)

$$L = \frac{\lambda}{2n_2}\left(m + \frac{1}{2}\right) = 110 \text{ nm}$$

for $n_2 = 1.25$ and $m = 0$.

(b) Now, we are dealing with strongly reflected light, so the condition is given by Eq. 35-36 (which would give no *transmission* in this scenario)

$$L = \frac{m\lambda}{2n_2} = 220 \text{ nm}$$

for $n_2 = 1.25$ and $m = 1$ (the $m = 0$ option is excluded in the problem statement).

102. We adapt the result of problem 21. Now, the phase difference in radians is

$$\frac{2\pi t}{\lambda}(n_2 - n_1) = 2m\pi.$$

The problem implies $m = 5$, so the thickness is

$$t = \frac{m\lambda}{n_2 - n_1} = \frac{5(480 \text{ nm})}{1.7 - 1.4} = 8.0 \times 10^3 \text{ nm} = 8.0 \mu\text{m}.$$

103. (a) Since $n_2 > n_3$, this case has no π-phase shift, and the condition for constructive interference is $m\lambda = 2Ln_2$. We solve for L:

$$L = \frac{m\lambda}{2n_2} = \frac{m(525 \text{ nm})}{2(1.55)} = (169 \text{ nm})m.$$

For the minimum value of L, let $m = 1$ to obtain $L_{min} = 169$ nm.

(b) The light of wavelength λ (other than 525 nm) that would also be preferentially transmitted satisfies $m'\lambda = 2n_2L$, or

$$\lambda = \frac{2n_2 L}{m'} = \frac{2(1.55)(169 \text{ nm})}{m'} = \frac{525 \text{ nm}}{m'}.$$

Here $m' = 2,3,4,\ldots$ (note that $m' = 1$ corresponds to the $\lambda = 525$ nm light, so it should not be included here). Since the minimum value of m' is 2, one can easily verify that no m' will give a value of λ which falls into the visible light range. So no other parts of the visible spectrum will be preferentially transmitted. They are, in fact, reflected.

(c) For a sharp reduction of transmission let

$$\lambda = \frac{2n_2 L}{m' + 1/2} = \frac{525 \text{ nm}}{m' + 1/2},$$

where $m' = 0, 1, 2, 3, \ldots$. In the visible light range $m' = 1$ and $\lambda = 350$ nm. This corresponds to the blue-violet light.

104. (a) Straightforward application of Eq. 35-3 and $v = \Delta x/\Delta t$ yields the result: film 1.

(b) The traversal time is equal to 4.0×10^{-15} s.

(c) Use of Eq. 35-9 leads to the number of wavelengths:

$$N = \frac{L_1 n_1 + L_2 n_2 + L_3 n_3}{\lambda} = 7.5.$$

105. (a) Following Sample Problem 35-1, we have

$$N_2 - N_1 = \frac{L}{\lambda}(n_2 - n_1) = 1.87$$

which represents a meaningful difference of 0.87 wavelength.

(b) The result in part (a) is closer to 1 wavelength (constructive interference) than it is to 1/2 wavelength (destructive interference) so the latter choice applies.

(c) This would insert a ±1/2 wavelength into the previous result — resulting in a meaningful difference (between the two rays) equal to 0.87 – 0.50 = 0.37 wavelength.

(d) The result in part (c) is closer to the destructive interference condition. Thus, there is intermediate illumination but closer to darkness.

106. (a) With $\lambda = 0.5$ μm, Eq. 35-14 leads to

$$\theta = \sin^{-1}\frac{(3)(0.5\,\mu\text{m})}{2.00\,\mu\text{m}} = 48.6°.$$

(b) Decreasing the frequency means increasing the wavelength — which implies y increases, and the third side bright fringe moves away from the center of the pattern. Qualitatively, this is easily seen with Eq. 35-17. One should exercise caution in appealing to Eq. 35-17 here, due to the fact the small angle approximation is not justified in this problem.

(c) The new wavelength is $0.5/0.9 = 0.556$ μm, which produces a new angle of

$$\theta = \sin^{-1}\frac{(3)(0.556\,\mu\text{m})}{2.00\,\mu\text{m}} = 56.4°.$$

Using $y = D\tan\theta$ for the old and new angles, and subtracting, we find

$$\Delta y = D(\tan 56.4° - \tan 48.6°) = 1.49\,\text{m}.$$

107. (a) A path length difference of $\lambda/2$ produces the first dark band, of $3\lambda/2$ produces the second dark band, and so on. Therefore, the fourth dark band corresponds to a path length difference of $7\lambda/2 = 1750$ nm $= 1.75$ μm.

(b) In the small angle approximation (which we assume holds here), the fringes are equally spaced, so that if Δy denotes the distance from one maximum to the next, then the distance from the middle of the pattern to the fourth dark band must be 16.8 mm $= 3.5\,\Delta y$. Therefore, we obtain $\Delta y = 16.8/3.5 = 4.8$ mm.

108. In the case of a distant screen the angle θ is close to zero so $\sin\theta \approx \theta$. Thus from Eq. 35-14,

$$\Delta\theta \approx \Delta\sin\theta = \Delta\left(\frac{m\lambda}{d}\right) = \frac{\lambda}{d}\Delta m = \frac{\lambda}{d},$$

or $d \approx \lambda/\Delta\theta = 589 \times 10^{-9}$ m/0.018 rad $= 3.3 \times 10^{-5}$ m $= 33\ \mu$m.

109. (a) Straightforward application of Eq. 35-3 $n = c/v$ and $v = \Delta x/\Delta t$ yields the result: pistol 1 with a time equal to $\Delta t = n\Delta x/c = 42.0 \times 10^{-12}$ s.

(b) For pistol 2, the travel time is equal to 42.3×10^{-12} s.

(c) For pistol 3, the travel time is equal to 43.2×10^{-12} s.

(d) For pistol 4 the travel time is equal to 41.8×10^{-12} s.

(e) We see that the blast from pistol 4 arrives first.

110. We use Eq. 35-36 for constructive interference: $2n_2L = (m + 1/2)\lambda$, or

$$\lambda = \frac{2n_2 L}{m+1/2} = \frac{2(1.50)(410\,\text{nm})}{m+1/2} = \frac{1230\ \text{nm}}{m+1/2},$$

where $m = 0, 1, 2, \ldots$. The only value of m which, when substituted into the equation above, would yield a wavelength which falls within the visible light range is $m = 1$. Therefore,

$$\lambda = \frac{1230\,\text{nm}}{1+1/2} = 492\ \text{nm}.$$

111. For the first maximum $m = 0$ and for the tenth one $m = 9$. The separation is $\Delta y = (D\lambda/d)\Delta m = 9D\lambda/d$. We solve for the wavelength:

$$\lambda = \frac{d\Delta y}{9D} = \frac{(0.15\times10^{-3}\,\text{m})(18\times10^{-3}\,\text{m})}{9(50\times10^{-2}\,\text{m})} = 6.0\times10^{-7}\,\text{m} = 600\ \text{nm}.$$

112. Light reflected from the upper oil surface (in contact with air) changes phase by π rad. Light reflected from the lower surface (in contact with glass) changes phase by π rad if the index of refraction of the oil is less than that of the glass and does not change phase if the index of refraction of the oil is greater than that of the glass.

• First, suppose the index of refraction of the oil is greater than the index of refraction of the glass. The condition for fully destructive interference is $2n_od = m\lambda$, where d is the thickness of the oil film, n_o is the index of refraction of the oil, λ is the wavelength in vacuum, and m is an integer. For the shorter wavelength, $2n_od = m_1\lambda_1$ and for the longer, $2n_od = m_2\lambda_2$. Since λ_1 is less than λ_2, m_1 is greater than m_2, and since fully destructive interference does not occur for any wavelengths between, $m_1 = m_2 + 1$. Solving $(m_2 + 1)\lambda_1 = m_2\lambda_2$ for m_2, we obtain

$$m_2 = \frac{\lambda_1}{\lambda_2 - \lambda_1} = \frac{500\,\text{nm}}{700\,\text{nm} - 500\,\text{nm}} = 2.50.$$

Since m_2 must be an integer, the oil cannot have an index of refraction that is greater than that of the glass.

• Now suppose the index of refraction of the oil is less than that of the glass. The condition for fully destructive interference is then $2n_o d = (2m + 1)\lambda$. For the shorter wavelength, $2m_o d = (2m_1 + 1)\lambda_1$, and for the longer, $2n_o d = (2m_2 + 1)\lambda_2$. Again, $m_1 = m_2 + 1$, so $(2m_2 + 3)\lambda_1 = (2m_2 + 1)\lambda_2$. This means the value of m_2 is

$$m_2 = \frac{3\lambda_1 - \lambda_2}{2(\lambda_2 - \lambda_1)} = \frac{3(500\,\text{nm}) - 700\,\text{nm}}{2(700\,\text{nm} - 500\,\text{nm})} = 2.00.$$

This is an integer. Thus, the index of refraction of the oil is less than that of the glass.

113. We use the formula obtained in Sample Problem 35-6:

$$L_{\text{min}} = \frac{\lambda}{4n_2} = \frac{\lambda}{4(1.25)} = 0.200\lambda \quad \Rightarrow \quad \frac{L_{\text{min}}}{\lambda} = 0.200.$$

114. We use Eq. 35-36:

$$L_{16} = \left(16 + \frac{1}{2}\right)\frac{\lambda}{2n_2}$$

$$L_6 = \left(6 + \frac{1}{2}\right)\frac{\lambda}{2n_2}$$

The difference between these, using the fact that $n_2 = n_{\text{air}} = 1.0$, is

$$L_{16} - L_6 = (10)\frac{480\,\text{nm}}{2(1.0)} = 2400\,\text{nm} = 2.4\,\mu\text{m}.$$

115. Let the position of the mirror measured from the point at which $d_1 = d_2$ be x. We assume the beam-splitting mechanism is such that the two waves interfere constructively for $x = 0$ (with some beam-splitters, this would not be the case). We can adapt Eq. 35-23 to this situation by incorporating a factor of 2 (since the interferometer utilizes directly reflected light in contrast to the double-slit experiment) and eliminating the sin θ factor. Thus, the phase difference between the two light paths is $\Delta\phi = 2(2\pi x/\lambda) = 4\pi x/\lambda$. Then from Eq. 35-22 (writing $4I_0$ as I_m) we find

$$I = I_m \cos^2\left(\frac{\Delta\phi}{2}\right) = I_m \cos^2\left(\frac{2\pi x}{\lambda}\right).$$

116. The index of refraction of fused quartz at $\lambda = 550$ nm is about 1.459, obtained from Fig. 34-19. Thus, from Eq. 35-3, we find

$$v = \frac{c}{n} = \frac{2.998 \times 10^8 \text{ m/s}}{1.459} = 2.06 \times 10^8 \text{ m/s} \approx 2.1 \times 10^8 \text{ m/s}.$$

117. (a) We use $\Delta y = D\lambda/d$ (see Sample Problem 35-2). Because of the placement of the mirror in the problem $D = 2(20.0 \text{ m}) = 40.0$ m, which we express in millimeters in the calculation below:

$$d = \frac{D\lambda}{\Delta y} = \frac{(4.00 \times 10^4 \text{ mm})(632.8 \times 10^{-6} \text{ mm})}{100 \text{ mm}} = 0.253 \text{ mm} .$$

(b) In this case the interference pattern will be shifted. At the location of the original central maximum, the effective phase difference is now $\frac{1}{2}$ wavelength, so there is now a minimum instead of a maximum.

118. (a) Dividing Eq. 35-12 by the wavelength, we obtain

$$N = \frac{\Delta L}{\lambda} = \frac{d}{\lambda}\sin\theta = 39.6$$

wavelengths.

(b) This is close to a half-integer value (destructive interference), so that the correct response is "intermediate illumination but closer to darkness."

119. We adapt Eq. 35-21 to the non-reflective coating on a glass lens: $I = I_{max} \cos^2 (\phi/2)$, where $\phi = (2\pi/\lambda)(2n_2 L) + \pi$.

(a) At $\lambda = 450$ nm

$$\frac{I}{I_{max}} = \cos^2\left(\frac{\phi}{2}\right) = \cos^2\left(\frac{2\pi n_2 L}{\lambda} + \frac{\pi}{2}\right) = \cos^2\left[\frac{2\pi(1.38)(99.6\text{nm})}{450\text{nm}} + \frac{\pi}{2}\right] = 0.883 \approx 88\%.$$

(b) At $\lambda = 650$ nm

$$\frac{I}{I_{max}} = \cos^2\left[\frac{2\pi(1.38)(99.6\text{nm})}{650\text{nm}} + \frac{\pi}{2}\right] = 0.942 \approx 94\%.$$

120. (a) Every time one more destructive (constructive) fringe appears the increase in thickness of the air gap is $\lambda/2$. Now that there are 6 more destructive fringes in addition to the one at point A, the thickness at B is $t_B = 6(\lambda/2) = 3(600 \text{ nm}) = 1.80 \ \mu\text{m}$.

(b) We must now replace λ by $\lambda' = \lambda/n_w$. Since t_B is unchanged $t_B = N(\lambda'/2) = N(\lambda/2n_w)$, or

$$N = \frac{2t_B n_w}{\lambda} = \frac{2(3\lambda)n_w}{\lambda} = 6n_w = 6(1.33) = 8 \ .$$

Counting the one at point A, a total of nine dark fringes will be observed.

121. We take the electric field of one wave, at the screen, to be

$$E_1 = E_0 \sin(\omega t)$$

and the electric field of the other to be

$$E_2 = 2E_0 \sin(\omega t + \phi) \ ,$$

where the phase difference is given by

$$\phi = \left(\frac{2\pi d}{\lambda} \right) \sin\theta.$$

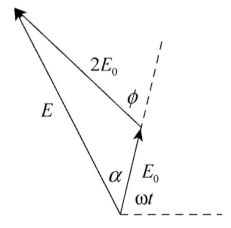

Here d is the center-to-center slit separation and λ is the wavelength. The resultant wave can be written $E = E_1 + E_2 = E\sin(\omega t + \alpha)$, where α is a phase constant. The phasor diagram is shown above. The resultant amplitude E is given by the trigonometric law of cosines:

$$E^2 = E_0^2 + (2E_0)^2 - 4E_0^2 \cos(180° - \phi) = E_0^2(5 + 4\cos\phi) \ .$$

The intensity is given by $I = I_0(5 + 4\cos\phi)$, where I_0 is the intensity that would be produced by the first wave if the second were not present. Since $\cos\phi = 2\cos^2(\phi/2) - 1$, this may also be written $I = I_0\left[1 + 8\cos^2(\phi/2)\right]$.

122. (a) To get to the detector, the wave from S_1 travels a distance x and the wave from S_2 travels a distance $\sqrt{d^2 + x^2}$. The phase difference (in terms of wavelengths) between the two waves is

$$\sqrt{d^2 + x^2} - x = m\lambda \quad m = 0, 1, 2, \ldots$$

where we are requiring constructive interference. The solution is

$$x = \frac{d^2 - m^2\lambda^2}{2m\lambda}.$$

We see that setting $m = 0$ in this expression produces $x = \infty$; hence, the phase difference between the waves when P is very far away is 0.

(b) The result of part (a) implies that the waves constructively interfere at P.

(c) As is particularly evident from our results in part (d), the phase difference increases as x decreases.

The condition for constructive interference is $\phi = 2\pi m$ or $\Delta L = m\lambda$ in, and the condition for destructive interference is $\phi = 2\pi(m + 1/2)$ or $\Delta L = (m + 1/2)\lambda$, with $m = 0, 1, 2, \ldots$

For parts (d) – (o), we can use our formula from part (a) for the 0.5λ, 1.50λ, etc. differences by allowing m in our formula to take on half-integer values. The half-integer values, though, correspond to destructive interference.

(d) When the phase difference is $\phi = 0$, the interference is fully constructive,

(e) and the interference occurs at $x = \infty$.

(f) When $\Delta L = 0.500\lambda$ $(m = 1/2)$, the interference is fully destructive.

(g) Using the values $\lambda = 0.500$ μm and $d = 2.00$ μm, we find $x = 7.88$ μm for $m = 1/2$.

(h) When $\Delta L = 1.00\lambda$ $(m = 1)$, the interference is fully constructive.

(i) Using the formula obtained in part (a), we have $x = 3.75$ μm for $m = 1$.

(j) When $\Delta L = 1.500\lambda$ $(m = 3/2)$, the interference is fully destructive.

(k) Using the formula obtained in part (a), we have $x = 2.29$ μm for $m = 3/2$.

(l) When $\Delta L = 2.00\lambda$ $(m = 2)$, the interference is fully constructive.

(m) Using the formula obtained in part (a), we have $x = 1.50$ μm for $m = 2$.

(n) When $\Delta L = 2.500\lambda$ $(m = 5/2)$, the interference is fully destructive.

(o) Using the formula obtained in part (a), we have $x = 0.975$ μm for $m = 5/2$.

123. (a) The binomial theorem (Appendix E) allows us to write

$$\sqrt{k(1+x)} = \sqrt{k}\left(1 + \frac{x}{2} + \frac{x^2}{8} + \frac{3x^3}{48} + \cdots\right) \approx \sqrt{k} + \frac{x}{2}\sqrt{k}$$

for $x \ll 1$. Thus, the end result from the solution of problem 49 yields

$$r_m = \sqrt{R\lambda m\left(1 + \frac{1}{2m}\right)} \approx \sqrt{R\lambda m} + \frac{1}{4m}\sqrt{R\lambda m}$$

and

$$r_{m+1} = \sqrt{R\lambda m\left(1 + \frac{3}{2m}\right)} \approx \sqrt{R\lambda m} + \frac{3}{4m}\sqrt{R\lambda m}$$

for very large values of m. Subtracting these, we obtain

$$\Delta r = \frac{3}{4m}\sqrt{R\lambda m} - \frac{1}{4m}\sqrt{R\lambda m} = \frac{1}{2}\sqrt{\frac{R\lambda}{m}} .$$

(b) We take the differential of the area: $dA = d(\pi r^2) = 2\pi r\, dr$, and replace dr with Δr in anticipation of using the result from part (a). Thus, the area between adjacent rings for large values of m is

$$2\pi r_m(\Delta r) \approx 2\pi\left(\sqrt{R\lambda m} + \frac{1}{4m}\sqrt{R\lambda m}\right)\left(\frac{1}{2}\sqrt{\frac{R\lambda}{m}}\right) \approx 2\pi\left(\sqrt{R\lambda m}\right)\left(\frac{1}{2}\sqrt{\frac{R\lambda}{m}}\right)$$

which simplifies to the desired result $(\pi\lambda R)$.

124. The *Hint* essentially answers the question, but we put in some algebraic details and arrive at the familiar analytic-geometry expression for a hyperbola. The distance $d/2$ is denoted a and the constant value for the path length difference is denoted $\phi = r_1 - r_2$, or

$$\sqrt{(a+x)^2 + y^2} - \sqrt{(a-x)^2 + y^2} = \phi$$

Rearranging and squaring, we have

$$\left(\sqrt{(a+x)^2 + y^2}\right)^2 = \left(\sqrt{(a-x)^2 + y^2} + \phi\right)^2$$

$$a^2 + 2ax + x^2 + y^2 = a^2 - 2ax + x^2 + y^2 + \phi^2 + 2\phi\sqrt{(a-x)^2 + y^2}$$

Many terms on both sides are identical and may be eliminated. This leaves us with

$$-2\phi\sqrt{(a-x)^2 + y^2} = \phi^2 - 4ax$$

at which point we square both sides again:

$$4\phi^2 a^2 - 8\phi^2 ax + 4\phi^2 x^2 + 4\phi^2 y^2 = \phi^4 - 8\phi^2 ax + 16a^2 x^2$$

We eliminate the $-8\phi^2 ax$ term from both sides and plug in $a = 2d$ to get back to the original notation used in the problem statement:

$$\phi^2 d^2 + 4\phi^2 x^2 + 4\phi^2 y^2 = \phi^4 + 4d^2 x^2$$

Then a simple rearrangement puts it in the familiar analytic geometry format:

$$\phi^2 d^2 - \phi^4 = 4(d^2 - \phi^2)x^2 - 4\phi^2 y^2$$

which can be further simplified by dividing through by $\phi^2 d^2 - \phi^4$.

Chapter 36

1. The condition for a minimum of a single-slit diffraction pattern is

$$a \sin \theta = m\lambda$$

where a is the slit width, λ is the wavelength, and m is an integer. The angle θ is measured from the forward direction, so for the situation described in the problem, it is 0.60° for $m = 1$. Thus

$$a = \frac{m\lambda}{\sin \theta} = \frac{633 \times 10^{-9}\,\text{m}}{\sin 0.60°} = 6.04 \times 10^{-5}\,\text{m} \; .$$

2. (a) $\theta = \sin^{-1}(1.50\,\text{cm}/2.00\,\text{m}) = 0.430°$.

(b) For the mth diffraction minimum $a \sin \theta = m\lambda$. We solve for the slit width:

$$a = \frac{m\lambda}{\sin \theta} = \frac{2(441\,\text{nm})}{\sin 0.430°} = 0.118\,\text{mm} \; .$$

3. (a) The condition for a minimum in a single-slit diffraction pattern is given by $a \sin \theta = m\lambda$, where a is the slit width, λ is the wavelength, and m is an integer. For $\lambda = \lambda_a$ and $m = 1$, the angle θ is the same as for $\lambda = \lambda_b$ and $m = 2$. Thus $\lambda_a = 2\lambda_b = 2(350\,\text{nm}) = 700\,\text{nm}$.

(b) Let m_a be the integer associated with a minimum in the pattern produced by light with wavelength λ_a, and let m_b be the integer associated with a minimum in the pattern produced by light with wavelength λ_b. A minimum in one pattern coincides with a minimum in the other if they occur at the same angle. This means $m_a\lambda_a = m_b\lambda_b$. Since $\lambda_a = 2\lambda_b$, the minima coincide if $2m_a = m_b$. Consequently, every other minimum of the λ_b pattern coincides with a minimum of the λ_a pattern. With $m_a = 2$, we have $m_b = 4$.

(c) With $m_a = 3$, we have $m_b = 6$.

4. (a) Eq. 36-3 and Eq. 36-12 imply smaller angles for diffraction for smaller wavelengths. This suggests that diffraction effects in general would decrease.

(b) Using Eq. 36-3 with $m = 1$ and solving for 2θ (the angular width of the central diffraction maximum), we find

$$2\theta = 2\sin^{-1}\left(\frac{\lambda}{a}\right) = 2\sin^{-1}\left(\frac{0.50\,\text{m}}{5.0\,\text{m}}\right) = 11°.$$

(c) A similar calculation yields 0.23° for $\lambda = 0.010$ m.

5. (a) A plane wave is incident on the lens so it is brought to focus in the focal plane of the lens, a distance of 70 cm from the lens.

(b) Waves leaving the lens at an angle θ to the forward direction interfere to produce an intensity minimum if $a \sin\theta = m\lambda$, where a is the slit width, λ is the wavelength, and m is an integer. The distance on the screen from the center of the pattern to the minimum is given by $y = D \tan\theta$, where D is the distance from the lens to the screen. For the conditions of this problem,

$$\sin\theta = \frac{m\lambda}{a} = \frac{(1)(590\times 10^{-9}\,\text{m})}{0.40\times 10^{-3}\,\text{m}} = 1.475\times 10^{-3}.$$

This means $\theta = 1.475 \times 10^{-3}$ rad and

$$y = (70\times 10^{-2}\,\text{m})\tan(1.475\times 10^{-3}\,\text{rad}) = 1.0\times 10^{-3}\,\text{m}.$$

6. (a) We use Eq. 36-3 to calculate the separation between the first ($m_1 = 1$) and fifth ($m_2 = 5$) minima:

$$\Delta y = D\Delta\sin\theta = D\Delta\left(\frac{m\lambda}{a}\right) = \frac{D\lambda}{a}\Delta m = \frac{D\lambda}{a}(m_2 - m_1).$$

Solving for the slit width, we obtain

$$a = \frac{D\lambda(m_2 - m_1)}{\Delta y} = \frac{(400\,\text{mm})(550\times 10^{-6}\,\text{mm})(5-1)}{0.35\,\text{mm}} = 2.5\,\text{mm}.$$

(b) For $m = 1$,

$$\sin\theta = \frac{m\lambda}{a} = \frac{(1)(550\times 10^{-6}\,\text{mm})}{2.5\,\text{mm}} = 2.2\times 10^{-4}.$$

The angle is $\theta = \sin^{-1}(2.2\times 10^{-4}) = 2.2\times 10^{-4}$ rad.

7. The condition for a minimum of intensity in a single-slit diffraction pattern is $a \sin\theta = m\lambda$, where a is the slit width, λ is the wavelength, and m is an integer. To find the angular position of the first minimum to one side of the central maximum, we set $m = 1$:

$$\theta_1 = \sin^{-1}\left(\frac{\lambda}{a}\right) = \sin^{-1}\left(\frac{589 \times 10^{-9}\,\text{m}}{1.00 \times 10^{-3}\,\text{m}}\right) = 5.89 \times 10^{-4}\,\text{rad}.$$

If D is the distance from the slit to the screen, the distance on the screen from the center of the pattern to the minimum is

$$y_1 = D\tan\theta_1 = (3.00\,\text{m})\tan(5.89 \times 10^{-4}\,\text{rad}) = 1.767 \times 10^{-3}\,\text{m}.$$

To find the second minimum, we set $m = 2$:

$$\theta_2 = \sin^{-1}\left(\frac{2(589 \times 10^{-9}\,\text{m})}{1.00 \times 10^{-3}\,\text{m}}\right) = 1.178 \times 10^{-3}\,\text{rad}.$$

The distance from the center of the pattern to this second minimum is

$$y_2 = D\tan\theta_2 = (3.00\,\text{m})\tan(1.178 \times 10^{-3}\,\text{rad}) = 3.534 \times 10^{-3}\,\text{m}.$$

The separation of the two minima is

$$\Delta y = y_2 - y_1 = 3.534\,\text{mm} - 1.767\,\text{mm} = 1.77\,\text{mm}.$$

8. From $y = m\lambda L/a$ we get

$$\Delta y = \Delta\left(\frac{m\lambda L}{a}\right) = \frac{\lambda L}{a}\Delta m = \frac{(632.8\,\text{nm})(2.60)}{1.37\,\text{mm}}[10-(-10)] = 24.0\,\text{mm}.$$

9. We note that nm $= 10^{-9}$ m $= 10^{-6}$ mm. From Eq. 36-4,

$$\Delta\phi = \left(\frac{2\pi}{\lambda}\right)(\Delta x \sin\theta) = \left(\frac{2\pi}{589 \times 10^{-6}\,\text{mm}}\right)\left(\frac{0.10\,\text{mm}}{2}\right)\sin 30° = 266.7\,\text{rad}.$$

This is equivalent to $266.7 - 84\pi = 2.8\,\text{rad} = 160°$.

10. (a) The slope of the plotted line is 12, and we see from Eq. 36-6 that this slope should correspond to

$$\frac{\pi a}{\lambda} = 12 \quad\Rightarrow\quad a = 2330\,\text{nm} = 2.33\,\mu\text{m}.$$

(b) Consider Eq. 36-3 with "continuously variable" m (of course, m should be an integer for diffraction minima, but for the moment we will solve for it as if it could be any real number):

$$m_{\text{max}} = \frac{a}{\lambda} (\sin \theta)_{\text{max}} \approx 3.8$$

which suggests that, on each side of the central maximum ($\theta_{\text{centr}} = 0$), there are three minima; considering both sides then implies there are six minima in the pattern.

(c) Setting $m = 1$ in Eq. 36-3 and solving for θ yields 15.2°.

(d) Setting $m = 3$ in Eq. 36-3 and solving for θ yields 51.8°.

11. (a) $\theta = \sin^{-1} (0.011 \text{ cm}/3.5 \text{ m}) = 0.18°$.

(b) We use Eq. 36-6:

$$\alpha = \left(\frac{\pi a}{\lambda} \right) \sin \theta = \frac{\pi (0.025 \text{ mm}) \sin 0.18°}{538 \times 10^{-6} \text{ mm}} = 0.46 \text{ rad} .$$

(c) Making sure our calculator is in radian mode, Eq. 36-5 yields

$$\frac{I(\theta)}{I_m} = \left(\frac{\sin \alpha}{\alpha} \right)^2 = 0.93 .$$

12. We will make use of arctangents and sines in our solution, even though they can be "shortcut" somewhat since the angles are small enough to justify the use of the small angle approximation.

(a) Given $y/D = 15/300$ (both expressed here in centimeters), then $\theta = \tan^{-1}(y/D) = 2.86°$. Use of Eq. 36-6 (with $a = 6000$ nm and $\lambda = 500$ nm) leads to

$$\alpha = \frac{\pi a}{\lambda} \sin \theta = 1.883 \text{ rad}$$

Thus,

$$\frac{I_P}{I_m} = \left(\frac{\sin \alpha}{\alpha} \right)^2 = 0.256 .$$

(b) Consider Eq. 36-3 with "continuously variable" m (of course, m should be an integer for diffraction minima, but for the moment we will solve for it as if it could be any real number):

$$m = \frac{a}{\lambda} \sin \theta \approx 0.6$$

which suggests that the angle takes us to a point between the central maximum ($\theta_{centr} = 0$) and the first minimum (which corresponds to $m = 1$ in Eq. 36-3).

13. (a) The intensity for a single-slit diffraction pattern is given by

$$I = I_m \frac{\sin^2 \alpha}{\alpha^2}$$

where $\alpha = (\pi a/\lambda) \sin \theta$, a is the slit width and λ is the wavelength. The angle θ is measured from the forward direction. We require $I = I_m/2$, so

$$\sin^2 \alpha = \frac{1}{2}\alpha^2 .$$

(b) We evaluate $\sin^2 \alpha$ and $\alpha^2/2$ for $\alpha = 1.39$ rad and compare the results. To be sure that 1.39 rad is closer to the correct value for α than any other value with three significant digits, we could also try 1.385 rad and 1.395 rad.

(c) Since $\alpha = (\pi a/\lambda) \sin \theta$,

$$\theta = \sin^{-1}\left(\frac{\alpha\lambda}{\pi a}\right).$$

Now $\alpha/\pi = 1.39/\pi = 0.442$, so

$$\theta = \sin^{-1}\left(\frac{0.442\lambda}{a}\right).$$

The angular separation of the two points of half intensity, one on either side of the center of the diffraction pattern, is

$$\Delta\theta = 2\theta = 2\sin^{-1}\left(\frac{0.442\lambda}{a}\right).$$

(d) For $a/\lambda = 1.0$,

$$\Delta\theta = 2\sin^{-1}(0.442/1.0) = 0.916\,\text{rad} = 52.5°.$$

(e) For $a/\lambda = 5.0$,

$$\Delta\theta = 2\sin^{-1}(0.442/5.0) = 0.177\,\text{rad} = 10.1°.$$

(f) For $a/\lambda = 10$, $\Delta\theta = 2\sin^{-1}(0.442/10) = 0.0884\,\text{rad} = 5.06°$.

14. Consider Huygens' explanation of diffraction phenomena. When A is in place only the Huygens' wavelets that pass through the hole get to point P. Suppose they produce a resultant electric field E_A. When B is in place, the light that was blocked by A gets to P and the light that passed through the hole in A is blocked. Suppose the electric field at P is now $\vec{E}_B$. The sum $\vec{E}_A + \vec{E}_B$ is the resultant of all waves that get to P when neither A nor B are present. Since P is in the geometric shadow, this is zero. Thus $\vec{E}_A = -\vec{E}_B$, and since the intensity is proportional to the square of the electric field, the intensity at P is the same when A is present as when B is present.

15. (a) The intensity for a single-slit diffraction pattern is given by

$$I = I_m \frac{\sin^2 \alpha}{\alpha^2}$$

where α is described in the text (see Eq. 36-6). To locate the extrema, we set the derivative of I with respect to α equal to zero and solve for α. The derivative is

$$\frac{dI}{d\alpha} = 2I_m \frac{\sin \alpha}{\alpha^3}(\alpha\cos\alpha - \sin\alpha) .$$

The derivative vanishes if $\alpha \neq 0$ but $\sin\alpha = 0$. This yields $\alpha = m\pi$, where m is a nonzero integer. These are the intensity minima: $I = 0$ for $\alpha = m\pi$. The derivative also vanishes for $\alpha\cos\alpha - \sin\alpha = 0$. This condition can be written $\tan\alpha = \alpha$. These implicitly locate the maxima.

(b) The values of α that satisfy $\tan\alpha = \alpha$ can be found by trial and error on a pocket calculator or computer. Each of them is slightly less than one of the values $(m + \frac{1}{2})\pi\,\text{rad}$, so we start with these values. They can also be found graphically. As in the diagram that follows, we plot $y = \tan\alpha$ and $y = \alpha$ on the same graph. The intersections of the line with the $\tan\alpha$ curves are the solutions.

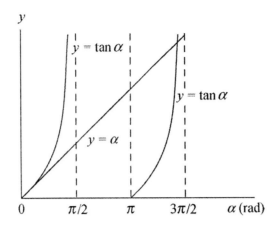

The smallest α is $\alpha = 0$.

(c) We write $\alpha = (m + \frac{1}{2})\pi$ for the maxima. For the central maximum, $\alpha = 0$ and $m = -1/2 = -0.500$.

(d) The next one can be found to be $\alpha = 4.493$ rad.

(e) For $\alpha = 4.4934$, $m = 0.930$.

(f) The next one can be found to be $\alpha = 7.725$ rad.

(g) For $\alpha = 7.7252$, $m = 1.96$.

16. We use Eq. 36-12 with $\theta = 2.5°/2 = 1.25°$. Thus,

$$ d = \frac{1.22\lambda}{\sin\theta} = \frac{1.22(550\,\text{nm})}{\sin 1.25°} = 31\,\mu\text{m} . $$

17. (a) We use the Rayleigh criteria. Thus, the angular separation (in radians) of the sources must be at least $\theta_R = 1.22\lambda/d$, where λ is the wavelength and d is the diameter of the aperture. For the headlights of this problem,

$$ \theta_R = \frac{1.22(550\times10^{-9}\,\text{m})}{5.0\times10^{-3}\,\text{m}} = 1.34\times10^{-4}\,\text{rad}, $$

or 1.3×10^{-4} rad , in two significant figures.

(b) If L is the distance from the headlights to the eye when the headlights are just resolvable and D is the separation of the headlights, then $D = L\theta_R$, where the small angle approximation is made. This is valid for θ_R in radians. Thus,

$$L = \frac{D}{\theta_R} = \frac{1.4\,\text{m}}{1.34 \times 10^{-4}\,\text{rad}} = 1.0 \times 10^4\,\text{m} = 10\,\text{km} \; .$$

18. (a) Using the notation of Sample Problem 36-3 (which is in the textbook supplement), the minimum separation is

$$D = L\theta_R = L\left(\frac{1.22\lambda}{d}\right) = \frac{(400 \times 10^3\,\text{m})(1.22)(550 \times 10^{-9}\,\text{m})}{(0.005\,\text{m})} \approx 50\,\text{m}.$$

(b) The Rayleigh criterion suggests that the astronaut will not be able to discern the Great Wall (see the result of part (a)).

(c) The signs of intelligent life would probably be, at most, ambiguous on the sunlit half of the planet. However, while passing over the half of the planet on the opposite side from the Sun, the astronaut would be able to notice the effects of artificial lighting.

19. Using the notation of Sample Problem 36-3 (which is in the textbook supplement), the minimum separation is

$$D = L\theta_R = L\left(1.22\frac{\lambda}{d}\right) = (3.82 \times 10^8\,\text{m})\frac{(1.22)(550 \times 10^{-9}\,\text{m})}{5.1\,\text{m}}$$

$$= 50\,\text{m} \; .$$

20. Using the notation of Sample Problem 36-3 (which is in the textbook supplement), the maximum distance is

$$L = \frac{D}{\theta_R} = \frac{D}{1.22\lambda/d} = \frac{(5.0 \times 10^{-3}\,\text{m})(4.0 \times 10^{-3}\,\text{m})}{1.22(550 \times 10^{-9}\,\text{m})} = 30\,\text{m} \; .$$

21. (a) We use the Rayleigh criteria. If L is the distance from the observer to the objects, then the smallest separation D they can have and still be resolvable is $D = L\theta_R$, where θ_R is measured in radians. The small angle approximation is made. Thus,

$$D = \frac{1.22\,L\lambda}{d} = \frac{1.22(8.0 \times 10^{10}\,\text{m})(550 \times 10^{-9}\,\text{m})}{5.0 \times 10^{-3}\,\text{m}} = 1.1 \times 10^7\,\text{m} = 1.1 \times 10^4\,\text{km} \; .$$

This distance is greater than the diameter of Mars; therefore, one part of the planet's surface cannot be resolved from another part.

(b) Now $d = 5.1$ m and

$$D = \frac{1.22(8.0 \times 10^{10}\,\text{m})(550 \times 10^{-9}\,\text{m})}{5.1\,\text{m}} = 1.1 \times 10^4\,\text{m} = 11\,\text{km}\ .$$

22. Using the notation of Sample Problem 36-3 (which is in the textbook supplement), the minimum separation is

$$D = L\theta_R = L\left(\frac{1.22\lambda}{d}\right) = \frac{(6.2 \times 10^3\,\text{m})(1.22)(1.6 \times 10^{-2}\,\text{m})}{2.3\,\text{m}} = 53\,\text{m}\ .$$

23. (a) Using the notation of Sample Problem 36-3,

$$L = \frac{D}{1.22\lambda/d} = \frac{2(50 \times 10^{-6}\,\text{m})(1.5 \times 10^{-3}\,\text{m})}{1.22(650 \times 10^{-9}\,\text{m})} = 0.19\ \text{m}\ .$$

(b) The wavelength of the blue light is shorter so $L_{\text{max}} \propto \lambda^{-1}$ will be larger.

24. Eq. 36-14 gives the Rayleigh angle (in radians):

$$\theta_R = \frac{1.22\lambda}{d} = \frac{D}{L}$$

where the rationale behind the second equality is given in Sample Problem 36-3.

(a) We are asked to solve for D and are given $\lambda = 1.40 \times 10^{-9}\,\text{m}$, $d = 0.200 \times 10^{-3}\,\text{m}$, and $L = 2000 \times 10^3\,\text{m}$. Consequently, we obtain $D = 17.1$ m.

(b) Intensity is power over area (with the area assumed spherical in this case, which means it is proportional to radius-squared), so the ratio of intensities is given by the square of a ratio of distances: $(d/D)^2 = 1.37 \times 10^{-10}$.

25. (a) The first minimum in the diffraction pattern is at an angular position θ, measured from the center of the pattern, such that $\sin\theta = 1.22\lambda/d$, where λ is the wavelength and d is the diameter of the antenna. If f is the frequency, then the wavelength is

$$\lambda = \frac{c}{f} = \frac{3.00 \times 10^8\,\text{m/s}}{220 \times 10^9\,\text{Hz}} = 1.36 \times 10^{-3}\,\text{m}\ .$$

Thus

$$\theta = \sin^{-1}\left(\frac{1.22\lambda}{d}\right) = \sin^{-1}\left(\frac{1.22(1.36 \times 10^{-3}\,\text{m})}{55.0 \times 10^{-2}\,\text{m}}\right) = 3.02 \times 10^{-3}\ \text{rad}\ .$$

The angular width of the central maximum is twice this, or 6.04×10^{-3} rad (0.346°).

(b) Now $\lambda = 1.6$ cm and $d = 2.3$ m, so

$$\theta = \sin^{-1}\left(\frac{1.22\left(1.6 \times 10^{-2} \text{ m}\right)}{2.3 \text{ m}}\right) = 8.5 \times 10^{-3} \text{ rad} .$$

The angular width of the central maximum is 1.7×10^{-2} rad (0.97°).

26. Eq. 36-14 gives $\theta_R = 1.22\lambda/d$, where in our case $\theta_R \approx D/L$, with $D = 60$ μm being the size of the object your eyes must resolve, and L being the maximum viewing distance in question. If $d = 3.00$ mm $= 3000$ μm is the diameter of your pupil, then

$$L = \frac{Dd}{1.22\lambda} = \frac{(60 \,\mu\text{m})(3000 \,\mu\text{m})}{1.22(0.55 \,\mu\text{m})} = 2.7 \times 10^5 \,\mu\text{m} = 27 \text{ cm} .$$

27. (a) Using Eq. 36-14, the angular separation is

$$\theta_R = \frac{1.22\lambda}{d} = \frac{(1.22)\left(550 \times 10^{-9} \text{ m}\right)}{0.76 \text{ m}} = 8.8 \times 10^{-7} \text{ rad} .$$

(b) Using the notation of Sample Problem 36-3 (which is in the textbook supplement), the distance between the stars is

$$D = L\theta_R = \frac{(10 \,\text{ly})\left(9.46 \times 10^{12} \text{ km/ly}\right)(0.18)\pi}{(3600)(180)} = 8.4 \times 10^7 \text{ km} .$$

(c) The diameter of the first dark ring is

$$d = 2\theta_R L = \frac{2(0.18)(\pi)(14 \text{ m})}{(3600)(180)} = 2.5 \times 10^{-5} \text{ m} = 0.025 \text{ mm} .$$

28. (a) Since $\theta = 1.22\lambda/d$, the larger the wavelength the larger the radius of the first minimum (and second maximum, etc). Therefore, the white pattern is outlined by red lights (with longer wavelength than blue lights).

(b) The diameter of a water drop is

$$d = \frac{1.22\lambda}{\theta} \approx \frac{1.22\left(7 \times 10^{-7} \text{ m}\right)}{1.5(0.50°)(\pi/180°)/2} = 1.3 \times 10^{-4} \text{ m} .$$

29. Bright interference fringes occur at angles θ given by $d \sin \theta = m\lambda$, where m is an integer. For the slits of this problem, $d = 11a/2$, so $a \sin \theta = 2m\lambda/11$ (see Sample Problem 36-5). The first minimum of the diffraction pattern occurs at the angle θ_1 given by $a \sin \theta_1 = \lambda$, and the second occurs at the angle θ_2 given by $a \sin \theta_2 = 2\lambda$, where a is the slit width. We should count the values of m for which $\theta_1 < \theta < \theta_2$, or, equivalently, the values of m for which $\sin \theta_1 < \sin \theta < \sin \theta_2$. This means $1 < (2m/11) < 2$. The values are $m = 6$, 7, 8, 9, and 10. There are five bright fringes in all.

30. In a manner similar to that discussed in Sample Problem 36-5, we find the number is $2(d/a) - 1 = 2(2a/a) - 1 = 3$.

31. (a) In a manner similar to that discussed in Sample Problem 36-5, we find the ratio should be $d/a = 4$. Our reasoning is, briefly, as follows: we let the location of the fourth bright fringe coincide with the first minimum of diffraction pattern, and then set $\sin \theta = 4\lambda/d = \lambda/a$ (so $d = 4a$).

(b) Any bright fringe which happens to be at the same location with a diffraction minimum will vanish. Thus, if we let

$$\sin \theta = m_1\lambda/d = m_2\lambda/a = m_1\lambda/4a,$$

or $m_1 = 4m_2$ where $m_2 = 1,2,3,\cdots$. The fringes missing are the 4th, 8th, 12th, and so on. Hence, every fourth fringe is missing.

32. The angular location of the mth bright fringe is given by $d \sin \theta = m\lambda$, so the linear separation between two adjacent fringes is

$$\Delta y = \Delta(D \sin \theta) = \Delta\left(\frac{D_m\lambda}{d}\right) = \frac{D\lambda}{d}\Delta m = \frac{D\lambda}{d}.$$

33. (a) The angular positions θ of the bright interference fringes are given by $d \sin \theta = m\lambda$, where d is the slit separation, λ is the wavelength, and m is an integer. The first diffraction minimum occurs at the angle θ_1 given by $a \sin \theta_1 = \lambda$, where a is the slit width. The diffraction peak extends from $-\theta_1$ to $+\theta_1$, so we should count the number of values of m for which $-\theta_1 < \theta < +\theta_1$, or, equivalently, the number of values of m for which $-\sin \theta_1 < \sin \theta < +\sin \theta_1$. This means $-1/a < m/d < 1/a$ or $-d/a < m < +d/a$. Now

$$d/a = (0.150 \times 10^{-3} \text{ m})/(30.0 \times 10^{-6} \text{ m}) = 5.00,$$

so the values of m are $m = -4, -3, -2, -1, 0, +1, +2, +3$, and $+4$. There are nine fringes.

(b) The intensity at the screen is given by

$$I = I_m \left(\cos^2 \beta\right) \left(\frac{\sin \alpha}{\alpha}\right)^2$$

where $\alpha = (\pi a/\lambda) \sin \theta$, $\beta = (\pi d/\lambda) \sin \theta$, and I_m is the intensity at the center of the pattern. For the third bright interference fringe, $d \sin \theta = 3\lambda$, so $\beta = 3\pi$ rad and $\cos^2 \beta = 1$. Similarly, $\alpha = 3\pi a/d = 3\pi/5.00 = 0.600\pi$ rad and

$$\left(\frac{\sin \alpha}{\alpha}\right)^2 = \left(\frac{\sin 0.600\pi}{0.600\pi}\right)^2 = 0.255 .$$

The intensity ratio is $I/I_m = 0.255$.

34. (a) We note that the slope of the graph is 80, and that Eq. 36-20 implies that the slope should correspond to

$$\frac{\pi d}{\lambda} = 80 \quad \Rightarrow \quad d = 11077 \text{ nm} = 11.1 \text{ } \mu\text{m} .$$

(b) Consider Eq. 36-25 with "continuously variable" m (of course, m should be an integer for interference maxima, but for the moment we will solve for it as if it could be any real number):

$$m_{max} = \frac{d}{\lambda} (\sin \theta)_{max} \approx 25.5$$

which indicates (on one side of the interference pattern) there are 25 bright fringes. Thus on the other side there are also 25 bright fringes. Including the one in the middle, then, means there are a total of 51 maxima in the interference pattern (assuming, as the problem remarks, that none of the interference maxima have been eliminated by diffraction minima).

(c) Clearly, the maximum closest to the axis is the middle fringe at $\theta = 0°$.

(d) If we set $m = 25$ in Eq. 36-25, we find

$$m\lambda = d \sin \theta \quad \Rightarrow \quad \theta = 79.0° .$$

35. (a) The first minimum of the diffraction pattern is at 5.00°, so

$$a = \frac{\lambda}{\sin \theta} = \frac{0.440 \text{ } \mu\text{m}}{\sin 5.00°} = 5.05 \mu\text{m} .$$

(b) Since the fourth bright fringe is missing, $d = 4a = 4(5.05 \text{ } \mu\text{m}) = 20.2 \text{ } \mu\text{m}$.

(c) For the $m = 1$ bright fringe,

$$\alpha = \frac{\pi a \sin\theta}{\lambda} = \frac{\pi(5.05\,\mu m)\sin 1.25°}{0.440\,\mu m} = 0.787\,\text{rad} \ .$$

Consequently, the intensity of the $m = 1$ fringe is

$$I = I_m\left(\frac{\sin\alpha}{\alpha}\right)^2 = \left(7.0\,\text{mW/cm}^2\right)\left(\frac{\sin 0.787\,\text{rad}}{0.787}\right)^2 = 5.7\,\text{mW/cm}^2 \ ,$$

which agrees with Fig. 36-43. Similarly for $m = 2$, the intensity is $I = 2.9$ mW/cm^2, also in agreement with Fig. 36-43.

36. We will make use of arctangents and sines in our solution, even though they can be "shortcut" somewhat since the angles are [almost] small enough to justify the use of the small angle approximation.

(a) Given $y/D = 70/400$ (both expressed here in centimeters), then

$$\theta = \tan^{-1}(y/D) = 0.173\,\text{rad}.$$

With d and λ in micrometers, Eq. 36-20 then gives

$$\beta = \frac{\pi d}{\lambda}\sin\theta = \frac{\pi(24)}{0.60}\sin(0.173\,\text{rad}) = 21.66\,\text{rad} \ .$$

Thus, use of Eq. 36-21 (with $a = 12$ µm and $\lambda = 0.60$ µm) leads to

$$\alpha = \frac{\pi a}{\lambda}\sin\theta = 10.83\,\text{rad} \ .$$

Thus,

$$\frac{I_P}{I_m} = \left(\frac{\sin\alpha}{\alpha}\right)^2 (\cos\beta)^2 = 0.00743 \ .$$

(b) Consider Eq. 36-25 with "continuously variable" m (of course, m should be an integer for interference maxima, but for the moment we will solve for it as if it could be any real number):

$$m = \frac{d}{\lambda}\sin\theta \approx 6.9$$

which suggests that the angle takes us to a point between the sixth minimum (which would have $m = 6.5$) and the seventh maximum (which corresponds to $m = 7$).

(c) Similarly, consider Eq. 36-3 with "continuously variable" m (of course, m should be an integer for diffraction minima, but for the moment we will solve for it as if it could be any real number):

$$m = \frac{a}{\lambda}\sin\theta \approx 3.4$$

which suggests that the angle takes us to a point between the third diffraction minimum ($m = 3$) and the fourth one ($m = 4$). The maxima (in the smaller peaks of the diffraction pattern) are not exactly midway between the minima; their location would make use of mathematics not covered in the prerequisites of the usual sophomore-level physics course.

37. The distance between adjacent rulings is

$$d = 20.0 \text{ mm}/6000 = 0.00333 \text{ mm} = 3.33 \ \mu\text{m}.$$

(a) Let $d\sin\theta = m\lambda \ (m = 0, \pm1, \pm2, \ldots)$. Since $|m|\lambda/d > 1$ for $|m| \geq 6$, the largest value of θ corresponds to $|m| = 5$, which yields

$$\theta = \sin^{-1}\left(|m|\lambda/d\right) = \sin^{-1}\left(\frac{5(0.589 \ \mu\text{m})}{3.33 \ \mu\text{m}}\right) = 62.1°$$

(b) The second largest value of θ corresponds to $|m| = 4$, which yields

$$\theta = \sin^{-1}\left(|m|\lambda/d\right) = \sin^{-1}\left(\frac{4(0.589 \ \mu\text{m})}{3.33 \ \mu\text{m}}\right) = 45.0°$$

(c) The third largest value of θ corresponds to $|m| = 3$, which yields

$$\theta = \sin^{-1}\left(|m|\lambda/d\right) = \sin^{-1}\left(\frac{3(0.589 \ \mu\text{m})}{3.33 \ \mu\text{m}}\right) = 32.0°$$

38. The angular location of the mth order diffraction maximum is given by $m\lambda = d\sin\theta$. To be able to observe the fifth-order maximum, we must let $\sin\theta|_{m=5} = 5\lambda/d < 1$, or

$$\lambda < \frac{d}{5} = \frac{1.00 \text{ nm}/315}{5} = 635 \text{ nm}.$$

Therefore, the longest wavelength that can be used is $\lambda = 635$ nm.

39. The ruling separation is $d = 1/(400 \text{ mm}^{-1}) = 2.5 \times 10^{-3}$ mm. Diffraction lines occur at angles θ such that $d\sin\theta = m\lambda$, where λ is the wavelength and m is an integer. Notice that for a given order, the line associated with a long wavelength is produced at a greater angle than the line associated with a shorter wavelength. We take λ to be the longest

wavelength in the visible spectrum (700 nm) and find the greatest integer value of m such that θ is less than 90°. That is, find the greatest integer value of m for which $m\lambda < d$. Since

$$d/\lambda = (2.5 \times 10^{-6}\text{ m})/(700 \times 10^{-9}\text{ m}) = 3.57,$$

that value is $m = 3$. There are three complete orders on each side of the $m = 0$ order. The second and third orders overlap.

40. We use Eq. 36-25 for diffraction maxima: $d \sin \theta = m\lambda$. In our case, since the angle between the $m = 1$ and $m = -1$ maxima is 26°, the angle θ corresponding to $m = 1$ is $\theta = 26°/2 = 13°$. We solve for the grating spacing:

$$d = \frac{m\lambda}{\sin \theta} = \frac{(1)(550\text{nm})}{\sin 13°} = 2.4\,\mu\text{m} \approx 2\,\mu\text{m}.$$

41. (a) Maxima of a diffraction grating pattern occur at angles θ given by $d \sin \theta = m\lambda$, where d is the slit separation, λ is the wavelength, and m is an integer. The two lines are adjacent, so their order numbers differ by unity. Let m be the order number for the line with $\sin \theta = 0.2$ and $m + 1$ be the order number for the line with $\sin \theta = 0.3$. Then, $0.2d = m\lambda$ and $0.3d = (m + 1)\lambda$. We subtract the first equation from the second to obtain $0.1d = \lambda$, or

$$d = \lambda/0.1 = (600 \times 10^{-9}\text{m})/0.1 = 6.0 \times 10^{-6}\text{ m}.$$

(b) Minima of the single-slit diffraction pattern occur at angles θ given by $a \sin \theta = m\lambda$, where a is the slit width. Since the fourth-order interference maximum is missing, it must fall at one of these angles. If a is the smallest slit width for which this order is missing, the angle must be given by $a \sin \theta = \lambda$. It is also given by $d \sin \theta = 4\lambda$, so

$$a = d/4 = (6.0 \times 10^{-6}\text{ m})/4 = 1.5 \times 10^{-6}\text{ m}.$$

(c) First, we set $\theta = 90°$ and find the largest value of m for which $m\lambda < d \sin \theta$. This is the highest order that is diffracted toward the screen. The condition is the same as $m < d/\lambda$ and since

$$d/\lambda = (6.0 \times 10^{-6}\text{ m})/(600 \times 10^{-9}\text{ m}) = 10.0,$$

the highest order seen is the $m = 9$ order. The fourth and eighth orders are missing, so the observable orders are $m = 0, 1, 2, 3, 5, 6, 7,$ and 9. Thus, the largest value of the order number is $m = 9$.

(d) Using the result obtained in (c), the second largest value of the order number is $m = 7$.

(e) Similarly, the third largest value of the order number is $m = 6$.

42. (a) For the maximum with the greatest value of m (= M) we have $M\lambda = a \sin \theta < d$, so $M < d/\lambda$ = 900 nm/600 nm = 1.5, or M = 1. Thus three maxima can be seen, with m = 0, ±1.

(b) From Eq. 36-28

$$\Delta\theta_{hw} = \frac{\lambda}{N d \cos\theta} = \frac{d \sin\theta}{N d \cos\theta} = \frac{\tan\theta}{N} = \frac{1}{N} \tan\left[\sin^{-1}\left(\frac{\lambda}{d}\right)\right]$$

$$= \frac{1}{1000} \tan\left[\sin^{-1}\left(\frac{600\,\text{nm}}{900\,\text{nm}}\right)\right] = 0.051°.$$

43. The angular positions of the first-order diffraction lines are given by $d \sin \theta = \lambda$. Let λ_1 be the shorter wavelength (430 nm) and θ be the angular position of the line associated with it. Let λ_2 be the longer wavelength (680 nm), and let $\theta + \Delta\theta$ be the angular position of the line associated with it. Here $\Delta\theta$ = 20°. Then, $d \sin \theta = \lambda_1$ and $d \sin (\theta + \Delta\theta) = \lambda_2$. We write

$$\sin(\theta + \Delta\theta) \text{ as } \sin\theta \cos\Delta\theta + \cos\theta \sin\Delta\theta,$$

then use the equation for the first line to replace $\sin \theta$ with λ_1/d, and $\cos \theta$ with $\sqrt{1 - \lambda_1^2/d^2}$. After multiplying by d, we obtain

$$\lambda_1 \cos\Delta\theta + \sqrt{d^2 - \lambda_1^2}\, \sin\Delta\theta = \lambda_2.$$

Solving for d, we find

$$d = \sqrt{\frac{(\lambda_2 - \lambda_1 \cos\Delta\theta)^2 + (\lambda_1 \sin\Delta\theta)^2}{\sin^2\Delta\theta}}$$

$$= \sqrt{\frac{[(680\,\text{nm}) - (430\,\text{nm})\cos 20°]^2 + [(430\,\text{nm})\sin 20°]^2}{\sin^2 20°}}$$

$$= 914\,\text{nm} = 9.14 \times 10^{-4}\,\text{mm}.$$

There are $1/d$ = 1/(9.14 × 10^{-4} mm) = 1.09 × 10^3 rulings per mm.

44. We use Eq. 36-25. For $m = \pm 1$

$$\lambda = \frac{d \sin\theta}{m} = \frac{(1.73\,\mu\text{m})\sin(\pm 17.6°)}{\pm 1} = 523\,\text{nm},$$

and for $m = \pm 2$

$$\lambda = \frac{(1.73\,\mu m)\sin(\pm 37.3°)}{\pm 2} = 524 \text{ nm.}$$

Similarly, we may compute the values of λ corresponding to the angles for $m = \pm 3$. The average value of these λ's is 523 nm.

45. At the point on the screen where we find the inner edge of the hole, we have tan θ = 5.0 cm/30 cm, which gives θ = 9.46°. We note that d for the grating is equal to 1.0 mm/350 = 1.0×10^6 nm/350.

(a) From $m\lambda = d \sin \theta$, we find

$$m = \frac{d\sin\theta}{\lambda} = \frac{\left(\frac{1.0\times10^6\ nm}{350}\right)(0.1644)}{\lambda} = \frac{470\,nm}{\lambda}.$$

Since for white light $\lambda > 400$ nm, the only integer m allowed here is $m = 1$. Thus, at one edge of the hole, $\lambda = 470$ nm. This is the shortest wavelength of the light that passes through the hole.

(b) At the other edge, we have tan $\theta' = 6.0$ cm/30 cm, which gives $\theta' = 11.31°$. This leads to

$$\lambda' = d\sin\theta' = \left(\frac{1.0\times10^6\ nm}{350}\right)\sin 11.31° = 560 \text{ nm.}$$

This corresponds to the longest wavelength of the light that passes through the hole.

46. We are given the "number of lines per millimeter" (which is a common way to express $1/d$ for diffraction gratings); thus,

$$\frac{1}{d} = 160 \text{ lines/mm} \Rightarrow d = 6.25 \times 10^{-6} \text{ m .}$$

(a) We solve Eq. 36-25 for θ with various values of m and λ. We show here the $m = 2$ and $\lambda = 460$ nm calculation:

$$\theta = \sin^{-1}\left(\frac{m\lambda}{d}\right) = \sin^{-1}\left(\frac{2(460 \times 10^{-9}\,m)}{6.25 \times 10^{-6}\,m}\right) = 8.46°$$

Similarly, we get 11.81° for $m = 2$ and $\lambda = 640$ nm, 12.75° for $m = 3$ and $\lambda = 460$ nm, and 17.89° for $m = 3$ and $\lambda = 640$ nm. The first indication of overlap occurs when we compute the angle for $m = 4$ and $\lambda = 460$ nm; the result is 17.12° which clearly shows overlap with the large-wavelength portion of the $m = 3$ spectrum.

(b) We solve Eq. 36-25 for *m* with $\theta = 90°$ and $\lambda = 640$ nm. In this case, we obtain *m* = 9.8 which means the largest order in which the full range (which must include that largest wavelength) is seen is ninth order.

(c) Now with *m* = 9, Eq. 36-25 gives $\theta = 41.5°$ for $\lambda = 460$ nm.

(d) It similarly gives $\theta = 67.2°$ for $\lambda = 640$ nm.

(e) We solve Eq. 36-25 for *m* with $\theta = 90°$ and $\lambda = 460$ nm. In this case, we obtain *m* = 13.6 which means the largest order in which that wavelength is seen is thirteenth order. Now with *m* = 13, Eq. 36-25 gives $\theta = 73.1°$ for $\lambda = 460$ nm.

47. Since the slit width is much less than the wavelength of the light, the central peak of the single-slit diffraction pattern is spread across the screen and the diffraction envelope can be ignored. Consider three waves, one from each slit. Since the slits are evenly spaced, the phase difference for waves from the first and second slits is the same as the phase difference for waves from the second and third slits. The electric fields of the waves at the screen can be written as

$$E_1 = E_0 \sin(\omega t),$$
$$E_2 = E_0 \sin(\omega t + \phi),$$
$$E_3 = E_0 \sin(\omega t + 2\phi),$$

where $\phi = (2\pi d/\lambda) \sin\theta$. Here *d* is the separation of adjacent slits and λ is the wavelength. The phasor diagram is shown below.

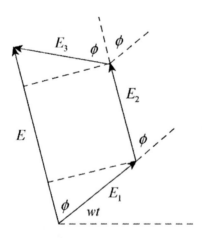

It yields

$$E = E_0 \cos\phi + E_0 \cos\phi = E_0(1 + 2\cos\phi).$$

for the amplitude of the resultant wave. Since the intensity of a wave is proportional to the square of the electric field, we may write $I = AE_0^2(1 + 2\cos\phi)^2$, where *A* is a constant of proportionality. If I_m is the intensity at the center of the pattern, for which $\phi = 0$, then $I_m = 9AE_0^2$. We take A to be $I_m/9E_0^2$ and obtain

$$I = \frac{I_m}{9}(1 + 2\cos\phi)^2 = \frac{I_m}{9}\left(1 + 4\cos\phi + 4\cos^2\phi\right).$$

48. (a) From $R = \lambda/\Delta\lambda = Nm$ we find

$$N = \frac{\lambda}{m\Delta\lambda} = \frac{(415.496\text{ nm} + 415.487\text{ nm})/2}{2(415.96\text{ nm} - 415.487\text{ nm})} = 23100.$$

(b) We note that $d = (4.0 \times 10^7\text{ nm})/23100 = 1732$ nm. The maxima are found at

$$\theta = \sin^{-1}\left(\frac{m\lambda}{d}\right) = \sin^{-1}\left[\frac{(2)(415.5\text{ nm})}{1732\text{ nm}}\right] = 28.7°.$$

49. (a) We note that $d = (76 \times 10^6\text{ nm})/40000 = 1900$ nm. For the first order maxima $\lambda = d\sin\theta$, which leads to

$$\theta = \sin^{-1}\left(\frac{\lambda}{d}\right) = \sin^{-1}\left(\frac{589\text{ nm}}{1900\text{ nm}}\right) = 18°.$$

Now, substituting $m = d\sin\theta/\lambda$ into Eq. 36-30 leads to

$$D = \tan\theta/\lambda = \tan 18°/589\text{ nm} = 5.5 \times 10^{-4}\text{ rad/nm} = 0.032°/\text{nm}.$$

(b) For $m = 1$, the resolving power is $R = Nm = 40000\ m = 40000 = 4.0 \times 10^4$.

(c) For $m = 2$ we have $\theta = 38°$, and the corresponding value of dispersion is $0.076°/\text{nm}$.

(d) For $m = 2$, the resolving power is $R = Nm = 40000\ m = (40000)2 = 8.0 \times 10^4$.

(e) Similarly for $m = 3$, we have $\theta = 68°$, and the corresponding value of dispersion is $0.24°/\text{nm}$.

(f) For $m = 3$, the resolving power is $R = Nm = 40000\ m = (40000)3 = 1.2 \times 10^5$.

50. Letting $R = \lambda/\Delta\lambda = Nm$, we solve for N:

$$N = \frac{\lambda}{m\Delta\lambda} = \frac{(589.6\text{ nm} + 589.0\text{ nm})/2}{2(589.6\text{ nm} - 589.0\text{ nm})} = 491.$$

51. If a grating just resolves two wavelengths whose average is λ_{avg} and whose separation is $\Delta\lambda$, then its resolving power is defined by $R = \lambda_{avg}/\Delta\lambda$. The text shows this is Nm, where N is the number of rulings in the grating and m is the order of the lines. Thus $\lambda_{avg}/\Delta\lambda = Nm$ and

$$N = \frac{\lambda_{avg}}{m\Delta\lambda} = \frac{656.3\,\text{nm}}{(1)(0.18\,\text{nm})} = 3.65 \times 10^3 \text{ rulings.}$$

52. (a) We find $\Delta\lambda$ from $R = \lambda/\Delta\lambda = Nm$:

$$\Delta\lambda = \frac{\lambda}{Nm} = \frac{500\,\text{nm}}{(600/\,\text{mm})(5.0\,\text{mm})(3)} = 0.056\,\text{nm} = 56\,\text{pm.}$$

(b) Since $\sin\theta = m_{max}\lambda/d < 1$,

$$m_{max} < \frac{d}{\lambda} = \frac{1}{(600/\,\text{mm})(500 \times 10^{-6}\,\text{mm})} = 3.3.$$

Therefore, $m_{max} = 3$. No higher orders of maxima can be seen.

53. (a) From $d \sin\theta = m\lambda$ we find

$$d = \frac{m\lambda_{avg}}{\sin\theta} = \frac{3(589.3\,\text{nm})}{\sin 10°} = 1.0 \times 10^4\,\text{nm} = 10\,\mu\text{m.}$$

(b) The total width of the ruling is

$$L = Nd = \left(\frac{R}{m}\right)d = \frac{\lambda_{avg}d}{m\Delta\lambda} = \frac{(589.3\,\text{nm})(10\,\mu\text{m})}{3(589.59\,\text{nm} - 589.00\,\text{nm})} = 3.3 \times 10^3\,\mu\text{m} = 3.3 \text{ mm.}$$

54. (a) From the expression for the half-width $\Delta\theta_{hw}$ (given by Eq. 36-28) and that for the resolving power R (given by Eq. 36-32), we find the product of $\Delta\theta_{hw}$ and R to be

$$\Delta\theta_{hw}R = \left(\frac{\lambda}{Nd\cos\theta}\right)Nm = \frac{m\lambda}{d\cos\theta} = \frac{d\sin\theta}{d\cos\theta} = \tan\theta,$$

where we used $m\lambda = d\sin\theta$ (see Eq. 36-25).

(b) For first order $m = 1$, so the corresponding angle θ_1 satisfies $d\sin\theta_1 = m\lambda = \lambda$. Thus the product in question is given by

$$\tan \theta_1 = \frac{\sin \theta_1}{\cos \theta_1} = \frac{\sin \theta_1}{\sqrt{1-\sin^2 \theta_1}} = \frac{1}{\sqrt{(1/\sin \theta_1)^2 - 1}} = \frac{1}{\sqrt{(d/\lambda)^2 - 1}}$$

$$= \frac{1}{\sqrt{(900 \text{nm}/600 \text{nm})^2 - 1}} = 0.89.$$

55. Bragg's law gives the condition for a diffraction maximum:

$$2d \sin \theta = m\lambda$$

where d is the spacing of the crystal planes and λ is the wavelength. The angle θ is measured from the surfaces of the planes. For a second-order reflection $m = 2$, so

$$d = \frac{m\lambda}{2\sin \theta} = \frac{2(0.12 \times 10^{-9} \text{m})}{2 \sin 28°} = 2.56 \times 10^{-10} \text{m} \approx 0.26 \text{nm}.$$

56. For x-ray ("Bragg") scattering, we have $2d \sin \theta_m = m\lambda$. This leads to

$$\frac{2d \sin \theta_2}{2d \sin \theta_1} = \frac{2\lambda}{1\lambda} \quad \Rightarrow \quad \sin \theta_2 = 2 \sin \theta_1.$$

Thus, with $\theta_1 = 3.4°$, this yields $\theta_2 = 6.8°$. The fact that θ_2 is very nearly twice the value of θ_1 is due to the small angles involved (when angles are small, $\sin \theta_2 / \sin \theta_1 = \theta_2/\theta_1$).

57. We use Eq. 36-34.

(a) From the peak on the left at angle 0.75° (estimated from Fig. 36-44), we have

$$\lambda_1 = 2d \sin \theta_1 = 2(0.94 \text{ nm}) \sin(0.75°) = 0.025 \text{nm} = 25 \text{pm}.$$

This is the shorter wavelength of the beam. Notice that the estimation should be viewed as reliable to within ±2 pm.

(b) We now consider the next peak:

$$\lambda_2 = 2d \sin \theta_2 = 2(0.94 \text{ nm}) \sin 1.15° = 0.038 \text{nm} = 38 \text{pm}.$$

This is the longer wavelength of the beam. One can check that the third peak from the left is the second-order one for λ_1.

58. The x-ray wavelength is $\lambda = 2d \sin \theta = 2(39.8 \text{ pm}) \sin 30.0° = 39.8$ pm.

59. (a) For the first beam $2d \sin \theta_1 = \lambda_A$ and for the second one $2d \sin \theta_2 = 3\lambda_B$. The values of d and λ_A can then be determined:

$$d = \frac{3\lambda_B}{2\sin\theta_2} = \frac{3(97\,\text{pm})}{2\sin 60°} = 1.7 \times 10^2\,\text{pm}.$$

(b) $\lambda_A = 2d\sin\theta_1 = 2(1.7 \times 10^2\,\text{pm})(\sin 23°) = 1.3 \times 10^2\,\text{pm}.$

60. The angle of incidence on the reflection planes is $\theta = 63.8° - 45.0° = 18.8°$, and the plane-plane separation is $d = a_0/\sqrt{2}$. Thus, using $2d\sin\theta = \lambda$, we get

$$a_0 = \sqrt{2}d = \frac{\sqrt{2}\lambda}{2\sin\theta} = \frac{0.260\,\text{nm}}{\sqrt{2}\sin 18.8°} = 0.570\,\text{nm}.$$

61. The sets of planes with the next five smaller interplanar spacings (after a_0) are shown in the diagram that follows.

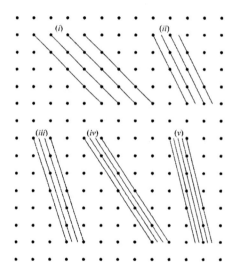

(a) In terms of a_0, the second largest interplanar spacing is $a_0/\sqrt{2} = 0.7071a_0$.

(b) The third largest interplanar spacing is $a_0/\sqrt{5} = 0.4472a_0$.

(c) The fourth largest interplanar spacing is $a_0/\sqrt{10} = 0.3162a_0$.

(d) The fifth largest interplanar spacing is $a_0/\sqrt{13} = 0.2774a_0$.

(e) The sixth largest interplanar spacing is $a_0/\sqrt{17} = 0.2425a_0$.

(f) Since a crystal plane passes through lattice points, its slope can be written as the ratio of two integers. Consider a set of planes with slope m/n, as shown in the diagram that follows. The first and last planes shown pass through adjacent lattice points along a horizontal line and there are $m - 1$ planes between. If h is the separation of the first and last planes, then the interplanar spacing is $d = h/m$. If the planes make the angle θ with the horizontal, then the normal to the planes (shown dashed) makes the angle $\phi = 90° - \theta$. The distance h is given by $h = a_0 \cos \phi$ and the interplanar spacing is $d = h/m = (a_0/m) \cos \phi$. Since $\tan \theta = m/n$, $\tan \phi = n/m$ and

$$\cos\phi = 1/\sqrt{1 + \tan^2 \phi} = m/\sqrt{n^2 + m^2}.$$

Thus,

$$d = \frac{h}{m} = \frac{a_0 \cos\phi}{m} = \frac{a_0}{\sqrt{n^2 + m^2}}.$$

62. The wavelengths satisfy

$$m\lambda = 2d \sin \theta = 2(275 \text{ pm})(\sin 45°) = 389 \text{ pm}.$$

In the range of wavelengths given, the allowed values of m are $m = 3, 4$.

(a) The longest wavelength is 389 pm/3 = 130 pm.

(b) The associated order number is $m = 3$.

(c) The shortest wavelength is 389 pm/4 = 97.2 pm.

(d) The associated order number is $m = 4$.

63. We want the reflections to obey the Bragg condition $2d \sin \theta = m\lambda$, where θ is the angle between the incoming rays and the reflecting planes, λ is the wavelength, and m is an integer. We solve for θ:

$$\theta = \sin^{-1}\left(\frac{m\lambda}{2d}\right) = \sin^{-1}\left(\frac{(0.125 \times 10^{-9}\,\text{m})m}{2(0.252 \times 10^{-9}\,\text{m})}\right) = 0.2480m.$$

(a) For $m = 2$ the above equation gives $\theta = 29.7°$. The crystal should be turned $\phi = 45° - 29.7° = 15.3°$ clockwise.

(b) For $m = 1$ the above equation gives $\theta = 14.4°$. The crystal should be turned $\phi = 45° - 14.4° = 30.6°$ clockwise.

(c) For $m = 3$ the above equation gives $\theta = 48.1°$. The crystal should be turned $\phi = 48.1° - 45° = 3.1°$ counterclockwise.

(d) For $m = 4$ the above equation gives $\theta = 82.8°$. The crystal should be turned $\phi = 82.8° - 45° = 37.8°$ counterclockwise.

Note that there are no intensity maxima for $m > 4$ as one can verify by noting that $m\lambda/2d$ is greater than 1 for m greater than 4.

64. Following the method of Sample Problem 36-5, we find

$$\frac{d}{a} = \frac{0.30 \times 10^{-3}\,\text{m}}{46 \times 10^{-6}\,\text{m}} = 6.52$$

which we interpret to mean that the first diffraction minimum occurs slightly farther "out" than the $m = 6$ interference maximum. This implies that the central diffraction envelope includes the central ($m = 0$) interference maximum as well as six interference maxima on each side of it. Therefore, there are $6 + 1 + 6 = 13$ bright fringes (interference maxima) in the central diffraction envelope.

65. Let the first minimum be a distance y from the central axis which is perpendicular to the speaker. Then

$$\sin\theta = y/\left(D^2 + y^2\right)^{1/2} = m\lambda/a = \lambda/a \ (\text{for } m = 1).$$

Therefore,

$$y = \frac{D}{\sqrt{(a/\lambda)^2 - 1}} = \frac{D}{\sqrt{(af/v_s)^2 - 1}} = \frac{100\,\text{m}}{\sqrt{\left[(0.300\,\text{m})(3000\,\text{Hz})/(343\,\text{m/s})\right]^2 - 1}} = 41.2\,\text{m}.$$

66. (a) We use Eq. 36-14:

$$\theta_R = 1.22\frac{\lambda}{d} = \frac{(1.22)(540\times10^{-6}\,\text{mm})}{5.0\,\text{mm}} = 1.3\times10^{-4}\,\text{rad}\ .$$

(b) The linear separation is $D = L\theta_R = (160 \times 10^3\,\text{m})\,(1.3 \times 10^{-4}\,\text{rad}) = 21\,\text{m}$.

67. Since we are considering the *diameter* of the central diffraction maximum, then we are working with *twice* the Rayleigh angle. Using notation similar to that in Sample Problem 36-3 (which is in the textbook supplement), we have $2(1.22\lambda/d) = D/L$. Therefore,

$$d = 2\frac{1.22\lambda L}{D} = 2\frac{(1.22)(500\times10^{-9}\,\text{m})(3.54\times10^5\,\text{m})}{9.1\,\text{m}} = 0.047\,\text{m}\ .$$

68. We denote the Earth-Moon separation as L. The energy of the beam of light which is projected onto the moon is concentrated in a circular spot of diameter d_1, where $d_1/L = 2\theta_R = 2(1.22\lambda/d_0)$, with d_0 the diameter of the mirror on Earth. The fraction of energy picked up by the reflector of diameter d_2 on the Moon is then $\eta' = (d_2/d_1)^2$. This reflected light, upon reaching the Earth, has a circular cross section of diameter d_3 satisfying

$$d_3/L = 2\theta_R = 2(1.22\lambda/d_2).$$

The fraction of the reflected energy that is picked up by the telescope is then $\eta'' = (d_0/d_3)^2$. Consequently, the fraction of the original energy picked up by the detector is

$$\eta = \eta'\eta'' = \left(\frac{d_0}{d_3}\right)^2\left(\frac{d_2}{d_1}\right)^2 = \left[\frac{d_0 d_2}{(2.44\lambda d_{em}/d_0)(2.44\lambda d_{em}/d_2)}\right]^2 = \left(\frac{d_0 d_2}{2.44\lambda d_{em}}\right)^4$$

$$= \left[\frac{(2.6\,\text{m})(0.10\,\text{m})}{2.44(0.69\times10^{-6}\,\text{m})(3.82\times10^8\,\text{m})}\right]^4 \approx 4\times10^{-13}\ .$$

69. Consider two of the rays shown in Fig. 36-48, one just above the other. The extra distance traveled by the lower one may be found by drawing perpendiculars from where the top ray changes direction (point P) to the incident and diffracted paths of the lower one. Where these perpendiculars intersect the lower ray's paths are here referred to as points A and C. Where the bottom ray changes direction is point B. We note that angle $\angle APB$ is the same as ψ, and angle BPC is the same as θ (see Fig. 36-48). The difference in path lengths between the two adjacent light rays is $\Delta x = |AB| + |BC| = d\sin\psi + d\sin\theta$. The condition for bright fringes to occur is therefore

$$\Delta x = d(\sin\psi + \sin\theta) = m\lambda$$

where $m = 0, 1, 2, \ldots$. If we set $\psi = 0$ then this reduces to Eq. 36-25.

70. Following Sample Problem 36-3, we use Eq. 36-17:

$$L = \frac{Dd}{1.22\lambda} = 164\,\text{m}\ .$$

71. (a) Employing Eq. 36-3 with the small angle approximation (sin $\theta \approx$ tan $\theta = y/D$ where y locates the minimum relative to the middle of the pattern), we find (with $m = 1$ and all lengths in mm)

$$D = \frac{ya}{m\lambda} = \frac{(0.9)(0.4)}{4.5\times10^{-4}} = 800$$

which places the screen 80 cm away from the slit.

(b) The above equation gives for the value of y (for $m = 3$)

$$y = \frac{(3)\lambda D}{a} = 2.7\,\text{mm}\ .$$

Subtracting this from the first minimum position $y = 0.9$ mm, we find the result $\Delta y = 1.8$ mm .

72. (a) We require that sin $\theta = m\lambda_{1,2}/d \le$ sin 30°, where $m = 1, 2$ and $\lambda_1 = 500$ nm. This gives

$$d \ge \frac{2\lambda_s}{\sin 30°} = \frac{2(600\,\text{nm})}{\sin 30°} = 2400\,\text{nm} = 2.4\,\mu\text{m}.$$

For a grating of given total width L we have $N = L/d \propto d^{-1}$, so we need to minimize d to maximize $R = mN \propto d^{-1}$. Thus we choose $d = 2400$ nm $= 2.4$ μm.

(b) Let the third-order maximum for $\lambda_2 = 600$ nm be the first minimum for the single-slit diffraction profile. This requires that d sin $\theta = 3\lambda_2 = a$ sin θ, or

$$a = d/3 = 2400\,\text{nm}/3 = 800\,\text{nm} = 0.80\ \mu\text{m}.$$

(c) Letting sin $\theta = m_{max}\lambda_2/d \le 1$, we obtain

$$m_{max} \le \frac{d}{\lambda_2} = \frac{2400\,\text{nm}}{800\,\text{nm}} = 3\ .$$

Since the third order is missing the only maxima present are the ones with $m = 0, 1$ and 2. Thus, the largest order of maxima produced by the grating is $m = 2$.

73. Letting $d \sin \theta = m\lambda$, we solve for λ:

$$\lambda = \frac{d \sin \theta}{m} = \frac{(1.0 \,\text{mm}/200)(\sin 30°)}{m} = \frac{2500 \,\text{nm}}{m}$$

where $m = 1, 2, 3 \ldots$. In the visible light range m can assume the following values: $m_1 = 4$, $m_2 = 5$ and $m_3 = 6$.

(a) The longest wavelength corresponds to $m_1 = 4$ with $\lambda_1 = 2500$ nm/4 = 625 nm.

(b) The second longest wavelength corresponds to $m_2 = 5$ with $\lambda_2 = 2500$ nm/5 = 500 nm.

(c) The third longest wavelength corresponds to $m_3 = 6$ with $\lambda_3 = 2500$ nm/6 = 416 nm.

74. Using the notation of Sample Problem 36-3,

$$L = \frac{D}{\theta_R} = \frac{D}{1.22\lambda/d} = \frac{(5.0 \times 10^{-2} \,\text{m})(4.0 \times 10^{-3} \,\text{m})}{1.22(0.10 \times 10^{-9} \,\text{m})} = 1.6 \times 10^{6} \,\text{m} = 1.6 \times 10^{3} \,\text{km} \,.$$

75. The condition for a minimum in a single-slit diffraction pattern is given by Eq. 36-3, which we solve for the wavelength:

$$\lambda = \frac{a \sin \theta}{m} = \frac{(0.022 \,\text{mm}) \sin 1.8°}{1} = 6.91 \times 10^{-4} \,\text{mm} = 691 \text{ nm} \,.$$

76. (a) We express all lengths in mm, and since $1/d = 180$, we write Eq. 36-25 as

$$\theta = \sin^{-1}\left(\frac{1}{d} m\lambda\right) = \sin^{-1}(180)(2)\lambda$$

where $\lambda_1 = 4 \times 10^{-4}$ and $\lambda_2 = 5 \times 10^{-4}$ (in mm). Thus, $\Delta\theta = \theta_2 - \theta_1 = 2.1°$.

(b) Use of Eq. 36-25 for each wavelength leads to the condition

$$m_1\lambda_1 = m_2\lambda_2$$

for which the smallest possible choices are $m_1 = 5$ and $m_2 = 4$. Returning to Eq. 36-25, then, we find

$$\theta = \sin^{-1}\left(\frac{1}{d} m_1\lambda_1\right) = 21°.$$

(c) There are no refraction angles greater than 90°, so we can solve for "m_{max}" (realizing it might not be an integer):

$$m_{max} = \frac{d \sin 90°}{\lambda_2} = 11$$

where we have rounded down. There are no values of m (for light of wavelength λ_2) greater than $m = 11$.

77. For $\lambda = 0.10$ nm, we have scattering for order m, and for $\lambda' = 0.075$ nm, we have scattering for order m'. From Eq. 36-34, we see that we must require

$$m\lambda = m'\lambda'$$

which suggests (looking for the smallest integer solutions) that $m = 3$ and $m' = 4$. Returning with this result and with $d = 0.25$ nm to Eq. 36-34, we obtain

$$\theta = \sin^{-1} \frac{m\lambda}{2d} = 37° .$$

Studying Figure 36-28, we conclude that the angle between incident and scattered beams is $180° - 2\theta = 106°$.

78. Letting $d \sin \theta = (L/N) \sin \theta = m\lambda$, we get

$$\lambda = \frac{(L/N) \sin \theta}{m} = \frac{(1.0 \times 10^7 \text{ nm})(\sin 30°)}{(1)(10000)} = 500 \text{ nm} .$$

79. As a slit is narrowed, the pattern spreads outward, so the question about "minimum width" suggests that we are looking at the lowest possible values of m (the label for the minimum produced by light $\lambda = 600$ nm) and m' (the label for the minimum produced by light $\lambda' = 500$ nm). Since the angles are the same, then Eq. 36-3 leads to

$$m\lambda = m'\lambda'$$

which leads to the choices $m = 5$ and $m' = 6$. We find the slit width from Eq. 36-3:

$$a = \frac{m\lambda}{\sin \theta} \approx \frac{m\lambda}{\theta}$$

which yields $a = 3.0$ mm.

80. The central diffraction envelope spans the range $-\theta_1 < \theta < +\theta_1$ where

$$\theta_1 = \sin^{-1} \frac{\lambda}{a}.$$

The maxima in the double-slit pattern are at

$$\theta_m = \sin^{-1}\frac{m\lambda}{d},$$

so that our range specification becomes

$$-\sin^{-1}\frac{\lambda}{a} < \sin^{-1}\frac{m\lambda}{d} < +\sin^{-1}\frac{\lambda}{a},$$

which we change (since sine is a monotonically increasing function in the fourth and first quadrants, where all these angles lie) to

$$-\frac{\lambda}{a} < \frac{m\lambda}{d} < +\frac{\lambda}{a}.$$

Rewriting this as $-d/a < m < +d/a$ we arrive at the result $m_{\max} < d/a \leq m_{\max} +1$. Due to the symmetry of the pattern, the multiplicity of the m values is $2m_{\max} + 1 = 17$ so that $m_{\max} = 8$, and the result becomes

$$8 < \frac{d}{a} \leq 9$$

where these numbers are as accurate as the experiment allows (that is, "9" means "9.000" if our measurements are that good).

81. (a) Use of Eq. 36-25 for the limit-wavelengths ($\lambda_1 = 700$ nm and $\lambda_2 = 550$ nm) leads to the condition

$$m_1\lambda_1 \geq m_2\lambda_2$$

for $m_1 + 1 = m_2$ (the low end of a high-order spectrum is what is overlapping with the high end of the next-lower-order spectrum). Assuming equality in the above equation, we can solve for "m_1" (realizing it might not be an integer) and obtain $m_1 \approx 4$ where we have rounded *up*. It is the fourth order spectrum that is the lowest-order spectrum to overlap with the next higher spectrum.

(b) The problem specifies $d = 1/200$ using the mm unit, and we note there are no refraction angles greater than 90°. We concentrate on the largest wavelength $\lambda = 700$ nm $= 7 \times 10^{-4}$ mm and solve Eq. 36-25 for "$m_{\max}$" (realizing it might not be an integer):

$$m_{\max} = \frac{d\sin 90°}{\lambda} = \frac{1}{(200)(7\times 10^{-4})} \approx 7$$

where we have rounded down. There are no values of m (for the appearance of the full spectrum) greater than $m = 7$.

82. From Eq. 36-3,

$$\frac{a}{\lambda} = \frac{m}{\sin\theta} = \frac{1}{\sin 45.0°} = 1.41.$$

83. (a) We use Eq. 36-12:

$$\theta = \sin^{-1}\left(\frac{1.22\lambda}{d}\right) = \sin^{-1}\left[\frac{1.22(v_s/f)}{d}\right]$$

$$= \sin^{-1}\left[\frac{(1.22)(1450\,\text{m/s})}{(25\times 10^3\,\text{Hz})(0.60\,\text{m})}\right] = 6.8°.$$

(b) Now $f = 1.0 \times 10^3$ Hz so

$$\frac{1.22\lambda}{d} = \frac{(1.22)(1450\,\text{m/s})}{(1.0\times 10^3\,\text{Hz})(0.60\,\text{m})} = 2.9 > 1.$$

Since $\sin\theta$ cannot exceed 1 there is no minimum.

84. We use Eq. 36-34. For smallest value of θ, we let $m = 1$. Thus,

$$\theta_{\min} = \sin^{-1}\left(\frac{m\lambda}{2d}\right) = \sin^{-1}\left[\frac{(1)(30\,\text{pm})}{2(0.30\times 10^3\,\text{pm})}\right] = 2.9°.$$

85. Employing Eq. 36-3, we find (with $m = 3$ and all lengths in μm)

$$\theta = \sin^{-1}\frac{m\lambda}{a} = \sin^{-1}\frac{(3)(0.5)}{2}$$

which yields $\theta = 48.6°$. Now, we use the experimental geometry ($\tan\theta = y/D$ where y locates the minimum relative to the middle of the pattern) to find

$$y = D\tan\theta = 2.27 \text{ m.}$$

86. The central diffraction envelope spans the range $-\theta_1 < \theta < +\theta_1$ where

$$\theta_1 = \sin^{-1}\frac{\lambda}{a}.$$

The maxima in the double-slit pattern are located at

$$\theta_m = \sin^{-1}\frac{m\lambda}{d},$$

so that our range specification becomes

$$-\sin^{-1}\frac{\lambda}{a} < \sin^{-1}\frac{m\lambda}{d} < +\sin^{-1}\frac{\lambda}{a},$$

which we change (since sine is a monotonically increasing function in the fourth and first quadrants, where all these angles lie) to

$$-\frac{\lambda}{a} < \frac{m\lambda}{d} < +\frac{\lambda}{a}.$$

Rewriting this as $-d/a < m < +d/a$, we find $-6 < m < +6$, or, since m is an integer, $-5 \le m \le +5$. Thus, we find eleven values of m that satisfy this requirement.

87. Assuming all $N = 2000$ lines are uniformly illuminated, we have

$$\frac{\lambda_{av}}{\Delta\lambda} = Nm$$

from Eq. 36-31 and Eq. 36-32. With $\lambda_{av} = 600$ nm and $m = 2$, we find $\Delta\lambda = 0.15$ nm.

88. Using the same notation found in Sample Problem 36-3,

$$\frac{D}{L} = \theta_R = 1.22\frac{\lambda}{d}$$

where we will assume a "typical" wavelength for visible light: $\lambda \approx 550 \times 10^{-9}$ m.

(a) With $L = 400 \times 10^3$ m and $D = 0.85$ m, the above relation leads to $d = 0.32$ m.

(b) Now with $D = 0.10$ m, the above relation leads to $d = 2.7$ m.

(c) The military satellites do not use Hubble Telescope-sized apertures. A great deal of very sophisticated optical filtering and digital signal processing techniques go into the final product, for which there is not space for us to describe here.

89. Although the angles in this problem are not particularly big (so that the small angle approximation could be used with little error), we show the solution appropriate for large as well as small angles (that is, we do not use the small angle approximation here). Eq. 36-3 gives

$$m\lambda = a\sin\theta \;\Rightarrow\; \theta = \sin^{-1}(m\lambda/a) = \sin^{-1}[2(0.42\ \mu m)/(5.1\ \mu m)] = 9.48°.$$

The geometry of Figure 35-8(a) is a useful reference (even though it shows a double slit instead of the single slit that we are concerned with here). We see in that figure the relation between y, D and θ:

$$y = D\tan\theta = (3.2\text{ m})\tan(9.48°) = 0.534\text{ m}.$$

90. The problem specifies $d = 12/8900$ using the mm unit, and we note there are no refraction angles greater than 90°. We convert $\lambda = 500$ nm to 5×10^{-4} mm and solve Eq. 36-25 for "m_{max}" (realizing it might not be an integer):

$$m_{max} = \frac{d\sin 90°}{\lambda} = \frac{12}{(8900)(5 \times 10^{-4})} \approx 2$$

where we have rounded down. There are no values of m (for light of wavelength λ) greater than $m = 2$.

91. (a) The central diffraction envelope spans the range $-\theta_1 < \theta < +\theta_1$ where

$$\theta_1 = \sin^{-1}\frac{\lambda}{a}$$

which could be further simplified *if* the small-angle approximation were justified (which it is *not*, since a is so small). The maxima in the double-slit pattern are at

$$\theta_m = \sin^{-1}\frac{m\lambda}{d}$$

so that our range specification becomes

$$-\sin^{-1}\frac{\lambda}{a} < \sin^{-1}\frac{m\lambda}{d} < +\sin^{-1}\frac{\lambda}{a}$$

which we change (since sine is a monotonically increasing function in the fourth and first quadrants, where all these angles lie) to

$$-\frac{\lambda}{a} < \frac{m\lambda}{d} < +\frac{\lambda}{a}.$$

Rewriting this as $-d/a < m < +d/a$ we arrive at the result $m_{max} < d/a \leq m_{max} + 1$. Due to the symmetry of the pattern, the multiplicity of the m values is $2m_{max} + 1 = 17$ so that $m_{max} = 8$, and the result becomes

$$8 < \frac{d}{a} \leq 9$$

where these numbers are as accurate as the experiment allows (that is, "9" means "9.000" if our measurements are that good).

92. We see that the total number of lines on the grating is $(1.8\ cm)(1400/cm) = 2520 = N$. Combining Eq. 36-31 and Eq. 36-32, we find

$$\Delta\lambda = \frac{\lambda_{avg}}{Nm} = \frac{450\ nm}{(2520)(3)} = 0.0595\ nm = 59.5\ pm.$$

93. (a) The central diffraction envelope spans the range $-\theta_1 < \theta < +\theta_1$ where

$$\theta_1 = \sin^{-1}\frac{\lambda}{a}$$

which could be further simplified if the small-angle approximation were justified (which it is not, since a is so small). The maxima in the double-slit pattern are at

$$\theta_m = \sin^{-1}\frac{m\lambda}{d}$$

so that our range specification becomes

$$-\sin^{-1}\frac{\lambda}{a} < \sin^{-1}\frac{m\lambda}{d} < +\sin^{-1}\frac{\lambda}{a}$$

which we change (since sine is a monotonically increasing function in the fourth and first quadrants, where all these angles lie) to

$$-\frac{\lambda}{a} < \frac{m\lambda}{d} < +\frac{\lambda}{a}.$$

Rewriting this as $-d/a < m < +d/a$ we arrive at the result $-7 < m < +7$ which implies (since m must be an integer) $-6 \leq m \leq +6$ which amounts to 13 distinct values for m. Thus, thirteen maxima are within the central envelope.

(b) The range (within one of the first-order envelopes) is now

$$-\sin^{-1}\frac{\lambda}{a} < \sin^{-1}\frac{m\lambda}{d} < +\sin^{-1}\frac{2\lambda}{a}$$

which leads to $d/a < m < 2d/a$ or $7 < m < 14$. Since m is an integer, this means $8 \leq m \leq 13$ which includes ^ distinct values for m in that one envelope. If we were to include the total from both first-order envelopes, the result would be twelve, but the wording of the problem implies six should be the answer (just one envelope).

94. Use of Eq. 36-21 leads to:

$$D = \frac{1.22 \lambda L}{d} = 6.1 \text{ mm.}$$

95. We refer (somewhat sloppily) to the 400 nm wavelength as "blue" and the 700 nm wavelength as "red." Consider Eq. 36-25 ($m\lambda = d \sin\theta$), for the 3rd order blue, and also for the 2nd order red:

$$(3) \; \lambda_{blue} = 1200 \text{ nm} = d \sin(\theta_{blue})$$

$$(2) \; \lambda_{red} = 1400 \text{ nm} = d \sin(\theta_{red}).$$

Since sine is an increasing function of angle (in the first quadrant) then the above set of values make clear that $\theta_{red \; (second \; order)} > \theta_{blue \; (third \; order)}$ which shows that the spectrums overlap (regardless of the value of d).

96. We note that the central diffraction envelope contains the central bright interference fringe (corresponding to $m = 0$ in Eq. 36-25) plus ten on either side of it. Since the eleventh order bright interference fringe is not seen in the central envelope, then we conclude the first diffraction minimum (satisfying $\sin\theta = \lambda/a$) coincides with the $m = 11$ instantiation of Eq. 36-25:

$$d = \frac{m\lambda}{\sin\theta} = \frac{11\lambda}{\lambda/a} = 11\,a.$$

Thus, the ratio d/a is equal to 11.

97. Following the method of Sample Problem 36-3, we have

$$\frac{1.22\lambda}{d} = \frac{D}{L}$$

where $\lambda = 550 \times 10^{-9}$ m, $D = 0.60$ m, and $d = 0.0055$ m. Thus we get $L = 4.9 \times 10^3$ m.

98. We use Eq. 36-3 for $m = 2$:

$$m\lambda = a \sin\theta \quad \Rightarrow \quad \frac{a}{\lambda} = \frac{m}{\sin\theta} = \frac{2}{\sin 37°} = 3.3.$$

99. We solve Eq. 36-25 for d:

$$d = \frac{m\lambda}{\sin\theta} = \frac{2(600 \times 10^{-9}\,\text{m})}{\sin(33°)} = 2.203 \times 10^{-6}\,\text{m} = 2.203 \times 10^{-4}\,\text{cm}$$

which is typically expressed in reciprocal form as the "number of lines per centimeter" (or per millimeter, or per inch):

$$\frac{1}{d} = 4539\,\text{lines/cm}\,.$$

The full width is 3.00 cm, so the number of lines is $(4539\,/\text{cm})(3.00\,\text{cm}) = 1.36 \times 10^4$.

100. We combine Eq. 36-31 ($R = \lambda_{avg}/\Delta\lambda$) with Eq. 36-32 ($R = Nm$) and solve for N:

$$N = \frac{\lambda_{avg}}{m\,\Delta\lambda} = \frac{590.2\,\text{nm}}{2\,(0.061\,\text{nm})} = 4.84 \times 10^3\,.$$

101. Eq. 36-14 gives the Rayleigh angle (in radians):

$$\theta_R = \frac{1.22\lambda}{d} = \frac{D}{L}$$

where the rationale behind the second equality is given in Sample Problem 36-3. We are asked to solve for d and are given $\lambda = 550 \times 10^{-9}$ m, $D = 30 \times 10^{-2}$ m, and $L = 160 \times 10^3$ m. Consequently, we obtain $d = 0.358$ m ≈ 36 cm.

102. Eq. 36-14 gives the Rayleigh angle (in radians):

$$\theta_R = \frac{1.22\lambda}{d} = \frac{D}{L}$$

where the rationale behind the second equality is given in Sample Problem 36-3. We are asked to solve for D and are given $\lambda = 500 \times 10^{-9}$ m, $d = 5.00 \times 10^{-3}$ m, and $L = 0.250$ m. Consequently, we obtain $D = 3.05 \times 10^{-5}$ m.

103. The dispersion of a grating is given by $D = d\theta/d\lambda$, where θ is the angular position of a line associated with wavelength λ. The angular position and wavelength are related by $\mathbf{d}\sin\theta = m\lambda$, where $\mathbf{d}$ is the slit separation (which we made boldfaced in order not to confuse it with the d used in the derivative, below) and m is an integer. We differentiate this expression with respect to θ to obtain

$$\frac{d\theta}{d\lambda}\mathbf{d}\cos\theta = m,$$

or

$$D = \frac{d\theta}{d\lambda} = \frac{m}{\mathbf{d}\cos\theta}.$$

Now $m = (\mathbf{d}/\lambda) \sin\theta$, so

$$D = \frac{\mathbf{d}\sin\theta}{\mathbf{d}\lambda\cos\theta} = \frac{\tan\theta}{\lambda}.$$

104. One strategy is to divide Eq. 36-25 by Eq. 36-3, assuming the same angle (a point we'll come back to, later) and the same light wavelength for both:

$$\frac{m}{m'} = \frac{m\lambda}{m'\lambda} = \frac{d\sin\theta}{a\sin\theta} = \frac{d}{a}.$$

We recall that d is measured from middle of transparent strip to the middle of the next transparent strip, which in this particular setup means $d = 2a$. Thus, $m/m' = 2$, or $m = 2m'$.

Now we interpret our result. First, the division of the equations is not valid when $m = 0$ (which corresponds to $\theta = 0$), so our remarks do not apply to the $m = 0$ maximum. Second, Eq. 36-25 gives the "bright" interference results, and Eq. 36-3 gives the "dark" diffraction results (where the latter overrules the former in places where they coincide – see Figure 36-16 in the textbook). For $m' =$ any nonzero integer, the relation $m = 2m'$ implies that $m =$ any nonzero *even* integer. As mentioned above, these are occurring at the same angle, so the even integer interference maxima are eliminated by the diffraction minima.

105. We imagine dividing the original slit into N strips and represent the light from each strip, when it reaches the screen, by a phasor. Then, at the central maximum in the diffraction pattern, we would add the N phasors, all in the same direction and each with the same amplitude. We would find that the intensity there is proportional to N^2. If we double the slit width, we need $2N$ phasors if they are each to have the amplitude of the phasors we used for the narrow slit. The intensity at the central maximum is proportional to $(2N)^2$ and is, therefore, four times the intensity for the narrow slit. The energy reaching the screen per unit time, however, is only twice the energy reaching it per unit time when the narrow slit is in place. The energy is simply redistributed. For example, the central peak is now half as wide and the integral of the intensity over the peak is only twice the analogous integral for the narrow slit.

106. The problem specifies $d = 1/500$ using the mm unit, and we note there are no refraction angles greater than 90°. We concentrate on the largest wavelength $\lambda = 700$ nm $= 7 \times 10^{-4}$ mm and solve Eq. 36-25 for "m_{max}" (realizing it might not be an integer):

$$m_{max} = \frac{d \sin 90°}{\lambda} = \frac{1}{(500)(7 \times 10^{-4})} \approx 2$$

where we have rounded down. There are no values of m (for appearance of the full spectrum) greater than $m = 2$.

107. The derivation is similar to that used to obtain Eq. 36-27. At the first minimum beyond the mth principal maximum, two waves from adjacent slits have a phase difference of $\Delta\phi = 2\pi m + (2\pi/N)$, where N is the number of slits. This implies a difference in path length of

$$\Delta L = (\Delta\phi/2\pi)\lambda = m\lambda + (\lambda/N).$$

If θ_m is the angular position of the mth maximum, then the difference in path length is also given by $\Delta L = d \sin(\theta_m + \Delta\theta)$. Thus

$$d \sin (\theta_m + \Delta\theta) = m\lambda + (\lambda/N).$$

We use the trigonometric identity

$$\sin(\theta_m + \Delta\theta) = \sin \theta_m \cos \Delta\theta + \cos \theta_m \sin \Delta\theta.$$

Since $\Delta\theta$ is small, we may approximate $\sin \Delta\theta$ by $\Delta\theta$ in radians and $\cos \Delta\theta$ by unity. Thus,

$$d \sin \theta_m + d \Delta\theta \cos \theta_m = m\lambda + (\lambda/N).$$

We use the condition $d \sin \theta_m = m\lambda$ to obtain $d \Delta\theta \cos \theta_m = \lambda/N$ and

$$\Delta\theta = \frac{\lambda}{N d \cos\theta_m}.$$

108. Referring to problem 69, we note that the angular deviation of a diffracted ray (the angle between the forward extrapolation of the incident ray and its diffracted ray) is $\psi' = \psi + \theta$. For $m = 1$, this becomes

$$\psi' = \psi + \theta = \psi + \sin^{-1}\left(\frac{\lambda}{d} - \sin \psi\right)$$

where the ratio $\lambda/d = 0.40$ using the values given in the problem statement. The graph of this is shown below (with radians used along both axes).

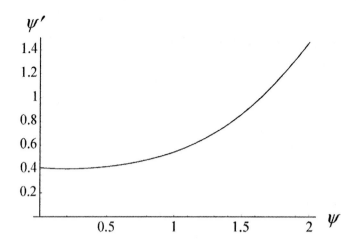

109. (a) Since the resolving power of a grating is given by $R = \lambda/\Delta\lambda$ and by Nm, the range of wavelengths that can just be resolved in order m is $\Delta\lambda = \lambda/Nm$. Here N is the number of rulings in the grating and λ is the average wavelength. The frequency f is related to the wavelength by $f\lambda = c$, where c is the speed of light. This means $f\Delta\lambda + \lambda\Delta f = 0$, so

$$\Delta\lambda = -\frac{\lambda}{f}\Delta f = -\frac{\lambda^2}{c}\Delta f$$

where $f = c/\lambda$ is used. The negative sign means that an increase in frequency corresponds to a decrease in wavelength. We may interpret Δf as the range of frequencies that can be resolved and take it to be positive. Then,

$$\frac{\lambda^2}{c}\Delta f = \frac{\lambda}{Nm}$$

and

$$\Delta f = \frac{c}{Nm\lambda}.$$

(b) The difference in travel time for waves traveling along the two extreme rays is $\Delta t = \Delta L/c$, where ΔL is the difference in path length. The waves originate at slits that are separated by $(N-1)d$, where d is the slit separation and N is the number of slits, so the path difference is $\Delta L = (N-1)d\sin\theta$ and the time difference is

$$\Delta t = \frac{(N-1)d\sin\theta}{c}.$$

If N is large, this may be approximated by $\Delta t = (Nd/c)\sin\theta$. The lens does not affect the travel time.

(c) Substituting the expressions we derived for Δt and Δf, we obtain

$$\Delta f \, \Delta t = \left(\frac{c}{Nm\lambda}\right)\left(\frac{Nd\sin\theta}{c}\right) = \frac{d\sin\theta}{m\lambda} = 1.$$

The condition $d \sin \theta = m\lambda$ for a diffraction line is used to obtain the last result.

110. There are two unknowns, the x-ray wavelength λ and the plane separation d, so data for scattering at two angles from the same planes should suffice. The observations obey Bragg's law, so

$$2d \sin \theta_1 = m_1 \lambda$$

and

$$2d \sin \theta_2 = m_2 \lambda.$$

However, these cannot be solved for the unknowns. For example, we can use the first equation to eliminate λ from the second. We obtain

$$m_2 \sin \theta_1 = m_1 \sin \theta_2,$$

an equation that does not contain either of the unknowns.

111. The key trigonometric identity used in this proof is $\sin(2\theta) = 2\sin\theta\cos\theta$. Now, we wish to show that Eq. 36-19 becomes (when $d = a$) the pattern for a single slit of width $2a$ (see Eq. 36-5 and Eq. 36-6):

$$I(\theta) = I_m \left(\frac{\sin(2\pi a\sin\theta/\lambda)}{2\pi a\sin\theta/\lambda}\right)^2 .$$

We note from Eq. 36-20 and Eq. 36-21, that the parameters β and α are identical in this case (when $d = a$), so that Eq. 36-19 becomes

$$I(\theta) = I_m \left(\frac{\cos(\pi a\sin\theta/\lambda)\sin(\pi a\sin\theta/\lambda)}{\pi a\sin\theta/\lambda}\right)^2 .$$

Multiplying numerator and denominator by 2 and using the trig identity mentioned above, we obtain

$$I(\theta) = I_m \left(\frac{2\cos(\pi a\sin\theta/\lambda)\sin(\pi a\sin\theta/\lambda)}{2\pi a\sin\theta/\lambda}\right)^2 = I_m \left(\frac{\sin(2\pi a\sin\theta/\lambda)}{2\pi a\sin\theta/\lambda}\right)^2$$

which is what we set out to show.

112. When the speaker phase difference is π rad (180°), we expect to see the "reverse" of Fig. 36-14 [translated into the acoustic context, so that "bright" becomes "loud" and "dark" becomes "quiet"]. That is, with 180° phase difference, all the peaks in Fig. 36-14 become valleys and all the valleys become peaks. As the phase changes from zero to 180° (and similarly for the change from 180° back to 360° = original pattern), the peaks should shift (and change height) in a continuous fashion – with the most dramatic feature being a large "dip" in the center diffraction envelope which deepens until it seems to split the central maximum into smaller diffraction maxima which (once the phase difference reaches π rad) will be located at angles given by $a\sin\theta = \pm\lambda$. How many interference fringes would actually "be inside" each of these smaller diffraction maxima would, of course, depend on the particular values of a, λ and d.

113. We equate Eq. 36-29 ($D = \Delta\theta/\Delta\lambda$) and Eq. 36-30 ($D = m/d\cos\theta$), and use the fact that $\sin^2\theta + \cos^2\theta = 1$, to obtain

$$\frac{\Delta\theta}{\Delta\lambda} = \frac{m}{d\sqrt{1 - \sin^2\theta}} = \frac{m}{\sqrt{d^2 - d^2\sin^2\theta}} = \frac{m}{\sqrt{d^2 - m^2\lambda^2}}$$

where we use Eq. 36-25 in that last step. Multiplying through by $\Delta\lambda$ and "simplifying" the right-hand side readily yields the final formula shown in the problem statement.

114. Among the many computer-based approaches that could be shown here, we chose a simple MAPLE program, where the details of searching for the maximum near 0.35 rad are given in the last step:

```
> restart:
> Digits:=20;
> lambda:=500; a:=5000;   #nanometers
> N:=200;  Delta[x]:=a/(N-1);  phi:=2*Pi/lambda*Delta[x]*sin(theta);
> E[h]:=Sum(cos(i*phi),i= 1 .. N);   E[v]:=Sum(sin(i*phi),i= 1 .. N);
> plot((E[h]^2 + E[v]^2)/N^2,theta= 0 .. 4);
> for inc to 9 do [theta = .35 + inc/1000,evalf(subs(theta = .35 + inc/1000,E[h]^2 +
  E[v]^2)/N^2)] od;
```

This seemed to give the maximum at $\theta = 0.353$ rad with an intensity ratio of $I/I_m = 0.00835$. A more exact treatment would give $\theta = 0.354$ rad and of $I/I_m = 0.00834$. Other maxima found in the computer-search manner indicated above were: $I/I_m = 0.0472$ at $\theta = 0.143$ rad, and $I/I_m = 0.0165$ at $\theta = 0.247$ rad.

Chapter 37

1. From the time dilation equation $\Delta t = \gamma \Delta t_0$ (where Δt_0 is the proper time interval, $\gamma = 1/\sqrt{1-\beta^2}$, and $\beta = v/c$), we obtain

$$\beta = \sqrt{1 - \left(\frac{\Delta t_0}{\Delta t}\right)^2}.$$

The proper time interval is measured by a clock at rest relative to the muon. Specifically, $\Delta t_0 = 2.2000$ μs. We are also told that Earth observers (measuring the decays of moving muons) find $\Delta t = 16.000$ μs. Therefore,

$$\beta = \sqrt{1 - \left(\frac{2.2000 \ \mu s}{16.000 \ \mu s}\right)^2} = 0.99050.$$

2. (a) We find β from $\gamma = 1/\sqrt{1-\beta^2}$:

$$\beta = \sqrt{1 - \frac{1}{\gamma^2}} = \sqrt{1 - \frac{1}{(1.0100000)^2}} = 0.14037076.$$

(b) Similarly, $\beta = \sqrt{1 - (10.000000)^{-2}} = 0.99498744$.

(c) In this case, $\beta = \sqrt{1 - (100.00000)^{-2}} = 0.99995000$.

(d) The result is $\beta = \sqrt{1 - (1000.0000)^{-2}} = 0.99999950$.

3. We solve the time dilation equation for the time elapsed (as measured by Earth observers):

$$\Delta t = \frac{\Delta t_0}{\sqrt{1 - (0.9990)^2}}$$

where $\Delta t_0 = 120$ y. This yields $\Delta t = 2684$ y $\approx 2.68 \times 10^3$ y.

4. Due to the time-dilation effect, the time between initial and final ages for the daughter is longer than the four years experienced by her father:

$$t_{f\,\text{daughter}} - t_{i\,\text{daughter}} = \gamma(4.000\text{ y})$$

where γ is Lorentz factor (Eq. 37-8). Letting T denote the age of the father, then the conditions of the problem require

$$T_i = t_{i\,\text{daughter}} + 20.00\text{ y} \quad \text{and} \quad T_f = t_{f\,\text{daughter}} - 20.00\text{ y} \ .$$

Since $T_f - T_i = 4.000$ y, then these three equations combine to give a single condition from which γ can be determined (and consequently v):

$$44 = \gamma 4 \implies \gamma = 11 \implies \beta = \frac{2\sqrt{30}}{11} = 0.9959.$$

5. In the laboratory, it travels a distance $d = 0.00105$ m $= vt$, where $v = 0.992c$ and t is the time measured on the laboratory clocks. We can use Eq. 37-7 to relate t to the proper lifetime of the particle t_0:

$$t = \frac{t_0}{\sqrt{1-(v/c)^2}} \implies t_0 = t\sqrt{1-\left(\frac{v}{c}\right)^2} = \frac{d}{0.992c}\sqrt{1-0.992^2}$$

which yields $t_0 = 4.46 \times 10^{-13}$ s $= 0.446$ ps.

6. From the value of Δt in the graph when $\beta = 0$, we infer than Δt_0 in Eq. 37-9 is 8.0 s. Thus, that equation (which describes the curve in Fig. 37-23) becomes

$$\Delta t = \frac{\Delta t_0}{\sqrt{1-(v/c)^2}} = \frac{8.0\text{ s}}{\sqrt{1-\beta^2}}$$

If we set $\beta = 0.98$ in this expression, we obtain approximately 40 s for Δt.

7. (a) The round-trip (discounting the time needed to "turn around") should be one year according to the clock you are carrying (this is your proper time interval Δt_0) and 1000 years according to the clocks on Earth which measure Δt. We solve Eq. 37-7 for β:

$$\beta = \sqrt{1-\left(\frac{\Delta t_0}{\Delta t}\right)^2} = \sqrt{1-\left(\frac{1\text{y}}{1000\text{y}}\right)^2} = 0.99999950.$$

(b) The equations do not show a dependence on acceleration (or on the direction of the velocity vector), which suggests that a circular journey (with its constant magnitude centripetal acceleration) would give the same result (if the speed is the same) as the one described in the problem. A more careful argument can be given to support this, but it should be admitted that this is a fairly subtle question which has occasionally precipitated debates among professional physicists.

8. The contracted length of the tube would be

$$L = L_0\sqrt{1-\beta^2} = (3.00\,\text{m})\sqrt{1-0.999987^2} = 0.0153\,\text{m}.$$

9. (a) The rest length L_0 = 130 m of the spaceship and its length L as measured by the timing station are related by Eq. 37-13. Therefore, $L = (130\,\text{m})\sqrt{1-(0.740)^2} = 87.4\,\text{m}.$

(b) The time interval for the passage of the spaceship is

$$\Delta t = \frac{L}{v} = \frac{87.4\,\text{m}}{(0.740)(3.00\times10^8\,\text{m/s})} = 3.94\times10^{-7}\,\text{s}.$$

10. Only the "component" of the length in the x direction contracts, so its y component stays

$$\ell'_y = \ell_y = \ell\sin30° = 0.5000\,\text{m}$$

while its x component becomes

$$\ell'_x = \ell_x\sqrt{1-\beta^2} = \ell\cos30°\sqrt{1-0.90^2} = 0.3775\,\text{m}.$$

Therefore, using the Pythagorean theorem, the length measured from S' is

$$\ell' = \sqrt{(\ell'_x)^2 + (\ell'_y)^2} = 0.63\,\text{m}.$$

11. The length L of the rod, as measured in a frame in which it is moving with speed v parallel to its length, is related to its rest length L_0 by $L = L_0/\gamma$, where $\gamma = 1/\sqrt{1-\beta^2}$ and $\beta = v/c$. Since γ must be greater than 1, L is less than L_0. For this problem, L_0 = 1.70 m and $\beta = 0.630$, so $L = (1.70\,\text{m})\sqrt{1-(0.630)^2} = 1.32\,\text{m}.$

12. (a) We solve Eq. 37-13 for v and then plug in:

$$\beta = \sqrt{1-\left(\frac{L}{L_0}\right)^2} = \sqrt{1-\left(\frac{1}{2}\right)^2} = 0.866.$$

(b) The Lorentz factor in this case is $\gamma = \dfrac{1}{\sqrt{1-(v/c)^2}} = 2.00$.

13. (a) The speed of the traveler is $v = 0.99c$, which may be equivalently expressed as 0.99 ly/y. Let d be the distance traveled. Then, the time for the trip as measured in the frame of Earth is

$$\Delta t = d/v = (26 \text{ ly})/(0.99 \text{ ly/y}) = 26.3 \text{ y} \approx 26 \text{ y}.$$

(b) The signal, presumed to be a radio wave, travels with speed c and so takes 26.0 y to reach Earth. The total time elapsed, in the frame of Earth, is

$$26.3 \text{ y} + 26.0 \text{ y} = 52.3 \text{ y} \approx 52 \text{ y}.$$

(c) The proper time interval is measured by a clock in the spaceship, so $\Delta t_0 = \Delta t/\gamma$. Now

$$\gamma = 1/\sqrt{1-\beta^2} = 1/\sqrt{1-(0.99)^2} = 7.09.$$

Thus, $\Delta t_0 = (26.3 \text{ y})/(7.09) = 3.7 \text{ y}$.

14. From the value of L in the graph when $\beta = 0$, we infer than L_0 in Eq. 37-13 is 0.80 m. Thus, that equation (which describes the curve in Fig. 37-24) with SI units understood becomes

$$L = L_0\sqrt{1 - (v/c)^2} = 0.80\sqrt{1-\beta^2}.$$

If we set $\beta = 0.95$ in this expression, we obtain approximately 0.25 m for L.

15. (a) Let $d = 23000 \text{ ly} = 23000 \, c \text{ y}$, which would give the distance in meters if we included a conversion factor for years $\rightarrow$ seconds. With $\Delta t_0 = 30$ y and $\Delta t = d/v$ (see Eq. 37-10), we wish to solve for v from Eq. 37-7. Our first step is as follows:

$$\Delta t = \frac{d}{v} = \frac{\Delta t_0}{\sqrt{1-\beta^2}} \quad \Rightarrow \quad \frac{23000 \text{ y}}{\beta} = \frac{30 \text{ y}}{\sqrt{1-\beta^2}},$$

at which point we can cancel the unit year and manipulate the equation to solve for the speed parameter β. This yields

$$\beta = \frac{1}{\sqrt{1+(30/23000)^2}} = 0.99999915.$$

(b) The Lorentz factor is $\gamma = 1/\sqrt{1-\beta^2} = 766.6680752$. Thus, the length of the galaxy measured in the traveler's frame is

$$L = \frac{L_0}{\gamma} = \frac{23000 \text{ ly}}{766.6680752} = 29.99999 \text{ ly} \approx 30 \text{ ly}.$$

16. The "coincidence" of $x = x' = 0$ at $t = t' = 0$ is important for Eq. 37-21 to apply without additional terms. In part (a), we apply these equations directly with $v = +0.400c = 1.199 \times 10^8$ m/s, and in part (c) we simply change $v \to -v$ and recalculate the primed values.

(a) The position coordinate measured in the S' frame is

$$x' = \gamma(x - vt) = \frac{x - vt}{\sqrt{1 - \beta^2}} = \frac{3.00 \times 10^8 \text{ m} - (1.199 \times 10^8 \text{m/s})(2.50 \text{s})}{\sqrt{1 - (0.400)^2}}$$

$$= 2.7 \times 10^5 \text{ m/s} \approx 0,$$

where we conclude that the numerical result (2.7×10^5 or 2.3×10^5 depending on how precise a value of v is used) is not meaningful (in the significant figures sense) and should be set equal to zero (that is, it is "consistent with zero" in view of the statistical uncertainties involved).

(b) The time coordinate measured in the S' frame is

$$t' = \gamma\left(t - \frac{vx}{c^2}\right) = \frac{t - \beta x/c}{\sqrt{1 - \beta^2}} = \frac{2.50 \text{s} - (0.400)(3.00 \times 10^8 \text{ m})/2.998 \times 10^8 \text{ m/s}}{\sqrt{1 - (0.400)^2}} = 2.29 \text{s}.$$

(c) Now, we obtain

$$x' = \frac{x + vt}{\sqrt{1 - \beta^2}} = \frac{3.00 \times 10^8 \text{ m} + (1.199 \times 10^8 \text{ m/s})(2.50 \text{ s})}{\sqrt{1 - (0.400)^2}} = 6.54 \times 10^8 \text{ m}.$$

(d) Similarly,

$$t' = \gamma\left(t + \frac{vx}{c^2}\right) = \frac{2.50 \text{s} + (0.400)(3.00 \times 10^8 \text{ m})/2.998 \times 10^8 \text{ m/s}}{\sqrt{1 - (0.400)^2}} = 3.16 \text{s}.$$

17. The proper time is not measured by clocks in either frame S or frame S' since a single clock at rest in either frame cannot be present at the origin and at the event. The full Lorentz transformation must be used:

$$x' = \gamma(x - vt) \quad \text{and} \quad t' = \gamma(t - \beta x/c)$$

where $\beta = v/c = 0.950$ and $\gamma = 1/\sqrt{1-\beta^2} = 1/\sqrt{1-(0.950)^2} = 3.20256$. Thus,

$$x' = (3.20256)\left(100\times10^3\,\text{m} - (0.950)(2.998\times10^8\,\text{m/s})(200\times10^{-6}\text{s})\right)$$
$$= 1.38\times10^5\,\text{m} = 138\,\text{km}.$$

(b) The temporal coordinate in S' is

$$t' = (3.20256)\left[200\times10^{-6}\text{s} - \frac{(0.950)(100\times10^3\,\text{m})}{2.998\times10^8\,\text{m/s}}\right] = -3.74\times10^{-4}\,\text{s} = -374\,\mu s.$$

18. The "coincidence" of $x = x' = 0$ at $t = t' = 0$ is important for Eq. 37-21 to apply without additional terms. We label the event coordinates with subscripts: $(x_1, t_1) = (0, 0)$ and $(x_2, t_2) = (3000, 4.0\times10^{-6})$ with SI units understood.

(a) We expect $(x'_1, t'_1) = (0, 0)$, and this may be verified using Eq. 37-21.

(b) We now compute (x'_2, t'_2), assuming $v = +0.60c = +1.799\times10^8$ m/s (the sign of v is not made clear in the problem statement, but the Figure referred to, Fig. 37-9, shows the motion in the positive x direction).

$$x'_2 = \frac{x - vt}{\sqrt{1-\beta^2}} = \frac{3000 - (1.799\times10^8)(4.0\times10^{-6})}{\sqrt{1-(0.60)^2}} = 2.85\times10^3$$

$$t'_2 = \frac{t - \beta x/c}{\sqrt{1-\beta^2}} = \frac{4.0\times10^{-6} - (0.60)(3000)/(2.998\times10^8)}{\sqrt{1-(0.60)^2}} = -2.5\times10^{-6}$$

(c) The two events in frame S occur in the order: first 1, then 2. However, in frame S' where $t'_2 < 0$, they occur in the reverse order: first 2, then 1. So the two observers see the two events in the reverse sequence.

We note that the distances $x_2 - x_1$ and $x'_2 - x'_1$ are larger than how far light can travel during the respective times $(c(t_2 - t_1) = 1.2$ km and $c\,|t'_2 - t'_1| \approx 750\,\text{m})$, so that no inconsistencies arise as a result of the order reversal (that is, no signal from event 1 could arrive at event 2 or vice versa).

19. (a) We take the flashbulbs to be at rest in frame S, and let frame S' be the rest frame of the second observer. Clocks in neither frame measure the proper time interval between the flashes, so the full Lorentz transformation (Eq. 37-21) must be used. Let t_s be the time and x_s be the coordinate of the small flash, as measured in frame S. Then, the time of the small flash, as measured in frame S', is

$$t'_s = \gamma \left(t_s - \frac{\beta\, x_s}{c} \right)$$

where $\beta = v/c = 0.250$ and

$$\gamma = 1/\sqrt{1-\beta^2} = 1/\sqrt{1-(0.250)^2} = 1.0328.$$

Similarly, let t_b be the time and x_b be the coordinate of the big flash, as measured in frame S. Then, the time of the big flash, as measured in frame S', is

$$t'_b = \gamma \left(t_b - \frac{\beta\, x_b}{c} \right).$$

Subtracting the second Lorentz transformation equation from the first and recognizing that $t_s = t_b$ (since the flashes are simultaneous in S), we find

$$\Delta t' = \frac{\gamma \beta (x_s - x_b)}{c} = \frac{(1.0328)(0.250)(30\times10^3\ \text{m})}{3.00\times10^8\ \text{m/s}} = 2.58\times10^{-5}\ \text{s}$$

where $\Delta t' = t'_b - t'_s$.

(b) Since $\Delta t'$ is negative, t'_b is greater than t'_s . The small flash occurs first in S'.

20. We refer to the solution of problem 18. We wish to adjust Δt so that

$$\Delta x' = 0 = \gamma(-720\,\text{m} - v\Delta t)$$

in the limiting case of $|v| \to c$. Thus,

$$\Delta t = \frac{720\,\text{m}}{2.998\times10^8\ \text{m/s}} = 2.40\times10^{-6}\ \text{s}.$$

21. (a) The Lorentz factor is

$$\gamma = \frac{1}{\sqrt{1-\beta^2}} = \frac{1}{\sqrt{1-(0.600)^2}} = 1.25.$$

(b) In the unprimed frame, the time for the clock to travel from the origin to $x = 180$ m is

$$t = \frac{x}{v} = \frac{180\,\text{m}}{(0.600)(3.00\times10^8\ \text{m/s})} = 1.00\times10^{-6}\ \text{s}.$$

The proper time interval between the two events (at the origin and at $x = 180$ m) is measured by the clock itself. The reading on the clock at the beginning of the interval is zero, so the reading at the end is

$$t' = \frac{t}{\gamma} = \frac{1.00 \times 10^{-6}\,\text{s}}{1.25} = 8.00 \times 10^{-7}\,\text{s} .$$

22. The time-dilation information in the problem (particularly, the 15 s on "his wristwatch… which takes 30.0 s according to you") reveals Lorentz factor is $\gamma = 2.00$ (see Eq. 37-9), which implies his speed is $v = 0.866c$.

(a) With $\gamma = 2.00$, Eq. 37-13 implies the contracted length is 0.500 m.

(b) There is no contraction along direction perpendicular to the direction of motion (or "boost" direction), so meter stick 2 still measures 1.00 m long.

(c) As in part (b), the answer is 1.00 m.

(d) Eq. 1′ in Table 37-2 gives

$$x_2' - x_1' = \gamma(\Delta x' - v\Delta t') = (2.00)[20.0 - (0.866)(2.998 \times 10^8)(40.0 \times 10^{-9})] = 19.2\ \text{m} .$$

(e) Eq. 2′ in Table 37-2 gives

$$t_2' - t_1' = \gamma(\Delta t' - v\Delta x'/c^2) = (2.00)[\ 40.0 \times 10^{-9} - (0.866)(20.0)/(2.998 \times 10^8)^2]$$

which yields −35.5 ns. In absolute value, the two events are separated by 35.5 ns.

(f) The negative sign obtained in part (e) implies event 2 occurred before event 1.

23. (a) In frame S, our coordinates are such that $x_1 = +1200$ m for the big flash, and $x_2 = 1200 - 720 = 480$ m for the small flash (which occurred later). Thus,

$$\Delta x = x_2 - x_1 = -720\ \text{m}.$$

If we set $\Delta x' = 0$ in Eq. 37-25, we find

$$0 = \gamma(\Delta x - v\Delta t) = \gamma\left(-720\,\text{m} - v(5.00 \times 10^{-6}\,\text{s})\right)$$

which yields $v = -1.44 \times 10^8$ m/s, or $\beta = v/c = 0.480$.

(b) The negative sign in part (a) implies that frame S' must be moving in the $-x$ direction.

(c) Eq. 37-28 leads to

$$\Delta t' = \gamma \left(\Delta t - \frac{v \Delta x}{c^2} \right) = \gamma \left(5.00 \times 10^{-6}\, s - \frac{(-1.44 \times 10^8\, m/s)(-720\, m)}{(2.998 \times 10^8\, m/s)^2} \right)$$

which turns out to be positive (regardless of the specific value of γ). Thus, the order of the flashes is the same in the S' frame as it is in the S frame (where Δt is also positive). Thus, the big flash occurs first, and the small flash occurs later.

(d) Finishing the computation begun in part (c), we obtain

$$\Delta t' = \frac{5.00 \times 10^{-6}\, s - (-1.44 \times 10^8\, m/s)(-720\, m)/(2.998 \times 10^8\, m/s)^2}{\sqrt{1 - 0.480^2}} = 4.39 \times 10^{-6}\, s .$$

24. From Eq. 2 in Table 37-2, we have $\Delta t = v\, \gamma\, \Delta x'/c^2 + \gamma\, \Delta t'$. The coefficient of $\Delta x'$ is the slope (4.0 µs / 400 m) of the graph, and the last term involving $\Delta t'$ is the "y-intercept" of the graph. From the first observation, we can solve for $\beta = v/c = 0.949$ and consequently $\gamma = 3.16$. Then, from the second observation, we find

$$\Delta t' = (2.0\ \text{µs})/\gamma = 0.63\ \text{µs}.$$

25. (a) Eq. 2′ of Table 37-2, with time in microseconds, becomes

$$\Delta t' = \gamma(\Delta t - \beta \Delta x/c) = \gamma[1.00 - \beta(400/299.8)]$$

where the Lorentz factor is itself a function of β (see Eq. 37-8).

(b) A plot of $\Delta t'$ as a function of β in the range $0 < \beta < 0.01$ is shown below:

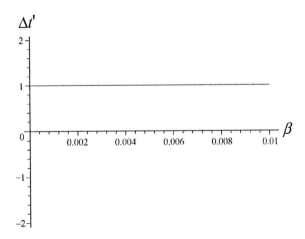

Note the limits of the vertical axis are +2 µs and −2 µs. We note how "flat" the curve is in this graph; the reason is that for low values of β, Bullwinkle's measure of the temporal separation between the two events is approximately our measure, namely +1.0 µs. There are no non-intuitive relativistic effects in this case.

(c) A plot of $\Delta t'$ as a function of β in the range $0.1 < \beta < 1$ is shown below:

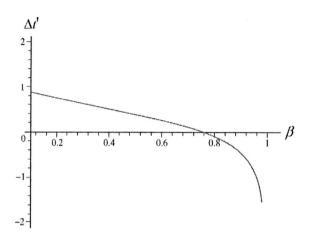

(d) Setting

$$\Delta t' = \gamma(\Delta t - \beta\Delta x/c) = \gamma[1.00 - \beta(400/299.8)] = 0,$$

leads to $\beta = 299.8/400 \approx 0.750$.

(e) For the graph shown in part (c), that as we increase the speed, the temporal separation according to Bullwinkle is positive for the lower values and then goes to zero and finally (as the speed approaches that of light) becomes progressively more negative. For the lower speeds with $\Delta t' > 0 \Rightarrow t_A' < t_B'$, or $0 < \beta < 0.750$, according to Bullwinkle event A occurs before event B just as we observe.

(f) For the higher speeds with $\Delta t' < 0 \Rightarrow t_A' > t_B'$, or $0.750 < \beta < 1$, according to Bullwinkle event B occurs before event A (the opposite of what we observe).

(g) No, event A cannot cause event B or vice versa. We note that

$$\Delta x/\Delta t = (400 \text{ m})/(1.00 \text{ µs}) = 4.00 \times 10^8 \text{ m/s} > c.$$

A signal cannot travel from event A to event B without exceeding c, so causal influences cannot originate at A and thus affect what happens at B, or vice versa.

26. (a) From Table 37-2, we find

$$\Delta x' = \gamma(\Delta x - \beta c\Delta t) = \gamma[400 \text{ m} - \beta c(1.00 \text{ µs})] = \frac{400 \text{ m} - (299.8 \text{ m})\beta}{\sqrt{1 - \beta^2}}$$

(b) A plot of $\Delta x'$ as a function of β with $0 < \beta < 0.01$ is shown below:

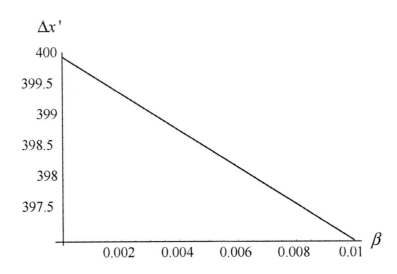

(c) A plot of $\Delta x'$ as a function of β with $0.1 < \beta < 1$ is shown below:

(d) To find the minimum, we can take a derivative of $\Delta x'$ with respect to β, simplify, and then set equal to zero:

$$\frac{d\,\Delta x'}{d\,\beta} = \gamma^3(\beta\Delta x - c\Delta t) = 0 \quad \Rightarrow \quad \beta = \frac{c\Delta t}{\Delta x} = 0.7495 \approx 0.750.$$

(e) Substituting this value of β into the part (a) expression yields $\Delta x' = 264.8$ m ≈ 265 m for its minimum value.

27. We assume S' is moving in the $+x$ direction. With $u' = +0.40c$ and $v = +0.60c$, Eq. 37-29 yields

$$u = \frac{u'+v}{1+u'v/c^2} = \frac{0.40c + 0.60c}{1+(0.40c)(+0.60c)/c^2} = 0.81c .$$

28. Using the notation of Eq. 37-29 and taking "away" (from us) as the positive direction, the problem indicates $v = +0.4c$ and $u = +0.8c$ (with 3 significant figures understood). We solve for the velocity of Q_2 relative to Q_1 (in multiple of c):

$$\frac{u'}{c} = \frac{u/c - v/c}{1 - uv/c^2} = \frac{0.8 - 0.4}{1 - (0.8)(0.4)} = 0.588$$

in a direction away from Earth.

29. (a) One thing Einstein's relativity has in common with the more familiar (Galilean) relativity is the reciprocity of relative velocity. If Joe sees Fred moving at 20 m/s eastward away from him (Joe), then Fred should see Joe moving at 20 m/s westward away from him (Fred). Similarly, if we see Galaxy A moving away from us at $0.35c$ then an observer in Galaxy A should see our galaxy move away from him at $0.35c$, or 0.35 in multiple of c.

(b) We take the positive axis to be in the direction of motion of Galaxy A, as seen by us. Using the notation of Eq. 37-29, the problem indicates $v = +0.35c$ (velocity of Galaxy A relative to Earth) and $u = -0.35c$ (velocity of Galaxy B relative to Earth). We solve for the velocity of B relative to A:

$$\frac{u'}{c} = \frac{u/c - v/c}{1 - uv/c^2} = \frac{(-0.35) - 0.35}{1 - (-0.35)(0.35)} = -0.62,$$

or $|u'/c| = 0.62$.

30. (a) We use Eq. 37-29:

$$v = \frac{v' + u}{1 + uv'/c^2} = \frac{0.47c + 0.62c}{1 + (0.47)(0.62)} = 0.84c,$$

in the direction of increasing x (since $v > 0$). In unit-vector notation, we have $\vec{v} = (0.84c)\hat{i}$.

(b) The classical theory predicts that $v = 0.47c + 0.62c = 1.1c$, or $\vec{v} = (1.1c)\hat{i}$

(c) Now $v' = -0.47c\,\hat{i}$ so

$$v = \frac{v' + u}{1 + uv'/c^2} = \frac{-0.47c + 0.62c}{1 + (-0.47)(0.62)} = 0.21c,$$

or $\vec{v} = (0.21c)\hat{i}$

(d) By contrast, the classical prediction is $v = 0.62c - 0.47c = 0.15c$, or $\vec{v} = (0.15c)\hat{i}$

31. Using the notation of Eq. 37-29 and taking the micrometeorite motion as the positive direction, the problem indicates $v = -0.82c$ (spaceship velocity) and $u = +0.82c$ (micrometeorite velocity). We solve for the velocity of the micrometeorite relative to the spaceship:

$$u' = \frac{u-v}{1-uv/c^2} = \frac{0.82c-(-0.82c)}{1-(0.82)(-0.82)} = 0.98c$$

or 2.94×10^8 m/s. Using Eq. 37-10, we conclude that observers on the ship measure a transit time for the micrometeorite (as it passes along the length of the ship) equal to

$$\Delta t = \frac{d}{u'} = \frac{350\,\text{m}}{2.94 \times 10^8 \,\text{m/s}} = 1.2 \times 10^{-6}\,\text{s} .$$

32. The Figure shows that $u' = 0.80c$ when $v = 0$. We therefore infer (using the notation of Eq. 37-29) that $u = 0.80c$. Now, u is a fixed value and v is variable, so u' as a function of v is given by

$$u' = \frac{0.80c - v}{1 - 0.80\,v/c}$$

which is Eq. 37-29 rearranged so that u' is isolated on the left-hand side. We use this expression to answer parts (a) and (b).

(a) Substituting $v = 0.90c$ in the expression above leads to $u' = -0.357c \approx -0.36c$.

(b) Substituting $v = c$ in the expression above leads to $u' = -c$ (regardless of the value of u).

33. (a) In the messenger's rest system (called S_m), the velocity of the armada is

$$v' = \frac{v - v_m}{1 - vv_m/c^2} = \frac{0.80c - 0.95c}{1 - (0.80c)(0.95c)/c^2} = -0.625c .$$

The length of the armada as measured in S_m is

$$L_1 = \frac{L_0}{\gamma_{v'}} = (1.0\,\text{ly})\sqrt{1-(-0.625)^2} = 0.781\,\text{ly} .$$

Thus, the length of the trip is

$$t' = \frac{L'}{|v'|} = \frac{0.781\,\mathrm{ly}}{0.625c} = 1.25\ \mathrm{y}\ .$$

(b) In the armada's rest frame (called S_a), the velocity of the messenger is

$$v' = \frac{v - v_a}{1 - vv_a / c^2} = \frac{0.95c - 0.80c}{1 - (0.95c)(0.80c)/c^2} = 0.625c\ .$$

Now, the length of the trip is

$$t' = \frac{L_0}{v'} = \frac{1.0\,\mathrm{ly}}{0.625c} = 1.60\ \mathrm{y}\ .$$

(c) Measured in system S, the length of the armada is

$$L = \frac{L_0}{\gamma} = 1.0\,\mathrm{ly}\sqrt{1 - (0.80)^2} = 0.60\ \mathrm{ly}\ ,$$

so the length of the trip is

$$t = \frac{L}{v_m - v_a} = \frac{0.60\,\mathrm{ly}}{0.95c - 0.80c} = 4.00\ \mathrm{y}\ .$$

34. (a) Eq. 37-34 leads to

$$v = \frac{\Delta\lambda}{\lambda}c = \frac{12.00\,\mathrm{nm}}{513.0\,\mathrm{nm}}(2.998 \times 10^8\,\mathrm{m/s}) = 7.000 \times 10^6\,\mathrm{m/s}.$$

(b) The line is shifted to a larger wavelength, which means shorter frequency. Recalling Eq. 37-31 and the discussion that follows it, this means galaxy NGC is moving away from Earth.

35. The spaceship is moving away from Earth, so the frequency received is given directly by Eq. 37-31. Thus,

$$f = f_0\sqrt{\frac{1-\beta}{1+\beta}} = (100\,\mathrm{MHz})\sqrt{\frac{1-0.9000}{1+0.9000}} = 22.9\,\mathrm{MHz}\ .$$

36. (a) Eq. 37-34 leads to a speed of

$$v = \frac{\Delta\lambda}{\lambda}c = (0.004)(3.0 \times 10^8\,\mathrm{m/s}) = 1.2 \times 10^6\,\mathrm{m/s} \approx 1 \times 10^6\,\mathrm{m/s}.$$

(b) The galaxy is receding.

37. We obtain

$$v = \frac{\Delta\lambda}{\lambda}c = \left(\frac{620-540}{620}\right)c = 0.13c.$$

38. We use the transverse Doppler shift formula, Eq. 37-37: $f = f_0\sqrt{1-\beta^2}$, or

$$\frac{1}{\lambda} = \frac{1}{\lambda_0}\sqrt{1-\beta^2}.$$

We solve for $\lambda - \lambda_0$:

$$\lambda - \lambda_0 = \lambda_0\left(\frac{1}{\sqrt{1-\beta^2}}-1\right) = (589.00\,\text{mm})\left[\frac{1}{\sqrt{1-(0.100)^2}}-1\right] = +2.97\,\text{nm}.$$

39. (a) The frequency received is given by

$$f = f_0\sqrt{\frac{1-\beta}{1+\beta}} \quad\Rightarrow\quad \frac{c}{\lambda} = \frac{c}{\lambda_0}\sqrt{\frac{1-0.20}{1+0.20}}$$

which implies

$$\lambda = (450\,\text{nm})\sqrt{\frac{1+0.20}{1-0.20}} = 550\,\text{nm}.$$

(b) This is in the yellow portion of the visible spectrum.

40. (a) The work-kinetic energy theorem applies as well to Einsteinian physics as to Newtonian; the only difference is the specific formula for kinetic energy. Thus, we use $W = \Delta K = m_e c^2(\gamma - 1)$ (Eq. 37-52) and $m_e c^2 = 511$ keV $= 0.511$ MeV (Table 37-3), and obtain

$$W = m_e c^2\left(\frac{1}{\sqrt{1-\beta^2}}-1\right) = (511\,\text{keV})\left[\frac{1}{\sqrt{1-(0.500)^2}}-1\right] = 79.1\ \text{keV}.$$

(b) $W = (0.511\,\text{MeV})\left(\frac{1}{\sqrt{1-(0.990)^2}}-1\right) = 3.11$ MeV.

(c) $W = (0.511\,\text{MeV})\left(\dfrac{1}{\sqrt{1-(0.990)^2}} - 1\right) = 10.9\,\text{MeV}.$

41. (a) From Eq. 37-52, $\gamma = (K/mc^2) + 1$, and from Eq. 37-8, the speed parameter is $\beta = \sqrt{1-(1/\gamma)^2}$. Table 37-3 gives $m_e c^2 = 511\,\text{keV} = 0.511\,\text{MeV}$, so the Lorentz factor is

$$\gamma = \frac{100\,\text{MeV}}{0.511\,\text{MeV}} + 1 = 196.695.$$

(b) The speed parameter is

$$\beta = \sqrt{1 - \frac{1}{(196.695)^2}} = 0.999987.$$

Thus, the speed of the electron is $0.999987c$, or 99.9987% of the speed of light.

42. The mass change is

$$\Delta M = (4.002603\,\text{u} + 15.994915\,\text{u}) - (1.007825\,\text{u} + 18.998405\,\text{u}) = -0.008712\,\text{u}.$$

Using Eq. 37-50 and Eq. 37-46, this leads to

$$Q = -\Delta M\, c^2 = -(-0.008712\,\text{u})(931.5\,\text{MeV}\,/\,\text{u}) = 8.12\,\text{MeV}.$$

43. (a) The work-kinetic energy theorem applies as well to Einsteinian physics as to Newtonian; the only difference is the specific formula for kinetic energy. Thus, we use $W = \Delta K$ where $K = m_e c^2 (\gamma - 1)$ (Eq. 37-52), and $m_e c^2 = 511\,\text{keV} = 0.511\,\text{MeV}$ (Table 37-3). Noting that $\Delta K = m_e c^2 (\gamma_f - \gamma_i)$, we obtain

$$W = m_e c^2 \left(\frac{1}{\sqrt{1-\beta_f^2}} - \frac{1}{\sqrt{1-\beta_i^2}}\right) = (511\,\text{keV})\left(\frac{1}{\sqrt{1-(0.19)^2}} - \frac{1}{\sqrt{1-(0.18)^2}}\right)$$

$$= 0.996\ \text{keV} \approx 1.0\ \text{keV}.$$

(b) Similarly,

$$W = (511\,\text{keV})\left(\frac{1}{\sqrt{1-(0.99)^2}} - \frac{1}{\sqrt{1-(0.98)^2}}\right) = 1055\,\text{keV} \approx 1.1\ \text{MeV}.$$

We see the dramatic increase in difficulty in trying to accelerate a particle when its initial speed is very close to the speed of light.

44. From Eq. 28-37, we have

$$Q = -\Delta Mc^2 = -(3(4.00151u) - 11.99671u)c^2 = -(0.00782u)(931.5 \text{MeV/u})$$
$$= -7.28 \text{Mev}.$$

Thus, it takes a minimum of 7.28 MeV supplied to the system to cause this reaction. We note that the masses given in this problem are strictly for the nuclei involved; they are not the "atomic" masses which are quoted in several of the other problems in this chapter.

45. (a) The strategy is to find the γ factor from $E = 14.24 \times 10^{-9}$ J and $m_p c^2 = 1.5033 \times 10^{-10}$ J and from that find the contracted length. From the energy relation (Eq. 37-48), we obtain

$$\gamma = \frac{E}{mc^2} = 94.73.$$

Consequently, Eq. 37-13 yields

$$L = \frac{L_0}{\gamma} = 0.222 \text{ cm} = 2.22 \times 10^{-3} \text{ m}.$$

(b) The time dilation formula (Eq. 37-7) leads to

$$\Delta t = \gamma \Delta t_0 = 7.01 \times 10^{-10} \text{ s}$$

which can be checked using $\Delta t = L_0/v$ in our frame of reference.

(c) From the γ factor, we find the speed:

$$v = c\sqrt{1 - \left(\frac{1}{\gamma}\right)^2} = 0.99994c.$$

Therefore, the trip (according to the proton) took

$$\Delta t_0 = 2.22 \times 10^{-3}/0.99994c = 7.40 \times 10^{-12} \text{ s}.$$

46. (a) From the information in the problem, we see that each kilogram of TNT releases $(3.40 \times 10^6 \text{ J/mol})/(0.227 \text{ kg/mol}) = 1.50 \times 10^7$ J. Thus,

$$(1.80 \times 10^{14} \text{ J})/(1.50 \times 10^7 \text{ J/kg}) = 1.20 \times 10^7 \text{ kg}$$

of TNT are needed. This is equivalent to a weight of $\approx 1.2 \times 10^8$ N.

(b) This is certainly more than can be carried in a backpack. Presumably, a train would be required.

(c) We have $0.00080mc^2 = 1.80 \times 10^{14}$ J, and find $m = 2.50$ kg of fissionable material is needed. This is equivalent to a weight of about 25 N, or 5.5 pounds.

(d) This can be carried in a backpack.

47. We set Eq. 37-55 equal to $(3.00mc^2)^2$, as required by the problem, and solve for the speed. Thus,

$$(pc)^2 + (mc^2)^2 = 9.00(mc^2)^2$$

leads to $p = mc\sqrt{8} \approx 2.83mc$.

48. (a) Using $K = m_e c^2 (\gamma - 1)$ (Eq. 37-52) and

$$m_e c^2 = 510.9989 \text{ keV} = 0.5109989 \text{ MeV},$$

we obtain

$$\gamma = \frac{K}{m_e c^2} + 1 = \frac{1.0000000 \text{ keV}}{510.9989 \text{ keV}} + 1 = 1.00195695 \approx 1.0019570.$$

(b) Therefore, the speed parameter is

$$\beta = \sqrt{1 - \frac{1}{\gamma^2}} = \sqrt{1 - \frac{1}{(1.0019570)^2}} = 0.062469542.$$

(c) For $K = 1.0000000$ MeV, we have

$$\gamma = \frac{K}{m_e c^2} + 1 = \frac{1.0000000 \text{ MeV}}{0.5109989 \text{ MeV}} + 1 = 2.956951375 \approx 2.9569514.$$

(d) The corresponding speed parameter is

$$\beta = \sqrt{1 - \gamma^{-2}} = 0.941079236 \approx 0.94107924.$$

(e) For $K = 1.0000000$ GeV, we have

$$\gamma = \frac{K}{m_e c^2} + 1 = \frac{1000.0000 \text{ MeV}}{0.5109989 \text{ MeV}} + 1 = 1957.951375 \approx 1957.9514.$$

(f) The corresponding speed parameter is

$$\beta = \sqrt{1-\gamma^{-2}} = 0.99999987$$

49. Since the rest energy E_0 and the mass m of the quasar are related by $E_0 = mc^2$, the rate P of energy radiation and the rate of mass loss are related by $P = dE_0/dt = (dm/dt)c^2$. Thus,

$$\frac{dm}{dt} = \frac{P}{c^2} = \frac{1 \times 10^{41}\,\text{W}}{\left(2.998 \times 10^8\,\text{m/s}\right)^2} = 1.11 \times 10^{24}\,\text{kg/s.}$$

Since a solar mass is 2.0×10^{30} kg and a year is 3.156×10^7 s,

$$\frac{dm}{dt} = \left(1.11 \times 10^{24}\,\text{kg/s}\right)\left(\frac{3.156 \times 10^7\,\text{s/y}}{2.0 \times 10^{30}\,\text{kg/smu}}\right) \approx 18\,\text{smu/y.}$$

50. From Eq. 37-52, $\gamma = (K/mc^2) + 1$, and from Eq. 37-8, the speed parameter is $\beta = \sqrt{1-(1/\gamma)^2}$.

(a) Table 37-3 gives $m_e c^2 = 511$ keV $= 0.511$ MeV, so the Lorentz factor is

$$\gamma = \frac{10.00\,\text{MeV}}{0.5110\,\text{MeV}} + 1 = 20.57,$$

(b) and the speed parameter is

$$\beta = \sqrt{1 - \frac{1}{(20.57)^2}} = 0.9988.$$

(c) Using $m_p c^2 = 938.272$ MeV, the Lorentz factor is

$$\gamma = 1 + 10.00\,\text{MeV}/938.272\,\text{MeV} = 1.01065 \approx 1.011.$$

(d) The speed parameter is

$$\beta = \sqrt{1-\gamma^{-2}} = 0.144844 \approx 0.1448.$$

(e) With $m_\alpha c^2 = 3727.40$ MeV, we obtain $\gamma = 10.00/3727.4 + 1 = 1.00268 \approx 1.003$.

(f) The speed parameter is

$$\beta = \sqrt{1 - \gamma^{-2}} = 0.0731037 \approx 0.07310 .$$

51. (a) We set Eq. 37-41 equal to mc, as required by the problem, and solve for the speed. Thus,

$$\frac{mv}{\sqrt{1 - v^2 / c^2}} = mc$$

leads to $\beta = 1/\sqrt{2} = 0.707$.

(b) Substituting $\beta = 1/\sqrt{2}$ into the definition of γ, we obtain

$$\gamma = \frac{1}{\sqrt{1 - v^2 / c^2}} = \frac{1}{\sqrt{1 - (1/2)}} = \sqrt{2} \approx 1.41.$$

(c) The kinetic energy is

$$K = (\gamma - 1) mc^2 = (\sqrt{2} - 1) mc^2 = 0.414 mc^2 = 0.414 E_0 .$$

which implies $K / E_0 = 0.414$.

52. (a) We set Eq. 37-52 equal to $2mc^2$, as required by the problem, and solve for the speed. Thus,

$$mc^2 \left(\frac{1}{\sqrt{1 - \beta^2}} - 1 \right) = 2mc^2$$

leads to $\beta = 2\sqrt{2}/3 \approx 0.943$.

(b) We now set Eq. 37-48 equal to $2mc^2$ and solve for the speed. In this case,

$$\frac{mc^2}{\sqrt{1 - \beta^2}} = 2mc^2$$

leads to $\beta = \sqrt{3}/2 \approx 0.866$.

53. The energy equivalent of one tablet is

$$mc^2 = (320 \times 10^{-6} \text{ kg}) (3.00 \times 10^8 \text{ m/s})^2 = 2.88 \times 10^{13} \text{ J.}$$

This provides the same energy as

$$(2.88 \times 10^{13} \text{ J})/(3.65 \times 10^7 \text{ J/L}) = 7.89 \times 10^5 \text{ L}$$

of gasoline. The distance the car can go is

$$d = (7.89 \times 10^5 \text{ L})(12.75 \text{ km/L}) = 1.01 \times 10^7 \text{ km}.$$

This is roughly 250 times larger than the circumference of Earth (see Appendix C).

54. (a) Squaring Eq. 37-47 gives

$$E^2 = \left(mc^2\right)^2 + 2mc^2 K + K^2$$

which we set equal to Eq. 37-55. Thus,

$$\left(mc^2\right)^2 + 2mc^2 K + K^2 = \left(pc\right)^2 + \left(mc^2\right)^2 \Rightarrow m = \frac{\left(pc\right)^2 - K^2}{2Kc^2}.$$

(b) At low speeds, the pre-Einsteinian expressions $p = mv$ and $K = \frac{1}{2}mv^2$ apply. We note that $pc \gg K$ at low speeds since $c \gg v$ in this regime. Thus,

$$m \to \frac{\left(mvc\right)^2 - \left(\frac{1}{2}mv^2\right)^2}{2\left(\frac{1}{2}mv^2\right)c^2} \approx \frac{\left(mvc\right)^2}{2\left(\frac{1}{2}mv^2\right)c^2} = m.$$

(c) Here, $pc = 121$ MeV, so

$$m = \frac{121^2 - 55^2}{2(55)c^2} = 105.6 \text{ MeV}/\text{c}^2.$$

Now, the mass of the electron (see Table 37-3) is $m_e = 0.511$ MeV/c^2, so our result is roughly 207 times bigger than an electron mass, i.e., $m/m_e \approx 207$. The particle is a muon.

55. The distance traveled by the pion in the frame of Earth is (using Eq. 37-12) $d = v\Delta t$. The proper lifetime Δt_0 is related to Δt by the time-dilation formula: $\Delta t = \gamma \Delta t_0$. To use this equation, we must first find the Lorentz factor γ (using Eq. 37-48). Since the total energy of the pion is given by $E = 1.35 \times 10^5$ MeV and its mc^2 value is 139.6 MeV, then

$$\gamma = \frac{E}{mc^2} = \frac{1.35 \times 10^5 \text{ MeV}}{139.6 \text{ MeV}} = 967.05.$$

Therefore, the lifetime of the moving pion as measured by Earth observers is

$$\Delta t = \gamma \Delta t_0 = (967.1)(35.0 \times 10^{-9} \text{ s}) = 3.385 \times 10^{-5} \text{ s},$$

and the distance it travels is

$$d \approx c\Delta t = (2.998 \times 10^8 \text{ m/s})(3.385 \times 10^{-5} \text{ s}) = 1.015 \times 10^4 \text{ m} = 10.15 \text{ km}$$

where we have approximated its speed as c (note: its speed can be found by solving Eq. 37-8, which gives $v = 0.9999995c$; this more precise value for v would not significantly alter our final result). Thus, the altitude at which the pion decays is 120 km − 10.15 km = 110 km.

56. (a) The binomial theorem tells us that, for x small,

$$(1 + x)^{\nu} \approx 1 + \nu x + \tfrac{1}{2} \nu(\nu - 1) x^2$$

if we ignore terms involving x^3 and higher powers (this is reasonable since if x is small, say $x = 0.1$, then x^3 is much smaller: $x^3 = 0.001$). The relativistic kinetic energy formula, when the speed v is much smaller than c, has a term that we can apply the binomial theorem to; identifying $-\beta^2$ as x and $-1/2$ as ν, we have

$$\gamma = (1 - \beta^2)^{-1/2} \approx 1 + (-\tfrac{1}{2})(-\beta^2) + \tfrac{1}{2}(-\tfrac{1}{2})((-\tfrac{1}{2}) - 1)(-\beta^2)^2.$$

Substituting this into Eq. 37-52 leads to

$$K = mc^2(\gamma - 1) \approx mc^2((-\tfrac{1}{2})(-\beta^2) + \tfrac{1}{2}(-\tfrac{1}{2})((-\tfrac{1}{2}) - 1)(-\beta^2)^2)$$

which simplifies to

$$K \approx \tfrac{1}{2} mc^2 \beta^2 + \tfrac{3}{8} mc^2 \beta^4 = \tfrac{1}{2} mv^2 + \tfrac{3}{8} mv^4/c^2 .$$

(b) If we use the mc^2 value for the electron found in Table 37-3, then for $\beta = 1/20$, the classical expression for kinetic energy gives

$$K_{\text{classical}} = \tfrac{1}{2} mv^2 = \tfrac{1}{2} mc^2 \beta^2 = \tfrac{1}{2}(8.19 \times 10^{-14} \text{ J})(1/20)^2 = 1.0 \times 10^{-16} \text{ J} .$$

(c) The first-order correction becomes

$$K_{\text{first-order}} = \tfrac{3}{8} mv^4/c^2 = \tfrac{3}{8} mc^2 \beta^4 = \tfrac{3}{8}(8.19 \times 10^{-14} \text{ J})(1/20)^4 = 1.9 \times 10^{-19} \text{ J}$$

which we note is much smaller than the classical result.

(d) In this case, $\beta = 0.80 = 4/5$, and the classical expression yields

$$K_{\text{classical}} = \tfrac{1}{2}mv^2 = \tfrac{1}{2}mc^2\beta^2 = \tfrac{1}{2}(8.19 \times 10^{-14}\,\text{J})(4/5)^2 = 2.6 \times 10^{-14}\,\text{J}.$$

(e) And the first-order correction is

$$K_{\text{first-order}} = \tfrac{3}{8}mv^4/c^2 = \tfrac{3}{8}mc^2\beta^4 = \tfrac{3}{8}(8.19 \times 10^{-14}\,\text{J})(4/5)^4 = 1.3 \times 10^{-14}\,\text{J}$$

which is comparable to the classical result. This is a signal that ignoring the higher order terms in the binomial expansion becomes less reliable the closer the speed gets to c.

(f) We set the first-order term equal to one-tenth of the classical term and solve for β:

$$\tfrac{3}{8}mc^2\beta^4 = \tfrac{1}{10}(\tfrac{1}{2}mc^2\beta^2)$$

and obtain $\beta = \sqrt{2/15} \approx 0.37$.

57. Using the classical orbital radius formula $r_0 = mv/|q|B$, the period is $T_0 = 2\pi r_0/v = 2\pi m/|q|B$. In the relativistic limit, we must use

$$r = \frac{p}{|q|B} = \frac{\gamma mv}{|q|B} = \gamma r_0$$

which yields

$$T = \frac{2\pi r}{v} = \gamma \frac{2\pi m}{|q|B} = \gamma T_0$$

(b) The period T is not independent of v.

(c) We interpret the given 10.0 MeV to be the kinetic energy of the electron. In order to make use of the mc^2 value for the electron given in Table 37-3 (511 keV = 0.511 MeV) we write the classical kinetic energy formula as

$$K_{\text{classical}} = \frac{1}{2}mv^2 = \frac{1}{2}(mc^2)\left(\frac{v^2}{c^2}\right) = \frac{1}{2}(mc^2)\beta^2.$$

If $K_{\text{classical}} = 10.0$ MeV, then

$$\beta = \sqrt{\frac{2K_{\text{classical}}}{mc^2}} = \sqrt{\frac{2(10.0\,\text{MeV})}{0.511\,\text{MeV}}} = 6.256,$$

which, of course, is impossible (see the Ultimate Speed subsection of §37-2). If we use this value anyway, then the classical orbital radius formula yields

$$r = \frac{mv}{|q|B} = \frac{m\beta c}{eB} = \frac{(9.11\times10^{-31}\,\text{kg})\,(6.256)\,(2.998\times10^{8}\,\text{m/s})}{(1.6\times10^{-19}\,\text{C})\,(2.20\text{T})} = 4.85\times10^{-3}\,\text{m}.$$

(d) Before using the relativistically correct orbital radius formula, we must compute β in a relativistically correct way:

$$K = mc^2(\gamma - 1) \Rightarrow \gamma = \frac{10.0\ \text{MeV}}{0.511\ \text{MeV}} + 1 = 20.57$$

which implies (from Eq. 37-8)

$$\beta = \sqrt{1 - \frac{1}{\gamma^2}} = 0.99882.$$

Therefore,

$$r = \frac{\gamma mv}{|q|B} = \frac{\gamma m\beta c}{eB} = \frac{(20.57)(9.11\times10^{-31}\,\text{kg})(0.99882)(2.998\times10^{8}\,\text{m/s})}{(1.6\times10^{-19}\,\text{C})(2.20\text{T})}$$
$$= 1.59\times10^{-2}\,\text{m}.$$

(e) The classical period is

$$T = \frac{2\pi r}{\beta c} = \frac{2\pi(4.85\times10^{-3}\ \text{m})}{(6.256)(2.998\times10^{8}\ \text{m/s})} = 1.63\times10^{-11}\ \text{s}.$$

(f) The period obtained with relativistic correction is

$$T = \frac{2\pi r}{\beta c} = \frac{2\pi(0.0159\ \text{m})}{(0.99882)(2.998\times10^{8}\ \text{m/s})} = 3.34\times10^{-10}\ \text{s}.$$

58. (a) The proper lifetime Δt_0 is 2.20 μs, and the lifetime measured by clocks in the laboratory (through which the muon is moving at high speed) is $\Delta t = 6.90$ μs. We use Eq. 37-7 to solve for the speed parameter:

$$\beta = \sqrt{1 - \left(\frac{\Delta t_0}{\Delta t}\right)^2} = 0.948\,.$$

(b) From the answer to part (a), we find $\gamma = 3.136$. Thus, with (see Table 37-3)

$$m_\mu c^2 = 207 m_e c^2 = 105.8 \text{ MeV},$$

Eq. 37-52 yields

$$K = m_\mu c^2 (\gamma - 1) = 226 \text{ MeV}.$$

(c) We write $m_\mu c = 105.8$ MeV/c and apply Eq. 37-41:

$$p = \gamma m_\mu v = \gamma m_\mu c \beta = (3.136)(105.8 \text{ MeV} / c)(0.9478) = 314 \text{ MeV} / c$$

which can also be expressed in SI units ($p = 1.7 \times 10^{-19}$ kg·m/s).

59. (a) Before looking at our solution to part (a) (which uses momentum conservation), it might be advisable to look at our solution (and accompanying remarks) for part (b) (where a very different approach is used). Since momentum is a vector, its conservation involves two equations (along the original direction of alpha particle motion, the x direction, as well as along the final proton direction of motion, the y direction). The problem states that all speeds are much less than the speed of light, which allows us to use the classical formulas for kinetic energy and momentum ($K = \frac{1}{2}mv^2$ and $\vec{p} = m\vec{v}$, respectively). Along the x and y axes, momentum conservation gives (for the components of $\vec{v}_{oxy}$):

$$m_\alpha v_\alpha = m_{oxy} v_{oxy,x} \Rightarrow v_{oxy,x} = \frac{m_\alpha}{m_{oxy}} v_\alpha \approx \frac{4}{17} v_\alpha$$

$$0 = m_{oxy} v_{oxy,y} + m_p v_p \Rightarrow v_{oxy,y} = -\frac{m_p}{m_{oxy}} v_p \approx -\frac{1}{17} v_p.$$

To complete these determinations, we need values (inferred from the kinetic energies given in the problem) for the initial speed of the alpha particle (v_α) and the final speed of the proton (v_p). One way to do this is to rewrite the classical kinetic energy expression as $K = \frac{1}{2}(mc^2)\beta^2$ and solve for β (using Table 37-3 and/or Eq. 37-46). Thus, for the proton, we obtain

$$\beta_p = \sqrt{\frac{2K_p}{m_p c^2}} = \sqrt{\frac{2(4.44 \text{ MeV})}{938 \text{ MeV}}} = 0.0973.$$

This is almost 10% the speed of light, so one might worry that the relativistic expression (Eq. 37-52) should be used. If one does so, one finds $\beta_p = 0.969$, which is reasonably close to our previous result based on the classical formula. For the alpha particle, we write

$$m_\alpha c^2 = (4.0026\,\text{u})(931.5\,\text{MeV/u}) = 3728\,\text{MeV}$$

(which is actually an overestimate due to the use of the "atomic mass" value in our calculation, but this does not cause significant error in our result), and obtain

$$\beta_\alpha = \sqrt{\frac{2K_\alpha}{m_\alpha c^2}} = \sqrt{\frac{2(7.70\,\text{MeV})}{3728\,\text{MeV}}} = 0.064.$$

Returning to our oxygen nucleus velocity components, we are now able to conclude:

$$v_{\text{oxy},x} \approx \frac{4}{17}v_\alpha \Rightarrow \beta_{\text{oxy},x} \approx \frac{4}{17}\beta_\alpha = \frac{4}{17}(0.064) = 0.015$$

$$|v_{\text{oxy},y}| \approx \frac{1}{17}v_p \Rightarrow \beta_{\text{oxy},y} \approx \frac{1}{17}\beta_p = \frac{1}{17}(0.097) = 0.0057$$

Consequently, with $m_{\text{oxy}}c^2 \approx (17\,\text{u})(931.5\,\text{MeV/u}) = 1.58 \times 10^4\,\text{MeV}$, we obtain

$$K_{\text{oxy}} = \frac{1}{2}(m_{\text{oxy}}c^2)(\beta_{\text{oxy},x}^2 + \beta_{\text{oxy},y}^2) = \frac{1}{2}(1.58\times10^4\,\text{MeV})(0.015^2 + 0.0057^2) \approx 2.08\,\text{MeV}.$$

(b) Using Eq. 37-50 and Eq. 37-46,

$$Q = -(1.007825\,\text{u} + 16.99914\,\text{u} - 4.00260\,\text{u} - 14.00307\,\text{u})c^2$$
$$= -(0.001295\,\text{u})(931.5\,\text{MeV/u})$$

which yields $Q = -1.206\,\text{MeV} \approx -1.21\,\text{MeV}$. Incidentally, this provides an alternate way to obtain the answer (and a more accurate one at that!) to part (a). Eq. 37-49 leads to

$$K_{\text{oxy}} = K_\alpha + Q - K_p = 7.70\,\text{MeV} - 1206\,\text{MeV} - 4.44\,\text{MeV}$$
$$= 2.05\,\text{MeV}.$$

This approach to finding K_{oxy} avoids the many computational steps and approximations made in part (a).

60. (a) Eq. 2′ of Table 37-2, becomes

$$\Delta t' = \gamma(\Delta t - \beta\Delta x/c) = \gamma[1.00\,\mu\text{s} - \beta(240\,\text{m})/(2.998 \times 10^2\,\text{m/}\mu\text{s})]$$
$$= \gamma(1.00 - 0.800\beta)\,\mu\text{s}$$

where the Lorentz factor is itself a function of β (see Eq. 37-8).

(b) A plot of $\Delta t'$ is shown for the range $0 < \beta < 0.01$:

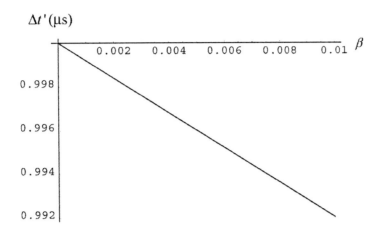

(c) A plot of $\Delta t'$ is shown for the range $0.1 < \beta < 1$:

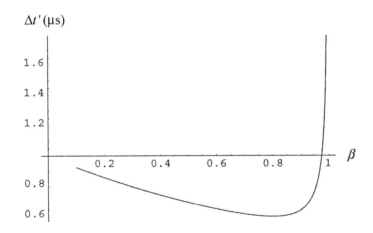

(d) The minimum for the $\Delta t'$ curve can be found from by taking the derivative and simplifying and then setting equal to zero:

$$\frac{d \Delta t'}{d \beta} = \gamma^3 (\beta \Delta t - \Delta x/c) = 0 .$$

Thus, the value of β for which the curve is minimum is $\beta = \Delta x/c\Delta t = 240/299.8$, or $\beta = 0.801$.

(e) Substituting the value of β from part (d) into the part (a) expression yields the minimum value $\Delta t' = 0.599$ μs.

(f) Yes. We note that $\Delta x/\Delta t = 2.4 \times 10^8$ m/s $< c$. A signal can indeed travel from event A to event B without exceeding c, so causal influences can originate at A and thus affect what happens at B. Such events are often described as being "time-like separated" – and we see in this problem that it is (always) possible in such a situation for us to find a frame of reference (here with $\beta \approx 0.801$) where the two events will seem to be at the same location (though at different times).

61. (a) Eq. 1' of Table 37-2 becomes $\Delta x' = \gamma(\Delta x - \beta c \Delta t) = \gamma[(240 \text{ m}) - \beta(299.8 \text{ m})]$.

(b) A plot of $\Delta x'$ for $0 < \beta < 0.01$ is shown below:

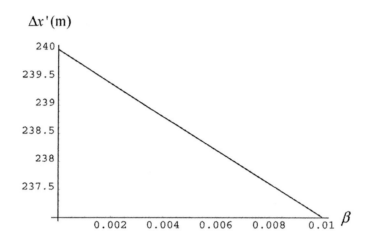

(c) A plot of $\Delta x'$ for $0.1 < \beta < 1$ is shown below:

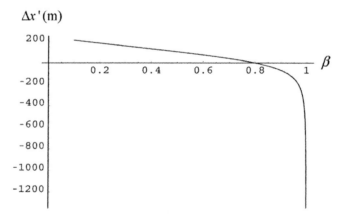

We see that $\Delta x'$ decreases from its $\beta = 0$ value (where it is equal to $\Delta x = 240$ m) to its zero value (at $\beta \approx 0.8$), and continues (without bound) downward in the graph (where it is negative – implying event B has a *smaller* value of x' than event A!).

(d) The zero value for $\Delta x'$ is easily seen (from the expression in part (b)) to come from the condition $\Delta x - \beta c \Delta t = 0$. Thus $\beta = 0.801$ provides the zero value of $\Delta x'$.

62. The line in the graph is described by Eq. 1 in Table 37-2:

$$\Delta x = v\gamma \Delta t' + \gamma \Delta x' = (\text{"slope"})\Delta t' + \text{"y-intercept"}$$

where the "slope" is 7.0×10^8 m/s. Setting this value equal to $v\gamma$ leads to $v = 2.8 \times 10^8$ m/s and $\gamma = 2.54$. Since the "y-intercept" is 2.0 m, we see that dividing this by γ leads to $\Delta x' = 0.79$ m.

63. (a) The spatial separation between the two bursts is vt. We project this length onto the direction perpendicular to the light rays headed to Earth and obtain $D_{app} = vt \sin \theta$.

(b) Burst 1 is emitted a time t ahead of burst 2. Also, burst 1 has to travel an extra distance L more than burst 2 before reaching the Earth, where $L = vt \cos \theta$ (see Fig. 37-30); this requires an additional time $t' = L/c$. Thus, the apparent time is given by

$$T_{app} = t - t' = t - \frac{vt \cos \theta}{c} = t \left[1 - \left(\frac{v}{c} \right) \cos \theta \right].$$

(c) We obtain

$$V_{app} = \frac{D_{app}}{T_{app}} = \left[\frac{(v/c) \sin \theta}{1 - (v/c) \cos \theta} \right] c = \left[\frac{(0.980) \sin 30.0°}{1 - (0.980) \cos 30.0°} \right] c = 3.24 \, c.$$

64. By examining the value of u' when $v = 0$ on the graph, we infer $u = -0.20c$. Solving Eq. 37-29 for u' and inserting this value for u, we obtain

$$u' = \frac{u - v}{1 - uv/c^2} = \frac{-0.20c - v}{1 + 0.20v/c}$$

for the equation of the curve shown in the figure.

(a) With $v = 0.80c$, the above expression yields $u' = -0.86c$.

(b) As expected, setting $v = c$ in this expression leads to $u' = -c$.

65. (a) From the length contraction equation, the length L'_c of the car according to Garageman is

$$L'_c = \frac{L_c}{\gamma} = L_c \sqrt{1 - \beta^2} = (30.5 \text{ m}) \sqrt{1 - (0.9980)^2} = 1.93 \text{ m}.$$

(b) Since the x_g axis is fixed to the garage $x_{g2} = L_g = 6.00$ m.

(c) As for t_{g2}, note from Fig. 37-32 (b) that, at $t_g = t_{g1} = 0$ the coordinate of the front bumper of the limo in the x_g frame is L'_c, meaning that the front of the limo is still a distance $L_g - L_c'$ from the back door of the garage. Since the limo travels at a speed v, the time it takes for the front of the limo to reach the back door of the garage is given by

$$\Delta t_g = t_{g2} - t_{g1} = \frac{L_g - L_c'}{v} = \frac{6.00 \text{ m} - 1.93 \text{ m}}{0.9980(2.998 \times 10^8 \text{ m/s})} = 1.36 \times 10^{-8} \text{ s.}$$

Thus $t_{g2} = t_{g1} + \Delta t_g = 0 + 1.36 \times 10^{-8} \text{ s} = 1.36 \times 10^{-8}$ s.

(d) The limo is inside the garage between times t_{g1} and t_{g2}, so the time duration is $t_{g2} - t_{g1}$ $= 1.36 \times 10^{-8}$ s.

(e) Again from Eq. 37-13, the length L_g' of the garage according to Carman is

$$L_g' = \frac{L_g}{\gamma} = L_g\sqrt{1-\beta^2} = (6.00 \text{ m})\sqrt{1-(0.9980)^2} = 0.379 \text{ m.}$$

(f) Again, since the x_c axis is fixed to the limo $x_{c2} = L_c = 30.5$ m.

(g) Now, from the two diagrams described in part (h) below, we know that at $t_c = t_{c2}$ (when event 2 takes place), the distance between the rear bumper of the limo and the back door of the garage is given by $L_c - L_g'$. Since the garage travels at a speed v, the front door of the garage will reach the rear bumper of the limo a time Δt_c later, where Δt_c satisfies

$$\Delta t_c = t_{c1} - t_{c2} = \frac{L_c - L_g'}{v} = \frac{30.5 \text{ m} - 0.379 \text{ m}}{0.9980(2.998 \times 10^8 \text{ m/s})} = 1.01 \times 10^{-7} \text{ s.}$$

Thus $t_{c2} = t_{c1} - \Delta t_c = 0 - 1.01 \times 10^{-7} \text{ s} = -1.01 \times 10^{-7}$ s.

(h) From Carman's point of view, the answer is clearly no.

(i) Event 2 occurs first according to Carman, since $t_{c2} < t_{c1}$.

(j) We describe the essential features of the two pictures. For event 2, the front of the limo coincides with the back door, and the garage itself seems very short (perhaps failing to reach as far as the front window of the limo). For event 1, the rear of the car coincides with the front door and the front of the limo has traveled a significant distance beyond the back door. In this picture, as in the other, the garage seems very short compared to the limo.

(k) No, the limo cannot be in the garage with both doors shut.

(l) Both Carman and Garageman are correct in their respective reference frames. But, in a sense, Carman should lose the bet since he dropped his physics course before reaching the Theory of Special Relativity!

66. (a) According to ship observers, the duration of proton flight is $\Delta t' = (760 \text{ m})/0.980c = 2.59\ \mu s$ (assuming it travels the entire length of the ship).

(b) To transform to our point of view, we use Eq. 2 in Table 37-2. Thus, with $\Delta x' = -750$ m, we have

$$\Delta t = \gamma \left(\Delta t' + (0.950c)\Delta x'/c^2 \right) = 0.572\,\mu s.$$

(c) For the ship observers, firing the proton from back to front makes no difference, and $\Delta t' = 2.59\ \mu s$ as before.

(d) For us, the fact that now $\Delta x' = +750$ m is a significant change.

$$\Delta t = \gamma \left(\Delta t' + (0.950c)\Delta x'/c^2 \right) = 16.0\,\mu s.$$

67. Interpreting v_{AB} as the x-component of the velocity of A relative to B, and defining the corresponding speed parameter $\beta_{AB} = v_{AB}/c$, then the result of part (a) is a straightforward rewriting of Eq. 37-29 (after dividing both sides by c). To make the correspondence with Fig. 37-11 clear, the particle in that picture can be labeled A, frame S' (or an observer at rest in that frame) can be labeled B, and frame S (or an observer at rest in it) can be labeled C. The result of part (b) is less obvious, and we show here some of the algebra steps:

$$M_{AC} = M_{AB}\, M_{BC}$$

$$\frac{1-\beta_{AC}}{1+\beta_{AC}} = \frac{1-\beta_{AB}}{1+\beta_{AB}}\frac{1-\beta_{BC}}{1+\beta_{BC}}$$

We multiply both sides by factors to get rid of the denominators

$$(1-\beta_{AC})\,(1+\beta_{AB})\,(1+\beta_{BC}) = (1-\beta_{AB})\,(1-\beta_{BC})\,(1+\beta_{AC})$$

and expand:

$$1 - \beta_{AC} + \beta_{AB} + \beta_{BC} - \beta_{AC}\beta_{AB} - \beta_{AC}\beta_{BC} + \beta_{AB}\beta_{BC} - \beta_{AB}\beta_{BC}\beta_{AC} =$$
$$1 + \beta_{AC} - \beta_{AB} - \beta_{BC} - \beta_{AC}\beta_{AB} - \beta_{AC}\beta_{BC} + \beta_{AB}\beta_{BC} + \beta_{AB}\beta_{BC}\beta_{AC}$$

We note that several terms are identical on both sides of the equals sign, and thus cancel, which leaves us with

$$-\beta_{AC} + \beta_{AB} + \beta_{BC} - \beta_{AB}\,\beta_{BC}\,\beta_{AC} = \beta_{AC} - \beta_{AB} - \beta_{BC} + \beta_{AB}\,\beta_{BC}\,\beta_{AC}$$

which can be rearranged to produce

$$2\beta_{AB} + 2\beta_{BC} = 2\beta_{AC} + 2\beta_{AB}\,\beta_{BC}\,\beta_{AC}$$

The left-hand side can be written as $2\beta_{AC}\,(1 + \beta_{AB}\,\beta_{BC})$ in which case it becomes clear how to obtain the result from part (a) [just divide both sides by $2(1 + \beta_{AB}\,\beta_{BC})$].

68. We note, because it is a pretty symmetry and because it makes the part (b) computation move along more quickly, that

$$M = \frac{1-\beta}{1+\beta} \quad \Rightarrow \quad \beta = \frac{1-M}{1+M} \quad .$$

Here, with β_{AB} given as 1/2 (see problem statement), then M_{AB} is seen to be 1/3 (which is $(1 - 1/2)$ divided by $(1 + 1/2)$). Similarly for β_{BC} .

(a) Thus,

$$M_{AC} = M_{AB}\,M_{BC} = \frac{1}{3} \cdot \frac{1}{3} = \frac{1}{9} \quad .$$

(b) Consequently,

$$\beta_{AC} = \frac{1 - M_{AC}}{1 + M_{AC}} = \frac{1 - 1/9}{1 + 1/9} = \frac{8}{10} = \frac{4}{5} = 0.80.$$

(c) By the definition of the speed parameter, we finally obtain $v_{AC} = 0.80c$.

69. We note, for use later in the problem, that

$$M = \frac{1-\beta}{1+\beta} \quad \Rightarrow \quad \beta = \frac{1-M}{1+M} \quad .$$

Now, with β_{AB} given as 1/5 (see problem statement), then M_{AB} is seen to be 2/3 (which is $(1 - 1/5)$ divided by $(1 + 1/5)$). With $\beta_{BC} = - 2/5$ we similarly find $M_{BC} = 7/3$, and for $\beta_{CD} = 3/5$ we get $M_{CD} = 1/4$. Thus,

$$M_{AD} = M_{AB}\,M_{BC}\,M_{CD} = \frac{2}{3} \cdot \frac{7}{3} \cdot \frac{1}{4} = \frac{7}{18} \quad .$$

Consequently,

$$\beta_{AD} = \frac{1 - M_{AD}}{1 + M_{AD}} = \frac{1 - 7/18}{1 + 7/18} = \frac{11}{25} = 0.44.$$

By the definition of the speed parameter, we obtain $v_{AD} = 0.44c$.

70. We are asked to solve Eq. 37-48 for the speed v. Algebraically, we find

$$\beta = \sqrt{1 - \left(\frac{mc^2}{E}\right)^2}.$$

Using $E = 10.611 \times 10^{-9}$ J and the very accurate values for c and m (in SI units) found in Appendix B, we obtain $\beta = 0.99990$.

71. Using Appendix C, we find that the contraction is

$$|\Delta L| = L_0 - L = L_0\left(1 - \frac{1}{\gamma}\right) = L_0\left(1 - \sqrt{1 - \beta^2}\right)$$

$$= 2(6.370 \times 10^6 \text{ m})\left(1 - \sqrt{1 - \left(\frac{3.0 \times 10^4 \text{ m/s}}{2.998 \times 10^8 \text{ m/s}}\right)^2}\right)$$

$$= 0.064 \text{ m}.$$

72. The speed of the spaceship after the first increment is $v_1 = 0.5c$. After the second one, it becomes

$$v_2 = \frac{v' + v_1}{1 + v'v_1/c^2} = \frac{0.50c + 0.50c}{1 + (0.50c)^2/c^2} = 0.80c,$$

and after the third one, the speed is

$$v_3 = \frac{v' + v_2}{1 + v'v_2/c^2} = \frac{0.50c + 0.50c}{1 + (0.50c)(0.80c)/c^2} = 0.929c.$$

Continuing with this process, we get $v_4 = 0.976c$, $v_5 = 0.992c$, $v_6 = 0.997c$ and $v_7 = 0.999c$. Thus, seven increments are needed.

73. The mean lifetime of a pion measured by observers on the Earth is $\Delta t = \gamma \Delta t_0$, so the distance it can travel (using Eq. 37-12) is

$$d = v\Delta t = \gamma v \Delta t_0 = \frac{(0.99)(2.998 \times 10^8 \text{ m/s})(26 \times 10^{-9} \text{ s})}{\sqrt{1 - (0.99)^2}} = 55 \text{ m}.$$

74. (a) For a proton (using Table 37-3), we have

$$E = \gamma m_p c^2 = \frac{938\,\text{MeV}}{\sqrt{1-(0.990)^2}} = 6.65\,\text{GeV}$$

which gives

$$K = E - m_p c^2 = 6.65\,\text{GeV} - 938\,\text{MeV} = 5.71\,\text{GeV}.$$

(b) From part (a), $E = 6.65\,\text{GeV}$.

(c) Similarly, we have $p = \gamma m_p v = \gamma (m_p c^2)\beta / c = \dfrac{(938\,\text{MeV})(0.990)/c}{\sqrt{1-(0.990)^2}} = 6.58\,\text{GeV}/c$

(d) For an electron, we have

$$E = \gamma m_e c^2 = \frac{0.511\,\text{MeV}}{\sqrt{1-(0.990)^2}} = 3.62\,\text{MeV}$$

which yields

$$K = E - m_e c^2 = 3.625\,\text{MeV} - 0.511\,\text{MeV} = 3.11\,\text{MeV}.$$

(e) From part (d), $E = 3.62\,\text{MeV}$.

(f) $p = \gamma m_e v = \gamma (m_e c^2)\beta / c = \dfrac{(0.511\,\text{MeV})(0.990)/c}{\sqrt{1-(0.990)^2}} = 3.59\,\text{MeV}/c.$

75. The strategy is to find the speed from $E = 1533$ MeV and $mc^2 = 0.511$ MeV (see Table 37-3) and from that find the time. From the energy relation (Eq. 37-48), we obtain

$$v = c\sqrt{1 - \left(\frac{mc^2}{E}\right)^2} = 0.99999994c \approx c$$

so that we conclude it took the electron 26 y to reach us. In order to transform to its own "clock" it's useful to compute γ directly from Eq. 37-48:

$$\gamma = \frac{E}{mc^2} = 3000$$

though if one is careful one can also get this result from $\gamma = 1/\sqrt{1-(v/c)^2}$. Then, Eq. 37-7 leads to

$$\Delta t_0 = \frac{26\,\mathrm{y}}{\gamma} = 0.0087\,\mathrm{y}$$

so that the electron "concludes" the distance he traveled is 0.0087 light-years (stated differently, the Earth, which is rushing towards him at very nearly the speed of light, seemed to start its journey from a distance of 0.0087 light-years away).

76. (a) Using Eq. 37-7, we expect the dilated time intervals to be

$$\tau = \gamma\tau_0 = \frac{\tau_0}{\sqrt{1-(v/c)^2}}\,.$$

(b) We rewrite Eq. 37-31 using the fact that period is the reciprocal of frequency ($f_R = \tau_R^{-1}$ and $f_0 = \tau_0^{-1}$):

$$\tau_R = \frac{1}{f_R} = \left(f_0\sqrt{\frac{1-\beta}{1+\beta}}\right)^{-1} = \tau_0\sqrt{\frac{1+\beta}{1-\beta}} = \tau_0\sqrt{\frac{c+v}{c-v}}\,.$$

(c) The Doppler shift combines two physical effects: the time dilation of the moving source *and* the travel-time differences involved in periodic emission (like a sine wave or a series of pulses) from a traveling source to a "stationary" receiver). To isolate the purely time-dilation effect, it's useful to consider "local" measurements (say, comparing the readings on a moving clock to those of two of your clocks, spaced some distance apart, such that the moving clock and each of your clocks can make a close-comparison of readings at the moment of passage).

77. We use the relative velocity formula (Eq. 37-29) with the primed measurements being those of the scout ship. We note that $v = -0.900c$ since the velocity of the scout ship relative to the cruiser is opposite to that of the cruiser relative to the scout ship.

$$u = \frac{u'+v}{1+u'v/c^2} = \frac{0.980c - 0.900c}{1-(0.980)(0.900)} = 0.678c\,.$$

78. (a) The relative contraction is

$$\frac{|\Delta L|}{L_0} = \frac{L_0(1-\gamma^{-1})}{L_0} = 1-\sqrt{1-\beta^2} \approx 1-\left(1-\frac{1}{2}\beta^2\right) = \frac{1}{2}\beta^2 = \frac{1}{2}\left(\frac{630\,\mathrm{m/s}}{3.00\times10^8\,\mathrm{m/s}}\right)^2$$
$$= 2.21\times10^{-12}\,.$$

(b) Letting $|\Delta t - \Delta t_0| = \Delta t_0(\gamma - 1) = \tau = 1.00\mu s$, we solve for Δt_0:

$$\Delta t_0 = \frac{\tau}{\gamma - 1} = \frac{\tau}{(1 - \beta^2)^{-1/2} - 1} \approx \frac{\tau}{1 + \frac{1}{2}\beta^2 - 1} = \frac{2\tau}{\beta^2}$$

$$= \frac{2(1.00 \times 10^{-6} \, s)(1 \, d / 86400 \, s)}{[(630 \, m/s)/(2.998 \times 10^8 \, m/s)]^2}$$

$$= 5.25 \, d \, .$$

79. Let the reference frame be S in which the particle (approaching the South Pole) is at rest, and let the frame that is fixed on Earth be S'. Then $v = 0.60c$ and $u' = 0.80c$ (calling "downwards" [in the sense of Fig. 37-35] positive). The relative speed is now the speed of the other particle as measured in S:

$$u = \frac{u' + v}{1 + u'v/c^2} = \frac{0.80c + 0.60c}{1 + (0.80c)(0.60c)/c^2} = 0.95c \, .$$

80. We refer to the particle in the first sentence of the problem statement as particle 2. Since the total momentum of the two particles is zero in S', it must be that the velocities of these two particles are equal in magnitude and opposite in direction in S'. Letting the velocity of the S' frame be v relative to S, then the particle which is at rest in S must have a velocity of $u'_1 = -v$ as measured in S', while the velocity of the other particle is given by solving Eq. 37-29 for u':

$$u'_2 = \frac{u_2 - v}{1 - u_2 v/c^2} = \frac{(c/2) - v}{1 - (c/2)(v/c^2)} \, .$$

Letting $u'_2 = -u'_1 = v$, we obtain

$$\frac{(c/2) - v}{1 - (c/2)(v/c^2)} = v \quad \Rightarrow \quad v = c(2 \pm \sqrt{3}) \approx 0.27c$$

where the quadratic formula has been used (with the smaller of the two roots chosen so that $v \le c$).

81. We use Eq. 37-54 with $mc^2 = 0.511$ MeV (see Table 37-3):

$$pc = \sqrt{K^2 + 2Kmc^2} = \sqrt{(2.00)^2 + 2(2.00)(0.511)}$$

This readily yields $p = 2.46$ MeV/c.

82. (a) Our lab-based measurement of its lifetime is figured simply from

$$t = L/v = 7.99 \times 10^{-13} \, s.$$

Use of the time-dilation relation (Eq. 37-7) leads to

$$\Delta t_0 = (7.99 \times 10^{-13} \text{ s})\sqrt{1-(0.960)^2} = 2.24 \times 10^{-13} \text{ s}.$$

(b) The length contraction formula can be used, or we can use the simple speed-distance relation (from the point of view of the particle, who watches the lab and all its meter sticks rushing past him at $0.960c$ until he expires): $L = v\Delta t_0 = 6.44 \times 10^{-5}$ m.

83. When $\beta = 0.9860$, we have $\gamma = 5.9972$, and when $\beta = 0.9850$, we have $\gamma = 5.7953$. Thus, $\Delta\gamma = 0.202$ and the change in kinetic energy (equal to the work) becomes (using Eq. 37-52)

$$W = \Delta K = mc^2\Delta\gamma = 189 \text{ MeV}$$

where $mc^2 = 938$ MeV has been used (see Table 37-3).

84. (a) Eq. 37-37 yields

$$\frac{\lambda_0}{\lambda} = \sqrt{\frac{1-\beta}{1+\beta}} \quad \Rightarrow \quad \beta = \frac{1-(\lambda_0/\lambda)^2}{1+(\lambda_0/\lambda)^2}.$$

With $\lambda_0/\lambda = 434/462$, we obtain $\beta = 0.062439$, or $v = 1.87 \times 10^7$ m/s.

(b) Since it is shifted "towards the red" (towards longer wavelengths) then the galaxy is moving away from us (receding).

85. (a) $\Delta E = \Delta mc^2 = (3.0 \text{ kg})(0.0010)(2.998 \times 10^8 \text{ m/s})^2 = 2.7 \times 10^{14}$ J.

(b) The mass of TNT is

$$m_{\text{TNT}} = \frac{(2.7 \times 10^{14} \text{ J})(0.227 \text{ kg/mol})}{3.4 \times 10^6 \text{ J}} = 1.8 \times 10^7 \text{ kg}.$$

(c) The fraction of mass converted in the TNT case is

$$\frac{\Delta m_{\text{TNT}}}{m_{\text{TNT}}} = \frac{(3.0 \text{ kg})(0.0010)}{1.8 \times 10^7 \text{ kg}} = 1.6 \times 10^{-9},$$

Therefore, the fraction is $0.0010/1.6 \times 10^{-9} = 6.0 \times 10^6$.

86. (a) We assume the electron starts from rest. The classical formula for kinetic energy is Eq. 37-51, so if $v = c$ then this (for an electron) would be $\frac{1}{2}mc^2 = \frac{1}{2}(511 \text{ ke V}) = 255.5$ ke V (using Table 37-3). Setting this equal to the potential energy loss (which is responsible for its acceleration), we find (using Eq. 25-7)

$$V = \frac{255.5 \text{ keV}}{|q|} = \frac{255 \text{ keV}}{e} = 255.5 \text{ kV} \approx 256 \text{ keV}.$$

(b) Setting this amount of potential energy loss ($|\Delta U| = 255.5$ keV) equal to the correct relativistic kinetic energy, we obtain (using Eq. 37-52)

$$mc^2 \left(\frac{1}{\sqrt{1-(v/c)^2}} - 1 \right) = |\Delta U| \Rightarrow v = c \sqrt{1 + \left(\frac{1}{1 - \Delta U / mc^2} \right)^2}$$

which yields $v = 0.745c = 2.23 \times 10^8$ m/s.

87. (a) $v_r = 2v = 2(27000 \text{ km/h}) = 5.4 \times 10^4$ km/h.

(b) We can express c in these units by multiplying by 3.6: $c = 1.08 \times 10^9$ km/h. The correct formula for v_r is $v_r = 2v/(1 + v^2/c^2)$, so the fractional error is

$$1 - \frac{1}{1 + v^2/c^2} = 1 - \frac{1}{1 + \left[(27000 \text{ km/h}) / (1.08 \times 10^9 \text{ km/h}) \right]^2} = 6.3 \times 10^{-10}.$$

88. Using Eq. 37-10,

$$\beta = \frac{v}{c} = \frac{d/c}{t} = \frac{6.0 \text{ y}}{2.0 \text{ y} + 6.0 \text{ y}} = 0.75.$$

Chapter 38

1. (a) Let $E = 1240$ eV·nm/$\lambda_{min} = 0.6$ eV to get $\lambda = 2.1 \times 10^3$ nm $= 2.1$ μm.

(b) It is in the infrared region.

2. The energy of a photon is given by $E = hf$, where h is the Planck constant and f is the frequency. The wavelength λ is related to the frequency by $\lambda f = c$, so $E = hc/\lambda$. Since $h = 6.626 \times 10^{-34}$ J·s and $c = 2.998 \times 10^8$ m/s,

$$hc = \frac{\left(6.626 \times 10^{-34} \text{ J·s}\right)\left(2.998 \times 10^8 \text{ m/s}\right)}{\left(1.602 \times 10^{-19} \text{ J/eV}\right)\left(10^{-9} \text{ m/nm}\right)} = 1240 \text{ eV·nm}.$$

Thus,

$$E = \frac{1240 \text{ eV·nm}}{\lambda}.$$

With $\lambda = 589$ nm, we obtain

$$E = \frac{hc}{\lambda} = \frac{1240 \text{ eV·nm}}{589 \text{ nm}} = 2.11 \text{ eV}.$$

3. Let R be the rate of photon emission (number of photons emitted per unit time) of the Sun and let E be the energy of a single photon. Then the power output of the Sun is given by $P = RE$. Now $E = hf = hc/\lambda$, where h is the Planck constant, f is the frequency of the light emitted, and λ is the wavelength. Thus $P = Rhc/\lambda$ and

$$R = \frac{\lambda P}{hc} = \frac{\left(550 \text{ nm}\right)\left(3.9 \times 10^{26} \text{ W}\right)}{\left(6.63 \times 10^{-34} \text{ J·s}\right)\left(2.998 \times 10^8 \text{ m/s}\right)} = 1.0 \times 10^{45} \text{ photons/s}.$$

4. We denote the diameter of the laser beam as d. The cross-sectional area of the beam is $A = \pi d^2/4$. From the formula obtained in problem 3, the rate is given by

$$\frac{R}{A} = \frac{\lambda P}{hc\left(\pi d^2/4\right)} = \frac{4\left(633 \text{ nm}\right)\left(5.0 \times 10^{-3} \text{ W}\right)}{\pi\left(6.63 \times 10^{-34} \text{ J·s}\right)\left(2.998 \times 10^8 \text{ m/s}\right)\left(3.5 \times 10^{-3} \text{ m}\right)^2}$$

$$= 1.7 \times 10^{21} \frac{\text{photons}}{\text{m}^2 \cdot \text{s}}.$$

5. Since

$$\lambda = (1, 650, 763.73)^{-1} \text{ m} = 6.0578021 \times 10^{-7} \text{ m} = 605.78021 \text{ nm},$$

the energy is (using the fact that $hc = 1240 \text{ eV} \cdot \text{nm}$),

$$E = \frac{hc}{\lambda} = \frac{1240 \text{ eV} \cdot \text{nm}}{605.78021 \text{ nm}} = 2.047 \text{ eV}.$$

6. Let

$$\frac{1}{2} m_e v^2 = E_{\text{photon}} = \frac{hc}{\lambda}$$

and solve for v:

$$v = \sqrt{\frac{2hc}{\lambda m_e}} = \sqrt{\frac{2hc}{\lambda m_e c^2} c^2} = c \sqrt{\frac{2hc}{\lambda (m_e c^2)}}$$

$$= (2.998 \times 10^8 \text{ m/s}) \sqrt{\frac{2(1240 \text{ eV} \cdot \text{nm})}{(590 \text{ nm})(511 \times 10^3 \text{ eV})}} = 8.6 \times 10^5 \text{ m/s}.$$

Since $v \ll c$, the non-relativistic formula $K = \frac{1}{2} m v^2$ may be used. The $m_e c^2$ value of Table 38-3 and $hc = 1240 \text{ eV} \cdot \text{nm}$ are used in our calculation.

7. The total energy emitted by the bulb is $E = 0.93 Pt$, where $P = 60$ W and

$$t = 730 \text{ h} = (730 \text{ h})(3600 \text{ s/h}) = 2.628 \times 10^6 \text{ s}.$$

The energy of each photon emitted is $E_{\text{ph}} = hc/\lambda$. Therefore, the number of photons emitted is

$$N = \frac{E}{E_{\text{ph}}} = \frac{0.93 Pt}{hc/\lambda} = \frac{(0.93)(60 \text{ W})(2.628 \times 10^6 \text{ s})}{(6.63 \times 10^{-34} \text{ J} \cdot \text{s})(2.998 \times 10^8 \text{ m/s})/(630 \times 10^{-9} \text{ m})} = 4.7 \times 10^{26}.$$

8. Following Sample Problem 38-1, we have

$$P = \frac{Rhc}{\lambda} = \frac{(100/\text{s})(6.63 \times 10^{-34} \text{ J} \cdot \text{s})(2.998 \times 10^8 \text{ m/s})}{550 \times 10^{-9} \text{ m}} = 3.6 \times 10^{-17} \text{ W}.$$

9. (a) Let R be the rate of photon emission (number of photons emitted per unit time) and let E be the energy of a single photon. Then, the power output of a lamp is given by $P = RE$ if all the power goes into photon production. Now, $E = hf = hc/\lambda$, where h is the Planck constant, f is the frequency of the light emitted, and λ is the wavelength. Thus $P = Rhc/\lambda$ and $R = \lambda P/hc$. The lamp emitting light with the longer wavelength (the 700 nm lamp) emits more photons per unit time. The energy of each photon is less, so it must emit photons at a greater rate.

(b) Let R be the rate of photon production for the 700 nm lamp. Then,

$$R = \frac{\lambda P}{hc} = \frac{(700\,\text{nm})(400\,\text{J/s})}{(1.60 \times 10^{-19}\,\text{J/eV})(1240\,\text{eV}\cdot\text{nm})} = 1.41 \times 10^{21}\ \text{photon/s}.$$

10. (a) The rate at which solar energy strikes the panel is

$$P = (1.39\ \text{kW/m}^2)(2.60\ \text{m}^2) = 3.61\ \text{kW}.$$

(b) The rate at which solar photons are absorbed by the panel is

$$R = \frac{P}{E_{\text{ph}}} = \frac{3.61 \times 10^3\ \text{W}}{(6.63 \times 10^{-34}\,\text{J}\cdot\text{s})(2.998 \times 10^8\ \text{m/s})/(550 \times 10^{-9}\ \text{m})} = 1.00 \times 10^{22}\ /\text{s}.$$

(c) The time in question is given by

$$t = \frac{N_A}{R} = \frac{6.02 \times 10^{23}}{1.00 \times 10^{22}\ /\text{s}} = 60.2\ \text{s}.$$

11. (a) We assume all the power results in photon production at the wavelength $\lambda = 589\ \text{nm}$. Let R be the rate of photon production and E be the energy of a single photon. Then, $P = RE = Rhc/\lambda$, where $E = hf$ and $f = c/\lambda$ are used. Here h is the Planck constant, f is the frequency of the emitted light, and λ is its wavelength. Thus,

$$R = \frac{\lambda P}{hc} = \frac{(589 \times 10^{-9}\ \text{m})(100\ \text{W})}{(6.63 \times 10^{-34}\,\text{J}\cdot\text{s})(3.00 \times 10^8\ \text{m/s})} = 2.96 \times 10^{20}\ \text{photon/s}.$$

(b) Let I be the photon flux a distance r from the source. Since photons are emitted uniformly in all directions, $R = 4\pi r^2 I$ and

$$r = \sqrt{\frac{R}{4\pi I}} = \sqrt{\frac{2.96 \times 10^{20}\ \text{photon/s}}{4\pi (1.00 \times 10^4\ \text{photon/m}^2\cdot\text{s})}} = 4.86 \times 10^7\ \text{m}.$$

(c) The photon flux is

$$I = \frac{R}{4\pi r^2} = \frac{2.96 \times 10^{20} \text{ photon}/\text{s}}{4\pi(2.00\,\text{m})^2} = 5.89 \times 10^{18} \frac{\text{photon}}{\text{m}^2 \cdot \text{s}}.$$

12. The rate at which photons are emitted from the argon laser source is given by $R = P/E_{\text{ph}}$, where $P = 1.5$ W is the power of the laser beam and $E_{\text{ph}} = hc/\lambda$ is the energy of each photon of wavelength λ. Since $\alpha = 84\%$ of the energy of the laser beam falls within the central disk, the rate of photon absorption of the central disk is

$$R' = \alpha R = \frac{\alpha P}{hc/\lambda} = \frac{(0.84)(1.5\,\text{W})}{(6.63 \times 10^{-34}\,\text{J} \cdot \text{s})(2.998 \times 10^8\,\text{m}/\text{s})/(515 \times 10^{-9}\,\text{m})}$$

$$= 3.3 \times 10^{18} \text{ photons}/\text{s}.$$

13. The energy of an incident photon is $E = hf = hc/\lambda$, where h is the Planck constant, f is the frequency of the electromagnetic radiation, and λ is its wavelength. The kinetic energy of the most energetic electron emitted is

$$K_m = E - \Phi = (hc/\lambda) - \Phi,$$

where Φ is the work function for sodium. The stopping potential V_0 is related to the maximum kinetic energy by $eV_0 = K_m$, so $eV_0 = (hc/\lambda) - \Phi$ and

$$\lambda = \frac{hc}{eV_0 + \Phi} = \frac{1240\,\text{eV} \cdot \text{nm}}{5.0\,\text{eV} + 2.2\,\text{eV}} = 170\,\text{nm}.$$

Here $eV_0 = 5.0$ eV and $hc = 1240$ eV·nm are used.

14. The energy of the most energetic photon in the visible light range (with wavelength of about 400 nm) is about $E = (1240 \text{ eV·nm}/400 \text{ nm}) = 3.1$ eV (using the fact that $hc = 1240$ eV·nm). Consequently, barium and lithium can be used, since their work functions are both lower than 3.1 eV.

15. The speed v of the electron satisfies

$$K_{\text{max}} = \tfrac{1}{2}m_e v^2 = \tfrac{1}{2}(m_e c^2)(v/c)^2 = E_{\text{photon}} - \Phi.$$

Using Table 38-3, we find

$$v = c\sqrt{\frac{2(E_{\text{photon}} - \Phi)}{m_e c^2}} = (2.998 \times 10^8\,\text{m}/\text{s})\sqrt{\frac{2(5.80\,\text{eV} - 4.50\,\text{eV})}{511 \times 10^3\,\text{eV}}} = 6.76 \times 10^5\,\text{m}/\text{s}.$$

16. We use Eq. 38-5 to find the maximum kinetic energy of the ejected electrons:

$$K_{max} = hf - \Phi = (4.14 \times 10^{-15}\,eV \cdot s)(3.0 \times 10^{15}\,Hz) - 2.3\,eV = 10\,eV.$$

17. (a) We use Eq. 38-6:

$$V_{stop} = \frac{hf - \Phi}{e} = \frac{hc/\lambda - \Phi}{e} = \frac{(1240\,eV \cdot nm / 400\,nm) - 1.8\,eV}{e} = 1.3\,V.$$

(b) We use the formula obtained in the solution of problem 15:

$$v = \sqrt{\frac{2(E_{photon} - \Phi)}{m_e}} = \sqrt{\frac{2eV_{stop}}{m_e}} = c\sqrt{\frac{2eV_{stop}}{m_e c^2}} = (2.998 \times 10^8\,m/s)\sqrt{\frac{2e(1.3\,V)}{511 \times 10^3\,eV}}$$
$$= 6.8 \times 10^5\,m/s.$$

18. To find the longest possible wavelength λ_{max} (corresponding to the lowest possible energy) of a photon which can produce a photoelectric effect in platinum, we set $K_{max} = 0$ in Eq. 38-5 and use $hf = hc/\lambda$. Thus $hc/\lambda_{max} = \Phi$. We solve for λ_{max}:

$$\lambda_{max} = \frac{hc}{\Phi} = \frac{1240\,eV \cdot nm}{5.32\,nm} = 233\,nm.$$

19. (a) The kinetic energy K_m of the fastest electron emitted is given by

$$K_m = hf - \Phi = (hc/\lambda) - \Phi,$$

where Φ is the work function of aluminum, f is the frequency of the incident radiation, and λ is its wavelength. The relationship $f = c/\lambda$ was used to obtain the second form. Thus,

$$K_m = \frac{1240\,eV \cdot nm}{200\,nm} - 4.20\,eV = 2.00\,eV.$$

Where we have used $hc = 1240$ eV·nm.

(b) The slowest electron just breaks free of the surface and so has zero kinetic energy.

(c) The stopping potential V_0 is given by $K_m = eV_0$, so $V_0 = K_m/e = (2.00\,eV)/e = 2.00$ V.

(d) The value of the cutoff wavelength is such that $K_m = 0$. Thus $hc/\lambda = \Phi$ or

$$\lambda = hc/\Phi = (1240\,eV \cdot nm)/(4.2\,eV) = 295\,nm.$$

If the wavelength is longer, the photon energy is less and a photon does not have sufficient energy to knock even the most energetic electron out of the aluminum sample.

20. We use Eq. 38-6 and the fact that $hc = 1240$ eV·nm:

$$K_{max} = E_{photon} - \Phi = \frac{hc}{\lambda} - \frac{hc}{\lambda_{max}} = \frac{1240\,\text{eV}\cdot\text{nm}}{254\,\text{nm}} - \frac{1240\,\text{eV}\cdot\text{nm}}{325\,\text{nm}} = 1.07\,\text{eV}.$$

21. (a) We use the photoelectric effect equation (Eq. 38-5) in the form $hc/\lambda = \Phi + K_m$. The work function depends only on the material and the condition of the surface, and not on the wavelength of the incident light. Let λ_1 be the first wavelength described and λ_2 be the second. Let $K_{m1} = 0.710$ eV be the maximum kinetic energy of electrons ejected by light with the first wavelength, and $K_{m2} = 1.43$ eV be the maximum kinetic energy of electrons ejected by light with the second wavelength. Then,

$$\frac{hc}{\lambda_1} = \Phi + K_{m1} \quad \text{and} \quad \frac{hc}{\lambda_2} = \Phi + K_{m2}.$$

The first equation yields $\Phi = (hc/\lambda_1) - K_{m1}$. When this is used to substitute for Φ in the second equation, the result is

$$(hc/\lambda_2) = (hc/\lambda_1) - K_{m1} + K_{m2}.$$

The solution for λ_2 is

$$\lambda_2 = \frac{hc\lambda_1}{hc + \lambda_1(K_{m2} - K_{m1})} = \frac{(1240\,\text{V}\cdot\text{nm})(491\,\text{nm})}{1240\,\text{eV}\cdot\text{nm} + (491\,\text{nm})(1.43\,\text{eV} - 0.710\,\text{eV})}$$
$$= 382\,\text{nm}.$$

Here $hc = 1240$ eV·nm has been used.

(b) The first equation displayed above yields

$$\Phi = \frac{hc}{\lambda_1} - K_{m1} = \frac{1240\,\text{eV}\cdot\text{nm}}{491\,\text{nm}} - 0.710\,\text{eV} = 1.82\,\text{eV}.$$

22. (a) For the first and second case (labeled 1 and 2) we have $eV_{01} = hc/\lambda_1 - \Phi$ and $eV_{02} = hc/\lambda_2 - \Phi$, from which h and Φ can be determined. Thus,

$$h = \frac{e(V_1 - V_2)}{c(\lambda_1^{-1} - \lambda_2^{-1})} = \frac{1.85\,\text{eV} - 0.820\,\text{eV}}{(3.00 \times 10^{17}\,\text{nm/s})\left[(300\,\text{nm})^{-1} - (400\,\text{nm})^{-1}\right]} = 4.12 \times 10^{-15}\,\text{eV}\cdot\text{s}.$$

(b) The work function is

$$\Phi = \frac{3(V_2\lambda_2 - V_1\lambda_1)}{\lambda_1 - \lambda_2} = \frac{(0.820 \text{ eV})(400 \text{ nm}) - (1.85 \text{ eV})(300 \text{ nm})}{300 \text{ nm} - 400 \text{ nm}} = 2.27 \text{ eV}.$$

(c) Let $\Phi = hc/\lambda_{max}$ to obtain

$$\lambda_{max} = \frac{hc}{\Phi} = \frac{1240 \text{ eV} \cdot \text{nm}}{2.27 \text{ eV}} = 545 \text{ nm}.$$

23. (a) Find the speed v of the electron from $r = m_e v/eB$: $v = rBe/m_e$. Thus

$$K_{max} = \frac{1}{2}m_e v^2 = \frac{1}{2}m_e\left(\frac{rBe}{m_e}\right)^2 = \frac{(rB)^2 e^2}{2m_e} = \frac{(1.88\times10^{-4} \text{ T}\cdot\text{m})^2(1.60\times10^{-19} \text{ C})^2}{2(9.11\times10^{-31} \text{ kg})(1.60\times10^{-19} \text{ J/eV})}$$
$$= 3.1 \text{ keV}.$$

(b) Using the fact that $hc = 1240$ eV·nm, the work done is

$$W = E_{photon} - K_{max} = \frac{1240 \text{ eV}\cdot\text{nm}}{71\times10^{-3} \text{ nm}} - 3.10 \text{ keV} = 14 \text{ keV}.$$

24. Using the fact that $hc = 1240$ eV·nm, the number of photons emitted from the laser per unit time is

$$R = \frac{P}{E_{ph}} = \frac{2.00\times10^{-3} \text{ W}}{(1240 \text{ eV}\cdot\text{nm}/600 \text{ nm})(1.60\times10^{-19} \text{ J/eV})} = 6.05\times10^{15}/\text{s},$$

of which $(1.0 \times 10^{-16})(6.05 \times 10^{15}/\text{s}) = 0.605/\text{s}$ actually cause photoelectric emissions. Thus the current is

$$i = (0.605/\text{s})(1.60 \times 10^{-19} \text{ C}) = 9.68 \times 10^{-20} \text{ A}.$$

25. (a) When a photon scatters from an electron initially at rest, the change in wavelength is given by $\Delta\lambda = (h/mc)(1 - \cos\phi)$, where m is the mass of an electron and ϕ is the scattering angle. Now, $h/mc = 2.43 \times 10^{-12}$ m = 2.43 pm, so

$$\Delta\lambda = (2.43 \text{ pm})(1 - \cos 30°) = 0.326 \text{ pm}.$$

The final wavelength is

$$\lambda' = \lambda + \Delta\lambda = 2.4 \text{ pm} + 0.326 \text{ pm} = 2.73 \text{ pm}.$$

(b) Now, $\Delta\lambda = (2.43 \text{ pm})(1 - \cos 120°) = 3.645$ pm and

$$\lambda' = 2.4 \text{ pm} + 3.645 \text{ pm} = 6.05 \text{ pm.}$$

26. (a) The rest energy of an electron is given by $E = m_e c^2$. Thus the momentum of the photon in question is given by

$$p = \frac{E}{c} = \frac{m_e c^2}{c} = m_e c = (9.11 \times 10^{-31} \text{ kg})(2.998 \times 10^8 \text{ m/s}) = 2.73 \times 10^{-22} \text{ kg} \cdot \text{m/s}$$
$$= 0.511 \text{ MeV}/c.$$

(b) From Eq. 38-7,

$$\lambda = \frac{h}{p} = \frac{6.63 \times 10^{-34} \text{ J} \cdot \text{s}}{2.73 \times 10^{-22} \text{ kg} \cdot \text{m/s}} = 2.43 \times 10^{-12} \text{ m} = 2.43 \text{ pm.}$$

(c) Using Eq. 38-1,

$$f = \frac{c}{\lambda} = \frac{2.998 \times 10^8 \text{ m/s}}{2.43 \times 10^{-12} \text{ m}} = 1.24 \times 10^{20} \text{ Hz.}$$

27. (a) The x-ray frequency is

$$f = \frac{c}{\lambda} = \frac{2.998 \times 10^8 \text{ m/s}}{35.0 \times 10^{-12} \text{ m}} = 8.57 \times 10^{18} \text{ Hz.}$$

(b) The x-ray photon energy is

$$E = hf = (4.14 \times 10^{-15} \text{ eV} \cdot \text{s})(8.57 \times 10^{18} \text{ Hz}) = 3.55 \times 10^4 \text{ eV.}$$

(c) From Eq. 38-7,

$$p = \frac{h}{\lambda} = \frac{6.63 \times 10^{-34} \text{ J} \cdot \text{s}}{35.0 \times 10^{-12} \text{ m}} = 1.89 \times 10^{-23} \text{ kg} \cdot \text{m/s} = 35.4 \text{ keV}/c.$$

28. (a) Eq. 38-11 yields

$$\Delta\lambda = \frac{h}{m_e c}(1 - \cos\phi) = (2.43 \text{ pm})(1 - \cos 180°) = +4.86 \text{ pm.}$$

(b) Using the fact that $hc = 1240$ eV·nm, the change in photon energy is

$$\Delta E = \frac{hc}{\lambda'} - \frac{hc}{\lambda} = (1240 \text{ eV·nm})\left(\frac{1}{0.01 \text{ nm} + 4.86 \text{ pm}} - \frac{1}{0.01 \text{ nm}}\right) = -40.6 \text{ keV.}$$

(c) From conservation of energy, $\Delta K = -\Delta E = 40.6$ keV.

(d) The electron will move straight ahead after the collision, since it has acquired some of the forward linear momentum from the photon. Thus, the angle between $+x$ and the direction of the electron's motion is zero.

29. (a) Since the mass of an electron is $m = 9.109 \times 10^{-31}$ kg, its Compton wavelength is

$$\lambda_C = \frac{h}{mc} = \frac{6.626 \times 10^{-34} \text{ J} \cdot \text{s}}{(9.109 \times 10^{-31} \text{ kg})(2.998 \times 10^8 \text{ m/s})} = 2.426 \times 10^{-12} \text{ m} = 2.43 \text{ pm.}$$

(b) Since the mass of a proton is $m = 1.673 \times 10^{-27}$ kg, its Compton wavelength is

$$\lambda_C = \frac{6.626 \times 10^{-34} \text{ J} \cdot \text{s}}{(1.673 \times 10^{-27} \text{ kg})(2.998 \times 10^8 \text{ m/s})} = 1.321 \times 10^{-15} \text{ m} = 1.32 \text{ fm.}$$

(c) We use the formula $hc = 1240$ eV·nm, which gives $E = (1240 \text{ eV·nm})/\lambda$, where E is the energy and λ is the wavelength. Thus for the electron,

$$E = (1240 \text{ eV·nm})/(2.426 \times 10^{-3} \text{ nm}) = 5.11 \times 10^5 \text{ eV} = 0.511 \text{ MeV.}$$

(d) For the proton,

$$E = (1240 \text{ eV·nm})/(1.321 \times 10^{-6} \text{ nm}) = 9.39 \times 10^8 \text{ eV} = 939 \text{ MeV.}$$

30. (a) Using the fact that $hc = 1240$ eV·nm, we find

$$\lambda = \frac{hc}{E} = \frac{1240 \text{ nm} \cdot \text{eV}}{0.511 \text{ MeV}} = 2.43 \times 10^{-3} \text{ nm} = 2.43 \text{ pm.}$$

(b) Now, Eq. 38-11 leads to

$$\lambda' = \lambda + \Delta\lambda = \lambda + \frac{h}{m_e c}(1 - \cos\phi) = 2.43 \text{ pm} + (2.43 \text{ pm})(1 - \cos 90.0°)$$

$$= 4.86 \text{ pm.}$$

(c) The scattered photons have energy equal to

$$E' = E\left(\frac{\lambda}{\lambda'}\right) = (0.511 \text{ MeV})\left(\frac{2.43 \text{ pm}}{4.86 \text{ pm}}\right) = 0.255 \text{ MeV.}$$

31. (a) The fractional change is

$$\frac{\Delta E}{E} = \frac{\Delta(hc/\lambda)}{hc/\lambda} = \lambda \Delta\left(\frac{1}{\lambda}\right) = \lambda\left(\frac{1}{\lambda'} - \frac{1}{\lambda}\right) = \frac{\lambda}{\lambda'} - 1 = \frac{\lambda}{\lambda + \Delta\lambda} - 1$$

$$= -\frac{1}{\lambda/\Delta\lambda + 1} = -\frac{1}{(\lambda/\lambda_C)(1-\cos\phi)^{-1} + 1}.$$

If $\lambda = 3.0$ cm $= 3.0 \times 10^{10}$ pm and $\phi = 90°$, the result is

$$\frac{\Delta E}{E} = -\frac{1}{(3.0\times10^{10}\,\text{pm}/2.43\,\text{pm})(1-\cos90°)^{-1}+1} = -8.1\times10^{-11} = -8.1\times10^{-9}\ \%.$$

(b) Now $\lambda = 500$ nm $= 5.00 \times 10^5$ pm and $\phi = 90°$, so

$$\frac{\Delta E}{E} = -\frac{1}{(5.00\times10^5\,\text{pm}/2.43\,\text{pm})(1-\cos90°)^{-1}+1} = -4.9\times10^{-6} = -4.9\times10^{-4}\ \%.$$

(c) With $\lambda = 25$ pm and $\phi = 90°$, we find

$$\frac{\Delta E}{E} = -\frac{1}{(25\,\text{pm}/2.43\,\text{pm})(1-\cos90°)^{-1}+1} = -8.9\times10^{-2} = -8.9\ \%.$$

(d) In this case, $\lambda = hc/E = 1240$ nm·eV$/1.0$ MeV $= 1.24 \times 10^{-3}$ nm $= 1.24$ pm, so

$$\frac{\Delta E}{E} = -\frac{1}{(1.24\,\text{pm}/2.43\,\text{pm})(1-\cos90°)^{-1}+1} = -0.66 = -66\ \%.$$

(e) From the calculation above, we see that the shorter the wavelength the greater the fractional energy change for the photon as a result of the Compton scattering. Since $\Delta E/E$ is virtually zero for microwave and visible light, the Compton effect is significant only in the x-ray to gamma ray range of the electromagnetic spectrum.

32. The $(1 - \cos\phi)$ factor in Eq. 38-11 is largest when $\phi = 180°$. Thus, using Table 38-3, we obtain

$$\Delta\lambda_{\text{max}} = \frac{hc}{m_p c^2}(1-\cos180°) = \frac{1240\,\text{MeV}\cdot\text{fm}}{938\,\text{MeV}}(1-(-1)) = 2.64\ \text{fm}$$

where we have used that fact that $hc = 1240$ eV·nm $= 1240$ MeV·fm.

33. If E is the original energy of the photon and E' is the energy after scattering, then the fractional energy loss is

$$\frac{\Delta E}{E} = \frac{E - E'}{E} = \frac{\Delta\lambda}{\lambda + \Delta\lambda}$$

using the result from Sample Problem 38-4. Thus

$$\frac{\Delta\lambda}{\lambda} = \frac{\Delta E / E}{1 - \Delta E / E} = \frac{0.75}{1 - 0.75} = 3 = 300\ \%.$$

A 300% increase in the wavelength leads to a 75% decrease in the energy of the photon.

34. The initial wavelength of the photon is (using $hc = 1240$ eV·nm)

$$\lambda = \frac{hc}{E} = \frac{1240\,\text{eV}\cdot\text{nm}}{17500\,\text{eV}} = 0.07086\ \text{nm}$$

or 70.86 pm. The maximum Compton shift occurs for $\phi = 180°$, in which case Eq. 38-11 (applied to an electron) yields

$$\Delta\lambda = \left(\frac{hc}{m_e c^2}\right)(1 - \cos 180°) = \left(\frac{1240\,\text{eV}\cdot\text{nm}}{511 \times 10^3\,\text{eV}}\right)(1 - (-1)) = 0.00485\ \text{nm}$$

where Table 38-3 is used. Therefore, the new photon wavelength is

$$\lambda' = 0.07086\ \text{nm} + 0.00485\ \text{nm} = 0.0757\ \text{nm}.$$

Consequently, the new photon energy is

$$E' = \frac{hc}{\lambda'} = \frac{1240\,\text{eV}\cdot\text{nm}}{0.0757\,\text{nm}} = 1.64 \times 10^4\ \text{eV} = 16.4\ \text{keV} \ .$$

By energy conservation, then, the kinetic energy of the electron must equal

$$E' - E = 17.5\ \text{keV} - 16.4\ \text{keV} = 1.1\ \text{keV}.$$

35. (a) From Eq. 38-11

$$\Delta\lambda = \frac{h}{m_e c}(1 - \cos\phi) = (2.43\,\text{pm})(1 - \cos 90°) = 2.43\,\text{pm}\ .$$

(b) The fractional shift should be interpreted as $\Delta\lambda$ divided by the original wavelength:

$$\frac{\Delta\lambda}{\lambda} = \frac{2.425\,\text{pm}}{590\,\text{nm}} = 4.11 \times 10^{-6}.$$

(c) The change in energy for a photon with $\lambda = 590$ nm is given by

$$\Delta E_{\text{ph}} = \Delta\left(\frac{hc}{\lambda}\right) \approx -\frac{hc\Delta\lambda}{\lambda^2}$$

$$= -\frac{(4.14\times10^{-15}\,\text{eV}\cdot\text{s})(2.998\times10^8\,\text{m/s})(2.43\,\text{pm})}{(590\,\text{nm})^2}$$

$$= -8.67\times10^{-6}\,\text{eV} .$$

(d) For an x-ray photon of energy $E_{\text{ph}} = 50$ keV, $\Delta\lambda$ remains the same (2.43 pm), since it is independent of E_{ph}.

(e) The fractional change in wavelength is now

$$\frac{\Delta\lambda}{\lambda} = \frac{\Delta\lambda}{hc/E_{\text{ph}}} = \frac{(50\times10^3\,\text{eV})(2.43\,\text{pm})}{(4.14\times10^{-15}\,\text{eV}\cdot\text{s})(2.998\times10^8\,\text{m/s})} = 9.78\times10^{-2} .$$

(f) The change in photon energy is now

$$\Delta E_{\text{ph}} = hc\left(\frac{1}{\lambda+\Delta\lambda} - \frac{1}{\lambda}\right) = -\left(\frac{hc}{\lambda}\right)\frac{\Delta\lambda}{\lambda+\Delta\lambda} = -E_{\text{ph}}\left(\frac{\alpha}{1+\alpha}\right)$$

where $\alpha = \Delta\lambda/\lambda$. With $E_{\text{ph}} = 50$ keV and $\alpha = 9.78 \times 10^{-2}$, we obtain $\Delta E_{\text{ph}} = -4.45$ keV. (Note that in this case $\alpha \approx 0.1$ is not close enough to zero so the approximation $\Delta E_{\text{ph}} \approx hc\Delta\lambda/\lambda^2$ is not as accurate as in the first case, in which $\alpha = 4.12 \times 10^{-6}$. In fact if one were to use this approximation here, one would get $\Delta E_{\text{ph}} \approx -4.89$ keV, which does not amount to a satisfactory approximation.)

36. Referring to Sample Problem 38-4, we see that the fractional change in photon energy is

$$\frac{E-E_n}{E} = \frac{\Delta\lambda}{\lambda+\Delta\lambda} = \frac{(h/mc)(1-\cos\phi)}{(hc/E)+(h/mc)(1-\cos\phi)} .$$

Energy conservation demands that $E - E' = K$, the kinetic energy of the electron. In the maximal case, $\phi = 180°$, and we find

$$\frac{K}{E} = \frac{(h/mc)(1-\cos180°)}{(hc/E)+(h/mc)(1-\cos180°)} = \frac{2h/mc}{(hc/E)+(2h/mc)} .$$

Multiplying both sides by E and simplifying the fraction on the right-hand side leads to

$$K = E\left(\frac{2/mc}{c/E + 2/mc}\right) = \frac{E^2}{mc^2/2 + E} .$$

37. (a) From Eq. 38-11, $\Delta\lambda = (h/m_e c)(1 - \cos\phi)$. In this case $\phi = 180°$ (so $\cos\phi = -1$), and the change in wavelength for the photon is given by $\Delta\lambda = 2h/m_e c$. The energy E' of the scattered photon (whose initial energy is $E = hc/\lambda$) is then

$$E' = \frac{hc}{\lambda + \Delta\lambda} = \frac{E}{1 + \Delta\lambda/\lambda} = \frac{E}{1 + (2h/m_e c)(E/hc)} = \frac{E}{1 + 2E/m_e c^2}$$

$$= \frac{50.0\,\text{keV}}{1 + 2(50.0\,\text{keV})/0.511\,\text{MeV}} = 41.8\,\text{keV} .$$

(b) From conservation of energy the kinetic energy K of the electron is given by

$$K = E - E' = 50.0\ \text{keV} - 41.8\ \text{keV} = 8.2\ \text{keV}.$$

38. The magnitude of the fractional energy change for the photon is given by

$$\left|\frac{\Delta E_{ph}}{E_{ph}}\right| = \left|\frac{\Delta(hc/\lambda)}{hc/\lambda}\right| = \left|\lambda\Delta\left(\frac{1}{\lambda}\right)\right| = \lambda\left(\frac{1}{\lambda} - \frac{1}{\lambda + \Delta\lambda}\right) = \frac{\Delta\lambda}{\lambda + \Delta\lambda} = \beta$$

where $\beta = 0.10$. Thus $\Delta\lambda = \lambda\beta/(1 - \beta)$. We substitute this expression for $\Delta\lambda$ in Eq. 38-11 and solve for $\cos\phi$:

$$\cos\phi = 1 - \frac{mc}{h}\Delta\lambda = 1 - \frac{mc\lambda\beta}{h(1 - \beta)} = 1 - \frac{\beta(mc^2)}{(1 - \beta)E_{ph}}$$

$$= 1 - \frac{(0.10)(511\ \text{keV})}{(1 - 0.10)(200\,\text{keV})} = 0.716 .$$

This leads to an angle of $\phi = 44°$.

39. We start with the result of Exercise 49: $\lambda = h/\sqrt{2mK}$. Replacing K with eV, where V is the accelerating potential and e is the fundamental charge, we obtain

$$\lambda = \frac{h}{\sqrt{2meV}} = \frac{6.626\times10^{-34}\,\text{J}\cdot\text{s}}{\sqrt{2(9.109\times10^{-31}\text{kg})(1.602\times10^{-19}\,\text{C})(25.0\times10^3\,\text{V})}}$$

$$= 7.75\times10^{-12}\ \text{m} = 7.75\,\text{pm}.$$

40. (a) Using Table 38-3 and the fact that $hc = 1240$ eV·nm, we obtain

$$\lambda = \frac{h}{p} = \frac{h}{\sqrt{2m_e K}} = \frac{hc}{\sqrt{2m_e c^2 K}} = \frac{1240\,\text{eV} \cdot \text{nm}}{\sqrt{2(511000\,\text{eV})(1000\,\text{eV})}} = 0.0388\,\text{nm}.$$

(b) A photon's de Broglie wavelength is equal to its familiar wave-relationship value. Using the fact that $hc = 1240$ eV·nm,

$$\lambda = \frac{hc}{E} = \frac{1240\,\text{eV} \cdot \text{nm}}{1.00\,\text{keV}} = 1.24\,\text{nm}.$$

(c) The neutron mass may be found in Appendix B. Using the conversion from electronvolts to Joules, we obtain

$$\lambda = \frac{h}{\sqrt{2m_n K}} = \frac{6.63 \times 10^{-34}\,\text{J} \cdot \text{s}}{\sqrt{2(1.675 \times 10^{-27}\,\text{kg})(1.6 \times 10^{-16}\,\text{J})}} = 9.06 \times 10^{-13}\,\text{m}.$$

41. If K is given in electron volts, then

$$\lambda = \frac{6.626 \times 10^{-34}\,\text{J} \cdot \text{s}}{\sqrt{2(9.109 \times 10^{-31}\,\text{kg})(1.602 \times 10^{-19}\,\text{J/eV})K}} = \frac{1.226 \times 10^{-9}\,\text{m} \cdot \text{eV}^{1/2}}{\sqrt{K}} = \frac{1.226\,\text{nm} \cdot \text{eV}^{1/2}}{\sqrt{K}},$$

where K is the kinetic energy. Thus

$$K = \left(\frac{1.226\,\text{nm} \cdot \text{eV}^{1/2}}{\lambda}\right)^2 = \left(\frac{1.226\,\text{nm} \cdot \text{eV}^{1/2}}{590\,\text{nm}}\right)^2 = 4.32 \times 10^{-6}\,\text{eV}.$$

42. (a) We solve v from $\lambda = h/p = h/(m_p v)$:

$$v = \frac{h}{m_p \lambda} = \frac{6.63 \times 10^{-34}\,\text{J} \cdot \text{s}}{(1.675 \times 10^{-27}\,\text{kg})(0.100 \times 10^{-12}\,\text{m})} = 3.96 \times 10^6\,\text{m/s}.$$

(b) We set $eV = K = \frac{1}{2}m_p v^2$ and solve for the voltage:

$$V = \frac{m_p v^2}{2e} = \frac{(1.67 \times 10^{-27}\,\text{kg})(3.96 \times 10^6\,\text{m/s})^2}{2(1.60 \times 10^{-19}\,\text{C})} = 8.18 \times 10^4\,\text{V} = 81.8\,\text{kV}.$$

43. (a) The momentum of the photon is given by $p = E/c$, where E is its energy. Its wavelength is

$$\lambda = \frac{h}{p} = \frac{hc}{E} = \frac{1240\,\text{eV} \cdot \text{nm}}{1.00\,\text{eV}} = 1240\,\text{nm}.$$

(b) The momentum of the electron is given by $p = \sqrt{2mK}$, where K is its kinetic energy and m is its mass. Its wavelength is

$$\lambda = \frac{h}{p} = \frac{h}{\sqrt{2mK}}.$$

If K is given in electron volts, then

$$\lambda = \frac{6.626 \times 10^{-34} \, \text{J} \cdot \text{s}}{\sqrt{2(9.109 \times 10^{-31} \, \text{kg})(1.602 \times 10^{-19} \, \text{J/eV})K}} = \frac{1.226 \times 10^{-9} \, \text{m} \cdot \text{eV}^{1/2}}{\sqrt{K}} = \frac{1.226 \, \text{nm} \cdot \text{eV}^{1/2}}{\sqrt{K}}.$$

For $K = 1.00 \, \text{eV}$, we have

$$\lambda = \frac{1.226 \, \text{nm} \cdot \text{eV}^{1/2}}{\sqrt{1.00 \, \text{eV}}} = 1.23 \, \text{nm}.$$

(c) For the photon,

$$\lambda = \frac{hc}{E} = \frac{1240 \, \text{eV} \cdot \text{nm}}{1.00 \times 10^9 \, \text{eV}} = 1.24 \times 10^{-6} \, \text{nm} = 1.24 \, \text{fm}.$$

(d) Relativity theory must be used to calculate the wavelength for the electron. According to Eq. 38-51, the momentum p and kinetic energy K are related by $(pc)^2 = K^2 + 2Kmc^2$. Thus,

$$pc = \sqrt{K^2 + 2Kmc^2} = \sqrt{\left(1.00 \times 10^9 \, \text{eV}\right)^2 + 2\left(1.00 \times 10^9 \, \text{eV}\right)\left(0.511 \times 10^6 \, \text{eV}\right)}$$
$$= 1.00 \times 10^9 \, \text{eV}.$$

The wavelength is

$$\lambda = \frac{h}{p} = \frac{hc}{pc} = \frac{1240 \, \text{eV} \cdot \text{nm}}{1.00 \times 10^9 \, \text{eV}} = 1.24 \times 10^{-6} \, \text{nm} = 1.24 \, \text{fm}.$$

44. (a) The momentum of the electron is

$$p = \frac{h}{\lambda} = \frac{6.63 \times 10^{-34} \, \text{J} \cdot \text{s}}{0.20 \times 10^{-9} \, \text{m}} = 3.3 \times 10^{-24} \, \text{kg} \cdot \text{m/s}.$$

(b) The momentum of the photon is the same as that of the electron: $p = 3.3 \times 10^{-24} \, \text{kg} \cdot \text{m/s}$.

(c) The kinetic energy of the electron is

$$K_e = \frac{p^2}{2m_e} = \frac{\left(3.3\times10^{-24}\,\text{kg}\cdot\text{m/s}\right)^2}{2\left(9.11\times10^{-31}\,\text{kg}\right)} = 6.0\times10^{-18}\,\text{J} = 38\,\text{eV}.$$

(d) The kinetic energy of the photon is

$$K_{\text{ph}} = pc = \left(3.3\times10^{-24}\,\text{kg}\cdot\text{m/s}\right)\left(2.998\times10^{8}\,\text{m/s}\right) = 9.9\times10^{-16}\,\text{J} = 6.2\,\text{keV}.$$

45. (a) The kinetic energy acquired is $K = qV$, where q is the charge on an ion and V is the accelerating potential. Thus

$$K = (1.60\times10^{-19}\,\text{C})(300\,\text{V}) = 4.80\times10^{-17}\,\text{J}.$$

The mass of a single sodium atom is, from Appendix F,

$$m = (22.9898\,\text{g/mol})/(6.02\times10^{23}\,\text{atom/mol}) = 3.819\times10^{-23}\,\text{g} = 3.819\times10^{-26}\,\text{kg}.$$

Thus, the momentum of an ion is

$$p = \sqrt{2mK} = \sqrt{2\left(3.819\times10^{-26}\,\text{kg}\right)\left(4.80\times10^{-17}\,\text{J}\right)} = 1.91\times10^{-21}\,\text{kg}\cdot\text{m/s}.$$

(b) The de Broglie wavelength is

$$\lambda = \frac{h}{p} = \frac{6.63\times10^{-34}\,\text{J}\cdot\text{s}}{1.91\times10^{-21}\,\text{kg}\cdot\text{m/s}} = 3.46\times10^{-13}\,\text{m}.$$

46. (a) We use the fact that $hc = 1240\,\text{nm}\cdot\text{eV}$:

$$E_{\text{photon}} = \frac{hc}{\lambda} = \frac{1240\,\text{nm}\cdot\text{eV}}{1.00\,\text{nm}} = 1.24\,\text{keV}.$$

(b) For the electron, we have

$$K = \frac{p^2}{2m_e} = \frac{(h/\lambda)^2}{2m_e} = \frac{(hc/\lambda)^2}{2m_e c^2} = \frac{1}{2(0.511\,\text{MeV})}\left(\frac{1240\,\text{eV}\cdot\text{nm}}{1.00\,\text{nm}}\right)^2 = 1.50\,\text{eV}.$$

(c) In this case, we find

$$E_{photon} = \frac{1240 \, nm \cdot eV}{1.00 \times 10^{-6} \, nm} = 1.24 \times 10^9 \, eV = 1.24 \, GeV.$$

(d) For the electron (recognizing that 1240 eV·nm = 1240 MeV·fm)

$$K = \sqrt{p^2 c^2 + \left(m_e c^2\right)^2} - m_e c^2 = \sqrt{(hc/\lambda)^2 + \left(m_e c^2\right)^2} - m_e c^2$$

$$= \sqrt{\left(\frac{1240 \, MeV \cdot fm}{1.00 \, fm}\right)^2 + \left(0.511 \, MeV\right)^2} - 0.511 \, MeV$$

$$= 1.24 \times 10^3 \, MeV = 1.24 \, GeV.$$

We note that at short λ (large K) the kinetic energy of the electron, calculated with the relativistic formula, is about the same as that of the photon. This is expected since now $K \approx E \approx pc$ for the electron, which is the same as $E = pc$ for the photon.

47. (a) We need to use the relativistic formula $p = \sqrt{(E/c)^2 - m_e^2 c^2} \approx E/c \approx K/c$ $\left(\text{since } E \gg m_e c^2\right)$. So

$$\lambda = \frac{h}{p} \approx \frac{hc}{K} = \frac{1240 \, eV \cdot nm}{50 \times 10^9 \, eV} = 2.5 \times 10^{-8} \, nm = 0.025 \, fm.$$

(b) With $R = 5.0 \, fm$, we obtain $R/\lambda = 2.0 \times 10^2$.

48. (a) Since $K = 7.5 \, MeV \ll m_\alpha c^2 = 4(932 \, MeV)$, we may use the non-relativistic formula $p = \sqrt{2m_\alpha K}$. Using Eq. 38-43 (and noting that 1240 eV·nm = 1240 MeV·fm), we obtain

$$\lambda = \frac{h}{p} = \frac{hc}{\sqrt{2m_\alpha c^2 K}} = \frac{1240 \, MeV \cdot fm}{\sqrt{2(4u)(931.5 \, MeV/u)(7.5 \, MeV)}} = 5.2 \, fm.$$

(b) Since $\lambda = 5.2 \, fm \ll 30 \, fm$, to a fairly good approximation, the wave nature of the α particle does not need to be taken into consideration.

49. The wavelength associated with the unknown particle is $\lambda_p = h/p_p = h/(m_p v_p)$, where p_p is its momentum, m_p is its mass, and v_p is its speed. The classical relationship $p_p = m_p v_p$ was used. Similarly, the wavelength associated with the electron is $\lambda_e = h/(m_e v_e)$, where m_e is its mass and v_e is its speed. The ratio of the wavelengths is $\lambda_p/\lambda_e = (m_e v_e)/(m_p v_p)$, so

$$m_p = \frac{v_e \lambda_e}{v_p \lambda_p} m_e = \frac{9.109 \times 10^{-31}\,\text{kg}}{3(1.813 \times 10^{-4})} = 1.675 \times 10^{-27}\,\text{kg}.$$

According to Appendix B, this is the mass of a neutron.

50. (a) Setting $\lambda = h/p = h/\sqrt{(E/c)^2 - m_e^2 c^2}$, we solve for $K = E - m_e c^2$:

$$K = \sqrt{\left(\frac{hc}{\lambda}\right)^2 + m_e^2 c^4} - m_e c^2 = \sqrt{\left(\frac{1240\,\text{eV}\cdot\text{nm}}{10 \times 10^{-3}\,\text{nm}}\right)^2 + (0.511\,\text{MeV})^2} - 0.511\,\text{MeV}$$

$$= 0.015\,\text{MeV} = 15\,\text{keV}.$$

(b) Using the fact that $hc = 1240\,\text{eV}\cdot\text{nm}$

$$E = \frac{hc}{\lambda} = \frac{1240\,\text{eV}\cdot\text{nm}}{10 \times 10^{-3}\,\text{nm}} = 1.2 \times 10^5\,\text{eV} = 120\,\text{keV}.$$

(c) The electron microscope is more suitable, as the required energy of the electrons is much less than that of the photons.

51. The same resolution requires the same wavelength, and since the wavelength and particle momentum are related by $p = h/\lambda$, we see that the same particle momentum is required. The momentum of a 100 keV photon is

$$p = E/c = (100 \times 10^3\,\text{eV})(1.60 \times 10^{-19}\,\text{J/eV})/(3.00 \times 10^8\,\text{m/s}) = 5.33 \times 10^{-23}\,\text{kg}\cdot\text{m/s}.$$

This is also the magnitude of the momentum of the electron. The kinetic energy of the electron is

$$K = \frac{p^2}{2m} = \frac{(5.33 \times 10^{-23}\,\text{kg}\cdot\text{m/s})^2}{2(9.11 \times 10^{-31}\,\text{kg})} = 1.56 \times 10^{-15}\,\text{J}.$$

The accelerating potential is

$$V = \frac{K}{e} = \frac{1.56 \times 10^{-15}\,\text{J}}{1.60 \times 10^{-19}\,\text{C}} = 9.76 \times 10^3\,\text{V}.$$

52. (a)

$$nn^* = (a + ib)(a + ib)^* = (a + ib)(a^* + i^* b^*) = (a + ib)(a - ib)$$

$$= a^2 + iba - iab + (ib)(-ib) = a^2 + b^2,$$

which is always real since both *a* and *b* are real.

(b)

$$|nm| = |(a+ib)(c+id)| = |ac+iad+ibc+(-i)^2 bd| = |(ac-bd)+i(ad+bc)|$$
$$= \sqrt{(ac-bd)^2+(ad+bc)^2} = \sqrt{a^2c^2+b^2d^2+a^2d^2+b^2c^2}.$$

However, since

$$|n||m| = |a+ib||c+id| = \sqrt{a^2+b^2}\sqrt{c^2+d^2}$$
$$= \sqrt{a^2c^2+b^2d^2+a^2d^2+b^2c^2},$$

we conclude that $|nm| = |n|\,|m|$.

53. We plug Eq. 38-17 into Eq. 38-16, and note that

$$\frac{d\psi}{dx} = \frac{d}{dx}\left(Ae^{ikx}+Be^{-ikx}\right) = ikAe^{ikx}-ikBe^{-ikx}.$$

Also,

$$\frac{d^2\psi}{dx^2} = \frac{d}{dx}\left(ikAe^{ikx}-ikBe^{-ikx}\right) = -k^2 Ae^{ikx}-k^2 Be^{ikx}.$$

Thus,

$$\frac{d^2\psi}{dx^2}+k^2\psi = -k^2 Ae^{ikx}-k^2 Be^{ikx}+k^2\left(Ae^{ikx}+Be^{-ikx}\right) = 0.$$

54. (a) We use Euler's formula $e^{i\phi} = \cos\phi + i\sin\phi$ to re-write $\psi(x)$ as

$$\psi(x) = \psi_0 e^{ikx} = \psi_0\left(\cos kx + i\sin kx\right) = \left(\psi_0\cos kx\right)+i\left(\psi_0\sin kx\right) = a+ib,$$

where $a = \psi_0\cos kx$ and $b = \psi_0\sin kx$ are both real quantities.

(b)

$$\psi(x,t) = \psi(x)e^{-i\omega t} = \psi_0 e^{ikx}e^{-i\omega t} = \psi_0 e^{i(kx-\omega t)}$$
$$= [\psi_0\cos(kx-\omega t)]+i[\psi_0\sin(kx-\omega t)].$$

55. The angular wave number k is related to the wavelength λ by $k = 2\pi/\lambda$ and the wavelength is related to the particle momentum p by $\lambda = h/p$, so $k = 2\pi p/h$. Now, the

kinetic energy K and the momentum are related by $K = p^2/2m$, where m is the mass of the particle. Thus $p = \sqrt{2mK}$ and

$$k = \frac{2\pi\sqrt{2mK}}{h}.$$

56. The wave function is now given by

$$\Psi(x,t) = \psi_0 e^{-i(kx+\omega t)}.$$

This function describes a plane matter wave traveling in the negative x direction. An example of the actual particles that fit this description is a free electron with linear momentum $\vec{p} = -(hk/2\pi)\hat{i}$ and kinetic energy $K = p^2/2m_e = h^2k^2/8\pi^2 m_e$.

57. For $U = U_0$, Schrödinger's equation becomes

$$\frac{d^2\psi}{dx^2} + \frac{8\pi^2 m}{h^2}[E - U_0]\psi = 0.$$

We substitute $\psi = \psi_0 e^{ikx}$. The second derivative is $d^2\psi/dx^2 = -k^2\psi_0 e^{ikx} = -k^2\psi$. The result is

$$-k^2\psi + \frac{8\pi^2 m}{h^2}[E - U_0]\psi = 0.$$

Solving for k, we obtain

$$k = \sqrt{\frac{8\pi^2 m}{h^2}[E - U_0]} = \frac{2\pi}{h}\sqrt{2m[E - U_0]}.$$

58. (a) The wave function is now given by

$$\Psi(x,t) = \psi_0\left[e^{i(kx-\omega t)} + e^{-i(kx+\omega t)}\right] = \psi_0 e^{-i\omega t}(e^{ikx} + e^{-ikx}).$$

Thus,

$$|\Psi(x,t)|^2 = \left|\psi_0 e^{-i\omega t}(e^{ikx} + e^{-ikx})\right|^2 = \left|\psi_0 e^{-i\omega t}\right|^2 \left|e^{ikx} + e^{-ikx}\right|^2 = \psi_0^2\left|e^{ikx} + e^{-ikx}\right|^2$$

$$= \psi_0^2\,|(\cos kx + i\sin kx) + (\cos kx - i\sin kx)|^2 = 4\psi_0^2(\cos kx)^2$$

$$= 2\psi_0^2(1 + \cos 2kx).$$

(b) Consider two plane matter waves, each with the same amplitude $\psi_0/\sqrt{2}$ and traveling in opposite directions along the x axis. The combined wave Ψ is a standing wave:

$$\Psi(x,t) = \psi_0 e^{i(kx-\omega t)} + \psi_0 e^{-i(kx+\omega t)} = \psi_0(e^{ikx} + e^{-ikx})e^{-i\omega t} = (2\psi_0 \cos kx)e^{-i\omega t}.$$

Thus, the squared amplitude of the matter wave is

$$|\Psi(x,t)|^2 = (2\psi_0 \cos kx)^2 |e^{-i\omega t}|^2 = 2\psi_0^2(1 + \cos 2kx),$$

which is shown below.

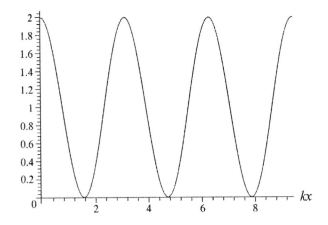

(c) We set $|\Psi(x,t)|^2 = 2\psi_0^2(1 + \cos 2kx) = 0$ to obtain $\cos(2kx) = -1$. This gives

$$2kx = 2\left(\frac{2\pi}{\lambda}\right) = (2n+1)\pi, \quad (n = 0, 1, 2, 3,...)$$

We solve for x:

$$x = \frac{1}{4}(2n+1)\lambda \ .$$

(d) The most probable positions for finding the particle are where $|\Psi(x,t)| \propto (1 + \cos 2kx)$ reaches its maximum. Thus $\cos 2kx = 1$, or

$$2kx = 2\left(\frac{2\pi}{\lambda}\right) = 2n\pi, \quad (n = 0, 1, 2, 3,...)$$

We solve for x:

$$x = \frac{1}{2}n\lambda \,.$$

59. If the momentum is measured at the same time as the position, then

$$\Delta p \approx \frac{\hbar}{\Delta x} = \frac{6.63 \times 10^{-34}\,\text{J} \cdot \text{s}}{2\pi(50\,\text{pm})} = 2.1 \times 10^{-24}\,\text{kg} \cdot \text{m/s}\,.$$

60. (a) Using the fact that $hc = 1240\,\text{nm} \cdot \text{eV}$, we have

$$E = \frac{hc}{\lambda} = \frac{1240\,\text{nm} \cdot \text{eV}}{10.0 \times 10^{-3}\,\text{nm}} = 124\,\text{keV}\,.$$

(b) The kinetic energy gained by the electron is equal to the energy decrease of the photon:

$$\Delta E = \Delta\!\left(\frac{hc}{\lambda}\right) = hc\!\left(\frac{1}{\lambda} - \frac{1}{\lambda + \Delta\lambda}\right) = \left(\frac{hc}{\lambda}\right)\!\left(\frac{\Delta\lambda}{\lambda + \Delta\lambda}\right) = \frac{E}{1 + \dfrac{\lambda}{\Delta\lambda}}$$

$$= \frac{E}{1 + \dfrac{\lambda}{\lambda_c(1 - \cos\phi)}} = \frac{124\,\text{keV}}{1 + \dfrac{10.0\,\text{pm}}{(2.43\,\text{pm})(1 - \cos 180^\circ)}}$$

$$= 40.5\,\text{keV}\,.$$

(c) It is impossible to "view" an atomic electron with such a high-energy photon, because with the energy imparted to the electron the photon would have knocked the electron out of its orbit.

61. We use the uncertainty relationship $\Delta x \Delta p \geq \hbar$. Letting $\Delta x = \lambda$, the de Broglie wavelength, we solve for the minimum uncertainty in p:

$$\Delta p = \frac{\hbar}{\Delta x} = \frac{h}{2\pi\lambda} = \frac{p}{2\pi}$$

where the de Broglie relationship $p = h/\lambda$ is used. We use $1/2\pi = 0.080$ to obtain $\Delta p = 0.080p$. We would expect the measured value of the momentum to lie between $0.92p$ and $1.08p$. Measured values of zero, $0.5p$, and $2p$ would all be surprising.

62. With

$$T \approx e^{-2bL} = \exp\!\left(-2L\sqrt{\frac{8\pi^2 m(U_b - E)}{h^2}}\right),$$

we have

$$E = U_b - \frac{1}{2m}\left(\frac{h\ln T}{4\pi L}\right)^2 = 6.0\,\text{eV} - \frac{1}{2(0.511\,\text{MeV})}\left[\frac{(1240\,\text{eV}\cdot\text{nm})(\ln 0.001)}{4\pi(0.70\,\text{nm})}\right]^2$$

$$= 5.1\,\text{eV}.$$

63. (a) The transmission coefficient T for a particle of mass m and energy E that is incident on a barrier of height U_b and width L is given by

$$T = e^{-2bL},$$

where

$$b = \sqrt{\frac{8\pi^2 m(U_b - E)}{h^2}}.$$

For the proton, we have

$$b = \sqrt{\frac{8\pi^2\left(1.6726\times10^{-27}\,\text{kg}\right)(10\,\text{MeV} - 3.0\,\text{MeV})\left(1.6022\times10^{-13}\,\text{J/MeV}\right)}{\left(6.6261\times10^{-34}\,\text{J}\cdot\text{s}\right)^2}}$$

$$= 5.8082\times10^{14}\,\text{m}^{-1}.$$

This gives $bL = \left(5.8082\times10^{14}\,\text{m}^{-1}\right)\left(10\times10^{-15}\,\text{m}\right) = 5.8082,$ and

$$T = e^{-2(5.8082)} = 9.02\times10^{-6}.$$

The value of b was computed to a greater number of significant digits than usual because an exponential is quite sensitive to the value of the exponent.

(b) Mechanical energy is conserved. Before the proton reaches the barrier, it has a kinetic energy of 3.0 MeV and a potential energy of zero. After passing through the barrier, the proton again has a potential energy of zero, thus a kinetic energy of 3.0 MeV.

(c) Energy is also conserved for the reflection process. After reflection, the proton has a potential energy of zero, and thus a kinetic energy of 3.0 MeV.

(d) The mass of a deuteron is 2.0141 u $= 3.3454 \times 10^{-27}$ kg, so

$$b = \sqrt{\frac{8\pi^2 \left(3.3454 \times 10^{-27}\, \text{kg}\right)\left(10\,\text{MeV} - 3.0\,\text{MeV}\right)\left(1.6022 \times 10^{-13}\, \text{J/MeV}\right)}{\left(6.6261 \times 10^{-34}\, \text{J} \cdot \text{s}\right)^2}}$$

$$= 8.2143 \times 10^{14}\, \text{m}^{-1}.$$

This gives $bL = \left(8.2143 \times 10^{14}\, \text{m}^{-1}\right)\left(10 \times 10^{-15}\, \text{m}\right) = 8.2143$, and

$$T = e^{-2(8.2143)} = 7.33 \times 10^{-8}.$$

(e) As in the case of a proton, mechanical energy is conserved. Before the deuteron reaches the barrier, it has a kinetic energy of 3.0 MeV and a potential energy of zero. After passing through the barrier, the deuteron again has a potential energy of zero, thus a kinetic energy of 3.0 MeV.

(f) Energy is also conserved for the reflection process. After reflection, the deuteron has a potential energy of zero, and thus a kinetic energy of 3.0 MeV.

64. (a) The rate at which incident protons arrive at the barrier is

$$n = 1.0\,\text{kA} / 1.60 \times 10^{-19}\, \text{C} = 6.25 \times 10^{23}/\text{s}.$$

Letting $nTt = 1$, we find the waiting time t:

$$t = (nT)^{-1} = \frac{1}{n} \exp\left(2L \sqrt{\frac{8\pi^2 m_p \left(U_b - E\right)}{h^2}} \right)$$

$$= \left(\frac{1}{6.25 \times 10^{23}/\text{s}} \right) \exp\left(\frac{2\pi (0.70\,\text{nm})}{1240\,\text{eV} \cdot \text{nm}} \sqrt{8(938\,\text{MeV})(6.0\,\text{eV} - 5.0\,\text{eV})} \right)$$

$$= 3.37 \times 10^{111}\, \text{s} \approx 10^{104}\, \text{y},$$

which is much longer than the age of the universe.

(b) Replacing the mass of the proton with that of the electron, we obtain the corresponding waiting time for an electron:

$$t = (nT)^{-1} = \frac{1}{n} \exp\left[2L \sqrt{\frac{8\pi^2 m_e \left(U_b - E\right)}{h^2}} \right]$$

$$= \left(\frac{1}{6.25 \times 10^{23}/\text{s}} \right) \exp\left[\frac{2\pi (0.70\,\text{nm})}{1240\,\text{eV} \cdot \text{nm}} \sqrt{8(0.511\,\text{MeV})(6.0\,\text{eV} - 5.0\,\text{eV})} \right]$$

$$= 2.1 \times 10^{-9}\, \text{s}.$$

The enormous difference between the two waiting times is the result of the difference between the masses of the two kinds of particles.

65. (a) If m is the mass of the particle and E is its energy, then the transmission coefficient for a barrier of height U_b and width L is given by

$$T = e^{-2bL},$$

where

$$b = \sqrt{\frac{8\pi^2 m \left(U_b - E\right)}{h^2}}.$$

If the change ΔU_b in U_b is small (as it is), the change in the transmission coefficient is given by

$$\Delta T = \frac{dT}{dU_b}\Delta U_b = -2LT\frac{db}{dU_b}\Delta U_b.$$

Now,

$$\frac{db}{dU_b} = \frac{1}{2\sqrt{U_b - E}}\sqrt{\frac{8\pi^2 m}{h^2}} = \frac{1}{2\left(U_b - E\right)}\sqrt{\frac{8\pi^2 m \left(U_b - E\right)}{h^2}} = \frac{b}{2\left(U_b - E\right)}.$$

Thus,

$$\Delta T = -LTb\frac{\Delta U_b}{U_b - E}.$$

For the data of Sample Problem 38-7, $2bL = 10.0$, so $bL = 5.0$ and

$$\frac{\Delta T}{T} = -bL\frac{\Delta U_b}{U_b - E} = -(5.0)\frac{(0.010)(6.8\,\mathrm{eV})}{6.8\,\mathrm{eV} - 5.1\,\mathrm{eV}} = -0.20.$$

There is a 20% decrease in the transmission coefficient.

(b) The change in the transmission coefficient is given by

$$\Delta T = \frac{dT}{dL}\Delta L = -2be^{-2bL}\Delta L = -2bT\Delta L$$

and

$$\frac{\Delta T}{T} = -2b\Delta L = -2\left(6.67\times10^{9}\ \text{m}^{-1}\right)(0.010)\left(750\times10^{-12}\ \text{m}\right) = -0.10\ .$$

There is a 10% decrease in the transmission coefficient.

(c) The change in the transmission coefficient is given by

$$\Delta T = \frac{dT}{dE}\Delta E = -2Le^{-2bL}\frac{db}{dE}\Delta E = -2LT\frac{db}{dE}\Delta E.$$

Now, $db/dE = -db/dU_b = -b/2(U_b - E)$, so

$$\frac{\Delta T}{T} = bL\frac{\Delta E}{U_b - E} = (5.0)\frac{(0.010)(5.1\text{eV})}{6.8\text{eV} - 5.1\text{eV}} = 0.15\ .$$

There is a 15% increase in the transmission coefficient.

66. (a) We calculate frequencies from the wavelengths (expressed in SI units) using Eq. 38-1. Our plot of the points and the line which gives the least squares fit to the data is shown below. The vertical axis is in volts and the horizontal axis, when multiplied by 10^{14}, gives the frequencies in Hertz.

From our least squares fit procedure, we determine the slope to be 4.14×10^{-15} V·s, which is in very good agreement with the value given in Eq. 38-3 (once it has been multiplied by e).

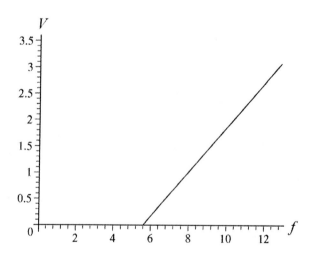

(b) Our least squares fit procedure can also determine the y-intercept for that line. The y-intercept is the negative of the photoelectric work function. In this way, we find $\Phi = 2.31$ eV.

67. Using the fact that $hc = 1240\text{eV}\cdot\text{nm}$, we obtain

$$E = \frac{hc}{\lambda} = \frac{1240\,\text{eV}\cdot\text{nm}}{21\times10^7\,\text{nm}} = 5.9\times10^{-6}\,\text{eV} = 5.9\,\mu\text{eV}.$$

68. (a) Since $E_{ph} = h/\lambda = 1240$ eV·nm/680 nm = 1.82 eV < Φ = 2.28 eV, there is no photoelectric emission.

(b) The cutoff wavelength is the longest wavelength of photons which will cause photoelectric emission. In sodium, this is given by $E_{ph} = hc/\lambda_{max} = \Phi$, or

$$\lambda_{max} = hc/\Phi = (1240 \text{ eV·nm})/2.28 \text{ eV} = 544 \text{ nm}.$$

(c) This corresponds to the color green.

69. (a) The average de Broglie wavelength is

$$\lambda_{avg} = \frac{h}{p_{avg}} = \frac{h}{\sqrt{2mK_{avg}}} = \frac{h}{\sqrt{2m(3kT/2)}} = \frac{hc}{\sqrt{2(mc^2)kT}}$$

$$= \frac{1240\,\text{eV}\cdot\text{nm}}{\sqrt{3(4)(938\,\text{MeV})(8.62\times10^{-5}\,\text{eV}/\text{K})(300\,\text{K})}}$$

$$= 7.3\times10^{-11}\,\text{m} = 73\,\text{pm}.$$

(b) The average separation is

$$d_{avg} = \frac{1}{\sqrt[3]{n}} = \frac{1}{\sqrt[3]{p/kT}} = \sqrt[3]{\frac{(1.38\times10^{-23}\,\text{J/K})(300\text{K})}{1.01\times10^5\,\text{Pa}}} = 3.4\,\text{nm}.$$

(c) Yes, since $\lambda_{avg} \ll d_{avg}$.

70. (a) The average kinetic energy is

$$K = \frac{3}{2}kT = \frac{3}{2}(1.38\times10^{-23}\,\text{J}/\text{K})(300\text{K}) = 6.21\times10^{-21}\,\text{J} = 3.88\times10^{-2}\,\text{eV}.$$

(b) The de Broglie wavelength is

$$\lambda = \frac{h}{\sqrt{2m_n K}} = \frac{6.63\times10^{-34}\,\text{J}\cdot\text{s}}{\sqrt{2(1.675\times10^{-27}\,\text{kg})(6.21\times10^{-21}\,\text{J})}} = 1.46\times10^{-10}\,\text{m}.$$

71. We rewrite Eq. 38-9 as

$$\frac{h}{m\lambda} - \frac{h}{m\lambda'}\cos\phi = \frac{v}{\sqrt{1-(v/c)^2}}\cos\theta,$$

and Eq. 38-10 as

$$\frac{h}{m\lambda'}\sin\phi = \frac{v}{\sqrt{1-(v/c)^2}}\sin\theta.$$

We square both equations and add up the two sides:

$$\left(\frac{h}{m}\right)^2\left[\left(\frac{1}{\lambda} - \frac{1}{\lambda'}\cos\phi\right)^2 + \left(\frac{1}{\lambda'}\sin\phi\right)^2\right] = \frac{v^2}{1-(v/c)^2},$$

where we use $\sin^2\theta + \cos^2\theta = 1$ to eliminate θ. Now the right-hand side can be written as

$$\frac{v^2}{1-(v/c)^2} = -c^2\left[1 - \frac{1}{1-(v/c)^2}\right],$$

so

$$\frac{1}{1-(v/c)^2} = \left(\frac{h}{mc}\right)^2\left[\left(\frac{1}{\lambda} - \frac{1}{\lambda'}\cos\phi\right)^2 + \left(\frac{1}{\lambda'}\sin\phi\right)^2\right] + 1.$$

Now we rewrite Eq. 38-8 as

$$\frac{h}{mc}\left(\frac{1}{\lambda} - \frac{1}{\lambda'}\right) + 1 = \frac{1}{\sqrt{1-(v/c)^2}}.$$

If we square this, then it can be directly compared with the previous equation we obtained for $[1 - (v/c)^2]^{-1}$. This yields

$$\left[\frac{h}{mc}\left(\frac{1}{\lambda} - \frac{1}{\lambda'}\right) + 1\right]^2 = \left(\frac{h}{mc}\right)^2\left[\left(\frac{1}{\lambda} - \frac{1}{\lambda'}\cos\phi\right)^2 + \left(\frac{1}{\lambda'}\sin\phi\right)^2\right] + 1.$$

We have so far eliminated θ and v. Working out the squares on both sides and noting that $\sin^2\phi + \cos^2\phi = 1$, we get

$$\lambda' - \lambda = \Delta\lambda = \frac{h}{mc}(1 - \cos\phi).$$

The enormous difference between the two waiting times is the result of the difference between the masses of the two kinds of particles.

65. (a) If m is the mass of the particle and E is its energy, then the transmission coefficient for a barrier of height U_b and width L is given by

$$T = e^{-2bL},$$

where

$$b = \sqrt{\frac{8\pi^2 m(U_b - E)}{h^2}}.$$

If the change ΔU_b in U_b is small (as it is), the change in the transmission coefficient is given by

$$\Delta T = \frac{dT}{dU_b}\Delta U_b = -2LT\frac{db}{dU_b}\Delta U_b.$$

Now,

$$\frac{db}{dU_b} = \frac{1}{2\sqrt{U_b - E}}\sqrt{\frac{8\pi^2 m}{h^2}} = \frac{1}{2(U_b - E)}\sqrt{\frac{8\pi^2 m(U_b - E)}{h^2}} = \frac{b}{2(U_b - E)}.$$

Thus,

$$\Delta T = -LTb\frac{\Delta U_b}{U_b - E}.$$

For the data of Sample Problem 38-7, $2bL = 10.0$, so $bL = 5.0$ and

$$\frac{\Delta T}{T} = -bL\frac{\Delta U_b}{U_b - E} = -(5.0)\frac{(0.010)(6.8\,\text{eV})}{6.8\,\text{eV} - 5.1\,\text{eV}} = -0.20.$$

There is a 20% decrease in the transmission coefficient.

(b) The change in the transmission coefficient is given by

$$\Delta T = \frac{dT}{dL}\Delta L = -2be^{-2bL}\Delta L = -2bT\Delta L$$

and

$$\frac{\Delta T}{T} = -2b\Delta L = -2\left(6.67\times10^{9}\ \mathrm{m}^{-1}\right)(0.010)\left(750\times10^{-12}\ \mathrm{m}\right) = -0.10\ .$$

There is a 10% decrease in the transmission coefficient.

(c) The change in the transmission coefficient is given by

$$\Delta T = \frac{dT}{dE}\Delta E = -2Le^{-2bL}\frac{db}{dE}\Delta E = -2LT\frac{db}{dE}\Delta E.$$

Now, $db/dE = -db/dU_b = -b/2(U_b - E)$, so

$$\frac{\Delta T}{T} = bL\frac{\Delta E}{U_b - E} = (5.0)\frac{(0.010)(5.1\mathrm{eV})}{6.8\mathrm{eV} - 5.1\mathrm{eV}} = 0.15\ .$$

There is a 15% increase in the transmission coefficient.

66. (a) We calculate frequencies from the wavelengths (expressed in SI units) using Eq. 38-1. Our plot of the points and the line which gives the least squares fit to the data is shown below. The vertical axis is in volts and the horizontal axis, when multiplied by 10^{14}, gives the frequencies in Hertz.

From our least squares fit procedure, we determine the slope to be 4.14×10^{-15} V·s, which is in very good agreement with the value given in Eq. 38-3 (once it has been multiplied by e).

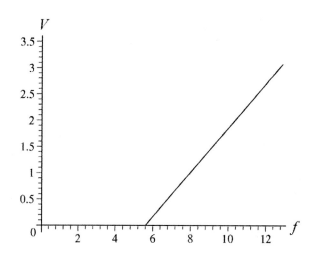

(b) Our least squares fit procedure can also determine the y-intercept for that line. The y-intercept is the negative of the photoelectric work function. In this way, we find $\Phi = 2.31$ eV.

67. Using the fact that $hc = 1240\mathrm{eV}\cdot\mathrm{nm}$, we obtain

72. The kinetic energy of the car of mass m moving at speed v is given by $E = \frac{1}{2}mv^2$, while the potential barrier it has to tunnel through is $U_b = mgh$, where $h = 24$ m. According to Eq. 38-21 and 38-22 the tunneling probability is given by $T \approx e^{-2bL}$, where

$$b = \sqrt{\frac{8\pi^2 m(U_b - E)}{h^2}} = \sqrt{\frac{8\pi^2 m\left(mgh - \frac{1}{2}mv^2\right)}{h^2}}$$

$$= \frac{2\pi(1500\,\text{kg})}{6.63\times10^{-34}\,\text{J}\cdot\text{s}} \sqrt{2\left[(9.8\,\text{m/s}^2)(24\text{m}) - \frac{1}{2}(20\,\text{m/s})^2\right]}$$

$$= 1.2\times10^{38}\,\text{m}^{-1}.$$

Thus, $2bL = 2(1.2\times10^{38}\,\text{m}^{-1})(30\text{m}) = 7.2\times10^{39}$. One can see that $T \approx e^{-2bL}$ is essentially zero.

73. The uncertainty in the momentum is

$$\Delta p = m\,\Delta v = (0.50\text{ kg})(1.0\text{ m/s}) = 0.50\text{ kg·m/s},$$

where Δv is the uncertainty in the velocity. Solving the uncertainty relationship $\Delta x \Delta p \geq \hbar$ for the minimum uncertainty in the coordinate x, we obtain

$$\Delta x = \frac{\hbar}{\Delta p} = \frac{0.60\,\text{J}\cdot\text{s}}{2\pi(0.50\,\text{kg}\cdot\text{m/s})} = 0.19\text{ m}.$$

74. (a) Since $p_x = p_y = 0$, $\Delta p_x = \Delta p_y = 0$. Thus from Eq. 38-20 both Δx and Δy are infinite. It is therefore impossible to assign a y or z coordinate to the position of an electron.

(b) Since it is independent of y and z the wave function $\Psi(x)$ should describe a plane wave that extends infinitely in both the y and z directions. Also from Fig. 38-12 we see that $|\Psi(x)|^2$ extends infinitely along the x axis. Thus the matter wave described by $\Psi(x)$ extends throughout the entire three-dimensional space.

75. The de Broglie wavelength for the bullet is

$$\lambda = \frac{h}{p} = \frac{h}{mv} = \frac{6.63\times10^{-34}\,\text{J.s}}{(40\times10^{-3}\,\text{kg})(1000\,\text{m/s})} = 1.7\times10^{-35}\text{ m}.$$

76. We substitute the classical relationship between momentum p and velocity v, $v = p/m$ into the classical definition of kinetic energy, $K = \frac{1}{2}mv^2$ to obtain $K = p^2/2m$. Here m is

the mass of an electron. Thus $p = \sqrt{2mK}$. The relationship between the momentum and the de Broglie wavelength λ is $\lambda = h/p$, where h is the Planck constant. Thus,

$$\lambda = \frac{h}{\sqrt{2mK}} \ .$$

If K is given in electron volts, then

$$\lambda = \frac{6.626 \times 10^{-34}\,\text{J} \cdot \text{s}}{\sqrt{2(9.109 \times 10^{-31}\,\text{kg})(1.602 \times 10^{-19}\,\text{J}/\text{eV})K}} = \frac{1.226 \times 10^{-9}\,\text{m} \cdot \text{eV}^{1/2}}{\sqrt{K}}$$

$$= \frac{1.226\,\text{nm} \cdot \text{eV}^{1/2}}{\sqrt{K}}\ .$$

77. We note that

$$|e^{ikx}|^2 = (e^{ikx})^*(e^{ikx}) = e^{-ikx}e^{ikx} = 1.$$

Referring to Eq. 38-14, we see therefore that $|\psi|^2 = |\Psi|^2$.

78. From Sample Problem 38-4, we have

$$\frac{\Delta E}{E} = \frac{\Delta \lambda}{\lambda + \Delta \lambda} = \frac{(h/mc)(1 - \cos\phi)}{\lambda'} = \frac{hf'}{mc^2}(1 - \cos\phi)$$

where we use the fact that $\lambda + \Delta\lambda = \lambda' = c/f'$.

79. With no loss of generality, we assume the electron is initially at rest (which simply means we are analyzing the collision from its initial rest frame). If the photon gave all its momentum and energy to the (free) electron, then the momentum and the kinetic energy of the electron would become

$$p = \frac{hf}{c} \quad \text{and} \quad K = hf,$$

respectively. Plugging these expressions into Eq. 38-51 (with m referring to the mass of the electron) leads to

$$(pc)^2 = K^2 + 2Kmc^2$$
$$(hf)^2 = (hf)^2 + 2hfmc^2$$

which is clearly impossible, since the last term ($2hf/mc^2$) is not zero. We have shown that considering total momentum and energy absorption of a photon by a free electron leads to an inconsistency in the mathematics, and thus cannot be expected to happen in nature.

80. The difference between the electron-photon scattering process in this problem and the one studied in the text (the Compton shift, see Eq. 38-11) is that the electron is in motion relative with speed v to the laboratory frame. To utilize the result in Eq. 38-11, shift to a new reference frame in which the electron is at rest before the scattering. Denote the quantities measured in this new frame with a prime ($'$), and apply Eq. 38-11 to yield

$$\Delta\lambda' = \lambda' - \lambda'_0 = \frac{h}{m_e c}(1 - \cos\pi) = \frac{2h}{m_e c},$$

where we note that $\phi = \pi$ (since the photon is scattered back in the direction of incidence). Now, from the Doppler shift formula (Eq. 38-25) the frequency f'_0 of the photon prior to the scattering in the new reference frame satisfies

$$f'_0 = \frac{c}{\lambda'_0} = f_0\sqrt{\frac{1+\beta}{1-\beta}},$$

where $\beta = v/c$. Also, as we switch back from the new reference frame to the original one after the scattering

$$f = f'\sqrt{\frac{1-\beta}{1+\beta}} = \frac{c}{\lambda'}\sqrt{\frac{1-\beta}{1+\beta}}.$$

We solve the two Doppler-shift equations above for λ' and λ'_0 and substitute the results into the Compton shift formula for $\Delta\lambda'$:

$$\Delta\lambda' = \frac{1}{f}\sqrt{\frac{1-\beta}{1+\beta}} - \frac{1}{f_0}\sqrt{\frac{1-\beta}{1+\beta}} = \frac{2h}{m_e c^2}.$$

Some simple algebra then leads to

$$E = hf = hf_0\left(1 + \frac{2h}{m_e c^2}\sqrt{\frac{1+\beta}{1-\beta}}\right)^{-1}.$$

81. (a) For $\lambda = 565$ nm

$$hf = \frac{hc}{\lambda} = \frac{1240\,\text{eV}\cdot\text{nm}}{565\,\text{nm}} = 2.20 \text{ eV}.$$

Since $\Phi_{potassium} > hf > \Phi_{cesium}$, the photoelectric effect can occur in cesium but not in potassium at this wavelength. The result hc = 1240 eV·nm is used in our calculation.

(b) Now λ = 518 nm so

$$hf = \frac{hc}{\lambda} = \frac{1240\,\text{eV}\cdot\text{nm}}{518\,\text{m}} = 2.40\,\text{eV}.$$

This is greater than both Φ_{cesium} and $\Phi_{potassium}$, so the photoelectric effect can now occur for both metals.

82. Eq. 38-3 gives h = 4.14 × 10^{-15} eV·s, but the metric prefix which stands for 10^{-15} is femto (f). Thus, h = 4.14 eV·fs.

83. The energy of a photon is given by $E = hf$, where h is the Planck constant and f is the frequency. The wavelength λ is related to the frequency by $\lambda f = c$, so $E = hc/\lambda$. Since h = 6.626 × 10^{-34} J·s and c = 2.998 × 10^8 m/s,

$$hc = \frac{\left(6.626 \times 10^{-34}\,\text{J}\cdot\text{s}\right)\left(2.998 \times 10^8\,\text{m/s}\right)}{\left(1.602 \times 10^{-19}\,\text{J/eV}\right)\left(10^{-9}\,\text{m/nm}\right)} = 1240\,\text{eV}\cdot\text{nm}.$$

Thus,

$$E = \frac{1240\,\text{eV}\cdot\text{nm}}{\lambda}.$$

Chapter 39

1. According to Eq. 39-4 $E_n \propto L^{-2}$. As a consequence, the new energy level E'_n satisfies

$$\frac{E'_n}{E_n} = \left(\frac{L'}{L}\right)^{-2} = \left(\frac{L}{L'}\right)^2 = \frac{1}{2},$$

which gives $L' = \sqrt{2}L$. Thus, the ratio is $L'/L = \sqrt{2} = 1.41$.

2. (a) The ground-state energy is

$$E_1 = \left(\frac{h^2}{8m_e L^2}\right) n^2 = \left(\frac{\left(6.63\times10^{-34}\,\text{J}\cdot\text{s}\right)^2}{8m_e\left(200\times10^{-12}\,\text{m}\right)^2}\right)(1)^2 = 1.51\times10^{-18}\,\text{J} = 9.42\,\text{eV}.$$

(b) With $m_p = 1.67 \times 10^{-27}$ kg, we obtain

$$E_1 = \left(\frac{h^2}{8m_p L^2}\right) n^2 = \left(\frac{\left(6.63\times10^{-34}\,\text{J}\cdot\text{s}\right)^2}{8m_p\left(200\times10^{-12}\,\text{m}\right)^2}\right)(1)^2 = 8.225\times10^{-22}\,\text{J} = 5.13\times10^{-3}\,\text{eV}.$$

3. To estimate the energy, we use Eq. 39-4, with $n = 1$, L equal to the atomic diameter, and m equal to the mass of an electron:

$$E = n^2 \frac{h^2}{8mL^2} = \frac{(1)^2\left(6.63\times10^{-34}\,\text{J}\cdot\text{s}\right)^2}{8\left(9.11\times10^{-31}\,\text{kg}\right)\left(1.4\times10^{-14}\,\text{m}\right)^2} = 3.07\times10^{-10}\,\text{J} = 1920\,\text{MeV} \approx 1.9\,\text{GeV}.$$

4. With $m_p = 1.67 \times 10^{-27}$ kg, we obtain

$$E_1 = \left(\frac{h^2}{8mL^2}\right) n^2 = \left(\frac{\left(6.63\times10^{-34}\,\text{J.s}\right)^2}{8m_p\left(100\times10^{12}\,\text{m}\right)^2}\right)(1)^2 = 3.29\times10^{-21}\,\text{J} = 0.0206\,\text{eV}.$$

Alternatively, we can use the mc^2 value for a proton from Table 37-3 (938×10^6 eV) and the $hc = 1240$ eV · nm value developed in problem 83 of Chapter 38 by writing Eq. 39-4 as

$$E_n = \frac{n^2 h^2}{8mL^2} = \frac{n^2 (hc)^2}{8(m_p c^2)L^2}.$$

This alternative approach is perhaps easier to plug into, but it is recommended that both approaches be tried to find which is most convenient.

5. We can use the mc^2 value for an electron from Table 37-3 (511×10^3 eV) and the $hc = 1240\ eV \cdot nm$ value developed in problem 83 of Chapter 38 by writing Eq. 39-4 as

$$E_n = \frac{n^2 h^2}{8mL^2} = \frac{n^2 (hc)^2}{8(mc^2)L^2}.$$

For $n = 3$, we set this expression equal to 4.7 eV and solve for L:

$$L = \frac{n(hc)}{\sqrt{8(mc^2)E_n}} = \frac{3(1240\,eV \cdot nm)}{\sqrt{8(511 \times 10^3\ eV)(4.7\,eV)}} = 0.85\ nm.$$

6. We can use the mc^2 value for an electron from Table 37-3 (511×10^3 eV) and the $hc = 1240\ eV \cdot nm$ value developed in problem 83 of Chapter 38 by writing Eq. 39-4 as

$$E_n = \frac{n^2 h^2}{8mL^2} = \frac{n^2 (hc)^2}{8(mc^2)L^2}.$$

The energy to be absorbed is therefore

$$\Delta E = E_4 - E_1 = \frac{(4^2 - 1^2)h^2}{8m_e L^2} = \frac{15(hc)^2}{8(m_e c^2)L^2} = \frac{15(1240\,eV \cdot nm)^2}{8(511 \times 10^3\,eV)(0.250\,nm)^2} = 90.3\,eV.$$

7. Since $E_n \propto L^{-2}$ in Eq. 39-4, we see that if L is doubled, then E_1 becomes $(2.6\ eV)(2)^{-2} = 0.65$ eV.

8. Let the quantum numbers of the pair in question be n and $n + 1$, respectively. We note that

$$E_{n+1} - E_n = \frac{(n+1)^2 h^2}{8mL^2} - \frac{n^2 h^2}{8mL^2} = \frac{(2n+1)h^2}{8mL^2}$$

Therefore, $E_{n+1} - E_n = (2n + 1)E_1$. Now

$$E_{n+1} - E_n = E_5 = 5^2 E_1 = 25E_1 = (2n+1)E_1,$$

which leads to $2n + 1 = 25$, or $n = 12$. Thus,

(a) the higher quantum number is $n+1 = 12+1 = 13$, and

(b) the lower quantum number is $n = 12$.

(c) Now let

$$E_{n+1} - E_n = E_6 = 6^2 E_1 = 36E_1 = (2n+1)E_1,$$

which gives $2n + 1 = 36$, or $n = 17.5$. This is not an integer, so it is impossible to find the pair that fits the requirement.

9. The energy levels are given by $E_n = n^2h^2/8mL^2$, where h is the Planck constant, m is the mass of an electron, and L is the width of the well. The frequency of the light that will excite the electron from the state with quantum number n_i to the state with quantum number n_f is $f = \Delta E/h = (h/8mL^2)(n_f^2 - n_i^2)$ and the wavelength of the light is

$$\lambda = \frac{c}{f} = \frac{8mL^2c}{h(n_f^2 - n_i^2)}.$$

We evaluate this expression for $n_i = 1$ and $n_f = 2, 3, 4$, and 5, in turn. We use $h = 6.626 \times 10^{-34}$ J $\cdot$ s, $m = 9.109 \times 10^{-31}$ kg, and $L = 250 \times 10^{-12}$ m, and obtain the following results:

(a) 6.87×10^{-8} m for $n_f = 2$, (the longest wavelength).

(b) 2.58×10^{-8} m for $n_f = 3$, (the second longest wavelength).

(c) 1.37×10^{-8} m for $n_f = 4$, (the third longest wavelength).

10. Let the quantum numbers of the pair in question be n and $n + 1$, respectively. Then $E_{n+1} - E_n = E_1 (n + 1)^2 - E_1n^2 = (2n + 1)E_1$. Letting

$$E_{n+1} - E_n = (2n+1)E_1 = 3(E_4 - E_3) = 3(4^2 E_1 - 3^2 E_1) = 21E_1,$$

we get $2n + 1 = 21$, or $n = 10$. Thus,

(a) the higher quantum number is $n + 1 = 10 + 1 = 11$, and

(b) the lower quantum number is $n = 10$.

(c) Now letting

$$E_{n+1} - E_n = (2n+1)E_1 = 2(E_4 - E_3) = 2(4^2 E_1 - 3^2 E_1) = 14E_1,$$

we get $2n + 1 = 14$, which does not have an integer-valued solution. So it is impossible to find the pair of energy levels that fits the requirement.

11. We can use the mc^2 value for an electron from Table 37-3 (511×10^3 eV) and the $hc = 1240$ eV · nm value developed in problem 83 of Chapter 38 by rewriting Eq. 39-4 as

$$E_n = \frac{n^2 h^2}{8mL^2} = \frac{n^2(hc)^2}{8(mc^2)L^2}.$$

(a) The first excited state is characterized by $n = 2$, and the third by $n' = 4$. Thus,

$$\Delta E = \frac{(hc)^2}{8(mc^2)L^2}(n'^2 - n^2) = \frac{(1240\text{eV}\cdot\text{nm})^2}{8(511\times10^3\,\text{eV})(0.250\text{nm})^2}(4^2 - 2^2) = (6.02\text{eV})(16-4)$$

which yields $\Delta E = 72.2$ eV.

Now that the electron is in the $n' = 4$ level, it can "drop" to a lower level (n'') in a variety of ways. Each of these drops is presumed to cause a photon to be emitted of wavelength

$$\lambda = \frac{hc}{E_{n'} - E_{n''}} = \frac{8(mc^2)L^2}{hc(n'^2 - n''^2)}.$$

For example, for the transition $n' = 4$ to $n'' = 3$, the photon emitted would have wavelength

$$\lambda = \frac{8(511\times10^3\,\text{eV})(0.250\,\text{nm})^2}{(1240\,\text{eV}\cdot\text{nm})(4^2 - 3^2)} = 29.4\ \text{nm},$$

and once it is then in level $n'' = 3$ it might fall to level $n''' = 2$ emitting another photon. Calculating in this way all the possible photons emitted during the de-excitation of this system, we obtain the following results:

(b) The shortest wavelength that can be emitted is $\lambda_{4\to1} = 13.7$ nm.

(c) The second shortest wavelength that can be emitted is $\lambda_{4\to2} = 17.2$ nm.

(d) The longest wavelength that can be emitted is $\lambda_{2\to1} = 68.7$ nm.

(e) The second longest wavelength that can be emitted is $\lambda_{3\to2} = 41.2\,\text{nm}$.

(f) The possible transitions are shown next. The energy levels are not drawn to scale.

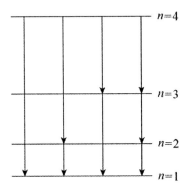

(g) A wavelength of 29.4 nm corresponds to $4\to3$ transition. Thus, it could make either the $3\to1$ transition or the pair of transitions: $3\to2$ and $2\to1$. The longest wavelength that can be emitted is $\lambda_{2\to1} = 68.7\,\text{nm}$.

(h) The shortest wavelength that can next be emitted is $\lambda_{3\to1} = 25.8\,\text{nm}$.

12. The frequency of the light that will excite the electron from the state with quantum number n_i to the state with quantum number n_f is $f = \Delta E/h = \left(h/8mL^2\right)\left(n_f^2 - n_i^2\right)$ and the wavelength of the light is

$$\lambda = \frac{c}{f} = \frac{8mL^2c}{h\left(n_f^2 - n_i^2\right)}.$$

The width of the well is

$$L = \sqrt{\frac{\lambda hc(n_f^2 - n_i^2)}{8mc^2}}$$

The longest wavelength shown in Figure 39-28 is $\lambda = 80.78$ nm which corresponds to a jump from $n_i = 2$ to $n_f = 3$. Thus, the width of the well is

$$L = \sqrt{\frac{(80.78\text{ nm})(1240\,\text{eV}\cdot\text{nm})(3^2 - 2^2)}{8(511\times10^3\,\text{eV})}} = 0.350\,\text{nm} = 350\text{ pm}.$$

13. The probability that the electron is found in any interval is given by $P = \int|\psi|^2\,dx$, where the integral is over the interval. If the interval width Δx is small, the probability can be approximated by $P = |\psi|^2\,\Delta x$, where the wave function is evaluated for the center

of the interval, say. For an electron trapped in an infinite well of width L, the ground state probability density is

$$|\psi|^2 = \frac{2}{L}\sin^2\left(\frac{\pi x}{L}\right),$$

so

$$P = \left(\frac{2\Delta x}{L}\right)\sin^2\left(\frac{\pi x}{L}\right).$$

(a) We take $L = 100$ pm, $x = 25$ pm, and $\Delta x = 5.0$ pm. Then,

$$P = \left[\frac{2(5.0\,\text{pm})}{100\,\text{pm}}\right]\sin^2\left[\frac{\pi(25\,\text{pm})}{100\,\text{pm}}\right] = 0.050.$$

(b) We take $L = 100$ pm, $x = 50$ pm, and $\Delta x = 5.0$ pm. Then,

$$P = \left[\frac{2(5.0\,\text{pm})}{100\,\text{pm}}\right]\sin^2\left[\frac{\pi(50\,\text{pm})}{100\,\text{pm}}\right] = 0.10.$$

(c) We take $L = 100$ pm, $x = 90$ pm, and $\Delta x = 5.0$ pm. Then,

$$P = \left[\frac{2(5.0\,\text{pm})}{100\,\text{pm}}\right]\sin^2\left[\frac{\pi(90\,\text{pm})}{100\,\text{pm}}\right] = 0.0095.$$

14. We follow Sample Problem 39-3 in the presentation of this solution. The integration result quoted below is discussed in a little more detail in that Sample Problem. We note that the arguments of the sine functions used below are in radians.

(a) The probability of detecting the particle in the region $0 \le x \le \frac{L}{4}$ is

$$\left(\frac{2}{L}\right)\left(\frac{L}{\pi}\right)\int_0^{\pi/4}\sin^2 y\,dy = \frac{2}{\pi}\left(\frac{y}{2} - \frac{\sin 2y}{4}\right)\Bigg|_0^{\pi/4} = 0.091.$$

(b) As expected from symmetry,

$$\left(\frac{2}{L}\right)\left(\frac{L}{\pi}\right)\int_{\pi/4}^{\pi}\sin^2 y\,dy = \frac{2}{\pi}\left(\frac{y}{2} - \frac{\sin 2y}{4}\right)\Bigg|_{\pi/4}^{\pi} = 0.091.$$

(c) For the region $\frac{L}{4} \le x \le \frac{3L}{4}$, we obtain

$$\left(\frac{2}{L}\right)\left(\frac{L}{\pi}\right)\int_{\pi/4}^{3\pi/4}\sin^2 y\,dy = \frac{2}{\pi}\left(\frac{y}{2} - \frac{\sin 2y}{4}\right)_{\pi/4}^{3\pi/4} = 0.82$$

which we could also have gotten by subtracting the results of part (a) and (b) from 1; that is, $1 - 2(0.091) = 0.82$.

15. According to Fig. 39-9, the electron's initial energy is 109 eV. After the additional energy is absorbed, the total energy of the electron is 109 eV + 400 eV = 509 eV. Since it is in the region $x > L$, its potential energy is 450 eV (see Section 39-5), so its kinetic energy must be 509 eV – 450 eV = 59 eV.

16. From Fig. 39-9, we see that the sum of the kinetic and potential energies in that particular finite well is 280 eV. The potential energy is zero in the region $0 < x < L$. If the kinetic energy of the electron is detected while it is in that region (which is the only region where this is likely to happen), we should find $K = 280$ eV.

17. Schrödinger's equation for the region $x > L$ is

$$\frac{d^2\psi}{dx^2} + \frac{8\pi^2 m}{h^2}[E - U_0]\psi = 0.$$

If $\psi = De^{2kx}$, then $d^2\psi/dx^2 = 4k^2 De^{2kx} = 4k^2\psi$ and

$$\frac{d^2\psi}{dx^2} + \frac{8\pi^2 m}{h^2}[E - U_0]\psi = 4k^2\psi + \frac{8\pi^2 m}{h^2}[E - U_0]\psi.$$

This is zero provided

$$k = \frac{\pi}{h}\sqrt{2m(U_0 - E)}.$$

The proposed function satisfies Schrödinger's equation provided k has this value. Since U_0 is greater than E in the region $x > L$, the quantity under the radical is positive. This means k is real. If k is positive, however, the proposed function is physically unrealistic. It increases exponentially with x and becomes large without bound. The integral of the probability density over the entire x axis must be unity. This is impossible if ψ is the proposed function.

18. We can use the mc^2 value for an electron from Table 37-3 (511×10^3 eV) and the $hc = 1240$ eV · nm value developed in problem 83 of Chapter 38 by writing Eq. 39-20 as

$$E_{nx,ny} = \frac{2h^2}{8m}\left(\frac{n_x^2}{L_x^2} + \frac{n_y^2}{L_y^2}\right) = \frac{(hc)^2}{8(mc^2)}\left(\frac{n_x^2}{L_x^2} + \frac{n_y^2}{L_y^2}\right).$$

For $n_x = n_y = 1$, we obtain

$$E_{1,1} = \frac{(1240\,\text{eV}\cdot\text{nm})^2}{8(511\times10^3\,\text{eV})}\left(\frac{1}{(0.800\,\text{nm})^2}+\frac{1}{(1.600\,\text{nm})^2}\right) = 0.734\text{ eV}.$$

19. We can use the mc^2 value for an electron from Table 37-3 (511×10^3 eV) and the $hc =$ 1240 eV $\cdot$ nm value developed in problem 83 of Chapter 38 by writing Eq. 39-21 as

$$E_{nx,ny,nz} = \frac{2h^2}{8m}\left(\frac{n_x^2}{L_x^2}+\frac{n_y^2}{L_y^2}+\frac{n_z^2}{L_z^2}\right) = \frac{(hc)^2}{8(mc^2)}\left(\frac{n_x^2}{L_x^2}+\frac{n_y^2}{L_y^2}+\frac{n_z^2}{L_z^2}\right).$$

For $n_x = n_y = n_z = 1$, we obtain

$$E_{1,1} = \frac{(1240\,\text{eV}\cdot\text{nm})^2}{8(511\times10^3\,\text{eV})}\left(\frac{1}{(0.800\,\text{nm})^2}+\frac{1}{(1.600\,\text{nm})^2}+\frac{1}{(0.400\,\text{nm})^2}\right) = 3.08\text{ eV}.$$

20. We are looking for the values of the ratio

$$\frac{E_{nx,ny}}{h^2/8mL^2} = L^2\left(\frac{n_x^2}{L_x^2}+\frac{n_y^2}{L_y^2}\right) = \left(n_x^2 + \frac{1}{4}n_y^2\right)$$

and the corresponding differences.

(a) For $n_x = n_y = 1$, the ratio becomes $1 + \frac{1}{4} = 1.25$.

(b) For $n_x = 1$ and $n_y = 2$, the ratio becomes $1 + \frac{1}{4}(4) = 2.00$. One can check (by computing other (n_x, n_y) values) that this is the next to lowest energy in the system.

(c) The lowest set of states that are degenerate are (n_x, n_y) = (1, 4) and (2, 2). Both of these states have that ratio equal to $1 + \frac{1}{4}(16) = 5.00$.

(d) For $n_x = 1$ and $n_y = 3$, the ratio becomes $1 + \frac{1}{4}(9) = 3.25$. One can check (by computing other (n_x, n_y) values) that this is the lowest energy greater than that computed in part (b). The next higher energy comes from (n_x, n_y) = (2, 1) for which the ratio is $4 + \frac{1}{4}(1) = 4.25$. The difference between these two values is $4.25 - 3.25 = 1.00$.

21. The energy levels are given by

$$E_{n_x,n_y} = \frac{h^2}{8m}\left[\frac{n_x^2}{L_x^2} + \frac{n_y^2}{L_y^2}\right] = \frac{h^2}{8mL^2}\left[n_x^2 + \frac{n_y^2}{4}\right]$$

where the substitutions $L_x = L$ and $L_y = 2L$ were made. In units of $h^2/8mL^2$, the energy levels are given by $n_x^2 + n_y^2/4$. The lowest five levels are $E_{1,1} = 1.25$, $E_{1,2} = 2.00$, $E_{1,3} = 3.25$, $E_{2,1} = 4.25$, and $E_{2,2} = E_{1,4} = 5.00$. It is clear that there are no other possible values for the energy less than 5. The frequency of the light emitted or absorbed when the electron goes from an initial state i to a final state f is $f = (E_f - E_i)/h$, and in units of $h/8mL^2$ is simply the difference in the values of $n_x^2 + n_y^2/4$ for the two states. The possible frequencies are as follows: $0.75(1,2 \rightarrow 1,1), 2.00(1,3 \rightarrow 1,1), 3.00(2,1 \rightarrow 1,1)$, $3.75(2,2 \rightarrow 1,1), 1.25(1,3 \rightarrow 1,2), 2.25(2,1 \rightarrow 1,2), 3.00(2,2 \rightarrow 1,2), 1.00(2,1 \rightarrow 1,3)$, $1.75(2,2 \rightarrow 1,3), 0.75(2,2 \rightarrow 2,1)$, all in units of $h/8mL^2$.

(a) From the above, we see that there are 8 different frequencies.

(b) The lowest frequency is, in units of $h/8mL^2$, 0.75 $(2, 2 \rightarrow 2,1)$.

(c) The second lowest frequency is, in units of $h/8mL^2$, 1.00 $(2, 1 \rightarrow 1,3)$.

(d) The third lowest frequency is, in units of $h/8mL^2$, 1.25 $(1, 3 \rightarrow 1,2)$.

(e) The highest frequency is, in units of $h/8mL^2$, 3.75 $(2, 2 \rightarrow 1,1)$.

(f) The second highest frequency is, in units of $h/8mL^2$, 3.00 $(2, 2 \rightarrow 1,2)$ or $(2, 1 \rightarrow 1,1)$.

(g) The third highest frequency is, in units of $h/8mL^2$, 2.25 $(2, 1 \rightarrow 1,2)$.

22. We are looking for the values of the ratio

$$\frac{E_{n_x,n_y,n_z}}{h^2/8mL^2} = L^2\left(\frac{n_x^2}{L_x^2} + \frac{n_y^2}{L_y^2} + \frac{n_z^2}{L_z^2}\right) = \left(n_x^2 + n_y^2 + n_z^2\right)$$

and the corresponding differences.

(a) For $n_x = n_y = n_z = 1$, the ratio becomes $1 + 1 + 1 = 3.00$.

(b) For $n_x = n_y = 2$ and $n_z = 1$, the ratio becomes $4 + 4 + 1 = 9.00$. One can check (by computing other (n_x, n_y, n_z) values) that this is the third lowest energy in the system. One can also check that this same ratio is obtained for $(n_x, n_y, n_z) = (2, 1, 2)$ and $(1, 2, 2)$.

(c) For $n_x = n_y = 1$ and $n_z = 3$, the ratio becomes $1 + 1 + 9 = 11.00$. One can check (by computing other (n_x, n_y, n_z) values) that this is three "steps" up from the lowest energy in the system. One can also check that this same ratio is obtained for $(n_x, n_y, n_z) = (1, 3, 1)$ and $(3, 1, 1)$. If we take the difference between this and the result of part (b), we obtain $11.0 - 9.00 = 2.00$.

(d) For $n_x = n_y = 1$ and $n_z = 2$, the ratio becomes $1 + 1 + 4 = 6.00$. One can check (by computing other (n_x, n_y, n_z) values) that this is the next to the lowest energy in the system. One can also check that this same ratio is obtained for $(n_x, n_y, n_z) = (2, 1, 1)$ and $(1, 2, 1)$. Thus, three states (three arrangements of (n_x, n_y, n_z) values) have this energy.

(e) For $n_x = 1$, $n_y = 2$ and $n_z = 3$, the ratio becomes $1 + 4 + 9 = 14.0$. One can check (by computing other (n_x, n_y, n_z) values) that this is five "steps" up from the lowest energy in the system. One can also check that this same ratio is obtained for $(n_x, n_y, n_z) = (1, 3, 2)$, $(2, 3, 1)$, $(2, 1, 3)$, $(3, 1, 2)$ and $(3, 2, 1)$. Thus, six states (six arrangements of (n_x, n_y, n_z) values) have this energy.

23. The ratios computed in problem 22 can be related to the frequencies emitted using $f = \Delta E/h$, where each level E is equal to one of those ratios multiplied by $h^2/8mL^2$. This effectively involves no more than a cancellation of one of the factors of h. Thus, for a transition from the second excited state (see part (b) of problem 22) to the ground state (treated in part (a) of that problem), we find

$$f = (9.00 - 3.00)\left(\frac{h}{8mL^2}\right) = (6.00)\left(\frac{h}{8mL^2}\right).$$

In the following, we omit the $h/8mL^2$ factors. For a transition between the fourth excited state and the ground state, we have $f = 12.00 - 3.00 = 9.00$. For a transition between the third excited state and the ground state, we have $f = 11.00 - 3.00 = 8.00$. For a transition between the third excited state and the first excited state, we have $f = 11.00 - 6.00 = 5.00$. For a transition between the fourth excited state and the third excited state, we have $f = 12.00 - 11.00 = 1.00$. For a transition between the third excited state and the second excited state, we have $f = 11.00 - 9.00 = 2.00$. For a transition between the second excited state and the first excited state, we have $f = 9.00 - 6.00 = 3.00$, which also results from some other transitions.

(a) From the above, we see that there are 7 frequencies.

(b) The lowest frequency is, in units of $h/8mL^2$, 1.00.

(c) The second lowest frequency is, in units of $h/8mL^2$, 2.00.

(d) The third lowest frequency is, in units of $h/8mL^2$, 3.00.

(e) The highest frequency is, in units of $h/8mL^2$, 9.00.

(f) The second highest frequency is, in units of $h/8mL^2$, 8.00.

(g) The third highest frequency is, in units of $h/8mL^2$, 6.00.

24. The difference between the energy absorbed and the energy emitted is

$$E_{\text{photon absorbed}} - E_{\text{photon emitted}} = \frac{hc}{\lambda_{\text{absorbed}}} - \frac{hc}{\lambda_{\text{emitted}}} .$$

Thus, using the result of problem 83 in Chapter 38 (hc = 1240 eV · nm), the net energy absorbed is

$$hc\Delta\left(\frac{1}{\lambda}\right) = (1240\,\text{eV}\cdot\text{nm})\left(\frac{1}{375\,\text{nm}} - \frac{1}{580\,\text{nm}}\right) = 1.17\,\text{eV} .$$

25. The energy E of the photon emitted when a hydrogen atom jumps from a state with principal quantum number u to a state with principal quantum number ℓ is given by

$$E = A\left(\frac{1}{\ell^2} - \frac{1}{u^2}\right)$$

where A = 13.6 eV. The frequency f of the electromagnetic wave is given by $f = E/h$ and the wavelength is given by $\lambda = c/f$. Thus,

$$\frac{1}{\lambda} = \frac{f}{c} = \frac{E}{hc} = \frac{A}{hc}\left(\frac{1}{\ell^2} - \frac{1}{u^2}\right) .$$

The shortest wavelength occurs at the series limit, for which $u = \infty$. For the Balmer series, $\ell = 2$ and the shortest wavelength is $\lambda_B = 4hc/A$. For the Lyman series, $\ell = 1$ and the shortest wavelength is $\lambda_L = hc/A$. The ratio is $\lambda_B/\lambda_L = 4.0$.

26. From Eq. 39-6,

$$\Delta E = hf = (4.14\times10^{-15}\,\text{eV}\cdot\text{s})(6.2\times10^{14}\,\text{Hz}) = 2.6\,\text{eV} .$$

27. (a) Since energy is conserved, the energy E of the photon is given by $E = E_i - E_f$, where E_i is the initial energy of the hydrogen atom and E_f is the final energy. The electron energy is given by $(-13.6\text{ eV})/n^2$, where n is the principal quantum number. Thus,

$$E = E_i - E_f = \frac{-13.6\,\text{eV}}{(3)^2} - \frac{-13.6\,\text{eV}}{(1)^2} = 12.1\,\text{eV} .$$

(b) The photon momentum is given by

$$p = \frac{E}{c} = \frac{(12.1\,\text{eV})(1.60 \times 10^{-19}\,\text{J/eV})}{3.00 \times 10^{8}\,\text{m/s}} = 6.45 \times 10^{-27}\,\text{kg} \cdot \text{m/s} \,.$$

(c) Using the result of problem 83 in Chapter 38 ($hc = 1240$ eV · nm), the wavelength is

$$\lambda = \frac{1240\,\text{eV} \cdot \text{nm}}{12.1\,\text{eV}} = 102\,\text{nm} \,.$$

28. (a) The energy level corresponding to the probability density distribution shown in Fig. 39-22 is the $n = 2$ level. Its energy is given by

$$E_{2} = -\frac{13.6\,\text{eV}}{2^{2}} = -3.4\,\text{eV} \,.$$

(b) As the electron is removed from the hydrogen atom the final energy of the proton-electron system is zero. Therefore, one needs to supply at least 3.4 eV of energy to the system in order to bring its energy up from $E_2 = -3.4$ eV to zero. (If more energy is supplied, then the electron will retain some kinetic energy after it is removed from the atom.)

29. If kinetic energy is not conserved, some of the neutron's initial kinetic energy is used to excite the hydrogen atom. The least energy that the hydrogen atom can accept is the difference between the first excited state ($n = 2$) and the ground state ($n = 1$). Since the energy of a state with principal quantum number n is $-(13.6\ \text{eV})/n^2$, the smallest excitation energy is 13.6 eV $-$ (13.6 eV)/(2)^2 = 10.2 eV. The neutron does not have sufficient kinetic energy to excite the hydrogen atom, so the hydrogen atom is left in its ground state and all the initial kinetic energy of the neutron ends up as the final kinetic energies of the neutron and atom. The collision must be elastic.

30. (a) We use Eq. 39-44. At $r = 0$, $P(r) \propto r^2 = 0$.

(b) At $r = a$

$$P(r) = \frac{4}{a^3}a^2 e^{-2a/a} = \frac{4e^{-2}}{a} = \frac{4e^{-2}}{5.29 \times 10^{-2}\,\text{nm}} = 10.2\,\text{nm}^{-1} \,.$$

(c) At $r = 2a$

$$P(r) = \frac{4}{a^3}(2a)^2 e^{-4a/a} = \frac{16e^{-4}}{a} = \frac{16e^{-4}}{5.29 \times 10^{-2}\,\text{nm}} = 5.54\,\text{nm}^{-1} \,.$$

31. (a) We use Eq. 39-39. At $r = a$

$$\psi^2(r) = \left(\frac{1}{\sqrt{\pi} a^{3/2}} e^{-a/a}\right)^2 = \frac{1}{\pi a^3} e^{-2} = \frac{1}{\pi \left(5.29 \times 10^{-2}\,\text{nm}\right)^3} e^{-2} = 291\,\text{nm}^{-3}\ .$$

(b) We use Eq. 39-44. At $r = a$

$$P(r) = \frac{4}{a^3} a^2 e^{-2a/a} = \frac{4e^{-2}}{a} = \frac{4e^{-2}}{5.29 \times 10^{-2}\,\text{nm}} = 10.2\,\text{nm}^{-1}\ .$$

32. (a) $\Delta E = -(13.6\ \text{eV})(4^{-2} - 1^{-2}) = 12.8\ \text{eV}.$

(b) There are 6 possible energies associated with the transitions $4 \to 3$, $4 \to 2$, $4 \to 1$, $3 \to 2$, $3 \to 1$ and $2 \to 1$.

(c) The greatest energy is $E_{4 \to 1} = 12.8\text{eV}$.

(d) The second greatest energy is $E_{3 \to 1} = -(13.6\text{eV})(3^{-2} - 1^{-2}) = 12.1\text{eV}$.

(e) The third greatest energy is $E_{2 \to 1} = -(13.6\text{eV})(2^{-2} - 1^{-2}) = 10.2\text{eV}$.

(f) The smallest energy is $E_{4 \to 3} = -(13.6\text{eV})(4^{-2} - 3^{-2}) = 0.661\ \text{eV}$.

(g) The second smallest energy is $E_{3 \to 2} = -(13.6\text{eV})(3^{-2} - 2^{-2}) = 1.89\text{eV}$.

(h) The third smallest energy is $E_{4 \to 2} = -(13.6\text{eV})(4^{-2} - 2^{-2}) = 2.55\text{eV}$.

33. (a) We take the electrostatic potential energy to be zero when the electron and proton are far removed from each other. Then, the final energy of the atom is zero and the work done in pulling it apart is $W = -E_i$, where E_i is the energy of the initial state. The energy of the initial state is given by $E_i = (-13.6\ \text{eV})/n^2$, where n is the principal quantum number of the state. For the ground state, $n = 1$ and $W = 13.6\ \text{eV}$.

(b) For the state with $n = 2$, $W = (13.6\ \text{eV})/(2)^2 = 3.40\ \text{eV}$.

34. Conservation of linear momentum of the atom-photon system requires that

$$p_{\text{recoil}} = p_{\text{photon}} \Rightarrow m_p v_{\text{recoil}} = \frac{hf}{c}$$

where we use Eq. 39-7 for the photon and use the classical momentum formula for the atom (since we expect its speed to be much less than c). Thus, from Eq. 39-6 and Table 38-3,

$$v_{recoil} = \frac{\Delta E}{m_p c} = \frac{E_4 - E_1}{\left(m_p c^2\right)/c} = \frac{(-13.6\text{eV})\left(4^{-2} - 1^{-2}\right)}{\left(938 \times 10^6 \text{ eV}\right)/\left(2.998 \times 10^8 \text{ m/s}\right)} = 4.1 \text{ m/s}.$$

35. (a) and (b) Letting $a = 5.292 \times 10^{-11}$ m be the Bohr radius, the potential energy becomes

$$U = -\frac{e^2}{4\pi\varepsilon_0 a} = \frac{\left(8.99 \times 10^9 \text{ N} \cdot \text{m}^2/\text{C}^2\right)\left(1.602 \times 10^{-19} \text{ C}\right)^2}{5.292 \times 10^{-11} \text{ m}} = -4.36 \times 10^{-18} \text{ J} = -27.2 \text{ eV}.$$

The kinetic energy is $K = E - U = (-13.6 \text{ eV}) - (-27.2 \text{ eV}) = 13.6 \text{ eV}$.

36. (a) The calculation is shown in Sample Problem 39-6. The difference in the values obtained in parts (a) and (b) of that Sample Problem is 122 nm – 91.4 nm ≈ 31 nm.

(b) We use Eq. 39-1. For the Lyman series,

$$\Delta f = \frac{2.998 \times 10^8 \text{ m/s}}{91.4 \times 10^{-9} \text{ m}} - \frac{2.998 \times 10^8 \text{ m/s}}{122 \times 10^{-9} \text{ m}} = 8.2 \times 10^{14} \text{ Hz}.$$

(c) Fig. 39-19 shows that the width of the Balmer series is 656.3 nm – 364.6 nm ≈ 292 nm ≈ 0.29 μm.

(d) The series limit can be obtained from the $\infty \to 2$ transition:

$$\Delta f = \frac{2.998 \times 10^8 \text{ m/s}}{364.6 \times 10^{-9} \text{ m}} - \frac{2.998 \times 10^8 \text{ m/s}}{656.3 \times 10^{-9} \text{ m}} = 3.65 \times 10^{14} \text{ Hz} \approx 3.7 \times 10^{14} \text{ Hz}.$$

37. The proposed wave function is

$$\psi = \frac{1}{\sqrt{\pi} a^{3/2}} e^{-r/a}$$

where a is the Bohr radius. Substituting this into the right side of Schrödinger's equation, our goal is to show that the result is zero. The derivative is

$$\frac{d\psi}{dr} = -\frac{1}{\sqrt{\pi} a^{5/2}} e^{-r/a}$$

so

$$r^2 \frac{d\psi}{dr} = -\frac{r^2}{\sqrt{\pi a^{5/2}}} e^{-r/a}$$

and

$$\frac{1}{r^2}\frac{d}{dr}\left(r^2\frac{d\psi}{dr}\right) = \frac{1}{\sqrt{\pi a^{5/2}}}\left[-\frac{2}{r}+\frac{1}{a}\right]e^{-r/a} = \frac{1}{a}\left[-\frac{2}{r}+\frac{1}{a}\right]\psi.$$

The energy of the ground state is given by $E = -me^4/8\varepsilon_0^2 h^2$ and the Bohr radius is given by $a = h^2\varepsilon_0/\pi me^2$, so $E = -e^2/8\pi\varepsilon_0 a$. The potential energy is given by $U = -e^2/4\pi\varepsilon_0 r$, so

$$\frac{8\pi^2 m}{h^2}[E-U]\psi = \frac{8\pi^2 m}{h^2}\left[-\frac{e^2}{8\pi\varepsilon_0 a}+\frac{e^2}{4\pi\varepsilon_0 r}\right]\psi = \frac{8\pi^2 m}{h^2}\frac{e^2}{8\pi\varepsilon_0}\left[-\frac{1}{a}+\frac{2}{r}\right]\psi$$

$$= \frac{\pi me^2}{h^2\varepsilon_0}\left[-\frac{1}{a}+\frac{2}{r}\right]\psi = \frac{1}{a}\left[-\frac{1}{a}+\frac{2}{r}\right]\psi.$$

The two terms in Schrödinger's equation cancel, and the proposed function ψ satisfies that equation.

38. Using Eq. 39-6 and the result of problem 83 in Chapter 38 ($hc = 1240$ eV · nm), we find

$$\Delta E = E_{photon} = \frac{hc}{\lambda} = \frac{1240\,\text{eV}\cdot\text{nm}}{121.6\,\text{nm}} = 10.2\,\text{eV}.$$

Therefore, $n_{low} = 1$, but what precisely is n_{high}?

$$E_{high} = E_{low} + \Delta E \quad\Rightarrow\quad -\frac{13.6\,\text{eV}}{n^2} = -\frac{13.6\,\text{eV}}{1^2} + 10.2\,\text{eV}$$

which yields $n = 2$ (this is confirmed by the calculation found from Sample Problem 39-6). Thus, the transition is from the $n = 2$ to the $n = 1$ state.

(a) The higher quantum number is $n = 2$.

(b) The lower quantum number is $n = 1$.

(c) Referring to Fig. 39-18, we see that this must be one of the Lyman series transitions.

39. According to Sample Problem 39-8, the probability the electron in the ground state of a hydrogen atom can be found inside a sphere of radius r is given by

$$p(r) = 1 - e^{-2x}\left(1 + 2x + 2x^2\right)$$

where $x = r/a$ and a is the Bohr radius. We want $r = a$, so $x = 1$ and

$$p(a) = 1 - e^{-2}(1 + 2 + 2) = 1 - 5e^{-2} = 0.323.$$

The probability that the electron can be found outside this sphere is $1 - 0.323 = 0.677$. It can be found outside about 68% of the time.

40. (a) Since $E_2 = -0.85$ eV and $E_1 = -13.6$ eV $+ 10.2$ eV $= -3.4$ eV, the photon energy is $E_{\text{photon}} = E_2 - E_1 = -0.85$ eV $- (-3.4$ eV$) = 2.6$ eV.

(b) From

$$E_2 - E_1 = (-13.6\,\text{eV})\left(\frac{1}{n_2^2} - \frac{1}{n_1^2}\right) = 2.6\text{ eV}$$

we obtain

$$\frac{1}{n_2^2} - \frac{1}{n_1^2} = \frac{2.6\text{ eV}}{13.6\,\text{eV}} \approx -\frac{3}{16} = \frac{1}{4^2} - \frac{1}{2^2}.$$

Thus, $n_2 = 4$ and $n_1 = 2$. So the transition is from the $n = 4$ state to the $n = 2$ state. One can easily verify this by inspecting the energy level diagram of Fig. 39-18. Thus, the higher quantum number is $n_2 = 4$.

(c) The lower quantum number is $n_1 = 2$.

41. The radial probability function for the ground state of hydrogen is $P(r) = (4r^2/a^3)e^{-2r/a}$, where a is the Bohr radius. (See Eq. 39-44). We want to evaluate the integral $\int_0^\infty P(r)\,dr$. Eq. 15 in the integral table of Appendix E is an integral of this form. We set $n = 2$ and replace a in the given formula with $2/a$ and x with r. Then

$$\int_0^\infty P(r)dr = \frac{4}{a^3}\int_0^\infty r^2 e^{-2r/a}dr = \frac{4}{a^3}\frac{2}{(2/a)^3} = 1.$$

42. From Sample Problem 39-8, we know that the probability of finding the electron in the ground state of the hydrogen atom inside a sphere of radius r is given by

$$p(r) = 1 - e^{-2x}\left(1 + 2x + 2x^2\right)$$

where $x = r/a$. Thus the probability of finding the electron between the two shells indicated in this problem is given by

$$p(a < r < 2a) = p(2a) - p(a) = \left[1 - e^{-2x}\left(1 + 2x + 2x^2\right)\right]_{x=2} - \left[1 - e^{-2x}\left(1 + 2x + 2x^2\right)\right]_{x=1}$$

$$= 0.439.$$

43. (a) ψ_{210} is real. Squaring it, we obtain the probability density:

$$P_{210}(r) = |\psi_{210}|^2 (4\pi r^2) = \left(\frac{r^2}{32\pi a^5} e^{-r/a} \cos^2 \theta\right)(4\pi r^2) = \frac{r^4}{8a^5} e^{-r/a} \cos^2 \theta.$$

(b) Each of the other functions is multiplied by its complex conjugate, obtained by replacing i with $-i$ in the function. Since $e^{i\phi} e^{-i\phi} = e^0 = 1$, the result is the square of the function without the exponential factor:

$$|\psi_{21+1}|^2 = \frac{r^2}{64\pi a^5} e^{-r/a} \sin^2 \theta$$

and

$$|\psi_{21-1}|^2 = \frac{r^2}{64\pi a^5} e^{-r/a} \sin^2 \theta.$$

The last two functions lead to the same probability density:

$$P_{21\pm1}(r) = |\psi_{21\pm1}|^2 (4\pi r^2) = \left(\frac{r^2}{64\pi a^5} e^{-r/a} \sin^2 \theta\right)(4\pi r^2) = \frac{r^4}{16a^5} e^{-r/a} \sin^2 \theta.$$

(c) The total probability density for the three states is the sum:

$$P_{210}(r) + P_{21+1}(r) + P_{21-1}(r) = \left(|\psi_{210}|^2 + |\psi_{21+1}|^2 + |\psi_{21-1}|^2\right)(4\pi r^2)$$

$$= \frac{r^4}{8a^5} e^{-r/a} \left[\cos^2 \theta + \frac{1}{2}\sin^2 \theta + \frac{1}{2}\sin^2 \theta\right] = \frac{r^4}{8a^5} e^{-r/a}.$$

The trigonometric identity $\cos^2 \theta + \sin^2 \theta = 1$ is used. We note that the total probability density does not depend on θ or ϕ; it is spherically symmetric.

44. Using Eq. 39-6 and the result of problem 83 in Chapter 38 ($hc = 1240$ eV · nm), we find

$$\Delta E = E_{\text{photon}} = \frac{hc}{\lambda} = \frac{1240 \text{ eV} \cdot \text{nm}}{106.6 \text{ nm}} = 12.09 \text{ eV}.$$

Therefore, $n_{low} = 1$, but what precisely is n_{high}?

$$E_{high} = E_{low} + \Delta E \quad \Rightarrow \quad -\frac{13.6 \text{eV}}{n^2} = -\frac{13.6 \text{eV}}{1^2} + 12.09 \text{eV}$$

which yields $n = 3$. Thus, the transition is from the $n = 3$ to the $n = 1$ state.

(a) The higher quantum number is $n = 3$.

(b) The lower quantum number is $n = 1$.

(c) Referring to Fig. 39-18, we see that this must be one of the Lyman series transitions.

45. Since Δr is small, we may calculate the probability using $p = P(r) \Delta r$, where $P(r)$ is the radial probability density. The radial probability density for the ground state of hydrogen is given by Eq. 39-44:

$$P(r) = \left(\frac{4r^2}{a^3}\right) e^{-2r/a}$$

where a is the Bohr radius.

(a) Here, $r = 0.500a$ and $\Delta r = 0.010a$. Then,

$$P = \left(\frac{4r^2 \Delta r}{a^3}\right) e^{-2r/a} = 4(0.500)^2 (0.010) e^{-1} = 3.68 \times 10^{-3} \approx 3.7 \times 10^{-3}.$$

(b) We set $r = 1.00a$ and $\Delta r = 0.010a$. Then,

$$P = \left(\frac{4r^2 \Delta r}{a^3}\right) e^{-2r/a} = 4(1.00)^2 (0.010) e^{-2} = 5.41 \times 10^{-3} \approx 5.4 \times 10^{-3}.$$

46. According to Fig. 39-25, the quantum number n in question satisfies $r = n^2 a$. Letting $r = 1.0$ mm, we solve for n:

$$n = \sqrt{\frac{r}{a}} = \sqrt{\frac{1.0 \times 10^{-3} \text{ m}}{5.29 \times 10^{-11} \text{ m}}} \approx 4.3 \times 10^3.$$

47. The radial probability function for the ground state of hydrogen is $P(r) = (4r^2/a^3) e^{-2r/a}$, where a is the Bohr radius. (See Eq. 39-44.) The integral table of Appendix E may be used to evaluate the integral $r_{avg} = \int_0^\infty r P(r) \, dr$. Setting $n = 3$ and replacing a in the given formula with $2/a$ (and x with r), we obtain

$$r_{\text{avg}} = \int_0^\infty r P(r)\, dr = \frac{4}{a^3} \int_0^\infty r^3 e^{-2r/a}\, dr = \frac{4}{a^3} \frac{6}{(2/a)^4} = 1.5a.$$

48. (a) The plot shown below for $|\psi_{200}(r)|^2$ is to be compared with the dot plot of Fig. 39-22. We note that the horizontal axis of our graph is labeled "r," but it is actually r/a (that is, it is in units of the parameter a). Now, in the plot below there is a high central peak between $r = 0$ and $r \sim 2a$, corresponding to the densely dotted region around the center of the dot plot of Fig. 39-22. Outside this peak is a region of near-zero values centered at $r = 2a$, where $\psi_{200} = 0$. This is represented in the dot plot by the empty ring surrounding the central peak. Further outside is a broader, flatter, low peak which reaches its maximum value at $r = 4a$. This corresponds to the outer ring with near-uniform dot density which is lower than that of the central peak.

(b) The extrema of $\psi^2(r)$ for $0 < r < \infty$ may be found by squaring the given function, differentiating with respect to r, and setting the result equal to zero:

$$-\frac{1}{32} \frac{(r - 2a)(r - 4a)}{a^6 \pi} e^{-r/a} = 0$$

which has roots at $r = 2a$ and $r = 4a$. We can verify directly from the plot above that $r = 4a$ is indeed a local maximum of $\psi_{200}^2(r)$. As discussed in part (a), the other root ($r = 2a$) is a local minimum.

(c) Using Eq. 39-43 and Eq. 39-41, the radial probability is

$$P_{200}(r) = 4\pi r^2 \psi_{200}^2(r) = \frac{r^2}{8a^3}\left(2 - \frac{r}{a}\right)^2 e^{-r/a}.$$

(d) Let $x = r/a$. Then

$$\int_0^\infty P_{200}(r)\,dr = \int_0^\infty \frac{r^2}{8a^3}\left(2-\frac{r}{a}\right)^2 e^{-r/a}\,dr = \frac{1}{8}\int_0^\infty x^2(2-x)^2 e^{-x}\,dx = \int_0^\infty (x^4 - 4x^3 + 4x^2)e^{-x}\,dx$$

$$= \frac{1}{8}[4! - 4(3!) + 4(2!)] = 1$$

where we have used the integral formula $\int_0^\infty x^n e^{-x}\,dx = n!$.

49. From Eq. 39-4,

$$E_{n+2} - E_n = \left(\frac{h^2}{8mL^2}\right)(n+2)^2 - \left(\frac{h^2}{8mL^2}\right)n^2 = \left(\frac{h^2}{2mL^2}\right)(n+1).$$

50. We can use the mc^2 value for an electron from Table 37-3 (511×10^3 eV) and the $hc = 1240$ eV · nm value developed in problem 83 of Chapter 38 by writing Eq. 39-4 as

$$E_n = \frac{n^2 h^2}{8mL^2} = \frac{n^2(hc)^2}{8(mc^2)L^2}.$$

(a) With $L = 3.0 \times 10^9$ nm, the energy difference is

$$E_2 - E_1 = \frac{1240^2}{8(511\times 10^3)(3.0\times 10^9)^2}(2^2 - 1^2) = 1.3 \times 10^{-19} \text{ eV}.$$

(b) Since $(n+1)^2 - n^2 = 2n + 1$, we have

$$\Delta E = E_{n+1} - E_n = \frac{h^2}{8mL^2}(2n+1) = \frac{(hc)^2}{8(mc^2)L^2}(2n+1).$$

Setting this equal to 1.0 eV, we solve for n:

$$n = \frac{4(mc^2)L^2\Delta E}{(hc)^2} - \frac{1}{2} = \frac{4(511\times 10^3 \text{ eV})(3.0\times 10^9 \text{ nm})^2 (1.0\text{eV})}{(1240\text{eV}\cdot\text{nm})^2} - \frac{1}{2} \approx 1.2\times 10^{19}.$$

(c) At this value of n, the energy is

$$E_n = \frac{1240^2}{8(511\times 10^3)(3.0\times 10^9)^2}(6\times 10^{18})^2 \approx 6\times 10^{18} \text{ eV}.$$

Thus

$$\frac{E_n}{mc^2} = \frac{6 \times 10^{18} \text{ eV}}{511 \times 10^3 \text{ eV}} = 1.2 \times 10^{13}.$$

(d) Since $E_n / mc^2 \gg 1$, the energy is indeed in the relativistic range.

51. (a) The allowed values of ℓ for a given n are 0, 1, 2, ..., $n - 1$. Thus there are n different values of ℓ.

(b) The allowed values of m_ℓ for a given ℓ are $-\ell$, $-\ell + 1$, ..., ℓ. Thus there are $2\ell + 1$ different values of m_ℓ.

(c) According to part (a) above, for a given n there are n different values of ℓ. Also, each of these ℓ's can have $2\ell + 1$ different values of m_ℓ [see part (b) above]. Thus, the total number of m_ℓ's is

$$\sum_{\ell=0}^{n-1} (2\ell + 1) = n^2.$$

52. (a) and (b) In the region $0 < x < L$, $U_0 = 0$, so Schrödinger's equation for the region is

$$\frac{d^2\psi}{dx^2} + \frac{8\pi^2 m}{h^2} E\psi = 0$$

where $E > 0$. If $\psi^2(x) = B \sin^2 kx$, then $\psi(x) = B' \sin kx$, where B' is another constant satisfying $B'^2 = B$. Thus $d^2\psi/dx^2 = -k^2 B' \sin kx = -k^2 \psi(x)$ and

$$\frac{d^2\psi}{dx^2} + \frac{8\pi^2 m}{h^2} E\psi = -k^2\psi + \frac{8\pi^2 m}{h^2} E\psi.$$

This is zero provided that

$$k^2 = \frac{8\pi^2 mE}{h^2}.$$

The quantity on the right-hand side is positive, so k is real and the proposed function satisfies Schrödinger's equation. In this case, there exists no physical restriction as to the sign of k. It can assume either positive or negative values. Thus, $k = \pm \dfrac{2\pi}{h}\sqrt{2mE}$.

53. (a) and (b) Schrödinger's equation for the region $x > L$ is

$$\frac{d^2\psi}{dx^2} + \frac{8\pi^2 m}{h^2}[E - U_0]\psi = 0,$$

where $E - U_0 < 0$. If $\psi^2(x) = Ce^{-2kx}$, then $\psi(x) = C'e^{-kx}$, where C' is another constant satisfying $C'^2 = C$. Thus $d^2\psi/dx^2 = 4k^2 C'e^{-kx} = 4k^2\psi$ and

$$\frac{d^2\psi}{dx^2} + \frac{8\pi^2 m}{h^2}[E - U_0]\psi = k^2\psi + \frac{8\pi^2 m}{h^2}[E - U_0]\psi.$$

This is zero provided that $k^2 = \frac{8\pi^2 m}{h^2}[U_0 - E]$.

The quantity on the right-hand side is positive, so k is real and the proposed function satisfies Schrödinger's equation. If k is negative, however, the proposed function would be physically unrealistic. It would increase exponentially with x. Since the integral of the probability density over the entire x axis must be finite, ψ diverging as $x \to \infty$ would be unacceptable. Therefore, we choose

$$k = \frac{2\pi}{h}\sqrt{2m(U_0 - E)} > 0.$$

54. (a) The allowed energy values are given by $E_n = n^2 h^2/8mL^2$. The difference in energy between the state n and the state $n + 1$ is

$$\Delta E_{adj} = E_{n+1} - E_n = \left[(n+1)^2 - n^2\right]\frac{h^2}{8mL^2} = \frac{(2n+1)h^2}{8mL^2}$$

and

$$\frac{\Delta E_{adj}}{E} = \left[\frac{(2n+1)h^2}{8mL^2}\right]\left(\frac{8mL^2}{n^2 h^2}\right) = \frac{2n+1}{n^2}.$$

As n becomes large, $2n+1 \to 2n$ and $(2n+1)/n^2 \to 2n/n^2 = 2/n$.

(b) No. As $n \to \infty$, ΔE_{adj} and E do not approach 0, but $\Delta E_{adj}/E$ does.

(c) No. See part (b).

(d) Yes. See part (b).

(e) $\Delta E_{adj}/E$ is a better measure than either ΔE_{adj} or E alone of the extent to which the quantum result is approximated by the classical result.

55. (a) We recall that a derivative with respect to a dimensional quantity carries the (reciprocal) units of that quantity. Thus, the first term in Eq. 39-18 has dimensions of ψ multiplied by dimensions of x^{-2}. The second term contains no derivatives, does contain ψ, and involves several other factors that turn out to have dimensions of x^{-2}:

$$\frac{8\pi^2 m}{h^2}[E - U(x)] \Rightarrow \frac{\text{kg}}{(\text{J} \cdot \text{s})^2}[\text{J}]$$

assuming SI units. Recalling from Eq. 7-9 that $\text{J} = \text{kg·m}^2/\text{s}^2$, then we see the above is indeed in units of m^{-2} (which means dimensions of x^{-2}).

(b) In one-dimensional Quantum Physics, the wavefunction has units of $\text{m}^{-1/2}$ as Sample Problem 39-2 shows. Thus, since each term in Eq. 39-18 has units of ψ multiplied by units of x^{-2}, then those units are $\text{m}^{-1/2} \cdot \text{m}^{-2} = \text{m}^{-2.5}$.

56. For $n = 1$

$$E_1 = -\frac{m_e e^4}{8\varepsilon_0^2 h^2} = -\frac{(9.11 \times 10^{-31}\,\text{kg})(1.6 \times 10^{-19}\,\text{C})^4}{8(8.85 \times 10^{-12}\,\text{F/m})^2 (6.63 \times 10^{-34}\,\text{J·s})^2 (1.60 \times 10^{-19}\,\text{J/eV})} = -13.6\,\text{eV} .$$

57. (a) and (b) Using Eq. 39-6 and the result of problem 83 in Chapter 38 ($hc = 1240$ eV · nm), we find

$$\Delta E = E_{\text{photon}} = \frac{hc}{\lambda} = \frac{1240\,\text{eV} \cdot \text{nm}}{486.1\,\text{nm}} = 2.55\,\text{eV} .$$

Therefore, $n_{\text{low}} = 2$, but what precisely is n_{high}?

$$E_{\text{high}} = E_{\text{low}} + \Delta E \quad \Rightarrow \quad -\frac{13.6\,\text{eV}}{n^2} = -\frac{13.6\,\text{eV}}{2^2} + 2.55\,\text{eV}$$

which yields $n = 4$. Thus, the transition is from the $n = 4$ to the $n = 2$ state.

(a) The higher quantum number is $n = 4$.

(b) The lower quantum number is $n = 2$.

(c) Referring to Fig. 39-18, we see that this must be one of the Balmer series transitions (this fact could also be found from Fig. 39-19).

58. (a) The "home-base" energy level for the Balmer series is $n = 2$. Thus the transition with the least energetic photon is the one from the $n = 3$ level to the $n = 2$ level. The energy difference for this transition is

$$\Delta E = E_3 - E_2 = -(13.6\,\text{eV})\left(\frac{1}{3^2} - \frac{1}{2^2}\right) = 1.889\,\text{eV} \ .$$

Using the result of problem 83 in Chapter 38 (hc = 1240 eV · nm), the corresponding wavelength is

$$\lambda = \frac{hc}{\Delta E} = \frac{1240\,\text{eV} \cdot \text{nm}}{1.889\,\text{eV}} = 658\,\text{nm} \ .$$

(b) For the series limit, the energy difference is

$$\Delta E = E_\infty - E_2 = -(13.6\,\text{eV})\left(\frac{1}{\infty^2} - \frac{1}{2^2}\right) = 3.40\,\text{eV} \ .$$

The corresponding wavelength is then $\lambda = \dfrac{hc}{\Delta E} = \dfrac{1240\,\text{eV} \cdot \text{nm}}{3.40\,\text{eV}} = 366\,\text{nm}$.

59. The wavelength λ of the photon emitted in a transition belonging to the Balmer series satisfies

$$E_{\text{ph}} = \frac{hc}{\lambda} = E_n - E_2 = -(13.6\,\text{eV})\left(\frac{1}{n^2} - \frac{1}{2^2}\right) \text{ where } n = 3,4,5,...$$

Using the result of problem 83 in Chapter 38 (hc = 1240 eV · nm), we find

$$\lambda = \frac{4hcn^2}{(13.6\ \text{eV})(n^2 - 4)} = \frac{4(1240\ \text{eV} \cdot \text{nm})}{13.6\ \text{eV}}\left(\frac{n^2}{n^2 - 4}\right).$$

Plugging in the various values of n, we obtain these values of the wavelength: $\lambda = 656$ nm (for $n = 3$), $\lambda = 486$ nm (for $n = 4$), $\lambda = 434$ nm (for $n = 5$), $\lambda = 410$ nm (for $n = 6$), $\lambda = 397$ nm (for $n = 7$), $\lambda = 389$ nm (for $n = 8$), etc. Finally for $n = \infty$, $\lambda = 365$ nm. These values agree well with the data found in Fig. 39-19. [One can also find λ beyond three significant figures by using the more accurate values for m_e, e and h listed in Appendix B when calculating E_n in Eq. 39-33. Another factor that contributes to the error is the motion of the atomic nucleus. It can be shown that this effect can be accounted for by replacing the mass of the electron m_e by $m_e m_p/(m_p + m_e)$ in Eq. 39-33, where m_p is the mass of the proton. Since $m_p \gg m_e$, this is not a major effect.]

Chapter 40

1. (a) For a given value of the principal quantum number n, the orbital quantum number ℓ ranges from 0 to $n - 1$. For $n = 3$, there are three possible values: 0, 1, and 2.

(b) For a given value of ℓ, the magnetic quantum number m_ℓ ranges from $-\ell$ to $+\ell$. For $\ell = 1$, there are three possible values: -1, 0, and $+1$.

2. For a given quantum number ℓ there are $(2\ell + 1)$ different values of m_ℓ. For each given m_ℓ the electron can also have two different spin orientations. Thus, the total number of electron states for a given ℓ is given by $N_\ell = 2(2\ell + 1)$.

(a) Now $\ell = 3$, so $N_\ell = 2(2 \times 3 + 1) = 14$.

(b) In this case, $\ell = 1$, which means $N_\ell = 2(2 \times 1 + 1) = 6$.

(c) Here $\ell = 1$, so $N_\ell = 2(2 \times 1 + 1) = 6$.

(d) Now $\ell = 0$, so $N_\ell = 2(2 \times 0 + 1) = 2$.

3. (a) We use Eq. 40-2:

$$L = \sqrt{\ell(\ell+1)}\hbar = \sqrt{3(3+1)}\left(1.055\times10^{-34}\,\text{J·s}\right) = 3.65\times10^{-34}\,\text{J·s}.$$

(b) We use Eq. 40-7: $L_z = m_\ell \hbar$. For the maximum value of L_z set $m_\ell = \ell$. Thus

$$\left[L_z\right]_{\text{max}} = \ell\hbar = 3\left(1.055\times10^{-34}\,\text{J·s}\right) = 3.16\times10^{-34}\,\text{J·s}.$$

4. For a given quantum number n there are n possible values of ℓ, ranging from 0 to $n - 1$. For each ℓ the number of possible electron states is $N_\ell = 2(2\ell + 1)$. Thus, the total number of possible electron states for a given n is

$$N_n = \sum_{l=0}^{n-1} N_\ell = 2\sum_{l=0}^{n-1} (2\ell+1) = 2n^2.$$

(a) In this case $n = 4$, which implies $N_n = 2(4^2) = 32$.

(b) Now $n = 1$, so $N_n = 2(1^2) = 2$.

(c) Here $n = 3$, and we obtain $N_n = 2(3^2) = 18$.

(d) Finally, $n = 2 \rightarrow N_n = 2(2^2) = 8$.

5. The magnitude L of the orbital angular momentum $\vec{L}$ is given by Eq. 40-2: $L = \sqrt{\ell(\ell+1)}\hbar$. On the other hand, the components L_z are $L_z = m_\ell \hbar$, where $m_\ell = -\ell, ... + \ell$. Thus, the semi-classical angle is $\cos\theta = L_z / L$. The angle is the smallest when $m = \ell$, or

$$\cos\theta = \frac{\ell\hbar}{\sqrt{\ell(\ell+1)}\hbar} \quad \Rightarrow \quad \theta = \cos^{-1}\left(\frac{\ell}{\sqrt{\ell(\ell+1)}}\right)$$

With $\ell = 5$, we have $\theta = 24.1°$.

6. (a) For $\ell = 3$, the greatest value of m_ℓ is $m_\ell = 3$.

(b) Two states ($m_s = \pm\frac{1}{2}$) are available for $m_\ell = 3$.

(c) Since there are 7 possible values for m_ℓ: $+3, +2, +1, 0, -1, -2, -3$, and two possible values for m_s, the total number of state available in the subshell $\ell = 3$ is 14.

7. (a) Using Table 40-1, we find $\ell = [m_\ell]_{max} = 4$.

(b) The smallest possible value of n is $n = \ell_{max} + 1 \geq \ell + 1 = 5$.

(c) As usual, $m_s = \pm\frac{1}{2}$, so two possible values.

8. For a given quantum number n there are n possible values of ℓ, ranging from 0 to $n-1$. For each ℓ the number of possible electron states is $N_\ell = 2(2\ell + 1)$. Thus the total number of possible electron states for a given n is

$$N_n = \sum_{\ell=0}^{n-1} N_\ell = 2\sum_{\ell=0}^{n-1} (2\ell+1) = 2n^2.$$

Thus, in this problem, the total number of electron states is $N_n = 2n^2 = 2(5)^2 = 50$.

9. (a) For $\ell = 3$, the magnitude of the orbital angular momentum is $L = \sqrt{\ell(\ell+1)}\hbar = \sqrt{3(3+1)}\hbar = \sqrt{12}\hbar$. So the multiple is $\sqrt{12} \approx 3.46$.

(b) The magnitude of the orbital dipole moment is $\mu_{orb} = \sqrt{\ell(\ell+1)}\mu_B = \sqrt{12}\mu_B$. So the multiple is $\sqrt{12} \approx 3.46$.

(c) The largest possible value of m_ℓ is $m_\ell = \ell = 3$.

(d) We use $L_z = m_\ell \hbar$ to calculate the z component of the orbital angular momentum. The multiple is $m_\ell = 3$.

(e) We use $\mu_z = -m_\ell \mu_B$ to calculate the z component of the orbital magnetic dipole moment. The multiple is $-m_\ell = -3$.

(f) We use $\cos\theta = m_\ell / \sqrt{\ell(\ell+1)}$ to calculate the angle between the orbital angular momentum vector and the z axis. For $\ell = 3$ and $m_\ell = 3$, we have $\cos\theta = 3/\sqrt{12} = \sqrt{3}/2$, or $\theta = 30.0°$.

(g) For $\ell = 3$ and $m_\ell = 2$, we have $\cos\theta = 2/\sqrt{12} = 1/\sqrt{3}$, or $\theta = 54.7°$.

(h) For $\ell = 3$ and $m_\ell = -3$, $\cos\theta = -3/\sqrt{12} = -\sqrt{3}/2$, or $\theta = 150°$.

10. (a) For $n = 3$ there are 3 possible values of ℓ: 0, 1, and 2.

(b) We interpret this as asking for the number of distinct values for m_ℓ (this ignores the multiplicity of any particular value). For each ℓ there are $2\ell + 1$ possible values of m_ℓ. Thus the number of possible m_ℓ's for $\ell = 2$ is $(2\ell + 1) = 5$. Examining the $\ell = 1$ and $\ell = 0$ cases cannot lead to any new (distinct) values for m_ℓ, so the answer is 5.

(c) Regardless of the values of n, ℓ and m_ℓ, for an electron there are always two possible values of $m_s : \pm\frac{1}{2}$.

(d) The population in the $n = 3$ shell is equal to the number of electron states in the shell, or $2n^2 = 2(3^2) = 18$.

(e) Each subshell has its own value of ℓ. Since there are three different values of ℓ for $n = 3$, there are three subshells in the $n = 3$ shell.

11. Since $L^2 = L_x^2 + L_y^2 + L_z^2$, $\sqrt{L_x^2 + L_y^2} = \sqrt{L^2 - L_z^2}$. Replacing L^2 with $\ell(\ell+1)\hbar^2$ and L_z with $m_\ell \hbar$, we obtain $\sqrt{L_x^2 + L_y^2} = \hbar\sqrt{\ell(\ell+1) - m_\ell^2}$. For a given value of ℓ, the greatest that m_ℓ can be is ℓ, so the smallest that $\sqrt{L_x^2 + L_y^2}$ can be is $\hbar\sqrt{\ell(\ell+1) - \ell^2} = \hbar\sqrt{\ell}$. The

smallest possible magnitude of m_ℓ is zero, so the largest $\sqrt{L_x^2 + L_y^2}$ can be is $\hbar\sqrt{\ell(\ell+1)}$. Thus,

$$\hbar\sqrt{\ell} \leq \sqrt{L_x^2 + L_y^2} \leq \hbar\sqrt{\ell(\ell+1)} \; .$$

12. (a) From Fig. 40-10 and Eq. 40-18,

$$\Delta E = 2\mu_B B = \frac{2(9.27 \times 10^{-24} \text{ J/T})(0.50 \text{ T})}{1.60 \times 10^{-19} \text{ J/eV}} = 58 \; \mu\text{eV} \; .$$

(b) From $\Delta E = hf$ we get

$$f = \frac{\Delta E}{h} = \frac{9.27 \times 10^{-24} \text{ J}}{6.63 \times 10^{-34} \text{ J} \cdot \text{s}} = 1.4 \times 10^{10} \text{ Hz} = 14 \text{ GHz} \; .$$

(c) The wavelength is

$$\lambda = \frac{c}{f} = \frac{2.998 \times 10^8 \text{ m/s}}{1.4 \times 10^{10} \text{ Hz}} = 2.1 \text{cm}.$$

(d) The wave is in the short radio wave region.

13. The magnitude of the spin angular momentum is $S = \sqrt{s(s+1)}\hbar = (\sqrt{3}/2)\hbar$, where $s = \frac{1}{2}$ is used. The z component is either $S_z = \hbar/2$ or $-\hbar/2$.

(a) If $S_z = +\hbar/2$ the angle θ between the spin angular momentum vector and the positive z axis is

$$\theta = \cos^{-1}\left(\frac{S_z}{S}\right) = \cos^{-1}\left(\frac{1}{\sqrt{3}}\right) = 54.7° \; .$$

(b) If $S_z = -\hbar/2$, the angle is $\theta = 180° - 54.7° = 125.3° \approx 125°$.

14. (a) From Eq. 40-19,

$$F = \mu_B \left|\frac{dB}{dz}\right| = (9.27 \times 10^{-24} \text{ J/T})(1.6 \times 10^2 \text{ T/m}) = 1.5 \times 10^{-21} \text{ N} \; .$$

(b) The vertical displacement is

$$\Delta x = \frac{1}{2}at^2 = \frac{1}{2}\left(\frac{F}{m}\right)\left(\frac{l}{v}\right)^2 = \frac{1}{2}\left(\frac{1.5\times10^{-21}\,\text{N}}{1.67\times10^{-27}\,\text{kg}}\right)\left(\frac{0.80\,\text{m}}{1.2\times10^5\,\text{m/s}}\right)^2 = 2.0\times10^{-5}\,\text{m}.$$

15. The acceleration is

$$a = \frac{F}{M} = \frac{(\mu\cos\theta)(dB/dz)}{M},$$

where M is the mass of a silver atom, μ is its magnetic dipole moment, B is the magnetic field, and θ is the angle between the dipole moment and the magnetic field. We take the moment and the field to be parallel ($\cos\theta = 1$) and use the data given in Sample Problem 40-1 to obtain

$$a = \frac{(9.27\times10^{-24}\,\text{J/T})(1.4\times10^3\,\text{T/m})}{1.8\times10^{-25}\,\text{kg}} = 7.2\times10^4\,\text{m/s}^2.$$

16. We let $\Delta E = 2\mu_B B_{\text{eff}}$ (based on Fig. 40-10 and Eq. 40-18) and solve for B_{eff}:

$$B_{\text{eff}} = \frac{\Delta E}{2\mu_B} = \frac{hc}{2\lambda\mu_B} = \frac{1240\,\text{nm}\cdot\text{eV}}{2(21\times10^{-7}\,\text{nm})(5.788\times10^{-5}\,\text{eV/T})} = 51\,\text{mT}.$$

17. The energy of a magnetic dipole in an external magnetic field $\vec{B}$ is $U = -\vec{\mu}\cdot\vec{B} = -\mu_z B$, where $\vec{\mu}$ is the magnetic dipole moment and μ_z is its component along the field. The energy required to change the moment direction from parallel to antiparallel is $\Delta E = \Delta U = 2\mu_z B$. Since the z component of the spin magnetic moment of an electron is the Bohr magneton μ_B,

$$\Delta E = 2\mu_B B = 2(9.274\times10^{-24}\,\text{J/T})(0.200\,\text{T}) = 3.71\times10^{-24}\,\text{J}.$$

The photon wavelength is

$$\lambda = \frac{c}{f} = \frac{hc}{\Delta E} = \frac{(6.63\times10^{-34}\,\text{J}\cdot\text{s})(3.00\times10^8\,\text{m/s})}{3.71\times10^{-24}\,\text{J}} = 5.36\times10^{-2}\,\text{m}.$$

18. The total magnetic field, $B = B_{\text{local}} + B_{\text{ext}}$, satisfies $\Delta E = hf = 2\mu B$ (see Eq. 40-22). Thus,

$$B_{\text{local}} = \frac{hf}{2\mu} - B_{\text{ext}} = \frac{(6.63\times10^{-34}\,\text{J}\cdot\text{s})(34\times10^6\,\text{Hz})}{2(1.41\times10^{-26}\,\text{J/T})} - 0.78\,\text{T} = 19\,\text{mT}.$$

19. Because of the Pauli principle (and the requirement that we construct a state of lowest possible total energy), two electrons fill the $n = 1, 2, 3$ levels and one electron occupies the $n = 4$ level. Thus, using Eq. 39-4,

$$E_{ground} = 2E_1 + 2E_2 + 2E_3 + E_4$$

$$= 2\left(\frac{h^2}{8mL^2}\right)(1)^2 + 2\left(\frac{h^2}{8mL^2}\right)(2)^2 + 2\left(\frac{h^2}{8mL^2}\right)(3)^2 + \left(\frac{h^2}{8mL^2}\right)(4)^2$$

$$= (2 + 8 + 18 + 16)\left(\frac{h^2}{8mL^2}\right) = 44\left(\frac{h^2}{8mL^2}\right).$$

Thus, the multiple of $h^2 / 8mL^2$ is 44.

20. Using Eq. 39-20 we find that the lowest four levels of the rectangular corral (with this specific "aspect ratio") are non-degenerate, with energies $E_{1,1} = 1.25$, $E_{1,2} = 2.00$, $E_{1,3} = 3.25$, and $E_{2,1} = 4.25$ (all of these understood to be in "units" of $h^2/8mL^2$). Therefore, obeying the Pauli principle, we have

$$E_{ground} = 2E_{1,1} + 2E_{1,2} + 2E_{1,3} + E_{2,1} = 2(1.25) + 2(2.00) + 2(3.25) + 4.25$$

which means (putting the "unit" factor back in) that the lowest possible energy of the system is $E_{ground} = 17.25(h^2/8mL^2)$. Thus, the multiple of $h^2 / 8mL^2$ is 17.25.

21. (a) Promoting one of the electrons (described in problem 19) to a not-fully occupied higher level, we find that the configuration with the least total energy greater than that of the ground state has the $n = 1$ and 2 levels still filled, but now has only one electron in the $n = 3$ level; the remaining two electrons are in the $n = 4$ level. Thus,

$$E_{first\ excited} = 2E_1 + 2E_2 + E_3 + 2E_4$$

$$= 2\left(\frac{h^2}{8mL^2}\right)(1)^2 + 2\left(\frac{h^2}{8mL^2}\right)(2)^2 + \left(\frac{h^2}{8mL^2}\right)(3)^2 + 2\left(\frac{h^2}{8mL^2}\right)(4)^2$$

$$= (2 + 8 + 9 + 32)\left(\frac{h^2}{8mL^2}\right) = 51\left(\frac{h^2}{8mL^2}\right).$$

Thus, the multiple of $h^2 / 8mL^2$ is 51.

(b) Now, the configuration which provides the next higher total energy, above that found in part (a), has the bottom three levels filled (just as in the ground state configuration) and has the seventh electron occupying the $n = 5$ level:

$$E_{\text{second excited}} = 2E_1 + 2E_2 + 2E_3 + E_5$$

$$= 2\left(\frac{h^2}{8mL^2}\right)(1)^2 + 2\left(\frac{h^2}{8mL^2}\right)(2)^2 + 2\left(\frac{h^2}{8mL^2}\right)(3)^2 + \left(\frac{h^2}{8mL^2}\right)(5)^2$$

$$= (2 + 8 + 18 + 25)\left(\frac{h^2}{8mL^2}\right) = 53\left(\frac{h^2}{8mL^2}\right).$$

Thus, the multiple of $h^2/8mL^2$ is 53.

(c) The third excited state has the $n = 1, 3, 4$ levels filled, and the $n = 2$ level half-filled:

$$E_{\text{third excited}} = 2E_1 + E_2 + 2E_3 + 2E_4$$

$$= 2\left(\frac{h^2}{8mL^2}\right)(1)^2 + \left(\frac{h^2}{8mL^2}\right)(2)^2 + 2\left(\frac{h^2}{8mL^2}\right)(3)^2 + 2\left(\frac{h^2}{8mL^2}\right)(4)^2$$

$$= (2 + 4 + 18 + 32)\left(\frac{h^2}{8mL^2}\right) = 56\left(\frac{h^2}{8mL^2}\right).$$

Thus, the multiple of $h^2/8mL^2$ is 56.

(d) The energy states of this problem and problem 19 are suggested in the sketch below:

_____ third excited $56(h^2/8mL^2)$

_____ second excited $53(h^2/8mL^2)$

_____ first excited $51(h^2/8mL^2)$

_____ ground state $44(h^2/8mL^2)$

22. (a) Using Eq. 39-20 we find that the lowest five levels of the rectangular corral (with this specific "aspect ratio") have energies $E_{1,1} = 1.25$, $E_{1,2} = 2.00$, $E_{1,3} = 3.25$, $E_{2,1} = 4.25$, and $E_{2,2} = 5.00$ (all of these understood to be in "units" of $h^2/8mL^2$). It should be noted that the energy level we denote $E_{2,2}$ actually corresponds to two energy levels ($E_{2,2}$ and $E_{1,4}$; they are degenerate), but that will not affect our calculations in this problem. The configuration which provides the lowest system energy higher than that of the ground state has the first three levels filled, the fourth one empty, and the fifth one half-filled:

$$E_{\text{first excited}} = 2E_{1,1} + 2E_{1,2} + 2E_{1,3} + E_{2,2} = 2(1.25) + 2(2.00) + 2(3.25) + 5.00$$

which means (putting the "unit" factor back in) the energy of the first excited state is $E_{\text{first excited}} = 18.00(h^2/8mL^2)$. Thus, the multiple of $h^2/8mL^2$ is 18.00.

(b) The configuration which provides the next higher system energy has the first two levels filled, the third one half-filled, and the fourth one filled:

$$E_{\text{second excited}} = 2E_{1,1} + 2E_{1,2} + E_{1,3} + 2E_{2,1} = 2(1.25) + 2(2.00) + 3.25 + 2(4.25)$$

which means (putting the "unit" factor back in) the energy of the second excited state is $E_{\text{second excited}} = 18.25(h^2/8mL^2)$. Thus, the multiple of $h^2/8mL^2$ is 18.25.

(c) Now, the configuration which provides the *next* higher system energy has the first two levels filled, with the next three levels half-filled:

$$E_{\text{third excited}} = 2E_{1,1} + 2E_{1,2} + E_{1,3} + E_{2,1} + E_{2,2} = 2(1.25) + 2(2.00) + 3.25 + 4.25 + 5.00$$

which means (putting the "unit" factor back in) the energy of the third excited state is $E_{\text{third excited}} = 19.00(h^2/8mL^2)$. Thus, the multiple of $h^2/8mL^2$ is 19.00.

(d) The energy states of this problem and problem 20 are suggested in the sketch below:

——————————————— third excited $19.00(h^2/8mL^2)$

——————————————— second excited $18.25(h^2/8mL^2)$

——————————————— first excited $18.00(h^2/8mL^2)$

——————————————— ground state $17.25(h^2/8mL^2)$

23. In terms of the quantum numbers n_x, n_y, and n_z, the single-particle energy levels are given by

$$E_{n_x,n_y,n_z} = \frac{h^2}{8mL^2}\left(n_x^2 + n_y^2 + n_z^2\right).$$

The lowest single-particle level corresponds to $n_x = 1$, $n_y = 1$, and $n_z = 1$ and is $E_{1,1,1} = 3(h^2/8mL^2)$. There are two electrons with this energy, one with spin up and one with spin down. The next lowest single-particle level is three-fold degenerate in the three integer quantum numbers. The energy is

$$E_{1,1,2} = E_{1,2,1} = E_{2,1,1} = 6(h^2/8mL^2).$$

Each of these states can be occupied by a spin up and a spin down electron, so six electrons in all can occupy the states. This completes the assignment of the eight electrons to single-particle states. The ground state energy of the system is

$$E_{gr} = (2)(3)(h^2/8mL^2) + (6)(6)(h^2/8mL^2) = 42(h^2/8mL^2).$$

Thus, the multiple of $h^2/8mL^2$ is 42.

24. We use the results of problem 22 in Chapter 39. The Pauli principle requires that no more than two electrons be in the lowest energy level (at $E_{1,1,1} = 3(h^2/8mL^2)$), but — due to their degeneracies — as many as six electrons can be in the next three levels

$$E' = E_{1,1,2} = E_{1,2,1} = E_{2,1,1} = 6(h^2/8mL^2)$$

$$E'' = E_{1,2,2} = E_{2,2,1} = E_{2,1,2} = 9(h^2/8mL^2)$$

$$E''' = E_{1,1,3} = E_{1,3,1} = E_{3,1,1} = 11(h^2/8mL^2).$$

Using Eq. 39-21, the level above those can only hold two electrons:

$$E_{2,2,2} = (2^2 + 2^2 + 2^2)(h^2/8mL^2) = 12(h^2/8mL^2).$$

And the next higher level can hold as much as twelve electrons (see part (e) of problem 22 in Chapter 39) and has energy $E'''' = 14(h^2/8mL^2)$.

(a) The configuration which provides the lowest system energy higher than that of the ground state has the first level filled, the second one with one vacancy, and the third one with one occupant:

$$E_{first\ excited} = 2E_{1,1,1} + 5E' + E'' = 2(3) + 5(6) + 9$$

which means (putting the "unit" factor back in) the energy of the first excited state is $E_{first\ excited} = 45(h^2/8mL^2)$. Thus, the multiple of $h^2/8mL^2$ is 45.

(b) The configuration which provides the next higher system energy has the first level filled, the second one with one vacancy, the third one empty, and the fourth one with one occupant:

$$E_{second\ excited} = 2E_{1,1,1} + 5E' + E'' = 2(3) + 5(6) + 11$$

which means (putting the "unit" factor back in) the energy of the second excited state is $E_{second\ excited} = 47(h^2/8mL^2)$. Thus, the multiple of $h^2/8mL^2$ is 47.

(c) Now, there are a couple of configurations which provide the *next* higher system energy. One has the first level filled, the second one with one vacancy, the third and fourth ones empty, and the fifth one with one occupant:

$$E_{\text{third excited}} = 2E_{1,1,1} + 5E' + E''' = 2(3) + 5(6) + 12$$

which means (putting the "unit" factor back in) the energy of the third excited state is $E_{\text{third excited}} = 48(h^2/8mL^2)$. Thus, the multiple of $h^2/8mL^2$ is 48. The other configuration with this same total energy has the first level filled, the second one with two vacancies, and the third one with one occupant.

(d) The energy states of this problem and problem 23 are suggested in the following sketch:

_____ third excited $48(h^2/8mL^2)$

_____ second excited $47(h^2/8mL^2)$

_____ first excited $45(h^2/8mL^2)$

_____ ground state $42(h^2/8mL^2)$

25. The first three shells ($n = 1$ through 3), which can accommodate a total of $2 + 8 + 18 = 28$ electrons, are completely filled. For selenium ($Z = 34$) there are still $34 - 28 = 6$ electrons left. Two of them go to the $4s$ subshell, leaving the remaining four in the highest occupied subshell, the $4p$ subshell. Thus,

(a) the highest occupied subshell is $4p$,

(b) and there are four electrons in the subshell.

For bromine ($Z = 35$) the highest occupied subshell is also the $4p$ subshell, which contains five electrons.

(c) Thus, the highest occupied subshell is $4p$, and

(d) there are five electrons in the subshell.

For krypton ($Z = 36$) the highest occupied subshell is also the $4p$ subshell, which now accommodates six electrons.

(e) Thus, the highest occupied subshell is $4p$, and

(f) there are six electrons in the subshell.

26. When a helium atom is in its ground state, both of its electrons are in the 1s state. Thus, for each of the electrons, $n = 1$, $\ell = 0$, and $m_\ell = 0$. One of the electrons is spin up $\left(m_s = +\frac{1}{2}\right)$ while the other is spin down $\left(m_s = -\frac{1}{2}\right)$. Thus,

(a) the quantum numbers (n, ℓ, m_ℓ, m_s) for the spin-up electron is $(1,0,0,+1/2)$, and

(b) the quantum numbers (n, ℓ, m_ℓ, m_s) for the spin-down electron is $(1,0,0,-1/2)$.

27. (a) All states with principal quantum number $n = 1$ are filled. The next lowest states have $n = 2$. The orbital quantum number can have the values $\ell = 0$ or 1 and of these, the $\ell = 0$ states have the lowest energy. The magnetic quantum number must be $m_\ell = 0$ since this is the only possibility if $\ell = 0$. The spin quantum number can have either of the values $m_s = -\frac{1}{2}$ or $+\frac{1}{2}$. Since there is no external magnetic field, the energies of these two states are the same. Therefore, in the ground state, the quantum numbers of the third electron are either $n = 2, \ell = 0, m_\ell = 0, m_s = -\frac{1}{2}$ or $n = 2, \ell = 0, m_\ell = 0, m_s = +\frac{1}{2}$. That is, $(n, \ell, m_\ell, m_s) = (2,0,0, +1/2)$ and $(2,0,0, -1/2)$.

(b) The next lowest state in energy is an $n = 2$, $\ell = 1$ state. All $n = 3$ states are higher in energy. The magnetic quantum number can be $m_\ell = -1, 0,$ or $+1$; the spin quantum number can be $m_s = -\frac{1}{2}$ or $+\frac{1}{2}$. Thus, $(n, \ell, m_\ell, m_s) = (2,1,1, +1/2)$ and $(2,1,1, -1/2)$, $(2,1,0,+1/2)$, $(2,1,0,-1/2)$, $(2,1,-1,+1/2)$ and $(2,1,-1,-1/2)$.

28. (a) The number of different m_ℓ's is $2\ell + 1 = 3$, ($m_\ell = 1,0,-1$) and the number of different m_s's is 2, which we denote as $+1/2$ and $-1/2$. The allowed states are $(m_{\ell 1}, m_{s1}, m_{\ell 2}, m_{s2}) = (1,+ 1/2,1,-1/2), (1,+ 1/2,0,+ 1/2), (1,+ 1/2,0,-1/2), (1,+ 1/2, -1,+ 1/2), (1,+ 1/2, -1, -1/2), (1, -1/2,0,+ 1/2), (1, -1/2,0, -1/2), (1, -1/2, -1,+ 1/2), (1, -1/2, -1, -1/2), (0,+ 1/2,0, -1/2), (0,+ 1/2, -1,+ 1/2), (0,+ 1/2, -1, -1/2), (0, -1/2, -1,+ 1/2), (0, -1/2, -1, -1/2), (-1,+ 1/2, -1, -1/2)$. So, there are 15 states.

(b) There are six states disallowed by the exclusion principle, in which both electrons share the quantum numbers: $(m_{\ell 1}, m_{s1}, m_{\ell 2}, m_{s2}) = (1,+ 1/2,1,+1/2), (1,-1/2,1, -1/2), (0,+ 1/2,0,+1/2), (0, -1/2, 0, -1/2), (-1,+ 1/2, -1, +1/2), (-1, -1/2, -1, -1/2)$. So, if Pauli exclusion principle is not applied, then there would be $15 + 6 = 21$ allowed states.

29. For a given value of the principal quantum number n, there are n possible values of the orbital quantum number ℓ, ranging from 0 to $n - 1$. For any value of ℓ, there are $2\ell + 1$ possible values of the magnetic quantum number m_ℓ, ranging from $-\ell$ to $+\ell$. Finally, for each set of values of ℓ and m_ℓ, there are two states, one corresponding to the spin quantum number $m_s = -\frac{1}{2}$ and the other corresponding to $m_s = +\frac{1}{2}$. Hence, the total number of states with principal quantum number n is

$$N = 2\sum_{0}^{n-1}(2\ell+1).$$

Now

$$\sum_{0}^{n-1}2\ell = 2\sum_{0}^{n-1}\ell = 2\frac{n}{2}(n-1) = n(n-1),$$

since there are n terms in the sum and the average term is $(n-1)/2$. Furthermore,

$$\sum_{0}^{n-1}1 = n .$$

Thus $N = 2[n(n-1)+n] = 2n^2$.

30. The kinetic energy gained by the electron is eV, where V is the accelerating potential difference. A photon with the minimum wavelength (which, because of $E = hc/\lambda$, corresponds to maximum photon energy) is produced when all of the electron's kinetic energy goes to a single photon in an event of the kind depicted in Fig. 40-15. Thus, using the result of problem 83 in Chapter 38,

$$eV = \frac{hc}{\lambda_{min}} = \frac{1240\,\text{eV}\cdot\text{nm}}{0.10\,\text{nm}} = 1.24\times10^4\,\text{eV} .$$

Therefore, the accelerating potential difference is $V = 1.24\times10^4$ V $= 12.4$ kV.

31. The initial kinetic energy of the electron is $K_0 = 50.0$ keV. After the first collision, the kinetic energy is $K_1 = 25$ keV; after the second, it is $K_2 = 12.5$ keV; and after the third, it is zero.

(a) The energy of the photon produced in the first collision is 50.0 keV $-$ 25.0 keV $=$ 25.0 keV. The wavelength associated with this photon is

$$\lambda = \frac{1240\,\text{eV}\cdot\text{nm}}{25.0\times10^3\,\text{eV}} = 4.96\times10^{-2}\,\text{nm} = 49.6\,\text{pm}$$

where the result of problem 83 of Chapter 38 is used.

(b) The energies of the photons produced in the second and third collisions are each 12.5 keV and their wavelengths are

$$\lambda = \frac{1240\,\text{eV}\cdot\text{nm}}{12.5\times10^3\,\text{eV}} = 9.92\times10^{-2}\,\text{nm} = 99.2\,\text{pm} .$$

32. (a) and (b) Let the wavelength of the two photons be λ_1 and $\lambda_2 = \lambda_1 + \Delta\lambda$. Then,

$$eV = \frac{hc}{\lambda_1} + \frac{hc}{\lambda_1 + \Delta\lambda} \quad \Rightarrow \quad \lambda_1 = \frac{-(\Delta\lambda/\lambda_0 - 2) \pm \sqrt{(\Delta\lambda/\lambda_0)^2 + 4}}{2/\Delta\lambda}.$$

Here, $\Delta\lambda = 130$ pm and $\lambda_0 = hc/eV = 1240$ keV·pm/20 keV = 62 pm. The result of problem 83 in Chapter 38 is adapted to these units (hc = 1240 eV·nm = 1240 keV·pm). We choose the plus sign in the expression for λ_1 (since $\lambda_1 > 0$) and obtain

$$\lambda_1 = \frac{-(130\,\text{pm}/62\,\text{pm} - 2) + \sqrt{(130\,\text{pm}/62\,\text{pm})^2 + 4}}{2/62\,\text{pm}} = 87\,\text{pm}.$$

The energy of the electron after its first deceleration is

$$K = K_i - \frac{hc}{\lambda_1} = 20\,\text{keV} - \frac{1240\,\text{keV}\cdot\text{pm}}{87\,\text{pm}} = 5.7\,\text{keV}.$$

(c) The energy of the first photon is

$$E_1 = \frac{hc}{\lambda_1} = \frac{1240\,\text{keV}\cdot\text{pm}}{87\,\text{pm}} = 14\,\text{keV}.$$

(d) The wavelength associated with the second photon is

$$\lambda_2 = \lambda_1 + \Delta\lambda = 87\,\text{pm} + 130\,\text{pm} = 2.2 \times 10^2\,\text{pm}.$$

(e) The energy of the second photon is

$$E_2 = \frac{hc}{\lambda_2} = \frac{1240\,\text{keV}\cdot\text{pm}}{2.2 \times 10^2\,\text{pm}} = 5.7\,\text{keV}.$$

33. (a) The cut-off wavelength $\lambda_{\min}$ is characteristic of the incident electrons, not of the target material. This wavelength is the wavelength of a photon with energy equal to the kinetic energy of an incident electron. According to the result of problem 83 of Chapter 38,

$$\lambda_{\min} = \frac{1240\,\text{eV}\cdot\text{nm}}{35 \times 10^3\,\text{eV}} = 3.54 \times 10^{-2}\,\text{nm} = 35.4\,\text{pm}.$$

(b) A K_α photon results when an electron in a target atom jumps from the L-shell to the K-shell. The energy of this photon is 25.51 keV – 3.56 keV = 21.95 keV and its wavelength is

$$\lambda_{K\alpha} = (1240\ eV \cdot nm)/(21.95 \times 10^3\ eV) = 5.65 \times 10^{-2}\ nm = 56.5\ pm.$$

(c) A K_β photon results when an electron in a target atom jumps from the *M*-shell to the *K*-shell. The energy of this photon is 25.51 keV – 0.53 keV = 24.98 keV and its wavelength is

$$\lambda_{K\beta} = (1240\ eV \cdot nm)/(24.98 \times 10^3\ eV) = 4.96 \times 10^{-2}\ nm = 49.6\ pm.$$

34. The result of problem 83 in Chapter 38 is adapted to these units (hc = 1240 eV·nm = 1240 keV·pm). For the K_α line from iron

$$\Delta E = \frac{hc}{\lambda} = \frac{1240\,\text{keV} \cdot \text{pm}}{193\,\text{pm}} = 6.42\ keV.$$

We remark that for the hydrogen atom the corresponding energy difference is

$$\Delta E_{12} = -(13.6\,\text{eV})\left(\frac{1}{2^2} - \frac{1}{1^1}\right) = 10\,\text{eV}\ .$$

That this difference is much greater in iron is due to the fact that its atomic nucleus contains 26 protons, exerting a much greater force on the *K*- and *L*-shell electrons than that provided by the single proton in hydrogen.

35. Suppose an electron with total energy *E* and momentum **p** spontaneously changes into a photon. If energy is conserved, the energy of the photon is *E* and its momentum has magnitude *E*/*c*. Now the energy and momentum of the electron are related by $E^2 = (pc)^2 + (mc^2)^2$, so $pc = \sqrt{E^2 - (mc^2)^2}$. Since the electron has non-zero mass, *E*/*c* and *p* cannot have the same value. Hence, momentum cannot be conserved. A third particle must participate in the interaction, primarily to conserve momentum. It does, however, carry off some energy.

36. (a) We use $eV = hc/\lambda_{min}$ (see Eq. 40-23 and Eq. 38-4). The result of problem 83 in Chapter 38 is adapted to these units (hc = 1240 eV·nm = 1240 keV·pm).

$$\lambda_{min} = \frac{hc}{eV} = \frac{1240\,\text{keV} \cdot \text{pm}}{50.0\,\text{keV}} = 24.8\,\text{pm}\ .$$

(b) The values of λ for the K_α and K_β lines do not depend on the external potential and are therefore unchanged.

37. Since the frequency of an x-ray emission is proportional to $(Z-1)^2$, where *Z* is the atomic number of the target atom, the ratio of the wavelength λ_{Nb} for the K_α line of

niobium to the wavelength λ_{Ga} for the K_α line of gallium is given by $\lambda_{Nb}/\lambda_{Ga} = (Z_{Ga} - 1)^2/(Z_{Nb} - 1)^2$, where Z_{Nb} is the atomic number of niobium (41) and Z_{Ga} is the atomic number of gallium (31). Thus $\lambda_{Nb}/\lambda_{Ga} = (30)^2/(40)^2 = 9/16 \approx 0.563$.

38. The result of problem 83 in Chapter 38 is adapted to these units (hc = 1240 eV·nm = 1240 keV·pm). The energy difference $E_L - E_M$ for the x-ray atomic energy levels of molybdenum is

$$\Delta E = E_L - E_M = \frac{hc}{\lambda_L} - \frac{hc}{\lambda_M} = \frac{1240\,\text{keV}\cdot\text{pm}}{63.0\,\text{pm}} - \frac{1240\,\text{keV}\cdot\text{pm}}{71.0\,\text{pm}} = 2.2\,\text{keV}.$$

39. (a) An electron must be removed from the K-shell, so that an electron from a higher energy shell can drop. This requires an energy of 69.5 keV. The accelerating potential must be at least 69.5 kV.

(b) After it is accelerated, the kinetic energy of the bombarding electron is 69.5 keV. The energy of a photon associated with the minimum wavelength is 69.5 keV, so its wavelength is

$$\lambda_{min} = \frac{1240\,\text{eV}\cdot\text{nm}}{69.5 \times 10^3\,\text{eV}} = 1.78 \times 10^{-2}\,\text{nm} = 17.8\,\text{pm}.$$

(c) The energy of a photon associated with the K_α line is 69.5 keV − 11.3 keV = 58.2 keV and its wavelength is

$$\lambda_{K\alpha} = (1240\ \text{eV·nm})/(58.2 \times 10^3\ \text{eV}) = 2.13 \times 10^{-2}\ \text{nm} = 21.3\ \text{pm}.$$

(d) The energy of a photon associated with the K_β line is 69.5 keV − 2.30 keV = 67.2 keV and its wavelength is

$$\lambda_{K\beta} = (1240\ \text{eV·nm})/(67.2 \times 10^3\ \text{eV}) = 1.85 \times 10^{-2}\ \text{nm} = 18.5\ \text{pm}.$$

The result of problem 83 of Chapter 38 is used.

40. From the data given in the problem, we calculate frequencies (using Eq. 38-1), take their square roots, look up the atomic numbers (see Appendix F), and do a least-squares fit to find the slope: the result is 5.02×10^7 with the odd-sounding unit of a square root of a Hertz. We remark that the least squares procedure also returns a value for the y-intercept of this statistically determined "best-fit" line; that result is negative and would appear on a graph like Fig. 40-17 to be at about − 0.06 on the vertical axis. Also, we can estimate the slope of the Moseley line shown in Fig. 40-17:

$$\frac{(1.95 - 0.50)10^9\,\text{Hz}^{1/2}}{40 - 11} \approx 5.0 \times 10^7\,\text{Hz}^{1/2}.$$

These are in agreement with the discussion in § 40-10.

41. We use Eq. 36-31, Eq. 39-6, and the result of problem 83 in Chapter 38, adapted to these units (hc = 1240 eV·nm = 1240 keV·pm). Letting $2d \sin \theta = m\lambda = mhc / \Delta E$, where $\theta = 74.1°$, we solve for d:

$$d = \frac{mhc}{2\Delta E \sin \theta} = \frac{(1)(1240\,\text{keV}\cdot\text{nm})}{2(8.979\,\text{keV} - 0.951\,\text{keV})(\sin 74.1°)} = 80.3\,\text{pm} .$$

42. (a) According to Eq. 40-26, $f \propto (Z-1)^2$, so the ratio of energies is (using Eq. 38-2) $f / f' = [(Z-1)/(Z'-1)]^2$.

(b) We refer to Appendix F. Applying the formula from part (a) to Z = 92 and Z' = 13, we obtain

$$\frac{E}{E'} = \frac{f}{f'} = \left(\frac{Z-1}{Z'-1}\right)^2 = \left(\frac{92-1}{13-1}\right)^2 = 57.5 .$$

(c) Applying this to Z = 92 and Z' = 3, we obtain

$$\frac{E}{E\text{n}} = \left(\frac{92-1}{3-1}\right)^2 = 2.07 \times 10^3 .$$

43. The transition is from n = 2 to n = 1, so Eq. 40-26 combined with Eq. 40-24 yields

$$f = \left(\frac{m_e e^4}{8\varepsilon_0^2 h^3}\right)\left(\frac{1}{1^2} - \frac{1}{2^2}\right)(Z-1)^2$$

so that the constant in Eq. 40-27 is

$$C = \sqrt{\frac{3m_e e^4}{32\varepsilon_0^2 h^3}} = 4.9673 \times 10^7 \text{ Hz}^{1/2}$$

using the values in the next-to-last column in the Table in Appendix B (but note that the power of ten is given in the middle column).

We are asked to compare the results of Eq. 40-27 (squared, then multiplied by the accurate values of h/e found in Appendix B to convert to x-ray energies) with those in the table of K_α energies (in eV) given at the end of the problem. We look up the corresponding atomic numbers in Appendix F.

(a) For Li, with $Z=3$, we have

$$E_{\text{theory}} = \frac{h}{e}C^2(Z-1)^2 = \frac{6.6260688\times10^{-34}\,\text{J}\cdot\text{s}}{1.6021765\times10^{-19}\,\text{J/eV}}\left(4.9673\times10^7\,\text{Hz}^{1/2}\right)^2(3-1)^2 = 40.817\,\text{eV}.$$

The percentage deviation is

$$\text{percentage deviation} = 100\left(\frac{E_{\text{theory}}-E_{\text{exp}}}{E_{\text{exp}}}\right) = 100\left(\frac{40.817-54.3}{54.3}\right) = -24.8\% \approx -25\%.$$

(b) For Be, with $Z=4$, using the steps outlined in (b), the percentage deviation is -15%.

(c) For B, with $Z=5$, using the steps outlined in (b), the percentage deviation is -11%.

(d) For C, with $Z=6$, using the steps outlined in (b), the percentage deviation is -7.9%.

(e) For N, with $Z=7$, using the steps outlined in (b), the percentage deviation is -6.4%.

(f) For O, with $Z=8$, using the steps outlined in (b), the percentage deviation is -4.7%.

(g) For F, with $Z=9$, using the steps outlined in (b), the percentage deviation is -3.5%.

(h) For Ne, with $Z=10$, using the steps outlined in (b), the percentage deviation is -2.6%.

(i) For Na, with $Z=11$, using the steps outlined in (b), the percentage deviation is -2.0%.

(j) For Mg, with $Z=12$, using the steps outlined in (b), the percentage deviation is -1.5%.

Note that the trend is clear from the list given above: the agreement between theory and experiment becomes better as Z increases. One might argue that the most questionable step in §40-10 is the replacement $e^4 \to (Z-1)^2 e^4$ and ask why this could not equally well be $e^4 \to (Z-.9)^2 e^4$ or $e^4 \to (Z-.8)^2 e^4$? For large Z, these subtleties would not matter so much as they do for small Z, since $Z - \xi \approx Z$ for $Z >> \xi$.

44. According to Sample Problem 40-6, $N_x/N_0 = 1.3 \times 10^{-38}$. Let the number of moles of the lasing material needed be n; then $N_0 = nN_A$, where N_A is the Avogadro constant. Also $N_x = 10$. We solve for n:

$$n = \frac{N_x}{(1.3\times10^{-38})N_A} = \frac{10}{(1.3\times10^{-38})(6.02\times10^{23})} = 1.3\times10^{15}\,\text{mol}.$$

45. (a) If t is the time interval over which the pulse is emitted, the length of the pulse is

$$L = ct = (3.00 \times 10^8 \text{ m/s})(1.20 \times 10^{-11} \text{ s}) = 3.60 \times 10^{-3} \text{ m.}$$

(b) If E_p is the energy of the pulse, E is the energy of a single photon in the pulse, and N is the number of photons in the pulse, then $E_p = NE$. The energy of the pulse is

$$E_p = (0.150 \text{ J})/(1.602 \times 10^{-19} \text{ J/eV}) = 9.36 \times 10^{17} \text{ eV}$$

and the energy of a single photon is $E = (1240 \text{ eV·nm})/(694.4 \text{ nm}) = 1.786 \text{ eV}$. Hence,

$$N = \frac{E_p}{E} = \frac{9.36 \times 10^{17} \text{ eV}}{1.786 \text{ eV}} = 5.24 \times 10^{17} \text{ photons.}$$

46. From Eq. 40-29, $N_2/N_1 = e^{-(E_2-E_1)/kT}$. We solve for T:

$$T = \frac{E_2 - E_1}{k \ln(N_1/N_2)} = \frac{3.2 \text{ eV}}{(1.38 \times 10^{-23} \text{ J/K}) \ln(2.5 \times 10^{15}/6.1 \times 10^{13})} = 1.0 \times 10^4 \text{ K.}$$

47. The number of atoms in a state with energy E is proportional to $e^{-E/kT}$, where T is the temperature on the Kelvin scale and k is the Boltzmann constant. Thus the ratio of the number of atoms in the thirteenth excited state to the number in the eleventh excited state is $n_{13}/n_{11} = e^{-\Delta E/kT}$, where ΔE is the difference in the energies: $\Delta E = E_{13} - E_{11} = 2(1.2 \text{ eV})$ = 2.4 eV. For the given temperature, $kT = (8.62 \times 10^{-2} \text{ eV/K})(2000 \text{ K}) = 0.1724 \text{ eV}$. Hence,

$$\frac{n_{13}}{n_{11}} = e^{-2.4/0.1724} = 9.0 \times 10^{-7}.$$

48. Consider two levels, labeled 1 and 2, with $E_2 > E_1$. Since $T = -|T| < 0$,

$$\frac{N_2}{N_1} = e^{-(E_2-E_1)/kT} = e^{-|E_2-E_1|/(-k|T|)} = e^{|E_2-E_1|/k|T|} > 1.$$

Thus, $N_2 > N_1$; this is population inversion. We solve for T:

$$T = -|T| = -\frac{E_2 - E_1}{k \ln(N_2/N_1)} = -\frac{2.26 \text{ eV}}{(8.62 \times 10^{-5} \text{ eV/K}) \ln(1 + 0.100)} = -2.75 \times 10^5 \text{ K.}$$

49. Let the power of the laser beam be P and the energy of each photon emitted be E. Then, the rate of photon emission is

$$R = \frac{P}{E} = \frac{P}{hc/\lambda} = \frac{P\lambda}{hc} = \frac{(2.3 \times 10^{-3} \text{ W})(632.8 \times 10^{-9} \text{ m})}{(6.63 \times 10^{-34} \text{ J·s})(2.998 \times 10^8 \text{ m/s})} = 7.3 \times 10^{15} \text{ s}^{-1}.$$

50. The Moon is a distance $R = 3.82 \times 10^8$ m from Earth (see Appendix C). We note that the "cone" of light has apex angle equal to 2θ. If we make the small angle approximation (equivalent to using Eq. 37-14), then the diameter D of the spot on the Moon is

$$D = 2R\theta = 2R\left(\frac{1.22\lambda}{d}\right) = \frac{2(3.82 \times 10^8 \text{ m})(1.22)(600 \times 10^{-9} \text{ m})}{0.12 \text{ m}} = 4.7 \times 10^3 \text{ m} = 4.7 \text{ km}.$$

51. Let the range of frequency of the microwave be Δf. Then the number of channels that could be accommodated is

$$N = \frac{\Delta f}{10 \text{ MHz}} = \frac{(2.998 \times 10^8 \text{ m/s})\left[(450 \text{ nm})^{-1} - (650 \text{ nm})^{-1}\right]}{10 \text{ MHz}} = 2.1 \times 10^7.$$

The higher frequencies of visible light would allow many more channels to be carried compared with using the microwave.

52. Let the power of the laser beam be P and the energy of each photon emitted be E. Then, the rate of photon emission is

$$R = \frac{P}{E} = \frac{P}{hc/\lambda} = \frac{P\lambda}{hc} = \frac{(5.0 \times 10^{-3} \text{ W})(0.80 \times 10^{-6} \text{ m})}{(6.63 \times 10^{-34} \text{ J} \cdot \text{s})(2.998 \times 10^8 \text{ m/s})} = 2.0 \times 10^{16} \text{ s}^{-1}.$$

53. (a) If both mirrors are perfectly reflecting, there is a node at each end of the crystal. With one end partially silvered, there is a node very close to that end. We assume nodes at both ends, so there are an integer number of half-wavelengths in the length of the crystal. The wavelength in the crystal is $\lambda_c = \lambda/n$, where λ is the wavelength in a vacuum and n is the index of refraction of ruby. Thus $N(\lambda/2n) = L$, where N is the number of standing wave nodes, so

$$N = \frac{2nL}{\lambda} = \frac{2(1.75)(0.0600 \text{ m})}{694 \times 10^{-9} \text{ m}} = 3.03 \times 10^5.$$

(b) Since $\lambda = c/f$, where f is the frequency, $N = 2nLf/c$ and $\Delta N = (2nL/c)\Delta f$. Hence,

$$\Delta f = \frac{c\Delta N}{2nL} = \frac{(2.998 \times 10^8 \text{ m/s})(1)}{2(1.75)(0.0600 \text{ m})} = 1.43 \times 10^9 \text{ Hz}.$$

(c) The speed of light in the crystal is c/n and the round-trip distance is $2L$, so the round-trip travel time is $2nL/c$. This is the same as the reciprocal of the change in frequency.

(d) The frequency is $f = c/\lambda = (2.998 \times 10^8 \text{ m/s})/(694 \times 10^{-9} \text{ m}) = 4.32 \times 10^{14}$ Hz and the fractional change in the frequency is

$$\Delta f/f = (1.43 \times 10^9 \text{ Hz})/(4.32 \times 10^{14} \text{ Hz}) = 3.31 \times 10^{-6}.$$

54. For the nth harmonic of the standing wave of wavelength λ in the cavity of width L we have $n\lambda = 2L$, so $n\Delta\lambda + \lambda\Delta n = 0$. Let $\Delta n = \pm 1$ and use $\lambda = 2L/n$ to obtain

$$|\Delta\lambda| = \frac{\lambda|\Delta n|}{n} = \frac{\lambda}{n} = \lambda\left(\frac{\lambda}{2L}\right) = \frac{(533 \text{ nm})^2}{2(8.0 \times 10^7 \text{ nm})} = 1.8 \times 10^{-12} \text{ m} = 1.8 \text{ pm}.$$

55. (a) We denote the upper level as level 1 and the lower one as level 2. From $N_1/N_2 = e^{-(E_2 - E_1)/kT}$ we get (using the result of problem 83 in Chapter 38)

$$N_1 = N_2 e^{-(E_1 - E_2)/kT} = N_2 e^{-hc/\lambda kT} = \left(4.0 \times 10^{20}\right) e^{-(1240 \text{ eV·nm})/[(580\,\text{nm})(8.62 \times 10^{-5} \text{ eV/K})(300\,\text{K})]}$$
$$= 5.0 \times 10^{-16} \ll 1,$$

so practically no electron occupies the upper level.

(b) With $N_1 = 3.0 \times 10^{20}$ atoms emitting photons and $N_2 = 1.0 \times 10^{20}$ atoms absorbing photons, then the net energy output is

$$E = \left(N_1 - N_2\right) E_{\text{photon}} = \left(N_1 - N_2\right) \frac{hc}{\lambda} = \left(2.0 \times 10^{20}\right) \frac{\left(6.63 \times 10^{-34} \text{ J·s}\right)\left(2.998 \times 10^8 \text{ m/s}\right)}{580 \times 10^{-9} \text{ m}}$$
$$= 68 \text{ J}.$$

56. (a) The radius of the central disk is

$$R = \frac{1.22 f \lambda}{d} = \frac{(1.22)(3.50 \text{ cm})(515 \text{ nm})}{3.00 \text{ mm}} = 7.33 \ \mu\text{m}.$$

(b) The average power flux density in the incident beam is

$$\frac{P}{\pi d^2/4} = \frac{4(5.00 \text{ W})}{\pi(3.00 \text{ mm})^2} = 7.07 \times 10^5 \text{ W/m}^2.$$

(c) The average power flux density in the central disk is

$$\frac{(0.84)P}{\pi R^2} = \frac{(0.84)(5.00 \text{ W})}{\pi(7.33 \mu\text{m})^2} = 2.49 \times 10^{10} \text{ W/m}^2.$$

57. (a) Using the result of problem 83 in Chapter 38,

$$\Delta E = hc\left(\frac{1}{\lambda_1} - \frac{1}{\lambda_2}\right) = (1240\,\text{eV}\cdot\text{nm})\left(\frac{1}{588.995\,\text{nm}} - \frac{1}{589.592\,\text{nm}}\right) = 2.13\,\text{meV} .$$

(b) From $\Delta E = 2\mu_B B$ (see Fig. 40-10 and Eq. 40-18), we get

$$B = \frac{\Delta E}{2\mu_B} = \frac{2.13\times10^{-3}\,\text{eV}}{2(5.788\times10^{-5}\,\text{eV/T})} = 18\,\text{T} .$$

58. (a) In the lasing action the molecules are excited from energy level E_0 to energy level E_2. Thus the wavelength λ of the sunlight that causes this excitation satisfies

$$\Delta E = E_2 - E_0 = \frac{hc}{\lambda},$$

which gives (using the result of problem 83 in Chapter 38)

$$\lambda = \frac{hc}{E_2 - E_0} = \frac{1240\,\text{eV}\cdot\text{nm}}{0.289\,\text{eV} - 0} = 4.29\times10^3\,\text{nm} = 4.29\,\mu\text{m}.$$

(b) Lasing occurs as electrons jump down from the higher energy level E_2 to the lower level E_1. Thus the lasing wavelength λ' satisfies

$$\Delta E' = E_2 - E_1 = \frac{hc}{\lambda'},$$

which gives

$$\lambda' = \frac{hc}{E_2 - E_1} = \frac{1240\,\text{eV}\cdot\text{nm}}{0.289\,\text{eV} - 0.165\,\text{eV}} = 1.00\times10^4\,\text{nm} = 10.0\,\mu\text{m}.$$

(c) Both λ and λ' belong to the infrared region of the electromagnetic spectrum.

59. (a) The intensity at the target is given by $I = P/A$, where P is the power output of the source and A is the area of the beam at the target. We want to compute I and compare the result with 10^8 W/m^2. The beam spreads because diffraction occurs at the aperture of the laser. Consider the part of the beam that is within the central diffraction maximum. The angular position of the edge is given by $\sin\theta = 1.22\lambda/d$, where λ is the wavelength and d is the diameter of the aperture (see Exercise 61). At the target, a distance D away, the radius of the beam is $r = D\tan\theta$. Since θ is small, we may approximate both $\sin\theta$ and $\tan\theta$ by θ, in radians. Then, $r = D\theta = 1.22D\lambda/d$ and

$$I = \frac{P}{\pi r^2} = \frac{Pd^2}{\pi(1.22D\lambda)^2} = \frac{(5.0\times10^6 \text{ W})(4.0\text{m})^2}{\pi\left[1.22(3000\times10^3 \text{ m})(3.0\times10^{-6} \text{ m})\right]^2} = 2.1\times10^5 \text{ W/m}^2,$$

not great enough to destroy the missile.

(b) We solve for the wavelength in terms of the intensity and substitute $I = 1.0 \times 10^8$ W/m²:

$$\lambda = \frac{d}{1.22D}\sqrt{\frac{P}{\pi I}} = \frac{4.0\text{m}}{1.22(3000\times10^3\text{m})}\sqrt{\frac{5.0\times10^6 \text{ W}}{\pi(1.0\times10^8 \text{ W/m}^2)}} = 1.40\times10^{-7} \text{ m} = 140\,\text{nm}.$$

60. (a) The energy difference between the two states 1 and 2 was equal to the energy of the photon emitted. Since the photon frequency was $f = 1666$ MHz, its energy was given by $hf = (4.14 \times 10^{-15} \text{ eV·s})(1666 \text{ MHz}) = 6.90 \times 10^{-6}$ eV. Thus,

$$E_2 - E_1 = hf = 6.9\times10^{-6} \text{ eV} = 6.9\ \mu\text{eV}.$$

(b) The emission was in the *radio* region of the electromagnetic spectrum.

61. Letting $eV = hc/\lambda_{\text{min}}$ (see Eq. 40-23 and Eq. 38-4), we get

$$\lambda_{\text{min}} = \frac{hc}{eV} = \frac{1240\,\text{nm}\cdot\text{eV}}{eV} = \frac{1240\,\text{pm}\cdot\text{keV}}{eV} = \frac{1240\,\text{pm}}{V}$$

where V is measured in kV.

62. (a) From Fig. 40-14 we estimate the wavelengths corresponding to the K_β line to be $\lambda_\beta = 63.0$ pm. Using the result of problem 83 in Chapter 38, adapted to these units ($hc = 1240$ eV·nm = 1240 keV·pm),

$$E_\beta = (1240 \text{ keV·nm})/(63.0 \text{ pm}) = 19.7 \text{ keV} \approx 20 \text{ keV}.$$

(b) For K_α, with $\lambda_\alpha = 70.0$ pm,

$$E_\alpha = \frac{hc}{\lambda_\alpha} = \frac{1240\text{keV}\cdot\text{pm}}{70.0\text{pm}} = 17.7\,\text{keV} \approx 18 \text{ keV}.$$

(c) Both Zr and Nb can be used, since $E_\alpha < 18.00$ eV $< E_\beta$ and $E_\alpha < 18.99$ eV $< E_\beta$. According to the hint given in the problem statement, Zr is the best choice.

(d) Nb is the second best choice.

63. (a) The length of the pulse's wave train is given by

$$L = c\Delta t = (2.998 \times 10^8 \text{ m/s})(10 \times 10^{-15} \text{ s}) = 3.0 \times 10^{-6} \text{ m}.$$

Thus, the number of wavelengths contained in the pulse is

$$N = \frac{L}{\lambda} = \frac{3.0 \times 10^{-6} \text{ m}}{500 \times 10^{-9} \text{ m}} = 6.0.$$

(b) We solve for X from 10 fm/1 m = 1 s/X:

$$X = \frac{(1 \text{ s})(1 \text{ m})}{10 \times 10^{-15} \text{ m}} = \frac{1 \text{ s}}{(10 \times 10^{-15})(3.15 \times 10^7 \text{ s/y})} = 3.2 \times 10^6 \text{ y}.$$

64. (a) The distance from the Earth to the Moon is $d_{em} = 3.82 \times 10^8$ m (see Appendix C). Thus, the time required is given by

$$t = \frac{2d_{em}}{c} = \frac{2(3.82 \times 10^8 \text{ m})}{2.998 \times 10^8 \text{ m/s}} = 2.55 \text{ s}.$$

(b) We denote the uncertainty in time measurement as δt and let $2\delta d_{es} = 15$ cm. Then, since $d_{em} \propto t$, $\delta t/t = \delta d_{em}/d_{em}$. We solve for δt:

$$\delta t = \frac{t \delta d_{em}}{d_{em}} = \frac{(2.55 \text{ s})(0.15 \text{ m})}{2(3.82 \times 10^8 \text{ m})} = 5.0 \times 10^{-10} \text{ s}.$$

(c) The angular divergence of the beam is

$$\theta = 2 \tan^{-1}\left(\frac{1.5 \times 10^3}{d_{em}}\right) = 2 \tan^{-1}\left(\frac{1.5 \times 10^3}{3.82 \times 10^8}\right) = (4.5 \times 10^{-4})^{\circ}$$

65. We use $eV = hc/\lambda_{min}$ (see Eq. 40-23 and Eq. 38-4):

$$h = \frac{eV\lambda_{min}}{c} = \frac{(1.60 \times 10^{-19} \text{ C})(40.0 \times 10^3 \text{ eV})(31.1 \times 10^{-12} \text{ m})}{2.998 \times 10^8 \text{ m/s}} = 6.63 \times 10^{-34} \text{ J} \cdot \text{s}.$$

66. For a given shell with quantum number n the total number of available electron states is $2n^2$. Thus, for the first four shells ($n = 1$ through 4) the number of available states are 2, 8, 18, and 32 (see Appendix G). Since $2 + 8 + 18 + 32 = 60 < 63$, according to the "logical" sequence the first four shells would be completely filled in an europium atom, leaving $63 - 60 = 3$ electrons to partially occupy the $n = 5$ shell. Two of these three electrons would fill up the 5s subshell, leaving only one remaining electron in the only

partially filled subshell (the 5*p* subshell). In chemical reactions this electron would have the tendency to be transferred to another element, leaving the remaining 62 electrons in chemically stable, completely filled subshells. This situation is very similar to the case of sodium, which also has only one electron in a partially filled shell (the 3*s* shell).

67. Without the spin degree of freedom the number of available electron states for each shell would be reduced by half. So the values of Z for the noble gas elements would become half of what they are now: $Z = 1, 5, 9, 18, 27,$ and 43. Of this set of numbers, the only one which coincides with one of the familiar noble gas atomic numbers ($Z = 2, 10, 18, 36, 54,$ and 86) is 18. Thus, argon would be the only one that would remain "noble."

68. (a) The value of ℓ satisfies $\sqrt{\ell(\ell+1)}\hbar \approx \sqrt{\ell^2}\hbar = \ell\hbar = L$, so $\ell \approx L/\hbar \approx 3 \times 10^{74}$.

(b) The number is $2\ell + 1 \approx 2(3 \times 10^{74}) = 6 \times 10^{74}$.

(c) Since

$$\cos\theta_{min} = \frac{m_{\ell\,max}\hbar}{\sqrt{\ell(\ell+1)}\hbar} = \frac{1}{\sqrt{\ell(\ell+1)}} \approx 1 - \frac{1}{2\ell} = 1 - \frac{1}{2(3\times10^{74})}$$

or $\cos\theta_{min} \approx 1 - \theta_{min}^2/2 \approx 1 - 10^{-74}/6$, we have $\theta_{min} \approx \sqrt{10^{-74}/3} = 6 \times 10^{-38}$ rad . The correspondence principle requires that all the quantum effects vanish as $\hbar \to 0$. In this case $\hbar/L$ is extremely small so the quantization effects are barely existent, with $\theta_{min} \approx 10^{-38}$ rad ≈ 0.

69. The principal quantum number n must be greater than 3. The magnetic quantum number m_ℓ can have any of the values $-3, -2, -1, 0, +1, +2,$ or $+3$. The spin quantum number can have either of the values $-\frac{1}{2}$ or $+\frac{1}{2}$.

70. One way to think of the units of h is that, because of the equation $E = hf$ and the fact that f is in cycles/second, then the "explicit" units for h should be J·s/cycle. Then, since 2π rad/cycle is a conversion factor for cycles $\to$ radians, $\hbar = h/2\pi$ can be thought of as the Planck constant expressed in terms of radians instead of cycles. Using the precise values stated in Appendix B,

$$\hbar = \frac{h}{2\pi} = \frac{6.62606876 \times 10^{-34}\,\text{J}\cdot\text{s}}{2\pi} = 1.05457 \times 10^{-34}\,\text{J}\cdot\text{s} = \frac{1.05457 \times 10^{-34}\,\text{J}\cdot\text{s}}{1.6021765 \times 10^{-19}\,\text{J/eV}}$$

$$= 6.582 \times 10^{-16}\,\text{eV}\cdot\text{s}.$$

Chapter 41

1. The number of atoms per unit volume is given by $n = d / M$, where d is the mass density of copper and M is the mass of a single copper atom. Since each atom contributes one conduction electron, n is also the number of conduction electrons per unit volume. Since the molar mass of copper is $A = 63.54\,g/mol$,

$$M = A / N_A = (63.54\,g/mol)/(6.022 \times 10^{23}\,mol^{-1}) = 1.055 \times 10^{-22}\,g.$$

Thus,

$$n = \frac{8.96\,g/cm^3}{1.055 \times 10^{-22}\,g} = 8.49 \times 10^{22}\,cm^{-3} = 8.49 \times 10^{28}\,m^{-3}.$$

2. We note that $n = 8.43 \times 10^{28}\,m^{-3} = 84.3\,nm^{-3}$. From Eq. 41-9,

$$E_F = \frac{0.121(hc)^2}{m_e c^2} n^{2/3} = \frac{0.121(1240\,eV \cdot nm)^2}{511 \times 10^3\,eV}(84.3\,nm^{-3})^{2/3} = 7.0\,eV$$

where the result of problem 83 in Chapter 38 is used.

3. (a) Eq. 41-5 gives

$$N(E) = \frac{8\sqrt{2}\pi m^{3/2}}{h^3} E^{1/2}$$

for the density of states associated with the conduction electrons of a metal. This can be written

$$n(E) = CE^{1/2}$$

where

$$C = \frac{8\sqrt{2}\pi m^{3/2}}{h^3} = \frac{8\sqrt{2}\pi(9.109 \times 10^{-31}\,kg)^{3/2}}{(6.626 \times 10^{-34}\,J \cdot s)^3} = 1.062 \times 10^{56}\,kg^{3/2}/J^3 \cdot s^3.$$

(b) Now, $1\,J = 1\,kg \cdot m^2/s^2$ (think of the equation for kinetic energy $K = \frac{1}{2}mv^2$), so $1\,kg = 1\,J \cdot s^2 \cdot m^{-2}$. Thus, the units of C can be written $(J \cdot s^2)^{3/2} \cdot (m^{-2})^{3/2} \cdot J^{-3} \cdot s^{-3} = J^{-3/2} \cdot m^{-3}$. This means

$$C = (1.062 \times 10^{56} \, \text{J}^{-3/2} \cdot \text{m}^{-3})(1.602 \times 10^{-19} \, \text{J}/\text{eV})^{3/2} = 6.81 \times 10^{27} \, \text{m}^{-3} \cdot \text{eV}^{-3/2} \, .$$

(c) If $E = 5.00$ eV, then

$$n(E) = (6.81 \times 10^{27} \, \text{m}^{-3} \cdot \text{eV}^{-3/2})(5.00 \, \text{eV})^{1/2} = 1.52 \times 10^{28} \, \text{eV}^{-1} \cdot \text{m}^{-3} \, .$$

4. We note that there is one conduction electron per atom and that the molar mass of gold is $197 \, \text{g}/\text{mol}$. Therefore, combining Eqs. 41-2, 41-3 and 41-4 leads to

$$n = \frac{(19.3 \, \text{g}/\text{cm}^3)(10^6 \, \text{cm}^3/\text{m}^3)}{(197 \, \text{g}/\text{mol})/(6.02 \times 10^{23} \, \text{mol}^{-1})} = 5.90 \times 10^{28} \, \text{m}^{-3} \, .$$

5. (a) At absolute temperature $T = 0$, the probability is zero that any state with energy above the Fermi energy is occupied.

(b) The probability that a state with energy E is occupied at temperature T is given by

$$P(E) = \frac{1}{e^{(E-E_F)/kT} + 1}$$

where k is the Boltzmann constant and E_F is the Fermi energy. Now, $E - E_F = 0.0620$ eV and

$$(E - E_F)/kT = (0.0620 \, \text{eV})/(8.62 \times 10^{-5} \, \text{eV}/\text{K})(320 \text{K}) = 2.248 \, ,$$

so

$$P(E) = \frac{1}{e^{2.248} + 1} = 0.0955 \, .$$

See Appendix B or Sample Problem 41-1 for the value of k.

6. We use the result of problem 3:

$$n(E) = CE^{1/2} = \left[6.81 \times 10^{27} \, \text{m}^{-3} \cdot (\text{eV})^{-2/3}\right](8.0 \, \text{eV})^{1/2} = 1.9 \times 10^{28} \, \text{m}^{-3} \cdot \text{eV}^{-1} \, .$$

This is consistent with Fig. 41-5.

7. According to Eq. 41-9, the Fermi energy is given by

$$E_F = \left(\frac{3}{16\sqrt{2}\pi}\right)^{2/3} \frac{h^2}{m} n^{2/3}$$

where n is the number of conduction electrons per unit volume, m is the mass of an electron, and h is the Planck constant. This can be written $E_F = An^{2/3}$, where

$$A = \left(\frac{3}{16\sqrt{2}\pi}\right)^{2/3}\frac{h^2}{m} = \left(\frac{3}{16\sqrt{2}\pi}\right)^{2/3}\frac{(6.626 \times 10^{-34}\,\text{J}\cdot\text{s})^2}{9.109 \times 10^{-31}\,\text{kg}} = 5.842 \times 10^{-38}\,\text{J}^2\cdot\text{s}^2\,/\,\text{kg}\,.$$

Since $1\,\text{J} = 1\,\text{kg}\cdot\text{m}^2\,/\,\text{s}^2$, the units of A can be taken to be $\text{m}^2\cdot\text{J}$. Dividing by $1.602 \times 10^{-19}\,\text{J}\,/\,\text{eV}$, we obtain $A = 3.65 \times 10^{-19}\,\text{m}^2\cdot\text{eV}$.

8. Let $E_1 = 63$ meV $+ E_F$ and $E_2 = -63$ meV $+ E_F$. Then according to Eq. 41-6,

$$P_1 = \frac{1}{e^{(E_1-E_F)/kT}+1} = \frac{1}{e^x+1}$$

where $x = (E_1 - E_F)\,/\,kT$. We solve for e^x:

$$e^x = \frac{1}{P_1} - 1 = \frac{1}{0.090} - 1 = \frac{91}{9}\,.$$

Thus,

$$P_2 = \frac{1}{e^{(E_2-E_F)/kT}+1} = \frac{1}{e^{-(E_1-E_F)/kT}+1} = \frac{1}{e^{-x}+1} = \frac{1}{(91/9)^{-1}+1} = 0.91,$$

where we use $E_2 - E_F = -63$ meV $= E_F - E_1 = -(E_1 - E_F)$.

9. The Fermi-Dirac occupation probability is given by $P_{FD} = 1/(e^{\Delta E/kT}+1)$, and the Boltzmann occupation probability is given by $P_B = e^{-\Delta E/kT}$. Let f be the fractional difference. Then

$$f = \frac{P_B - P_{FD}}{P_B} = \frac{e^{-\Delta E/kT} - \frac{1}{e^{\Delta E/kT}+1}}{e^{-\Delta E/kT}}\,.$$

Using a common denominator and a little algebra yields

$$f = \frac{e^{-\Delta E/kT}}{e^{-\Delta E/kT}+1}\,.$$

The solution for $e^{-\Delta E/kT}$ is

$$e^{-\Delta E/kT} = \frac{f}{1-f} \, .$$

We take the natural logarithm of both sides and solve for T. The result is

$$T = \frac{\Delta E}{k \ln\left(\dfrac{f}{1-f}\right)} \, .$$

(a) Letting f equal 0.01, we evaluate the expression for T:

$$T = \frac{(1.00\,\text{eV})(1.60\times10^{-19}\,\text{J/eV})}{(1.38\times10^{-23}\,\text{J/K})\ln\left(\dfrac{0.010}{1-0.010}\right)} = 2.50\times10^{3}\,\text{K}.$$

(b) We set f equal to 0.10 and evaluate the expression for T:

$$T = \frac{(1.00\,\text{eV})(1.60\times10^{-19}\,\text{J/eV})}{(1.38\times10^{-23}\,\text{J/K})\ln\left(\dfrac{0.10}{1-0.10}\right)} = 5.30\times10^{3}\,\text{K}.$$

10. We reproduce the calculation of Problem 4: Combining Eqs. 41-2, 41-3 and 41-4, the number density of conduction electrons in gold is

$$n = \frac{(19.3\,\text{g}/\text{cm}^{3})(6.02\times10^{23}\,/\,\text{mol})}{(197\,\text{g}/\text{mol})} = 5.90\times10^{22}\,\text{cm}^{-3} = 59.0\,\text{nm}^{-3} \, .$$

Now, using the result of Problem 83 in Chapter 38, Eq. 41-9 leads to

$$E_F = \frac{0.121(hc)^{2}}{(m_e c^{2})}n^{2/3} = \frac{0.121(1240\,\text{eV}\cdot\text{nm})^{2}}{511\times10^{3}\,\text{eV}}(59.0\,\text{nm}^{-3})^{2/3} = 5.52\,\text{eV} \, .$$

11. (a) Eq. 41-6 leads to

$$E = E_F + kT \ln (P^{-1}-1) = 7.00\,\text{eV} + (8.62\times10^{-5}\,\text{eV}/\text{K})(1000\,\text{K})\ln\left(\frac{1}{0.900}-1\right)$$

$$= 6.81\,\text{eV}.$$

(b) $n(E) = CE^{1/2} = \left(6.81\times10^{27}\,\text{m}^{-3}\cdot\text{eV}^{-3/2}\right)(6.81\,\text{eV})^{1/2} = 1.77\times10^{28}\,\text{m}^{-3}\cdot\text{eV}^{-1}.$

(c) $n_0(E) = P(E)n(E) = (0.900)(1.77\times10^{28}\,\text{m}^{-3}\cdot\text{eV}^{-1}) = 1.59\times10^{28}\,\text{m}^{-3}\cdot\text{eV}^{-1}.$

12. (a) The volume per cubic meter of sodium occupied by the sodium ions is

$$V_{Na} = \frac{(971\,kg)(6.022\times10^{23}\,/\,mol)(4\pi/3)(98.0\times10^{-12}\,m)^3}{(23.0\,g/mol)} = 0.100\,m^3,$$

so the fraction available for conduction electrons is $1-(V_{Na}/1.00\,m^3) = 1-0.100 = 0.900$, or 90.0%.

(b) For copper,

$$V_{Cu} = \frac{(8960\,kg)(6.022\times10^{23}\,/\,mol)(4\pi/3)(135\times10^{-12}\,m)^3}{(63.5\,g/mol)} = 0.1876\,m^{-3}.$$

Thus, the fraction is $1-(V_{Cu}/1.00\,m^3) = 1-0.876 = 0.124$, or 12.4%.

(c) Sodium, because the electrons occupy a greater portion of the space available.

13. We use

$$N_0 = N(E)P(E) = CE^{1/2}\left[e^{(E-E_F)/kT}+1\right]^{-1},$$

where C is given in problem 3(b).

(a) At $E = 4.00$ eV,

$$n_0 = \frac{\left(6.81\times10^{27}\,m^{-3}\cdot(eV)^{-3/2}\right)(4.00\,eV)^{1/2}}{e^{(4.00\,eV-7.00\,eV)/[(8.62\times10^{-5}\,eV/K)(1000\,K)]}+1} = 1.36\times10^{28}\,m^{-3}\cdot eV^{-1}.$$

(b) At $E = 6.75$ eV,

$$n_0 = \frac{\left(6.81\times10^{27}\,m^{-3}\cdot(eV)^{-3/2}\right)(6.75\,eV)^{1/2}}{e^{(6.75\,eV-7.00\,eV)/[(8.62\times10^{-5}\,eV/K)(1000\,K)]}+1} = 1.68\times10^{28}\,m^{-3}\cdot eV^{-1}.$$

(c) Similarly, at $E = 7.00$ eV, the value of $n_0(E)$ is $9.01 \times 10^{27}\,m^{-3}\cdot eV^{-1}$.

(d) At $E = 7.25$ eV, the value of $n_0(E)$ is $9.56 \times 10^{26}\,m^{-3}\cdot eV^{-1}$.

(e) At $E = 9.00$ eV, the value of $n_0(E)$ is $1.71 \times 10^{18}\,m^{-3}\cdot eV^{-1}$.

14. The probability P_h that a state is occupied by a hole is the same as the probability the state is *unoccupied* by an electron. Since the total probability that a state is either occupied or unoccupied is 1, we have $P_h + P = 1$. Thus,

$$P_h = 1 - \frac{1}{e^{(E-E_F)/kT} + 1} = \frac{e^{(E-E_F)/kT}}{1 + e^{(E-E_F)/kT}} = \frac{1}{e^{-(E-E_F)/kT} + 1}.$$

15. (a) We evaluate $P(E) = 1/\left(e^{(E-E_F)/kT} + 1\right)$ for the given value of E, using

$$kT = \frac{(1.381 \times 10^{-23} \text{ J} / \text{K})(273 \text{ K})}{1.602 \times 10^{-19} \text{ J} / \text{eV}} = 0.02353 \text{ eV}.$$

For $E = 4.4$ eV, $(E - E_F)/kT = (4.4 \text{ eV} - 5.5 \text{ eV})/(0.02353 \text{ eV}) = -46.25$ and

$$P(E) = \frac{1}{e^{-46.25} + 1} = 1.0.$$

(b) Similarly, for $E = 5.4$ eV, $P(E) = 0.986 \approx 0.99$.

(c) For $E = 5.5$ eV, $P(E) = 0.50$.

(d) For $E = 5.6$ eV, $P(E) = 0.014$.

(e) For $E = 6.4$ eV, $P(E) = 2.447 \times 10^{-17} \approx 2.4 \times 10^{-17}$.

(f) Solving $P = 1/(e^{\Delta E/kT} + 1)$ for $e^{\Delta E/kT}$, we get

$$e^{\Delta E/kT} = \frac{1}{P} - 1.$$

Now, we take the natural logarithm of both sides and solve for T. The result is

$$T = \frac{\Delta E}{k \ln\left(\frac{1}{P} - 1\right)} = \frac{(5.6 \text{eV} - 5.5 \text{eV})(1.602 \times 10^{-19} \text{ J/eV})}{(1.381 \times 10^{-23} \text{ J/K}) \ln\left(\frac{1}{0.16} - 1\right)} = 699 \text{ K} \approx 7.0 \times 10^2 \text{ K}.$$

16. The molar mass of carbon is $m = 12.01115$ g/mol and the mass of the Earth is $M_e = 5.98 \times 10^{24}$ kg. Thus, the number of carbon atoms in a diamond as massive as the Earth is $N = (M_e/m)N_A$, where N_A is the Avogadro constant. From the result of Sample Problem 41-1, the probability in question is given by

$$P = N_e^{-E_g/kT} = \left(\frac{M_e}{m}\right)N_A e^{-E_g/kT} = \left(\frac{5.98 \times 10^{24} \text{ kg}}{12.01115 \text{ g/mol}}\right)(6.02 \times 10^{23} \text{ /mol})(3 \times 10^{-93})$$

$$= 9 \times 10^{-43} \approx 10^{-42}.$$

17. Let N be the number of atoms per unit volume and n be the number of free electrons per unit volume. Then, the number of free electrons per atom is n/N. We use the result of Exercise 11 to find n: $E_F = An^{2/3}$, where $A = 3.65 \times 10^{-19} \text{ m}^2 \cdot \text{eV}$. Thus,

$$n = \left(\frac{E_F}{A}\right)^{3/2} = \left(\frac{11.6 \text{ eV}}{3.65 \times 10^{-19} \text{ m}^2 \cdot \text{eV}}\right)^{3/2} = 1.79 \times 10^{29} \text{ m}^{-3}.$$

If M is the mass of a single aluminum atom and d is the mass density of aluminum, then $N = d/M$. Now,

$$M = (27.0 \text{ g/mol})/(6.022 \times 10^{23} \text{ mol}^{-1}) = 4.48 \times 10^{-23} \text{ g},$$

so

$$N = (2.70 \text{ g/cm}^3)/(4.48 \times 10^{-23} \text{ g}) = 6.03 \times 10^{22} \text{ cm}^{-3} = 6.03 \times 10^{28} \text{ m}^{-3}.$$

Thus, the number of free electrons per atom is

$$\frac{n}{N} = \frac{1.79 \times 10^{29} \text{ m}^{-3}}{6.03 \times 10^{28} \text{ m}^{-3}} = 2.97 \approx 3.$$

18. (a) The ideal gas law in the form of Eq. 20-9 leads to $p = NkT/V = nkT$. Thus, we solve for the molecules per cubic meter:

$$n = \frac{p}{kT} = \frac{(1.0 \text{ atm})(1.0 \times 10^5 \text{ Pa/atm})}{(1.38 \times 10^{-23} \text{ J/K})(273 \text{ K})} = 2.7 \times 10^{25} \text{ m}^{-3}.$$

(b) Combining Eqs. 41-2, 41-3 and 41-4 leads to the conduction electrons per cubic meter in copper:

$$n = \frac{8.96 \times 10^3 \text{ kg/m}^3}{(63.54)(1.67 \times 10^{-27} \text{ kg})} = 8.43 \times 10^{28} \text{ m}^{-3}.$$

(c) The ratio is $(8.43 \times 10^{28} \text{ m}^{-3})/(2.7 \times 10^{25} \text{ m}^{-3}) = 3.1 \times 10^3$.

(d) We use $d_{avg} = n^{-1/3}$. For case (a), $d_{avg} = (2.7 \times 10^{25} \text{ m}^{-3})^{-1/3}$ which equals 3.3 nm.

(e) For case (b), $d_{avg} = (8.43 \times 10^{28} \text{ m}^{-3})^{-1/3} = 0.23$ nm.

19. (a) According to Appendix F the molar mass of silver is 107.870 g/mol and the density is 10.49 g/cm^3. The mass of a silver atom is

$$\frac{107.870 \times 10^{-3} \text{ kg / mol}}{6.022 \times 10^{23} \text{ mol}^{-1}} = 1.791 \times 10^{-25} \text{ kg .}$$

We note that silver is monovalent, so there is one valence electron per atom (see Eq. 41-2). Thus, Eqs. 41-4 and 41-3 lead to

$$n = \frac{\rho}{M} = \frac{10.49 \times 10^{-3} \text{ kg/m}^3}{1.791 \times 10^{25} \text{ kg}} = 5.86 \times 10^{28} \text{ m}^{-3} .$$

(b) The Fermi energy is

$$E_F = \frac{0.121 h^2}{m} n^{2/3} = \frac{(0.121)(6.626 \times 10^{-34} \text{ J} \cdot \text{s})^2}{9.109 \times 10^{-31} \text{ kg}} = (5.86 \times 10^{28} \text{ m}^{-3})^{2/3}$$

$$= 8.80 \times 10^{-19} \text{ J} = 5.49 \text{eV} .$$

(c) Since $E_F = \frac{1}{2} m v_F^2$,

$$v_F = \sqrt{\frac{2 E_F}{m}} = \sqrt{\frac{2(8.80 \times 10^{-19} \text{ J})}{9.109 \times 10^{-31} \text{ kg}}} = 1.39 \times 10^6 \text{ m / s .}$$

(d) The de Broglie wavelength is

$$\lambda = \frac{h}{m v_F} = \frac{6.626 \times 10^{-34} \text{ J} \cdot \text{s}}{(9.109 \times 10^{-31} \text{ kg})(1.39 \times 10^6 \text{ m/s})} = 5.22 \times 10^{-10} \text{ m.}$$

20. Let the energy of the state in question be an amount ΔE above the Fermi energy E_F. Then, Eq. 41-6 gives the occupancy probability of the state as

$$P = \frac{1}{e^{(E_F + \Delta E - E_F)/kT} + 1} = \frac{1}{e^{\Delta E/kT} + 1} .$$

We solve for ΔE to obtain

$$\Delta E = kT \ln\left(\frac{1}{P} - 1\right) = (1.38 \times 10^{23} \text{ J / K})(300 \text{ K}) \ln\left(\frac{1}{0.10} - 1\right) = 9.1 \times 10^{-21} \text{ J} ,$$

which is equivalent to 5.7×10^{-2} eV = 57 meV.

21. The average energy of the conduction electrons is given by

$$E_{\text{avg}} = \frac{1}{n} \int_0^{\infty} E N(E) P(E) dE$$

where n is the number of free electrons per unit volume, $N(E)$ is the density of states, and $P(E)$ is the occupation probability. The density of states is proportional to $E^{1/2}$, so we may write $N(E) = CE^{1/2}$, where C is a constant of proportionality. The occupation probability is one for energies below the Fermi energy and zero for energies above. Thus,

$$E_{\text{avg}} = \frac{C}{n} \int_0^{E_F} E^{3/2} dE = \frac{2C}{5n} E_F^{5/2} .$$

Now

$$n = \int_0^{\infty} N(E) P(E) dE = C \int_0^{E_F} E^{1/2} dE = \frac{2C}{3} E_F^{3/2} .$$

We substitute this expression into the formula for the average energy and obtain

$$E_{\text{avg}} = \left(\frac{2C}{5} \right) E_F^{5/2} \left(\frac{3}{2CE_F^{3/2}} \right) = \frac{3}{5} E_F .$$

22. (a) Combining Eqs. 41-2, 41-3 and 41-4 leads to the conduction electrons per cubic meter in zinc:

$$n = \frac{2(7.133 \, \text{g} / \text{cm}^3)}{(65.37 \, \text{g} / \text{mol}) / (6.02 \times 10^{23} \, \text{mol})} = 1.31 \times 10^{23} \, \text{cm}^{-3} = 1.31 \times 10^{29} \, \text{m}^{-3} .$$

(b) From Eq. 41-9,

$$E_F = \frac{0.121 h^2}{m_e} n^{2/3} = \frac{0.121(6.63 \times 10^{-34} \, \text{J} \cdot \text{s})^2 (1.31 \times 10^{29} \, \text{m}^{-3})^{2/3}}{(9.11 \times 10^{-31} \, \text{kg})(1.60 \times 10^{-19} \, \text{J} / \text{eV})} = 9.43 \, \text{eV}.$$

(c) Equating the Fermi energy to $\frac{1}{2} m_e v_F^2$ we find (using the $m_e c^2$ value in Table 37-3)

$$v_F = \sqrt{\frac{2 E_F c^2}{m_e c^2}} = \sqrt{\frac{2(9.43 \, \text{eV})(2.998 \times 10^8 \, \text{m} / \text{s})^2}{511 \times 10^3 \, \text{eV}}} = 1.82 \times 10^6 \, \text{m} / \text{s} .$$

(d) The de Broglie wavelength is

$$\lambda = \frac{h}{m_e v_F} = \frac{6.63 \times 10^{-34} \, \text{J} \cdot \text{s}}{(9.11 \times 10^{-31} \, \text{kg})(1.82 \times 10^6 \, \text{m} / \text{s})} = 0.40 \, \text{nm} .$$

23. Let the volume be $v = 1.00 \times 10^{-6} \, \text{m}^3$. Then,

$$K_{\text{total}} = N E_{\text{avg}} = nv E_{\text{avg}} = (8.43 \times 10^{28} \, \text{m}^{-3})(1.00 \times 10^{-6} \, \text{m}^3)\left(\frac{3}{5}\right)(7.00 \, \text{eV})(1.60 \times 10^{-19} \, \text{J/eV})$$

$$= 5.71 \times 10^4 \, \text{J} = 57.1 \, \text{kJ}.$$

24. (a) At $T = 300$ K

$$f = \frac{3kT}{2E_F} = \frac{3(8.62 \times 10^{-5} \, \text{eV} / \text{K})(300 \, \text{K})}{2(7.0 \, \text{eV})} = 5.5 \times 10^{-3} \, .$$

(b) At $T = 1000$ K,

$$f = \frac{3kT}{2E_F} = \frac{3(8.62 \times 10^{-5} \, \text{eV} / \text{K})(1000 \, \text{K})}{2(7.0 \, \text{eV})} = 1.8 \times 10^{-2} \, .$$

(c) Many calculators and most math software packages (here we use MAPLE) have built-in numerical integration routines. Setting up ratios of integrals of Eq. 41-7 and canceling common factors, we obtain

$$frac = \frac{\displaystyle\int_{E_F}^{\infty} \sqrt{E} \, / \, (e^{(E-E_F)/kT} + 1) dE}{\displaystyle\int_{0}^{\infty} \sqrt{E} \, / \, (e^{(E-E_F)/kT} + 1) dE}$$

where $k = 8.62 \times 10^{-5}$ eV/K. We use the Fermi energy value for copper ($E_F = 7.0$ eV) and evaluate this for $T = 300$ K and $T = 1000$ K; we find *frac* = 0.00385 and *frac* = 0.0129, respectively.

25. The fraction f of electrons with energies greater than the Fermi energy is (approximately) given in Problem 41-24:

$$f = \frac{3kT/2}{E_F}$$

where T is the temperature on the Kelvin scale, k is the Boltzmann constant, and E_F is the Fermi energy. We solve for T:

$$T = \frac{2fE_F}{3k} = \frac{2(0.013)(4.70 \, \text{eV})}{3(8.62 \times 10^{-5} \, \text{eV} / \text{K})} = 472 \, \text{K}.$$

26. (a) Using Eq. 41-4, the energy released would be

$$E = NE_{avg} = \frac{(3.1g)}{(63.54\,g/mol)/(6.02\times10^{23}/mol)}\left(\frac{3}{5}\right)(7.0\,eV)(1.6\times10^{-19}\,J/eV)$$

$$= 1.97\times10^{4}\,J.$$

(b) Keeping in mind that a Watt is a Joule per second, we have

$$\frac{1.97\times10^{4}\,J}{100\,J/s} = 197\,s.$$

27. (a) Since the electron jumps from the conduction band to the valence band, the energy of the photon equals the energy gap between those two bands. The photon energy is given by $hf = hc/\lambda$, where f is the frequency of the electromagnetic wave and λ is its wavelength. Thus, $E_g = hc/\lambda$ and

$$\lambda = \frac{hc}{E_g} = \frac{(6.63\times10^{-34}\,J\cdot s)(2.998\times10^{8}\,m/s)}{(5.5\,eV)(1.60\times10^{-19}\,J/eV)} = 2.26\times10^{-7}\,m = 226\,nm.$$

Photons from other transitions have a greater energy, so their waves have shorter wavelengths.

(b) These photons are in the ultraviolet portion of the electromagnetic spectrum.

28. Each Arsenic atom is connected (by covalent bonding) to four Gallium atoms, and each Gallium atom is similarly connected to four Arsenic atoms.

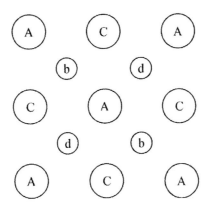

The "depth" of their very non-trivial lattice structure is, of course, not evident in a flattened-out representation such as shown for Silicon in Fig. 41-9. Still we try to convey some sense of this (in the [1, 0, 0] view shown — for those who might be familiar with Miller indices) by using letters to indicate the depth: A for the closest atoms (to the observer), b for the next layer deep, C for further into the page, d for the last layer seen, and E (not shown) for the atoms that are at the deepest layer (and are behind the A's) needed for our description of the structure. The capital letters are used for the Gallium atoms, and the small letters for the Arsenic.

Consider the Arsenic atom (with the letter b) near the upper left; it has covalent bonds with the two A's and the two C's near it. Now consider the Arsenic atom (with the letter d) near the upper right; it has covalent bonds with the two C's which are near it and with the two E's (which are behind the A's which are near :+).

(a) The 3p, 3d and 4s subshells of both Arsenic and Gallium are filled. They both have partially filled 4p subshells. An isolated, neutral Arsenic atom has three electrons in the 4p subshell, and an isolated, neutral Gallium atom has one electron in the 4p subshell. To supply the total of eight shared electrons (for the four bonds connected to each ion in the lattice), not only the electrons from 4p must be shared but also the electrons from 4s. The core of the Gallium ion has charge $q = +3e$ (due to the "loss" of its single 4p and two 4s electrons).

(b) The core of the Arsenic ion has charge $q = +5e$ (due to the "loss" of the three 4p and two 4s electrons).

(c) As remarked in part (a), there are two electrons shared in each of the covalent bonds. This is the same situation that one finds for Silicon (see Fig. 41-9).

29. (a) At the bottom of the conduction band $E = 0.67$ eV. Also $E_F = 0.67$ eV/2 = 0.335 eV. So the probability that the bottom of the conduction band is occupied is

$$P(E) = \frac{1}{e^{(E-E_F)/kT}+1} = \frac{1}{e^{(0.67\,\text{eV}-0.335\,\text{eV})/[(8.62\times10^{-5}\,\text{eV/K})(290\,\text{K})]}+1} = 1.5\times10^{-6}.$$

(b) At the top of the valence band $E = 0$, so the probability that the state is *unoccupied* is given by

$$1-P(E) = 1 - \frac{1}{e^{(E-E_F)/kT}+1} = \frac{1}{e^{-(E-E_F)/kT}+1} = \frac{1}{e^{-(0-0.335\,\text{eV})/[(8.62\times10^{-5}\,\text{eV/K})(290\,\text{K})]}+1}$$

$$= 1.5\times10^{-6}.$$

30. (a) The number of electrons in the valence band is

$$N_{ev} = N_v P(E_v) = \frac{N_v}{e^{(E_v-E_F)/kT}+1}.$$

Since there are a total of N_v states in the valence band, the number of holes in the valence band is

$$N_{hv} = N_v - N_{ev} = N_v\left[1 - \frac{1}{e^{(E_v-E_F)/kT}+1}\right] = \frac{N_v}{e^{-(E_v-E_F)/kT}+1}.$$

Now, the number of electrons in the conduction band is

$$N_{ec} = N_c P(E_c) = \frac{N_c}{e^{(E_c - E_F)/kT} + 1},$$

Hence, from $N_{ev} = N_{hc}$, we get

$$\frac{N_v}{e^{-(E_v - E_F)/kT} + 1} = \frac{N_c}{e^{(E_c - E_F)/kT} + 1}.$$

(b) In this case, $e^{(E_c - E_F)/kT} \gg 1$ and $e^{-(E_v - E_F)/kT} \gg 1$. Thus, from the result of part (a),

$$\frac{N_c}{e^{(E_c - E_F)/kT}} \approx \frac{N_v}{e^{-(E_v - E_F)/kT}},$$

or $e^{(E_v - E_c + 2E_F)/kT} \approx N_v / N_c$. We solve for E_F:

$$E_F \approx \frac{1}{2}(E_c + E_v) + \frac{1}{2}kT \ln\left(\frac{N_v}{N_c}\right).$$

31. Sample Problem 41-6 gives the fraction of silicon atoms that must be replaced by phosphorus atoms. We find the number the silicon atoms in 1.0 g, then the number that must be replaced, and finally the mass of the replacement phosphorus atoms. The molar mass of silicon is 28.086 g/mol, so the mass of one silicon atom is

$$(28.086 \text{ g/mol})/(6.022 \times 10^{23} \text{ mol}^{-1}) = 4.66 \times 10^{-23} \text{ g}$$

and the number of atoms in 1.0 g is $(1.0 \text{ g})/(4.66 \times 10^{-23} \text{ g}) = 2.14 \times 10^{22}$. According to Sample Problem 41-6 one of every 5×10^6 silicon atoms is replaced with a phosphorus atom. This means there will be $(2.14 \times 10^{22})/(5 \times 10^6) = 4.29 \times 10^{15}$ phosphorus atoms in 1.0 g of silicon. The molar mass of phosphorus is 30.9758 g/mol so the mass of a phosphorus atom is

$$(30.9758 \text{ g/mol})/(6.022 \times 10^{-23} \text{ mol}^{-1}) = 5.14 \times 10^{-23} \text{ g}.$$

The mass of phosphorus that must be added to 1.0 g of silicon is

$$(4.29 \times 10^{15})(5.14 \times 10^{-23} \text{ g}) = 2.2 \times 10^{-7} \text{ g}.$$

32. (a) n-type, since each phosphorus atom has one more valence electron than a silicon atom.

(b) The added charge carrier density is

$$n_P = 10^{-7}\, n_{Si} = 10^{-7}\, (5 \times 10^{28}\text{ m}^{-3}) = 5 \times 10^{21}\text{ m}^{-3}.$$

(c) The ratio is $(5 \times 10^{21}\text{ m}^{-3})/[2(5 \times 10^{15}\text{ m}^{-3})] = 5 \times 10^5$. Here the factor of 2 in the denominator reflects the contribution to the charge carrier density from *both* the electrons in the conduction band *and* the holes in the valence band.

33. (a) The probability that a state with energy E is occupied is given by

$$P(E) = \frac{1}{e^{(E-E_F)/kT} + 1}$$

where E_F is the Fermi energy, T is the temperature on the Kelvin scale, and k is the Boltzmann constant. If energies are measured from the top of the valence band, then the energy associated with a state at the bottom of the conduction band is $E = 1.11$ eV. Furthermore,

$$kT = (8.62 \times 10^{-5}\text{ eV/K})(300\text{ K}) = 0.02586\text{ eV}.$$

For pure silicon,

$$E_F = 0.555\text{ eV and } (E - E_F)/kT = (0.555\text{ eV})/(0.02586\text{ eV}) = 21.46.$$

Thus,

$$P(E) = \frac{1}{e^{21.46} + 1} = 4.79 \times 10^{-10}.$$

(b) For the doped semiconductor, $(E - E_F)/kT = (0.11\text{ eV})/(0.02586\text{ eV}) = 4.254$ and

$$P(E) = \frac{1}{e^{4.254} + 1} = 1.40 \times 10^{-2}.$$

(c) The energy of the donor state, relative to the top of the valence band, is 1.11 eV $- 0.15$ eV $= 0.96$ eV. The Fermi energy is 1.11 eV $- 0.11$ eV $= 1.00$ eV. Hence,

$$(E - E_F)/kT = (0.96\text{ eV} - 1.00\text{ eV})/(0.02586\text{ eV}) = -1.547$$

and

$$P(E) = \frac{1}{e^{-1.547} + 1} = 0.824.$$

34. (a) Measured from the top of the valence band, the energy of the donor state is $E = 1.11$ eV $- 0.11$ eV $= 1.0$ eV. We solve E_F from Eq. 41-6:

$$E_F = E - kT \ln\left[P^{-1} - 1 \right] = 1.0\text{eV} - \left(8.62 \times 10^{-5}\ \text{eV/K} \right) \left(300\text{K} \right) \ln\left[\left(5.00 \times 10^{-5} \right)^{-1} - 1 \right]$$

$$= 0.744\text{eV}.$$

(b) Now $E = 1.11$ eV, so

$$P(E) = \frac{1}{e^{(E-E_F)/kT} + 1} = \frac{1}{e^{(1.11\text{eV}-0.744\text{eV})/\left[\left(8.62 \times 10^{-5}\ \text{eV/K} \right) \left(300\text{K} \right) \right]} + 1} = 7.13 \times 10^{-7}.$$

35. The energy received by each electron is exactly the difference in energy between the bottom of the conduction band and the top of the valence band (1.1 eV). The number of electrons that can be excited across the gap by a single 662-keV photon is

$$N = (662 \times 10^3\ \text{eV})/(1.1\ \text{eV}) = 6.0 \times 10^5.$$

Since each electron that jumps the gap leaves a hole behind, this is also the number of electron-hole pairs that can be created.

36. (a) The vertical axis in the graph below is the current in nanoamperes:

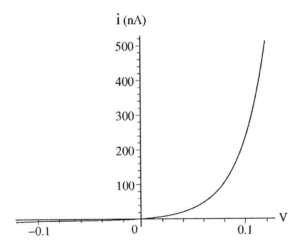

(b) The ratio is

$$\frac{i\big|_{v=+0.50\,\text{V}}}{i\big|_{v=-0.50\,\text{V}}} = \frac{i_0\left[e^{+0.50\text{eV}/\left[\left(8.62 \times 10^{-5}\ \text{eV/K} \right) \left(300\text{K} \right) \right]} - 1 \right]}{i_0\left[e^{-0.50\text{eV}/\left[\left(8.62 \times 10^{-5}\ \text{eV/K} \right) \left(300\text{K} \right) \right]} - 1 \right]} = 2.5 \times 10^8.$$

37. The valence band is essentially filled and the conduction band is essentially empty. If an electron in the valence band is to absorb a photon, the energy it receives must be sufficient to excite it across the band gap. Photons with energies less than the gap width are not absorbed and the semiconductor is transparent to this radiation. Photons with

energies greater than the gap width are absorbed and the semiconductor is opaque to this radiation. Thus, the width of the band gap is the same as the energy of a photon associated with a wavelength of 295 nm. We use the result of Problem 83 of Chapter 38 to obtain

$$E_{gap} = \frac{1240\,\text{eV}\cdot\text{nm}}{\lambda} = \frac{1240\,\text{eV}\cdot\text{nm}}{295\,\text{nm}} = 4.20\,\text{eV}.$$

38. Since (using the result of problem 83 in Chapter 38)

$$E_{photon} = \frac{hc}{\lambda} = \frac{1240\,\text{eV}\cdot\text{nm}}{140\,\text{nm}} = 8.86\,\text{eV} > 7.6\,\text{eV},$$

the light will be absorbed by the KCI crystal. Thus, the crystal is opaque to this light.

39. We denote the maximum dimension (side length) of each transistor as ℓ_{max}, the size of the chip as A, and the number of transistors on the chip as N. Then $A = N\ell_{max}^2$. Therefore,

$$\ell_{max} = \sqrt{\frac{A}{N}} = \sqrt{\frac{(1.0\,\text{in.} \times 0.875\,\text{in.})(2.54\times 10^{-2}\,\text{m/in.})^2}{3.5\times 10^6}} = 1.3\times 10^{-5}\,\text{m} = 13\,\mu\text{m}.$$

40. (a) According to Chapter 25, the capacitance is $C = \kappa\varepsilon_0 A/d$. In our case $\kappa = 4.5$, $A = (0.50\,\mu\text{m})^2$, and $d = 0.20\,\mu\text{m}$, so

$$C = \frac{\kappa\varepsilon_0 A}{d} = \frac{(4.5)(8.85\times 10^{-12}\,\text{F/m})(0.50\,\mu\text{m})^2}{0.20\,\mu\text{m}} = 5.0\times 10^{-17}\,\text{F}.$$

(b) Let the number of elementary charges in question be N. Then, the total amount of charges that appear in the gate is $q = Ne$. Thus, $q = Ne = CV$, which gives

$$N = \frac{CV}{e} = \frac{(5.0\times 10^{-17}\,\text{F})(1.0\,\text{V})}{1.6\times 10^{-19}\,\text{C}} = 3.1\times 10^2.$$

41. (a) Setting $E = E_F$ (see Eq. 41-9), Eq. 41-5 becomes

$$N(E_F) = \frac{8\pi m\sqrt{2m}}{h^3}\left(\frac{3}{16\pi\sqrt{2}}\right)^{1/3}\frac{h}{\sqrt{m}}n^{1/3}.$$

Noting that $16\sqrt{2} = 2^4 2^{1/2} = 2^{9/2}$ so that the cube root of this is $2^{3/2} = 2\sqrt{2}$, we are able to simplify the above expression and obtain

$$N(E_F) = \frac{4m}{h^2}\sqrt[3]{3\pi^2 n}$$

which is equivalent to the result shown in the problem statement. Since the desired numerical answer uses eV units, we multiply numerator and denominator of our result by c^2 and make use of the mc^2 value for an electron in Table 38-3 as well as the hc value found in problem 83 of Chapter 38:

$$N(E_F) = \left(\frac{4mc^2}{(hc)^2}\sqrt[3]{3\pi^2}\right)n^{1/3} = \left(\frac{4(511\times10^3\,\mathrm{eV})}{(1240\,\mathrm{eV\cdot nm})^2}\sqrt[3]{3\pi^2}\right)n^{1/3} = (4.11\,\mathrm{nm}^{-2}\cdot\mathrm{eV}^{-1})n^{1/3}$$

which is equivalent to the value indicated in the problem statement.

(b) Since there are 10^{27} cubic nanometers in a cubic meter, then the result of problem 1 may be written

$$n = 8.49\times10^{28}\,\mathrm{m}^{-3} = 84.9\,\mathrm{nm}^{-3}\ .$$

The cube root of this is $n^{1/3} \approx 4.4$/nm. Hence, the expression in part (a) leads to

$$N(E_F) = (4.11\,\mathrm{nm}^{-2}\cdot\mathrm{eV}^{-1})(4.4\,\mathrm{nm}^{-1}) = 18\,\mathrm{nm}^{-3}\cdot\mathrm{eV}^{-1} = 1.8\times10^{28}\,\mathrm{m}^{-3}\cdot\mathrm{eV}^{-1}.$$

If we multiply this by 10^{27} m^3/nm^3, we see this compares very well with the curve in Fig. 41-5 evaluated at 7.0 eV.

42. If we use the approximate formula discussed in problem 41-24, we obtain

$$frac = \frac{3(8.62\times10^{-5}\,\mathrm{eV/K})(961+273\,\mathrm{K})}{2(5.5\,\mathrm{eV})} \approx 0.03\ .$$

The numerical approach is briefly discussed in part (c) of problem 32. Although the problem does not ask for it here, we remark that numerical integration leads to a fraction closer to 0.02.

43. The description in the problem statement implies that an atom is at the center point C of the regular tetrahedron, since its four *neighbors* are at the four vertices. The side length for the tetrahedron is given as $a = 388$ pm. Since each face is an equilateral triangle, the "altitude" of each of those triangles (which is not to be confused with the altitude of the tetrahedron itself) is $h' = \frac{1}{2}a\sqrt{3}$ (this is generally referred to as the "slant height" in the solid geometry literature). At a certain location along the line segment representing "slant height" of each face is the center C' of the face. Imagine this line segment starting at atom A and ending at the midpoint of one of the sides. Knowing that this line segment bisects the $60°$ angle of the equilateral face, then it is easy to see that C' is a distance

$AC' = a/\sqrt{3}$. If we draw a line from C' all the way to the farthest point on the tetrahedron (this will land on an atom we label B), then this new line is the altitude h of the tetrahedron. Using the Pythagorean theorem,

$$h = \sqrt{a^2 - (AC')^2} = \sqrt{a^2 - \left(\frac{a}{\sqrt{3}}\right)^2} = a\sqrt{\frac{2}{3}}.$$

Now we include coordinates: imagine atom B is on the $+y$ axis at $y_b = h = a\sqrt{2/3}$, and atom A is on the $+x$ axis at $x_a = AC' = a/\sqrt{3}$. Then point C' is the origin. The tetrahedron center point C is on the y axis at some value y_c which we find as follows: C must be equidistant from A and B, so

$$y_b - y_c = \sqrt{x_a^2 + y_c^2}$$

$$a\sqrt{\frac{2}{3}} - y_c = \sqrt{\left(\frac{a}{\sqrt{3}}\right)^2 + y_c^2}$$

which yields $y_c = a/2\sqrt{6}$.

(a) In unit vector notation, using the information found above, we express the vector starting at C and going to A as

$$\vec{r}_{ac} = x_a \hat{i} + (-y_c)\hat{j} = \frac{a}{\sqrt{3}}\hat{i} - \frac{a}{2\sqrt{6}}\hat{j}.$$

Similarly, the vector starting at C and going to B is

$$\vec{r}_{bc} = (y_b - y_c)\hat{j} = \tfrac{a}{2}\sqrt{3/2}\hat{j}.$$

Therefore, using Eq. 3-20,

$$\theta = \cos^{-1}\left(\frac{\vec{r}_{ac} \cdot \vec{r}_{bc}}{|\vec{r}_{ac}||\vec{r}_{bc}|}\right) = \cos^{-1}\left(-\frac{1}{3}\right)$$

which yields $\theta = 109.5°$ for the angle between adjacent bonds.

(b) The length of vector $\vec{r}_{bc}$ (which is, of course, the same as the length of $\vec{r}_{ac}$) is

$$|\vec{r}_{bc}| = \frac{a}{2}\sqrt{\frac{3}{2}} = \frac{388\,\text{pm}}{2}\sqrt{\frac{3}{2}} = 237.6 \text{ pm} \approx 238 \text{ pm}.$$

We note that in the solid geometry literature, the distance $\frac{a}{2}\sqrt{\frac{3}{2}}$ is known as the circumradius of the regular tetrahedron.

44. According to Eq. 41-6,

$$P(E_F + \Delta E) = \frac{1}{e^{(E_F + \Delta E - E_F)/kT} + 1} = \frac{1}{e^{\Delta E/kT} + 1} = \frac{1}{e^x + 1}$$

where $x = \Delta E / kT$. Also,

$$P(E_F + \Delta E) = \frac{1}{e^{(E_F - \Delta E - E_F)/kT} + 1} = \frac{1}{e^{-\Delta E/kT} + 1} = \frac{1}{e^{-x} + 1}.$$

Thus,

$$P(E_F + \Delta E) + P(E_F - \Delta E) = \frac{1}{e^x + 1} + \frac{1}{e^{-x} + 1} = \frac{e^x + 1 + e^{-x} + 1}{(e^{-x} + 1)(e^x + 1)} = 1.$$

A special case of this general result can be found in problem 13, where $\Delta E = 63$ meV and

$$P(E_F + 63 \text{ meV}) + P(E_F - 63 \text{ meV}) = 0.090 + 0.91 = 1.0.$$

45. (a) The derivative of $P(E)$ is

$$\left(\frac{-1}{\left(e^{(E-E_F)/kT} + 1\right)^2}\right) \frac{d}{dE} e^{(E-E_F)/kT} = \left(\frac{-1}{\left(e^{(E-E_F)/kT} + 1\right)^2}\right) \frac{1}{kT} e^{(E-E_F)/kT}.$$

Evaluating this at $E = E_F$ we readily obtain the desired result.

(b) The equation of a line may be written $y = m(x - x_o)$ where m is the slope (here: equal to $-1/kT$, from part (a)) and x_o is the x-intercept (which is what we are asked to solve for). It is clear that $P(E_F) = 2$, so our equation of the line, evaluated at $x = E_F$, becomes

$$2 = (-1/kT)(E_F - x_o),$$

which leads to $x_o = E_F + 2kT$.

46. (a) For copper, Eq. 41-10 leads to

$$\frac{d\rho}{dT} = [\rho\alpha]_{Cu} = (2 \times 10^{-8}\ \Omega \cdot m)(4 \times 10^{-3} K^{-1}) = 8 \times 10^{-11}\ \Omega \cdot m / K.$$

(b) For silicon,

$$\frac{d\rho}{dT} = [\rho\alpha]_{\text{Si}} = (3\times10^3\,\Omega\cdot\text{m})(-70\times10^{-3}\,\text{K}^{-1}) = -2.1\times10^2\,\Omega\cdot\text{m}/\text{K}.$$

47. We use the ideal gas law in the form of Eq. 20-9:

$$p = nkT = (8.43\times10^{28}\,\text{m}^{-3})(1.38\times10^{-23}\,\text{J/K})(300\text{ K}) = 3.49\times10^8\,\text{Pa} = 3.49\times10^3\text{ atm}.$$

48. We equate E_F with $\frac{1}{2}m_e v_F^2$ and write our expressions in such a way that we can make use of the electron mc^2 value found in Table 37-3:

$$v_F = \sqrt{\frac{2E_F}{m}} = c\sqrt{\frac{2E_F}{mc^2}} = (3.0\times10^5\text{ km/s})\sqrt{\frac{2(7.0\,\text{eV})}{5.11\times10^5\,\text{eV}}} = 1.6\times10^3\text{ km/s}.$$

49. We compute $\left(\frac{3}{16\sqrt{2\pi}}\right)^{2/3} \approx 0.121.$

Chapter 42

1. Our calculation is similar to that shown in Sample Problem 42-1. We set $K = 5.30 \, \text{MeV} = U = (1/4\pi\varepsilon_0)(q_\alpha q_{Cu}/r_{min})$ and solve for the closest separation, r_{min}:

$$r_{min} = \frac{q_\alpha q_{Cu}}{4\pi\varepsilon_0 K} = \frac{kq_\alpha q_{Cu}}{4\pi\varepsilon_0 K} = \frac{(2e)(29)(1.60\times10^{-19}\,\text{C})(8.99\times10^9\,\text{V}\cdot\text{m/C})}{5.30\times10^6\,\text{eV}}$$

$$= 1.58\times10^{-14}\,\text{m} = 15.8\,\text{fm}.$$

We note that the factor of e in $q_\alpha = 2e$ was not set equal to 1.60×10^{-19} C, but was instead allowed to cancel the "e" in the non-SI energy unit, electronvolt.

2. Kinetic energy (we use the classical formula since v is much less than c) is converted into potential energy (see Eq. 24-43). From Appendix F or G, we find $Z = 3$ for Lithium and $Z = 90$ for Thorium; the charges on those nuclei are therefore $3e$ and $90e$, respectively. We manipulate the terms so that one of the factors of e cancels the "e" in the kinetic energy unit MeV, and the other factor of e is set equal to its SI value 1.6×10^{-19} C. We note that $k = 1/4\pi\varepsilon_0$ can be written as 8.99×10^9 V·m/C. Thus, from energy conservation, we have

$$K = U \Rightarrow r = \frac{kq_1q_2}{K} = \frac{(8.99\times10^9\,\frac{\text{V}\cdot\text{m}}{\text{C}})(3\times1.6\times10^{-19}\,\text{C})(90e)}{3.00\times10^6\,\text{eV}}$$

which yields $r = 1.3 \times 10^{-13}$ m (or about 130 fm).

3. The conservation laws of (classical kinetic) energy and (linear) momentum determine the outcome of the collision (see Chapter 9). The final speed of the α particle is

$$v_{\alpha f} = \frac{m_\alpha - m_{Au}}{m_\alpha + m_{Au}} v_{\alpha i},$$

and that of the recoiling gold nucleus is

$$v_{Au,f} = \frac{2m_\alpha}{m_\alpha + m_{Au}} v_{\alpha i}.$$

(a) Therefore, the kinetic energy of the recoiling nucleus is

$$K_{Au,f} = \frac{1}{2} m_{Au} v_{Au,f}^2 = \frac{1}{2} m_{Au} \left(\frac{2m_\alpha}{m_\alpha + m_{Au}} \right)^2 v_{\alpha i}^2 = K_{\alpha i} \frac{4m_{Au} m_\alpha}{\left(m_\alpha + m_{Au} \right)^2}$$

$$= \left(5.00 \text{MeV} \right) \frac{4(197 \text{u})(4.00 \text{u})}{\left(4.00 \text{u} + 197 \text{u} \right)^2}$$

$$= 0.390 \text{MeV}.$$

(b) The final kinetic energy of the alpha particle is

$$K_{\alpha f} = \frac{1}{2} m_\alpha v_{\alpha f}^2 = \frac{1}{2} m_\alpha \left(\frac{m_\alpha - m_{Au}}{m_\alpha + m_{Au}} \right)^2 v_{\alpha i}^2 = K_{\alpha i} \left(\frac{m_\alpha - m_{Au}}{m_\alpha + m_{Au}} \right)^2$$

$$= \left(5.00 \text{MeV} \right) \left(\frac{4.00 \text{u} - 197 \text{u}}{4.00 \text{u} + 197 \text{u}} \right)^2$$

$$= 4.61 \text{MeV}.$$

We note that $K_{\alpha f} + K_{Au,f} = K_{\alpha i}$ is indeed satisfied.

4. (a) 6 protons, since $Z = 6$ for carbon (see Appendix F).

(b) 8 neutrons, since $A - Z = 14 - 6 = 8$ (see Eq. 42-1).

5. (a) Table 42-1 gives the atomic mass of ^{1}H as $m = 1.007825$ u. Therefore, the *mass excess* for ^{1}H is $\Delta = (1.007825 \text{ u} - 1.000000 \text{ u}) = 0.007825$ u.

(b) In the unit MeV/c^2, $\Delta = (1.007825 \text{ u} - 1.000000 \text{ u})(931.5 \text{ MeV}/c^2 \cdot \text{u}) = +7.290 \text{ MeV}/c^2$.

(c) The mass of the neutron is given in Sample Problem 42-3. Thus, for the neutron,

$$\Delta = (1.008665 \text{ u} - 1.000000 \text{ u}) = 0.008665 \text{ u}.$$

(d) In the unit MeV/c^2, $\Delta = (1.008665 \text{ u} - 1.000000 \text{ u})(931.5 \text{ MeV}/ c^2 \cdot \text{u}) = +8.071$ MeV/c^2.

(e) Appealing again to Table 42-1, we obtain, for ^{120}Sn,

$$\Delta = (119.902199 \text{ u} - 120.000000 \text{ u}) = -0.09780 \text{ u}.$$

(f) In the unit MeV/c^2,

$$\Delta = (119.902199 \text{ u} - 120.000000 \text{ u})(931.5 \text{ MeV}/ c^2 \cdot \text{u}) = -91.10 \text{ MeV}/c^2.$$

6. (a) The atomic number $Z = 39$ corresponds to the element Yttrium (see Appendix F and/or Appendix G).

(b) The atomic number $Z = 53$ corresponds to Iodine.

(c) A detailed listing of stable nuclides (such as the website http://nucleardata. nuclear.lu.se/nucleardata) shows that the stable isotope of Yttrium has 50 neutrons (this can also be inferred from the Molar Mass values listed in Appendix F).

(d) Similarly, the stable isotope of Iodine has 74 neutrons

(e) The number of neutrons left over is $235 - 127 - 89 = 19$.

7. We note that the mean density and mean radius for the Sun are given in Appendix C. Since $\rho = M/V$ where $V \propto r^3$, we get $r \propto \rho^{-1/3}$. Thus, the new radius would be

$$r = R_s \left(\frac{\rho_s}{\rho} \right)^{1/3} = (6.96 \times 10^8 \text{ m}) \left(\frac{1410 \text{ kg} / \text{m}^3}{2 \times 10^{17} \text{ kg} / \text{m}^3} \right)^{1/3} = 1.3 \times 10^4 \text{ m.}$$

8. (a) Since $U > 0$, the energy represents a tendency for the sphere to blow apart.

(b) For ^{239}Pu, $Q = 94e$ and $R = 6.64$ fm. Including a conversion factor for $J \rightarrow eV$ we obtain

$$U = \frac{3Q^2}{20\pi\varepsilon_0 r} = \frac{3\left[94(1.60 \times 10^{-19} \text{ C})\right]^2 (8.99 \times 10^9 \text{ N} \cdot \text{m}^2 / \text{C}^2)}{5(6.64 \times 10^{-15} \text{ m})} \left(\frac{1 \text{ eV}}{1.60 \times 10^{-19} \text{ J}} \right)$$

$$= 1.15 \times 10^9 \text{ eV} = 1.15 \text{ GeV.}$$

(c) Since $Z = 94$, the electrostatic potential per proton is 1.15 GeV/94 = 12.2 MeV/proton.

(d) Since $A = 239$, the electrostatic potential per nucleon is 1.15 GeV/239 = 4.81 MeV/nucleon.

(e) The strong force that binds the nucleus is very strong.

9. (a) For ^{55}Mn the mass density is

$$\rho_m = \frac{M}{V} = \frac{0.055 \text{kg/mol}}{(4\pi/3)\left[(1.2 \times 10^{-15} \text{ m})(55)^{1/3} \right]^3 (6.02 \times 10^{23} / \text{mol})} = 2.3 \times 10^{17} \text{ kg/m}^3.$$

(b) For ^{209}Bi,

$$\rho_m = \frac{M}{V} = \frac{0.209 \, \text{kg/mol}}{(4\pi/3)\left[\left(1.2 \times 10^{-15} \, \text{m}\right)(209)^{1/3}\right]^3 \left(6.02 \times 10^{23} \, / \, \text{mol}\right)} = 2.3 \times 10^{17} \, \text{kg/m}^3.$$

(c) Since $V \propto r^3 = \left(r_0 A^{1/3}\right)^3 \propto A$, we expect $\rho_m \propto A / V \propto A / A \approx$ const. for all nuclides.

(d) For ^{55}Mn the charge density is

$$\rho_q = \frac{Ze}{V} = \frac{(25)\left(1.6 \times 10^{-19} \, \text{C}\right)}{(4\pi/3)\left[\left(1.2 \times 10^{-15} \, \text{m}\right)(55)^{1/3}\right]^3} = 1.0 \times 10^{25} \, \text{C/m}^3.$$

(e) For ^{209}Bi

$$\rho_q = \frac{Ze}{V} = \frac{(83)\left(1.6 \times 10^{-19} \, \text{C}\right)}{(4\pi/3)\left[\left(1.2 \times 10^{-15} \, \text{m}\right)(209)^{1/3}\right]^3} = 8.8 \times 10^{24} \, \text{C/m}^3.$$

Note that $\rho_q \propto Z / V \propto Z / A$ should gradually decrease since $A > 2Z$ for large nuclides.

10. (a) The mass number A is the number of nucleons in an atomic nucleus. Since $m_p \approx m_n$ the mass of the nucleus is approximately Am_p. Also, the mass of the electrons is negligible since it is much less than that of the nucleus. So $M \approx Am_p$.

(b) For ^{1}H, the approximate formula gives

$$M \approx Am_p = (1)(1.007276 \, \text{u}) = 1.007276 \, \text{u}.$$

The actual mass is (see Table 42-1) 1.007825 u. The percentage deviation committed is then

$$\delta = (1.007825 \, \text{u} - 1.007276 \, \text{u})/1.007825 \, \text{u} = 0.054\% \approx 0.05\%.$$

(c) Similarly, for ^{31}P, $\delta = 0.81\%$.

(d) For ^{120}Sn, $\delta = 0.81\%$.

(e) For ^{197}Au, $\delta = 0.74\%$.

(f) For ^{239}Pu, $\delta = 0.71\%$.

(g) No. In a typical nucleus the binding energy per nucleon is several MeV, which is a bit less than 1% of the nucleon mass times c^2. This is comparable with the percent error calculated in parts (b) – (f) , so we need to use a more accurate method to calculate the nuclear mass.

11. (a) The de Broglie wavelength is given by $\lambda = h/p$, where p is the magnitude of the momentum. The kinetic energy K and momentum are related by Eq. 38-51, which yields

$$pc = \sqrt{K^2 + 2Kmc^2} = \sqrt{(200\,\text{MeV})^2 + 2(200\,\text{MeV})(0.511\,\text{MeV})} = 200.5\,\text{MeV}.$$

Thus,

$$\lambda = \frac{hc}{pc} = \frac{1240\,\text{eV} \cdot \text{nm}}{200.5 \times 10^6\,\text{eV}} = 6.18 \times 10^{-6}\,\text{nm} \approx 6.2\ \text{fm}.$$

(b) The diameter of a copper nucleus, for example, is about 8.6 fm, just a little larger than the de Broglie wavelength of a 200-MeV electron. To resolve detail, the wavelength should be smaller than the target, ideally a tenth of the diameter or less. 200-MeV electrons are perhaps at the lower limit in energy for useful probes.

12. From Appendix F and/or G, we find $Z = 107$ for Bohrium, so this isotope has $N = A - Z = 262 - 107 = 155$ neutrons. Thus,

$$\Delta E_{\text{ben}} = \frac{\left(Zm_{\text{H}} + Nm_n - m_{\text{Bh}}\right)c^2}{A}$$

$$= \frac{\left((107)(1.007825\,\text{u}) + (155)(1.008665\,\text{u}) - 262.1231\,\text{u}\right)(931.5\,\text{MeV/u})}{262}$$

which yields 7.31 MeV per nucleon.

13. Let f_{24} be the abundance of ^{24}Mg, let f_{25} be the abundance of ^{25}Mg, and let f_{26} be the abundance of ^{26}Mg. Then, the entry in the periodic table for Mg is

$$24.312 = 23.98504 f_{24} + 24.98584 f_{25} + 25.98259 f_{26}.$$

Since there are only three isotopes, $f_{24} + f_{25} + f_{26} = 1$. We solve for f_{25} and f_{26}. The second equation gives $f_{26} = 1 - f_{24} - f_{25}$. We substitute this expression and $f_{24} = 0.7899$ into the first equation to obtain

$24.312 = (23.98504)(0.7899) + 24.98584 f_{25} + 25.98259 - (25.98259)(0.7899) - 25.98259 f_{25}.$

The solution is $f_{25} = 0.09303$. Then,

$$f_{26} = 1 - 0.7899 - 0.09303 = 0.1171.\ 78.99\%$$

of naturally occurring magnesium is ^{24}Mg.

(a) Thus, 9.303% is ^{25}Mg.

(b) 11.71% is ^{26}Mg.

14. (a) The first step is to add energy to produce $^4\text{He} \rightarrow p + ^3\text{H}$, which — to make the electrons "balance" — may be rewritten as $^4\text{He} \rightarrow ^1\text{H} + ^3\text{H}$. The energy needed is

$$\Delta E_1 = \left(m_{^3\text{H}} + m_{^1\text{H}} - m_{^4\text{He}}\right)c^2 = \left(3.01605\,\text{u} + 1.00783\,\text{u} - 4.00260\,\text{u}\right)\left(931.5\,\text{MeV/u}\right)$$
$$= 19.8\,\text{MeV}.$$

(b) The second step is to add energy to produce $^3\text{H} \rightarrow n + ^2\text{H}$. The energy needed is

$$\Delta E_2 = \left(m_{^2\text{H}} + m_n - m_{^3\text{H}}\right)c^2 = \left(2.01410\,\text{u} + 1.00867\,\text{u} - 3.01605\,\text{u}\right)\left(931.5\,\text{MeV/u}\right)$$
$$= 6.26\,\text{MeV}.$$

(c) The third step: $^2\text{H} \rightarrow p + n$, which — to make the electrons "balance" — may be rewritten as $^2\text{H} \rightarrow ^1\text{H} + n$. The work required is

$$\Delta E_3 = \left(m_{^1\text{H}} + m_n - m_{^2\text{H}}\right)c^2 = \left(1.00783\,\text{u} + 1.00867\,\text{u} - 2.01410\,\text{u}\right)\left(931.5\,\text{MeV/u}\right)$$
$$= 2.23\,\text{MeV}.$$

(d) The total binding energy is

$$\Delta E_{\text{be}} = \Delta E_1 + \Delta E_2 + \Delta E_3 = 19.8\,\text{MeV} + 6.26\,\text{MeV} + 2.23\,\text{MeV} = 28.3\,\text{MeV}.$$

(e) The binding energy per nucleon is $\Delta E_{\text{ben}} = \Delta E_{\text{be}} / A = 28.3\,\text{MeV} / 4 = 7.07\,\text{MeV}$.

(f) No, the answers do not match.

15. The binding energy is given by $\Delta E_{\text{be}} = \left[Zm_H + (A - Z)m_n - M_{\text{Pu}}\right]c^2$, where Z is the atomic number (number of protons), A is the mass number (number of nucleons), m_H is the mass of a hydrogen atom, m_n is the mass of a neutron, and M_{Pu} is the mass of a $^{239}_{94}\text{Pu}$ atom. In principle, nuclear masses should be used, but the mass of the Z electrons included in ZM_H is canceled by the mass of the Z electrons included in M_{Pu}, so the result is the same. First, we calculate the mass difference in atomic mass units:

$$\Delta m = (94)(1.00783\,\text{u}) + (239 - 94)(1.00867\,\text{u}) - (239.05216\,\text{u}) = 1.94101\,\text{u}.$$

Since 1 u is equivalent to 931.5 MeV,

$$\Delta E_{\text{be}} = (1.94101\,\text{u})(931.5\,\text{MeV/u}) = 1808\,\text{MeV}.$$

Since there are 239 nucleons, the binding energy per nucleon is

$$\Delta E_{ben} = E/A = (1808 \text{ MeV})/239 = 7.56 \text{ MeV}.$$

16. We first "separate" all the nucleons in one copper nucleus (which amounts to simply calculating the nuclear binding energy) and then figure the number of nuclei in the penny (so that we can multiply the two numbers and obtain the result). To begin, we note that (using Eq. 42-1 with Appendix F and/or G) the copper-63 nucleus has 29 protons and 34 neutrons. We use the more accurate values given in Sample Problem 42-3:

$$\Delta E_{be} = (29(1.007825 \text{ u}) + 34(1.008665 \text{ u}) - 62.92960 \text{ u})(931.5 \text{ MeV} / \text{u}) = 551.4 \text{ MeV}.$$

To figure the number of nuclei (or, equivalently, the number of atoms), we adapt Eq. 42-21:

$$N_{Cu} = \left(\frac{3.0 \text{ g}}{62.92960 \text{ g} / \text{mol}} \right)(6.02 \times 10^{23} \text{ atoms} / \text{mol}) \approx 2.9 \times 10^{22} \text{ atoms}.$$

Therefore, the total energy needed is

$$N_{Cu} \Delta E_{be} = (551.4 \text{ MeV})(2.9 \times 10^{22}) = 1.6 \times 10^{25} \text{ MeV}.$$

17. (a) Since the nuclear force has a short range, any nucleon interacts only with its nearest neighbors, not with more distant nucleons in the nucleus. Let N be the number of neighbors that interact with any nucleon. It is independent of the number A of nucleons in the nucleus. The number of interactions in a nucleus is approximately $N A$, so the energy associated with the strong nuclear force is proportional to $N A$ and, therefore, proportional to A itself.

(b) Each proton in a nucleus interacts electrically with every other proton. The number of pairs of protons is $Z(Z- 1)/2$, where Z is the number of protons. The Coulomb energy is, therefore, proportional to $Z(Z- 1)$.

(c) As A increases, Z increases at a slightly slower rate but Z^2 increases at a faster rate than A and the energy associated with Coulomb interactions increases faster than the energy associated with strong nuclear interactions.

18. It should be noted that when the problem statement says the "masses of the proton and the deuteron are …" they are actually referring to the corresponding atomic masses (given to very high precision). That is, the given masses include the "orbital" electrons. As in many computations in this chapter, this circumstance (of implicitly including electron masses in what should be a purely nuclear calculation) does not cause extra difficulty in the calculation (see remarks in Sample Problems 42-4, 42-6, and 42-7). Setting the gamma ray energy equal to ΔE_{be}, we solve for the neutron mass (with each term understood to be in u units):

$$m_{\mathrm{n}} = M_{\mathrm{d}} - m_{\mathrm{H}} + \frac{E_{\gamma}}{c^2}$$

$$= 2.013553212 - 1.007276467 + \frac{2.2233}{931.502}$$

$$= 1.0062769 + 0.0023868$$

which yields $m_{\mathrm{n}} = 1.0086637$ u ≈ 1.0087 u.

19. If a nucleus contains Z protons and N neutrons, its binding energy is $\Delta E_{\mathrm{be}} = (Zm_H + Nm_n - m)c^2$, where m_H is the mass of a hydrogen atom, m_n is the mass of a neutron, and m is the mass of the atom containing the nucleus of interest. If the masses are given in atomic mass units, then mass excesses are defined by $\Delta_H = (m_H - 1)c^2$, $\Delta_n = (m_n - 1)c^2$, and $\Delta = (m - A)c^2$. This means $m_H c^2 = \Delta_H + c^2$, $m_n c^2 = \Delta_n + c^2$, and $mc^2 = \Delta + Ac^2$. Thus,

$$E = (Z\Delta_H + N\Delta_n - \Delta) + (Z + N - A)c^2 = Z\Delta_H + N\Delta_n - \Delta,$$

where $A = Z + N$ is used. For $^{197}_{79}\mathrm{Au}$, $Z = 79$ and $N = 197 - 79 = 118$. Hence,

$$\Delta E_{\mathrm{be}} = (79)(7.29\,\mathrm{MeV}) + (118)(8.07\,\mathrm{MeV}) - (-31.2\,\mathrm{MeV}) = 1560\,\mathrm{MeV}.$$

This means the binding energy per nucleon is $\Delta E_{\mathrm{ben}} = (1560\,\mathrm{MeV}) / 197 = 7.92\,\mathrm{MeV}$.

20. Using Eq. 42-15 with Eq. 42-18, we find the fraction remaining:

$$\frac{N}{N_0} = e^{-t\ln 2/T_{1/2}} = e^{-30\ln 2/29} = 0.49.$$

21. (a) Since $60\,\mathrm{y} = 2(30\,\mathrm{y}) = 2T_{1/2}$, the fraction left is $2^{-2} = 1/4 = 0.250$.

(b) Since $90\,\mathrm{y} = 3(30\,\mathrm{y}) = 3T_{1/2}$, the fraction that remains is $2^{-3} = 1/8 = 0.125$.

22. By the definition of half-life, the same has reduced to $\frac{1}{2}$ its initial amount after 140 d. Thus, reducing it to $\frac{1}{4} = \left(\frac{1}{2}\right)^2$ of its initial number requires that two half-lives have passed: $t = 2T_{1/2} = 280$ d.

23. (a) The decay rate is given by $R = \lambda N$, where λ is the disintegration constant and N is the number of undecayed nuclei. Initially, $R = R_0 = \lambda N_0$, where N_0 is the number of undecayed nuclei at that time. One must find values for both N_0 and λ. The disintegration constant is related to the half-life $T_{1/2}$ by

$$\lambda = (\ln 2)/T_{1/2} = (\ln 2)/(78\text{h}) = 8.89 \times 10^{-3} \, \text{h}^{-1}.$$

If M is the mass of the sample and m is the mass of a single atom of gallium, then $N_0 = M/m$. Now, $m = (67 \text{ u})(1.661 \times 10^{-24} \text{ g/u}) = 1.113 \times 10^{-22}$ g and

$$N_0 = (3.4 \text{ g})/(1.113 \times 10^{-22} \text{ g}) = 3.05 \times 10^{22}.$$

Thus

$$R_0 = (8.89 \times 10^{-3} \text{ h}^{-1}) (3.05 \times 10^{22}) = 2.71 \times 10^{20} \text{ h}^{-1} = 7.53 \times 10^{16} \text{ s}^{-1}.$$

(b) The decay rate at any time t is given by

$$R = R_0 \, e^{-\lambda t}$$

where R_0 is the decay rate at $t = 0$. At $t = 48$ h, $\lambda t = (8.89 \times 10^{-3} \text{ h}^{-1}) (48 \text{ h}) = 0.427$ and

$$R = (7.53 \times 10^{16} \text{ s}^{-1}) e^{-0.427} = 4.91 \times 10^{16} \text{ s}^{-1}.$$

24. We note that $t = 24$ h is four times $T_{1/2} = 6.5$ h. Thus, it has reduced by half, four-fold:

$$\left(\frac{1}{2} \right)^4 (48 \times 10^{19}) = 3.0 \times 10^{19}.$$

25. (a) The half-life $T_{1/2}$ and the disintegration constant are related by $T_{1/2} = (\ln 2)/\lambda$, so $T_{1/2} = (\ln 2)/(0.0108 \text{ h}^{-1}) = 64.2$ h.

(b) At time t, the number of undecayed nuclei remaining is given by

$$N = N_0 \, e^{-\lambda t} = N_0 \, e^{-(\ln 2)t/T_{1/2}}.$$

We substitute $t = 3T_{1/2}$ to obtain

$$\frac{N}{N_0} = e^{-3 \ln 2} = 0.125.$$

In each half-life, the number of undecayed nuclei is reduced by half. At the end of one half-life, $N = N_0/2$, at the end of two half-lives, $N = N_0/4$, and at the end of three half-lives, $N = N_0/8 = 0.125 N_0$.

(c) We use

$$N = N_0 \, e^{-\lambda t}.$$

10.0 d is 240 h, so $\lambda t = (0.0108\ \text{h}^{-1})(240\ \text{h}) = 2.592$ and

$$\frac{N}{N_0} = e^{-2.592} = 0.0749.$$

26. (a) We adapt Eq. 42-21:

$$N_{\text{Pu}} = \left(\frac{0.002\,\text{g}}{239\,\text{g/mol}}\right)(6.02 \times 10^{23}\ \text{nuclei/mol}) \approx 5.04 \times 10^{18}\ \text{nuclei}.$$

(b) Eq. 42-20 leads to

$$R = \frac{N \ln 2}{T_{1/2}} = \frac{5 \times 10^{18}\ \ln 2}{2.41 \times 10^4\ \text{y}} = 1.4 \times 10^{14}\ /\,\text{y}$$

which is equivalent to $4.60 \times 10^6/\text{s} = 4.60 \times 10^6$ Bq (the unit becquerel is defined in §42-3).

27. The rate of decay is given by $R = \lambda N$, where λ is the disintegration constant and N is the number of undecayed nuclei. In terms of the half-life $T_{1/2}$, the disintegration constant is $\lambda = (\ln 2)/T_{1/2}$, so

$$N = \frac{R}{\lambda} = \frac{R T_{1/2}}{\ln 2} = \frac{(6000\ \text{Ci})(3.7 \times 10^{10}\ \text{s}^{-1}/\ \text{Ci})(5.27\,\text{y})(3.16 \times 10^7\ \text{s}/\text{y})}{\ln 2}$$

$$= 5.33 \times 10^{22}\ \text{nuclei}.$$

28. We note that 3.82 days is 330048 s, and that a becquerel is a disintegration per second (see §42-3). From Eq. 34-19, we have

$$\frac{N}{V} = \frac{R}{V}\frac{T_{1/2}}{\ln 2} = \left(1.55 \times 10^5\ \frac{\text{Bq}}{\text{m}^3}\right)\frac{330048\,\text{s}}{\ln 2} = 7.4 \times 10^{10}\ \frac{\text{atoms}}{\text{m}^3}$$

where we have divided by volume v. We estimate v (the volume breathed in 48 h = 2880 min) as follows:

$$\left(2\frac{\text{Liters}}{\text{breath}}\right)\left(\frac{1\,\text{m}^3}{1000\,\text{L}}\right)\left(40\frac{\text{breaths}}{\text{min}}\right)(2880\,\text{min})$$

which yields $v \approx 200\ \text{m}^3$. Thus, the order of magnitude of N is

$$\left(\frac{N}{V}\right)(V) \approx \left(7 \times 10^{10}\ \frac{\text{atoms}}{\text{m}^3}\right)(200\,\text{m}^3) \approx 1 \times 10^{13}\ \text{atoms}.$$

29. Using Eq. 42-16 with Eq. 42-18, we find the initial activity:

$$R_0 = Re^{t \ln 2 / T_{1/2}} = \left(7.4 \times 10^8 \text{ Bq}\right) e^{24 \ln 2 / 83.61} = 9.0 \times 10^8 \text{ Bq}.$$

30. (a) Eq. 42-20 leads to

$$R = \frac{\ln 2}{T_{1/2}} N = \frac{\ln 2}{30.2 \text{y}} \left(\frac{M_{sam}}{m_{atom}} \right) = \frac{\ln 2}{9.53 \times 10^8 \text{ s}} \left(\frac{0.0010 \text{kg}}{137 \times 1.661 \times 10^{-27} \text{ kg}} \right)$$
$$= 3.2 \times 10^{12} \text{ Bq}.$$

(b) Using the conversion factor $1 \text{ Ci} = 3.7 \times 10^{10}$ Bq, $R = 3.2 \times 10^{12}$ Bq.=86Ci.

31. (a) We assume that the chlorine in the sample had the naturally occurring isotopic mixture, so the average mass number was 35.453, as given in Appendix F. Then, the mass of ^{226}Ra was

$$m = \frac{226}{226 + 2(35.453)} (0.10 \text{ g}) = 76.1 \times 10^{-3} \text{ g}.$$

The mass of a ^{226}Ra nucleus is $(226 \text{ u})(1.661 \times 10^{-24} \text{ g/u}) = 3.75 \times 10^{-22}$ g, so the number of ^{226}Ra nuclei present was

$$N = (76.1 \times 10^{-3} \text{ g})/(3.75 \times 10^{-22} \text{ g}) = 2.03 \times 10^{20}.$$

(b) The decay rate is given by $R = N\lambda = (N \ln 2)/T_{1/2}$, where λ is the disintegration constant, $T_{1/2}$ is the half-life, and N is the number of nuclei. The relationship $\lambda = (\ln 2)/T_{1/2}$ is used. Thus,

$$R = \frac{\left(2.03 \times 10^{20}\right) \ln 2}{(1600 \text{ y})(3.156 \times 10^7 \text{ s/y})} = 2.79 \times 10^9 \text{ s}^{-1}.$$

32. (a) Molybdenum beta decays into Technetium:

$$^{99}_{42}\text{Mo} \rightarrow ^{99}_{43}\text{Tc} + e^- + \nu$$

(b) Each decay corresponds to a photon produced when the Technetium nucleus de-excites [note that the de-excitation half-life is much less than the beta decay half-life]. Thus, the gamma rate is the same as the decay rate: 8.2×10^7/s.

(c) Eq. 42-20 leads to

$$N = \frac{RT_{1/2}}{\ln 2} = \frac{(38/s)(6.0\,h)(3600\,s/h)}{\ln 2} = 1.2 \times 10^6.$$

33. Using Eq. 42-15 and Eq. 42-18 (and the fact that mass is proportional to the number of atoms), the amount decayed is

$$|\Delta m| = m\Big|_{t_f = 16.0\,h} - m\Big|_{t_f = 14.0\,h} = m_0\left(1 - e^{-t_i \ln 2/T_{1/2}}\right) - m_0\left(1 - e^{-t_f \ln 2/T_{1/2}}\right)$$

$$= m_0\left(e^{-t_f \ln 2/T_{1/2}} - e^{-t_i \ln 2/T_{1/2}}\right) = (5.50\,g)\left[e^{-(16.0\,h/12.7\,h)\ln 2} - e^{-(14.0\,h/12.7\,h)\ln 2}\right]$$

$$= 0.265\,g.$$

34. We label the two isotopes with subscripts 1 (for ^{32}P) and 2 (for ^{33}P). Initially, 10% of the decays come from ^{33}P, which implies that the initial rate $R_{02} = 9R_{01}$. Using Eq. 42-17, this means

$$R_{01} = \lambda_1 N_{01} = \frac{1}{9}R_{02} = \frac{1}{9}\lambda_2 N_{02}.$$

At time t, we have $R_1 = R_{01}e^{-\lambda_1 t}$ and $R_2 = R_{02}e^{-\lambda_2 t}$. We seek the value of t for which $R_1 = 9R_2$ (which means 90% of the decays arise from ^{33}P). We divide equations to obtain $\left(R_{01}/R_{02}\right)e^{-(\lambda_1-\lambda_2)t} = 9$, and solve for t:

$$t = \frac{1}{\lambda_1 - \lambda_2}\ln\left(\frac{R_{01}}{9R_{02}}\right) = \frac{\ln\left(R_{01}/9R_{02}\right)}{\ln 2/T_{1/2_1} - \ln 2/T_{1/2_2}} = \frac{\ln\left[(1/9)^2\right]}{\ln 2\left[(14.3\,d)^{-1} - (25.3\,d)^{-1}\right]}$$

$$= 209\,d.$$

35. The number N of undecayed nuclei present at any time and the rate of decay R at that time are related by $R = \lambda N$, where λ is the disintegration constant. The disintegration constant is related to the half-life $T_{1/2}$ by $\lambda = (\ln 2)/T_{1/2}$, so $R = (N \ln 2)/T_{1/2}$ and $T_{1/2} = (N \ln 2)/R$. Since 15.0% by mass of the sample is ^{147}Sm, the number of ^{147}Sm nuclei present in the sample is

$$N = \frac{(0.150)(1.00\,g)}{(147\,u)(1.661 \times 10^{-24}\,g/u)} = 6.143 \times 10^{20}.$$

Thus

$$T_{1/2} = \frac{(6.143 \times 10^{20})\ln 2}{120\,s^{-1}} = 3.55 \times 10^{18}\,s = 1.12 \times 10^{11}\,y.$$

36. We have one alpha particle (helium nucleus) produced for every plutonium nucleus that decays. To find the number that have decayed, we use Eq. 42-15, Eq. 42-18, and adapt Eq. 42-21:

$$N_0 - N = N_0\left(1 - e^{-t\ln 2/T_{1/2}}\right) = N_A \frac{12.0\,\text{g}/\text{mol}}{239\,\text{g}/\text{mol}}\left(1 - e^{-20000\ln 2/24100}\right)$$

where N_A is the Avogadro constant. This yields 1.32×10^{22} alpha particles produced. In terms of the amount of helium gas produced (assuming the α particles slow down and capture the appropriate number of electrons), this corresponds to

$$m_{\text{He}} = \left(\frac{1.32 \times 10^{22}}{6.02 \times 10^{23}/\text{mol}}\right)(4.0\,\text{g}/\text{mol}) = 87.9 \times 10^{-3}\,\text{g}.$$

37. If N is the number of undecayed nuclei present at time t, then

$$\frac{dN}{dt} = R - \lambda N$$

where R is the rate of production by the cyclotron and λ is the disintegration constant. The second term gives the rate of decay. Rearrange the equation slightly and integrate:

$$\int_{N_0}^{N} \frac{dN}{R - \lambda N} = \int_0^t dt$$

where N_0 is the number of undecayed nuclei present at time $t = 0$. This yields

$$-\frac{1}{\lambda}\ln\frac{R - \lambda N}{R - \lambda N_0} = t.$$

We solve for N:

$$N = \frac{R}{\lambda} + \left(N_0 - \frac{R}{\lambda}\right)e^{-\lambda t}.$$

After many half-lives, the exponential is small and the second term can be neglected. Then, $N = R/\lambda$, regardless of the initial value N_0. At times that are long compared to the half-life, the rate of production equals the rate of decay and N is a constant.

38. Combining Eqs. 42-20 and 42-21, we obtain

$$M_{\text{sam}} = N\frac{M_K}{M_A} = \left(\frac{RT_{1/2}}{\ln 2}\right)\left(\frac{40\,\text{g}/\text{mol}}{6.02 \times 10^{23}/\text{mol}}\right)$$

which gives 0.66 g for the mass of the sample once we plug in $1.7 \times 10^5/s$ for the decay rate and 1.28×10^9 y $= 4.04 \times 10^{16}$ s for the half-life.

39. (a) The sample is in secular equilibrium with the source and the decay rate equals the production rate. Let R be the rate of production of ^{56}Mn and let λ be the disintegration constant. According to the result of problem 41, $R = \lambda N$ after a long time has passed. Now, $\lambda N = 8.88 \times 10^{10}$ s^{-1}, so $R = 8.88 \times 10^{10}$ s^{-1}.

(b) We use $N = R/\lambda$. If $T_{1/2}$ is the half-life, then the disintegration constant is

$$\lambda = (\ln 2)/T_{1/2} = (\ln 2)/(2.58 \text{ h}) = 0.269 \text{ h}^{-1} = 7.46 \times 10^{-5} \text{ s}^{-1},$$

so $N = (8.88 \times 10^{10} \text{ s}^{-1})/(7.46 \times 10^{-5} \text{ s}^{-1}) = 1.19 \times 10^{15}$.

(c) The mass of a ^{56}Mn nucleus is $(56 \text{ u}) (1.661 \times 10^{-24} \text{ g/u}) = 9.30 \times 10^{-23}$ g and the total mass of ^{56}Mn in the sample at the end of the bombardment is

$$Nm = (1.19 \times 10^{15})(9.30 \times 10^{-23} \text{ g}) = 1.11 \times 10^{-7} \text{ g}.$$

40. (a) The nuclear reaction is written as ^{238}U$\rightarrow^{234}$Th$+$ ^{4}He. The energy released is

$$\Delta E_1 = \left(m_U - m_{He} - m_{Th}\right)c^2$$
$$= (238.05079 \text{ u} - 4.00260 \text{ u} - 234.04363 \text{ u})(931.5 \text{ MeV}/\text{u})$$
$$= 4.25 \text{ MeV}.$$

(b) The reaction series consists of ^{238}U$\rightarrow^{237}$U$+n$, followed by

$$^{237}\text{U}\rightarrow^{236}\text{Pa}+p$$
$$^{236}\text{Pa}\rightarrow^{235}\text{Pa}+n$$
$$^{235}\text{Pa}\rightarrow^{234}\text{Th}+p$$

The net energy released is then

$$\Delta E_2 = \left(m_{238_U} - m_{237_U} - m_n\right)c^2 + \left(m_{237_U} - m_{236_{Pa}} - m_p\right)c^2$$
$$+ \left(m_{236_{Pa}} - m_{235_{Pa}} - m_n\right)c^2 + \left(m_{235_{Pa}} - m_{234_{Th}} - m_p\right)c^2$$
$$= \left(m_{238_U} - 2m_n - 2m_p - m_{234_{Th}}\right)c^2$$
$$= \left[238.05079 \text{ u} - 2(1.00867 \text{ u}) - 2(1.00783 \text{ u}) - 234.04363 \text{ u}\right](931.5 \text{ MeV}/\text{u})$$
$$= -24.1 \text{ MeV}.$$

(c) This leads us to conclude that the binding energy of the α particle is

$$\left| \left(2m_n + 2m_p - m_{He}\right)c^2 \right| = \left| -24.1\,\text{MeV} - 4.25\,\text{MeV} \right| = 28.3\,\text{MeV}.$$

41. The fraction of undecayed nuclei remaining after time t is given by

$$\frac{N}{N_0} = e^{-\lambda t} = e^{-(\ln 2)t/T_{1/2}}$$

where λ is the disintegration constant and $T_{1/2}$ ($= (\ln 2)/\lambda$) is the half-life. The time for half the original ^{238}U nuclei to decay is 4.5×10^9 y.

(a) For ^{244}Pu at that time,

$$\frac{(\ln 2)t}{T_{1/2}} = \frac{(\ln 2)\left(4.5 \times 10^9\,\text{y}\right)}{8.0 \times 10^7\,\text{y}} = 39$$

and

$$\frac{N}{N_0} = e^{-39.0} \approx 1.2 \times 10^{-17}.$$

(b) For ^{248}Cm at that time,

$$\frac{(\ln 2)t}{T_{1/2}} = \frac{(\ln 2)\left(4.5 \times 10^9\,\text{y}\right)}{3.4 \times 10^5\,\text{y}} = 9170$$

and

$$\frac{N}{N_0} = e^{-9170} = 3.31 \times 10^{-3983}.$$

For any reasonably sized sample this is less than one nucleus and may be taken to be zero. A standard calculator probably cannot evaluate e^{-9170} directly. Our recommendation is to treat it as $(e^{-91.70})^{100}$.

42. (a) The disintegration energy for uranium-235 "decaying" into thorium-232 is

$$\begin{aligned} Q_3 &= \left(m_{235_U} - m_{232_{Th}} - m_{3_{He}}\right)c^2 \\ &= \left(235.0439\,\text{u} - 232.0381\,\text{u} - 3.0160\,\text{u}\right)\left(931.5\,\text{MeV}/\text{u}\right) \\ &= -9.50\,\text{MeV}. \end{aligned}$$

(b) Similarly, the disintegration energy for uranium-235 decaying into thorium-231 is

$$Q_4 = \left(m_{235\text{U}} - m_{231\text{Th}} - m_{4\text{He}}\right)c^2$$
$$= \left(235.0439\,\text{u} - 231.0363\,\text{u} - 4.0026\,\text{u}\right)\left(931.5\,\text{MeV/u}\right)$$
$$= 4.66\,\text{MeV}.$$

(c) Finally, the considered transmutation of uranium-235 into thorium-230 has a Q-value of

$$Q_5 = \left(m_{235\text{U}} - m_{230\text{Th}} - m_{5\text{He}}\right)c^2$$
$$= \left(235.0439\,\text{u} - 230.0331\,\text{u} - 5.0122\,\text{u}\right)\left(931.5\,\text{MeV}/\text{u}\right)$$
$$= -1.30\,\text{MeV}.$$

Only the second decay process (the α decay) is spontaneous, as it releases energy.

43. Energy and momentum are conserved. We assume the residual thorium nucleus is in its ground state. Let K_α be the kinetic energy of the alpha particle and K_{Th} be the kinetic energy of the thorium nucleus. Then, $Q = K_\alpha + K_{\text{Th}}$. We assume the uranium nucleus is initially at rest. Then, conservation of momentum yields $0 = p_\alpha + p_{\text{Th}}$, where p_α is the momentum of the alpha particle and p_{Th} is the momentum of the thorium nucleus. Both particles travel slowly enough that the classical relationship between momentum and energy can be used. Thus $K_{\text{Th}} = p_{\text{Th}}^2 / 2m_{\text{Th}}$, where m_{Th} is the mass of the thorium nucleus. We substitute $p_{\text{Th}} = -p_\alpha$ and use $K_\alpha = p_\alpha^2 / 2m_\alpha$ to obtain $K_{\text{Th}} = (m_\alpha/m_{\text{Th}})K_\alpha$. Consequently,

$$Q = K_\alpha + \frac{m_\alpha}{m_{\text{Th}}}K_\alpha = \left(1 + \frac{m_\alpha}{m_{\text{Th}}}\right)K_\alpha = \left(1 + \frac{4.00\,\text{u}}{234\,\text{u}}\right)\left(4.196\,\text{MeV}\right) = 4.269\,\text{MeV}.$$

44. (a) For the first reaction

$$Q_1 = \left(m_{\text{Ra}} - m_{\text{Pb}} - m_{\text{C}}\right)c^2 = \left(223.01850\,\text{u} - 208.98107\,\text{u} - 14.00324\,\text{u}\right)\left(931.5\,\text{MeV/u}\right)$$
$$= 31.8\,\text{MeV}.$$

(b) For the second one

$$Q_2 = \left(m_{\text{Ra}} - m_{\text{Rn}} - m_{\text{He}}\right)c^2 = \left(223.01850\,\text{u} - 219.00948\,\text{u} - 4.00260\,\text{u}\right)\left(931.5\,\text{MeV/u}\right)$$
$$= 5.98\,\text{MeV}.$$

(c) From $U \propto q_1 q_2/r$, we get

$$U_1 \approx U_2\left(\frac{q_{Pb}\, q_C}{q_{Rn}\, q_{He}}\right) = (30.0\,\mathrm{MeV})\frac{(82e)(6.0e)}{(86e)(2.0e)} = 86\,\mathrm{MeV}.$$

45. Let M_{Cs} be the mass of one atom of $^{137}_{55}\mathrm{Cs}$ and M_{Ba} be the mass of one atom of $^{137}_{56}\mathrm{Ba}$. To obtain the nuclear masses, we must subtract the mass of 55 electrons from M_{Cs} and the mass of 56 electrons from M_{Ba}. The energy released is

$$Q = [(M_{Cs} - 55m) - (M_{Ba} - 56m) - m]\,c^2,$$

where m is the mass of an electron. Once cancellations have been made, $Q = (M_{Cs} - M_{Ba})c^2$ is obtained. Therefore,

$$Q = [136.9071\,\mathrm{u} - 136.9058\,\mathrm{u}]c^2 = (0.0013\,\mathrm{u})c^2 = (0.0013\,\mathrm{u})(931.5\,\mathrm{MeV}/\mathrm{u}) = 1.21\,\mathrm{MeV}.$$

46. (a) We recall that $mc^2 = 0.511$ MeV from Table 38-3, and note that the result of problem 3 in Chapter 39 can be written as $hc = 1240$ MeV·fm. Using Eq. 38-51 and Eq. 39-13, we obtain

$$\lambda = \frac{h}{p} = \frac{hc}{\sqrt{K^2 + 2Kmc^2}}$$

$$= \frac{1240\,\mathrm{MeV}\cdot\mathrm{fm}}{\sqrt{(1.0\,\mathrm{MeV})^2 + 2(1.0\,\mathrm{MeV})(0.511\,\mathrm{MeV})}} = 9.0\times10^2\,\mathrm{fm}.$$

(b) $r = r_0 A^{1/3} = (1.2\,\mathrm{fm})(150)^{1/3} = 6.4$ fm.

(c) Since $\lambda \gg r$ the electron cannot be confined in the nuclide. We recall from Chapters 40 and 41 that at least $\lambda/2$ was needed in any particular direction, to support a standing wave in an "infinite well." A finite well is able to support *slightly* less than $\lambda/2$ (as one can infer from the ground state wavefunction in Fig. 40-8), but in the present case λ/r is far too big to be supported.

(d) A strong case can be made on the basis of the remarks in part (c), above.

47. The decay scheme is $n \to p + e^- + \nu$. The electron kinetic energy is a maximum if no neutrino is emitted. Then,

$$K_{max} = (m_n - m_p - m_e)c^2,$$

where m_n is the mass of a neutron, m_p is the mass of a proton, and m_e is the mass of an electron. Since $m_p + m_e = m_H$, where m_H is the mass of a hydrogen atom, this can be written $K_{max} = (m_n - m_H)c^2$. Hence,

$$K_{max} = (840 \times 10^{-6}\,\mathrm{u})c^2 = (840 \times 10^{-6}\,\mathrm{u})(931.5\,\mathrm{MeV}/\mathrm{u}) = 0.783\,\mathrm{MeV}.$$

48. Assuming the neutrino has negligible mass, then

$$\Delta mc^2 = \left(\mathbf{m}_{Ti} - \mathbf{m}_V - m_e\right)c^2.$$

Now, since Vanadium has 23 electrons (see Appendix F and/or G) and Titanium has 22 electrons, we can add and subtract $22m_e$ to the above expression and obtain

$$\Delta mc^2 = \left(\mathbf{m}_{Ti} + 22m_e - \mathbf{m}_V - 23m_e\right)c^2 = \left(m_{Ti} - m_V\right)c^2.$$

We note that our final expression for Δmc^2 involves the *atomic* masses, and that this assumes (due to the way they are usually tabulated) the atoms are in the ground states (which is certainly not the case here, as we discuss below). The question now is: do we set $Q = -\Delta mc^2$ as in Sample Problem 42-7? The answer is "no." The atom is left in an excited (high energy) state due to the fact that an electron was captured from the lowest shell (where the absolute value of the energy, E_K, is quite large for large Z — see Eq. 39-25). To a very good approximation, the energy of the K-shell electron in Vanadium is equal to that in Titanium (where there is now a "vacancy" that must be filled by a readjustment of the whole electron cloud), and we write $Q = -\Delta mc^2 - E_K$ so that Eq. 42-26 still holds. Thus,

$$Q = \left(m_V - m_{Ti}\right)c^2 - E_K.$$

49. (a) Since the positron has the same mass as an electron, and the neutrino has negligible mass, then

$$\Delta mc^2 = \left(\mathbf{m}_B + m_e - \mathbf{m}_C\right)c^2.$$

Now, since Carbon has 6 electrons (see Appendix F and/or G) and Boron has 5 electrons, we can add and subtract $6m_e$ to the above expression and obtain

$$\Delta mc^2 = \left(\mathbf{m}_B + 7m_e - \mathbf{m}_C - 6m_e\right)c^2 = \left(m_B + 2m_e - m_C\right)c^2.$$

We note that our final expression for Δmc^2 involves the *atomic* masses, as well an "extra" term corresponding to two electron masses. From Eq. 38-47 and Table 38-3, we obtain

$$Q = \left(m_C - m_B - 2m_e\right)c^2 = \left(m_C - m_B\right)c^2 - 2\left(0.511\,\text{MeV}\right).$$

(b) The disintegration energy for the positron decay of Carbon-11 is

$$Q = \left(11.011434\,\text{u} - 11.009305\,\text{u}\right)\left(931.5\,\text{MeV / u}\right) - 1.022\,\text{MeV} = 0.961\,\text{MeV}.$$

50. (a) The rate of heat production is

$$\frac{dE}{dt} = \sum_{i=1}^{3} R_i Q_i = \sum_{i=1}^{3} \lambda_1 N_i Q_i = \sum_{i=1}^{3} \left(\frac{\ln 2}{T_{1/2_i}} \right) \frac{(1.00\,\text{kg}) f_i}{m_i} Q_i$$

$$= \frac{(1.00\,\text{kg})(\ln 2)(1.60 \times 10^{-13}\,\text{J}/\text{MeV})}{(3.15 \times 10^{7}\,\text{s}/\text{y})(1.661 \times 10^{-27}\,\text{kg}/\text{u})} \left[\frac{(4 \times 10^{-6})(51.7\,\text{MeV})}{(238\,\text{u})(4.47 \times 10^{9}\,\text{y})} \right.$$

$$\left. + \frac{(13 \times 10^{-6})(42.7\,\text{MeV})}{(232\,\text{u})(1.41 \times 10^{10}\,\text{y})} + \frac{(4 \times 10^{-6})(1.31\,\text{MeV})}{(40\,\text{u})(1.28 \times 10^{9}\,\text{y})} \right]$$

$$= 1.0 \times 10^{-9}\,\text{W}.$$

(b) The contribution to heating, due to radioactivity, is

$$P = (2.7 \times 10^{22}\,\text{kg})(1.0 \times 10^{-9}\,\text{W/kg}) = 2.7 \times 10^{13}\,\text{W},$$

which is very small compared to what is received from the Sun.

51. Since the electron has the maximum possible kinetic energy, no neutrino is emitted. Since momentum is conserved, the momentum of the electron and the momentum of the residual sulfur nucleus are equal in magnitude and opposite in direction. If p_e is the momentum of the electron and p_S is the momentum of the sulfur nucleus, then $p_S = -p_e$. The kinetic energy K_S of the sulfur nucleus is $K_S = p_S^2 / 2M_S = p_e^2 / 2M_S$, where M_S is the mass of the sulfur nucleus. Now, the electron's kinetic energy K_e is related to its momentum by the relativistic equation $(p_e c)^2 = K_e^2 + 2K_e mc^2$, where m is the mass of an electron. See Eq. 38-51. Thus,

$$K_S = \frac{(p_e c)^2}{2M_S c^2} = \frac{K_e^2 + 2K_e mc^2}{2M_S c^2} = \frac{(1.71\,\text{MeV})^2 + 2(1.71\,\text{MeV})(0.511\,\text{MeV})}{2(32\,\text{u})(931.5\,\text{MeV}/\text{u})}$$

$$= 7.83 \times 10^{-5}\,\text{MeV} = 78.3\,\text{eV}$$

where $mc^2 = 0.511$ MeV is used (see Table 38-3).

52. We solve for t from $R = R_0 e^{-\lambda t}$:

$$t = \frac{1}{\lambda} \ln \frac{R_0}{R} = \left(\frac{5730\,\text{y}}{\ln 2} \right) \ln \left[\left(\frac{15.3}{63.0} \right) \left(\frac{5.00}{1.00} \right) \right] = 1.61 \times 10^{3}\,\text{y}.$$

53. (a) The mass of a ^{238}U atom is $(238\,\text{u})(1.661 \times 10^{-24}\,\text{g}/\text{u}) = 3.95 \times 10^{-22}$ g, so the number of uranium atoms in the rock is

$$N_U = (4.20 \times 10^{-3}\,\text{g})/(3.95 \times 10^{-22}\,\text{g}) = 1.06 \times 10^{19}.$$

(b) The mass of a ^{206}Pb atom is $(206\ u)(1.661 \times 10^{-24}\ g) = 3.42 \times 10^{-22}\ g$, so the number of lead atoms in the rock is

$$N_{Pb} = (2.135 \times 10^{-3}\ g)/(3.42 \times 10^{-22}\ g) = 6.24 \times 10^{18}.$$

(c) If no lead was lost, there was originally one uranium atom for each lead atom formed by decay, in addition to the uranium atoms that did not yet decay. Thus, the original number of uranium atoms was

$$N_{U0} = N_U + N_{Pb} = 1.06 \times 10^{19} + 6.24 \times 10^{18} = 1.68 \times 10^{19}.$$

(d) We use

$$N_U = N_{U0}e^{-\lambda t}$$

where λ is the disintegration constant for the decay. It is related to the half-life $T_{1/2}$ by $\lambda = (\ln 2)/T_{1/2}$. Thus

$$t = -\frac{1}{\lambda}\ln\left(\frac{N_U}{N_{U0}}\right) = -\frac{T_{1/2}}{\ln 2}\ln\left(\frac{N_U}{N_{U0}}\right) = -\frac{4.47 \times 10^9\ y}{\ln 2}\ln\left(\frac{1.06 \times 10^{19}}{1.68 \times 10^{19}}\right) = 2.97 \times 10^9\ y.$$

54. The original amount of ^{238}U the rock contains is given by

$$m_0 = me^{\lambda t} = (3.70\ mg)e^{(\ln 2)\left(260 \times 10^6\ y\right)/\left(4.47 \times 10^9\ y\right)} = 3.85\ mg.$$

Thus, the amount of lead produced is

$$m' = (m_0 - m)\left(\frac{m_{206}}{m_{238}}\right) = (3.85\ mg - 3.70\ mg)\left(\frac{206}{238}\right) = 0.132\ mg.$$

55. We can find the age t of the rock from the masses of ^{238}U and ^{206}Pb. The initial mass of ^{238}U is

$$m_{U0} = m_U + \frac{238}{206}m_{Pb}.$$

Therefore, $m_U = m_{U0}e^{-\lambda_U t} = \left(m_U + m_{238\,Pb}/206\right)e^{-(t\ln 2)/T_{1/2U}}.$ We solve for t:

$$t = \frac{T_{1/2_U}}{\ln 2}\ln\left(\frac{m_U + (238/206)m_{Pb}}{m_U}\right) = \frac{4.47 \times 10^9\ y}{\ln 2}\ln\left[1 + \left(\frac{238}{206}\right)\left(\frac{0.15\,mg}{0.86\,mg}\right)\right]$$

$$= 1.18 \times 10^9\ y.$$

For the β decay of ^{40}K, the initial mass of ^{40}K is

$$m_{K0} = m_K + (40/40)m_{Ar} = m_K + m_{Ar},$$

so

$$m_K = m_{K0}e^{-\lambda_K t} = (m_K + m_{Ar})e^{-\lambda_K t}.$$

We solve for m_K:

$$m_K = \frac{m_{Ar}e^{-\lambda_K t}}{1-e^{-\lambda_K t}} = \frac{m_{Ar}}{e^{\lambda_K t}-1} = \frac{1.6\,\text{mg}}{e^{(\ln 2)(1.18\times10^9\,\text{y})/(1.25\times10^9\,\text{y})}-1} = 1.7\,\text{mg}.$$

56. We note that every Calcium-40 atom and Krypton-40 atom found now in the sample was once one of the original number of Potassium atoms. Thus, using Eq. 42-14 and Eq. 42-18, we find

$$\ln\left(\frac{N_K}{N_K + N_{Ar} + N_{Ca}}\right) = -\lambda t \quad \Rightarrow \quad \ln\left(\frac{1}{1+1+8.54}\right) = -\frac{\ln 2}{T_{1/2}}t$$

which (with $T_{1/2} = 1.26 \times 10^9$ y) yields $t = 4.28 \times 10^9$ y.

57. The decay rate R is related to the number of nuclei N by $R = \lambda N$, where λ is the disintegration constant. The disintegration constant is related to the half-life $T_{1/2}$ by $\lambda = (\ln 2)/T_{1/2}$, so $N = R/\lambda = RT_{1/2}/\ln 2$. Since 1 Ci $= 3.7 \times 10^{10}$ disintegrations/s,

$$N = \frac{(250\,\text{Ci})(3.7\times10^{10}\,\text{s}^{-1}/\text{Ci})(2.7\,\text{d})(8.64\times10^4\,\text{s}/\text{d})}{\ln 2} = 3.11\times10^{18}.$$

The mass of a ^{198}Au atom is $M = (198\,\text{u})(1.661 \times 10^{-24}\,\text{g/u}) = 3.29 \times 10^{-22}$ g, so the mass required is

$$N M = (3.11 \times 10^{18})(3.29 \times 10^{-22}\,\text{g}) = 1.02 \times 10^{-3}\,\text{g} = 1.02\,\text{mg}.$$

58. The becquerel (Bq) and curie (Ci) are defined in §42-3.

(a) $R = 8700/60 = 145$ Bq.

(b) $R = \dfrac{145\,\text{Bq}}{3.7\times10^{10}\,\text{Bq}/\text{Ci}} = 3.92 \times 10^{-9}$ Ci.

59. The dose equivalent is the product of the absorbed dose and the RBE factor, so the absorbed dose is

$$(\text{dose equivalent})/(\text{RBE}) = (250 \times 10^{-6}\ \text{Sv})/(0.85) = 2.94 \times 10^{-4}\ \text{Gy}.$$

But 1 Gy = 1 J/kg, so the absorbed dose is

$$\left(2.94 \times 10^{-4}\ \text{Gy}\right)\left(1\frac{\text{J}}{\text{kg} \cdot \text{Gy}}\right) = 2.94 \times 10^{-4}\ \text{J} / \text{kg}.$$

To obtain the total energy received, we multiply this by the mass receiving the energy:

$$E = (2.94 \times 10^{-4}\ \text{J/kg})(44\ \text{kg}) = 1.29 \times 10^{-2}\ \text{J} \approx 1.3 \times 10^{-2}\ \text{J}.$$

60. (a) Using Eq. 42-32, the energy absorbed is

$$\left(2.4 \times 10^{-4}\ \text{Gy}\right)\left(75\,\text{kg}\right) = 18\,\text{mJ}.$$

(b) The dose equivalent is

$$\left(2.4 \times 10^{-4}\ \text{Gy}\right)\left(12\right) = 2.9 \times 10^{-3}\ \text{Sv}.$$

(c) Using Eq. 42-33, we have $2.9 \times 10^{-3}\ \text{Sv} = 0.29\,\text{rem}$

61. (a) Adapting Eq. 42-21, we find

$$N_0 = \frac{\left(2.5 \times 10^{-3}\ \text{g}\right)\left(6.02 \times 10^{23}\ / \text{mol}\right)}{239\,\text{g} / \text{mol}} = 6.3 \times 10^{18}.$$

(b) From Eq. 42-15 and Eq. 42-18,

$$|\Delta N| = N_0\left[1 - e^{-t\ln 2/T_{1/2}}\right] = \left(6.3 \times 10^{18}\right)\left[1 - e^{-(12\,\text{h})\ln 2/(24,100\,\text{y})(8760\,\text{h/y})}\right] = 2.5 \times 10^{11}.$$

(c) The energy absorbed by the body is

$$\left(0.95\right)E_\alpha|\Delta N| = \left(0.95\right)\left(5.2\,\text{MeV}\right)\left(2.5 \times 10^{11}\right)\left(1.6 \times 10^{-13}\ \text{J} / \text{MeV}\right) = 0.20\,\text{J}.$$

(d) On a per unit mass basis, the previous result becomes (according to Eq. 42-32)

$$\frac{0.20\,\text{mJ}}{85\,\text{kg}} = 2.3 \times 10^{-3}\ \text{J} / \text{kg} = 2.3\,\text{mGy}.$$

(e) Using Eq. 42-31, $(2.3 \text{ mGy})(13) = 30 \text{ mSv}$.

62. From Eq. 19-24, we obtain

$$T = \frac{2}{3}\left(\frac{K_{avg}}{k}\right) = \frac{2}{3}\left(\frac{5.00 \times 10^6 \text{ eV}}{8.62 \times 10^{-5} \text{ eV/K}}\right) = 3.87 \times 10^{10} \text{ K}.$$

63. (a) Following Sample Problem 42-10, we compute

$$\Delta E \approx \frac{\hbar}{t_{avg}} = \frac{\left(4.14 \times 10^{-15} \text{ eV} \cdot \text{fs}\right)/2\pi}{1.0 \times 10^{-22} \text{ s}} = 6.6 \times 10^6 \text{ eV}.$$

(b) In order to fully distribute the energy in a fairly large nucleus, and create a "compound nucleus" equilibrium configuration, about 10^{-15} s is typically required. A reaction state that exists no more than about 10^{-22} s does not qualify as a compound nucleus.

64. (a) We compare both the proton numbers (atomic numbers, which can be found in Appendix F and/or G) and the neutron numbers (see Eq. 42-1) with the magic nucleon numbers (special values of either Z or N) listed in §42-8. We find that ^{18}O, ^{60}Ni, ^{92}Mo, ^{144}Sm, and ^{207}Pb each have a filled shell for either the protons or the neutrons (two of these, ^{18}O and ^{92}Mo, are explicitly discussed in that section).

(b) Consider ^{40}K, which has $Z = 19$ protons (which is one less than the magic number 20). It has $N = 21$ neutrons, so it has one neutron outside a closed shell for neutrons, and thus qualifies for this list. Others in this list include ^{91}Zr, ^{121}Sb, and ^{143}Nd.

(c) Consider ^{13}C, which has $Z = 6$ and $N = 13 - 6 = 7$ neutrons. Since 8 is a magic number, then ^{13}C has a vacancy in an otherwise filled shell for neutrons. Similar arguments lead to inclusion of ^{40}K, ^{49}Ti, ^{205}Tl, and ^{207}Pb in this list.

65. A generalized formation reaction can be written $X + x \rightarrow Y$, where X is the target nucleus, x is the incident light particle, and Y is the excited compound nucleus (^{20}Ne). We assume X is initially at rest. Then, conservation of energy yields

$$m_X c^2 + m_x c^2 + K_x = m_Y c^2 + K_Y + E_Y$$

where m_X, m_x, and m_Y are masses, K_x and K_Y are kinetic energies, and E_Y is the excitation energy of Y. Conservation of momentum yields

$$p_x = p_Y.$$

Now, $K_Y = p_Y^2/2m_Y = p_x^2/2m_Y = (m_x/m_Y)K_x$, so

$$m_x c^2 + m_x c^2 + K_x = m_y c^2 + (m_x / m_y) K_x + E_Y$$

and

$$K_x = \frac{m_Y}{m_Y - m_x} \left[(m_Y - m_X - m_x) c^2 + E_Y \right].$$

(a) Let x represent the alpha particle and X represent the ^{16}O nucleus. Then,

$$(m_Y - m_X - m_x)c^2 = (19.99244 \text{ u} - 15.99491 \text{ u} - 4.00260 \text{ u})(931.5 \text{ MeV/u}) = -4.722 \text{ MeV}$$

and

$$K_\alpha = \frac{19.99244 \text{ u}}{19.99244 \text{ u} - 4.00260 \text{ u}} (-4.722 \text{ MeV} + 25.0 \text{ MeV}) = 25.35 \text{ MeV} \approx 25.4 \text{ MeV}.$$

(b) Let x represent the proton and X represent the ^{19}F nucleus. Then,

$$(m_Y - m_X - m_x)c^2 = (19.99244 \text{ u} - 18.99841 \text{ u} - 1.00783 \text{ u})(931.5 \text{ MeV/u}) = -12.85 \text{ MeV}$$

and

$$K_\alpha = \frac{19.99244 \text{ u}}{19.99244 \text{ u} - 1.00783 \text{ u}} (-12.85 \text{ MeV} + 25.0 \text{ MeV}) = 12.80 \text{ MeV}.$$

(c) Let x represent the photon and X represent the ^{20}Ne nucleus. Since the mass of the photon is zero, we must rewrite the conservation of energy equation: if E_γ is the energy of the photon, then $E_\gamma + m_X c^2 = m_Y c^2 + K_Y + E_Y$. Since $m_X = m_Y$, this equation becomes $E_\gamma = K_Y + E_Y$. Since the momentum and energy of a photon are related by $p_\gamma = E_\gamma / c$, the conservation of momentum equation becomes $E_\gamma / c = p_Y$. The kinetic energy of the compound nucleus is $K_Y = p_Y^2 / 2m_Y = E_\gamma^2 / 2m_Y c^2$. We substitute this result into the conservation of energy equation to obtain

$$E_\gamma = \frac{E_\gamma^2}{2m_Y c^2} + E_Y.$$

This quadratic equation has the solutions

$$E_\gamma = m_Y c^2 \pm \sqrt{\left(m_Y c^2 \right)^2 - 2 m_Y c^2 E_Y}.$$

If the problem is solved using the relativistic relationship between the energy and momentum of the compound nucleus, only one solution would be obtained, the one corresponding to the negative sign above. Since

$$m_Y c^2 = (19.99244 \text{ u})(931.5 \text{ MeV/u}) = 1.862 \times 10^4 \text{ MeV},$$

we have

$$E_\gamma = \left(1.862 \times 10^4 \text{ MeV}\right) - \sqrt{\left(1.862 \times 10^4 \text{ MeV}\right)^2 - 2\left(1.862 \times 10^4 \text{ MeV}\right)\left(25.0 \text{ MeV}\right)}$$

$$= 25.0 \text{ MeV}.$$

The kinetic energy of the compound nucleus is very small; essentially all of the photon energy goes to excite the nucleus.

66. (a) From the decay series, we know that N_{210}, the amount of ^{210}Pb nuclei, changes because of two decays: the decay from ^{226}Ra into ^{210}Pb at the rate $R_{226} = \lambda_{226} N_{226}$, and the decay from ^{210}Pb into ^{206}Pb at the rate $R_{210} = \lambda_{210} N_{210}$. The first of these decays causes N_{210} to increase while the second one causes it to decrease. Thus,

$$\frac{dN_{210}}{dt} = R_{226} - R_{210} = \lambda_{226} N_{226} - \lambda_{210} N_{210}.$$

(b) We set $dN_{210}/dt = R_{226} - R_{210} = 0$ to obtain $R_{226}/R_{210} = 1.00$.

(c) From $R_{226} = \lambda_{226} N_{226} = R_{210} = \lambda_{210} N_{210}$, we obtain

$$\frac{N_{226}}{N_{210}} = \frac{\lambda_{210}}{\lambda_{226}} = \frac{T_{1/2\,226}}{T_{1/2\,210}} = \frac{1.60 \times 10^3 \text{ y}}{22.6 \text{ y}} = 70.8.$$

(d) Since only 1.00% of the ^{226}Ra remains, the ratio R_{226}/R_{210} is 0.00100 of that of the equilibrium state computed in part (b). Thus the ratio is $(0.0100)(1) = 0.0100$.

(e) This is similar to part (d) above. Since only 1.00% of the ^{226}Ra remains, the ratio N_{226}/N_{210} is 1.00% of that of the equilibrium state computed in part (c), or $(0.0100)(70.8) = 0.708$.

(f) Since the actual value of N_{226}/N_{210} is 0.09, which much closer to 0.0100 than to 1, the sample of the lead pigment cannot be 300 years old. So *Emmaus* is not a *Vermeer*.

67. (a) We use $R = R_0 e^{-\lambda t}$ to find t:

$$t = \frac{1}{\lambda} \ln \frac{R_0}{R} = \frac{T_{1/2}}{\ln 2} \ln \frac{R_0}{R} = \frac{14.28 \text{ d}}{\ln 2} \ln \frac{3050}{170} = 59.5 \text{ d}.$$

(b) The required factor is

$$\frac{R_0}{R} = e^{\lambda t} = e^{t \ln 2/T_{1/2}} = e^{(3.48d/14.28d)\ln 2} = 1.18.$$

68. (a) Assuming a "target" area of one square meter, we establish a ratio:

$$\frac{\text{rate through you}}{\text{total rate upward}} = \frac{1\,\text{m}^2}{(2.6 \times 10^5\,\text{km}^2)(1000\,\text{m/km})^2} = 3.8 \times 10^{-12}.$$

The SI unit becquerel is equivalent to a disintegration per second. With half the beta-decay electrons moving upward, we find

$$\text{rate through you} = \frac{1}{2}(1 \times 10^{16}/\text{s})(3.8 \times 10^{-12}) = 1.9 \times 10^4/\text{s}$$

which implies (converting s → h) the rate of electrons you would intercept is $R_0 = 7 \times 10^7/\text{h}$. So in one hour, 7×10^7 electrons would be intercepted.

(b) Let D indicate the current year (2003, 2004, etc). Combining Eq. 42-16 and Eq. 42-18, we find

$$R = R_0 e^{-t \ln 2/T_{1/2}} = (7 \times 10^7/\text{h}) e^{-(D-1996)\ln 2/(30.2\,\text{y})}.$$

69. Since the spreading is assumed uniform, the count rate $R = 74,000/\text{s}$ is given by $R = \lambda N = \lambda(M/m)(a/A)$, where $M = 400$ g, m is the mass of the ^{90}Sr nucleus, $A = 2000$ km^2, and a is the area in question. We solve for a:

$$a = A\left(\frac{m}{M}\right)\left(\frac{R}{\lambda}\right) = \frac{AmRT_{1/2}}{M \ln 2}$$

$$= \frac{(2000 \times 10^6\,\text{m}^2)(90\,\text{g/mol})(29\,\text{y})(3.15 \times 10^7\,\text{s/y})(74,000/\text{s})}{(400\,\text{g})(6.02 \times 10^{23}/\text{mol})(\ln 2)}$$

$$= 7.3 \times 10^{-2}\,\text{m}^{-2} = 730\,\text{cm}^2.$$

70. (a) The rate at which Radium-226 is decaying is

$$R = \lambda N = \left(\frac{\ln 2}{T_{1/2}}\right)\left(\frac{M}{m}\right) = \frac{(\ln 2)(1.00\,\text{mg})(6.02 \times 10^{23}/\text{mol})}{(1600\,\text{y})(3.15 \times 10^7\,\text{s/y})(226\,\text{g/mol})} = 3.66 \times 10^7\,\text{s}^{-1}.$$

The activity is 3.66×10^7 Bq.

(b) The activity of ^{222}Rn is also 3.66×10^7 Bq.

(c) From $R_{Ra} = R_{Rn}$ and $R = \lambda N = (\ln 2/T_{1/2})(M/m)$, we get

$$M_{Rn} = \left(\frac{T_{1/2_{Rn}}}{T_{1/2_{Ra}}}\right)\left(\frac{m_{Rn}}{m_{Ra}}\right)M_{Ra} = \frac{(3.82\,\mathrm{d})(1.00\times10^{-3}\,\mathrm{g})(222\,\mathrm{u})}{(1600\,\mathrm{y})(365\,\mathrm{d/y})(226\,\mathrm{u})} = 6.42\times10^{-9}\,\mathrm{g}.$$

71. We obtain

$$\begin{aligned} Q &= (m_V - m_{Ti})c^2 - E_K \\ &= (48.94852\,\mathrm{u} - 48.94787\,\mathrm{u})(931.5\,\mathrm{MeV/u}) - 0.00547\,\mathrm{MeV} \\ &= 0.600\,\mathrm{MeV}. \end{aligned}$$

72. In order for the α particle to penetrate the gold nucleus, the separation between the centers of mass of the two particles must be no greater than

$$r = r_{Cu} + r_\alpha = 6.23\ \mathrm{fm} + 1.80\ \mathrm{fm} = 8.03\ \mathrm{fm}.$$

Thus, the minimum energy K_α is given by

$$\begin{aligned} K_\alpha = U &= \frac{1}{4\pi\varepsilon_0}\frac{q_\alpha q_{Au}}{r} = \frac{kq_\alpha q_{Au}}{r} \\ &= \frac{(8.99\times10^9\ \mathrm{V\cdot m/C})(2e)(79)(1.60\times10^{-19}\ \mathrm{C})}{8.03\times10^{-15}\ \mathrm{m}} = 28.3\times10^6\ \mathrm{eV}. \end{aligned}$$

We note that the factor of e in $q_\alpha = 2e$ was not set equal to 1.60×10^{-19} C, but was instead carried through to become part of the final units.

73. We note that $hc = 1240$ MeV·fm (see problem 83 of Chapter 38), and that the classical kinetic energy $\frac{1}{2}mv^2$ can be written directly in terms of the classical momentum $p = mv$ (see below). Letting $p \simeq \Delta p \simeq \Delta h/\Delta x \simeq h/r$, we get

$$E = \frac{p^2}{2m} \simeq \frac{(hc)^2}{2(mc^2)r^2} = \frac{(1240\,\mathrm{MeV\cdot fm})^2}{2(938\,\mathrm{MeV})\left[(1.2\,\mathrm{fm})(100)^{1/3}\right]^2} \simeq 30\,\mathrm{MeV}.$$

74. Adapting Eq. 42-21, we have

$$N_{Kr} = \frac{M_{sam}}{M_{Kr}}N_A = \left(\frac{20\times10^{-9}\,\mathrm{g}}{92\,\mathrm{g/mol}}\right)(6.02\times10^{23}\ \mathrm{atoms/mol}) = 1.3\times10^{14}\ \mathrm{atoms}.$$

Consequently, Eq. 42-20 leads to

$$R = \frac{N \ln 2}{T_{1/2}} = \frac{\left(1.3 \times 10^{14}\right) \ln 2}{1.84\,\text{s}} = 4.9 \times 10^{13}\,\text{Bq}.$$

75. Since R is proportional to N (see Eq. 42-17) then $N/N_0 = R/R_0$. Combining Eq. 42-14 and Eq. 42-18 leads to

$$t = -\frac{T_{1/2}}{\ln 2} \ln\left(\frac{R}{R_0}\right) = -\frac{5730\,\text{y}}{\ln 2} \ln(0.020) = 3.2 \times 10^4\,\text{y}.$$

76. (a) The mass number A of a radionuclide changes by 4 in an α decay and is unchanged in a β decay. If the mass numbers of two radionuclides are given by $4n + k$ and $4n' + k$ (where $k = 0, 1, 2, 3$), then the heavier one can decay into the lighter one by a series of α (and β) decays, as their mass numbers differ by only an integer times 4. If $A = 4n + k$, then after α-decaying for m times, its mass number becomes

$$A = 4n + k - 4m = 4(n - m) + k,$$

still in the same chain.

(b) For ^{235}U, $235 = 58 \times 4 + 3 = 4n + 3$.

(c) For ^{236}U, $236 = 59 \times 4 = 4n$.

(d) For ^{238}U, $238 = 59 \times 4 + 2 = 4n + 2$.

(e) For ^{239}Pu, $239 = 59 \times 4 + 3 = 4n + 3$.

(f) For ^{240}Pu, $240 = 60 \times 4 = 4n$.

(g) For ^{245}Cm, $245 = 61 \times 4 + 1 = 4n + 1$.

(h) For ^{246}Cm, $246 = 61 \times 4 + 2 = 4n + 2$.

(i) For ^{249}Cf, $249 = 62 \times 4 + 1 = 4n + 1$.

(j) For ^{253}Fm, $253 = 63 \times 4 + 1 = 4n + 1$.

77. Let $^{A}_{Z}X$ represent the unknown nuclide. The reaction equation is

$$^{A}_{Z}X + ^{1}_{0}\text{n} \rightarrow ^{0}_{-1}\text{e} + 2^{4}_{2}\text{He}.$$

Conservation of charge yields $Z + 0 = -1 + 4$ or $Z = 3$. Conservation of mass number yields $A + 1 = 0 + 8$ or $A = 7$. According to the periodic table in Appendix G (also see Appendix F), lithium has atomic number 3, so the nuclide must be ^7_3Li.

78. We note that 2.42 min = 145.2 s. We are asked to plot (with SI units understood)

$$\ln R = \ln\left(R_0 e^{-\lambda t} + R_0' e^{-\lambda' t}\right)$$

where $R_0 = 3.1 \times 10^5$, $R_0' = 4.1 \times 10^6$, $\lambda = \ln 2/145.2$ and $\lambda' = \ln 2/24.6$. Our plot is shown below.

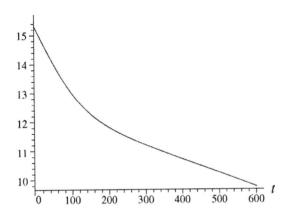

We note that the magnitude of the slope for small t is λ' (the disintegration constant for ^{110}Ag), and for large t is λ (the disintegration constant for ^{108}Ag).

79. The lines that lead toward the lower left are alpha decays, involving an atomic number change of $\Delta Z_\alpha = -2$ and a mass number change of $\Delta A_\alpha = -4$. The short horizontal lines toward the right are beta decays (involving electrons, not positrons) in which case A stays the same but the change in atomic number is $\Delta Z_\beta = +1$. Fig. 42-16 shows three alpha decays and two beta decays; thus,

$$Z_f = Z_i + 3\Delta Z_\alpha + 2\Delta Z_\beta \text{ and } A_f = A_i + 3\Delta A_\alpha.$$

Referring to Appendix F or G, we find $Z_i = 93$ for Neptunium, so

$$Z_f = 93 + 3(-2) + 2(1) = 89,$$

which indicates the element Actinium. We are given $A_i = 237$, so $A_f = 237 + 3(-4) = 225$. Therefore, the final isotope is ^{225}Ac.

80. (a) Replacing differentials with deltas in Eq. 42-12, we use the fact that $\Delta N = -12$ during $\Delta t = 1.0$ s to obtain

$$\frac{\Delta N}{N} = -\lambda \Delta t \Rightarrow \lambda = 4.8 \times 10^{-18} / \text{s}$$

where $N = 2.5 \times 10^{18}$, mentioned at the second paragraph of §42-3, is used.

(b) Eq. 42-18 yields $T_{1/2} = \ln 2/\lambda = 1.4 \times 10^{17}$ s, or about 4.6 billion years.

81. Eq. 24-43 gives the electrostatic potential energy between two uniformly charged spherical charges (in this case $q_1 = 2e$ and $q_2 = 90e$) with r being the distance between their centers. Assuming the "uniformly charged spheres" condition is met in this instance, we write the equation in such a way that we can make use of $k = 1/4 \pi \varepsilon_0$ and the electronvolt unit:

$$U = k\frac{(2e)(90e)}{r} = \left(8.99 \times 10^9 \ \frac{\text{V} \cdot \text{m}}{\text{C}}\right)\frac{(3.2 \times 10^{-19} \ \text{C})(90e)}{r} = \frac{2.59 \times 10^{-7}}{r} \text{eV}$$

with r understood to be in meters. It is convenient to write this for r in femtometers, in which case $U = 259/r$ MeV. This is shown plotted below.

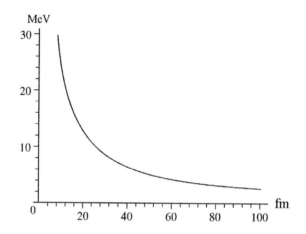

82. We locate a nuclide from Table 42-1 by finding the coordinate (N, Z) of the corresponding point in Fig. 42-4. It is clear that all the nuclides listed in Table 42-1 are stable except the last two, ^{227}Ac and ^{239}Pu.

83. Although we haven't drawn the requested lines in the following table, we can indicate their slopes: lines of constant A would have $- 45°$ slopes, and those of constant $N - Z$ would have 45°. As an example of the latter, the $N - Z = 20$ line (which is one of "eighteen-neutron excess") would pass through Cd-114 at the lower left corner up through Te-122 at the upper right corner. The first column corresponds to $N = 66$, and the bottom row to $Z = 48$. The last column corresponds to $N = 70$, and the top row to $Z = 52$. Much of the information below (regarding values of $T_{1/2}$ particularly) was obtained from the websites http://nucleardata.nuclear.lu.se/nucleardata and http://www.nndc.bnl.gov/nndc/ensdf (we refer the reader to the remarks we made in the solution to problem 8).

^{118}Te	^{119}Te	^{120}Te	^{121}Te	^{122}Te
6.0 days	16.0 h	0.1%	19.4 days	2.6%
^{117}Sb	^{118}Sb	^{119}Sb	^{120}Sb	^{121}Sb
2.8 h	3.6 min	38.2 s	15.9 min	57.2%
^{116}Sn	^{117}Sn	^{118}Sn	^{119}Sn	^{120}Sn
14.5%	7.7%	24.2%	8.6%	32.6%
^{115}In	^{116}In	^{117}In	^{118}In	^{119}In
95.7%	14.1 s	43.2 min	5.0 s	2.4 min
^{114}Cd	^{115}Cd	^{116}Cd	^{117}Cd	^{118}Cd
28.7%	53.5 h	7.5%	2.5 h	50.3 min

84. The problem with Web-based services is that there are no guarantees of accuracy or that the webpage addresses will not change from the time this solution is written to the time someone reads this. Still, it is worth mentioning that a very accessible website for a wide variety of periodic table and isotope-related information is http://www.webelements.com. Two websites aimed more towards the nuclear professional are http://nucleardata.nuclear.lu.se/nucleardata and http://www.nndc.bnl.gov/nndc/ensdf, which are where some of the information mentioned below was obtained.

(a) According to Appendix F, the atomic number 60 corresponds to the element Neodymium (Nd). The first website mentioned above gives ^{142}Nd, ^{143}Nd, ^{144}Nd, ^{145}Nd, ^{146}Nd, ^{148}Nd, and ^{150}Nd in its list of naturally occurring isotopes. Two of these, ^{144}Nd and ^{150}Nd, are not perfectly stable, but their half-lives are much longer than the age of the universe (detailed information on their half-lives, modes of decay, etc are available at the last two websites referred to, above).

(b) In this list, we are asked to put the nuclides which contain 60 neutrons and which are recognized to exist but not stable nuclei (this is why, for example, ^{108}Cd is not included here). Although the problem does not ask for it, we include the half-lives of the nuclides in our list, though it must be admitted that not all reference sources agree on those values (we picked ones we regarded as "most reliable"). Thus, we have ^{97}Rb (0.2 s), ^{98}Sr (0.7 s), ^{99}Y (2 s), ^{100}Zr (7 s), ^{101}Nb (7 s), ^{102}Mo (11 minutes), ^{103}Tc (54 s), ^{105}Rh (35 hours), ^{109}In (4 hours), ^{110}Sn (4 hours), ^{111}Sb (75 s), ^{112}Te (2 minutes), ^{113}I (7 s), ^{114}Xe (10 s), ^{115}Cs (1.4 s), and ^{116}Ba (1.4 s).

(c) We would include in this list: ^{60}Zn, ^{60}Cu, ^{60}Ni, ^{60}Co, ^{60}Fe, ^{60}Mn, ^{60}Cr, and ^{60}V.

85. (a) In terms of the original value of u, the newly defined u is greater by a factor of 1.007825. So the mass of ^{1}H would be 1.000000 u, the mass of ^{12}C would be (12.000000/1.007825) u = 11.90683 u.

(b) The mass of ^{238}U would be (238.050785/ 1.007825) u = 236.2025 u.

86. We take the speed to be constant, and apply the classical kinetic energy formula:

$$t = \frac{d}{v} = \frac{d}{\sqrt{2K/m}} = 2r\sqrt{\frac{m_n}{2K}} = \frac{r}{c}\sqrt{\frac{2mc^2}{K}}$$

$$\approx \frac{\left(1.2 \times 10^{-15}\,\text{m}\right)(100)^{1/3}}{3.0 \times 10^8\,\text{m/s}}\sqrt{\frac{2(938\,\text{MeV})}{5\,\text{MeV}}}$$

$$\approx 4 \times 10^{-22}\,\text{s}.$$

87. We solve for A from Eq. 42-3:

$$A = \left(\frac{r}{r_0}\right)^3 = \left(\frac{3.6\,\text{fm}}{1.2\,\text{fm}}\right)^3 = 27.$$

Chapter 43

1. If R is the fission rate, then the power output is $P = RQ$, where Q is the energy released in each fission event. Hence,

$$R = P/Q = (1.0 \text{ W})/(200 \times 10^6 \text{ eV})(1.60 \times 10^{-19} \text{ J/eV}) = 3.1 \times 10^{10} \text{ fissions/s}.$$

2. We note that the sum of superscripts (mass numbers A) must balance, as well as the sum of Z values (where reference to Appendix F or G is helpful). A neutron has $Z = 0$ and $A = 1$. Uranium has $Z = 92$.

(a) Since xenon has $Z = 54$, then "Y" must have $Z = 92 - 54 = 38$, which indicates the element Strontium. The mass number of "Y" is $235 + 1 - 140 - 1 = 95$, so "Y" is ^{95}Sr.

(b) Iodine has $Z = 53$, so "Y" has $Z = 92 - 53 = 39$, corresponding to the element Yttrium (the symbol for which, coincidentally, is Y). Since $235 + 1 - 139 - 2 = 95$, then the unknown isotope is ^{95}Y.

(c) The atomic number of Zirconium is $Z = 40$. Thus, $92 - 40 - 2 = 52$, which means that "X" has $Z = 52$ (Tellurium). The mass number of "X" is $235 + 1 - 100 - 2 = 134$, so we obtain ^{134}Te.

(d) Examining the mass numbers, we find $b = 235 + 1 - 141 - 92 = 3$.

3. (a) The mass of a single atom of ^{235}U is $(235 \text{ u})(1.661 \times 10^{-27} \text{ kg/u}) = 3.90 \times 10^{-25} \text{ kg}$, so the number of atoms in 1.0 kg is

$$(1.0 \text{ kg})/(3.90 \times 10^{-25} \text{ kg}) = 2.56 \times 10^{24} \approx 2.6 \times 10^{24}.$$

An alternate approach (but essentially the same once the connection between the "u" unit and N_A is made) would be to adapt Eq. 42-21.

(b) The energy released by N fission events is given by $E = NQ$, where Q is the energy released in each event. For 1.0 kg of ^{235}U,

$$E = (2.56 \times 10^{24})(200 \times 10^6 \text{ eV})(1.60 \times 10^{-19} \text{ J/eV}) = 8.19 \times 10^{13} \text{ J} \approx 8.2 \times 10^{13} \text{ J}.$$

(c) If P is the power requirement of the lamp, then

$$t = E/P = (8.19 \times 10^{13} \text{ J})/(100 \text{ W}) = 8.19 \times 10^{11} \text{ s} = 2.6 \times 10^4 \text{ y}.$$

The conversion factor 3.156×10^7 s/y is used to obtain the last result.

4. Adapting Eq. 42-21, there are

$$N_{Pu} = \frac{M_{sam}}{M_{Pu}} NA = \left(\frac{1000 \text{ g}}{239 \text{ g/mol}} \right)(6.02 \times 10^{23} \text{/mol}) = 2.5 \times 10^{24}$$

plutonium nuclei in the sample. If they all fission (each releasing 180 MeV), then the total energy release is 4.54×10^{26} MeV.

5. If M_{Cr} is the mass of a ^{52}Cr nucleus and M_{Mg} is the mass of a ^{26}Mg nucleus, then the disintegration energy is

$$Q = (M_{Cr} - 2M_{Mg})c^2 = [51.94051 \text{ u} - 2(25.98259 \text{ u})](931.5 \text{ MeV/u}) = -23.0 \text{ MeV}.$$

6. (a) We consider the process ^{98}Mo $\rightarrow$ ^{49}Sc + ^{49}Sc. The disintegration energy is

$$Q = (m_{Mo} - 2m_{Sc})c^2 = [97.90541 \text{ u} - 2(48.95002 \text{ u})](931.5 \text{ MeV/u}) = +5.00 \text{ MeV}.$$

(b) The fact that it is positive does not necessarily mean we should expect to find a great deal of Molybdenum nuclei spontaneously fissioning; the energy barrier (see Fig. 43-3) is presumably higher and/or broader for Molybdenum than for Uranium.

7. (a) Using Eq. 42-20 and adapting Eq. 42-21 to this sample, the number of fission-events per second is

$$R_{fission} = \frac{N \ln 2}{T_{1/2_{fission}}} = \frac{M_{sam} N_A \ln 2}{M_U T_{1/2_{fission}}}$$

$$= \frac{(1.0 \text{ g})(6.02 \times 10^{23} \text{/mol}) \ln 2}{(235 \text{ g/mol})(3.0 \times 10^{17} \text{ y})(365 \text{ d/y})} = 16 \text{ fissions/day}.$$

(b) Since $R \propto \frac{1}{T_{1/2}}$ (see Eq. 42-20), the ratio of rates is

$$\frac{R_\alpha}{R_{fission}} = \frac{T_{1/2_{fission}}}{T_{1/2_\alpha}} = \frac{3.0 \times 10^{17} \text{ y}}{7.0 \times 10^{8} \text{ y}} = 4.3 \times 10^{8}.$$

8. The energy released is

$$Q = (m_U + m_n - m_{Cs} - m_{Rb} - 2m_n)c^2$$
$$= (235.04392 \text{ u} - 1.00867 \text{ u} - 140.91963 \text{ u} - 92.92157 \text{ u})(931.5 \text{ MeV/u})$$
$$= 181 \text{ MeV}.$$

9. (a) If X represents the unknown fragment, then the reaction can be written

$$\,_{92}^{235}\mathrm{U} + \,_{0}^{1}\mathrm{n} \rightarrow \,_{32}^{82}\mathrm{Ge} + \,_{Z}^{A}X$$

where A is the mass number and Z is the atomic number of the fragment. Conservation of charge yields $92 + 0 = 32 + Z$, so $Z = 60$. Conservation of mass number yields $235 + 1 = 83 + A$, so $A = 153$. Looking in Appendix F or G for nuclides with $Z = 60$, we find that the unknown fragment is $\,_{60}^{153}\mathrm{Nd}$.

(b) We neglect the small kinetic energy and momentum carried by the neutron that triggers the fission event. Then, $Q = K_{\mathrm{Ge}} + K_{\mathrm{Nd}}$, where K_{Ge} is the kinetic energy of the germanium nucleus and K_{Nd} is the kinetic energy of the neodymium nucleus. Conservation of momentum yields $\vec{p}_{\mathrm{Ge}} + \vec{p}_{\mathrm{Nd}} = 0$. Now, we can write the classical formula for kinetic energy in terms of the magnitude of the momentum vector:

$$K = \frac{1}{2}mv^2 = \frac{p^2}{2m}$$

which implies that $K_{\mathrm{Nd}} = (m_{\mathrm{Ge}}/m_{\mathrm{Nd}})K_{\mathrm{Ge}}$. Thus, the energy equation becomes

$$Q = K_{\mathrm{Ge}} + \frac{M_{\mathrm{Ge}}}{M_{\mathrm{Nd}}}K_{\mathrm{Ge}} = \frac{M_{\mathrm{Nd}} + M_{\mathrm{Ge}}}{M_{\mathrm{Nd}}}K\mathrm{Ge}$$

and

$$K_{\mathrm{Ge}} = \frac{M_{\mathrm{Nd}}}{M_{\mathrm{Nd}} + M_{\mathrm{Ge}}}Q = \frac{153 \text{ u}}{153 \text{ u} + 83 \text{ u}}(170 \text{ MeV}) = 110 \text{ MeV}.$$

(c) Similarly,

$$K_{\mathrm{Nd}} = \frac{M_{\mathrm{Ge}}}{M_{\mathrm{Nd}} + M_{\mathrm{Ge}}}Q = \frac{83 \text{ u}}{153 \text{ u} + 83 \text{ u}}(170 \text{ MeV}) = 60 \text{ MeV}.$$

(d) The initial speed of the germanium nucleus is

$$v_{\mathrm{Ge}} = \sqrt{\frac{2K_{\mathrm{Ge}}}{M_{\mathrm{Ge}}}} = \sqrt{\frac{2(110 \times 10^6 \text{ eV})(1.60 \times 10^{-19} \text{ J/eV})}{(83 \text{ u})(1.661 \times 10^{-27} \text{ kg/u})}} = 1.60 \times 10^7 \text{ m/s}.$$

(e) The initial speed of the neodymium nucleus is

$$v_{\mathrm{Nd}} = \sqrt{\frac{2K_{\mathrm{Nd}}}{M_{\mathrm{ND}}}} = \sqrt{\frac{2(60 \times 10^6 \text{ eV})(1.60 \times 10^{-19} \text{ J/eV})}{(153 \text{ u})(1.661 \times 10^{-27} \text{ kg/u})}} = 8.69 \times 10^6 \text{ m/s}.$$

10. (a) The surface area a of a nucleus is given by

$$a \simeq 4\pi R^2 \simeq 4\pi[R_0 A^{1/3}]^2 \propto A^{2/3}.$$

Thus, the fractional change in surface area is

$$\frac{\Delta a}{a_i} = \frac{a_f - a_i}{a_i} = \frac{(140)^{2/3} + (96)^{2/3}}{(236)^{2/3}} - 1 = +0.25.$$

(b) Since $V \propto R^3 \propto (A^{1/3})^3 = A$, we have

$$\frac{\Delta V}{V} = \frac{V_f}{V_i} - 1 = \frac{140 + 96}{236} - 1 = 0.$$

(c) The fractional change in potential energy is

$$\frac{\Delta U}{U} = \frac{U_f}{U_i} - 1 = \frac{Q_{Xe}^2 / R_{Xe} + Q_{Sr}^2 / R_{Sr}}{Q_U^2 / R_U} - 1 = \frac{(54)^2 (140)^{-1/3} + (38)^2 (96)^{-1/3}}{(92)^2 (236)^{-1/3}} - 1$$

$$= -0.36.$$

11. (a) The electrostatic potential energy is given by

$$U = \frac{1}{4\pi\varepsilon_0} \frac{Z_{Xe} Z_{Sr} e^2}{r_{Xe} + r_{Sr}}$$

where Z_{Xe} is the atomic number of xenon, Z_{Sr} is the atomic number of strontium, r_{Xe} is the radius of a xenon nucleus, and r_{Sr} is the radius of a strontium nucleus. Atomic numbers can be found either in Appendix F or Appendix G. The radii are given by $r = (1.2 \text{ fm})A^{1/3}$, where A is the mass number, also found in Appendix F. Thus,

$$r_{Xe} = (1.2 \text{ fm})(140)^{1/3} = 6.23 \text{ fm} = 6.23 \times 10^{-15} \text{ m}$$

and

$$r_{Sr} = (1.2 \text{ fm})(96)^{1/3} = 5.49 \text{ fm} = 5.49 \times 10^{-15} \text{ m}.$$

Hence, the potential energy is

$$U = (8.99 \times 10^9 \text{ V} \cdot \text{m} / \text{C}) \frac{(54)(38)(1.60 \times 10^{-19} \text{ C})^2}{6.23 \times 10^{-15} \text{ m} + 5.49 \times 10^{-15} \text{ m}} = 4.08 \times 10^{-11} \text{ J} = 251 \text{ MeV}.$$

(b) The energy released in a typical fission event is about 200 MeV, roughly the same as the electrostatic potential energy when the fragments are touching. The energy appears as kinetic energy of the fragments and neutrons produced by fission.

12. (a) Consider the process $^{239}U + n \rightarrow {}^{140}Ce + {}^{99}Ru + Ne$. We have

$$Z_f - Z_i = Z_{Ce} + Z_{Ru} - Z_U = 58 + 44 - 92 = 10.$$

Thus the number of beta-decay events is 10.

(b) Using Table 37-3, the energy released in this fission process is

$$Q = (m_U + m_n - m_{Ce} - m_{Ru} - 10m_e)c^2$$
$$= (238.05079 \ u + 1.00867 \ u - 139.90543 \ u - 98.90594 \ u)(931.5 \ MeV/u) - 10(0.511 \ MeV)$$
$$= 226 \ MeV.$$

13. If P is the power output, then the energy E produced in the time interval Δt (= 3 y) is

$$E = P \Delta t = (200 \times 10^6 \ W)(3 \ y)(3.156 \times 10^7 \ s/y) = 1.89 \times 10^{16} \ J$$
$$= (1.89 \times 10^{16} \ J)/(1.60 \times 10^{-19} \ J/eV) = 1.18 \times 10^{35} \ eV$$
$$= 1.18 \times 10^{29} \ MeV.$$

At 200 MeV per event, this means $(1.18 \times 10^{29})/200 = 5.90 \times 10^{26}$ fission events occurred. This must be half the number of fissionable nuclei originally available. Thus, there were $2(5.90 \times 10^{26}) = 1.18 \times 10^{27}$ nuclei. The mass of a ^{235}U nucleus is

$$(235 \ u)(1.661 \times 10^{-27} \ kg/u) = 3.90 \times 10^{-25} \ kg,$$

so the total mass of ^{235}U originally present was $(1.18 \times 10^{27})(3.90 \times 10^{-25} \ kg) = 462 \ kg$.

14. When a neutron is captured by ^{237}Np it gains 5.0 MeV, more than enough to offset the 4.2 MeV required for ^{238}Np to fission. Consequently, ^{237}Np is fissionable by thermal neutrons.

15. If R is the decay rate then the power output is $P = RQ$, where Q is the energy produced by each alpha decay. Now $R = \lambda N = N \ln 2/T_{1/2}$, where λ is the disintegration constant and $T_{1/2}$ is the half-life. The relationship $\lambda = (\ln 2)/T_{1/2}$ is used. If M is the total mass of material and m is the mass of a single ^{238}Pu nucleus, then

$$N = \frac{M}{m} = \frac{1.00 \ kg}{(238 \ u)(1.661 \times 10^{-27} \ kg/u)} = 2.53 \times 10^{24}.$$

Thus,

$$P = \frac{NQ\ln 2}{T_{1/2}} = \frac{(2.53\times10^{24})(5.50\times10^6 \text{ eV})(1.60\times10^{-19} \text{ J/eV})(\ln 2)}{(87.7\,\text{y})(3.156\times10^7\,\text{s/y})} = 557\,\text{W}.$$

16. (a) Using the result of problem 4, the TNT equivalent is

$$\frac{(2.50 \text{ kg})(4.54\times10^{26} \text{ MeV / kg})}{2.6\times10^{28} \text{ MeV} / 10^6 \text{ ton}} = 4.4\times10^4 \text{ ton} = 44 \text{ kton}.$$

(b) Assuming that this is a fairly inefficiently designed bomb, then much of the remaining 92.5 kg is probably "wasted" and was included perhaps to make sure the bomb did not "fizzle." There is also an argument for having more than just the critical mass based on the short assembly-time of the material during the implosion, but this so-called "super-critical mass," as generally quoted, is much less than 92.5 kg, and does not necessarily have to be purely Plutonium.

17. (a) We solve Q_{eff} from $P = RQ_{\text{eff}}$:

$$Q_{\text{eff}} = \frac{P}{R} = \frac{P}{N\lambda} = \frac{mPT_{1/2}}{M\ln 2}$$

$$= \frac{(90.0 \text{ u})(1.66\times10^{-27} \text{ kg / u})(0.93 \text{ W})(29 \text{ y})(3.15\times10^7 \text{ s/ y})}{(1.00\times10^{-3} \text{ kg})(\ln 2)(1.60\times10^{-13} \text{ J / MeV})}$$

$$= 1.2 \text{ MeV}.$$

(b) The amount of ^{90}Sr needed is

$$M = \frac{150 \text{ W}}{(0.050)(0.93 \text{ W / g})} = 3.2 \text{ kg}.$$

18. After each time interval t_{gen} the number of nuclides in the chain reaction gets multiplied by k. The number of such time intervals that has gone by at time t is t/t_{gen}. For example, if the multiplication factor is 5 and there were 12 nuclei involved in the reaction to start with, then after one interval 60 nuclei are involved. And after another interval 300 nuclei are involved. Thus, the number of nuclides engaged in the chain reaction at time t is $N(t) = N_0 k^{t/t_{\text{gen}}}$. Since $P \propto N$ we have

$$P(t) = P_0 k^{t/t_{\text{gen}}}.$$

19. (a) The energy yield of the bomb is

$$E = (66\times10^{-3} \text{ megaton})(2.6\times10^{28} \text{ MeV/ megaton}) = 1.72\times10^{27} \text{ MeV}.$$

At 200 MeV per fission event,

$$(1.72 \times 10^{27}\ \text{MeV})/(200\ \text{MeV}) = 8.58 \times 10^{24}$$

fission events take place. Since only 4.0% of the ^{235}U nuclei originally present undergo fission, there must have been $(8.58 \times 10^{24})/(0.040) = 2.14 \times 10^{26}$ nuclei originally present. The mass of ^{235}U originally present was

$$(2.14 \times 10^{26})(235\ \text{u})(1.661 \times 10^{-27}\ \text{kg/u}) = 83.7\ \text{kg} \approx 84\ \text{kg}.$$

(b) Two fragments are produced in each fission event, so the total number of fragments is

$$2(8.58 \times 10^{24}) = 1.72 \times 10^{25} \approx 1.7 \times 10^{25}.$$

(c) One neutron produced in a fission event is used to trigger the next fission event, so the average number of neutrons released to the environment in each event is 1.5. The total number released is

$$(8.58 \times 10^{24})(1.5) = 1.29 \times 10^{25} \approx 1.3 \times 10^{25}.$$

20. We use the formula from problem 22:

$$P(t) = P_0 k^{t/t_{\text{gen}}} = (400\,\text{MW})(1.0003)^{(5.00\,\text{min})(60\,\text{s/min})/(0.00300\,\text{s})} = 8.03 \times 10^3\,\text{MW}.$$

21. (a) Let v_{ni} be the initial velocity of the neutron, v_{nf} be its final velocity, and v_f be the final velocity of the target nucleus. Then, since the target nucleus is initially at rest, conservation of momentum yields $m_n v_{ni} = m_n v_{nf} + m v_f$ and conservation of energy yields $\frac{1}{2} m_n v_{ni}^2 = \frac{1}{2} m_n v_{nf}^2 + \frac{1}{2} m v_f^2$. We solve these two equations simultaneously for v_f. This can be done, for example, by using the conservation of momentum equation to obtain an expression for v_{nf} in terms of v_f and substituting the expression into the conservation of energy equation. We solve the resulting equation for v_f. We obtain $v_f = 2 m_n v_{ni}/(m + m_n)$. The energy lost by the neutron is the same as the energy gained by the target nucleus, so

$$\Delta K = \frac{1}{2} m v_f^2 = \frac{1}{2} \frac{4 m_n^2 m}{(m + m_n)^2} v_{ni}^2.$$

The initial kinetic energy of the neutron is $K = \frac{1}{2} m_n v_{ni}^2$, so

$$\frac{\Delta K}{K} = \frac{4 m_n m}{(m + m_n)_2}.$$

(b) The mass of a neutron is 1.0 u and the mass of a hydrogen atom is also 1.0 u. (Atomic masses can be found in Appendix G.) Thus,

$$\frac{\Delta K}{K} = \frac{4(1.0 \text{ u})(1.0 \text{ u})}{(1.0 \text{ u} + 1.0 \text{ u})^2} = 1.0.$$

(c) Similarly, the mass of a deuterium atom is 2.0 u, so

$$(\Delta K)/K = 4(1.0 \text{ u})(2.0 \text{ u})/(2.0 \text{ u} + 1.0 \text{ u})^2 = 0.89.$$

(d) The mass of a carbon atom is 12 u, so $(\Delta K)/K = 4(1.0 \text{ u})(12 \text{ u})/(12 \text{ u} + 1.0 \text{ u})^2 = 0.28.$

(e) The mass of a lead atom is 207 u, so

$$(\Delta K)/K = 4(1.0 \text{ u})(207 \text{ u})/(207 \text{ u} + 1.0 \text{ u})^2 = 0.019.$$

(f) During each collision, the energy of the neutron is reduced by the factor $1 - 0.89 = 0.11$. If E_i is the initial energy, then the energy after n collisions is given by $E = (0.11)^n E_i$. We take the natural logarithm of both sides and solve for n. The result is

$$n = \frac{\ln(E/E_i)}{\ln 0.11} = \frac{\ln(0.025 \text{ eV}/1.00 \text{ eV})}{\ln 0.11} = 7.9.$$

The energy first falls below 0.025 eV on the eighth collision.

22. We recall Eq. 43-6: $Q \approx 200 \text{ MeV} = 3.2 \times 10^{-11}$ J. It is important to bear in mind that Watts multiplied by seconds give Joules. From $E = Pt_{gen} = NQ$ we get the number of free neutrons:

$$N = \frac{Pt_{gen}}{Q} = \frac{(500 \times 10^6 \text{ W})(1.0 \times 10^{-3} \text{ s})}{3.2 \times 10^{-11} \text{ J}} = 1.6 \times 10^{16}.$$

23. Let P_0 be the initial power output, P be the final power output, k be the multiplication factor, t be the time for the power reduction, and t_{gen} be the neutron generation time. Then, according to the result of Problem 22,

$$P = P_0 k^{t/t_{gen}}.$$

We divide by P_0, take the natural logarithm of both sides of the equation and solve for $\ln k$:

$$\ln k = \frac{t_{gen}}{t} \ln \frac{P}{P_0} = \frac{1.3 \times 10^{-3} \text{ s}}{2.6 \text{ s}} \ln \frac{350 \text{ MW}}{1200 \text{ MW}} = -0.0006161.$$

Hence, $k = e^{-0.0006161} = 0.99938.$

24. (a) $P_{avg} = (15 \times 10^9 \text{ W·y})/(200,000 \text{ y}) = 7.5 \times 10^4 \text{ W} = 75 \text{ kW}.$

(b) Using the result of Eq. 43-6, we obtain

$$M = \frac{m_U E_{total}}{Q} = \frac{(235\,u)(1.66 \times 10^{-27}\,kg/u)(15 \times 10^9\,W \cdot y)(3.15 \times 10^7\,s/y)}{(200\,MeV)(1.6 \times 10^{-13}\,J/MeV)}$$

$$= 5.8 \times 10^3\,kg.$$

25. Our approach is the same as that shown in Sample Problem 43-3. We have

$$\frac{N_5(t)}{N_8(t)} = \frac{N_5(0)}{N_8(0)} e^{-(\lambda_5 - \lambda_8)t},$$

or

$$t = \frac{1}{\lambda_8 - \lambda_5} \ln \left[\left(\frac{N_5(t)}{N_8(t)} \right) \left(\frac{N_8(0)}{N_5(0)} \right) \right] = \frac{1}{(1.55 - 9.85)10^{-10}\,y^{-1}} \ln[(0.0072)(0.15)^{-1}]$$

$$= 3.6 \times 10^9\,y.$$

26. The nuclei of ^{238}U can capture neutrons and beta-decay. With a large amount of neutrons available due to the fission of ^{235}U, the probability for this process is substantially increased, resulting in a much higher decay rate for ^{238}U and causing the depletion of ^{238}U (and relative enrichment of ^{235}U).

27. Let t be the present time and $t = 0$ be the time when the ratio of ^{235}U to ^{238}U was 3.0%. Let N_{235} be the number of ^{235}U nuclei present in a sample now and $N_{235,0}$ be the number present at $t = 0$. Let N_{238} be the number of ^{238}U nuclei present in the sample now and $N_{238,0}$ be the number present at $t = 0$. The law of radioactive decay holds for each specie, so

$$N_{235} = N_{235,0} e^{-\lambda_{235}t}$$

and

$$N_{238} = N_{238,0} e^{-\lambda_{238}t}.$$

Dividing the first equation by the second, we obtain

$$r = r_0 e^{-(\lambda_{235} - \lambda_{238})t}$$

where $r = N_{235}/N_{238}$ ($= 0.0072$) and $r_0 = N_{235,0}/N_{238,0}$ ($= 0.030$). We solve for t:

$$t = -\frac{1}{\lambda_{235} - \lambda_{238}} \ln \frac{r}{r_0}.$$

Now we use $\lambda_{235} = (\ln 2) / T_{1/2_{235}}$ and $\lambda_{238} = (\ln 2) / T_{1/2_{238}}$ to obtain

$$t = \frac{T_{1/2_{235}} T_{1/2_{238}}}{(T_{1/2_{238}} - T_{1/2_{235}}) \ln 2} \ln \frac{r}{r_0} = -\frac{(7.0 \times 10^8 \text{ y})(4.5 \times 10^9 \text{ y})}{(4.5 \times 10^9 \text{ y} - 7.0 \times 10^8 \text{ y}) \ln 2} \ln \frac{0.0072}{0.030}$$

$$= 1.7 \times 10^9 \text{ y}.$$

28. We are given the energy release per fusion ($Q = 3.27$ MeV $= 5.24 \times 10^{-13}$ J) and that a pair of deuterium atoms are consumed in each fusion event. To find how many pairs of deuterium atoms are in the sample, we adapt Eq. 42-21:

$$N_{d \text{ pairs}} = \frac{M_{\text{sam}}}{2 M_d} N_A = \left(\frac{1000 \text{ g}}{2(2.0 \text{ g} / \text{mol})} \right)(6.02 \times 10^{23} / \text{mol}) = 1.5 \times 10^{26}.$$

Multiplying this by Q gives the total energy released: 7.9×10^{13} J. Keeping in mind that a Watt is a Joule per second, we have

$$t = \frac{7.9 \times 10^{13} \text{ J}}{100 \text{ W}} = 7.9 \times 10^{11} \text{ s} = 2.5 \times 10^4 \text{ y}.$$

29. The height of the Coulomb barrier is taken to be the value of the kinetic energy K each deuteron must initially have if they are to come to rest when their surfaces touch (see Sample Problem 43-4). If r is the radius of a deuteron, conservation of energy yields

$$2K = \frac{1}{4\pi\varepsilon_0} \frac{e^2}{2r},$$

so

$$K = \frac{1}{4\pi\varepsilon_0} \frac{e^2}{4r} = (8.99 \times 10^9 \text{ V} \cdot \text{m} / \text{C}) \frac{(1.60 \times 10^{-19} \text{ C})^2}{4(2.1 \times 10^{-15} \text{ m})} = 2.74 \times 10^{-14} \text{ J} = 170 \text{ keV}.$$

30. (a) Our calculation is identical to that in Sample Problem 43-4 except that we are now using R appropriate to two deuterons coming into "contact," as opposed to the $R = 1.0$ fm value used in the Sample Problem. If we use $R = 2.1$ fm for the deuterons (this is the value given in problem 33), then our K is simply the K calculated in Sample Problem 43-4, divided by 2.1:

$$K_{d+d} = \frac{K_{p+p}}{2.1} = \frac{360 \text{ keV}}{2.1} \approx 170 \text{ keV}.$$

Consequently, the voltage needed to accelerate each deuteron from rest to that value of K is 170 kV.

(b) Not all deuterons that are accelerated towards each other will come into "contact" and not all of those that do so will undergo nuclear fusion. Thus, a great many deuterons must be repeatedly encountering other deuterons in order to produce a macroscopic energy release. An accelerator needs a fairly good vacuum in its beam pipe, and a very large number flux is either impractical and/or very expensive. Regarding expense, there are other factors that have dissuaded researchers from using accelerators to build a controlled fusion "reactor," but those factors may become less important in the future — making the feasibility of accelerator "add-on's" to magnetic and inertial confinement schemes more cost-effective.

31. Our calculation is very similar to that in Sample Problem 43-4 except that we are now using R appropriate to two Lithium-7 nuclei coming into "contact," as opposed to the $R = 1.0$ fm value used in the Sample Problem. If we use

$$R = r = r_0 A^{1/3} = (1.2 \text{ fm})\sqrt[3]{7} = 2.3 \text{ fm}$$

and $q = Ze = 3e$, then our K is given by (see Sample Problem 43-4)

$$K = \frac{Z^2 e^2}{16\pi\varepsilon_0 r} = \frac{3^2 (1.6 \times 10^{-19} \text{ C})^2}{16\pi(8.85 \times 10^{-12} \text{ F/m})(2.3 \times 10^{15} \text{ m})}$$

which yields 2.25×10^{-13} J $= 1.41$ MeV. We interpret this as the answer to the problem, though the term "Coulomb barrier height" as used here may be open to other interpretations.

32. From the expression for $n(K)$ given we may write $n(K) \propto K^{1/2} e^{-K/kT}$. Thus, with

$$k = 8.62 \times 10^{-5} \text{ eV/K} = 8.62 \times 10^{-8} \text{ keV/K},$$

we have

$$\frac{n(K)}{n(K_{avg})} = \left(\frac{K}{K_{avg}}\right)^{1/2} e^{-(K-K_{avg})/kT} = \left(\frac{5.00 \text{ keV}}{1.94 \text{ keV}}\right)^{1/2} e^{-(5.00 \text{ keV}-1.94 \text{ keV})/[(8.62\times10^{-8} \text{ keV})(1.50\times10^7 \text{ K})]}$$

$$= 0.151.$$

33. If M_{He} is the mass of an atom of helium and M_C is the mass of an atom of carbon, then the energy released in a single fusion event is

$$Q = [3M_{He} - M_C]c^2 = [3(4.0026 \text{ u}) - (12.0000 \text{ u})](931.5 \text{ MeV/u}) = 7.27 \text{ MeV}.$$

Note that $3M_{He}$ contains the mass of six electrons and so does M_C. The electron masses cancel and the mass difference calculated is the same as the mass difference of the nuclei.

34. In Fig. 43-11, let $Q_1 = 0.42$ MeV, $Q_2 = 1.02$ MeV, $Q_3 = 5.49$ MeV and $Q_4 = 12.86$ MeV. For the overall proton-proton cycle

$$Q = 2Q_1 + 2Q_2 + 2Q_3 + Q_4$$
$$= 2(0.42 \text{ MeV} + 1.02 \text{ MeV} + 5.49 \text{ MeV}) + 12.86 \text{ MeV} = 26.7 \text{ MeV}.$$

35. (a) Let M be the mass of the Sun at time t and E be the energy radiated to that time. Then, the power output is $P = dE/dt = (dM/dt)c^2$, where $E = Mc^2$ is used. At the present time,

$$\frac{dM}{dt} = \frac{P}{c^2} = \frac{3.9 \times 10^{26} \text{ W}}{\left(2.998 \times 10^8 \text{ m/s}\right)^2} = 4.3 \times 10^9 \text{ kg/s}.$$

(b) We assume the rate of mass loss remained constant. Then, the total mass loss is

$$\Delta M = (dM/dt) \, \Delta t = (4.33 \times 10^9 \text{ kg/s}) \, (4.5 \times 10^9 \text{ y}) \, (3.156 \times 10^7 \text{ s/y}) = 6.15 \times 10^{26} \text{ kg}.$$

The fraction lost is

$$\frac{\Delta M}{M + \Delta M} = \frac{6.15 \times 10^{26} \text{ kg}}{2.0 \times 10^{30} \text{ kg} + 6.15 \times 10^{26} \text{ kg}} = 3.1 \times 10^{-4}.$$

36. We assume the neutrino has negligible mass. The photons, of course, are also taken to have zero mass.

$$Q_1 = \left(2m_p - m_2 - m_e\right)c^2 = \left[2(m_1 - m_e) - (m_2 - m_e) - m_e\right]c^2$$
$$= \left[2(1.007825 \text{ u}) - 2.014102 \text{ u} - 2(0.0005486 \text{ u})\right](931.5 \text{ MeV/u})$$
$$= 0.42 \text{ MeV}$$
$$Q_2 = \left(m_2 + m_p - m_3\right)c^2 = \left(m_2 + m_p - m_3\right)c^2$$
$$= (2.014102 \text{ u}) + 1.007825 \text{ u} - 3.016029 \text{ u})(931.5 \text{ MeV/u})$$
$$= 5.49 \text{ MeV}$$
$$Q_3 = \left(2m_3 - m_4 - 2m_p\right)c^2 = \left(2m_3 - m_4 - 2m_p\right)c^2$$
$$= \left[2(3.016029 \text{ u}) - 4.002603 \text{ u} - 2(1.007825 \text{ u})\right](931.5 \text{ MeV/u})$$
$$= 12.86 \text{ MeV}.$$

37. (a) Since two neutrinos are produced per proton-proton cycle (see Eq. 43-10 or Fig. 43-11), the rate of neutrino production R_ν satisfies

$$R_v = \frac{2P}{Q} = \frac{2(3.9\times10^{26}\,\text{W})}{(26.7\,\text{MeV})(1.6\times10^{-13}\,\text{J/MeV})} = 1.8\times10^{38}\,\text{s}^{-1}\ .$$

(b) Let d_{es} be the Earth to Sun distance, and R be the radius of Earth (see Appendix C). Earth represents a small cross section in the "sky" as viewed by a fictitious observer on the Sun. The rate of neutrinos intercepted by that area (very small, relative to the area of the full "sky") is

$$R_{v,\text{Earth}} = R_v\left(\frac{\pi R_e^2}{4\pi d_{es}^2}\right) = \frac{(1.8\times10^{38}\,\text{s}^{-1})}{4}\left(\frac{6.4\times10^6\,\text{m}}{1.5\times10^{11}\,\text{m}}\right)^2 = 8.2\times10^{28}\,\text{s}^{-1}\ .$$

38. (a) We are given the energy release per fusion (calculated in §43-7: $Q = 26.7$ MeV = 4.28×10^{-12} J) and that four protons are consumed in each fusion event. To find how many sets of four protons are in the sample, we adapt Eq. 42-21:

$$N_{4p} = \frac{M_{\text{sam}}}{4M_H}N_A = \left(\frac{1000\,\text{g}}{4(1.0\,\text{g/mol})}\right)(6.02\times10^{23}/\text{mol}) = 1.5\times10^{26}\ .$$

Multiplying this by Q gives the total energy released: 6.4×10^{14} J. It is not required that the answer be in SI units; we could have used MeV throughout (in which case the answer is 4.0×10^{27} MeV).

(b) The number of ^{235}U nuclei is

$$N_{235} = \left(\frac{1000\,\text{g}}{235\,\text{g/mol}}\right)(6.02\times10^{23}/\text{mol}) = 2.56\times10^{24}\ .$$

If all the U-235 nuclei fission, the energy release (using the result of Eq. 43-6) is

$$N_{235}Q_{\text{fission}} = (2.56\times10^{22})(200\,\text{MeV}) = 5.1\times10^{26}\,\text{MeV} = 8.2\times10^{13}\,\text{J}\ .$$

We see that the fusion process (with regard to a unit mass of fuel) produces a larger amount of energy (despite the fact that the Q value per event is smaller).

39. (a) The mass of a carbon atom is $(12.0\,\text{u})(1.661\times10^{-27}\,\text{kg/u}) = 1.99\times10^{-26}$ kg, so the number of carbon atoms in 1.00 kg of carbon is

$$(1.00\,\text{kg})/(1.99\times10^{-26}\,\text{kg}) = 5.02\times10^{25}.$$

The heat of combustion per atom is

$(3.3 \times 10^7 \text{ J/kg})/(5.02 \times 10^{25} \text{ atom/kg}) = 6.58 \times 10^{-19}$ J/atom.

This is 4.11 eV/atom.

(b) In each combustion event, two oxygen atoms combine with one carbon atom, so the total mass involved is $2(16.0 \text{ u}) + (12.0 \text{ u}) = 44$ u. This is

$$(44 \text{ u})(1.661 \times 10^{-27} \text{ kg/u}) = 7.31 \times 10^{-26} \text{ kg}.$$

Each combustion event produces 6.58×10^{-19} J so the energy produced per unit mass of reactants is $(6.58 \times 10^{-19} \text{ J})/(7.31 \times 10^{-26} \text{ kg}) = 9.00 \times 10^6$ J/kg.

(c) If the Sun were composed of the appropriate mixture of carbon and oxygen, the number of combustion events that could occur before the Sun burns out would be

$$(2.0 \times 10^{30} \text{ kg})/(7.31 \times 10^{-26} \text{ kg}) = 2.74 \times 10^{55}.$$

The total energy released would be

$$E = (2.74 \times 10^{55})(6.58 \times 10^{-19} \text{ J}) = 1.80 \times 10^{37} \text{ J}.$$

If P is the power output of the Sun, the burn time would be

$$t = \frac{E}{P} = \frac{1.80 \times 10^{37} \text{ J}}{3.9 \times 10^{26} \text{ W}} = 4.62 \times 10^{10} \text{ s} = 1.46 \times 10^3 \text{ y},$$

1.5×10^3 y, to two significant figures.

40. (a) The products of the carbon cycle are $2e^+ + 2\nu + {}^4\text{He}$, the same as that of the proton-proton cycle (see Eq. 43-10). The difference in the number of photons is not significant.

(b) $Q_{\text{carbon}} = Q_1 + Q_2 + \cdots + Q_6 = (1.95 \times 1.19 + 7.55 + 7.30 + 1.73 + 4.97) \text{ MeV} = 24.7 \text{ MeV}$, which is the same as that for the proton-proton cycle (once we subtract out the electron-positron annihilations; see Fig. 43-11):

$$Q_{p-p} = 26.7 \text{ MeV} - 2(1.02 \text{ MeV}) = 24.7 \text{ MeV}.$$

41. Since the mass of a helium atom is $(4.00 \text{ u})(1.661 \times 10^{-27} \text{ kg/u}) = 6.64 \times 10^{-27}$ kg, the number of helium nuclei originally in the star is

$$(4.6 \times 10^{32} \text{ kg})/(6.64 \times 10^{-27} \text{ kg}) = 6.92 \times 10^{58}.$$

Since each fusion event requires three helium nuclei, the number of fusion events that can take place is $N = 6.92 \times 10^{58}/3 = 2.31 \times 10^{58}$. If Q is the energy released in each event and t is the conversion time, then the power output is $P = NQ/t$ and

$$t = \frac{NQ}{P} = \frac{(2.31 \times 10^{58})(7.27 \times 10^6 \text{ eV})(1.60 \times 10^{-19} \text{ J/eV})}{5.3 \times 10^{30} \text{ W}} = 5.07 \times 10^{15} \text{ s} = 1.6 \times 10^8 \text{ y}.$$

42. The mass of the hydrogen in the Sun's core is $m_\text{H} = 0.35(\frac{1}{8} M_\text{Sun})$. The time it takes for the hydrogen to be entirely consumed is

$$t = \frac{M_\text{H}}{dm/dt} = \frac{(0.35)(\frac{1}{8})(2.0 \times 10^{30} \text{ kg})}{(6.2 \times 10^{11} \text{ kg/s})(3.15 \times 10^7 \text{ s/y})} = 5 \times 10^9 \text{ y}.$$

43. (a)

$$Q = \left(5m_{^2\text{H}} - m_{^3\text{He}} - m_{^4\text{He}} - m_{^1\text{H}} - 2m_n\right)c^2$$
$$= \left[5(2.014102\,\text{u}) - 3.016029\,\text{u} - 4.002603\,\text{u} - 1.007825\,\text{u} - 2(1.008665\,\text{u})\right](931.5\,\text{MeV/u})$$
$$= 24.9\,\text{MeV}.$$

(b) Assuming 30.0% of the deuterium undergoes fusion, the total energy released is

$$E = NQ = \left(\frac{0.300\,M}{5m_{^2\text{H}}}\right)Q.$$

Thus, the rating is

$$R = \frac{E}{2.6 \times 10^{28}\,\text{MeV/megaton TNT}}$$
$$= \frac{(0.300)(500\,\text{kg})(24.9\,\text{MeV})}{5(2.0\,\text{u})(1.66 \times 10^{-27}\,\text{kg/u})(2.6 \times 10^{28}\,\text{MeV/megaton TNT})}$$
$$= 8.65\,\text{megaton TNT}.$$

44. In Eq. 43-13,

$$Q = \left(2m_{^2\text{H}} - m_{^3\text{He}} - m_n\right)c^2$$
$$= \left[2(2.014102\,\text{u}) - 3.016049\,\text{u} - 1.008665\,\text{u}\right](931.5\,\text{MeV/u})$$
$$= 3.27\,\text{MeV}.$$

In Eq. 43-14,

$$Q = \left(2m_{^2H} - m_{^3H} - m_{^1H}\right)c^2$$
$$= \left[2(2.014102\,\text{u}) - 3.016049\,\text{u} - 1.007825\,\text{u}\right](931.5\,\text{MeV/u})$$
$$= 4.03\,\text{MeV} .$$

Finally, in Eq. 43-15,

$$Q = \left(m_{^2H} + m_{^3H} - m_{^4He} - m_n\right)c^2$$
$$= \left[2.014102\,\text{u} + 3.016049\,\text{u} - 4.002603\,\text{u} - 1.008665\,\text{u}\right](931.5\,\text{MeV/u})$$
$$= 17.59\,\text{MeV} .$$

45. Since 1.00 L of water has a mass of 1.00 kg, the mass of the heavy water in 1.00 L is 0.0150×10^{-2} kg $= 1.50 \times 10^{-4}$ kg. Since a heavy water molecule contains one oxygen atom, one hydrogen atom and one deuterium atom, its mass is

$$(16.0\,\text{u} + 1.00\,\text{u} + 2.00\,\text{u}) = 19.0\,\text{u} = (19.0\,\text{u})(1.661 \times 10^{-27}\,\text{kg/u}) = 3.16 \times 10^{-26}\,\text{kg.}$$

The number of heavy water molecules in a liter of water is

$$(1.50 \times 10^{-4}\,\text{kg})/(3.16 \times 10^{-26}\,\text{kg}) = 4.75 \times 10^{21}.$$

Since each fusion event requires two deuterium nuclei, the number of fusion events that can occur is $N = 4.75 \times 10^{21}/2 = 2.38 \times 10^{21}$. Each event releases energy

$$Q = (3.27 \times 10^6\,\text{eV})(1.60 \times 10^{-19}\,\text{J/eV}) = 5.23 \times 10^{-13}\,\text{J.}$$

Since all events take place in a day, which is 8.64×10^4 s, the power output is

$$P = \frac{NQ}{t} = \frac{(2.38 \times 10^{21})(5.23 \times 10^{-13}\,\text{J})}{8.64 \times 10^4\,\text{s}} = 1.44 \times 10^4\,\text{W} = 14.4\,\text{kW} .$$

46. (a) From $E = NQ = (M_{sam}/4m_p)Q$ we get the energy per kilogram of hydrogen consumed:

$$\frac{E}{M_{sam}} = \frac{Q}{4m_p} = \frac{(26.2\,\text{MeV})(1.60 \times 10^{-13}\,\text{J/MeV})}{4(1.67 \times 10^{-27}\,\text{kg})} = 6.3 \times 10^{14}\,\text{J/kg} .$$

(b) Keeping in mind that a Watt is a Joule per second, the rate is

$$\frac{dm}{dt} = \frac{3.9 \times 10^{26} \text{ W}}{6.3 \times 10^{14} \text{ J/kg}} = 6.2 \times 10^{11} \text{ kg/s} .$$

This agrees with the computation shown in Sample Problem 43-5.

(c) From the Einstein relation $E = Mc^2$ we get $P = dE/dt = c^2 dM/dt$, or

$$\frac{dM}{dt} = \frac{P}{c^2} = \frac{3.9 \times 10^{26} \text{ W}}{\left(3.0 \times 10^8 \text{ m/s}\right)^2} = 4.3 \times 10^9 \text{ kg/s} .$$

(d) This finding, that $dm/dt > dM/dt$, is in large part due to the fact that, as the protons are consumed, their mass is mostly turned into alpha particles (helium), which remain in the Sun.

(e) The time to lose 0.10% of its total mass is

$$t = \frac{0.0010 \, M}{dM/dt} = \frac{(0.0010)\left(2.0 \times 10^{30} \text{ kg}\right)}{\left(4.3 \times 10^9 \text{ kg/s}\right)\left(3.15 \times 10^7 \text{ s/y}\right)} = 1.5 \times 10^{10} \text{ y} .$$

47. (a) From $\rho_H = 0.35\rho = n_p m_p$, we get the proton number density n_p:

$$n_p = \frac{0.35\rho}{m_p} = \frac{(0.35)\left(1.5 \times 10^5 \text{ kg/m}^3\right)}{1.67 \times 10^{-27} \text{ kg}} = 3.1 \times 10^{31} \text{ m}^{-3} .$$

(b) From Chapter 19 (see Eq. 19-9), we have

$$\frac{N}{V} = \frac{p}{kT} = \frac{1.01 \times 10^5 \text{ Pa}}{\left(1.38 \times 10^{-23} \text{ J/K}\right)(273\,\text{K})} = 2.68 \times 10^{25} \text{ m}^{-3}$$

for an ideal gas under "standard conditions." Thus,

$$\frac{n_p}{(N/V)} = \frac{3.14 \times 10^{31} \text{ m}^{-3}}{2.44 \times 10^{25} \text{ m}^{-3}} = 1.2 \times 10^6 .$$

48. Conservation of energy gives $Q = K_\alpha + K_n$, and conservation of linear momentum (due to the assumption of negligible initial velocities) gives $|p_\alpha| = |p_n|$. We can write the classical formula for kinetic energy in terms of momentum:

$$K = \frac{1}{2} mv^2 = \frac{p^2}{2m}$$

which implies that $K_n = (m_\alpha/m_n)K_\alpha$.

(a) Consequently, conservation of energy and momentum allows us to solve for kinetic energy of the alpha particle which results from the fusion:

$$K_\alpha = \frac{Q}{1+\frac{m_\alpha}{m_n}} = \frac{17.59\,\text{MeV}}{1+\frac{4.0015\,\text{u}}{1.008665\,\text{u}}} = 3.541\,\text{MeV}$$

where we have found the mass of the alpha particle by subtracting two electron masses from the ^{4}He mass (quoted several times in this and the previous chapter).

(b) Then, $K_n = Q - K_\alpha$ yields 14.05 MeV for the neutron kinetic energy.

49. Since Plutonium has $Z = 94$ and Uranium has $Z = 92$, we see that (to conserve charge) two electrons must be emitted so that the nucleus can gain a $+2e$ charge. In the beta decay processes described in Chapter 42, electrons and neutrinos are emitted. The reaction series is as follows:

$$^{238}\text{U} + \text{n} \rightarrow {}^{239}\text{Np} + {}^{239}\text{U} + e + \nu$$
$$^{239}\text{Np} \rightarrow {}^{239}\text{Pu} + e + \nu$$

50. (a) Rather than use $P(v)$ as it is written in Eq. 19-27, we use the more convenient nK expression given in problem 32 of this chapter [43]. The $n(K)$ expression can be derived from Eq. 19-27, but we do not show that derivation here. To find the most probable energy, we take the derivative of $n(K)$ and set the result equal to zero:

$$\left.\frac{dn(K)}{dK}\right|_{K=K_p} = \frac{1.13n}{(kT)^{3/2}}\left(\frac{1}{2K^{1/2}} - \frac{K^{3/2}}{kT}\right)e^{-K/kT}\Bigg|_{K=K_p} = 0,$$

which gives $K_p = \frac{1}{2}kT$. Specifically, for $T = 1.5 \times 10^7$ K we find

$$K_p = \frac{1}{2}kT = \frac{1}{2}(8.62 \times 10^{-5}\,\text{eV/K})(1.5 \times 10^7\,\text{K}) = 6.5 \times 10^2\,\text{eV}$$

or 0.65 keV, in good agreement with Fig. 43-10.

(b) Eq. 19-35 gives the most probable speed in terms of the molar mass M, and indicates its derivation (see also Sample Problem 19-6). Since the mass m of the particle is related to M by the Avogadro constant, then using Eq. 19-7,

$$v_p = \sqrt{\frac{2RT}{M}} = \sqrt{\frac{2RT}{mN_A}} = \sqrt{\frac{2kT}{m}}.$$

With $T = 1.5 \times 10^7$ K and $m = 1.67 \times 10^{-27}$ kg, this yields $v_p = 5.0 \times 10^5$ m/s.

(c) The corresponding kinetic energy is

$$K_{v,p} = \frac{1}{2}mv_p^2 = \frac{1}{2}m\left(\sqrt{\frac{2kT}{m}}\right)^2 = kT$$

which is twice as large as that found in part (a). Thus, at $T = 1.5 \times 10^7$ K we have $K_{v,p} = 1.3$ keV, which is indicated in Fig. 43-10 by a single vertical line.

51. In Sample Problem 43-2, it is noted that the rate of consumption of U-235 by (nonfission) neutron capture is one-fourth as big as the rate of neutron-induced fission events. Consequently, the mass of ^{235}U should be larger than that computed in problem 15 by 25%: $(1.25)(462 \text{ kg}) = 5.8 \times 10^2 \text{ kg} \approx 6 \times 10^2$ kg. If appeal is to made to other sources (other than Sample Problem 43-2), then it might be possible to argue for a factor other than 1.25 (we found others in our brief search) and thus to a somewhat different result.

52. First, we figure out the mass of U-235 in the sample (assuming "3.0%" refers to the proportion by weight as opposed to proportion by number of atoms):

$$M_{U-235} = (3.0\%)M_{sam}\left(\frac{(97\%)m_{238} + (3.0\%)m_{235}}{(97\%)m_{238} + (3.0\%)m_{235} + 2m_{16}}\right)$$

$$= (0.030)(1000 \text{ g})\left(\frac{0.97(238) + 0.030(235)}{0.97(238) + 0.030(235) + 2(16.0)}\right)$$

$$= 26.4 \text{ g.}$$

Next, this uses some of the ideas illustrated in Sample Problem 42-5; our notation is similar to that used in that example. The number of ^{235}U nuclei is

$$N_{235} = \frac{(26.4 \text{ g})(6.02 \times 10^{23} / \text{mol})}{235 \text{ g} / \text{mol}} = 6.77 \times 10^{22}.$$

If all the U-235 nuclei fission, the energy release (using the result of Eq. 43-6) is

$$N_{235}Q_{fission} = (6.77 \times 10^{22})(200 \text{ MeV}) = 1.35 \times 10^{25} \text{ MeV} = 2.17 \times 10^{12} \text{ J.}$$

Keeping in mind that a Watt is a Joule per second, the time that this much energy can keep a 100-W lamp burning is found to be

$$t = \frac{2.17 \times 10^{12} \text{ J}}{100 \text{ W}} = 2.17 \times 10^{10} \text{ s} \approx 690 \text{ y}.$$

If we had instead used the $Q = 208$ MeV value from Sample Problem 43-1, then our result would have been 715 y, which perhaps suggests that our result is meaningful to just one significant figure ("roughly 700 years").

53. At $T = 300$ K, the average kinetic energy of the neutrons is (using Eq. 20-24)

$$K_{\text{avg}} = \frac{3}{2} KT = \frac{3}{2}(8.62 \times 10^{-5} \text{ eV / K})(300 \text{ K}) \approx 0.04 \text{ eV}.$$

54. (a) Fig. 42-9 shows the barrier height to be about 30 MeV.

(b) The potential barrier height listed in Table 43-2 is roughly 5 MeV. There is some model-dependence involved in arriving at this estimate, and other values can be found in the literature (6 MeV is frequently cited).

Chapter 44

1. Conservation of momentum requires that the gamma ray particles move in opposite directions with momenta of the same magnitude. Since the magnitude p of the momentum of a gamma ray particle is related to its energy by $p = E/c$, the particles have the same energy E. Conservation of energy yields $m_\pi c^2 = 2E$, where m_π is the mass of a neutral pion. The rest energy of a neutral pion is $m_\pi c^2 = 135.0$ MeV, according to Table 44-4. Hence, $E = (135.0 \text{ MeV})/2 = 67.5$ MeV. We use the result of Problem 83 of Chapter 38 to obtain the wavelength of the gamma rays:

$$\lambda = \frac{1240 \text{ eV} \cdot \text{nm}}{67.5 \times 10^6 \text{ eV}} = 1.84 \times 10^{-5} \text{ nm} = 18.4 \text{ fm}.$$

2. We establish a ratio, using Eq. 22-4 and Eq. 14-1:

$$\frac{F_{\text{gravity}}}{F_{\text{electric}}} = \frac{Gm_e^2/r^2}{ke^2/r^2} = \frac{4\pi\varepsilon_0 Gm_e^2}{e^2} = \frac{\left(6.67 \times 10^{-11} \text{ N} \cdot \text{m}^2/\text{C}^2\right)\left(9.11 \times 10^{-31} \text{ kg}\right)^2}{\left(9.0 \times 10^9 \text{ N} \cdot \text{m}^2/\text{C}^2\right)\left(1.60 \times 10^{-19} \text{ C}\right)^2}$$

$$= 2.4 \times 10^{-43}.$$

Since $F_{\text{gravity}} \ll F_{\text{electric}}$, we can neglect the gravitational force acting between particles in a bubble chamber.

3. Since the density of water is $\rho = 1000 \text{ kg/m}^3 = 1$ kg/L, then the total mass of the pool is $\rho V = 4.32 \times 10^5$ kg, where V is the given volume. Now, the fraction of that mass made up by the protons is 10/18 (by counting the protons versus total nucleons in a water molecule). Consequently, if we ignore the effects of neutron decay (neutrons can beta decay into protons) in the interest of making an order-of-magnitude calculation, then the number of particles susceptible to decay via this $T_{1/2} = 10^{32}$ y half-life is

$$N = \frac{\frac{10}{18} M_{\text{pool}}}{m_p} = \frac{\frac{10}{18}\left(4.32 \times 10^5 \text{ kg}\right)}{1.67 \times 10^{-27} \text{ kg}} = 1.44 \times 10^{32}.$$

Using Eq. 42-20, we obtain

$$R = \frac{N \ln 2}{T_{1/2}} = \frac{\left(1.44 \times 10^{32}\right)\ln 2}{10^{32} \text{ y}} \approx 1 \text{ decay/y}.$$

4. By charge conservation, it is clear that reversing the sign of the pion means we must reverse the sign of the muon. In effect, we are replacing the charged particles by their antiparticles. Less obvious is the fact that we should now put a "bar" over the neutrino (something we should also have done for some of the reactions and decays discussed in the previous two chapters, except that we had not yet learned about antiparticles). To understand the "bar" we refer the reader to the discussion in §44-4. The decay of the negative pion is $\pi^- \rightarrow \mu^- + \bar{v}$. A subscript can be added to the antineutrino to clarify what "type" it is, as discussed in §44-4.

5. From Eq. 37-45, the Lorentz factor would be

$$\gamma = \frac{E}{mc^2} = \frac{1.5 \times 10^6 \text{ eV}}{20 \text{ eV}} = 75000.$$

Solving Eq. 37-8 for the speed, we find

$$\gamma = \frac{1}{\sqrt{1-(v/c)^2}} \Rightarrow v = c\sqrt{1-\frac{1}{\gamma^2}}$$

which implies that the difference between v and c is

$$c - v = c\left(1 - \sqrt{1-\frac{1}{\gamma^2}}\right) \approx c\left(1 - \left(1 - \frac{1}{2\gamma^2} + \cdots\right)\right)$$

where we use the binomial expansion (see Appendix E) in the last step. Therefore,

$$c - v \approx c\left(\frac{1}{2\gamma^2}\right) = (299792458 \text{ m/s})\left(\frac{1}{2(75000)^2}\right) = 0.0266 \text{ m/s} \approx 2.7 \text{ cm/s}.$$

6. (a) In SI units,

$$K = (2200 \text{ MeV})(1.6 \times 10^{-13} \text{ J/MeV}) = 3.52 \times 10^{-10} \text{ J}.$$

Similarly, $mc^2 = 2.85 \times 10^{-10}$ J for the positive tau. Eq. 37-51 leads to the relativistic momentum:

$$p = \frac{1}{c}\sqrt{K^2 + 2Kmc^2} = \frac{1}{2.998 \times 10^8}\sqrt{(3.52 \times 10^{-10})^2 + 2(3.52 \times 10^{-10})(2.85 \times 10^{-10})}$$

which yields $p = 1.90 \times 10^{-18}$ kg·m/s.

(b) According to problem 57 in Chapter 37, the radius should be calculated with the relativistic momentum:

$$r = \frac{\gamma m v}{|q| B} = \frac{p}{eB}$$

where we use the fact that the positive tau has charge $e = 1.6 \times 10^{-19}$ C. With $B = 1.20$ T, this yields $r = 9.90$ m.

7. Table 44-4 gives the rest energy of each pion as 139.6 MeV. The magnitude of the momentum of each pion is $p_\pi = (358.3$ MeV$)/c$. We use the relativistic relationship between energy and momentum (Eq. 37-52) to find the total energy of each pion:

$$E_\pi = \sqrt{(p_\pi c)^2 + (m_\pi c^2)^2} = \sqrt{(358.3 \text{ MeV})^2 + (139.6 \text{ Mev})} = 384.5 \text{ MeV}.$$

Conservation of energy yields

$$m_\rho c^2 = 2E_\pi = 2(384.5 \text{ MeV}) = 769 \text{ MeV}.$$

8. From Eq. 37-49, the Lorentz factor is

$$\gamma = 1 + \frac{K}{mc^2} = 1 + \frac{80 \text{ MeV}}{135 \text{ MeV}} = 1.59.$$

Solving Eq. 37-8 for the speed, we find

$$\gamma = \frac{1}{\sqrt{1 - (v/c)^2}} \Rightarrow v = c\sqrt{1 - \frac{1}{\gamma^2}}$$

which yields $v = 0.778c$ or $v = 2.33 \times 10^8$ m/s. Now, in the reference frame of the laboratory, the lifetime of the pion is not the given τ value but is "dilated." Using Eq. 37-9, the time in the lab is

$$t = \gamma \tau = (1.59)(8.3 \times 10^{-17} \text{ s}) = 1.3 \times 10^{-16} \text{ s}.$$

Finally, using Eq. 37-10, we find the distance in the lab to be

$$x = vt = (2.33 \times 10^8 \text{ m/s}) (1.3 \times 10^{-16} \text{ s}) = 3.1 \times 10^{-8} \text{ m}.$$

9. (a) Conservation of energy gives

$$Q = K_2 + K_3 = E_1 - E_2 - E_3$$

where E refers here to the *rest* energies (mc^2) instead of the total energies of the particles. Writing this as $K_2 + E_2 - E_1 = -(K_3 + E_3)$ and squaring both sides yields

$$K_2^2 + 2K_2E_2 - 2K_2E_1 + (E_1 - E_2)^2 = K_3^2 + 2K_3E_3 + E_3^2.$$

Next, conservation of linear momentum (in a reference frame where particle 1 was at rest) gives $|p_2| = |p_3|$ (which implies $(p_2c)^2 = (p_3c)^2$). Therefore, Eq. 37-54 leads to

$$K_2^2 + 2K_2E_2 = K_3^2 + 2K_3E_3$$

which we subtract from the above expression to obtain

$$-2K_2E_1 + (E_1 - E_2)^2 = E_3^2.$$

This is now straightforward to solve for K_2 and yields the result stated in the problem.

(b) Setting $E_3 = 0$ in

$$K_2 = \frac{1}{2E_1}\left[(E_1 - E_2)^2 - E_3^2\right]$$

and using the rest energy values given in Table 44-1 readily gives the same result for K_μ as computed in Sample Problem 44-1.

10. (a) Noting that there are two positive pions created (so, in effect, its decay products are doubled), then we count up the electrons, positrons and neutrinos: $2e^+ + e^- + 5v + 4\bar{v}$.

(b) The final products are all leptons, so the baryon number of A_2^+ is zero. Both the pion and rho meson have integer-valued spins, so A_2^+ is a boson.

(c) A_2^+ is also a meson.

(d) As stated in (b), the baryon number of A_2^+ is zero.

11. (a) The conservation laws considered so far are associated with energy, momentum, angular momentum, charge, baryon number, and the three lepton numbers. The rest energy of the muon is 105.7 MeV, the rest energy of the electron is 0.511 MeV, and the rest energy of the neutrino is zero. Thus, the total rest energy before the decay is greater than the total rest energy after. The excess energy can be carried away as the kinetic energies of the decay products and energy can be conserved. Momentum is conserved if the electron and neutrino move away from the decay in opposite directions with equal magnitudes of momenta. Since the orbital angular momentum is zero, we consider only spin angular momentum. All the particles have spin $\hbar/2$. The total angular momentum after the decay must be either $\hbar$ (if the spins are aligned) or zero (if the spins are antialigned). Since the spin before the decay is $\hbar/2$ angular momentum cannot be conserved. The muon has charge $-e$, the electron has charge $-e$, and the neutrino has

charge zero, so the total charge before the decay is −e and the total charge after is −e. Charge is conserved. All particles have baryon number zero, so baryon number is conserved. The muon lepton number of the muon is +1, the muon lepton number of the muon neutrino is +1, and the muon lepton number of the electron is 0. Muon lepton number is conserved. The electron lepton numbers of the muon and muon neutrino are 0 and the electron lepton number of the electron is +1. Electron lepton number is not conserved. The laws of conservation of angular momentum and electron lepton number are not obeyed and this decay does not occur.

(b) We analyze the decay in the same way. We find that charge and the muon lepton number L_μ are not conserved.

(c) Here we find that energy and muon lepton number L_μ cannot be conserved.

12. (a) Referring to Tables 44-3 and 44-4, we find the strangeness of K^0 is +1, while it is zero for both π^+ and π^-. Consequently, strangeness is not conserved in this decay; $K^0 \rightarrow \pi^+ + \pi^-$ does not proceed via the strong interaction.

(b) The strangeness of each side is −1, which implies that the decay is governed by the strong interaction.

(c) The strangeness or Λ^0 is −1 while that of $p+\pi^-$ is zero, so the decay is not via the strong interaction.

(d) The strangeness of each side is −1; it proceeds via the strong interaction.

13. For purposes of deducing the properties of the antineutron, one may cancel a proton from each side of the reaction and write the equivalent reaction as

$$\pi^+ \rightarrow p = \bar{n}.$$

Particle properties can be found in Tables 44-3 and 44-4. The pion and proton each have charge +e, so the antineutron must be neutral. The pion has baryon number zero (it is a meson) and the proton has baryon number +1, so the baryon number of the antineutron must be −1. The pion and the proton each have strangeness zero, so the strangeness of the antineutron must also be zero. In summary, for the antineutron,

(a) $q = 0$,

(b) $B = -1$,

(c) and $S = 0$.

14. (a) From Eq. 37-50,

$$Q = -\Delta mc^2 = (m_{\Sigma^+} + m_{K^+} - m_{\pi^+} - m_p)c^2$$
$$= 1189.4\,\text{MeV} + 493.7\,\text{MeV} - 139.6\,\text{MeV} - 938.3\,\text{MeV}$$
$$= 605\,\text{MeV}.$$

(b) Similarly,

$$Q = -\Delta mc^2 = (m_{\Lambda^0} + m_{\pi^0} - m_{K^-} - m_p)c^2$$
$$= 1115.6\,\text{MeV} + 135.0\,\text{MeV} - 493.7\,\text{MeV} - 938.3\,\text{MeV}$$
$$= -181\,\text{MeV}.$$

15. (a) See the solution to Problem 11 for the quantities to be considered, adding strangeness to the list. The lambda has a rest energy of 1115.6 MeV, the proton has a rest energy of 938.3 MeV, and the kaon has a rest energy of 493.7 MeV. The rest energy before the decay is less than the total rest energy after, so energy cannot be conserved. Momentum can be conserved. The lambda and proton each have spin $\hbar/2$ and the kaon has spin zero, so angular momentum can be conserved. The lambda has charge zero, the proton has charge $+e$, and the kaon has charge $-e$, so charge is conserved. The lambda and proton each have baryon number $+1$, and the kaon has baryon number zero, so baryon number is conserved. The lambda and kaon each have strangeness -1 and the proton has strangeness zero, so strangeness is conserved. Only energy cannot be conserved.

(b) The omega has a rest energy of 1680 MeV, the sigma has a rest energy of 1197.3 MeV, and the pion has a rest energy of 135 MeV. The rest energy before the decay is greater than the total rest energy after, so energy can be conserved. Momentum can be conserved. The omega and sigma each have spin $\hbar/2$ and the pion has spin zero, so angular momentum can be conserved. The omega has charge $-e$, the sigma has charge $-e$, and the pion has charge zero, so charge is conserved. The omega and sigma have baryon number $+1$ and the pion has baryon number 0, so baryon number is conserved. The omega has strangeness -3, the sigma has strangeness -1, and the pion has strangeness zero, so strangeness is not conserved.

(c) The kaon and proton can bring kinetic energy to the reaction, so energy can be conserved even though the total rest energy after the collision is greater than the total rest energy before. Momentum can be conserved. The proton and lambda each have spin $\hbar/2$ and the kaon and pion each have spin zero, so angular momentum can be conserved. The kaon has charge $-e$, the proton has charge $+e$, the lambda has charge zero, and the pion has charge $+e$, so charge is not conserved. The proton and lambda each have baryon number $+1$, and the kaon and pion each have baryon number zero; baryon number is conserved. The kaon has strangeness -1, the proton and pion each have strangeness zero, and the lambda has strangeness -1, so strangeness is conserved. Only charge is not conserved.

16. The formula for T_z as it is usually written to include strange baryons is $T_z = q - (S + B)/2$. Also, we interpret the symbol q in the T_z formula in terms of elementary charge units; this is how q is listed in Table 44-3. In terms of charge q as we have used it in previous chapters, the formula is

$$T_z = \frac{q}{e} - \frac{1}{2}(B+S).$$

For instance, $T_z = +\frac{1}{2}$ for the proton (and the neutral Xi) and $T_z = -\frac{1}{2}$ for the neutron (and the negative Xi). The baryon number B is +1 for all the particles in Fig. 44-4(a). Rather than use a sloping axis as in Fig. 44-4 (there it is done for the q values), one reproduces (if one uses the "corrected" formula for T_z mentioned above) exactly the same pattern using regular rectangular axes (T_z values along the horizontal axis and Y values along the vertical) with the neutral lambda and sigma particles situated at the origin.

17. (a) As far as the conservation laws are concerned, we may cancel a proton from each side of the reaction equation and write the reaction as $p \rightarrow \Lambda^0 + x$. Since the proton and the lambda each have a spin angular momentum of $\hbar/2$, the spin angular momentum of x must be either zero or $\hbar$. Since the proton has charge $+e$ and the lambda is neutral, x must have charge $+e$. Since the proton and the lambda each have a baryon number of $+1$, the baryon number of x is zero. Since the strangeness of the proton is zero and the strangeness of the lambda is -1, the strangeness of x is $+1$. We take the unknown particle to be a spin zero meson with a charge of $+e$ and a strangeness of $+1$. Look at Table 44-4 to identify it as a K^+ particle.

(b) Similar analysis tells us that x is a spin-$\frac{1}{2}$ antibaryon ($B = -1$) with charge and strangeness both zero. Inspection of Table 44-3 reveals it is an antineutron.

(c) Here x is a spin-0 (or spin-1) meson with charge zero and strangeness -1. According to Table 44-4, it could be a $\overline{K}^0$ particle.

18. Conservation of energy (see Eq. 37-47) leads to

$$K_f = -\Delta mc^2 + K_i = (m_{\Sigma^-} - m_{\pi^-} - m_n)c^2 + K_i$$
$$= 1197.3\,\text{MeV} - 139.6\,\text{MeV} - 939.6\,\text{MeV} + 220\,\text{MeV}$$
$$= 338\,\text{MeV}.$$

19. (a) From Eq. 37-50,

$$Q = -\Delta mc^2 = (m_{\Lambda^0} - m_p - m_{\pi^-})c^2$$
$$= 1115.6\,\text{MeV} - 938.3\,\text{MeV} - 139.6\,\text{MeV} = 37.7\,\text{MeV}.$$

(b) We use the formula obtained in problem 44-9 (where it should be emphasized that E is used to mean the rest energy, not the total energy):

$$K_p = \frac{1}{2E_\Lambda}\left[\left(E_\Lambda - E_p\right)^2 - E_\pi^2\right]$$

$$= \frac{\left(1115.6\,\text{MeV} - 938.3\,\text{MeV}\right)^2 - \left(139.6\,\text{MeV}\right)^2}{2\left(1115.6\,\text{MeV}\right)} = 5.35\,\text{MeV}.$$

(c) By conservation of energy,

$$K_{\pi^-} = Q - K_p = 37.7\,\text{MeV} - 5.35\,\text{MeV} = 32.4\,\text{MeV}.$$

20. (a) The combination ddu has a total charge of $\left(-\frac{1}{3} - \frac{1}{3} + \frac{2}{3}\right) = 0$, and a total strangeness of zero. From Table 44-3, we find it to be a neutron (n).

(b) For the combination uus, we have $Q = +\frac{2}{3} + \frac{2}{3} - \frac{1}{3} = 1$ and $S = 0 + 0 - 1 = 1$. This is the Σ^+ particle.

(c) For the quark composition ssd, we have $Q = -\frac{1}{3} - \frac{1}{3} - \frac{1}{3} = -1$ and $S = -1 - 1 + 0 = -2$. This is a Ξ^-.

21. (a) We indicate the antiparticle nature of each quark with a "bar" over it. Thus, $\overline{u}\,\overline{u}\,\overline{d}$ represents an antiproton.

(b) Similarly, $\overline{u}\,\overline{d}\,\overline{d}$ represents an antineutron.

22. (a) Using Table 44-3, we find $q = 0$ and $S = -1$ for this particle (also, $B = 1$, since that is true for all particles in that table). From Table 44-5, we see it must therefore contain a strange quark (which has charge $-1/3$), so the other two quarks must have charges to add to zero. Assuming the others are among the lighter quarks (none of them being an antiquark, since $B = 1$), then the quark composition is $s\,\overline{u}\,d$.

(b) The reasoning is very similar to that of part (a). The main difference is that this particle must have two strange quarks. Its quark combination turns out to be $\overline{u}ss$.

23. (a) Looking at the first three lines of Table 44-5, since the particle is a baryon, we determine that it must consist of three quarks. To obtain a strangeness of -2, two of them must be s quarks. Each of these has a charge of $-e/3$, so the sum of their charges is $-2e/3$. To obtain a total charge of e, the charge on the third quark must be $5e/3$. There is no quark with this charge, so the particle cannot be constructed. In fact, such a particle has never been observed.

(b) Again the particle consists of three quarks (and no antiquarks). To obtain a strangeness of zero, none of them may be s quarks. We must find a combination of three u and d quarks with a total charge of $2e$. The only such combination consists of three u quarks.

24. If we were to use regular rectangular axes, then this would appear as a right triangle. Using the sloping q axis as the problem suggests, it is similar to an "upside down" equilateral triangle as we show below.

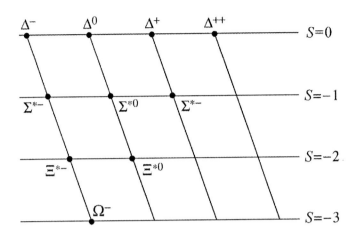

The leftmost slanted line is for the −1 charge, and the rightmost slanted line is for the +2 charge.

25. From $\gamma = 1 + K/mc^2$ (see Eq. 37-52) and $v = \beta c = c\sqrt{1-\gamma^{-2}}$ (see Eq. 37-8), we get

$$v = c\sqrt{1-\left(1+\frac{K}{mc^2}\right)^{-2}}.$$

(a) Therefore, for the Σ^{*0} particle,

$$v = (2.9979 \times 10^8 \text{ m/s})\sqrt{1-\left(1+\frac{1000\,\text{MeV}}{1385\,\text{MeV}}\right)^{-2}} = 2.4406 \times 10^8 \text{ m/s}.$$

For Σ^0,

$$v' = (2.9979 \times 10^8 \text{ m/s})\sqrt{1-\left(1+\frac{1000\,\text{MeV}}{1192.5\,\text{MeV}}\right)^{-2}} = 2.5157 \times 10^8 \text{ m/s}.$$

Thus Σ^0 moves faster than Σ^{*0}.

(b) The speed difference is

$$\Delta v = v' - v = (2.5157 - 2.4406)(10^8 \text{ m/s}) = 7.51 \times 10^6 \text{ m/s}.$$

26. Letting $v = Hr = c$, we obtain

$$r = \frac{c}{H} = \frac{3.0 \times 10^8 \text{ m/s}}{0.0218 \text{ m/s} \cdot \text{ly}} = 1.376 \times 10^{10} \text{ ly} \approx 1.4 \times 10^{10} \text{ ly}.$$

27. We apply Eq. 37-36 for the Doppler shift in wavelength:

$$\frac{\Delta\lambda}{\lambda} = \frac{v}{c}$$

where v is the recessional speed of the galaxy. We use Hubble's law to find the recessional speed: $v = Hr$, where r is the distance to the galaxy and H is the Hubble constant $\left(21.8 \times 10^{-3} \frac{\text{m}}{\text{s} \cdot \text{ly}}\right)$. Thus,

$$v = \left[21.8 \times 10^{-3} \tfrac{\text{m}}{\text{s} \cdot \text{ly}}\right] \left(2.40 \times 10^8 \text{ ly}\right) = 5.23 \times 10^6 \text{ m/s}$$

and

$$\Delta\lambda = \frac{v}{c}\lambda = \left(\frac{5.23 \times 10^6 \text{ m/s}}{3.00 \times 10^8 \text{ m/s}}\right)(656.3 \text{ nm}) = 11.4 \text{ nm}.$$

Since the galaxy is receding, the observed wavelength is longer than the wavelength in the rest frame of the galaxy. Its value is 656.3 nm + 11.4 nm = 667.7 nm ≈ 668 nm.

28. First, we find the speed of the receding galaxy from Eq. 37-31:

$$\beta = \frac{1 - (f/f_0)^2}{1 + (f/f_0)^2} = \frac{1 - (\lambda_0/\lambda)^2}{1 + (\lambda_0/\lambda)^2}$$

$$= \frac{1 - (590.0 \text{ nm}/602.0 \text{ nm})^2}{1 + (590.0 \text{ nm}/602.0 \text{ nm})^2} = 0.02013$$

where we use $f = c/\lambda$ and $f_0 = c/\lambda_0$. Then from Eq. 44-19,

$$r = \frac{v}{H} = \frac{\beta c}{H} = \frac{(0.02013)(2.998 \times 10^8 \text{ m/s})}{0.0218 \text{ m/s} \cdot \text{ly}} = 2.77 \times 10^8 \text{ ly}.$$

29. (a) From $f = c/\lambda$ and Eq. 37-31, we get

$$\lambda_0 = \lambda\sqrt{\frac{1-\beta}{1+\beta}} = (\lambda_0 + \Delta\lambda)\sqrt{\frac{1-\beta}{1+\beta}}.$$

Dividing both sides by λ_0 leads to

$$1 = (1+z)\sqrt{\frac{1-\beta}{1+\beta}}.$$

We solve for β:

$$\beta = \frac{(1+z)^2 - 1}{(1+z)^2 + 1} = \frac{z^2 + 2z}{z^2 + 2z + 2}.$$

(b) Now $z = 4.43$, so

$$\beta = \frac{(4.43)^2 + 2(4.43)}{(4.43)^2 + 2(4.43) + 2} = 0.934.$$

(c) From Eq. 44-19,

$$r = \frac{v}{H} = \frac{\beta c}{H} = \frac{(0.934)(3.0 \times 10^8 \text{ m/s})}{0.0218 \text{ m/s} \cdot \text{ly}} = 1.28 \times 10^{10} \text{ ly}.$$

30. (a) Letting $v(r) = Hr \le v_e = \sqrt{2G\,M/r}$, we get $M/r^3 \ge H^2/2G$. Thus,

$$\rho = \frac{M}{4\pi r^2/3} = \frac{3}{4\pi}\frac{M}{r^3} \ge \frac{3H^2}{8\pi G}.$$

(b) The density being expressed in H-atoms/m^3 is equivalent to expressing it in terms of $\rho_0 = m_H/\text{m}^3 = 1.67 \times 10^{-27}$ kg/m^3. Thus,

$$\rho = \frac{3H^2}{8\pi G \rho_0}\left(\text{H atoms/m}^3\right) = \frac{3(0.0218 \text{ m/s} \cdot \text{ly})^2 (1.00 \text{ly}/9.460 \times 10^{15} \text{ m})^2 \left(\text{H atoms/m}^3\right)}{8\pi(6.67 \times 10^{-11} \text{ m}^3/\text{kg} \cdot \text{s}^2)(1.67 \times 10^{-27} \text{ kg/m}^3)}$$

$$= 5.7 \text{ H atoms/m}^3.$$

31. (a) From Eq. 41-29, we know that $N_2/N_1 = e^{-\Delta E/kT}$. We solve for ΔE:

$$\Delta E = kT \ln\frac{N_1}{N_2} = (8.62 \times 10^{-5} \text{ eV/K})(2.7\text{K}) \ln\left(\frac{1-0.25}{0.25}\right)$$

$$= 2.56 \times 10^{-4} \text{ eV} \approx 0.26 \text{ meV}.$$

(b) Using the result of problem 83 in Chapter 38,

$$\lambda = \frac{hc}{\Delta E} = \frac{1240 \text{eV} \cdot \text{nm}}{2.56 \times 10^{-4} \text{eV}} = 4.84 \times 10^6 \text{ nm} \approx 4.8 \text{mm}.$$

32. From $F_{\text{grav}} = GMm/r^2 = mv^2/r$ we find $M \propto v^2$. Thus, the mass of the Sun would be

$$M'_s = \left(\frac{v_{\text{Mercury}}}{v_{\text{Pluto}}}\right)^2 M_s = \left(\frac{47.9 \text{ km/s}}{4.74 \text{ km/s}}\right)^2 M_s = 102 \, M_s \,.$$

33. (a) The mass M within Earth's orbit is used to calculate the gravitational force on Earth. If r is the radius of the orbit, R is the radius of the new Sun, and M_S is the mass of the Sun, then

$$M = \left(\frac{r}{R}\right)^3 M_s = \left(\frac{1.50 \times 10^{11} \text{ m}}{5.90 \times 10^{12} \text{ m}}\right)^3 (1.99 \times 10^{30} \text{ kg}) = 3.27 \times 10^{25} \text{ kg} \,.$$

The gravitational force on Earth is given by GMm/r^2, where m is the mass of Earth and G is the universal gravitational constant. Since the centripetal acceleration is given by v^2/r, where v is the speed of Earth, $GMm/r^2 = mv^2/r$ and

$$v = \sqrt{\frac{GM}{r}} = \sqrt{\frac{(6.67 \times 10^{-11} \text{ m}^3/\text{s}^2 \cdot \text{kg})(3.27 \times 10^{25} \text{ kg})}{1.50 \times 10^{11} \text{ m}}} = 1.21 \times 10^2 \text{ m/s} \,.$$

(b) The ratio is

$$\frac{1.21 \times 10^2 \text{ m/s}}{2.98 \times 10^4 \text{ m/s}} = 0.00406.$$

(c) The period of revolution is

$$T = \frac{2\pi r}{v} = \frac{2\pi (1.50 \times 10^{11} \text{ m})}{1.21 \times 10^2 \text{ m/s}} = 7.82 \times 10^9 \text{ s} = 248 \text{ y} \,.$$

34. (a) The mass of the portion of the galaxy within the radius r from its center is given by $M' = (r/R)^3 M$. Thus, from $GM'm/r^2 = mv^2/r$ (where m is the mass of the star) we get

$$v = \sqrt{\frac{GM'}{r}} = \sqrt{\frac{GM}{r}\left(\frac{r}{R}\right)^3} = r\sqrt{\frac{GM}{R^3}} \,.$$

(b) In the case where $M' = M$, we have

$$T = \frac{2\pi r}{v} = 2\pi r \sqrt{\frac{r}{GM}} = \frac{2\pi r^{3/2}}{\sqrt{GM}} \, .$$

35. (a) We substitute $\lambda = (2898 \ \mu m \cdot K)/T$ into the result of Problem 83 of Chapter 38: $E = (1240 \ eV \cdot nm)/\lambda$. First, we convert units: $2898 \ \mu m \cdot K = 2.898 \times 10^6 \ nm \cdot K$ and $1240 \ eV \cdot nm = 1.240 \times 10^{-3} \ MeV \cdot nm$. Hence,

$$E = \frac{\left(1.240 \times 10^{-3} \ MeV \cdot nm\right)T}{2.898 \times 10^6 \ nm \cdot K} = \left(4.28 \times 10^{-10} \ MeV/K\right)T \, .$$

(b) The minimum energy required to create an electron-positron pair is twice the rest energy of an electron, or $2(0.511 \ MeV) = 1.022 \ MeV$. Hence,

$$T = \frac{E}{4.28 \times 10^{-10} \ MeV/K} = \frac{1.022 \ MeV}{4.28 \times 10^{-10} \ MeV/K} = 2.39 \times 10^9 \ K \, .$$

36. (a) For the universal microwave background, Wien's law leads to

$$T = \frac{2898 \ \mu m \cdot K}{\lambda_{max}} = \frac{2898 \ mm \cdot K}{1.1 \ mm} = 2.6 \ K \, .$$

(b) At "decoupling" (when the universe became approximately "transparent"),

$$\lambda_{max} = \frac{2898 \ \mu m \cdot K}{T} = \frac{2898 \ \mu m \cdot K}{2970 K} = 0.976 \ \mu m = 976 \ nm.$$

37. The energy released would be twice the rest energy of Earth, or

$$E = 2mc^2 = 2(5.98 \times 10^{24} \ kg)(2.998 \times 10^8 \ m/s)^2 = 1.08 \times 10^{42} \ J.$$

The mass of Earth can be found in Appendix C.

38. We note from track 1, and the quantum numbers of the original particle (A), that positively charged particles move in counterclockwise curved paths, and — by inference — negatively charged ones move along clockwise arcs. This immediately shows that tracks 1, 2, 4, 6, and 7 belong to positively charged particles, and tracks 5, 8 and 9 belong to negatively charged ones. Looking at the fictitious particles in the table (and noting that each appears in the cloud chamber once [or not at all]), we see that this observation (about charged particle motion) greatly narrows the possibilities:

$$\text{tracks } 2,4,6,7, \ \leftrightarrow \ \text{particles } C,F,H,J$$
$$\text{tracks } 5,8,9 \ \leftrightarrow \ \text{particles } D,E,G$$

This tells us, too, that the particle that does not appear at all is either B or I (since only one neutral particle "appears"). By charge conservation, tracks 2, 4 and 6 are made by particles with a single unit of positive charge (note that track 5 is made by one with a single unit of negative charge), which implies (by elimination) that track 7 is made by particle H. This is confirmed by examining charge conservation at the end-point of track 6. Having exhausted the charge-related information, we turn now to the fictitious quantum numbers. Consider the vertex where tracks 2, 3 and 4 meet (the Whimsy number is listed here as a subscript):

$$\text{tracks } 2,4 \ \leftrightarrow \ \text{particles } C_2, F_0, J_{-6}$$
$$\text{tracks } 3 \ \leftrightarrow \ \text{particle } B_4 \text{ or } I_6$$

The requirement that the Whimsy quantum number of the particle making track 4 must equal the sum of the Whimsy values for the particles making tracks 2 and 3 places a powerful constraint (see the subscripts above). A fairly quick trial and error procedure leads to the assignments: particle F makes track 4, and particles J and I make tracks 2 and 3, respectively. Particle B, then, is irrelevant to this set of events. By elimination, the particle making track 6 (the only positively charged particle not yet assigned) must be C. At the vertex defined by

$$A \rightarrow F + C + (\text{track } 5)_- \ ,$$

where the charge of that particle is indicated by the subscript, we see that Cuteness number conservation requires that the particle making track 5 has Cuteness $= -1$, so this must be particle G. We have only one decision remaining:

$$\text{tracks } 8,9, \ \leftrightarrow \ \text{particles } D,E$$

Re-reading the problem, one finds that the particle making track 8 must be particle D since it is the one with seriousness $= 0$. Consequently, the particle making track 9 must be E.

Thus, we have the following:

(a) Particle A for track 1.

(b) Particle J for track 2.

(c) Particle I for track 3.

(d) Particle F for track 4.

(e) Particle G for track 5.

(f) Particle C for track 6.

(g) Particle H for track 7.

(h) Particle D for track 8.

(i) Particle E for track 9.

39. Since only the strange quark (s) has non-zero strangeness, in order to obtain $S = -1$ we need to combine s with some non-strange anti-quark (which would have the negative of the quantum numbers listed in Table 44-5). The difficulty is that the charge of the strange quark is $-1/3$, which means that (to obtain a total charge of $+1$) the anti-quark would have to have a charge of $+\frac{4}{3}$. Clearly, there are no such anti-quarks in our list. Thus, a meson with $S = -1$ and $q = +1$ cannot be formed with the quarks/anti-quarks of Table 44-5. Similarly, one can show that, since no quark has $q = -\frac{4}{3}$, there cannot be a meson with $S = +1$ and $q = -1$.

40. Assuming the line passes through the origin, its slope is $0.40c/(5.3 \times 10^9 \text{ ly})$. Then,

$$T = \frac{1}{H} = \frac{1}{\text{slope}} = \frac{5.3 \times 10^9 \text{ ly}}{0.40c} = \frac{5.3 \times 10^9 \text{ y}}{0.40} \approx 13 \times 10^9 \text{ y}.$$

41. (a) We use the relativistic relationship between speed and momentum:

$$p = \gamma mv = \frac{mv}{\sqrt{1-(v/c)^2}},$$

which we solve for the speed v:

$$\frac{v}{c} = \sqrt{1 - \frac{1}{\left(\frac{pc}{mc^2}\right)^2 + 1}}.$$

For an antiproton $mc^2 = 938.3$ MeV and $pc = 1.19$ GeV $= 1190$ MeV, so

$$v = c\sqrt{1 - \frac{1}{(1190\,\text{MeV}/938.3\,\text{MeV})^2 + 1}} = 0.785c.$$

(b) For the negative pion $mc^2 = 193.6$ MeV, and pc is the same. Therefore,

$$v = c\sqrt{1 - \frac{1}{(1190\,\text{MeV}/193.6\,\text{MeV})^2 + 1}} = 0.993\,c \;.$$

(c) Since the speed of the antiprotons is about $0.78c$ but not over $0.79c$, an antiproton will trigger C2.

(d) Since the speed of the negative pions exceeds $0.79c$, a negative pion will trigger C1.

(e) We use $\Delta t = d/v$, where $d = 12$ m. For an antiproton

$$\Delta t = \frac{1}{0.785(2.998\times10^8\,\text{m/s})} = 5.1\times10^{-8}\,\text{s} = 51\text{ns}.$$

(f) For a negative pion

$$\Delta t = \frac{12\,\text{m}}{0.993(2.998\times10^8\,\text{m/s})} = 4.0\times10^{-8}\,\text{s} = 40\,\text{ns}\;.$$

42. (a) Eq. 44-14 conserves charge since both the proton and the positron have $q = +e$ (and the neutrino is uncharged).

(b) Energy conservation is not violated since $m_p c^2 > m_e c^2 + m_\nu c^2$.

(c) We are free to view the decay from the rest frame of the proton. Both the positron and the neutrino are able to carry momentum, and so long as they travel in opposite directions with appropriate values of p (so that $\sum \vec{p} = 0$) then linear momentum is conserved.

(d) If we examine the spin angular momenta, there does seem to be a violation of angular momentum conservation (Eq. 44-14 shows a spin-one-half particle decaying into two spin-one-half particles).

43. (a) During the time interval Δt, the light emitted from galaxy A has traveled a distance $c\Delta t$. Meanwhile, the distance between Earth and the galaxy has expanded from r to $r' = r + r\alpha\,\Delta t$. Let $c\Delta t = r' = r + ra\Delta t$, which leads to

$$\Delta t = \frac{r}{c - r\alpha}\;.$$

(b) The detected wavelength λ' is longer than λ by $\lambda\alpha\Delta t$ due to the expansion of the universe: $\lambda' = \lambda + \lambda\alpha\Delta t$. Thus,

$$\frac{\Delta\lambda}{\lambda} = \frac{\lambda' - \lambda}{\lambda} = \alpha\Delta t = \frac{\alpha r}{c - \alpha r}\;.$$

(c) We use the binomial expansion formula (see Appendix E):

$$(1 \pm x)^n = 1 \pm \frac{nx}{1!} + \frac{n(n-1)x^2}{2!} + \cdots \quad (x^2 < 1)$$

to obtain

$$\frac{\Delta\lambda}{\lambda} = \frac{\alpha r}{c - \alpha r} = \frac{\alpha r}{c}\left(1 - \frac{\alpha r}{c}\right)^{-1} = \frac{\alpha r}{c}\left[1 + \frac{-1}{1!}\left(-\frac{\alpha r}{c}\right) + \frac{(-1)(-2)}{2!}\left(-\frac{\alpha r}{c}\right)^2 + \cdots\right]$$

$$\approx \frac{\alpha r}{c} + \left(\frac{\alpha r}{c}\right)^2 + \left(\frac{\alpha r}{c}\right)^3 .$$

(d) When only the first term in the expansion for $\Delta\lambda/\lambda$ is retained we have

$$\frac{\Delta\lambda}{\lambda} \approx \frac{\alpha r}{c} .$$

(e) We set

$$\frac{\Delta\lambda}{\lambda} = \frac{v}{c} = \frac{Hr}{c}$$

and compare with the result of part (d) to obtain $\alpha = H$.

(f) We use the formula $\Delta\lambda/\lambda = \alpha r/(c - \alpha r)$ to solve for r:

$$r = \frac{c(\Delta\lambda/\lambda)}{\alpha(1 + \Delta\lambda/\lambda)} = \frac{(2.998 \times 10^8 \text{ m/s})(0.050)}{(0.0218 \text{ m/s}\cdot\text{ly})(1 + 0.050)} = 6.548 \times 10^8 \text{ ly} \approx 6.5 \times 10^8 \text{ ly}.$$

(g) From the result of part (a),

$$\Delta t = \frac{r}{c - \alpha r} = \frac{(6.5 \times 10^8 \text{ ly})(9.46 \times 10^{15} \text{ m/ly})}{2.998 \times 10^8 \text{ m/s} - (0.0218 \text{ m/s}\cdot\text{ly})(6.5 \times 10^8 \text{ ly})} = 2.17 \times 10^{16} \text{ s},$$

which is equivalent to 6.9×10^8 y.

(h) Letting $r = c\Delta t$, we solve for Δt:

$$\Delta t = \frac{r}{c} = \frac{6.5 \times 10^8 \text{ ly}}{c} = 6.5 \times 10^8 \text{ y}.$$

(i) The distance is given by

$$r = c\Delta t = c\left(6.9\times10^8\,\text{y}\right) = 6.9\times10^8\,\text{ly}.$$

(j) From the result of part (f),

$$r_B = \frac{c\left(\Delta\lambda/\lambda\right)}{\alpha\left(1+\Delta\lambda/\lambda\right)} = \frac{\left(2.998\times10^8\ \text{m/s}\right)\left(0.080\right)}{\left(0.0218\ \text{mm/s}\cdot\text{ly}\right)\left(1+0.080\right)} = 1.018\times10^9\,\text{ly} \approx 1.0\times10^9\,\text{ly}.$$

(k) From the formula obtained in part (a),

$$\Delta t_B = \frac{r_B}{c - r_B\alpha} = \frac{\left(1.0\times10^9\ \text{ly}\right)\left(9.46\times10^{15}\ \text{m/ly}\right)}{2.998\times10^8\ \text{m/s} - \left(1.0\times10^9\ \text{ly}\right)\left(0.0218\,\text{m/s}\cdot\text{ly}\right)} = 3.4\times10^{16}\ \text{s},$$

which is equivalent to 1.1×10^9 y.

(l) At the present time, the separation between the two galaxies A and B is given by $r_{\text{now}} = c\Delta t_B - c\Delta t_A$. Since $r_{\text{now}} = r_{\text{then}} + r_{\text{then}}\alpha\Delta t$, we get

$$r_{\text{then}} = \frac{r_{\text{now}}}{1+\alpha\Delta t} = 3.9\times10^8\,\text{ly}.$$

44. Using Table 44-1, the difference in mass between the muon and the pion is

$$\Delta m = \left(139.6\,\tfrac{\text{MeV}}{c^2} - 105.7\,\tfrac{\text{MeV}}{c^2}\right) = \frac{\left(33.9\ \text{MeV}\right)\left(1.60\times10^{-13}\ \text{J/MeV}\right)}{\left(2.998\times10^8\ \text{m/s}\right)^2} = 6.03\times10^{-29}\ \text{kg}.$$

NOTES

NOTES

NOTES

NOTES

NOTES

9 780471 470